Turbo Coding,
Turbo Equalisation and
Space-Time Coding

for Transmission over Fading Channels

Turbo Coding, Turbo Equalisation and Space-Time Coding

for Transmission over Fading Channels

L. Hanzo
T. H. Liew
B.L. Yeap

All of University of Southampton, UK

IEEE PRESS

IEEE Communications Society, Sponsor

JOHN WILEY & SONS, LTD

Other Wiley Editorial Offices

John Wiley & Sons Inc., 111 River Street, Hoboken, NJ 07030, USA

Jossey-Bass, 989 Market Street, San Francisco, CA 94103-1741, USA

Wiley-VCH Verlag GmbH, Boschstr. 12, D-69469 Weinheim, Germany

John Wiley & Sons Australia Ltd, 33 Park Road, Milton, Queensland 4064, Australia

John Wiley & Sons (Asia) Pte Ltd, 2 Clementi Loop #02-01, Jin Xing Distripark, Singapore 129809

John Wiley & Sons Canada Ltd, 22 Worcester Road, Etobicoke, Ontario, Canada M9W 1L1

IEEE Communications Society, Sponsor
COMMS-S Liaison to IEEE Press, Mostafa Hashem Sherif

Library of Congress Cataloging-in-Publication Data

Hanzo, Lajos, 1952-
    Turbo coding, Turbo equalization, and space-time coding / by L. Hanzo, T.H. Liew, B.L. Yeap.
     p.cm.
    Includes bibliographical references and indexes.
    ISBN 0-470-84726-3
    1.   Signal processing–Mathematics. 2. Coding theory, 3. Iterative methods
    (Mathematics) I. Liew, T. H. (Tong Hooi) II. Yeap, B. L. (Bee Long) III. Title.

    TK5102.92 H36 2002
    003'.54–dc21   2002069162

British Library Cataloguing in Publication Data

A catalogue record for this book is available from the British Library

ISBN 0-470-84726-3

Produced from LaTeX files supplied by the author
Printed and bound in Great Britain by Biddles Ltd, Guildford and King's Lynn
This book is printed on acid-free paper responsibly manufactured from sustainable forestry
in which at least two trees are planted for each one used for paper production.

We dedicate this monograph to Rita, Ching Ching, Khar Yee
and to our parents as well as to the numerous contributors of this field,
many of whom are listed in the Author Index

Contents

IV   Space-Time Block and Space-Time Trellis Coding   393

10 Space-time Block Codes   395

Acknowledgments

We are indebted to our many colleagues who have enhanced our understanding of the subject, in particular to Prof. Emeritus Raymond Steele. These colleagues and valued friends, too numerous to be mentioned, have influenced our views concerning various aspects of wireless multimedia communications. We thank them for the enlightenment gained from our collaborations on various projects, papers and books. We are grateful to Steve Braithwaite, Jan Brecht, Jon Blogh, Marco Breiling, Marco del Buono, Sheng Chen, Peter Cherriman, Stanley Chia, Byoung Jo Choi, Joseph Cheung, Sheyam Lal Dhomeja, Dirk Didascalou, Lim Dongmin, Stephan Ernst, Peter Fortune, Eddie Green, David Greenwood, Hee Thong How, Thomas Keller, Ee Lin Kuan, W. H. Lam, C. C. Lee, Xiao Lin, Chee Siong Lee, Tong-Hooi Liew, Matthias Münster, Vincent Roger-Marchart, Jason Ng, Michael Ng, M. A. Nofal, Jeff Reeve, Redwan Salami, Clare Somerville, Rob Stedman, David Stewart, Jürgen Streit, Jeff Torrance, Spyros Vlahoyiannatos, William Webb, Stephan Weiss, John Williams, Jason Woodard, Choong Hin Wong, Henry Wong, James Wong, Lie-Liang Yang, Bee-Leong Yeap, Mong-Suan Yee, Kai Yen, Andy Yuen, and many others with whom we enjoyed an association.

We also acknowledge our valuable associations with the Virtual Centre of Excellence (VCE) in Mobile Communications, in particular with its chief executive, Dr Walter Tuttlebee, and other leading members of the VCE, namely Dr Keith Baughan, Prof. Hamid Aghvami, Prof. Ed Candy, Prof. John Dunlop, Prof. Barry Evans, Prof. Peter Grant, Dr Mike Barnard, Prof. Joseph McGeehan, Dr Steve McLaughlin and many other valued colleagues. Our sincere thanks are also due to the EPSRC, UK for supporting our research. We would also like to thank Dr Joao Da Silva, Dr Jorge Pereira, Dr Bartholome Arroyo, Dr Bernard Barani, Dr Demosthenes Ikonomou, Dr Fabrizio Sestini and other valued colleagues from the Commission of the European Communities, Brussels, Belgium, as well as Andy Aftelak, Mike Philips, Andy Wilton, Luis Lopes and Paul Crichton from Motorola ECID, Swindon, UK, for sponsoring some of our recent research. Further thanks are due to Tim Wilkinson and Ian Johnson at HP in Bristol, UK for funding some of our research efforts.

We feel particularly indebted to Rita Hanzo as well as Denise Harvey for their skilful assistance in typesetting the manuscript in LaTeX. Without the kind support of Mark Hammond, Sarah Hinton, Zöe Pinnock and their colleagues at the Wiley editorial office in Chichester, UK this monograph would never have materialised. Finally, our sincere gratitude is due to the numerous authors listed in the Author Index — as well as to those whose work was not cited due to space limitations — for their contributions to the state of the art, without whom this book would not have materialised.

Lajos Hanzo, Tong Hooi Liew and Bee Leong Yeap
Department of Electronics and Computer Science
University of Southampton

Contributors of the book:

Chapter 4
J.P. Woddard, L. Hanzo

Chapter 5
M. Breiling, L. Hanzo

Chapter 7
T.H. Liew, L.L. Yang, L. Hanzo

Chapter 8
S.X. Ng, L. Hanzo

Chapter 1

Historical Perspective, Motivation and Outline

1.1 A Historical Perspective on Channel Coding

The history of channel coding or Forward Error Correction (FEC) coding dates back to Shannon's pioneering work [1] in 1948, predicting that arbitrarily reliable communications are achievable with the aid of channel coding, upon adding redundant information to the transmitted messages. However, Shannon refrained from proposing explicit channel coding schemes for practical implementations. Furthermore, although the amount of redundancy added increases as the associated information delay increases, he did not specify the maximum delay that may have to be tolerated, in order to be able to communicate near the Shannonian limit. In recent years researchers have been endeavouring to reduce the amount of latency inflicted for example by a turbo codec's interleaver that has to be tolerated for the sake of attaining a given target performance.

Historically, one of the first practical FEC codes was the single error correcting Hamming code [2], which was a block code proposed in 1950. Convolutional FEC codes date back to 1955 [3], which were discovered by Elias, while Wozencraft and Reiffen [4,5], as well as Fano [6] and Massey [7], proposed various algorithms for their decoding. A major milestone in the history of convolutional error correction coding was the invention of a maximum likelihood sequence estimation algorithm by Viterbi [8] in 1967. A classic interpretation of the Viterbi Algorithm (VA) can be found, for example, in Forney's often-quoted paper [9]. One of the first practical applications of convolutional codes was proposed by Heller and Jacobs [10] during the 1970s.

We note here that the VA does not result in minimum Bit Error Rate (BER), rather it finds the most likely sequence of transmitted bits. However, it performs close to the minimum possible BER, which can be achieved only with the aid of an extremely complex full-search algorithm evaluating the probability of all possible 2^n binary strings of a k-bit message. The minimum BER decoding algorithm was proposed in 1974 by Bahl *et al.* [11], which was termed the Maximum A-Posteriori (MAP) algorithm. Although the MAP algorithm slightly outperforms the VA in BER terms, because of its significantly higher complexity it was rarely used in practice, until turbo codes were contrived by Berrou *et al.* in 1993 [12,13].

Focusing our attention on block codes, the single error correcting Hamming block code was too weak for practical applications. An important practical milestone was the discovery

1

of the family of multiple error correcting Bose–Chaudhuri–Hocquenghem (BCH) binary block codes [14] in 1959 and in 1960 [15, 16]. In 1960, Peterson [17] recognised that these codes exhibit a cyclic structure, implying that all cyclically shifted versions of a legitimate codeword are also legitimate codewords. The first method for constructing trellises for linear block codes was proposed by Wolf [18] in 1978. Owing to the associated high complexity, there was only limited research in trellis decoding of linear block codes [19, 20]. It was in 1988, when Forney [21] showed that some block codes have relatively simple trellis structures. Motivated by Forney's work, Honary, Markarian and Farrell *et al.* [19, 22–25] as well as Lin and Kasami *et al.* [20, 26, 27] proposed various methods for reducing the associated complexity. The Chase algorithm [28] is one of the most popular techniques proposed for near maximum likelihood decoding of block codes.

Furthermore, in 1961 Gorenstein and Zierler [29] extended the binary coding theory to treat non-binary codes as well, where code symbols were constituted by a number of bits, and this led to the birth of burst-error correcting codes. They also contrived a combination of algorithms, which is referred to as the Peterson–Gorenstein–Zierler (PGZ) algorithm. In 1960 a prominent non-binary subset of BCH codes was discovered by Reed and Solomon [30]; they were named Reed–Solomon (RS) codes after their inventors. These codes exhibit certain optimality properties, since their codewords have the highest possible minimum distance between the legitimate codewords for a given code rate. This, however, does not necessarily guarantee attaining the lowest possible BER. The PGZ decoder can also be invoked for decoding non-binary RS codes. A range of powerful decoding algorithms for RS codes was found by Berlekamp [31, 32] and Massey [33, 34]. Various soft-decision decoding algorithms were proposed for the soft decoding of RS codes by Sweeney [35–37] and Honary [19]. In recent years RS codes have found practical applications, for example, in Compact Disc (CD) players, in deep-space scenarios [38], and in the family of Digital Video Broadcasting (DVB) schemes [39], which were standardised by the European Telecommunications Standardisation Institute (ETSI).

Inspired by the ancient theory of Residue Number Systems (RNS) [40–42], which constitute a promising number system for supporting fast arithmetic operations [40, 41], a novel class of non-binary codes referred to as Redundant Residue Number System (RRNS) codes were introduced in 1967. An RRNS code is a maximum–minimum distance block code, exhibiting similar distance properties to RS codes. Watson and Hastings [42] as well as Krishna *et al.* [43, 44] exploited the properties of the RRNS for detecting or correcting a single error and also for detecting multiple errors. Recently, the soft decoding of RRNS codes was proposed in [45].

During the early 1970s, FEC codes were incorporated in various deep-space and satellite communications systems, and in the 1980s they also became common in virtually all cellular mobile radio systems. However, for a long time FEC codes and modulation have been treated as distinct subjects in communication systems. By integrating FEC and modulation, in 1987 Ungerboeck [46–48] proposed Trellis Coded Modulation (TCM), which is capable of achieving significant coding gains over power and band-limited transmission media. A further historic breakthrough was the invention of turbo codes by Berrou, Glavieux, and Thitimajshima [12, 13] in 1993, which facilitate the operation of communications systems near the Shannonian limits. Turbo coding is based on a composite codec constituted by two parallel concatenated codecs. Since its recent invention turbo coding has evolved at an unprecedented rate and has reached a state of maturity within just a few years due to the intensive research efforts of the turbo coding community. As a result of this dramatic evolution, turbo coding has also found its way into standardised systems, such as for example the recently ratified third-generation (3G) mobile radio systems [49]. Even more impressive performance gains can be attained with the aid of turbo coding in the context of video broadcast systems, where the associated system delay is less critical than in delay-sensitive interactive systems.

More specifically, in their proposed scheme Berrou *et al.* [12, 13] used a parallel concatenation of two Recursive Systematic Convolutional (RSC) codes, accommodating the turbo interleaver between the two encoders. At the decoder an iterative structure using a modified version of the classic minimum BER MAP invented by Bahl *et al.* [11] was invoked by Berrou *et al.*, in order to decode these parallel concatenated codes. Again, since 1993 a large amount of work has been carried out in the area, aiming for example to reduce the associated decoder complexity. Practical reduced-complexity decoders are for example the Max-Log-MAP algorithm proposed by Koch and Baier [50], as well as by Erfanian *et al.* [51], the Log-MAP algorithm suggested by Robertson, Villebrun and Hoeher [52], and the SOVA advocated by Hagenauer as well as Hoeher [53,54]. Le Goff, Glavieux and Berrou [55], Wachsmann and Huber [56] as well as Robertson and Worz [57] suggested the use of these codes in conjunction with bandwidth-efficient modulation schemes. Further advances in understanding the excellent performance of the codes are due, for example, to Benedetto and Montorsi [58, 59] and Perez, Seghers and Costello [60]. During the mid-1990s Hagenauer, Offer and Papke [61], as well as Pyndiah [62], extended the turbo concept to parallel concatenated block codes as well. Nickl *et al.* show in [63] that Shannon's limit can be approached within 0.27 dB by employing a simple turbo Hamming code. In [64] Acikel and Ryan proposed an efficient procedure for designing the puncturing patterns for high-rate turbo convolutional codes. Jung and Nasshan [65, 66] characterised the achievable turbo-coded performance under the constraints of short transmission frame lengths, which is characteristic of interactive speech systems. In collaboration with Blanz they also applied turbo codes to a CDMA system using joint detection and antenna diversity [67]. Barbulescu and Pietrobon addressed the issues of interleaver design [68]. The tutorial paper by Sklar [69] is also highly recommended as background reading.

Driven by the urge to support high data rates for a wide range of bearer services, Tarokh, Seshadri and Calderbank [70] proposed space-time trellis codes in 1998. By jointly designing the FEC, modulation, transmit diversity and optional receive diversity scheme, they increased the throughput of band-limited wireless channels. A few months later, Alamouti [71] invented a low-complexity space-time block code, which offers significantly lower complexity at the cost of a slight performance degradation. Alamouti's invention motivated Tarokh *et al.* [72, 73] to generalise Alamouti's scheme to an arbitrary number of transmitter antennas. Then, Tarokh *et al.*, Bauch *et al.* [74, 75], Agrawal *et al.* [76], Li *et al.* [77, 78] and Naguib *et al.* [79] extended the research of space-time codes from considering narrowband channels to dispersive channels [70, 71, 73, 79, 80].

In Figure 1.1, we show the evolution of channel coding research over the past 50 years since Shannon's legendary contribution [1]. These milestones have been incorporated also in the range of monographs and textbooks summarised in Figure 1.2. At the time of writing, the Shannon limit has been approached within 0.27 dB [63] over Gaussian channels. Also at the time of writing the challenge is to contrive FEC schemes which are capable of achieving a performance near the *capacity of wireless channels*.

1.2 Motivation of the Book

The design of an attractive channel coding and modulation scheme depends on a range of contradictory factors, which are portrayed in Figure 1.3. The message of this illustration is multi-fold. For example, given a certain transmission channel, it is always feasible to design a coding and modulation ('codulation') system, which can further reduce the BER achieved. This typically implies, however, further investments and/or penalties in terms of the required increased imple-

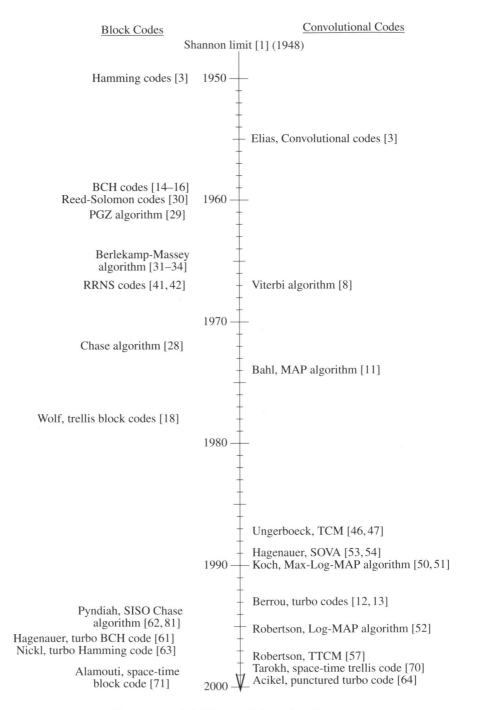

Figure 1.1: A brief history of channel coding.

mentational complexity and coding/interleaving delay as well as reduced effective throughput. Different solutions accrue when optimising different codec features. For example, in many applications the most important codec parameter is the achievable coding gain, which quantifies the amount of bit-energy reduction attained by a codec at a certain target BER. Naturally, transmitted power reduction is extremely important in battery-powered devices. This transmitted power reduction is only achievable at the cost of an increased implementational complexity, which itself typically increases the power consumption and hence erodes some of the power gain.

Viewing this system optimisation problem from a different perspective, it is feasible to transmit at a higher bit rate in a given fixed bandwidth by increasing the number of bits per modulated symbol. However, when aiming for a given target BER, the channel coding rate has to be reduced, in order to increase the transmission integrity. Naturally, this reduces the *effective throughput* of the system and results in an overall increased system complexity. When the channel's characteristic and the associated bit error statistics change, different solutions may become more attractive. This is because Gaussian channels, narrowband and wideband Rayleigh fading or various Nakagami fading channels inflict different impairments. These design trade-offs constitute the subject of this monograph.

Our intention with the book is multi-fold:

1) *First, we would like to pay tribute to all researchers, colleagues and valued friends who contributed to the field.* Hence this book is dedicated to them, since without their quest for better coding solutions to communications problems this monograph could not have been conceived. They are too numerous to name here, hence they appear in the author index of the book.

2) The invention of turbo coding not only assisted in attaining a performance approaching the Shannonian limits of channel coding for transmissions over **Gaussian channels**, but also revitalised channel coding research. In other words, turbo coding opened a new chapter in the design of iterative detection-assisted communications systems, such as turbo trellis coding schemes, turbo channel equalisers, etc. Similarly dramatic advances have been attained with the advent of space-time coding, when communicating over dispersive, fading **wireless channels**. *Recent trends indicate that better overall system performance may be attained by jointly optimising a number of system components, such as channel coding, channel equalisation, transmit and received diversity and the modulation scheme, than in case of individually optimising the system components. This is the main objective of this monograph.*

3) *Since at the time of writing no joint treatment of the subjects covered by this book exists, it is timely to compile the most recent advances in the field. Hence it is our hope that the conception of this monograph on the topic will present an adequate portrayal of the last decade of research and spur this innovation process by stimulating further research in the coding and communications community.*

1.3 Organisation of the Book

Below, we present the outline and rationale of the book:

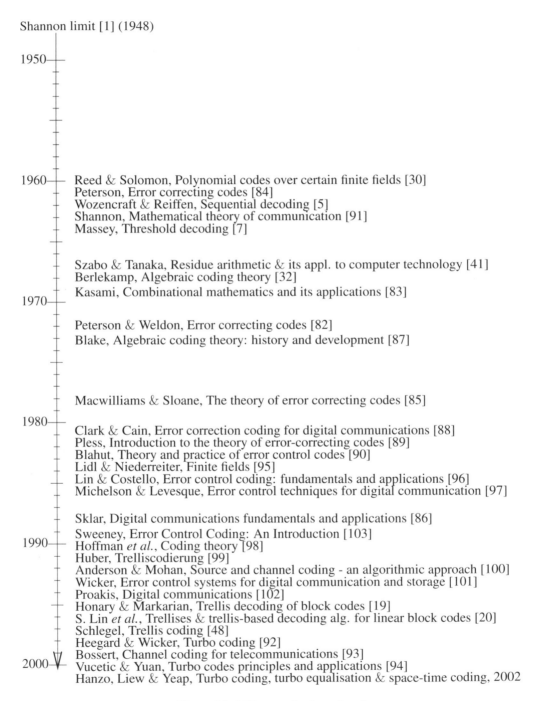

Figure 1.2: Milestones in channel coding.

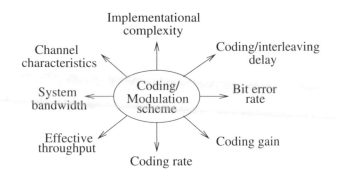

Figure 1.3: Factors affecting the design of channel coding and modulation scheme.

- **Chapter 2:** For the sake of completeness and wider reader appeal virtually no prior knowledge is assumed in the field of channel coding. Hence in Chapter 2 we commence our discourse by introducing the family of convolutional codes and the hard- as well as soft-decision Viterbi algorithm in simple conceptual terms with the aid of worked examples.

- **Chapter 3:** This chapter provides a rudimentary introduction to the most prominent classes of block codes, namely to Reed–Solomon (RS) and Bose–Chaudhuri–Hocquenghem (BCH) codes. A range of algebraic decoding techiques are also reviewed and worked examples are included.

- **Chapter 4:** Based on the simple Viterbi decoding concepts introduced in Chapter 2, in this chapter an overview of the family of conventional binary BCH codes is given, with special emphasis on their trellis decoding. In parallel to our elaborations in Chapter 2 on the context of convolutional codes, the Viterbi decoding of binary BCH codes is detailed with the aid of worked examples. These discussions are followed by the simulation-based performance characterisation of various BCH codes employing both hard-decision and soft-decision decoding methods. The classic Chase algorithm is introduced and its performance is investigated.

- **Chapter 5:** This chapter introduces the concept of turbo convolutional codes and gives a detailed discourse on the Maximum A-Posteriori (MAP) algorithm and its computationally less demanding counterparts, namely the Log-MAP and Max-Log-MAP algorithms. The Soft-Output Viterbi Algorithm (SOVA) is also highlighted and its concept is augmented with the aid of a detailed worked example. Then the effects of the various turbo codec parameters are investigated, namely that of the number of iterations, the puncturing patterns used, the component decoders, the influence of the interlever depth, which is related to the codeword length, etc. The various codecs' performance is studied also when communicating over Rayleigh fading channels.

- **Chapter 6:** While in Chapter 5 we invoked iterative turbo decoders, in this chapter a super-trellis is constructed from the two constituent convolutional codes' trellises and the maximum likelihood codeword is output in a single, but implementationally coplex decoding step, without iterations. The advantage of the associated super-trellis is that it allows us to explore the trellis describing the construction of turbo codes and also to relate turbo codes to high-constraint-length convolutional codes of the same decoding complexity.

- **Chapter 7:** The concept of turbo codes using BCH codes as component codes is introduced. A detailed derivation of the MAP algorithm is given, building on the concepts introduced in Chapter 5 in the context of convolutional turbo codes, but this time cast in the framework of turbo BCH codes. Then, the MAP algorithm is modified in order to highlight the concept of the Max-Log-MAP and Log-MAP algorithms, again, with reference to binary turbo BCH codes. Furthermore, the SOVA-based binary BCH decoding algorithm is introduced. Then a simple turbo decoding example is given, highlighting how iterative decoding assists in correcting multiple errors. We also describe a novel MAP algorithm for decoding extended BCH codes. Finally, we show the effects of the various coding parameters on the performance of turbo BCH codes.

- **Chapter 8:** The concept of Residue Number Systems (RNS) is introduced and extended to Redundant Residue Number Systems (RRNS), introducing the family of RRNS codes. Some coding-theoretic aspects of RRNS codes is investigated, demonstrating that RRNS codes exhibit similar distance properties to RS codes. A procedure for multiple-error correction is then given. Different bit-to-symbol mapping methods are highlighted, yielding non-systematic and systematic RRNS codes. A novel bit-to-symbol mapping method is introduced, which results in efficient systematic RRNS codes. The classic Chase algorithm is then modified in order to create a Soft-Input Soft-Output (SISO) RRNS decoder. This enables us to implement the iterative decoding of turbo RRNS codes. Finally, simulation results are given for various RRNS codes, employing hard-decision and soft-decision decoding methods. The performance of the RRNS codes is compared to that of RS codes and the performance of turbo RRNS codes is studied.

- **Chapter 9:** Our previous discussions on various channel coding schemes evolves to the family of joint coding and modulation-based arrangements, which are often referred to as coded modulation schemes. Specifically, Trellis-Coded Modulation (TCM), Turbo Trellis-Coded Modulation (TTCM), Bit-Interleaved Coded Modulation (BICM) as well as iterative joint decoding and demodulation-assisted BICM (BICM-ID) will be studied and compared under various narrowband and wideband propagation conditions.

- **Chapter 10:** Space-time block codes are introduced. The derivation of the MAP decoding of space-time block codes is then given. A system is proposed by concatenating space-time block codes and various channel codes. The complexity and memory requirements of various channel decoders are derived, enabling us to compare the performance of the proposed channel codes by considering their decoder complexity. Our simulation results related to space-time block codes using no channel coding are presented first. Then, we investigate the effect of mapping data and parity bits from binary channel codes to non-binary modulation schemes. Finally, we compare our simulation results for various channel codes concatenated with a simple space-time block code. Our performance comparisons are conducted by also considering the complexity of the associated channel decoder.

- **Chapter 11:** The encoding process of space-time trellis codes is highlighted. This is followed by employing an Orthogonal Frequency Division Multiplexing (OFDM) modem in conjunction with space-time codes over wideband channels. Turbo codes and RS codes are concatenated with space-time codes in order to improve their performance. Then, the performance of the advocated space-time block code and space-time trellis codes is compared. Their complexity is also considered in comparing both schemes. The effect of delay spread and maximum Doppler frequency on the performance of the space-time codes is investigated. A Signal to Interference Ratio (SIR) related term is defined in the context of dispersive channels for the advocated space-time block code, and we will show how the SIR affects the performance of the system. In our last section, we propose space-time-coded Adaptive OFDM (AOFDM). We then show by employing multiple antennas that with the advent of space-time coding, the wideband fading channels have been converted to AWGN-like channels.

- **Chapter 12:** The discussions of Chapters 10 and 11 were centred around the topic of employing multiple-transmitter, multiple-receiver (MIMO) based transmit and receive-diversity assisted space-time coding schemes. These arrangements have the potential of significantly mitigating the hostile fading wireless channel's near-instantaneous channel quality fluctuations. Hence these space-time codecs can be advantageously combined with powerful channel codecs originally designed for Gaussian channels. As a lower-complexity design alternative, this chapter introduces the concept of near-instantaneously Adaptive Quadrature Amplitude Modulation (AQAM), combined with near-instantaneously adaptive turbo channel coding. These adaptive schemes are capable of mitigating the wireless channel's quality fluctuations by near-instantaneously adapting both the modulation mode used as well as the coding rate of the channel codec invoked. The design and performance study of these novel schemes constitutes the topic of Chapter 12.

- **Chapter 13:** This chapter focuses on the portrayal of partial-response modulation schemes, which exhibit impressive performance gains in the context of joint iterative, joint channel equalisation and channel decoding. This joint iterative receiver principle is termed turbo equalisation. An overview of Soft-In/Soft-Out (SISO) algorithms, namely that of the MAP algorithm and Log-MAP algorithm, is presented in the context of GMSK channel equalisation, since these algorithms are used in the investigated joint channel equaliser and turbo decoder scheme.

- **Chapter 14:** Based on the introductory concepts of Chapter 13, in this chapter the detailed principles of iterative joint channel equalisation and channel decoding techniques known as turbo equalisation are introduced. This technique is invoked in order to overcome the unintentional Inter-Symbol Interference (ISI) and Controlled Inter-Symbol Interference (CISI) introduced by the channel and the modulator, respectively. Modifications of the SISO algorithms employed in the equaliser and decoder are also portrayed, in order to generate information related not only to the source bits but also to the parity bits of the codewords. The performance of coded systems employing turbo equalisation is analysed. Three classes of encoders are utilised, namely convolutional codes, convolutional-coding-based turbo codes and BCH-coding-based turbo codes.

- **Chapter 15:** Theoretical models are devised for the coded schemes in order to derive the maximum likelihood bound of the system. These models are based on the Serial Concatenated Convolutional Code (SCCC) analysis presented in reference [104]. Essentially, this analysis can be employed since the modulator could be represented accurately as a rate $R = 1$ convolutional encoder. Apart from convolutional-coded systems, turbo-coded schemes are also considered. Therefore the theoretical concept of Parallel Concatenated Convolutional Codes (PCCC) [59] is utilised in conjunction with the SCCC principles in order to determine the Maximum Likelihood (ML) bound of the turbo-coded systems, which are modelled as hybrid codes consisting of a parallel concatenated convolutional code, serially linked with another convolutional code. An abstract interleaver from reference [59] — termed the uniform interleaver — is also utilised, in order to reduce the complexity associated with determining all the possible interleaver permutations.

- **Chapter 16:** A comparative study of coded BPSK systems, employing high-rate channel encoders, is presented. The objective of this study is to investigate the performance of turbo equalisers in systems employing different classes of codes for high code rates of $R = \frac{3}{4}$ and $R = \frac{5}{6}$, since known turbo equalisation results have only been presented for turbo equalisers using convolutional codes and convolutional-based turbo codes for code rates of $R = \frac{1}{3}$ and $R = \frac{1}{2}$ [105, 106]. Specifically, convolutional codes, convolutional-coding-based turbo codes, and Bose–Chaudhuri–Hocquengham (BCH)-coding-based [14, 15] turbo codes are employed in this study.

- **Chapter 17:** A novel reduced-complexity trellis-based equaliser is presented. In each turbo equalisation iteration the decoder generates information which reflects the reliability of the source and parity bits of the codeword. With successive iteration, the reliability of this information improves. This decoder information is exploited in order to decompose the received signal such that each quadrature component consists of the in-phase or quadrature-phase component signals. Therefore, the equaliser only has to consider the possible in-phase or quadrature-phase components, which is a smaller set of signals than all of their possible combinations.

- **Chapter 18:** For transmissions over wideband fading channels and fast fading channels, space-time trellis coding (STTC) is a more appropriate diversity technique than space-time block coding. STTC [70] relies on the joint design of channel coding, modulation, transmit diversity and the optional receiver diversity schemes. The decoding operation is performed by using a maximum likelihood detector. This is an effective scheme, since it combines the benefits of Forward Error Correction (FEC) coding and transmit diversity, in order to obtain performance gains. However, the cost of this is the additional computational complexity, which increases as a function of bandwidth efficiency (bits/s/Hz) and the required diversity order. In this chapter STTC is investigated for transmission over wideband fading channels.

- **Chapter 19:** This chapter provides a brief summary of the book.

It is our hope that this book portrays the range of contradictory system design trade-offs associated with the conception of channel coding arrangements in an unbiased fashion and that readers will be able to glean information from it in order to solve their own particular channel coding and communications problem. Most of all, however, we hope that they will find it an enjoyable and informative read, providing them with intellectual stimulation.

Lajos Hanzo, Tong Hooi Liew and Bee Leong Yeap
Department of Electronics and Computer Science
University of Southampton

Part I

Convolutional and Block Coding

Chapter 2

Convolutional Channel Coding

2.1 Brief Channel Coding History

In this chapter a rudimentary introduction to convolutional coding is offered to those readers who are not familiar with the associated concepts. Readers who are familiar with the basic concepts of convolutional coding may proceed to the chapter of their immediate interest.

The history of channel coding or Forward Error Correction (FEC) coding dates back to Shannon's pioneering work in which he predicted that arbitrarily reliable communications are achievable by redundant FEC coding, although he refrained from proposing explicit schemes for practical implementations. Historically, one of the first practical codes was the single error correcting Hamming code [2], which was a block code proposed in 1950. Convolutional FEC codes date back to 1955 [3], which were discovered by Elias, whereas Wozencraft and Reiffen [4,5], as well as Fano [6] and Massey [7], proposed various algorithms for their decoding. A major milestone in the history of convolutional error correction coding was the invention of a maximum likelihood sequence estimation algorithm by Viterbi [8] in 1967. A classic interpretation of the Viterbi Algorithm (VA) can be found, for example, in Forney's often-quoted paper [9], and one of the first applications was proposed by Heller and Jacobs [107].

We note, however, that the VA does not result in minimum Bit Error Rate (BER). The minimum BER decoding algorithm was proposed in 1974 by Bahl *et al.* [11], which was termed the Maximum A-Posteriori (MAP) algorithm. Although the MAP algorithm slightly outperforms the VA in BER terms, because of its significantly higher complexity it was rarely used, until turbo codes were contrived [12].

During the early 1970s, FEC codes were incorporated in various deep-space and satellite communications systems, and in the 1980s they also became common in virtually all cellular mobile radio systems. A further historic breakthrough was the invention of the turbo codes by Berrou, Glavieux, and Thitimajshima [12] in 1993, which facilitates the operation of communications systems near the Shannonian limits.

Focusing our attention on block codes, the single error correcting Hamming block code was too weak, however, for practical applications. An important practical milestone was the discovery of the family of multiple error correcting Bose–Chaudhuri–Hocquenghem (BCH) binary block codes [14] in 1959 and in 1960 [15, 16]. In 1960, Peterson [17] recognised that these codes exhibit a cyclic structure, implying that all cyclically shifted versions of a legitimate codeword are also legitimate codewords. Furthermore, in 1961 Gorenstein and Zierler [29] extended the binary coding theory to treat non-binary codes as well, where code symbols were constituted by a number of bits, and this led to the birth of burst-error correcting codes. They also contrived a combination of algorithms, which are referred to as the Peterson–Gorenstein–Zierler (PGZ) algorithm. We will elaborate on this algorithm later in this chapter. In 1960 a prominent non-binary subset of BCH codes were discovered by Reed and Solomon [30]; they were named Reed–Solomon (RS) codes after their inventors. These codes exhibit certain optimality properties, and they will also be treated in more depth in this chapter. We will show that the PGZ decoder can also be invoked for decoding non-binary RS codes.

A range of powerful decoding algorithms for RS codes was found by Berlekamp [31,32] and Massey [33,34], which also constitutes the subject of this chapter. In recent years, these codes have found practical applications, for example, in Compact Disc (CD) players, in deep-space scenarios [38], and in the family of Digital Video Broadcasting (DVB) schemes, which were standardised by the European Telecommunications Standardization Institute (ETSI). We now consider the conceptually less complex class of convolutional codes, which will be followed by our discussions on block coding.

2.2 Convolutional Encoding

Both block codes and Convolutional Codes (CCs) can be classified as systematic or non-systematic codes, where the terminology suggests that in systematic codes the original information bits or symbols constitute part of the encoded codeword and hence they can be recognised explicitly at the output of the encoder. Their encoders can typically be implemented by the help of linear shift-register circuitries, an example of which can be seen in Figure 2.1. The figure will be explored in more depth after introducing some of the basic convolutional coding parameters.

Specifically, in general a k-bit information symbol is entered into the encoder, constituted by K shift-register stages. In our example of Figure 2.1, the corresponding two shift-register stages are s_1 and s_2. In general, the number of shift-register stages K is referred to as the constraint length of the code. An alternative terminology is to refer to this code as a memory three code, implying that the memory of the CC is given by $K + 1$. The current shift-register state s_1, s_2 plus the incoming bit b_i determine the next state of this state machine. The number of output bits is typically denoted by n, while the coding rate by $R = k/n$, implying that $R \leq 1$. In order to fully specify the code, we also have to stipulate the generator polynomial, which describes the topology of the modulo-2 gates generating the output bits of the convolutional encoder. For generating n bits, n generator polynomials are necessary. In general, a CC is denoted as a $CC(n, k, K)$ scheme, and given the n generator polynomials, the code is fully specified.

Once a specific bit enters the encoder's shift register in Figure 2.1, it has to traverse through the register, and hence the register's sequence of state transitions is not arbitrary. Furthermore, the modulo-2 gates impose additional constraints concerning the output bit-stream. Because of these constraints, the legitimate transmitted sequences are restricted to certain bit patterns, and if there are transmission errors, the decoder will conclude that such an encoded sequence could not have been generated by the encoder and that it must be due to channel errors. In this case, the decoder will attempt to choose the most resemblent legitimately encoded sequence and output the corresponding bit-stream as the decoded string. These processes will be elaborated on in more detail later in the chapter.

The n generator polynomials $g_1, g_2, \ldots, g_n$ are described by the specific connections to the register stages. Upon clocking the shift register, a new information bit is inserted in the register, while the bits constituting the states of this state machine move to the next register stage and the last bit is shifted out of the register. The generator polynomials are constituted by a binary

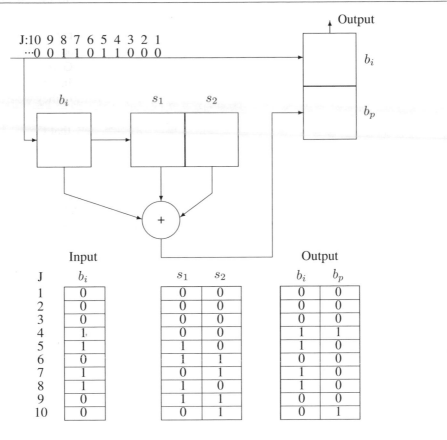

Figure 2.1: Systematic half-rate, constraint-length two convolutional encoder $CC(2,1,2)$.

pattern, indicating the presence or absence of a specific link from a shift register stage by a binary one or zero, respectively. For example, in Figure 2.1 we observe that the generator polynomials are constituted by:

$$g_1 = [1\,0\,0] \quad \text{and} \quad g_2 = [1\,1\,1]\,, \tag{2.1}$$

or, in an equivalent polynomial representation, as:

$$g_1(z) = 1 + 0 \cdot z^1 + 0 \cdot z^2 \quad \text{and} \quad g_2(z) = 1 + z + z^2\,. \tag{2.2}$$

We note that in a non-systematic CC, g_1 would also have more than one non-zero term. It is intuitively expected that the more constraints are imposed by the encoder, the more powerful the code becomes, facilitating the correction of a higher number of bits, which renders non-systematic CCs typically more powerful than their systematic counterparts.

Again, in a simple approach, we will demonstrate the encoding and decoding principles in the context of the systematic code specified as $(k = 1)$, half-rate $(R = k/n = 1/2)$, $CC(2, 1, 2)$, with a memory of three binary stages $(K = 2)$. These concepts can then be extended to arbitrary codecs. At the commencement of the encoding, the shift register is typically cleared by setting it to the all-zero state, before the information bits are input to it. Figure 2.1 demonstrates the encoder's operation for the duration of the first ten clock cycles, tabulating the input bits, the

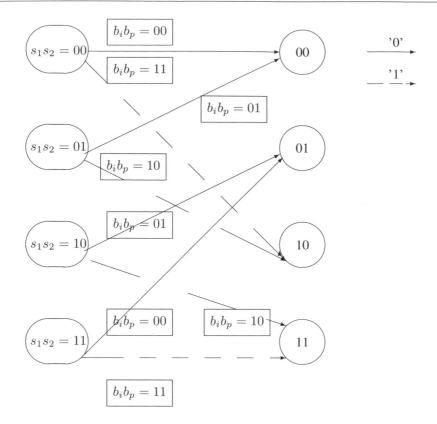

Figure 2.2: State transition diagram of the $CC(2, 1, 2)$ systematic code, where broken lines indicate transitions due to an input one, while continuous lines correspond to input zeros.

shift-register states s_1, s_2, and the corresponding output bits. At this stage, the uninitiated reader is requested to follow the operations summarised in the figure before proceeding to the next stage of operations.

2.3 State and Trellis Transitions

An often-used technique for characterising the operations of a state machine, such as our convolutional encoder, is to refer to the state transition diagram of Figure 2.2. Given that there are two bits in the shift register at any moment, there are four possible states in the state machine and the state transitions are governed by the incoming bit b_i. A state transition due to a logical zero is indicated by a continuous line in the figure, while a transition activated by a logical one is represented by a broken line. The inherent constraints imposed by the encoder manifest themselves here in that from any state there are only two legitimate state transitions, depending on the binary input bit. Similarly, in each state there are two merging paths. It is readily seen from the encoder circuit of Figure 2.1 that, for example, from state (s_1, s_2)=(1,1) a logical one input results in a transition to (1,1), while an input zero leads to state (0,1). The remaining transitions can also be readily checked by the reader. A further feature of this figure is that the associated encoded output bits are also plotted in the boxes associated with each of the transitions. Hence,

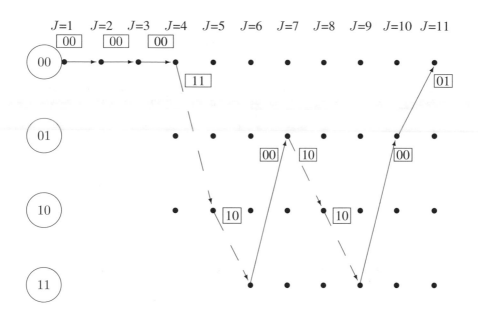

Figure 2.3: Trellis-diagram of the $CC(2, 1, 2)$ systematic code, where broken lines indicate transitions due to a binary one, while continuous lines correspond to input zeros.

this diagram fully describes the operations of the encoder.

Another simple way of characterising the encoder is to portray its trellis diagram, which is depicted in Figure 2.3. At the left of the figure, the four legitimate encoder states are portrayed. Commencing operations from the all-zero register state (0,0) allows us to mirror the encoder's actions seen in Figures 2.1 and 2.2 also in the trellis diagram, using the same input bit-stream. As before, the state transitions are governed by the incoming bits b_i and a state transition due to a logical zero is indicated by a continuous line, while a transition activated by a logical one is represented by a broken line.

Again, the inherent constraints imposed by the encoder manifest themselves here in that from any state there are only two legitimate state transitions, depending on the binary input bit, and in each state there are two merging paths. Given our specific input bit-stream, it is readily seen from the encoder circuit of Figure 2.1 and the state transition diagram of Figure 2.2 that, for example, from state (s_1, s_2)=(0,0) a logical zero input bit results in a transition to (0,0), while an input one leads to state (1,0). The remaining transitions shown in the figure are associated with our specific input bit-stream, which can be readily explored by the reader. As before, the associated output bits are indicated in the boxes along each of the transitions. Hence, the trellis diagram gives a similarly unambiguous description of the encoder's operations to the state diagram of Figure 2.2. Armed with the above description of CCs, we are now ready to give an informal description of the maximum likelihood Viterbi algorithm in the next section.

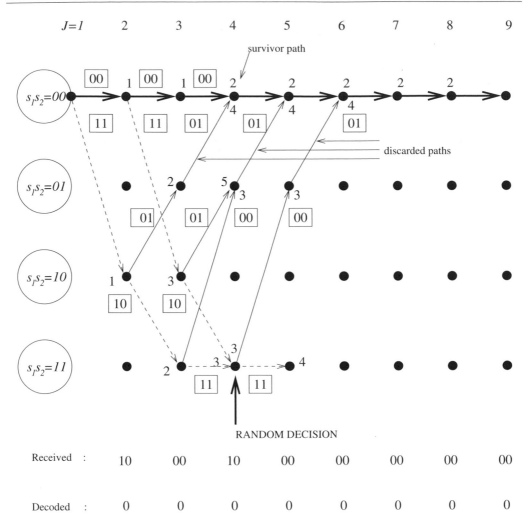

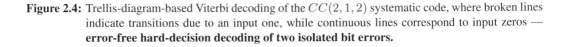

Figure 2.4: Trellis-diagram-based Viterbi decoding of the $CC(2, 1, 2)$ systematic code, where broken lines indicate transitions due to an input one, while continuous lines correspond to input zeros — **error-free hard-decision decoding of two isolated bit errors.**

2.4 The Viterbi Algorithm

2.4.1 Error-free Hard-decision Viterbi Decoding

Given the received bit-stream, the decoder has to arrive at the best possible estimate of the original uncoded information sequence. Hence, the previously mentioned constraints imposed by the encoder on the legitimate bit sequences have to be exploited in order to eradicate illegitimate sequences and thereby remove the transmission errors. For the sake of computational simplicity, let us assume that the all-zero bit-stream has been transmitted and the received sequence of Figure 2.4

has been detected by the demodulator, which has been passed to the FEC decoder. We note here that if the demodulator carries out a binary decision concerning the received bit, this operation is referred to as *hard-decision demodulation*. By contrast, if the demodulator refrains from making a binary decision and instead it outputs a more finely graded multilevel confidence measure concerning the probability of a binary one and a binary zero, then it is said to invoke *soft-decision demodulation*.

For the time being we will consider only hard-decision demodulation. The decoder now has to compare the received bits with all the legitimate sequences of the trellis diagram of Figure 2.4 and quantify the probability of each of the associated paths, which ultimately assigns a probability-related quantity to specific decoded sequences, as we will show below.

Referring to Figure 2.4 and beginning the decoding operations from the all-zero state, we compute the Hamming distance of the received two-bit symbol with respect to both of the legitimate encoded sequences of the trellis diagram for the first trellis section (i.e. for the first trellis transition), noting that the Hamming distance is given by the number of different bit positions between two binary sequences. For example, for the first two-bit received symbol 10, the associated Hamming distances are 1 with respect to both the 00 and the 11 encoded sequences. Thus, at this stage the decoder is unable to express any preference as to whether 00 or 11 was the more likely transmitted symbol. We also note these Hamming distances in the trellis diagram of Figure 2.4, indicated at the top of the nodes corresponding to the new encoder states we arrived at, as a consequence of the state transitions due to a logical one and zero, respectively. These Hamming distances are known in the context of Viterbi decoding as the *branch metric*. The power of the Viterbi decoding algorithm accrues from the fact that it carries out a maximum likelihood sequence estimation, as opposed to arriving at symbol-by-symbol decisions, and thereby exploits the constraints imposed by the encoder on the legitimate encoded sequences. Hence, the branch metrics will be accumulated over a number of consecutive trellis stages before a decision as to the most likely encoder path and information sequence can be released.

Proceeding to the next received two-bit symbol, namely 00, the operations of the decoder are identical; that is, the Hamming distance between the encoded symbols of all four legitimate paths and the received symbol is computed. These distances yield the new branch metrics associated with the second trellis stage. By now the encoded symbols of two original input bits have been received, and this is why there are now four possible trellis states in which the decoder may reside. The branch metrics computed for these four legitimate transitions from top to bottom are 0, 2, 1 and 1, respectively. These are now added to the previous branch metrics of 1 in order to generate the *path metrics* of 1, 2, 3 and 2, respectively, quantifying the probability of each legitimate trellis path in terms of the accumulated Hamming distance. A low Hamming distance indicates a high similarity between the received sequence and the encoded sequence concerned, which is characteristic of the most likely encoded sequence, since the probability of a high number of errors is exponentially decreasing with the number of errors.

Returning to Figure 2.4 again, the corresponding accumulated Hamming distances or branch metrics from top to bottom are 1, 2, 3 and 2, respectively. At this stage, we can observe that the top branch has the lowest branch metric and hence it is the most likely encountered encoder path. The reader *knows* this, but the decoder can only *quantify the probability* of the corresponding paths and thus it cannot be sure of the validity of its decision. The other three encoder paths and their associated information bits also have a finite probability.

Continuing the legitimate paths of Figure 2.4 further, at trellis stage three the received sequence of 10 is compared to the four legitimate two-bit encoded symbols and the associated path metrics now become dependent on the actual path followed, since at this stage there are merging paths. For example, at the top node we witness the merger of the 00, 00, 00 path with the 11, 01, 01 path, where the associated original information bits were 0,0,0 and 1,0,0, respectively. On the basis of the associated path metrics, the decoder may 'suspect' that the former one was the original information sequence, but it still refrains from carrying out a decision. Considering the two merging paths, future input bits would affect both of these in an identical fashion, resulting in an indistinguishable future set of transitions. Their path metrics will therefore also evolve identically, suggesting that it is pointless to keep track of both of the merging paths, since the one with the lower metric will always remain the more likely encoder path. This is reflected in Figure 2.4 by referring to the path exhibiting the lower metric as the *survivor path*, while the higher metric merging path will be discarded.

We also note that at the bottom node of trellis stage three we ended up with two identical path metrics, namely 3, and in this case a random decision must be made as to which one becomes the survivor. This event is indicated in Figure 2.4 by the arrow. In this particular example, this decision does not affect the final outcome of the decoder's decision, since the top path appears to be the lowest metric path. Nonetheless, in some situations such random decisions will influence the decoded bit sequence and may indeed determine whether decoding errors are encountered. It is plausible that the confidence in the decoder's decision is increased, as the accumulation of the branch metrics continues. Indeed, one may argue that the 'best' decision can be taken upon receiving the complete information sequence. However, deferring decisions for so long may not be acceptable in latency terms, in particular in delay-sensitive interactive speech or video communications. Nor is it necessary in practical terms, since experience shows that the decoder's bit error rate is virtually unaffected by curtailing the decision interval to about five times the encoder's memory, which was three in our example.

In our example, the received bit sequence does not contain any more transmission errors, and so it is plausible that the *winning path* remains the one at the top of Figure 2.4 and the associated branch metric of 2 reflects the actual number of transmission errors. We are now ready to release the error-free decoded sequence, namely the all-zero sequence, as seen explicitly in terms of the corresponding binary bits at the bottom of Figure 2.4. The corresponding winning path was drawn in bold in the figure.

2.4.2   Erroneous Hard-decision Viterbi Decoding

Following the above double-error correction scenario, below we consider another instructive example where the number of transmission errors remains two and even their separation is increased. Yet the decoder may become unable to correct the errors, depending on the outcome of a random decision. This is demonstrated in Figure 2.5 at stage four of the top path. Observe furthermore that the corresponding received bits and path metrics of Figure 2.4 are also indicated in Figure 2.5, but they are crossed out and superseded by the appropriately updated values according to the current received pattern. Depending on the actual choice of the survivor path at stage four, the first decoded bit may become a logical one, as indicated at the bottom of Figure 2.5. The accumulated Hamming distance becomes 2, regardless of the random choice of the survivor path, which indicates that in the case of decoding errors the path metric is not a reliable measure of the actual number of errors encountered. This will become even more evident in our next example.

Let us now consider a scenario in which there are more than two transmission errors in the received sequence, as seen in Figure 2.6. Furthermore, the bit errors are more concentrated, forming a burst of errors, rather than remaining isolated error events. In this example, we show that the decoder becomes 'overloaded' by the plethora of errors, and hence it will opt for an erroneous trellis path, associated with the wrong decoded bits.

Observe in Figure 2.4 that up to trellis stage three the lowest-metric path is the one at the

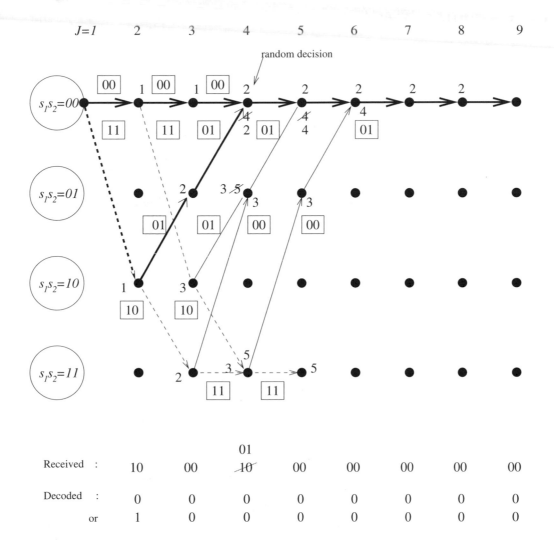

Figure 2.5: Trellis-diagram-based Viterbi decoding of the $CC(2, 1, 2)$ systematic code, where broken lines indicate transitions due to an input one, while continuous lines correspond to input zeros — **erroneous hard-decision decoding of two isolated bit errors.**

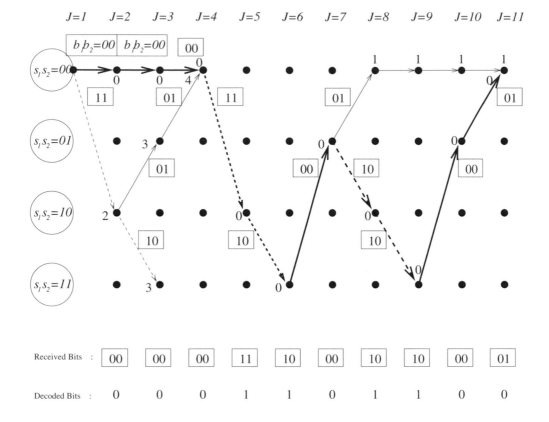

Figure 2.6: Trellis-diagram-based Viterbi decoding of the $CC(2, 1, 2)$ systematic code, where broken lines indicate transitions due to an input one, while continuous lines correspond to input zeros — **erroneous hard-decision decoding of burst errors.**

top, which is associated with the error-free all-zero sequence. However, the double error in the fourth symbol of Figure 2.6 results in a 'dramatic' turn of events, since the survivor path deviates from the error-free all-zero sequence. Because of the specific received sequence encountered, the path metric remains 0 up to trellis stage $J = 11$, the last stage, accumulating a total of zero Hamming distance, despite actually encountering a total of six transmission errors, resulting in four decoding errors at the bottom of Figure 2.6. Again, the winning path was drawn in bold in Figure 2.4.

From this experience we can infer two observations. First, the high-error rate scenario encountered is less likely than the previously considered double-error case, but it has a finite probability and hence it may be encountered in practice. Second, since the decoder carries out a maximum likelihood decision, in such cases it will opt for the wrong decoded sequence, in which case the accumulated path metric will not correctly reflect the number of transmission errors encountered. We therefore conclude that, in contrast to block codes, CCs do not possess an ability to monitor the number of transmission errors encountered.

2.4.3   Error-free Soft-decision Viterbi Decoding

Having considered a number of hard-decision decoding scenarios, let us now demonstrate the added power of soft-decision decoding. Recall from our earlier discussions that if the demodulator refrains from making a binary decision and instead it outputs a finely graded *soft-decision confidence measure* related to the probability of a binary one and a binary zero, respectively, then it is said to invoke soft-decision demodulation. As an example, we may invoke an eight-level soft-decision demodulator output. This provides a more accurate indication of whether the demodulator's decision concerning a certain demodulated bit is a high- or low-reliability one. This clearly supplies the Viterbi decoder with substantially more information than the previous binary zero/one decisions. Hence, a higher error correction capability will be achieved, as we will demonstrate in Figure 2.7.

Specifically, let us assume that on our eight-level confidence scale +4 indicates the highest possible confidence concerning the demodulator's decision for a binary one and -4 the lowest possible confidence. In fact, if the demodulator outputs -4, the low confidence in a logical one implies a high confidence in a logical zero, and conversely, the demodulator output of +4 implies a very low probability of a binary zero. Bearing this eight-level confidence scale of [-4, -3, ... +3, +4] in mind, the previously erroneously decoded double-error scenario of Figure 2.5 can now be revisited in Figure 2.7, where we will demonstrate that the more powerful soft-decision decoding allows us to remove all transmission errors.

Let us first consider the soft-decision metrics provided by the demodulator, which now replace the previously used hard-decision values at the bottom of Figure 2.5, appearing crossed out in Figure 2.7. For example, the first two values appear to be a high-confidence one and zero, respectively. The second two values are relatively high-confidence zeros, whereas the previously erroneously demodulated third symbol, namely 01, is now represented by the confidence values of -2,+1, indicating that these bits may well have been either one or zero. The rest of the soft-decision metrics appear to be of higher value, apart from the last-but-one.

The computation of the branch metrics and path metrics now has to be slightly modified. Previously, we were accumulating only 'penalties' in terms of the Hamming distances encountered. By contrast, in soft-decision decoding we will have to accumulate both penalties and credits, since we now consider the possibility of all demodulated values being both a binary one and a zero and quantify their associated probabilities using the soft-decision (SD) metrics. Explicitly, in Figure 2.7 we replace the crossed-out hard-decision metrics by the corresponding soft-decision metrics.

Considering trellis stage one and the 00 encoded symbol, the first SD metric of +3 does not tally well with the bit zero; rather, it indicates a strong probability of a one; hence, we accumulate a 'penalty' of -3. The second SD metric, however, indicates a strong probability of a zero, earning a credit of +3, which cancels the previous -3 penalty, yielding a SD branch metric of 0. Similar arguments are valid for the trellis branch from (0,0) to (1,0), which is associated with the encoded symbol 11, also yielding a SD branch metric of 0. During stage two, the received SD metrics of -2,-3 suggest a high probability of two zeros, earning an added credit of +5 along the branch (0,0) to (0,0). By contrast, these SD values do not tally well with the encoded symbol of 11 along the transition of (0,0) to (1,0), yielding a penalty of -5. During the third stage along the path from (0,0) to (0,0), we encounter a penalty of -1 and a credit of +2, bringing the total credits for the all-zero path to +6.

At this stage of the hard-decision decoder of Figure 2.7, we encountered a random decision, which now will not be necessary, since the merging path has a lower credit of +2. Clearly, at trellis stage nine we have a total credit of +29, allowing the decoder to release the correct original all-zero information sequence.

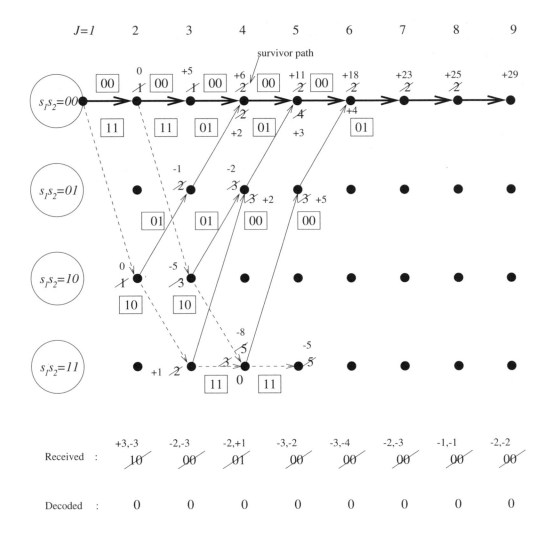

Figure 2.7: Trellis-diagram-based Viterbi decoding of the $CC(2, 1, 2)$ systematic code, where broken lines indicate transitions due to an input one, while continuous lines correspond to input zeros — **error-free soft-decision decoding of two isolated bit errors.**

2.5 Summary and Conclusions

This brief chapter commenced by providing a historical perspective on convolution coding. The convolutional encoder has been characterised with the aid of its state transition diagram and trellis diagram. Then the classic Viterbi algorithm has been introduced in the context of a simple decoding example, considering both hard- and soft-decision-based scenarios. In the next chapter, we focus our attention on the family of block codes, which have found numerous applications in both standard and proprietary wireless communications systems. A variety of video schemes characterised in this monograph have also opted for employing Bose–Chaudhuri–Hocquenghem codes.

Chapter 3

Block Coding

3.1 Introduction

Having studied the basics of convolutional coding in the previous chapter, we now focus our attention on methods of combatting transmission errors with the aid of block codes. The codes discussed in this chapter have found favour in numerous standard communications systems and hence they are historically important. While a basic understanding of the construction of these codes is important for equipping the reader with the necessary background for studying turbo Bose–Chaudhuri–Hocquenghem codes, for example, the full appreciation of the associated so-called algebraic decoding techniques discussed in this this chapter is not necessary.

After demodulating the received signal, transmission errors can be removed from the digital information if their number does not exceed the error correcting power of the error correction code used. The employment of Forward Error Correction (FEC) coding techniques becomes important for transmissions over hostile mobile radio channels, where the violent channel fading precipitates bursts of error. This is particularly true when using vulnerable multilevel modulation schemes, such as Quadrature Amplitude Modulation (QAM). This chapter addresses issues of Reed–Solomon (RS) and Bose–Chaudhuri–Hocquenghem (BCH) coding in order to provide a self-contained reference for readers who want to delve into the theory of FEC coding.

The theory and practice of FEC coding has been well documented in classic references [32, 82, 84, 87–90, 96, 97]; hence for an in-depth treatment, the reader is referred to these sources. Both convolutional and block FEC codes have been successfully utilised in various communications systems [108, 109]. In this brief overview, to a certain extent we will follow the philosophy of these references, which are also recommended for a more detailed discourse on the subject. Since the applications described in this book often use the family of RS and BCH block codes, in this chapter we concentrate on their characterisation. We will draw the reader's attention to any differences between them, as they arise during our discussions. Both RS and BCH codes are defined over the mathematical structure of finite fields; therefore we briefly consider their construction. For a more detailed discourse on finite fields, the reader is referred to Lidl's work [95].

+	0	1	2	3	4	5	6
0	0	1	2	3	4	5	6
1	1	2	3	4	5	6	0
2	2	3	4	5	6	0	1
3	3	4	5	6	0	1	2
4	4	5	6	0	1	2	3
5	5	6	0	1	2	3	4
6	6	0	1	2	3	4	5

×	0	1	2	3	4	5	6
0	0	0	0	0	0	0	0
1	0	1	2	3	4	5	6
2	0	2	4	6	1	3	5
3	0	3	6	2	5	1	4
4	0	4	1	5	2	6	3
5	0	5	3	1	6	4	2
6	0	6	5	4	3	2	1

Table 3.1: $Modulo - 7$ addition and multiplication tables for the prime field. $GF(7)$

3.2   Finite Fields

3.2.1   Definitions

Loosely speaking, an *algebraic field* is any arithmetic structure in which addition, subtraction, multiplication and division are defined and the associative, distributive and commutative laws apply. In conventional algebraic fields, like the rational, real or complex field, these operations are trivial, but in other fields they are somewhat different, such as *modulo* or *modulo polynomial* operations.

A more formal field definition can be formulated as follows [90]. An *algebraic field* is a set of elements that has addition and multiplication defined over it, such that the set is closed under both addition and multiplication; that is, the results of these operations are also elements in the same set. Furthermore, both addition and multiplication are associative and commutative. There is a zero element, such that $a + 0 = a$, and an additive inverse element, such that $a + (-a) = 0$. Subtraction is defined as $a - b = a + (-b)$. There is a one element, such that $1 \cdot a = a$, and a multiplicative inverse element for which we have $a \cdot (a^{-1}) = 1$. Division is defined as $a/b = a \cdot (b^{-1})$.

Digital Signal Processing (DSP) has been historically studied in the real algebraic field constituted by all finite and infinite decimals, or in the complex algebraic field with elements of the form $a + jb$, where a and b are real. In both the real and complex fields, the operations are fairly straightforward and well known, and the number of elements is infinite; consequently, they are *infinite algebraic fields*.

Finite algebraic fields or shortly finite fields are constituted by a finite number q of elements if there exists a field with q elements. They are also referred to as *Galois Fields GF(q)*. Every field must have a zero and a one element. Consequently, the smallest field is $GF(2)$ with mod 2 addition and multiplication, because the field must be closed under the operations. That is, their result must be a field element, too. If the number of elements in a field is a prime p, the $GF(p)$ constructed from the elements $\{0, 1, 2, \ldots, (p-1)\}$ is called a *prime field* with *modulo p* operations.

Example 1: `Galois prime-field modulo operation.`

As an illustrative example, let us briefly consider the basic operations, namely, addition and multiplication, over the prime field $GF(7)$. The prime field $GF(7)$ is constituted by the elements $\{0,1,2,3,4,5,6\}$, and the corresponding $modulo - 7$ addition and multiplication tables are summarised in Table 3.1.

Since every field must contain the unity element, we can define the *order n of a field element* α, such that $\alpha^n = 1$. In other words, to determine the order of a field element the exponent n yielding $\alpha^n = 1$ has to be found. Every $GF(p)$ has at least one element α of order $(p-1)$, which is called a *primitive element* and which always exists in a $GF(p)$ [32]. Hence, $\alpha^{(p-1)} = 1$ must apply. For practical FEC applications we must have sufficiently large finite fields, constituted by many field elements, since only this finite number of elements can be used in order to represent our codewords.

Example 2: Extension field construction example.

In order to augment the concept of *extension fields*, let us consider first the simple example of extending the real field to the complex field. Thus, we attempt to find a second-order polynomial over the real field that cannot be factorised over it, which is also referred to in finite field parlance as an *irreducible polynomial*. For example, it is well understood that the polynomial $x^2 + 1$ cannot be factorised over the real field because it has no real zeros. However, one can define a special zero j of the polynomial $x^2 + 1$, which we refer to as an imaginary number, for which $j^2 = -1$ and $x^2 + 1 = (x + j)(x - j)$. This way the real number system has been extended to the complex field, which has complex elements $\alpha_1 + j\alpha_2$, where α_1 and α_2 are real. The new number system derived from the real field fulfils all the criteria necessary for an extension field.

The elements $\alpha_1 + j\alpha_2$ of the complex field can also be interpreted as first-order polynomials $\alpha_1 x^0 + \alpha_2 x^1$ having real coefficients, if we replace j by x. In this context, multiplication of the field elements $(\alpha_1 + \alpha_2 x)$ and $(\alpha_3 + \alpha_4 x)$ is equivalent to modulo polynomial multiplication of the field elements:

$$
\begin{aligned}
\alpha_5 + \alpha_6 x &= (\alpha_1 + \alpha_2 x)(\alpha_3 + \alpha_4 x) \quad (\text{mod } x^2 + 1) \\
&= \alpha_1 \alpha_3 + \alpha_2 \alpha_3 x + \alpha_1 \alpha_4 x + \alpha_2 \alpha_4 x^2 \quad (\text{mod } x^2 + 1) \\
&= \alpha_1 \alpha_3 + (\alpha_2 \alpha_3 + \alpha_1 \alpha_4)x + \alpha_2 \alpha_4 x^2 \quad (\text{mod } x^2 + 1),
\end{aligned}
\tag{3.1}
$$

where $\alpha_1 \ldots \alpha_6$ are arbitrary elements of the original real field. But since $x^2 \ (\text{mod } x^2 + 1) = \text{Remainder}\{x^2 : (x^2 + 1)\} = -1$, this polynomial formulation delivers identical results to those given by the complex number formulation, where $j^2 = -1$. So we may write:

$$
\begin{aligned}
\alpha_5 + \alpha_6 x &= (\alpha_1 \alpha_3 - \alpha_2 \alpha_4) + (\alpha_2 \alpha_3 + \alpha_1 \alpha_4)x \\
\end{aligned}
\tag{3.2}
$$

$$
\alpha_5 = \alpha_1 \alpha_3 - \alpha_2 \alpha_4
\tag{3.3}
$$

$$
\alpha_6 = \alpha_2 \alpha_3 + \alpha_1 \alpha_4 .
\tag{3.4}
$$

Pursuing this polynomial representation, we find that the general framework for extending the finite prime field $GF(p)$ is to have m components rather than just two.

Explicitly, a convenient way of generating large algebraic fields is to extend the prime field $GF(p)$ to a so-called extension field $GF(p^m)$. In general, the elements of the extension field $GF(p^m)$ are all the possible m dimensional vectors, where all m vector coordinates are elements of the original prime field $GF(p)$, and m is an integer. For example, an extension field $GF(p^m)$ of the original prime field $GF(p)$ with $p = 2$ elements $\{0, 1\}$ contains all the possible combinations of m number of $GF(p)$ elements, which simply means that there are p^m number of extension field elements. Consequently, an element of the extension field $GF(p^m)$ can be written as a polynomial of order $m - 1$, with coefficients from $GF(p)$:

$$
GF(p^m) = \{(a_0 x^0 + a_1 x^1 + a_2 x^2 + \ldots + a_{m-1} x^{m-1})\}
$$

with

$$\{a_0, a_1 \ldots a_{m-1}\} \in \{0, 1 \ldots (p-1)\}.$$

The operations in the extension field $GF(p^m)$ are *modulo polynomial* operations rather than conventional modulo operations. Hence, the addition is carried out as the addition of two polynomials:

$$
\begin{aligned}
c(x) &= a(x) + b(x) & (3.5) \\
&= a_0 + a_1 x + a_2 x^2 + \ldots + a_{m-1} x^{m-1} + \\
&+ b_0 + b_1 x + b_2 x^2 + \ldots + b_{m-1} x^{m-1} & (3.6) \\
&= (a_0 + b_0) + (a_1 + b_1)x^1 + (a_2 + b_2)x^2 + \ldots \\
&+ (a_{m-1} + b_{m-1})x^{m-1}. & (3.7)
\end{aligned}
$$

Since

$$\{a_0, a_1, \ldots, a_{m-1}\} \in \{0, 1, \ldots, (p-1)\}$$

and

$$\{b_0, b_1, \ldots, b_{m-1}\} \in \{0, 1, \ldots, (p-1)\},$$

the component-wise addition of the polynomial coefficients must be $\bmod p$ addition, so that the coefficients of the result polynomial are also elements of $GF(p)$.

The definition of the modulo polynomial multiplication is somewhat more complicated. First, one has to find an irreducible polynomial $p(x)$ of degree m, which cannot be factorised over $GF(p)$. In other words, $p(x)$ must be a polynomial with coefficients from $GF(p)$, and it must not have any zeros; that is, it cannot be factorised into polynomials of lower order with coefficients from $GF(p)$. Recall, for example, that the polynomial $x^2 + 1$ was irreducible over the real field. More formally, the polynomial $p(x)$ is *irreducible* if and only if it is divisible exclusively by $\alpha.p(x)$ or by α, where α is an arbitrary field element in $GF(p)$. Whenever the highest-order coefficient of an irreducible polynomial is equal to one, the polynomial is called a *prime polynomial*.

Once we have found an appropriate prime polynomial $p(x)$, the multiplication of two extension field elements over the extension field is modulo polynomial multiplication of the polynomial representations of the field elements. Explicitly, the product of the two extension field elements given by their polynomial representations must be divided by $p(x)$, and the remainder must be retained as the final result. This is to ensure that the result is also an element of the extension field (i.e. a polynomial with an order less than p). Then the extension field is declared closed under the operations, yielding results that are elements of the original field. Accordingly, the nature of the extension field depends on the choice of the prime polynomial $p(x)$. Sometimes there exist several prime polynomials, and some of them result in somewhat more advantageous extension field construction than others. Sometimes there is no prime polynomial of a given degree m over a given prime field $GF(p)$; consequently, $GF(p)$ cannot be extended to $GF(p^m)$. At this stage, we note that we have constructed a mathematical environment necessary for representing our information carrying signals (i.e. codewords), and defined the operations over the extension field.

Since the source data to be encoded are usually binary, we represent the non-binary information symbols of the extension field as a sequence of m bits. Logically, our prime field is the binary field $GF(2)$ with elements $\{0,1\}$ and $p = 2$. Then we extend $GF(2)$ to $GF(2^m)$ in order to generate a sufficiently large working field for signal processing, where an m-bit non-binary information symbol constitutes a field element of $GF(2^m)$. If, for example, an information symbol is constituted by one byte of information (i.e. $m = 8$), the extension field $GF(2^m) = GF(256)$ contains 256 different field elements. The appropriate prime polynomial can be chosen, for example, from Table 4.7 on p. 409 of reference [109], where primitive polynomials of degree less than 25 over $GF(2)$ are listed. These prime polynomials allow us to construct any arbitrary extension field from $GF(2^2)$ to $GF(2^{25})$.

It is useful for our further discourse to represent the field elements of $GF(q = 2^m)$ each as a unique power of an element α, which we refer to as the *primitive element*. The primitive element α was earlier defined to be the one that fulfils the condition $\alpha^{q-1} = \alpha^{2^m-1} = 1$. With this notation the elements of the extension field $GF(q = 2^m)$ can be written as $\{0, 1, \alpha, \alpha^2, \alpha^3, \dots, \alpha^{q-2}\}$, and their polynomial representation is given by the remainder of x^n upon division by the prime polynomial $p(x)$:

$$\alpha^n = Remainder\ \{x^n/p(x)\}. \tag{3.8}$$

This relationship will become clearer in the practical example of the next subsection.

3.2.2   Galois Field Construction

Example 3: GF(16) construction example.

As an example, let us consider the construction of $GF(2^4) = GF(16)$ based on the prime polynomial $p(x) = x^4 + x + 1$ [109], p. 409, where $m = 4$. Each extension field element α can be represented as a polynomial of degree three with coefficients over $GF(2)$. Equivalently, each extension field element can be described by the help of the binary representation of the coefficients of the polynomial or by means of the decimal value of the binary representation.

Let us proceed with the derivation of all the representations mentioned above, which are summarised in Table 3.2. Since every field must contain the zero and one element, we have:

$$0 = 0$$

$$\alpha^0 = 1.$$

Further plausible assignments are:

$$\alpha^1 = x$$

$$\alpha^2 = x^2$$

$$\alpha^3 = x^3,$$

because the remainders of x, x^2 and x^3 upon division by the primitive polynomial $p(x) = x^4+x+1$ are themselves. However, the polynomial representation of x^4 cannot be derived without polynomial division:

$$\alpha^4 = Rem\ \{x^4 : (x^4 + x + 1)\}_{p(x)=x^4+x+1}$$

Exponential Represent.	Polynomial Represent.	Binary Represent.	Decimal Represent.
0	0	0000	0
α^0	1	0001	1
α^1	x	0010	2
α^2	x^2	0100	4
α^3	x^3	1000	8
α^4	$x+1$	0011	3
α^5	x^2+x	0110	6
α^6	x^3+x^2	1100	12
α^7	x^3+x+1	1011	11
α^8	x^2+1	0101	5
α^9	x^3+x	1010	10
α^{10}	x^2+x+1	0111	7
α^{11}	x^3+x^2+x	1110	14
α^{12}	x^3+x^2+x+1	1111	15
α^{13}	x^3+x^2+1	1101	13
α^{14}	x^3+1	1001	9

Table 3.2: Different representations of $GF(16)$ elements ($\alpha^{15} \equiv \alpha^0 \equiv 1$) generated using the prime polynomial $p(x) = x^4 + x + 1$

$$\text{The polynomial division:} \qquad x^4 : (x^4 + x + 1) = 1 \qquad (3.9)$$
$$\underline{-(x^4 + x + 1)} \qquad\qquad\quad (3.10)$$
$$\text{The remainder:} \qquad -(x + 1). \qquad\qquad\qquad\quad (3.11)$$

Hence we have:

$$\alpha^4 = x + 1.$$

Following these steps the first two columns of Table 3.2 can easily be filled in. The binary notation simply comprises the coefficients of the polynomial representation, whereas the decimal notation is the decimal number computed from the binary representation.

3.2.3   Galois Field Arithmetic

Multiplication in the extension field is carried out by multiplying the polynomial representations of the elements, dividing the result by the prime polynomial $p(x) = x^4 + x + 1$, and finally taking the remainder. For example, for the field elements:

$$\alpha^4 \equiv x + 1$$
$$\alpha^7 \equiv x^3 + x + 1$$

we have:

$$
\begin{aligned}
\alpha^4 \cdot \alpha^7 &= (x+1)(x^3+x+1)_{\mathrm{mod}\,p(x)=x^4+x+1} \\
&= Rem\left\{(x+1)(x^3+x+1):(x^4+x+1)\right\} \\
&= Rem\left\{(x^4+x^2+x+x^3+x+1):(x^4+x+1)\right\} \\
&= Rem\left\{(x^4+x^3+x^2+1):(x^4+x+1)\right\}.
\end{aligned}
$$

Then the polynomial division is computed as follows:

$$
(x^4+x^3+x^2+0+1):(x^4+x+1)=1
$$
$$
\underline{-(x^4 \qquad +x+1)}
$$

The remainder: $\quad x^3+x^2+x+0 = x^3+x^2+x.$

Therefore, the required product is given by:

$$
\alpha^4 \cdot \alpha^7 = x^3+x^2+x \equiv \alpha^{11}.
$$

From Table 3.2 it is readily recognised that the exponential representation of the field elements allows us to simply add the exponents without referring back to their polynomial representations. Whenever the exponent happens to be larger than $q-1 = 15$, it is simply collapsed back into the finite field by taking its value modulo $(q-1)$, that is modulo 15. For example:

$$
\alpha^{12} \cdot \alpha^6 = \alpha^{18} = \alpha^{15} \cdot \alpha^3 = 1 \cdot \alpha^3 = \alpha^3.
$$

The addition of two field elements is also carried out easily by referring to the polynomial representation by the help of a component-wise addition, as follows:

$$
\begin{aligned}
\alpha^{11} + \alpha^7 &\equiv (x^3+x^2+x+0) \\
&+ \underline{(x^3+0+x+1)} \\
& \quad 0+x^2+0+1 = x^2+1 \equiv \alpha^8.
\end{aligned}
$$

This is also equivalent to the modulo-2 addition of the binary representations:

$$
\begin{aligned}
\alpha^{11} + \alpha^7 &\equiv 1110 \\
&+ \underline{1011} \\
& \quad 0101 \equiv x^2+1 \equiv \alpha^8.
\end{aligned}
$$

The fastest way to compute $GF(16)$ addition of exponentially represented field elements is to use the precomputed $GF(16)$ addition table, namely Table 3.3. With our background in finite fields, we can now proceed to define and characterise the important so-called maximum–minimum distance family — a term to be clarified during our later discussion — of non-binary block codes referred to as RS codes and their binary subclass, BCH codes, which are often used in our prototype systems.

Expon. Repr.	Binary Repr.	Hexad. Repr.	α^0	α^1	α^4	α^2	α^8	α^5	α^{10}	α^3	α^{14}	α^9	α^7	α^6	α^{13}	α^{11}	α^{12}
			0001	0010	0011	0100	0101	0110	0111	1000	1001	1010	1011	1100	1101	1110	1111
			1	2	3	4	5	6	7	8	9	A	B	C	D	E	F
α^0	0001	1	0	α^4	α^1	α^8	α^2	α^{10}	α^5	α^{14}	α^3	α^7	α^9	α^{13}	α^6	α^{12}	α^{11}
α^1	0010	2	α^4	0	α^0	α^5	α^{10}	α^2	α^8	α^9	α^7	α^3	α^{14}	α^{11}	α^{12}	α^6	α^{13}
α^4	0011	3	α^1	α^0	0	α^{10}	α^5	α^8	α^2	α^7	α^9	α^{14}	α^3	α^{12}	α^{11}	α^{13}	α^6
α^2	0100	4	α^8	α^5	α^{10}	0	α^0	α^1	α^4	α^6	α^{13}	α^{11}	α^{12}	α^3	α^{14}	α^9	α^7
α^8	0101	5	α^2	α^{10}	α^5	α^0	0	α^4	α^1	α^{13}	α^6	α^{12}	α^{11}	α^{14}	α^3	α^7	α^9
α^5	0110	6	α^{10}	α^2	α^8	α^1	α^4	0	α^0	α^{11}	α^{12}	α^6	α^{13}	α^9	α^7	α^3	α^{14}
α^{10}	0111	7	α^5	α^8	α^2	α^4	α^1	α^0	0	α^{12}	α^{11}	α^{13}	α^6	α^7	α^9	α^{14}	α^3
α^3	1000	8	α^{14}	α^9	α^7	α^6	α^{13}	α^{11}	α^{12}	0	α^0	α^1	α^4	α^2	α^8	α^5	α^{10}
α^{14}	1001	9	α^3	α^7	α^9	α^{13}	α^6	α^{12}	α^{11}	α^0	0	α^4	α^1	α^8	α^2	α^{10}	α^5
α^9	1010	A	α^7	α^3	α^{14}	α^{11}	α^{12}	α^6	α^{13}	α^1	α^4	0	α^0	α^5	α^{10}	α^2	α^8
α^7	1011	B	α^9	α^{14}	α^3	α^{12}	α^{11}	α^{13}	α^6	α^4	α^1	α^0	0	α^{10}	α^5	α^8	α^2
α^6	1100	C	α^{13}	α^{11}	α^{12}	α^3	α^{14}	α^9	α^7	α^2	α^8	α^5	α^{10}	0	α^0	α^1	α^4
α^{13}	1101	D	α^6	α^{12}	α^{11}	α^{14}	α^3	α^7	α^9	α^8	α^2	α^{10}	α^5	α^0	0	α^4	α^1
α^{11}	1110	E	α^{12}	α^6	α^{13}	α^9	α^7	α^3	α^{14}	α^5	α^{10}	α^2	α^8	α^1	α^4	0	α^0
α^{12}	1111	F	α^{11}	α^{13}	α^6	α^7	α^9	α^{14}	α^3	α^{10}	α^5	α^8	α^2	α^4	α^1	α^0	0

Table 3.3: $GF(16)$ addition table, ©Wong 1989 [108], 1999 Wong, Hanzo [109].

3.3 Reed-Solomon and Bose-Chaudhuri-Hocquenghem Block Codes

3.3.1 Definitions

As mentioned above, RS codes represent a non-binary subclass of (BCH) multiple error correcting codes. Because of their non-binary nature, RS codes pack the information represented by m consecutive bits into non-binary symbols, which are elements in the extension field $GF(2^m)$ of $GF(2)$. In general, an RS code is defined as a block of n non-binary symbols over $GF(2^m)$, constructed from k original information symbols by adding $n - k = 2t$ number of redundancy symbols from the same extension field, giving $n = k + 2t$. This code is often written as $RS(n, k, t)$ over $GF(2^m)$.

RS codes are cyclic codes, implying that any cyclically shifted version of a codeword is also a codeword of the same code. A further property of cyclic codes is that all of the codewords can be generated as a linear combination of any codeword and its cyclically shifted versions. The distance properties of a code are crucial in assessing its error correction capability. The *minimum distance* of a code is the minimum number of positions in which two arbitrary codewords differ. If we define the weight of a codeword as the number of non-zero symbol positions, then the minimum distance of a code is actually the weight of its minimum weight codeword, since the 'all-zero' word is always a codeword in a linear code. It is plausible that if in an (n, k) code less than half the minimum distance number of symbols are in error, it can be uniquely recognised which k symbol's long information message has been sent. This is because the received message is still closer to the transmitted one than to any other one. On the other hand, if more than half the minimum distance number of symbols are in error, the decoder decides erroneously on the basis of the nearest legitimate codeword.

The *Singleton bound* imposes an upper limit on the maximum achievable minimum distance of a code on the following basis. If in a codeword one information symbol is changed, the highest possible distance for the newly computed codeword from the original one will be $d = (n-k)+1$, provided that all the $n-k$ parity symbols also changed as a result, which is an extreme assumption. Consequently, for the code's minimum distance we have $d_{min} \leq (n-k) + 1$. RS codes are one of the very few codes which reach the maximum possible minimum distance of $d = (n-k) + 1$. This is why they are referred to as maximum–minimum distance codes. In general, an $RS(n, k, t)$ code can correct up to $t = (n-k)/2$ symbol errors, or in other words a t error correcting code must contain $2t = (n-k)$ number of redundancy symbols. Therefore, the minimum distance between codewords must be $d_{min} \geq 2t + 1$. Substituting $(n-k) = 2t$ into $d_{min} \geq 2t + 1$ and taking into account that previously we showed that $d_{min} \leq (n-k) + 1$, we find that the minimum distance of RS codes is exactly $d_{min} = 2t + 1$.

Before we proceed to define the RS encoding rules, the *generator polynomial* has to be introduced, which is defined to be the polynomial of order $2t$, which has its zeros at any $2t$ out of the 2^m possible field elements. For the sake of simplicity, but without any loss of generality, we always use the GF elements $\alpha^1, \alpha^2, \ldots, \alpha^{2t}$ in order to determine the generator polynomial of an RS code:

$$g(x) = (x - \alpha)(x - \alpha^2) \ldots (x - \alpha^{2t}) = \prod_{j=1}^{2t} (x - \alpha^j) = \sum_{j=0}^{2t} g_j x^j. \qquad (3.12)$$

In the next subsection, we highlight two algorithms using the above-mentioned generator polynomial for the encoding of the information symbols in order to generate RS-coded codewords.

3.3.2   RS Encoding

Since RS codes are cyclic, any cyclic encoding method can be used for their coding. When opting for the *non-systematic encoding* rule, the information symbols are not explicitly recognisable in the encoded codeword — hence the terminology. Non-systematic cyclic encoders generate the encoded word by multiplying the information polynomial $i(x)$ by the generator polynomial $g(x)$ using modulo polynomial algebra as follows:

$$c(x) \;=\; i(x) \,.\, g(x) \tag{3.13}$$

$$i(x) \;=\; i_1 x + i_2 x^2 + \ldots + i_k x^k = \sum_{j=1}^{k} i_j x^j \tag{3.14}$$

$$c(x) \;=\; c_1 x + c_2 x^2 + \ldots + c_n x^n = \sum_{j=1}^{n} c_j x^j \tag{3.15}$$

where the information polynomial $i(x)$ is of order k and the encoded codeword polynomial $c(x)$ is of order $n = k + 2t$. The coefficients of the polynomials are the non-binary information carrying symbols, which are elements in $GF(2^m)$, and the powers of x can be thought of as place markers for the symbols in the codeword. Again, since the polynomial $c(x)$ does not contain explicitly the original k information symbols, this is called a non-systematic encoder. The codeword is then modulated and sent over a non-ideal transmission medium, where the code symbols or polynomial coefficients may become corrupted.

The channel effects can be modelled by the help of an additive error polynomial $e(x)$ of order n as follows:

$$
\begin{aligned}
r(x) = c(x) + e(x) \;=\;\; & c_1 x + c_2 x^2 + \ldots + c_n x^n \\
+ \;\; & e_1 x + e_2 x^2 + \ldots + e_n x^n \,,
\end{aligned}
\tag{3.16}
$$

where $r(x)$ is the corrupted received polynomial. The decoder has to determine up to t error positions and compute t error locations. In other words, the decoder has to compute $t + t = 2t$ unknowns from the $2t$ redundancy symbols and correct the errors to produce the error-free codeword $c(x)$. After error correction, the information symbols must be recovered by the help of the inverse operation of the encoding, using the following simple decoding rule:

$$i(x) = \frac{c(x)}{g(x)} \,. \tag{3.17}$$

However, if there are more than t transmission errors in the received polynomial $r(x)$, the decoder fails to produce the error-free codeword polynomial $c(x)$. This is plausible, since owing to its excessive corruption the received codeword will become more similar to another legitimate codeword rather than the transmitted one. In algebraic terms, this is equivalent to saying that we cannot determine more than t unknown error positions and error values, when using $2 \times t$ redundancy symbols. We note that in the case of binary BCH codes it is plausible that having determined the error positions, the corresponding bits are simply inverted in order to correct them, while for non-binary RS codes a more complicated procedure will have to be employed. These statements will be made more explicit during our further discourse.

The *systematic RS encoding* rule is somewhat more sophisticated than the non-systematic one, but the original information symbols are simply copied into the encoded word. This property is often attractive, because in some applications it is advantageous to know where the original information symbols reside in the encoded block. Explicitly, when a received codeword is deemed to be overwhelmed by transmission errors, a non-systematic RS or BCH code has no other option than to attempt to compute the error locations and 'magnitudes' for their correction, but this operation will be flawed by the plethora of transmission errors, and hence the decoding operation may actually corrupt more received symbols by carrying out a flawed decoding operation.

By contrast, in systematic RS or BCH codecs, instead of erroneously computing the error locations and 'magnitudes' for their correction, this 'code overload' condition can be detected. Hence, this flawed action can be avoided by simply extracting the k original information symbols from $c(x)$. Because of these differences, non-systematic codes usually result in a Bit Error Rate (BER) increase under hostile channel conditions, while powerful systematic codes can maintain a similar BER performance to the uncoded case under similar circumstances.

Again, in systematic RS codes the first k encoded symbols are chosen to be the original k information symbols. We simply multiply the information polynomial $i(x)$ by x^{n-k} in order to shift it into the highest-order position of the codeword $c(x)$. Then we choose the parity symbols constituted by the polynomial $p(x)$ according to the systematic encoding rule in order to result in a legitimate codeword. Legitimate in this sense means that the remainder of the encoded word $c(x)$ upon division by the generator polynomial $g(x)$ is zero. Using the codeword $c(x)$ hosting the shifted information word $i(x)$ plus the parity segment $p(x)$, we have:

$$c(x) = x^{(n-k)} \cdot i(x) + p(x), \tag{3.18}$$

and according to the above definition of $c(x)$, we have:

$$Rem \left\{ \frac{c(x)}{g(x)} \right\} = 0 \tag{3.19}$$

$$Rem \left\{ \frac{x^{(n-k)} \cdot i(x) + p(x)}{g(x)} \right\} = 0 \tag{3.20}$$

$$Rem \left\{ \frac{x^{(n-k)} \cdot i(x)}{g(x)} \right\} + Rem \left\{ \frac{p(x)}{g(x)} \right\} = 0. \tag{3.21}$$

Since the order of the parity polynomial $p(x)$ is less than $(n-k)$ and the order of $g(x)$ is $(n-k)$, we have:

$$Rem \left\{ \frac{p(x)}{g(x)} \right\} = p(x). \tag{3.22}$$

By substituting Equation 3.22 into Equation 3.21 and rearranging it, we get:

$$- Rem \left\{ \frac{x^{(n-k)} \cdot i(x)}{g(x)} \right\} = p(x). \tag{3.23}$$

Hence, if we substitute Equation 3.23 into Equation 3.18 and take into account that over GFs addition and subtraction are the same, the systematic encoding rule is as follows:

$$c(x) = x^{(n-k)} \cdot i(x) + Rem \left\{ \frac{x^{(n-k)} \cdot i(x)}{g(x)} \right\}. \tag{3.24}$$

The error correction ensues in a completely equivalent manner to that of the non-systematic decoder, which will be the subject of our later discussion, but recovering the information symbols from the corrected received codeword is simpler. Namely, the first k information symbols of an n-symbol long codeword are retained as corrected decoded symbols. Let us now revise these definitions and basic operations with reference to the following example.

3.3.3 RS Encoding Example

Example 4:
Systematic and non-systematic $RS(12, 8, 2)$ encoding example.

Let us consider a low-complexity double-error correcting RS code, namely the RS(12,8,2) code over GF(16), and demonstrate the operation of both the systematic and non-systematic encoder. We begin our example with the determination of the generator polynomial $g(x)$, which is a polynomial of order $2t = 4$, having zeros at the first four elements of $GF(16)$, namely at $\alpha^1, \alpha^2, \alpha^3$ and α^4. Recall, however, that we could have opted for the last four $GF(16)$ elements or any other four elements of it. Remember furthermore that multiplication is based on adding the exponents of the exponential representations, while addition is conveniently carried out using Table 3.3. Then the generator polynomial is arrived at as follows:

$$
\begin{aligned}
g(x) &= (x - \alpha^1)(x - \alpha^2)(x - \alpha^3)(x - \alpha^4) & (3.25)\\
&= (x^2 - \alpha^1 x - \alpha^2 x + \alpha^1 \alpha^2)(x^2 - \alpha^3 x - \alpha^4 x + \alpha^3 \alpha^4)\\
&= (x^2 - \underbrace{(\alpha^1 + \alpha^2)}_{\alpha^5} x + \alpha^3)(x^2 - \underbrace{(\alpha^3 + \alpha^4)}_{\alpha^7} x + \alpha^7)\\
&= (x^2 + \alpha^5 x + \alpha^3)(x^2 + \alpha^7 x + \alpha^7)\\
&= x^4 + \alpha^7 x^3 + \alpha^7 x^2 + \alpha^5 x^3 + \alpha^{12} x^2 + \alpha^{12} x + \alpha^3 x^2 + \alpha^{10} x + \alpha^{10}\\
&= x^4 + \underbrace{(\alpha^7 + \alpha^5)}_{\alpha^{13}} x^3 + \underbrace{(\alpha^7 + \alpha^{12} + \alpha^3)}_{\alpha^6} x^2 + \underbrace{(\alpha^{12} + \alpha^{10})}_{\alpha^3} x + \alpha^{10}\\
&= x^4 + \alpha^{13} . x^3 + \alpha^6 . x^2 + \alpha^3 . x + \alpha^{10}.
\end{aligned}
$$

Let us now compute the codeword polynomial $c(x)$ for the $RS(12, 8, 2)$ double-error correcting code, when the 'all-one' information polynomial $i(x) = 11 \ldots 11$ is to be encoded. In hexadecimal format, this can be expressed as $i(x) = FFFF\ FFFF\#H$. Since $GF(2^m) = GF(16)$ for $m = 4$, $8.4 = 32$ bits; that is, 8 hexadecimal symbols must be encoded into 12 hexadecimal symbols. The exponential representation of the $1111 = F\#H$ hexadecimal symbol is α^{12}, as seen in Table 3.2. Hence, the seventh-order information polynomial is given by:

$$
i(x) = \alpha^{12} x^7 + \alpha^{12} x^6 + \alpha^{12} x^5 + \alpha^{12} x^4 + \alpha^{12} x^3 + \alpha^{12} x^2 + \alpha^{12} x + \alpha^{12}.
$$

Then the non-systematic encoder simply computes the product of the information polynomial and the generator polynomial, as shown below:

$$
\begin{aligned}
c(x) &= g(x)i(x) &&(3.26)\\
c(x) &= (x^4 + \alpha^{13}x^3 + \alpha^6 x^2 + \alpha^3 x + \alpha^{10})\\
&\quad\cdot\ (\alpha^{12}x^7 + \alpha^{12}x^6 + \alpha^{12}x^5 + \alpha^{12}x^4\\
&\qquad + \alpha^{12}x^3 + \alpha^{12}x^2 + \alpha^{12}x + \alpha^{12})\\
&= \alpha^{12}x^{11} + \alpha^{12}x^{10} + \alpha^{12}x^9 + \alpha^{12}x^8 + \alpha^{12}x^7 + \alpha^{12}x^6 + \alpha^{12}x^5\\
&+ \alpha^{12}x^4 + \alpha^{10}x^{10} + \alpha^{10}x^9 + \alpha^{10}x^8 + \alpha^{10}x^7 + \alpha^{10}x^6 + \alpha^{10}x^5\\
&+ \alpha^{10}x^4 + \alpha^{10}x^3 + \alpha^3 x^9 + \alpha^3 x^8 + \alpha^3 x^7 + \alpha^3 x^6 + \alpha^3 x^5\\
&+ \alpha^3 x^4 + \alpha^3 x^3 + \alpha^3 x^2 + x^8 + x^7 + x^6 + x^5 + x^4 + x^3 + x^2 + x\\
&+ \alpha^7 x^7 + \alpha^7 x^6 + \alpha^7 x^5 + \alpha^7 x^4 + \alpha^7 x^3 + \alpha^7 x^2 + \alpha^7 x + \alpha^7\\
&= \alpha^{12}x^{11} + (\alpha^{12} + \alpha^{10})x^{10} + (\alpha^{12} + \alpha^{10} + \alpha^3)x^9\\
&+ (\alpha^{12} + \alpha^{10} + \alpha^3 + 1)x^8 + (\alpha^{12} + \alpha^{10} + \alpha^3 + 1 + \alpha^7)x^7\\
&+ (\alpha^{12} + \alpha^{10} + \alpha^3 + 1 + \alpha^7)x^6 + (\alpha^{12} + \alpha^{10} + \alpha^3 + 1 + \alpha^7)x^5\\
&+ (\alpha^{12} + \alpha^{10} + \alpha^3 + 1 + \alpha^7)x^4 + (\alpha^{10} + \alpha^3 + 1 + \alpha^7)x^3 &&(3.27)\\
&+ (\alpha^3 + 1 + \alpha^7)x^2 + (1 + \alpha^7)x + \alpha^7\\
&= \alpha^{12}x^{11} + \alpha^3 x^{10} + 0x^9 + \alpha^0 x^8 + \alpha^9 x^7 + \alpha^9 x^6 + \alpha^9 x^5\\
&+ \alpha^9 x^4 + \alpha^8 x^3 + \alpha^1 x^2 + \alpha^9 x + \alpha^7.
\end{aligned}
$$

Here it becomes clear that when using the non-systematic encoding rule, the original information polynomial $i(x)$ cannot be directly recognised in $c(x)$.

We argued above that in the case of systematic RS and BCH codes the BER can be kept lower than that of the non-systematic codes if a small additional signal processing complexity is tolerated. Thus, from now on we concentrate our attention on systematic codes. In order to compute the systematic codeword $c(x)$, Equations 3.18–3.22 must be used:

$$
c(x) = x^4 \cdot i(x) + p(x),
$$

where:

$$
p(x) = Rem\left\{ \frac{x^4\, i(x)}{g(x)} \right\}.
$$

The quotient $q(x)$ and remainder $p(x)$ of the above polynomial division are computed in Table 3.4, and the reader may find it beneficial at this stage to work through this example. Although the quotient polynomial is not necessary for our further elaborations, it is delivered by these operations, while the remainder appears in the bottom line of the table, which are given by:

$$
q(x) = \alpha^{12}x^7 + \alpha^3 x^6 + \alpha^8 x^5 + \alpha^3 x^4 + \alpha^4 x^3 + \alpha^1 x^2 + \alpha^8 x + \alpha^{11}
$$

$$
p(x) = \alpha^{14}x^3 + \alpha^2 x^2 + \alpha^0 x + \alpha^6,
$$

where:

$$
\frac{x^4 i(x)}{g(x)} = q(x)g(x) + p(x)
$$

$$\frac{x^4 \cdot i(x)}{g(x)} = \frac{\left.\begin{array}{l}\alpha^{12}.x^{11} + \alpha^{12}.x^{10} + \alpha^{12}.x^9 + \alpha^{12}.x^8 + \alpha^{12}.x^7 + \alpha^{12}.x^6 + \alpha^{12}.x^5 + \alpha^{12}.x^4\end{array}\right\}q(x)}{x^4 + \alpha^{13}.x^3 + \alpha^6.x^2 + \alpha^3.x + \alpha^{10}}$$

$$= \alpha^{12}.x^7 + \alpha^3.x^6 + \alpha^8.x^5 + \alpha^3.x^4 + \alpha^4.x^3 + \alpha^1.x^2 + \alpha^8.x + \alpha^{11}$$

x^{11}	x^{10}	x^9	x^8	x^7	x^6	x^5	x^4	x^3	x^2	x^1	x^0
$\alpha^{12}.x^{11}$	$+\alpha^{12}.x^{10}$	$+\alpha^{12}.x^9$	$+\alpha^{12}.x^8$	$+\alpha^{12}.x^7$	$+\alpha^{12}.x^6$	$+\alpha^{12}.x^5$	$+\alpha^{12}.x^4$				
$-(\alpha^{12}.x^{11}$	$+\alpha^{10}.x^{10}$	$+\alpha^3.x^9$	$+\alpha^0.x^8$	$+\alpha^7.x^7)$							
0	$+(\alpha^{12}+\alpha^{10}).x^{10}$	$+(\alpha^{12}+\alpha^3).x^9$	$+(\alpha^{12}+\alpha^0).x^8$	$+(\alpha^7+\alpha^{12}).x^7$	$+\alpha^{12}.x^6$						
	$-(\alpha^3.x^{10}$	$+\alpha^{10}.x^9$	$+\alpha^{11}.x^8$	$+\alpha^2.x^7$							
		$+\alpha^1.x^9$	$+\alpha^9.x^8$	$+\alpha^6.x^7$	$+\alpha^{13}.x^6)$						
	0	$+(\alpha^1+\alpha^{10}).x^9$	$+(\alpha^9+\alpha^{11}).x^8$	$+(\alpha^2+\alpha^6).x^7$	$+(\alpha^{12}+\alpha^{13}).x^6$						
		$-(\alpha^8.x^9$	$+\alpha^2.x^8$	$+\alpha^3.x^7$	$+\alpha^1.x^6$						
			$+\alpha^6.x^8$	$+\alpha^{14}.x^7$	$+\alpha^{11}.x^6$	$+\alpha^3.x^5)$					
		0	$+(\alpha^6+\alpha^2).x^8$	$+(\alpha^{14}+\alpha^3).x^7$	$+(\alpha^1+\alpha^{11}).x^6$	$+(\alpha^{12}+\alpha^3).x^5$					
			$-(\alpha^3.x^8$	$+\alpha^0.x^7$	$+\alpha^6.x^6$	$+\alpha^{10}.x^5$					
				$+\alpha^1.x^7$	$+\alpha^9.x^6$	$+\alpha^6.x^5$	$+\alpha^{13}.x^4)$				
			0	$+(\alpha^0+\alpha^1).x^7$	$+(\alpha^6+\alpha^9).x^6$	$+(\alpha^{10}+\alpha^6).x^5$	$+(\alpha^{12}+\alpha^{13}).x^4$				
				$-(\alpha^4.x^7$	$+\alpha^5.x^6$	$+\alpha^7.x^5$	$+\alpha^1.x^4$				
					$+\alpha^2.x^6$	$+\alpha^{10}.x^5$	$+\alpha^7.x^4$	$+\alpha^{14}.x^3)$			
				0	$+(\alpha^5+\alpha^2).x^6$	$+(\alpha^7+\alpha^{10}).x^5$	$+(\alpha^1+\alpha^7).x^4$	$+(0+\alpha^{14}).x^3$			
					$-(\alpha^1.x^6$	$+\alpha^6.x^5$	$+\alpha^{14}.x^4$	$+\alpha^{14}.x^3$			
						$+\alpha^{14}.x^5$	$+\alpha^7.x^4$	$+\alpha^4.x^3$	$+\alpha^{11}.x^2)$		
					0	$+(\alpha^6+\alpha^{14}).x^5$	$+(\alpha^{14}+\alpha^7).x^4$	$+(\alpha^{14}+\alpha^4).x^3$	$+(0+\alpha^{11}).x^2$		
						$-(\alpha^8.x^5$	$+\alpha^1.x^4$	$+\alpha^9.x^3$	$+\alpha^{11}.x^2$		
							$+\alpha^6.x^4$	$+\alpha^{14}.x^3$	$+\alpha^{11}.x^2$	$+\alpha^3.x)$	
						0	$+(\alpha^1+\alpha^6).x^4$	$+(\alpha^9+\alpha^{14}).x^3$	$+(\alpha^{11}+\alpha^{11}).x^2$	$+(0+\alpha^3).x$	
							$-(\alpha^{11}.x^4$	$+\alpha^4.x^3$	$+0.x^2$	$+\alpha^3.x$	
								$+\alpha^9.x^3$	$+\alpha^2.x^2$	$+\alpha^{14}.x$	$+\alpha^6)$
							0	$+(\alpha^9+\alpha^4).x^3$	$+(\alpha^2+0).x^2$	$+(\alpha^{14}+\alpha^3).x$	$+(0+\alpha^6)$
								$+\alpha^{14}.x^3$	$+\alpha^2.x^2$	$+\alpha^0.x$	$+\alpha^6 = p(x)$

Table 3.4: Systematic $RS(12,8,2)$ encoding example using polynomial division.

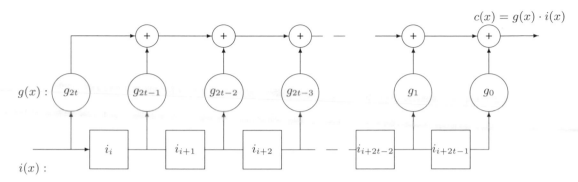

$$c(x) = g(x) \cdot i(x)$$

Figure 3.1: LSR circuit for multiplying polynomials in non-systematic RS and BCH encoders.

holds.

Since we now know the parity polynomial $p(x)$, the systematic codeword $c(x)$ is known from Equation 3.18 as well:

$$c(x) = x^4 i(x) + p(x) \tag{3.28}$$

$$
\begin{aligned}
c(x) &= \alpha^{12}x^{11} + \alpha^{12}x^{10} + \alpha^{12}x^9 + \alpha^{12}x^8 \\
&+ \alpha^{12}x^7 + \alpha^{12}x^6 + \alpha^{12}x^5 + \alpha^{12}x^4 \\
&+ \alpha^{14}x^3 + \alpha^2 x^2 + \alpha^0 x + \alpha^6.
\end{aligned}
\tag{3.29}
$$

If there are no transmission errors in the received polynomial $r(x)$, then we have $r(x) = c(x)$, and decoding simply ensues by taking the first $k = 8$ information symbols of $r(x)$, concluding Example 4. Let us continue our discourse by considering an implementation-ally attractive Linear Shift Register (LSR) encoder structure in the next subsection.

3.3.4 Linear Shift-register Circuits for Cyclic Encoders

3.3.4.1 Polynomial Multiplication

Since RS codes constitute a subclass of cyclic codes, any cyclic encoding circuitry can be used for RS and BCH encoding. The non-systematic RS or BCH encoder can be implemented as a Linear Shift Register (LSR) depicted in Figure 3.1, which multiplies the information polynomial $i(x)$ by the fixed generator polynomial $g(x)$ as follows:

$$c(x) = i(x) \cdot g(x),$$

where:

$$
\begin{aligned}
i(x) &= i_{k-1}x^{k-1} + i_{k-2}x^{k-2} + \ldots + i_1 x + i_0 \tag{3.30} \\
g(x) &= g_{2t}x^{2t} + g_{2t-1}x^{2t-1} + \ldots + g_1 x + g_0. \tag{3.31}
\end{aligned}
$$

The prerequisites for the circuit to carry out proper polynomial multiplications are as follows:

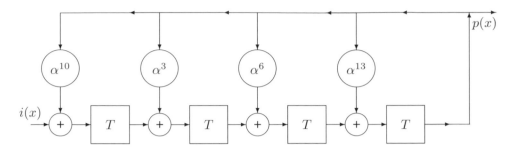

Figure 3.2: $RS(12, 8, 2)$ systematic encoder using polynomial division.

1) The LSR must be cleared before two polynomials are multiplied.

2) The k symbols of the information polynomial $i(x)$ must be followed by $2t$ zeros.

3) The highest-order $i(x)$ coefficients must be entered into the LSR first.

When the first information symbol i_{k-1} appears at the input of the LSR of Figure 3.1, its output is $i_{k-1} \cdot g_{2t} \cdot x^{(k-1+2t)}$, since there is no contribution from its internal stages because its contents were cleared before entering a new codeword. After one clockpulse, the output is $(i_{k-2} \cdot g_{2t} + i_{k-1} \cdot g_{2t-1})x^{(k-2+2t)}$, and so on. After $(2t + k - 1)$ clockpulses, the LSR contains $0, 0 \ldots 0, i_0, i_1$, and the output is $(i_0 \cdot g_1 + i_1 \cdot g_0)x$, while the last non-zero output is $i_0 \cdot g_0$. Consequently, the product polynomial at the LSR's output is given by:

$$
\begin{aligned}
c(x) = i(x).g(x) &= i_{k-1}.g_{2t}.x^{k-1+2t} \\
&+ (i_{k-2}.g_{2t} + i_{k-1}.g_{2t-1})x^{k-2+2t} \\
&+ (i_{k-3}.g_{2t} + i_{k-2}.g_{2t-1} + i_{k-3}.g_{2t-2})x^{k-3+2t} \\
&\quad\vdots \\
&+ (i_0.g_1 + i_1.g_0)x \\
&+ i_0.g_0. \quad\quad\quad\quad\quad\quad\quad\quad\quad\quad\quad\quad (3.32)
\end{aligned}
$$

3.3.4.2 Systematic Cyclic Shift-register Encoding Example

In this subsection we demonstrate how LSR circuits can be used in order to carry out the polynomial division necessary for computing the parity polynomial $p(x)$ of systematic RS or BCH codes. Let us attempt to highlight the operation of the LSR division circuit by referring to the example computed for the systematic $RS(12, 8)$ encoder previously.

Example 5:
LSR polynomial division example for the $RS(12,8)$ $GF(16)$ code used for the generation of the parity polynomial $p(x)$ in cyclic systematic encoders.

The corresponding division circuit is depicted in Figure 3.2, where the generator polynomial of Equation 3.25 was used. Note that the highest-order information polynomial coefficients must be entered in the LSR first. Hence, $i(x)$ has to be arranged with the high-order coefficients at its right, ready for entering the LSR of Figure 3.2 from the left.

No. of shifts j	LSR content after j shifts				Output symbol after j shifts	Feedback				Input symbol
0	0	0	0	0	0	–	–	–	–	–
1	$\alpha^{12}+0=\alpha^{12}$	0	0	0	0	$0.\alpha^{10}=0$	$0.\alpha^3=0$	$0.\alpha^6=0$	$0.\alpha^{13}=0$	α^{12}
2	$\alpha^{12}+0=\alpha^{12}$	$\alpha^{12}+0=\alpha^{12}$	$0+0=0$	$0+0=0$	0	$0.\alpha^{10}=0$	$0.\alpha^3=0$	$0.\alpha^6=0$	$0.\alpha^{13}=0$	α^{12}
3	$\alpha^{12}+0=\alpha^{12}$	$\alpha^{12}+0=\alpha^{12}$	$\alpha^{12}+0=\alpha^{12}$	$0+0=0$	0	$0.\alpha^{10}=0$	$0.\alpha^3=0$	$0.\alpha^6=0$	$0.\alpha^{13}=0$	α^{12}
4	$\alpha^{12}+0=\alpha^{12}$	$\alpha^{12}+0=\alpha^{12}$	$\alpha^{12}+0=\alpha^{12}$	$\alpha^{12}+0=\alpha^{12}$	α^{12}	$\alpha^{12}.\alpha^{10}=\alpha^7$	$\alpha^{12}.\alpha^3=\alpha^0$	$\alpha^{12}.\alpha^6=\alpha^3$	$\alpha^{12}.\alpha^{13}=\alpha^{10}$	α^{12}
5	$\alpha^{12}+\alpha^7=\alpha^2$	$\alpha^{12}+\alpha^0=\alpha^{11}$	$\alpha^{12}+\alpha^3=\alpha^{10}$	$\alpha^{12}+\alpha^3=\alpha^3$	α^3	$\alpha^3.\alpha^{10}=\alpha^{13}$	$\alpha^3.\alpha^3=\alpha^6$	$\alpha^3.\alpha^6=\alpha^9$	$\alpha^3.\alpha^{13}=\alpha^1$	α^{12}
6	$\alpha^{12}+\alpha^{13}=\alpha^1$	$\alpha^2+\alpha^6=\alpha^3$	$\alpha^{11}+\alpha^9=\alpha^2$	$\alpha^{10}+\alpha^1=\alpha^8$	α^8	$\alpha^8.\alpha^{10}=\alpha^3$	$\alpha^8.\alpha^3=\alpha^{11}$	$\alpha^8.\alpha^6=\alpha^{14}$	$\alpha^8.\alpha^{13}=\alpha^6$	α^{12}
7	$\alpha^{12}+\alpha^3=\alpha^{10}$	$\alpha^1+\alpha^{11}=\alpha^6$	$\alpha^3+\alpha^{14}=\alpha^0$	$\alpha^2+\alpha^6=\alpha^3$	α^3	$\alpha^3.\alpha^{10}=\alpha^{13}$	$\alpha^3.\alpha^3=\alpha^6$	$\alpha^3.\alpha^6=\alpha^9$	$\alpha^3.\alpha^{13}=\alpha^1$	α^{12}
8	$\alpha^{12}+\alpha^{13}=\alpha^1$	$\alpha^{10}+\alpha^6=\alpha^7$	$\alpha^6+\alpha^9=\alpha^5$	$\alpha^0+\alpha^1=\alpha^4$	α^4	$\alpha^4.\alpha^{10}=\alpha^{14}$	$\alpha^4.\alpha^3=\alpha^7$	$\alpha^4.\alpha^6=\alpha^{10}$	$\alpha^4.\alpha^{13}=\alpha^2$	α^{12}
9	$0+\alpha^{14}=\alpha^{14}$	$\alpha^1+\alpha^7=\alpha^{14}$	$\alpha^7+\alpha^{10}=\alpha^6$	$\alpha^5+\alpha^2=\alpha^1$	α^1	$\alpha^1.\alpha^{10}=\alpha^{11}$	$\alpha^1.\alpha^3=\alpha^4$	$\alpha^1.\alpha^6=\alpha^7$	$\alpha^1.\alpha^{13}=\alpha^{14}$	0
10	$0+\alpha^{11}=\alpha^{11}$	$\alpha^{14}+\alpha^4=\alpha^9$	$\alpha^{14}+\alpha^7=\alpha^1$	$\alpha^6+\alpha^{14}=\alpha^8$	α^8	$\alpha^8.\alpha^{10}=\alpha^3$	$\alpha^8.\alpha^3=\alpha^{11}$	$\alpha^8.\alpha^6=\alpha^{14}$	$\alpha^8.\alpha^{13}=\alpha^6$	0
11	$0+\alpha^3=\alpha^3$	$\alpha^{11}+\alpha^{11}=0$	$\alpha^9+\alpha^{14}=\alpha^4$	$\alpha^1+\alpha^6=\alpha^{11}$	α^{11}	$\alpha^{11}.\alpha^{10}=\alpha^6$	$\alpha^{11}.\alpha^3=\alpha^{14}$	$\alpha^{11}.\alpha^6=\alpha^2$	$\alpha^{11}.\alpha^{13}=\alpha^9$	0
12	$0+\alpha^6=\alpha^6$	$\alpha^3+\alpha^{14}=\alpha^0$	$0+\alpha^2=\alpha^2$	$\alpha^4+\alpha^9=\alpha^{14}$	α^{14}	$\alpha^{14}.\alpha^{10}=\alpha^9$	$\alpha^{14}.\alpha^3=\alpha^2$	$\alpha^{14}.\alpha^6=\alpha^5$	$\alpha^{14}.\alpha^{13}=\alpha^{12}$	0

Table 3.5: List of LSR internal states for $RS(12,8)$ systematic encoder.

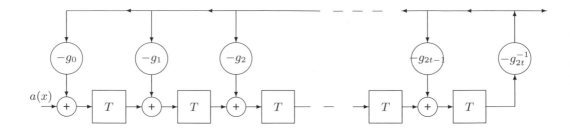

Figure 3.3: LSR circuit for dividing polynomials in non-systematic RS and BCH encoders.

By contrast, in the previously invoked polynomial division algorithm, $i(x)$ was arranged with the high-order coefficients at the left. In order to demonstrate the operation of the LSR encoder, we listed its contents during each clock cycle in Table 3.5, which mimics the operation carried out during the polynomial division portrayed in Table 3.4. Explicitly, the feedback loop of the schematic in Figure 3.2 carries out the required multiplications by the generator polynomial after each division step, while the modulo-2 additions correspond to subtracting the terms of the same order during the polynomial division, where subtraction and addition are identical in our modulo algebra. Careful inspection of Tables 3.4 and 3.5 reveals the inherent analogy of the associated operations. The quotient polynomial exits the encoder at its output after 12 clock cycles, which is seen in the central column of Table 3.5, while the remainder appears at the bottom left section of the table. Observe in the table that the first $2t = 4$ shifts of the LSR generate no output and consequently have no equivalent in the polynomial division algorithm. However, the following output symbols are identical to the quotient coefficients of $q(x)$ in the polynomial division algorithm of Table 3.4, and the LSR content after the 12th clockpulse or shift is exactly the remainder polynomial $p(x)$ computed by polynomial division. The field additions at the inputs of the LSR cells are equivalent to subtracting the terms of identical order in the polynomial division, which actually represent the updating of the dividend polynomial in the course of the polynomial division after computing a new quotient coefficient.

Following similar generalised arguments, an arbitrary systematic encoder can be implemented by the help of the *division circuit* of Figure 3.3. As in the case of the multiplier circuit, the LSR must again be cleared initially. Then the output of the LSR is zero for the first $2t$ clockpulses, and the first non-zero output is the first coefficient of the quotient, as we have seen in Table 3.5 for the $RS(12, 8, 2)$ encoder. Again, as we have demonstrated in our systematic $RS(12, 8)$ encoder example, for each quotient coefficient q_j the product polynomial $q_j.g(x)$ must be subtracted from the dividend polynomial, which is contained in the LSR. This is carried out by arranging for the current quotient coefficient q_j, multiplied by $g(x)$, to manipulate cells of the LSR by the help of the modulo gates in order to update its contents to hold the current dividend polynomial. After n clockpulses, the entire quotient has appeared at the LSR's output, and the current dividend polynomial in the LSR has an order which is lower than that of $g(x)$. Hence, it cannot be divided by $g(x)$; it is the remainder of the polynomial division operation. It therefore constitutes the required parity polynomial.

After this short overview of the analogies between GF field operations and their LSR implementations, we proceed with the outline of decoding and error correction algorithms for RS and BCH codes. We emphasise at this point that LSR implementations of GF arithmetics are usually computationally quite effective, though conceptually often a bit mysterious. However, one often accepts a conceptually more cumbersome solution for the sake of computational efficiency. Following the above portrayal of RS and BCH encoding techniques, let us now consider the corresponding decoding algorithms.

3.3.5 RS Decoding

3.3.5.1 Formulation of the Key Equations [32, 34, 82, 88, 90, 96, 97, 108, 109]

Since RS codes are non-binary BCH codes and BCH codes are cyclic codes, any cyclic decoding technique can be deployed for their decoding. In this section we follow the philosophies of Peterson [82, 84], Clark and Cain [88], Blahut [90], Berlekamp [32], Lin and Costello [96], and a number of other prestigious authors, who have contributed substantially to the better understanding of algebraic decoding techniques and derive a set of non-linear equations to compute the error locations and magnitudes, which can be solved by matrix inversion. This approach was originally suggested for binary BCH codes by Peterson [17], extended for non-binary codes by Gorenstein and Zierler [29], and refined again by Peterson [82, 84].

Let us now attempt to explore what information is given by the parity symbols represented by the coefficients of $p(x)$ about the location and magnitude of error symbols represented by the error polynomial $e(x)$, which are corrupting the encoded codeword $c(x)$ into the received codeword $r(x)$. Since the encoded codeword is given by:

$$c(x) = i(x) \cdot x^{n-k} + p(x)$$

and the error polynomial $e(x)$ is added component-wise in the GF:

$$
\begin{aligned}
r(x) &= c(x) + e(x) \\
&= c_{n-1}.x^{n-1} + c_{n-2}.x^{n-2} + \ldots + c_1.x + c_0 \\
&+ e_{n-1}.x^{n-1} + e_{n-2}.x^{n-2} + \ldots + e_1 + e_0.
\end{aligned}
\tag{3.33}
$$

May we remind the reader at this point that an $RS(n, k, t)$ code over $GF(2^m)$ contains in a codeword $c(x)$ n number of m-ary codeword symbols, constituted by k information and $(n-k) = 2t$ parity symbols of the polynomials $i(x)$ and $p(x)$, respectively. This code can correct up to t symbol errors, from which we suppose that $e(x)$ contains at most t non-zero coefficients. Otherwise, the error correction algorithm fails to compute the proper error positions and magnitudes. If we knew the error polynomial $e(x)$, we could remove its effect from the received codeword $r(x)$ resulting into $c(x)$, the error-free codeword, and we could recover the information $i(x)$ sent. Since effectively $e(x)$ carries merely $2t$ unknowns in the form of t unknown error positions and t unknown error magnitudes, it would be sufficient to know the value of $e(x)$ at $2t$ field elements to determine the $2t$ unknowns, required for error correction.

Recall that the generator polynomial has been defined as:

$$g(x) = \prod_{j=1}^{2t}(x - \alpha^j),$$

with zeros at the first $2t$ number of GF elements, and:

$$c(x) = i(x) \cdot g(x).$$

Therefore, $c(x) = 0$ at all zeros of $g(x)$ even in the case of systematic encoders. Hence, if we evaluate the received polynomial $r(x)$ at the zeros of $g(x)$, which are the first $2t$ GF elements, we arrive at:

$$
\begin{aligned}
r(x)|_{\alpha^1 \ldots \alpha^{2t}} &= [c(x) + e(x)]|_{\alpha^1 \ldots \alpha^{2t}} \\
&= [i(x).g(x) + e(x)]|_{\alpha^1 \ldots \alpha^{2t}} \\
&= [i(x).g(x)]|_{\alpha^1 \ldots \alpha^{2t}} + e(x)|_{\alpha^1 \ldots \alpha^{2t}} \\
&= 0 + e(x)|_{\alpha^1 \ldots \alpha^{2t}}.
\end{aligned}
\tag{3.34}
$$

Consequently, the received polynomial $r(x)$ evaluated at the zeros of $g(x)$, that is at the first $2t$ GF elements, provides us with the $2t$ values of the error polynomial $e(x)$, which are necessary for computation of the t error positions and t error magnitudes. These $2t$ characteristic values are called the *syndromes* $S_1 \ldots S_{2t}$, which can be viewed as the algebraic equivalents of the symptoms of an illness.

If we now assume that actually $v \leq t$ errors have occurred and the error positions are $P_1 \ldots P_v$, while the error magnitudes are $M_1 \ldots M_v$, then an actual transmission error, that is each non-zero component of the error polynomial $e(x)$, is characterised by the pair of GF elements (P_j, M_j). The $2t$ equations to be solved for error correction are as follows:

$$
\begin{aligned}
S_1 &= r(\alpha) = (e_{n-1}.x^{n-1} + e_{n-2}.x^{n-2} + \ldots + e_1.x + e_0)|_{x=\alpha} \\
S_2 &= r(\alpha^2) = (e_{n-1}.x^{n-1} + e_{n-2}.x^{n-2} + \ldots + e_1.x + e_0)|_{x=\alpha^2} \\
&\vdots \qquad \vdots \\
S_{2t} &= r(\alpha^{2t}) = (e_{n-1}.x^{n-1} + e_{n-2}.x^{n-2} + \ldots + e_1.x + e_0)|_{x=\alpha^{2t}}.
\end{aligned}
\tag{3.35}
$$

There are only $v \leq t$ non-zero error magnitudes $M_1 \ldots M_v$ with the corresponding positions $P_1 \ldots P_v$, which can be thought of as place markers for the error magnitudes, and all the positions $P_1 \ldots P_v$ and magnitudes $M_1 \ldots M_v$ are $GF(2^m)$ elements, both of which can be expressed as powers of the primitive element α. Wherever on the right-hand side (RHS) of the first equation in the set of Equations 3.35 there is a non-zero error magnitude $M_j \neq 0$, $j = 1 \ldots v$, $x = \alpha$ is substituted. In the corresponding terms of the second equation, $x = \alpha^2$ is substituted, and so forth, while in the last equation $x = \alpha^{2t}$ is used. Regardless of the actual error positions, the non-zero terms in the syndrome equations $S_1 \ldots S_{2t}$ are always ordered in corresponding columns above each other with appropriately increasing powers of the original non-zero error positions. This is so whether they are expressed as powers of the primitive element α or that of the error positions $P_1 \ldots P_v$. When formulating this equation in terms of the error positions, we arrive at:

$$
\begin{aligned}
S_1 &= r(\alpha) = M_1.P_1 + M_2.P_2 + \ldots + M_v.P_v = \sum_{i=1}^{v} M_i.P_i \\
S_2 &= r(\alpha^2) = M_1.P_1^2 + M_2.P_2^2 + \ldots + M_v.P_v^2 = \sum_{i=1}^{v} M_i.P_i^2 \\
&\vdots \qquad \vdots \\
S_{2t} &= r(\alpha^{2t}) = M_1.P_1^{2t} + M_2.P_2^{2t} + \ldots + M_v.P_v^{2t} = \sum_{i=1}^{v} M_i.P_i^{2t}.
\end{aligned}
\tag{3.36}
$$

Equation 3.36 can also be conveniently expressed in matrix form, as given below:

$$
\begin{bmatrix} S_1 \\ S_2 \\ \vdots \\ S_{2t} \end{bmatrix} = \begin{bmatrix} P_1 & P_2 & P_3 & \dots & P_v \\ P_1^2 & P_2^2 & P_3^2 & \dots & P_v^2 \\ \vdots & & & & \\ P_1^{2t} & P_2^{2t} & P_3^{2t} & \dots & P_v^{2t} \end{bmatrix} \begin{bmatrix} M_1 \\ M_2 \\ \vdots \\ M_v \end{bmatrix}
\tag{3.37}
$$

$$
\underline{S} = \underline{\underline{P}} \cdot \underline{M}.
$$

However, this set of equations is non-linear, and hence a direct solution appears to be too complicated, since in general there are many solutions. All the solutions must be found, and the most likely error pattern is the one with the lowest number of errors, which in fact minimises the probability of errors and the BER.

Peterson suggested a simple method [17, 87] for binary BCH codes, which has been generalised by Gorenstein and Zierler [29, 87] for non-binary RS codes. Hence, the corresponding algorithm is referred to as the Peterson–Gorenstein–Zierler (PGZ) decoder.

Their approach was based on the introduction of the *error locator polynomial L(x)*, which can be computed from the syndromes. The error locator polynomial can then be employed in order to linearise the set of equations, resulting in a more tractable solution for the error locations. Some authors define $L(x)$ as the polynomial with zeros at the error locations [97]. More frequently, however, [90, 96] it is supposed to have zeros at the multiplicative inverses of the error positions [90, 96], as suggested by the following equation:

$$
L(x) = (1 - x.P_1)(1 - x.P_2)\dots(1 - x.P_v) = \prod_{j=1}^{v}(1 - x.P_j)
\tag{3.38}
$$

$$
L(x) = L_v.x^v + L_{v-1}.x^{v-1} + \dots + L_1.x + 1 = \sum_{j=0}^{v} L_j.x^j.
\tag{3.39}
$$

Clearly, $L(x) = 0$ for $x = P_1^{-1}, P_2^{-1}, \dots, P_v^{-1}$. If $L(x)$, that is its coefficients or zeros, were known, we would know the error positions; therefore, we try to determine $L(x)$ from $S_1 \dots S_{2t}$ [82, 90]. Upon multiplying Equation 3.39 by $M_i.P_i^{(j+v)}$, we get:

$$
L(x).M_i.P_i^{j+v} = M_i.P_i^{j+v}(L_v.x^v + L_{v-1}.x^{v-1} + \dots + L_1.x + 1)
\tag{3.40}
$$

and on substituting $x = P_i^{-1}$ into Equation 3.40 we arrive at:

$$
0 = M_i.P_i^{j+v}(L_v.P_i^{-v} + L_{v-1}.P_i^{v-1} + \dots + L_1.P_i^{-1} + 1),
\tag{3.41}
$$

$$
0 = M_i(L_v.P_i^j + L_{v-1}.P_i^{j-1} + \dots + L_1.P_i^{j+v-1} + P_i^{j+v}).
\tag{3.42}
$$

There exists such an equation for all $i = 1 \dots v$ and all j. If we sum the equations for $i = 1 \dots v$, for each $j, j = 1 \dots 2t$, we get an equation of the form of Equation 3.43:

$$
\sum_{i=1}^{v} M_i(L_v.P_i^j + L_{v-1}.P_i^{j-1} + \dots + L_i.P_i^{j+v-1} + P_i^{j+v}) = 0.
\tag{3.43}
$$

Equivalently:

$$\sum_{i=1}^{v} M_i.L_v.P_i^j + \sum_{i=1}^{v} M_i.L_{v-1}.P_i^{j-1} + \ldots + \sum_{i=1}^{v} M_i.P_i^{j+v} = 0. \qquad (3.44)$$

If we compare Equation 3.44 with Equation 3.36, we can recognise the syndromes in the sums. Therefore, we arrive at:

$$L_v.S_j + L_{v-1}.S_{j+1} + \ldots + L_1.S_{j+v-1} + S_{j+v} = 0. \qquad (3.45)$$

The highest syndrome index is $j + v$, but since only the first $2 \cdot t$ syndromes $S_1 \ldots S_{2t}$ are specified, and since $v \leq t$, the condition $1 \leq j \leq t$ must be fulfilled. After rearranging Equation 3.45 we get a set of linear equations for the unknown coefficients $L_1 \ldots L_v$ as a function of the known syndromes $S_1 \ldots S_{2t}$, which is in fact the *key equation* for correcting RS or BCH codes. Any algorithm that delivers a solution to this set of equations can also be employed for correcting errors in RS codes. From Equation 3.45 we can also write:

$$L_v.S_j + L_{v-1}.S_{j+1} + \ldots + L_1.S_{j+v-1} = -S_{j+v} \quad \text{for } j = 1 \ldots v. \qquad (3.46)$$

The key equation is more easily understood in a matrix form:

$$\begin{bmatrix} S_1 & S_2 & S_3 & \ldots & S_{v-1} & S_v \\ S_2 & S_3 & S_4 & \ldots & S_v & S_{v-1} \\ \vdots & & & & & \\ S_v & S_{v+1} & S_{v+2} & \ldots & S_{2v-2} & S_{2v-1} \end{bmatrix} \begin{bmatrix} L_v \\ L_{v-1} \\ \vdots \\ L_1 \end{bmatrix} = \begin{bmatrix} -S_{v+1} \\ -S_{v+2} \\ \vdots \\ -S_{2v} \end{bmatrix} \qquad (3.47)$$

$$\underline{\underline{S}}.\underline{L} = \underline{S}. \qquad (3.48)$$

Equation 3.47 can be solved for the unknown coefficients of the error locator polynomial $L(x)$ (i.e. for the error positions $P_1 \ldots P_v$) if the matrix of syndromes is non-singular. Pless [89] p. 97 showed that the matrix $\underline{\underline{S}}$ has the form of the *Vandermonde matrix*, which plays a prominent role in the theory of error correction coding. $\underline{\underline{S}}$ can be shown to be non-singular if it is of dimension $v \times v$, where v is the actual rather than the maximum number of errors that occurred, while it is singular if the dimension of $\underline{\underline{S}}$ is greater than v [29, 82]. This theorem provides the basis for determining the actual number of errors v and determining the error positions $P_1 \ldots P_v$.

Before we proceed with our discourse on the various solutions to the key equation, we formulate it in different ways, from which the various solutions accrue. Notice that Equation 3.46 can also be interpreted in the form of a set of recursive formulae, which led to Massey's solution by the synthesis of an autoregressive filter [34]. Namely, if we assume that the coefficients $L_1 \ldots L_v$ are known for $j = 1 \ldots v$, Equation 3.46 generates recursively the next syndrome from the previous v number of syndromes as follows:

$$\begin{aligned} j = 1 \quad -S_{v+1} &= L_v.S_1 + L_{v-1}.S_2 + L_{v-2}.S_3 + \ldots + L_1.S_v \\ j = 2 \quad -S_{v+2} &= L_v.S_2 + L_{v-1}.S_3 + L_{v-2}.S_4 + \ldots + L_1.S_{v+1} \\ &\vdots \qquad \vdots \\ j = v \quad -S_{2v} &= L_v.S_v + L_{v-1}.S_{v+1} + L_{v-2}.S_{v+2} + \ldots + L_1.S_{2v-1}. \end{aligned}$$

$$(3.49)$$

This set of equations can also be written as:

$$S_j = - \sum_{n=1}^{v} L_n . S_{j-n} \quad \text{for } j = (v+1), (v+2), \dots, 2v, \qquad (3.50)$$

which is reminiscent of a convolutional expression and hence can be implemented by the help of a linear feedback shift register or *autoregressive filter* having taps $L_1 \dots L_v$, and with S_j fed back into the shift register. Massey's approach constitutes one of the most computationally effective alternatives to the solution of the key equations, in particular when the codeword length is high and the number of correctible errors is also high. Hence, we will demonstrate how it can be invoked after describing the conceptually simpler PGZ decoding algorithm.

A plethora of other solutions for the key equation in Equation 3.50 are also possible because in this form it is equivalent to the problem found, for example, in spectral estimation in speech coding, when solving a set of recursive equations of the same form for finding the prediction filter's coefficients. The most powerful solution to the set of equations accrues, if one recognises that the matrix of syndromes in Equation 3.47 can be shown to have both *Vandermonde* structure, and its values along the main diagonal are identical. In other words, it exhibits a symmetric Toeplitz structure. Efficient special algorithms for the solution of Toeplitz-type matrix equations have been proposed by Levinson, Robinson, Durbin, Berlekamp, Massey, Trench, Burg and Schur in the literature for various prediction and equalisation problems. All these techniques can be invoked for solving the key equations in Equation 3.50 when ensuring that all the operations are carried out over the previously described finite field $GF(2^m)$. Excellent discourses on the topic have been published in the literature by Makhoul [110], Blahut [111] pp. 352–387 and Schur [112].

Having derived various representations of the key equation for error correction (see Equations 3.46–3.50), it is instructive to continue with the description of the conceptually most simple solution. This was originally proposed by Peterson [17] for binary BCH codes, and it was extended by Gorenstein and Zierler [29] for non-binary RS codes on the basis of inverting the matrix of syndromes.

3.3.5.2 Peterson-Gorenstein-Zierler Decoder

As mentioned above, a number of solutions to the key equations have been suggested in [17, 29, 90, 97]. The following set of equations can be derived from Equation 3.48, whose solution is based on inverting the matrix of syndromes, as shown below:

$$\underline{L} = \underline{\underline{S}}^{-1} . \underline{S}. \qquad (3.51)$$

The solution is based on the following theorem. The *Vandermonde* matrix $\underline{\underline{S}}$ constituted by the syndromes is non-singular and can be inverted if its dimension is $v \times v$, but it is singular and cannot be inverted if its dimension is greater than v, where v is the actual number of errors that occurred.

We have to determine v to be able to invert $\underline{\underline{S}}$ in order to solve Equation 3.51. Initially, we set $v = t$, since t is the maximum possible number of errors, and we compute the determinant of $\underline{\underline{S}}$. If $det(\underline{\underline{S}}) = 0$, then $v = t$, and we can proceed with the matrix inversion. Otherwise, we decrement v by one and tentatively try $v = t - 1$, and so forth, down to $v = 0$, until $det(\underline{\underline{S}}) \neq 0$ is found. If we have found the specific v value, for which $det(\underline{\underline{S}}) \neq 0$, we compute $\underline{\underline{S}}^{-1}$ by matrix inversion and derive $\underline{L} = \underline{\underline{S}}^{-1} . \underline{S}$.

Now the zeros of the error locator polynomial are determined by trial and error. This is carried out by substituting all non-zero field elements into $L(x)$, and finding those for which we

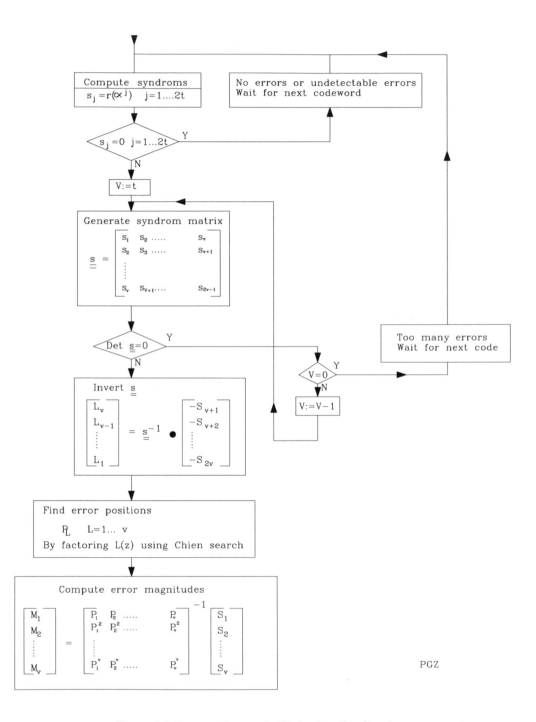

Figure 3.4: Peterson-Gorenstein-Zierler decoding flowchart.

have $L(x) = 0$. This method is called the *Chien search* [113]. The error positions $P_1 \ldots P_v$ can then be found by determining the multiplicative inverses of the zeros of $L(x)$. With this step, the solution is complete for binary BCH codes; the bits at the computed error positions must be inverted for error correction.

For non-binary RS codes the error magnitudes $M_1 \ldots M_v$ must be determined in a further step. This is relatively straightforward now from the syndrome Equation 3.37; simply the known matrix of error positions $\underline{\underline{P}}$ has to be inverted to compute the vector $\underline{M}$ of error magnitudes:

$$\underline{M} = \underline{\underline{P}}^{-1}.\underline{S}. \tag{3.52}$$

Similarly to the matrix of syndromes $\underline{S}$, $\underline{\underline{P}}$ can also be shown to have *Vandermonde* structure and consequently is non-singular and can be inverted, if exactly v errors have occurred. Although Equation 3.52 represents a set of $2t$ equations, it is sufficient to solve v of them for the v unknown error magnitudes $M_1 \ldots M_v$. The PGZ method of RS decoding is summarised in the flowchart of Figure 3.4. Observe in the figure that the error correction problem was converted into a problem of inverting two matrices.

Before providing a worked numerical example for PGZ decoding, we briefly allude to the *error detection capability* of the RS and BCH codes. Explicitly, it is a very attractive property of RS and BCH codes that after error correction the syndromes can be recomputed in order to check whether the code was capable of removing the transmission errors. This is necessary, because without it the received codeword could have been more similar to some other legitimate codeword, and hence the decoder might have concluded that this other, more similar codeword was transmitted. If the error correction action of the decoder was successful, all recomputed syndromes must be zero. If, however, there are errors after the error correction, a systematic decoder can at least separate the original information part of the codeword. This measure allows the decoder to minimize the 'damage', which would have been more catastrophic had the decoder attempted to correct the received symbols in the wrong positions, thereby actually corrupting potentially correct symbols.

In order to appreciate the details of the PGZ decoding algorithm, we now offer a worked example using the previously invoked $RS(12, 8, 2)$ code.

3.3.5.3 PGZ Decoding Example

Example 6:
Consider the double-error correcting $RS(12, 8, 2)$ code over $GF(2^4)$ and carry out error correction by matrix inversion, using the PGZ decoder.

Let us assume that the 'all-one' information sequence has been sent in the form of the systematically encoded codeword of Section 3.3.3:

$$
\begin{aligned}
i(x) &= \alpha^{12}.x^{11} + \alpha^{12}.x^{10} + \alpha^{12}.x^9 + \alpha^{12}.x^8 \\
&+ \alpha^{12}.x^7 + \alpha^{12}.x^6 + \alpha^{12}.x^5 + \alpha^{12}.x^4 \\
p(x) &= \alpha^{14}.x^3 + \alpha^2.x^2 + \alpha^0.x + \alpha^6.
\end{aligned}
$$

Let us also assume that two errors have occurred in $c(x)$ in positions 11 and 3, resulting in the following received polynomial:

$$\begin{aligned}
r(x) \;=\; & \underbrace{\alpha^4}_{error}.x^{11} + \alpha^{12}.x^{10} + \alpha^{12}.x^9 + \alpha^{12}.x^8 \\
+ \; & \alpha^{12}.x^7 + \alpha^{12}.x^6 + \alpha^{12}.x^5 + \alpha^{12}.x^4 \\
+ \; & \underbrace{\alpha^5}_{error}.x^3 + \alpha^2.x^2 + \alpha^0.x + \alpha^6.
\end{aligned}$$

Since $r(x) = c(x) + e(x)$, based on Table 3.3, we infer that $e(x)$ contains two non-zero terms, in position 11, where the systematic data symbol was α^{12} and hence the additive error was:

$$e_{11} = r_{11} - c_{11} = \alpha^4 + \alpha^{12} = \alpha^6$$

and in position 3, where the parity symbol was α^{14}, requiring an error magnitude of:

$$e_3 = r_3 - c_3 = \alpha^5 + \alpha^{14} = \alpha^{12}.$$

Hence:

$$e(x) = \alpha^6.x^{11} + 0 + 0 + 0 + 0 + 0 + 0 + 0 + \alpha^{12}.x^3 + 0 + 0 + 0.$$

This exponential representation is easily translated into bit patterns by referring to Table 3.2, which contains the various representations of $GF(2^4)$ elements.

Syndrome calculation: Since the syndromes are the received polynomials evaluated at the $GF(2^4)$ elements $\alpha^1 \ldots \alpha^{2t} = \alpha^1 \ldots \alpha^4$, we have:

$$\begin{aligned}
S_1 \;=\; & r(\alpha^1) \\
=\; & \alpha^4\alpha^{11} + \alpha^{12}\alpha^{10} + \alpha^{12}\alpha^9 + \alpha^{12}\alpha^8 + \alpha^{12}\alpha^7 + \alpha^{12}\alpha^6 \\
+\; & \alpha^{12}\alpha^5 + \alpha^{12}\alpha^4 + \alpha^5\alpha^3 + \alpha^2\alpha^2 + \alpha^0\alpha^1 + \alpha^6 \\
=\; & \underbrace{\alpha^0 + \alpha^7}_{\alpha^9} + \alpha^6 + \alpha^5 + \alpha^4 + \alpha^3 + \alpha^2 + \alpha^1 + \alpha^8 + \alpha^4 + \alpha^1 + \alpha^6 \\
=\; & \underbrace{\alpha^9 + \alpha^{11}}_{\alpha^2} + \alpha^0 \\
=\; & \alpha^2 + \alpha^0 \\
S_1 \;=\; & \alpha^8
\end{aligned}$$

$$
\begin{aligned}
S_2 &= r(\alpha^2) \\
&= \alpha^4\alpha^{22} + \alpha^{12}\alpha^{20} + \alpha^{12}\alpha^{18} + \alpha^{12}\alpha^{16} + \alpha^{12}\alpha^{14} \\
&+ \alpha^{12}\alpha^{12} + \alpha^{12}\alpha^{10} + \alpha^{12}\alpha^{8} + \alpha^{5}\alpha^{6} + \alpha^{2}\alpha^{4} + \alpha^{0}\alpha^{2} + \alpha^{6} \\
&= \alpha^{11} + \alpha^2 + \underbrace{\alpha^0 + \alpha^{13}}_{\alpha^6} + \alpha^{11} + \underbrace{\alpha^9 + \alpha^7}_{\alpha^0} + \underbrace{\alpha^5 + \alpha^{11}}_{\alpha^3} + \alpha^6 + \alpha^2 + \alpha^6 \\
&= \underbrace{\alpha^6 + \alpha^0}_{\alpha^{13}} + \alpha^3 \\
&= \alpha^{13} + \alpha^3 \\
S_2 &= \alpha^8
\end{aligned}
$$

$$
\begin{aligned}
S_3 &= r(\alpha^3) \\
&= \alpha^4\alpha^{33} + \alpha^{12}\alpha^{30} + \alpha^{12}\alpha^{27} + \alpha^{12}\alpha^{24} + \alpha^{12}\alpha^{21} \\
&+ \alpha^{12}\alpha^{18} + \alpha^{12}\alpha^{15} + \alpha^{12}\alpha^{12} + \alpha^{5}\alpha^{9} + \alpha^{2}\alpha^{6} + \alpha^{0}\alpha^{3} + \alpha^{6} \\
&= \alpha^7 + \alpha^{12} + \alpha^9 + \alpha^6 + \alpha^3 + \alpha^0 + \alpha^{12} + \alpha^9 + \underbrace{\alpha^{14} + \alpha^8}_{\alpha^6} + \alpha^3 + \alpha^6 \\
&= \alpha^9 + \alpha^6 \\
S_3 &= \alpha^5
\end{aligned}
$$

$$
\begin{aligned}
S_4 &= r(\alpha^4) \\
&= \alpha^4\alpha^{44} + \alpha^{12}\alpha^{40} + \alpha^{12}\alpha^{36} + \alpha^{12}\alpha^{32} + \alpha^{12}\alpha^{28} + \alpha^{12}\alpha^{24} \\
&+ \alpha^{12}\alpha^{20} + \alpha^{12}\alpha^{16} + \alpha^{5}\alpha^{12} + \alpha^{2}\alpha^{8} + \alpha^{0}\alpha^{4} + \alpha^{6} \\
&= \alpha^3 + \underbrace{\alpha^7 + \alpha^3 + \alpha^{14}}_{\alpha^1} + \alpha^{10} + \alpha^6 + \alpha^2 + \alpha^{13} + \alpha^2 + \alpha^{10} + \alpha^4 + \alpha^6 \\
&= \alpha^1 + \alpha^{11} \\
S_4 &= \alpha^6.
\end{aligned}
$$

Since we have no information on the actual number of errors v in the receiver, we must determine v first. Let us suppose initially $v = t = 2$. The key equation to be solved according to Equation 3.47 is as follows:

$$
\underbrace{\begin{bmatrix} S_1 & S_2 \\ S_2 & S_3 \end{bmatrix}}_{\underline{\underline{S}}} \underbrace{\begin{bmatrix} L_2 \\ L_1 \end{bmatrix}}_{\underline{L}} = - \underbrace{\begin{bmatrix} S_3 \\ S_4 \end{bmatrix}}_{\underline{S}}.
$$

Let us now compute the determinant of $\underline{\underline{S}}$, as portrayed in:

$$det(\underline{\underline{S}}) = det \begin{vmatrix} S_1 & S_2 \\ S_2 & S_3 \end{vmatrix} = (S_1 S_3 - S_2^2).$$

By substituting the syndromes and using Table 3.3 we arrive at:

$$S_1 = \alpha^8, \; S_2 = \alpha^8, \; S_3 = \alpha^5, \; S_4 = \alpha^6$$

$$det(\underline{\underline{S}}) = (\alpha^8.\alpha^5 - (\alpha^8)^2) = \alpha^{13} - \alpha^1 = \alpha^{12} \neq 0.$$

Since $det(\underline{\underline{S}}) \neq 0$, $v = 2$ errors occurred, and we can compute $\underline{\underline{S}}^{-1}$.

Let us use *Gauss–Jordan* elimination [114] for the computation of $\underline{\underline{S}}^{-1}$. This can be achieved by transforming both the original matrix $\underline{\underline{S}}$, which has to be inverted, and the unit matrix in the same way, until the original $\underline{\underline{S}}$ is transformed to a unit matrix. The transformed unit matrix then becomes inverted matrix $\underline{\underline{S}}^{-1}$. The operations are naturally $GF(2^4)$ operations, and we commence the transformations with the following matrices:

$$\underline{\underline{S}} = \begin{bmatrix} \alpha^8 & \alpha^8 \\ \alpha^8 & \alpha^5 \end{bmatrix} \quad \underline{\underline{I}} = \begin{bmatrix} \alpha^0 & 0 \\ 0 & \alpha^0 \end{bmatrix}.$$

After adding row 1 of both matrices to their second rows, we get:

$$\begin{bmatrix} \alpha^8 & \alpha^8 \\ \alpha^8 + \alpha^8 = 0 & \alpha^5 + \alpha^8 = \alpha^4 \end{bmatrix} \quad \begin{bmatrix} \alpha^0 & 0 \\ \alpha^0 & \alpha^0 \end{bmatrix}.$$

In a second step we carry out the assignment row 1 = row 1 . α^{11} + row 2, yielding:

$$\begin{bmatrix} \alpha^4 & 0 \\ 0 & \alpha^4 \end{bmatrix} \quad \begin{bmatrix} \alpha^{11} + \alpha^0 = \alpha^{12} & \alpha^0 \\ \alpha^0 & \alpha^0 \end{bmatrix}.$$

Finally, we compute row 1 = row 1 . α^{11} and row 2 = row 2 . α^{11} in order to render $\underline{\underline{S}}$ a unit matrix, as seen in:

$$\begin{bmatrix} \alpha^0 & 0 \\ 0 & \alpha^0 \end{bmatrix} \quad \begin{bmatrix} \alpha^8 & \alpha^{11} \\ \alpha^{11} & \alpha^{11} \end{bmatrix}.$$

Therefore, the inverted matrix becomes:

$$\underline{\underline{S}}^{-1} = \begin{bmatrix} \alpha^8 & \alpha^{11} \\ \alpha^{11} & \alpha^{11} \end{bmatrix}$$

and hence:

$$\underline{L} = \underline{\underline{S}}^{-1}.\underline{S}$$

which gives:

$$\begin{bmatrix} L_2 \\ L_1 \end{bmatrix} = \begin{bmatrix} \alpha^8 & \alpha^{11} \\ \alpha^{11} & \alpha^{11} \end{bmatrix} . \begin{bmatrix} \alpha^5 \\ \alpha^6 \end{bmatrix} = \begin{bmatrix} \alpha^{13} + \alpha^2 \\ \alpha^1 + \alpha^2 \end{bmatrix} = \begin{bmatrix} \alpha^{14} \\ \alpha^5 \end{bmatrix}.$$

Now we can compute the error positions from $L(x)$, where:

$$L(x) = \alpha^{14}.x^2 + \alpha^5.x + 1,$$

by trying all $GF(2^4)$ elements according to the Chien search:

$$
\begin{aligned}
L(\alpha^0) &= \alpha^{14}\alpha^0 + \alpha^5\alpha^0 + 1 = \alpha^{14} + \alpha^5 + \alpha^0 = \alpha^{11} \\
L(\alpha^1) &= \alpha^{14}\alpha^2 + \alpha^5\alpha^1 + 1 = \alpha^1 + \alpha^6 + \alpha^0 = \alpha^{11} + \alpha^0 = \alpha^{12} \\
L(\alpha^2) &= \alpha^{14}\alpha^4 + \alpha^5\alpha^2 + 1 = \alpha^3 + \alpha^7 + \alpha^0 = \alpha^4 + \alpha^0 = \alpha^1
\end{aligned}
$$

$$
\begin{aligned}
L(\alpha^3) &= \alpha^{14}\alpha^6 + \alpha^5\alpha^3 + 1 = \alpha^5 + \alpha^8 + \alpha^0 = \alpha^4 + \alpha^0 = \alpha^1 \\
L(\alpha^4) &= \alpha^{14}\alpha^8 + \alpha^5\alpha^4 + 1 = \alpha^7 + \alpha^9 + \alpha^0 = \alpha^0 + \alpha^0 = 0 \longleftarrow \\
L(\alpha^5) &= \alpha^{14}\alpha^{10} + \alpha^5\alpha^5 + 1 = \alpha^9 + \alpha^{10} + \alpha^0 = \alpha^{13} + \alpha^0 = \alpha^6
\end{aligned}
$$

$$
\begin{aligned}
L(\alpha^6) &= \alpha^{14}\alpha^{12} + \alpha^5\alpha^6 + 1 = \alpha^{11} + \alpha^{11} + \alpha^0 = 0 + \alpha^0 = \alpha^0 \\
L(\alpha^7) &= \alpha^{14}\alpha^{14} + \alpha^5\alpha^7 + 1 = \alpha^{13} + \alpha^{12} + \alpha^0 = \alpha^1 + \alpha^0 = \alpha^4 \\
L(\alpha^8) &= \alpha^{14}\alpha^{16} + \alpha^5\alpha^8 + 1 = \alpha^0 + \alpha^{13} + \alpha^0 = 0 + \alpha^{13} = \alpha^{13}
\end{aligned}
$$

$$
\begin{aligned}
L(\alpha^9) &= \alpha^{14}\alpha^{18} + \alpha^5\alpha^9 + 1 = \alpha^2 + \alpha^{14} + \alpha^0 = \alpha^{13} + \alpha^0 = \alpha^6 \\
L(\alpha^{10}) &= \alpha^{14}\alpha^{20} + \alpha^5\alpha^{10} + 1 = \alpha^4 + \alpha^0 + \alpha^0 = \alpha^4 + 0 = \alpha^4 \\
L(\alpha^{11}) &= \alpha^{14}\alpha^{22} + \alpha^5\alpha^{11} + 1 = \alpha^6 + \alpha^1 + \alpha^0 = \alpha^{11} + \alpha^0 = \alpha^{12}
\end{aligned}
$$

$$
\begin{aligned}
L(\alpha^{12}) &= \alpha^{14}\alpha^{24} + \alpha^5\alpha^{12} + 1 = \alpha^8 + \alpha^2 + \alpha^0 = \alpha^0 + \alpha^0 = 0 \longleftarrow \\
L(\alpha^{13}) &= \alpha^{14}\alpha^{26} + \alpha^5\alpha^{13} + 1 = \alpha^{10} + \alpha^3 + \alpha^0 = \alpha^{12} + \alpha^0 = \alpha^{11} \\
L(\alpha^{14}) &= \alpha^{14}\alpha^{28} + \alpha^5\alpha^{14} + 1 = \alpha^{12} + \alpha^4 + \alpha^0 = \alpha^6 + \alpha^0 = \alpha^{13}.
\end{aligned}
$$

Since the error locator polynomial has its zeros at the inverse error positions, their multiplicative inverses have to be found in order to determine the actual error positions:

$$
\begin{aligned}
(\alpha^4)^{-1} &= \alpha^{11} = P_1 \\
(\alpha^{12})^{-1} &= \alpha^3 = P_2,
\end{aligned}
$$

which are indeed the positions, where the $c(x)$ symbols have been corrupted by $e(x)$.

If we had a binary BCH code, the error correction would simply be the inversion of the bits in positions $P_1 = \alpha^{11}$ and $P_2 = \alpha^3$. For the non-binary $RS(12, 8, 2)$ code over $GF(2^4)$, the error magnitudes still have to be determined. Now we know that there are $v = 2$ errors and so the matrix $\underline{\underline{P}}$ of error positions in Equation 3.37 can be inverted and

Equation 3.52 can be solved for the error positions $M_1 \ldots M_v$. But since $v = 2$, there are only two equations to be solved for M_1 and M_2, so we simply substitute P_1 and P_2 into Equation 3.36, which gives:

$$
\begin{aligned}
S_1 &= M_1 P_1 + M_2 P_2 \\
S_2 &= M_1 P_1^2 + M_2 P_2^2 \\
\alpha^8 &= M_1 \alpha^{11} + M_2 \alpha^3 \\
\alpha^8 &= M_1 \alpha^7 + M_2 \alpha^6.
\end{aligned}
$$

From the first equation we can express M_1 and substitute it into the second one as follows:

$$
\begin{aligned}
M_1 &= \frac{\alpha^8 + M_2 \alpha^3}{\alpha^{11}} = (\alpha^8 + M_2 \alpha^3)\alpha^4 \\
M_1 &= \alpha^{12} + M_2.\alpha^7
\end{aligned}
$$

$$
\begin{aligned}
\alpha^8 &= (\alpha^{12} + M_2.\alpha^7)\alpha^7 + M_2.\alpha^6 \\
\alpha^8 &= \alpha^4 + M_2.\alpha^{14} + M_2.\alpha^6 \\
\alpha^8 - \alpha^4 &= M_2(\alpha^{14} + \alpha^6)
\end{aligned}
$$

$$
\begin{aligned}
M_2 &= \frac{\alpha^8 + \alpha^4}{\alpha^{14} + \alpha^6} = \frac{\alpha^5}{\alpha^8} = \alpha^5.\alpha^7 = \alpha^{12} \\
M_1 &= \alpha^{12} + M_2.\alpha^7 = \alpha^{12} + \alpha^{12}.\alpha^7 = \alpha^{12} + \alpha^4 = \alpha^6.
\end{aligned}
$$

Now we are able to correct the errors by the help of the pairs:

$$
\begin{aligned}
(P_1, M_1) &= (\alpha^{11}, \alpha^6) \\
(P_2, M_2) &= (\alpha^3, \alpha^{12}),
\end{aligned}
$$

if we simply add the error polynomial $e(x) = \alpha^6.x^{11} + \alpha^{12}.x^3$ to the received polynomial $r(x)$ to recover the error-free codeword polynomial $c(x)$, carrying the original information $i(x)$.

   As we have seen, the PGZ method involves the inversion of two matrices, one to compute the error positions $P_1 \ldots P_v$, and one to determine the error magnitudes $M_1 \ldots M_v$. For short codes over small GFs with low values of the n, k, t and m parameters, the computational complexity is relatively low. This is because the number of multiplications is proportional to t^3. However, for higher t values the matrix inversions have to be circumvented. The first matrix inversion can, for example, be substituted by the algorithms suggested by Berlekamp [32] and Massey [34, 87], which pursue similar approaches. For the efficient computation of the error magnitudes, Forney has proposed a method [115] that can conveniently substitute the second matrix inversion. Let us now embark on highlighting the principles of these techniques, noting that at this stage we can proceed to the next chapter without jeopardising the seamless flow of thought.

3.3.5.4 Berlekamp–Massey Algorithm [32, 34, 82, 88, 90, 96, 97, 108, 109]

In Section 3.3.5.1 we formulated the key equation first in matrix form and showed a solution based on matrix inversion. We also mentioned that by exploiting the Toeplitz structure of the

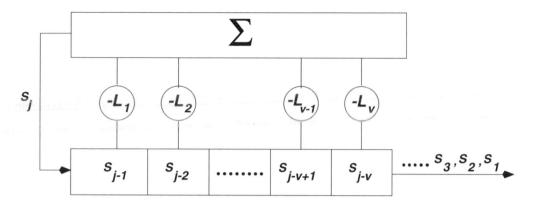

Figure 3.5: LFSR design for generating the syndromes $S_1 \ldots S_{2t}$ where $S_j = -\sum_{n=1}^{v} L_n \cdot S_{j-n}$ for $j = (v+1), \ldots, 2v$.

syndrome matrix $\underline{\underline{S}}$, more efficient methods can be contrived. Instead of pursuing this matrix approach, we now concentrate on the form of the key equation described by Equation 3.50, which suggests the design of an autoregressive filter or *Linear Feedback Shift Register (LFSR)* in order to generate the required sequence of syndromes $S_1 \ldots S_{2t}$. Berlekamp suggested an efficient iterative heuristic solution to the problem [32], p. 180, which can be better understood in Massey's original interpretation [34]. However, even Massey's approach is somewhat abstract, and several authors have attempted to give their own slants on its justification [82, 88, 90, 96, 97]. Some others find it instructive to introduce it after explaining a simpler algorithm, the Euclidean algorithm [116]. In this treatise, we will attempt to make the Berlekamp–Massey (BM) algorithm plausible by the help of a LFSR approach [90, 117], which shows that it is possible to design the taps of a LFSR, constituted in our case by the coefficients $L_0 \ldots L_v$ of the error locator polynomial $L(x)$ so that it produces a required sequence, which is represented by the syndromes $S_1 \ldots S_{2t}$ in our case. In a conventional signal processing context, this would be equivalent to designing the coefficients of a filter such that it would generate a required output sequence. However, in our current context all operations are over the finite GF.

According to this approach, Equation 3.50 can be modelled by the LFSR depicted in Figure 3.5. Clearly, our objective is to determine the length v and the feedback tap values L_n of the LFSR, such that it recursively generates the already known syndromes. Naturally, there is only a limited number of LFSR designs, which produce the required sequence $S_1 \ldots S_{2t}$, and an error polynomial is associated with each possible LFSR design. Note that the order of the error locator polynomial predetermines the number of errors per codeword. The probability of receiving a transmitted codeword with one symbol error is always lower than the joint probability of having a double error within the codeword. In general terms, the most likely error pattern is the one with the lowest number of errors per codeword, which is equivalent to saying that we are looking for the minimum length LFSR design; that is, for the minimum order error locator polynomial $L(x)$.

Since the BM algorithm is iterative, it is instructive to start from iteration one ($i = 1$), where the shortest possible LFSR length to produce the first syndrome S_1 is $l^{(i)} = l^{(1)} = 1$ and the corresponding LFSR connection polynomial, that is the error locator polynomial at $i = 1$, is $L^i(x) = L^1(x) = 1$. Let us now assume tentatively that this polynomial also produces the second syndrome S_2, and let us attempt to compute the first estimated syndrome S_{2e} produced by this tentative feedback connection polynomial $L^1(x) = 1$ at $i = 1$ by referring to Equation 3.50

and Figure 3.5:

$$S_{i+1} = -\sum_{n=1}^{l^{(i)}} L_n^{(i)} S_{i+1-n}. \tag{3.53}$$

Since at the current stage of iteration we have $l^{(i)} = 1$, $j = 2$, $L^{(1)} = 1$, the estimated syndrome is given by:

$$S_{2e} = -1 \cdot S_1 = -S_1,$$

which means that the second estimated syndrome S_{2e} is actually approximated by the first one. This might be true in some special cases, but not in general, and a so-called *discrepancy* or error term $d^{(i)}$ is generated in order to check whether the present LFSR design adequately produces the next syndrome. This discrepancy is logically the difference of the required precomputed syndrome $S_{(i+1)}$ and its estimate $S_{(i+1)e}$, which is given by:

$$d^{(i)} = S_{i+1} - S_{(i+1)e} = S_{i+1} + \sum_{n=1}^{l^{(i)}} L_n^{(i)} \cdot S_{i+1-n} = \sum_{n=0}^{l^{(i)}} L_n^{(i)} S_{i+1-n}. \tag{3.54}$$

Upon exploiting that $L_0 \neq 1$, for the iteration index $i = 1$ from Equation 3.54, we have:

$$d^{(1)} = L_0^{(1)} \cdot S_2 + L_1^{(1)} \cdot S_1 = S_2 + S_1.$$

If the current discrepancy or syndrome estimation error is $d^{(1)} = 0$, then the present LFSR design correctly produces the next syndrome S_2. This means that the internal variables $l^{(i)}, L^{(i)}(x), i$ of the iterative process must remain tentatively unaltered for the duration of the next iteration. In other words, the present LFSR design will be initially employed in the next iteration as a trial design in order to estimate the next precomputed syndrome. Therefore, the following 'update' has to take place:

$$\begin{aligned}
l^{(2)} &:= l^{(1)} = 1 \\
L^{(2)}(x) &:= L^{(1)}(x) \\
i &:= i+1 = 2.
\end{aligned}$$

The above statements are also valid at other stages of the syndrome iteration and hence in general, if $d^{(i)} = 0$, then the following assignments become effective:

$$\begin{aligned}
l^{(i+1)} &:= l^{(i)} \\
L^{(i+1)}(x) &:= L^{(i)}(x) \\
i &:= i+1.
\end{aligned}$$

However, if $d^{(i)} \neq 0$, the LFSR design must be modified until a solution is found, where $d^{(i)} = 0$. This means that the LFSR must be lengthened by the minimum possible number of stages and another connection polynomial has to be found, which produces the next desired syndrome S_{i+1} with an estimation error or discrepancy of $d_i = 0$. Furthermore, all the previous syndromes $S_1 \ldots S_i$ have to be also properly generated.

   One possible way of doing this is to remember the last case at iteration m, when the LFSR has failed to produce the next syndrome S_m, and use the mth LFSR design with its associated discrepancy of d_m in order to modify the present design as follows. The linearity of the LFSR circuitry allows us to invoke superposition, where the non-zero discrepancy of the current LFSR design can be cancelled by appropriately scaling the mth non-zero discrepancy and superimposing it on the current non-zero estimation error. This can be achieved by actually superimposing two LFSR designs exhibiting non-zero estimation errors. In terms of discrepancies we have:

$$d^{(i)} - d^{(i)} \cdot \frac{d^{(m)}}{d^{(m)}} = 0, \tag{3.55}$$

which suggests a solution for the choice of the proper connection polynomial in an equivalent form:

$$L^{(i+1)}(x) = L^{(i)}(x) - x^{i-m} \cdot \frac{d^{(i)}}{d^{(m)}} \cdot L^{(m)}(x). \tag{3.56}$$

In order to understand Equation 3.56, we emphasise again that two LFSRs can be superimposed to produce the required syndrome with zero discrepancy, since they are linear circuits, where superposition applies. Consequently, we can have a separate auxiliary LFSR of length $l^{(m)}$ with connection polynomial $L^{(m)}(x)$ and discrepancy d_m, the coefficients of which are scaled by the factor $\frac{d^{(i)}}{d^{(m)}}$ in order to compensate properly for the non-zero discrepancy. Note, however, that the auxiliary connection polynomial $L^{(m)}(x)$ has to be shifted to the required stage of the LFSR. This can be carried out using the multiplicative factor $x^{(i-m)}$. Having been shifted by $(i - m)$ positions, the output of the appropriately positioned auxiliary connection polynomial $x^{(i-m)} \cdot L^{(m)}(x)$ is added to that of the LFSR design $L^{(i)}(x)$, cancelling the undesirable non-zero discrepancy at this iteration.

We note that because of its shift by $x^{(i-m)}$ the connection polynomial $L^{(m)}(x)$ actually contributes to $L^{(i+1)}(x)$ only in coefficients with indices in excess of $(i-m)$, which implies that the first m number of $L^{(i)}(x)$ coefficients are not altered by this operation, while those above the index m must be altered to result in $d^{(i)} = 0$. Since we have chosen the most recent mth iteration, at which $d^{(m)} \neq 0$ occurred and consequently LFSR lengthening was necessary, the LFSR length $l^{(m+1)}$ is higher than $l^{(m)}$, but only by the minimum required number of stages. Hence, we have a minimum length LFSR design. Because of the linearity of the LFSR circuit, the auxiliary LFSR can be physically merged with the main LFSR. In order to assist in future modifications of the LFSR, when a non-zero discrepancy is produced, the most recent auxiliary LFSR with $d^{(m)} \neq 0$ must be stored as well.

Since we have made the inner workings of the BM algorithm plausible, we now attempt to summarise formally the set of iterative steps to be carried out to design the LFSR. The generation of the appropriate LFSR is synonymous with the determination of the minimum length error locator polynomial $L(x)$, which generates the required precomputed sequence of syndromes. The algorithm had originally been stated in the form of a number of theorems, lemmas and corollaries, for which the rigorous proofs can be found in Massey's original paper [34], but only its essence is presented here.

Theorem 1: An LFSR of length $l^{(i)}$, which generates the required syndromes $S_1, S_2, \ldots, S_{i-1}$ and the required syndrome sequence $S_1, S_2, \ldots, S_i$, does not have to be lengthened; hence, $l^{(i+1)} = l^{(i)}$ is satisfied. Conversely, the LFSR of length $l^{(i)}$ that generates $S_1, S_2, \ldots, S_{i-1}$, but fails to generate the required syndrome sequence $S_1, S_2, \ldots, S_i$, has to be lengthened. In general, the LFSR length has to obey $l^{(i+1)} := MAX\{l^{(i)}, (i + 1 - l^{(i)})\}$. Thus, the LFSR has to be lengthened if and only if:

$$i + 1 - l^{(i)} > l^{(i)} \tag{3.57}$$

$$\text{or} \quad i + 1 > 2l^{(i)}, \tag{3.58}$$

$$\text{i.e.,} \quad i \geq 2l^{(i)}, \tag{3.59}$$

and the increased LFSR length is given by:

$$l^{(i+1)} = i + 1 - l^{(i)}. \tag{3.60}$$

Following the above introductory elaborations, the BM algorithm can be more formally summarised following Blahut's interpretation [90] as seen in Figure 3.6. The various stages of processing are numbered in the flowchart, and these steps are listed below.

1 Set the initial conditions:
   Iteration index: $i = 0$
   LFSR length: $l^{(1)} = 0$
   Connection polynomial: $L^{(1)}(x) = 1$
   Auxiliary connection polynomial: $A^{(1)}(x) = 1$
   For $i := 0 \ldots 2t - 1$ apply the set of recursive relations specified by *Steps* $2 \ldots 11$.

2 Check whether we reached the end of iteration, that is whether $i = 2t$, and if so, branch to *Step* 13.

3 Compute the discrepancy or estimation error associated with the generation of the next syndrome from Equation 3.54:

$$d^{(i)} = \sum_{n=0}^{l^{(i)}} L_n^{(i)} . S_{i+1-n}$$

and go to *Step* 4.

4 Check whether the discrepancy is zero, and if $d^{(i)} = 0$, then the present LFSR design having a length of $l^{(i)}$ and connection polynomial $L^{(i)}(x)$ does produce the next syndrome S_{i+1}. Hence, go to *Step* 5; otherwise go to *Step* 6.

5 Simply shift the auxiliary LFSR by one position using the following operation:

$$A^{(i)}(x) := x . A^{(i)}(x) \tag{3.61}$$

and go to *Step* 12.

6 Since $d^{(i)} \neq 0$, we correct the temporary connection polynomial $T^{(i)}(x)$ by adding the properly shifted, normalised and scaled auxiliary connection polynomial $A^{(i)}(x)$ to the current connection polynomial $L^{(i)}(x)$, as seen below:

$$T^{(i)}(x) = L^{(i)}(x) - d^{(i)} . x . A^{(i)}(x). \tag{3.62}$$

7 Check whether the LFSR has to be lengthened, by using *Theorem 1* and Equation 3.59. If $2l^{(i)} \leq i$, go to *Step* 8; otherwise go to *Step* 9.

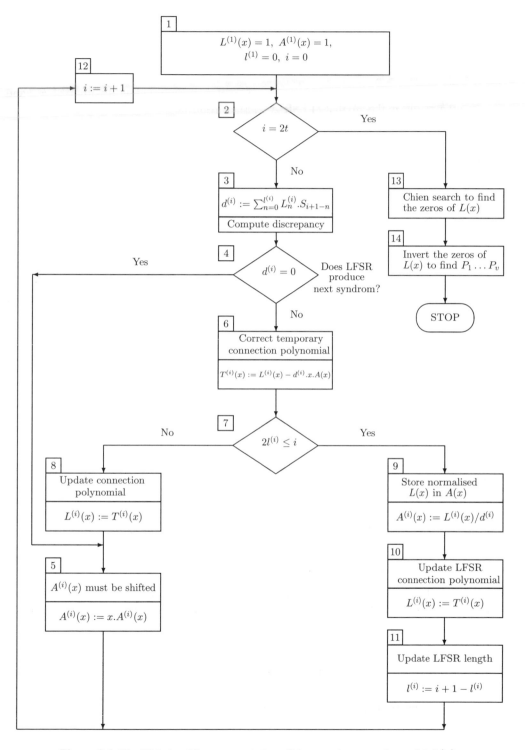

Figure 3.6: The BM algorithm: computation of the error locator polynomial $L(x)$.

8 Update the connection polynomial according to the following formula:

$$L^{(i)}(x) := T^{(i)}(x) \tag{3.63}$$

and go to *Step* 5.

9 Normalise the most recent connection polynomial $L(x)$ by dividing it with $d^{(i)} \neq 0$ and store the results in the auxiliary LFSR $A^{(i)}(x)$, as follows:

$$A^{(i)}(x) := L^{(i)}(x)/d^{(i)} \tag{3.64}$$

and go to *Step* 10.

10 Now one can overwrite $L^{(i)}(x)$, since it has been normalised and stored in $A^{(i)}(x)$ during *Step* 9; hence, update the connection polynomial $L^{(i)}(x)$ according to:

$$L^{(i)}(x) := T^{(i)}(x) \tag{3.65}$$

and go to *Step* 11.

11 Since according to *Step* 7 the LFSR must be lengthened, we update the LFSR length by using *Theorem 1* and Equation 3.60, yielding:

$$l^{(i+1)} = i + 1 - l^{(i)} \tag{3.66}$$

and go to *Step* 12.

12 Increment the iteration index i and go to *Step* 2.

13 Invoke the Chien search in order to find the zeros of $L(x)$ by tentatively substituting all possible *GF* elements into $L(x)$ and finding those that render $L(x)$ zero.

14 Invert the zeros of $L(x)$ over the given *GF*, since the zeros of $L(x)$ are the inverse error locations, and hence find the error positions $P_1 \dots P_v$.

Having formalised the BM algorithm, let us now augment our exposition by working through our $RS(12, 8, 2)$ example assuming the same error pattern, as in our PGZ decoder example.

3.3.5.5   Berlekamp–Massey Decoding Example

In order to familiarise ourselves with the somewhat heuristic nature of the BM algorithm, we return to our example in Section 3.3.5.3 and solve the same problem using the BM algorithm instead of the PGZ algortihm.

Example 7:
Computation of the error locator polynomial and error positions in the $RS(12, 8)$ $GF(2^m)$ code employing the BM algorithm.

The syndromes from Section 3.3.5.3 are as follows:

$$S_1 = \alpha^8, \quad S_2 = \alpha^8, \quad S_3 = \alpha^5, \quad S_4 = \alpha^6.$$

Let us now follow *Steps* 1-14 of the formally stated BM algorithm:

1 Initialisation:

$$i = 0, \ l = 0, \ A(x) = 1 \ L(x) = L_0 = 1.$$

Now we work through the flowchart of Figure 3.6 $2t = 4$ times:

2 End of iteration? No, because $0 \neq 4$; hence, go to *Step 3*.

3 Compute the discrepancy:

$$d = \sum_{n=0}^{l} L_n.S_{i+1-n} = L_0.S_1 = 1.\alpha^8 = \alpha^8.$$

4 Check if $d = 0$? Since $d = \alpha^8 \neq 0$, a new LFSR design is required; hence, go to *Step 6*.

6 The new temporary connection polynomial is given by:

$$T(x) = L(x) - d.x.A(x) = 1 - \alpha^8.x,$$

go to *Step 7*.

7 Check whether the LFSR has to be lengthened. Since $2l = 0$ and $i = 0$, $2l \leq i$ is true and the LFSR must be lenghtened; hence, go to *Step 9*.

9 Store the most recent connection polynomial $L(x)$ after normalising it by its discrepancy d for later use as auxiliary LFSR $A(x)$:

$$A(x) = \frac{L(x)}{d} = \frac{1}{\alpha^8} = \alpha^7$$

and go to *Step 10*.

10 Update the connection polynomial $L(x)$ with the contents of the temporary connection polynomial as follows:

$$L(x) = T(x) = 1 - \alpha^8.x$$

and go to *Step 11*.

11 Update the LFSR length according to:

$$l = i + 1 - l = 0 + 1 - 0 = 1$$

and go to *Step 12*.

12 Increment the iteration index by assigning $i := i + 1 = 0 + 1 = 1$ and go to *Step 2*:

$$i = 1, \ l = 1, \ L(x) = 1 - \alpha^8.x, \ A(x) = \alpha^7.$$

---

2 Continue iterating, and go to *Step 3*.

3

$$
\begin{aligned}
d &= \sum_{n=0}^{l} L_n.S_{i+1-n} = L_0.S_2 + L_1.S_1 \\
&= 1.\alpha^8 + (-\alpha^8).\alpha^8 = \alpha^8 + \alpha^1 = \alpha^{10}.
\end{aligned}
$$

4 Since $d \neq 0$, go to *Step 6*.

6 Compute the temporary corrected LFSR connection polynomial:

$$
\begin{aligned}
T(x) &= L(x) - d.x.A(x) = 1 - \alpha^8.x - \alpha^{10}.x.\alpha^7 \\
     &= 1 + (\alpha^8 + \alpha^2).x = 1 + \alpha^0.x
\end{aligned}
$$

and go to *Step 7*.

7 Since $2l = 2$ and $i = 1$, $2l \leq i$ is false and the LFSR does not have to be lengthened; therefore, go to *Step 8*.

8 Update the connection polynomial by the temporary connection polynomial:

$$
L(x) = T(x) = 1 + \alpha^0.x,
$$

and go to *Step 5*.

5 The contents of the auxiliary LFSR remain unchanged; they must simply be shifted:

$$
A(x) = A(x).x = \alpha^7.x.
$$

Go to *Step 12*.

12 $i := i + 1 = 2$ and go to *Step 2*:

$$
i = 2, \; l = 1, \; L(x) = 1 + \alpha^0.x, \; A(x) = \alpha^7.x.
$$

2 Carry on iterating, and go to *Step 3*.

3

$$
d = L_0.S_3 + L_1.S_2 = 1.\alpha^5 + \alpha^0.\alpha^8 = \alpha^5 + \alpha^8 = \alpha^4;
$$

go to *Step 4*.

4 Since $d = \alpha^4 \neq 0$, go to *Step 6*.

6

$$
T(x) = L(x) - d.x.A(x) = 1 + \alpha^0.x - \alpha^4.x.\alpha^7.x = 1 + \alpha^0.x + \alpha^{11}.x^2;
$$

go to *Step 7*.

7 Since $2l \leq 2$ is true, because $2 \leq 2$, go to *Step 9*.

9 The LFSR must be lengthened; hence, we store the most recent normalised connection polynomial into the auxiliary LFSR:

$$
A(x) = \frac{L(x)}{d} = \frac{1 + \alpha^0.x}{\alpha^4} = \alpha^{11} + \alpha^{11}.x;
$$

go to *Step 10*.

10 Update the LFSR connection polynomial:

$$L(x) = T(x) = 1 + \alpha^0.x + \alpha^{11}.x^2;$$

go to *Step 11*.

11 Update the LFSR length:

$$l = i + 1 - l = 2 + 1 - 1 = 2;$$

go to *Step 12*.

12 $i := i + 1 = 3$; go to *Step 2*:

$$i = 3, \quad l = 2, \quad L(x) = 1 + \alpha^0.x + \alpha^{11}.x^2, \quad A(x) = \alpha^{11} + \alpha^{11}.x.$$

2 Continue iterating, go to *Step 3*.

3

$$d = L_0.S_4 + L_1.S_3 + L_2.S_1 = 1.\alpha^6 + \alpha^0.\alpha^5 + \alpha^{11}.\alpha^8 = \alpha^6 + \alpha^5 + \alpha^4$$

$$d = \alpha^9 + \alpha^4 = \alpha^{14};$$

go to *Step 4*.

4 Since $d = \alpha^{14} \neq 0$, go to *Step 6*.

6 Compute the new temporary connection polynomial:

$$
\begin{aligned}
T(x) &= L(x) - d.x.A(x) \\
&= 1 + \alpha^0.x + \alpha^{11}.x^2 - \alpha^{14}.x.(\alpha^{11} + \alpha^{11}.x) \\
&= 1 + \alpha^0.x + \alpha^{11}.x^2 - \alpha^{10}.x - \alpha^{10}.x^2 \\
&= 1 + (\alpha^0 + \alpha^{10}).x + (\alpha^{11} + \alpha^{10}).x^2 \\
&= 1 + \alpha^5.x + \alpha^{14}.x^2;
\end{aligned}
$$

go to *Step 7*.

7 Because $2l \leq i$ is false, since $4 > 3$, the LFSR does not have to be lengthened; go to *Step 8*.

8 Simply the temporary connection polynomial is stored in $L(x)$:

$$L(x) = T(x) = 1 + \alpha^5.x + \alpha^{14}.x^2;$$

go to *Step 5*.

5

$$A(x) = x.A(x) = x.(\alpha^{11} + \alpha^{11}.x) = \alpha^{11}.x + \alpha^{11}.x^2;$$

go to *Step 12*.

12 $i := i + 1 = 3 + 1 = 4$; go to *Step 2*.

2 Since $i = 2t$ is true, we have computed the final error locator polynomial $L(x)$, which is identical to that computed in Section 3.3.5.3 using the PGZ decoder.

We can therefore use the error positions, determined previously by Chien search. The error positions computed from this second-order connection polynomial $L(x)$ are: $P_1 = \alpha^{11}$, $P_2 = \alpha^3$. Having determined the error positions, all we have to do for the sake of error correction now is to compute the error magnitudes. This was achieved by the PGZ decoder using matrix inversion, but for large matrices this becomes a computationally demanding operation. Fortunately, it can be circumvented, for example by the Forney algorithm, which will be the subject of the next subsection.

3.3.5.6 Computation of the Error Magnitudes by the Forney Algorithm [32, 82, 88, 90, 96, 97, 108, 109, 115]

The Forney algorithm has been described in a number of classic references [32, 34, 82, 88, 90, 96, 97, 108, 109], and here we follow their philosophy. Once the error locator polynomial has been computed to give the error positions $P_1 \ldots P_v$, we concentrate on the determination of the error magnitudes $M_1 \ldots M_v$, which can be computed from the so-called error evaluator polynomial $E(x)$, defined as follows:

$$E(x) = S(x).L(x), \tag{3.67}$$

where $L(x)$ is the error locator polynomial from Equation 3.68:

$$L(x) = \prod_{l=1}^{v}(1 - x.P_l) \tag{3.68}$$

and $S(x)$ is the so-called syndrome polynomial defined here as:

$$S(x) = \sum_{j=1}^{2t} S_j.x^j. \tag{3.69}$$

By substituting the syndromes $S_1 \ldots S_{2t}$ from Equation 3.36 in the above equation, we arrive at:

$$S(x) = \sum_{j=1}^{2t}(\sum_{i=1}^{v} M_i.P_i^j).x^j. \tag{3.70}$$

The error evaluator polynomial $E(x)$ depends on both the error positions and the error magnitudes, as opposed to the error locator $L(x)$, which only depends on the error positions. This fact also reflects the parallelism to the PGZ decoder, where we had two matrix inversions to carry out. The first one has been circumvented by the BM algorithm to determine the error locator polynomial $L(x)$, that is the error positions. The second matrix inversion is substituted by the Forney algorithm through the computation of the error evaluator polynomial by using the precomputed error positions $P_1 \ldots P_v$ as well.

Since the decoder is capable of computing only the first $2t$ syndromes $S_1 \ldots S_{2t}$ for a t error correcting code, the error evaluator polynomial has to be defined in $mod\ x^{2t}$. This is because, given the code construction, there are $2t$ parity symbols, allowing the determination of t error positions and t error magnitudes. Hence we have:

$$E(x) = S(x).L(x)\ \ mod\ x^{2t}, \tag{3.71}$$

which is the key equation to be solved for the unknown error evaluator polynomial in order to determine the error magnitudes $M_1 \ldots M_v$. Let us substitute the polynomials $S(x)$ from Equation 3.70 and $L(x)$ from Equation 3.68 into Equation 3.71:

$$E(x) = (\sum_{j=1}^{2t} \sum_{i=1}^{v} M_i.P_i^j.x^j)(\prod_{l=1}^{v}(1 - x.P_l)) \ (mod \ x^{2t}). \tag{3.72}$$

By changing the order of summations and rearranging, we get:

$$
\begin{aligned}
E(x) &= [\sum_{i=1}^{v} M_i \sum_{j=1}^{2t} P_i^j.x^j] \prod_{l=1}^{v}(1 - x.P_l) \ (mod \ x^{2t}) \\
&= [\sum_{i=1}^{v} M_i.P_i.x \sum_{j=1}^{2t}(P_i.x)^{j-1}](1 - x.P_i) \prod_{l \neq i}(1 - x.P_l) \ (mod \ x^{2t}) \\
&= x.\sum_{i=1}^{v} M_i.P_i.[(1 - x.P_i) \sum_{j=1}^{2t}(P_i.x)^{j-1}] \prod_{l \neq i}(1 - x.P_l) \ (mod \ x^{2t}). \tag{3.73}
\end{aligned}
$$

If we now expand the square bracketed term, we arrive at:

$$
\begin{aligned}
(1 - xP_i) \sum_{j=1}^{2t}(P_i x)^{j-1} &= (1 - xP_i)[1 + P_i x + (P_i x)^2 + \ldots + (P_i x)^{2t-1}] \\
&= 1 + P_i x + (P_i x)^2 + (P_i x)^3 + \ldots + (P_i x)^{2t-1} \\
&- P_i x - (P_i x)^2 - (P_i x)^3 - \ldots - (P_i x)^{2t-1} - (P_i x)^{2t} \\
&= 1 - (P_i x)^{2t}. \tag{3.74}
\end{aligned}
$$

If we substitute this simplification according to Equation 3.74 into Equation 3.73, we get:

$$E(x) = x.\sum_{i=1}^{v} M_i.P_i.[1 - (P_i x)^{2t}] \prod_{l \neq i}(1 - xP_l) \ (mod \ x^{2t}). \tag{3.75}$$

Since $(P_i x)^{2t}$ in the square brackets gives zero in $(mod \ x^{2t})$, Equation 3.75 yields:

$$E(x) = x.\sum_{i=1}^{v} M_i.P_i. \prod_{\substack{l=1 \\ l \neq i}}^{v}(1 - x.P_l), \tag{3.76}$$

which is a closed-form equation for the computation of the error evaluator polynomial $E(x)$ in terms of the error positions $P_1 \ldots P_v$ and error magnitudes $M_1 \ldots M_v$, enabling us to determine the error magnitudes by the help of the Forney algorithm [115].

Forney algorithm: If one evaluates the error evaluator polynomial at the inverse error positions $P_1^{-1} \ldots P_v^{-1}$, the error magnitudes are given by:

$$M_l = \frac{E(P_l^{-1})}{\prod_{j \neq l}(1 - P_j P_l^{-1})} = -\frac{E(P_l^{-1})}{P_l^{-1} L'(P_l^{-1})}, \tag{3.77}$$

where L' stands for the derivative of the error locator polynomial $L(x)$ with respect to x.

Proof of the Forney algorithm: Let us substitute $x = P_l^{-1}$ into Equation 3.76, which yields:

$$
\begin{aligned}
E(P_l^{-1}) &= \sum_{i=1}^{v} M_i.P_i.P_l^{-1} \prod_{\substack{j=1 \\ j \neq i}}^{v} (1 - P_l^{-1}.P_j) \\
&= M_1.P_1.P_l^{-1} \prod_{\substack{j=1 \\ j \neq 1}}^{v} (1 - P_l^{-1}.P_j) \\
&+ M_2.P_2.P_l^{-1} \prod_{\substack{j=1 \\ j \neq 2}}^{v} (1 - P_l^{-1}.P_j) \\
&\vdots \\
&+ M_v.P_v.P_l^{-1} \prod_{\substack{j=1 \\ j \neq v}}^{v} (1 - P_l^{-1}.P_j)
\end{aligned}
\tag{3.78}
$$

$$
\begin{aligned}
E(P_l^{-1}) &= M_1.P_1.P_l^{-1}[(1 - P_l^{-1}.P_2)(1 - P_l^{-1}.P_3)\ldots(1 - P_l^{-1}.P_v)] \\
&+ M_2.P_2.P_l^{-1}[(1 - P_l^{-1}.P_1)(1 - P_l^{-1}.P_3)\ldots(1 - P_l^{-1}.P_v)] \\
&\vdots \\
&+ M_v.P_v.P_l^{-1}[(1 - P_l^{-1}.P_1)(1 - P_l^{-1}.P_2)\ldots(1 - P_l^{-1}.P_{v-1})].
\end{aligned}
\tag{3.79}
$$

Observe in the above equation that all the square bracketed terms contain all but one combination of the factors $(1 - P_l^{-1}.P_j)$, including $(1 - P_l^{-1}.P_l)$, which is zero. There is, however, one such square bracketed term, where the missing factor happens to be $(1 - P_l^{-1}.P_l) = 0$. Taking this into account, we arrive at:

$$
E(P_l^{-1}) = M_l.P_l.P_l^{-1} \prod_{\substack{j=1 \\ j \neq l}}^{v} (1 - P_l^{-1}.P_j),
\tag{3.80}
$$

or after expressing the error magnitude M_l from the above equation we get:

$$
M_l = \frac{E(P_l^{-1})}{\prod_{\substack{j=1 \\ j \neq l}}^{v} (1 - P_l^{-1}.P_j)},
\tag{3.81}
$$

which is the first form of the Forney algorithm.

The second form of the Forney algorithm in Equation 3.78 can be proved by computing the derivative $L'(x)$ of the error locator polynomial $L(x)$, which ensues as follows:

$$
L(x) = (1 - xP_1)(1 - xP_2)\ldots(1 - xP_v) = \prod_{j=1}^{v}(1 - xP_j)
$$

$$L'(x) = \frac{d}{dx}L(x) = -P_1(1 - xP_2)\ldots(1 - xP_v) \tag{3.82}$$
$$- (1 - xP_1)(1 - xP_3)\ldots(1 - xP_v)$$
$$- (1 - xP_1)(1 - xP_2)\ldots(1 - xP_v)$$
$$\vdots$$
$$- (1 - xP_1)(1 - xP_2)\ldots(1 - xP_{v-1})$$

$$L'(x) = -\sum_{i=1}^{v} P_i \prod_{j=1}^{v}(1 - xP_j). \tag{3.83}$$

By evaluating $L'(x)$ from Equation 3.83 at P_l^{-1} we get:

$$L'(P_l^{-1}) = -\sum_{i=1}^{v} P_i \prod_{\substack{j=1 \\ j \neq i}}^{v}(1 - P_l^{-1}P_j), \tag{3.84}$$

where similarly to Equation 3.79 we have only one non-zero term left in the summation. Thus, we arrive at:

$$L'(P_l^{-1}) = -P_l \prod_{\substack{j=1 \\ j \neq l}}^{v}(1 - P_l^{-1}P_j). \tag{3.85}$$

Hence, the denominator of Equation 3.78 can be formulated as follows:

$$\prod_{\substack{j=1 \\ j \neq l}}^{v}(1 - P_l^{-1}P_j) = -P_l^{-1}.L'(P_l^{-1}), \tag{3.86}$$

which proves the second form of the algorithm in Equation 3.78.

Let us now compute the error magnitudes in our standard example of the $RS(12, 8)$ code.

3.3.5.7 Forney Algorithm Example

Example 8:

```
Computation of the error evaluator polynomial and error
magnitudes in an RS(12,8) GF(2⁴) code using the Forney
algorithm.
```

The error locator polynomial from Section 3.3.5.5 is given by:

$$S(x) = \sum_{j=1}^{2t} S_j x^j = S_1 x + S_2 x^2 + S_3 x^3 + S_4 x^4$$
$$S(x) = \alpha^8 x + \alpha^8 x^2 + \alpha^5 x^3 + \alpha^6 x^4.$$

The error evaluator polynomial $E(x) = L(x)S(x)$ from the definition of Equation 3.71 is computed as follows:

$$E(x) = (1 + \alpha^5 x + \alpha^{14} x^2)(\alpha^8 x + \alpha^8 x^2 + \alpha^5 x^3 + \alpha^6 x^4) \ (mod \ x^4)$$

$$
\begin{aligned}
= \ & \alpha^8 x && +\alpha^8 x^2 && +\alpha^5 x^3 && +\alpha^6 x^4 \\
& && +\alpha^{13} x^2 && +\alpha^{13} x^3 && +\alpha^{10} x^4 && +\alpha^{11} x^5 \\
& && && +\alpha^7 x^3 && +\alpha^7 x^4 && +\alpha^4 x^5 && +\alpha^5 x^6
\end{aligned}
$$

$$
\begin{aligned}
E(x) \ &= \ \alpha^8 x + (\alpha^8 + \alpha^{13})x^2 + (\alpha^5 + \alpha^{13} + \alpha^7)x^3 \\
&+ \ (\alpha^6 + \alpha^{10} + \alpha^7)x^4 + (\alpha^{11} + \alpha^4)x^5 + \alpha^5 x^6 \\
&= \ (\alpha^8 x + \alpha^3 x^2 + 0.x^3 + 0.x^4 + \alpha^{13} x^5 + \alpha^5 x^6) \ (mod \ x^4) \\
&= \ \alpha^3 x^2 + \alpha^8 x.
\end{aligned}
$$

The error magnitudes can be computed from $E(x)$ by Forney's algorithm using Equation 3.77:

$$M_l = \frac{E(P_l^{-1})}{\prod_{j \neq l}(1 - P_j P_l^{-1})}. \tag{3.87}$$

The error positions are: $P_1 = \alpha^{11}$, $P_1^{-1} = \alpha^4$, $P_2 = \alpha^3$, $P_2^{-1} = \alpha^{12}$; hence, we have:

$$M_1 = \frac{E(P_1^{-1})}{(1 - P_2 P_1^{-1})} = \frac{\alpha^3 (\alpha^4)^2 + \alpha^8 \alpha^4}{1 - \alpha^3 \alpha^4} = \frac{\alpha^{11} + \alpha^{12}}{1 - \alpha^7} = \frac{\alpha^0}{\alpha^9} = \alpha^6$$

$$M_2 = \frac{E(P_2^{-1})}{(1 - P_1 P_2^{-1})} = \frac{\alpha^3 (\alpha^{12})^2 + \alpha^8 \alpha^{12}}{1 - \alpha^{11} \alpha^{12}} = \frac{\alpha^{12} + \alpha^5}{1 - \alpha^8} = \frac{\alpha^{14}}{\alpha^2} = \alpha^{12}.$$

The error magnitudes computed by the Forney algorithm are identical to those computed by the PGZ decoder. Thus, the quantities required for error correction are given by:

$$(P_1, M_1) = (\alpha^{11}, \alpha^6)$$

$$(P_2, M_2) = (\alpha^3, \alpha^{12}).$$

In order to compute the error-free information symbols, one simply has to add the error magnitudes to the received symbols at the error positions computed, as we have shown in our PGZ decoding example.

Clark and Cain [88] have noted that since $E(x) = L(x).S(x)$, $E(x)$ can be computed in the same iterative loop as $L(x)$, which was portrayed using the BM algorithm in Figure 3.6. In order to achieve this, however, the original BM algorithm's flowchart in Figure 3.7 has to be properly initialised and slightly modified, as it will be elaborated on below. If we associate a separate normalised LFSR correction term $C(x)$ with $E(x)$, which is the counterpart of the auxiliary LFSR connection polynomial $A(x)$ in the BM algorithm, then the correction formula for $E(x)$ is identical to that of $A(x)$ in Equation 3.62, yielding:

$$E^{(i+1)}(x) = E^{(i)}(x) - d^{(i)}.x.C(x). \tag{3.88}$$

Then, by using a second temporary storage polynomial $T_2(x)$, *Steps 6, 8, 5, 9* and *10* of the flowchart in Figure 3.6 are extended by the corresponding operations for $T_2(x)$, $E(x)$ and $C(x)$, as follows:

Step 6b:

$$T_2^{(i)}(x) := E^{(i)}(x) - d^{(i)}.x.C(x) \tag{3.89}$$

Step 8b:

$$E^{(i)}(x) := T_2^{(i)}(x) \tag{3.90}$$

Step 5b:

$$C^{(i)}(x) := x.C^{(i)}(x) \tag{3.91}$$

Step 9b:

$$C^{(i)}(x) := E^{(i)}(x)/d^{(i)} \tag{3.92}$$

Step 10b:

$$E^{(i)}(x) := T_2^{(i)}(x). \tag{3.93}$$

The complete BM flowchart with these final refinements is depicted in Figure 3.7.

We now round off our discussion of the BM algorithm with the computation of the error evaluator polynomial $E(x)$ for our $RS(12,8)\ GF(2^4)$ example using the flowchart of Figure 3.7.

3.3.5.8 Error Evaluator Polynomial Computation

Example 9:

```
Computation of the error evaluator polynomial E(x) in an
RS(12,8) GF(2^4) systematic code.
```

1 $E(x) = 0$, $C(x) = 1$, $l = 0$, $i = 0$; go to *Step 2*.

2 $0 \neq 4$; go to *Step 3*.

3 $d = L_0.S_1 = 1.\alpha^8 = \alpha^8$; go to *Step 4*.

4 $d \neq 0$; go to *Step 6*.

6b $T_2(x) = E(x) - d.x.C(x) = 0 - \alpha^8.x.1 = \alpha^8.x$; go to *Step 7*.

7 $0 \leq 0$; go to *Step 9*.

9b $C(x) = E(x)/d = 0/\alpha^8 = 0$; go to *Step 10*.

10b $E(x) = T_2(x) = \alpha^8.x$; go to *Step 11*.

11 $l = i + 1 - l = 0 + 1 - 0 = 1$; go to *Step 12*.

12 $i = i + 1 = 0 + 1 = 1$; go to *Step 2*.

$$\boxed{i = 1,\ l = 1,\ E(x) = \alpha^8.x,\ C(x) = 0;}$$

2 $1 \neq 4$; go to *Step 3*.

3 $d = L_0.S_2 + L_1.S_1 = 1.\alpha^8 + \alpha^8\alpha^8 = \alpha^8 + \alpha^1 = \alpha^{10}$; go to *Step 4*.

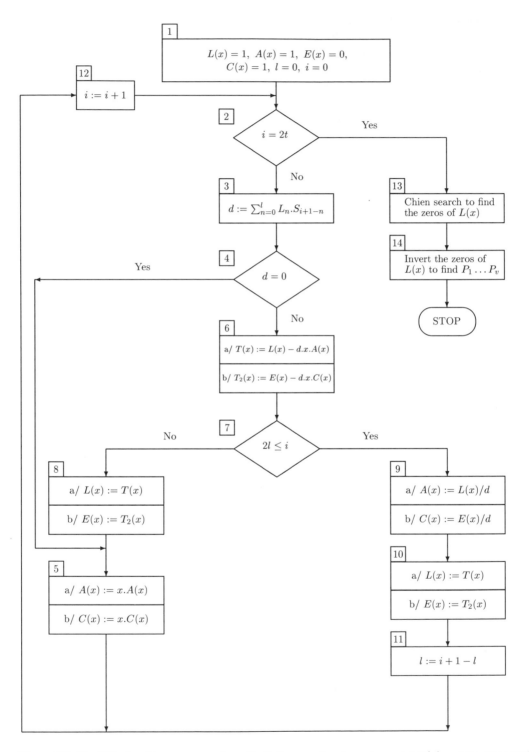

Figure 3.7: The BM algorithm: computation of both the error locator polynomial $L(x)$ and the error evaluator polynomial $E(x)$.

4 $d \neq 0$; go to *Step 6*.

6b $T_2(x) = E(x) - d.x.C(x) = \alpha^8.x - \alpha^{10}.x.0 = \alpha^8.x$; go to *Step 7*.

7 $2l \leq i$ is false; go to *Step 8*.

8b $E(x) = T_2(x) = \alpha^8.x$; go to *Step 5*.

5b $C(x) = x.C(x) = 0$; go to *Step 12*.

12 $i = i + 1 = 2$; go to *Step 2*.

$$\boxed{i = 2, \; l = 1, \; E(x) = \alpha^8.x, \; C(x) = 0;}$$

2 $2 \neq 4$; go to *Step 3*.

3 $d = L_0.S_3 + L_1.S_2 = 1.\alpha^5 + \alpha^0\alpha^8 = \alpha^5 + \alpha^8 = \alpha^4$; go to *Step 4*.

4 $d \neq 0$; go to *Step 6*.

6b $T_2(x) = E(x) - d.x.C(x) = \alpha^8.x - \alpha^4.x.0 = \alpha^8.x$; go to *Step 7*.

7 $2l \leq i$ is true; go to *Step 9*.

9b $C(x) = E(x)/d = \alpha^8.x/\alpha^4 = \alpha^4.x$; go to *Step 10*.

10b $E(x) = T_2(x) = \alpha^8.x$; go to *Step 11*.

11 $l = i + 1 - l = 2 + 1 - 1 = 2$; go to *Step 12*.

12 $i = i + 1 = 2 + 1 = 3$; go to *Step 2*.

$$\boxed{i = 3, \; l = 2, \; E(x) = \alpha^8.x, \; C(x) = \alpha^4.x;}$$

2 $3 \neq 4$; go to *Step 3*.

3 $d = L_0.S_4 + L_1.S_3 + L_2.S_1 = 1.\alpha^6 + \alpha^0\alpha^5 + \alpha^{11}\alpha^8 = \alpha^6 + \alpha^5 + \alpha^4, \quad d = \alpha^9 + \alpha^4 = \alpha^{14}$; go to *Step 4*.

4 $d \neq 0$; go to *Step 6*.

6b $T_2(x) = E(x) - d.x.C(x) = \alpha^8.x - \alpha^{14}.x. \, \alpha^4.x = \alpha^8 + \alpha^3.x^2;$ go to *Step 7*.

7 Since $2l \leq i$ is false, go to *Step 8*.

8b $E(x) = T_2(x) = \alpha^8.x + \alpha^3.x^2$; go to *Step 5*

5b $C(x) = x.C(x) = x.\alpha^4.x = \alpha^4.x^2$; go to *Step 12*.

12 $i = i + 1 = 4$; go to *Step 2*.

2 Since $i = 2t$; go to *Step 13*.

13 *Chien* search; go to *Step 14*.

14 Find P_1, P_2 by inverting the zeros of $L(x)$. Stop.

We have found the error evaluator polynomial $E(x) = \alpha^8.x + \alpha^3.x^2$, which is the same as that in our previous example, and consequently results in the same error magnitudes. Having considered some of the algorithmic issues of RS and BCH codecs, in closing in the next chapter we provide a range of simulation results for a variety of practical BCH codecs, in order to be able to gauge the BER reductions achieved. A more comprehensive set of results for a range of convolutional, block and concatenated codes can be found, for example, in [109].

3.4 Summary and Conclusions

In this chapter, after a rudimentary introduction to finite fields, we have considered two decoding algorithms for RS and BCH codes. The PGZ decoder, which is based on matrix inversion techniques, and a combination of the BM and Forney algorithm was considered. The latter technique constitutes a computationally more effective iterative solution to the determination of the error positions and error magnitudes. The number of multiplications is proportional to t^3 in the case of the matrix inversion method, as opposed to $6t^2$ in the case of the BM algorithm, which means that whenever $t > 6$, the BM algorithm requires less computation than the matrix inversion-based PGZ decoder. Although for the sake of completeness we have provided a rudimentary introduction to RS and BCH codes, as well as to their algebraic decoding, for more detailed discussions on the topic the interested reader is referred to the literature [31, 32, 49, 85, 88, 90, 96].

Chapter 4

Soft Decoding and Performance of BCH Codes

4.1 Introduction

In Chapter 3 we familiarised ourselves with the basic concepts of Reed–Solomon (RS) and Bose–Chaudhuri–Hocquenghem (BCH) codes and discussed their decoding. No deep knowledge concerning the algebraic decoding of these RS and BCH codes is required for following the discussions of this chapter, but again, the basic code construction principles will not be repeated here from the previous chapter. It is also an advantage if one is familiar with the underlying concepts of convolutional coding and the rudiments of Viterbi decoding, since in this chapter the Viterbi algorithm will be invoked again, although this time for the decoding of binary BCH codes.

As mentioned in the previous chapter, BCH codes were discovered by Hocquenghem [14] and independently by Bose and Chaudhuri [15] in 1959 and 1960, respectively. These codes constitute a prominent class of *cyclic block codes* that have multiple-error detection and correction capabilities.

In this chapter, we will commence with an introduction to BCH codes in Section 4.2. Their state and trellis diagrams are constructed in Section 4.2.2. The trellis decoding of BCH codes using the Viterbi Algorithm (VA) [8] is detailed in Section 4.3.2. Simulation results of various BCH codes employing the Berlekamp–Massey Algorithm (BMA) [31–34] and the Viterbi algorithm are given in Section 4.3.5. Finally, in Section 4.4 we investigate the low-complexity Chase algorithm [28] in the context of the soft decoding of BCH codes.

4.2 BCH codes

A BCH encoder accepts k information data bits and produces n coded bits. The minimum Hamming distance of the codewords is d_{min} and the BCH code concerned is denoted as $\text{BCH}(n, k, d_{min})$. Table 4.1 lists some commonly used code generators, $g(x)$, for the construction of BCH codes [118]. The coefficients of $g(x)$ are presented as octal numbers arranged so that when they are converted to binary digits, the rightmost digit corresponds to the zero-degree coefficient of $g(x)$.

n	k	d_{min}	$g(x)$
7	4	3	13
15	11	3	23
	7	5	721
31	26	3	45
	21	5	3551
	16	7	107657
63	57	3	103
	51	5	12471
	45	7	1701317
	39	9	166623567
	36	11	1033500423
127	120	3	211
	113	5	41567
	106	7	11554743
	99	9	3447023271
	92	11	624730022327
	85	13	130704476322273
	78	15	26230002166130115
	71	19	62550107132531 27753
	64	21	120653402557077310 0045
255	247	3	435
	239	5	267543
	231	7	156720665
	223	9	75626641375
	215	11	23157564726421
	207	13	16176560567636227
	199	15	7633031270420722341
	191	17	26634701761153337 14567
	187	19	5275531354000132223 6351
	179	21	2262471071734043241 6300455
	171	23	1541621421234235607 7061630637
	163	25	7500415510075602551 574724514601
	155	27	3757513005407665015 722506464677633
	147	29	1642130173537165525 304165305441011711
	139	31	4614017320601755615 70722730247453567445
	131	37	2157133314715101512 61250277442142024165471

Table 4.1: Table of generators for BCH codes [118] ©IEEE, Stenbit.

4.2.1 BCH Encoder

Since BCH codes are cyclic codes, their encoders can be implemented using shift-register circuits [119–121]. The codes can be encoded either non-systematically or systematically. However, systematic BCH codes were found to perform slightly better than their non-systematic counterparts. Hence, only systematic BCH codes will be discussed in this chapter.

For systematic codes, the generator polynomial, $g(x)$, is written as follows:

$$g(x) = g_0 + g_1 x + g_2 x^2 + \ldots + g_{n-k-1} x^{n-k-1} + g_{n-k} x^{n-k}. \tag{4.1}$$

The generator polynomial, $g(x)$, formulates n codeword bits by appending $(n-k)$ parity bits to the k information data bits. The encoder employs a shift register having $(n-k)$ stages as depicted in Figure 4.1, where $\otimes$ represents multiplication, whereas $\oplus$ is modulo-2 addition. In simple plausible terms the code exhibits error correction ability, since only certain encoded sequences obeying the encoding rules are legitimate and hence corrupted or illegitimate codewords can be recognised and corrected. The parity bits are computed from the information data bits according to the rules imposed by the generator polynomial.

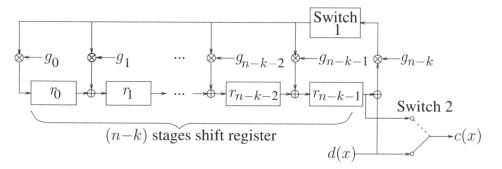

Figure 4.1: Systematic encoder for BCH codes having $(n-k)$ shift-register stages.

The following steps describe the encoding procedures:

1) Switch 1 is closed during the first k shifts, in order to allow the information data bits, $d(x)$, to shift into the $n-k$ stages of the shift register.

2) At the same time, Switch 2 is in the down position to allow the data bits, $d(x)$, to be copied directly to the codeword, $c(x)$.

3) After kth shifts, Switch 1 is opened and Switch 2 is moved to the upper position.

4) The remaining $n-k$ shifts clear the shift register by appending the parity bits to the codeword, $c(x)$.

Let us consider the BCH(7,4,3) code as an example for illustrating the process of encoding. From Table 4.1, the generator polynomial is:

$$
\begin{aligned}
g &= 13_{octal} \\
&= 1011_{bin} \\
g(x) &= x^3 + x + 1. \tag{4.2}
\end{aligned}
$$

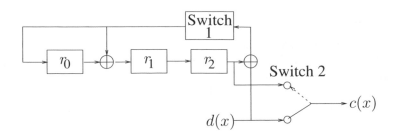

Figure 4.2: Systematic encoder for the BCH(7,4,3) code having $n - k = 3$ register stages.

Figure 4.2 shows the specific encoder, which is a derivative of Figure 4.1. Observe that all the multipliers illustrated in Figure 4.1 are absent in Figure 4.2. Explicitly, if the generator polynomial coefficient is 1, the multiplier is replaced by a direct hard-wire connection as shown in Figure 4.2, whereas if the coefficient is 0, no connection is made.

Let us use the shift register shown in Figure 4.2 for encoding four ($k = 4$) information data bits, $d = 1\ 0\ 1\ 1$ ($d(x) = 1 + x^2 + x^3$). The operational steps are as follows:

Input queue	Shift index	Shift register $r_0 r_1 r_2$	Codeword $c_0 c_1 c_2 c_3 c_4 c_5 c_6$
1 0 1 1	0	0 0 0	- - - - - - -
1 0 1	1	1 1 0	- - - - - - 1
1 0	2	1 0 1	- - - - - 1 1
1	3	1 0 0	- - - - 0 1 1
-	4	1 0 0	- - - 1 0 1 1
-	5	0 1 0	- - 0 1 0 1 1
-	6	0 0 1	- 0 0 1 0 1 1
-	7	0 0 0	1 0 0 1 0 1 1

The shift registers must be reset to zero before the encoding process starts. After the fourth shift, Switch 1 is opened and Switch 2 is moved to the upper position. The parity bits contained in the shift register are appended to the codeword. The codeword is $c = 1\ 0\ 0\ 1\ 0\ 1\ 1$ ($c(x) = 1 + x^3 + x^5 + x^6$). The binary representation of both $d(x)$ and $c(x)$ is shown in Figure 4.3.

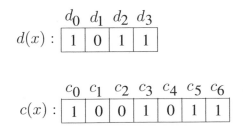

Figure 4.3: Binary representation of the uncoded data bits and coded bits.

4.2.2 State and Trellis Diagrams

Let us study Figure 4.2 in the context of the example outlined in Section 4.2.1. As the data bits are shifted into the register by one bit at a time, the parity bits, $\{r_0, r_1, r_2\}$, represent the state of the register. The corresponding operations are shown below:

Input queue	Shift index	Shift register $r_0 r_1 r_2$	State	Output bit
1 0 1 1	0	0 0 0	0	-
1 0 1	1	1 1 0	6	1
1 0	2	1 0 1	5	1
1	3	1 0 0	4	0
-	4	1 0 0	4	1
-	5	0 1 0	2	0
-	6	0 0 1	1	0
-	7	0 0 0	0	1

In the example above, there are a few points worth noting:

- The encoding process always starts at the all-zero state and ends at the all zero state.

- The number of the output bits is always one following a clockpulse.

- For the first k (which is four for this example) shifts, the output bit is the same as the input bit.

- After the kth shift, the parity bits of the shift register are shifted to the output.

- The number of states is equal to 2^{n-k} increasing exponentially, when $n - k$ increases.

For the BCH(7,4,3) code, $n - k$ is 3 and the total number of encoder states is $2^3 = 8$. By using the shift register shown in Figure 4.2, we can find all the subsequent states when the register is in a particular state. Figure 4.4 shows all possible state transitions at any encoder state for the BCH(7,4,3) code. The branch emanating from the present state to the next state indicates the state transition. The broken line branch is the transition initiated by a data bit of logical 0, whereas the continuous branch is due to the data bit being logical 1. The number of branches emanating from the present state is 2, which corresponds to the number of possible input bits, namely 1 and 0. As explained earlier, the output bit is the same as the data bit.

The state diagram corresponding to the state transition diagram is shown in Figure 4.5. It consists of a total of $2^{n-k} = 8$ states connected by all the possible transitions shown in the state transition diagram of Figure 4.4. By using the state diagram in Figure 4.5, we can encode the data bits, $d = 1\,0\,1\,1$, without using the shift register shown in Figure 4.2. The first data bit is a logical 1, hence the state changes from 000 to 110, as illustrated by the solid branch emanating from state 000 in Figure 4.5. The encoder output is the same as the input data bit, which is a logical 1. At the next instant, the present state becomes 110 and the data bit is logical 1. This causes the state transition from 110 to 101. The encoding cycle is repeated for subsequent data bits, which change the states. By following the change of states throughout the first k cycles of the encoding process, a particular path associated with states $000 \rightarrow 110 \rightarrow 101 \rightarrow 100 \rightarrow 100$ can be observed.

After the kth cycle, the state changes correspond to shifting out the parity bits from the shift register. In our example, the parity bits are 100 at the kth cycle. In the following cycle, the

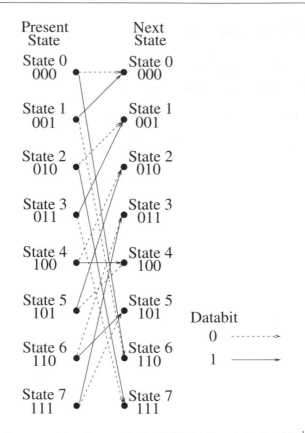

Figure 4.4: State transition diagram for the BCH(7,4,3) code having $2^{n-k} = 8$ states.

parity bits are shifted to the right. The rightmost bit of the parity bits is shifted out to become the output bit and the leftmost bit is filled with logical 0. As a result, the state changes are $100 \rightarrow 010 \rightarrow 001 \rightarrow 000$. The whole encoding process can be associated with state transitions of $000 \rightarrow 110 \rightarrow 101 \rightarrow 100 \rightarrow 100 \rightarrow 010 \rightarrow 001 \rightarrow 000$.

Another representation of the encoding process is the trellis diagram shown in Figure 4.6. This is formed by concatenating the consecutive instants of the state transition diagram of Figure 4.4 starting from the all-zero state. The diagram illustrates all the possible $2^k = 16$ paths for the BCH(7,4,3) code. The trellis has $2^{n-k} = 8$ rows (8 different states) and $n + 1 = 8$ columns. The nodes in the same row represent the same state, whereas the nodes in the same column illustrate all the possible states 000 (State a), 001 (State b), 010 (State c), ..., 111 (State h). The state transitions between adjacent columns are drawn either by a continuous line or broken line, according to whether the encoder output bit is logical 1 or 0, respectively.

Initially, there is only one state, which is the all-zero state (State a). The number of trellis states increases, as each new data bit is inserted into the encoder. The symbol signalling instants corresponding to the column positions in the trellis, shown in Figure 4.6, are indexed by the integer T. On inserting the first data bit into the encoder, $T = 0$, two different nodes are possible at the next instant. The arrival of the second data bit, when $T = 1$, causes the number of possible nodes at the next instant to increase to 2^2. The number of possible nodes continues to increase with T, until the maximum number of $2^{n-k} = 8$ is reached. The maximum number of states is

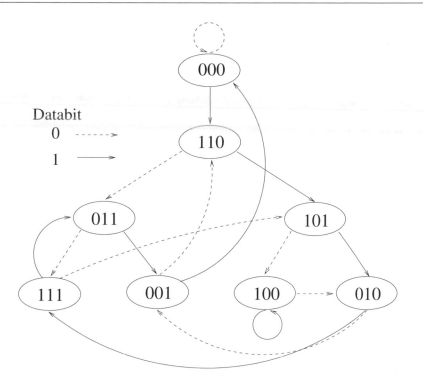

Figure 4.5: State diagram for the BCH(7,4,3) code having $2^{n-k} = 8$ states.

reached when $T = n - k = 3$, and from then on the number of possible states is constant. After $T = k$, the number of possible states is thus divided by two at every instant in the trellis merging towards the zero state, which is reached at $T = n$.

4.3 Trellis Decoding

4.3.1 Introduction

The trellis decoding of *linear block codes* was first introduced by Wolf [18] in 1978. However, this technique is only feasible for certain BCH codes, since the number of states increases exponentially when $n - k$ increases. The reason was outlined in the earlier sections.

4.3.2 Viterbi Algorithm

The Viterbi Algorithm (VA) was proposed by Viterbi in 1967 [8]. The algorithm searches all the possible paths in the trellis and their Hamming or Euclidean distances from the received sequence at the decoder's input are compared. The path exhibiting the smallest distance from the received sequence is selected as the most likely transmitted sequence and the associated information data bits are regenerated. This method is known as maximum likelihood sequence estimation, since the most likely path is selected from the set of all the paths in the trellis.

Figure 4.7 records the 'history' of the paths selected by the BCH(7,4,3) Viterbi decoder. Suppose that there are no channel errors and hence the input sequence of the decoder is the same

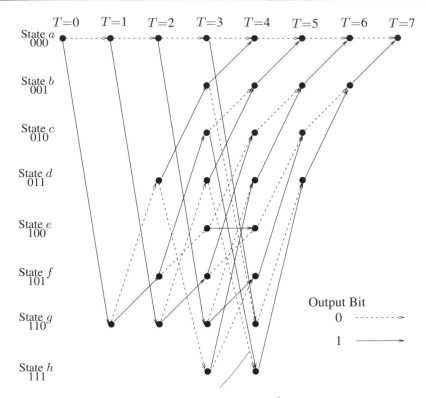

Figure 4.6: Trellis diagram for the BCH(7,4,3) code having $2^{n-k} = 8$ states and $n + 1 = 8$ consecutive stages.

as the encoded sequence, namely 0 0 0 0 0 0 0. At the first instant, $T = 1$, the received bit is logical 0, which is compared with the possible transmitted bits 0 and 1 of the branches from node a to a and from node a to g, respectively. The metrics of these two branches are their Hamming distances, namely the differences between the possible transmitted bits 0 or 1 and the received bit 0. Their Hamming distances are 0 and 1, respectively.

Now, we define the branch metric as the Hamming distance of an individual branch from the received bits, and the path metric at the Tth instant as the sum of the branch metrics at all of its branches from $T = 0$ to the Tth instant. Hence the path metrics, printed on top of each branch in Figure 4.7, at instant $T = 1$ are 0 and 1 for the paths $a \rightarrow a$ and $a \rightarrow g$, respectively. At the second instant $T = 2$, the received bit is 0 and the branch metrics are 0, 1, 0 and 1 for the branches $a \rightarrow a$, $a \rightarrow g$, $g \rightarrow d$ and $g \rightarrow f$, respectively. The path metrics[1] are 0, 1, 1 and 2 for the corresponding paths $a \rightarrow a \rightarrow a$, $a \rightarrow a \rightarrow g$, $a \rightarrow g \rightarrow d$ and $a \rightarrow g \rightarrow f$. At the third instant, the received bit is 0. There are eight possible branches and their path metrics, which are shown in Figure 4.7, are 0, 1, 2, 1, 3, 2, 1 and 2 for the paths $a \rightarrow a \rightarrow a \rightarrow a$, $a \rightarrow a \rightarrow a \rightarrow g$, $a \rightarrow g \rightarrow d \rightarrow b$, $a \rightarrow g \rightarrow d \rightarrow h$, $a \rightarrow g \rightarrow f \rightarrow c$, $a \rightarrow g \rightarrow f \rightarrow e$, $a \rightarrow a \rightarrow g \rightarrow d$ and $a \rightarrow a \rightarrow g \rightarrow f$, respectively.

Let α_1 and α_2 denote the corresponding paths $a \rightarrow a \rightarrow a \rightarrow a \rightarrow a$ and $a \rightarrow g \rightarrow d \rightarrow b \rightarrow a$ that begin at the initial node a and remerge in node a at $T = 4$. Their respective path metrics are 0 and 3. Any further branches associated with $T > 4$ stemming from node a at

---

[1]It is also known as the accumulated metric.

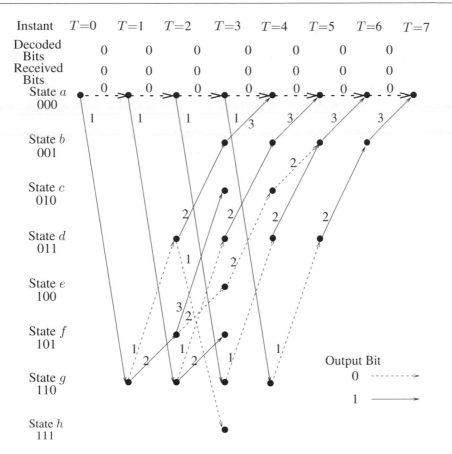

Figure 4.7: Example of Viterbi decoding of the BCH(7,4,3) code.

$T = 4$ will add identical branch metrics to the path metrics of both paths α_1 and α_2, and this means that the path metric of α_2 is larger at $T = 4$ and will remain larger for $T > 4$. The Viterbi decoder will select the path having the smallest metric, which is the all-zero state sequence, and therefore discards the path α_2. The path α_1 is referred to as the *survivor*. This procedure is also applied at the other nodes for $T \geq n - k = 3$. Notice that paths $a \rightarrow g \rightarrow f \rightarrow c$ and $a \rightarrow a \rightarrow g \rightarrow f$ etc. cannot survive, since their path metrics are larger than that of their counterparts of the merging pairs and they are therefore eliminated from the decoder's memory. Thus, there are only $2^{n-k} = 8$ paths that survive from instant $T = n - k$ to $T = k$. Following instant $T = k$, the number of surviving paths reduces by a factor of two for each instant.

Sometimes, two paths will merge, which have the same path metrics. At instant $T = 5$, the paths $a \rightarrow a \rightarrow a \rightarrow g \rightarrow d \rightarrow b$ and $a \rightarrow g \rightarrow f \rightarrow e \rightarrow c \rightarrow b$ remerge at node b. Both paths have the same path metric, which is 2. Normally, the Viterbi decoder will choose the survivor randomly and discard the other path. However, this situation never (or rarely) occurs in the *Soft-Decision Viterbi Algorithm* or *Soft-Output Viterbi Algorithm* (SOVA), which is the preferred algorithm in practical applications.

4.3.3 Hard-decision Viterbi Decoding

For hard-decision decoding, the demodulator provides only hard decisions (logical 1 or 0) when regenerating the transmitted sequence. In this case, the Hamming distances between the received bits and the estimated transmitted bits in the trellis are used as a metric, i.e. as a confidence measure.

4.3.3.1 Correct Hard-decision Decoding

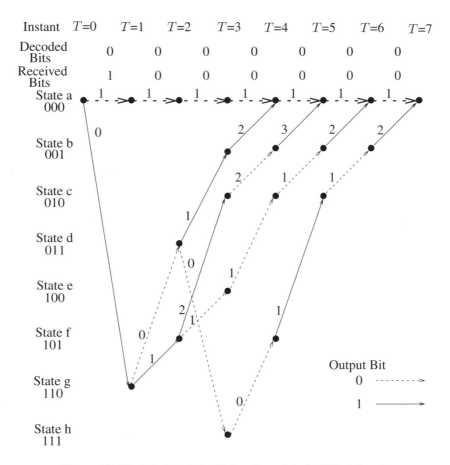

Figure 4.8: Hard-decision Viterbi decoding of the BCH(7,4,3) code.

Let us illustrate the philosophy of hard-decision decoding using the BCH(7,4,3) code that was previously used as an example. Assume that the transmitted sequence is 0 0 0 0 0 0 0. A channel error is introduced to the first transmitted bit and the received sequence provided by the demodulator is 1 0 0 0 0 0 0. The decoder compares the demodulator's output bit with both of the possible decoded bits indicated by the continuous and broken lines in Figure 4.8, which correspond to a binary one and zero, respectively. When the demodulator's output bit and the decoded bit are identical, their Hamming distance is zero. By contrast, when these two bits are different, a Hamming distance of one is added to the accumulated path metrics, which is written on the corresponding trellis transition. As we traverse through the trellis, the above-mentioned branch metrics are summed and at $T = 7$ the path having the lowest Hamming weight is deemed the survivor path. Hence the decoded sequence is the associated string of ones and zeros.

Again, Figure 4.8 demonstrates how the Viterbi decoder selects the survivor path (marked by the thick broken line) at the top of the figure, which has the smallest path metric, and hence decodes the received sequence correctly. Note that the path metric of the survivor is equivalent to the number of errors in the received sequence, as long as the decoder is capable of correcting the errors. This is not true when owing to channel errors the decoder diverges from the error-free trellis path.

4.3.3.2 Incorrect Hard-decision Decoding

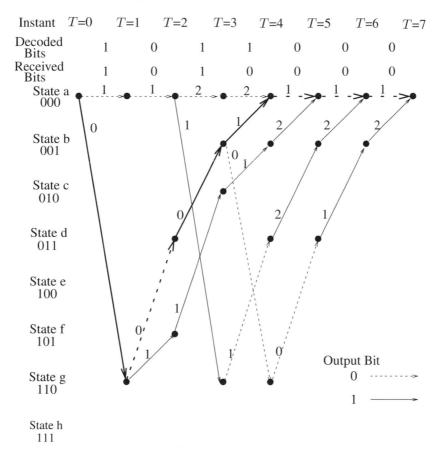

Figure 4.9: Incorrect hard-decision Viterbi decoding of the BCH(7,4,3) code.

When the number of channel errors exceeds the correcting capability of the code, incorrect decoding will occur, as illustrated in Figure 4.9. Two channel errors are introduced in the first and third position of the received sequence. The incorrect decoding occurs in the four initial branches (marked by a thick line), which results in the decoded sequence of 1 0 1 1 0 0 0.

This and the previous example of correct decoding are related to decisions that depend on whether the Hamming distance of the received sequence with respect to the correct path is smaller than the distance of the received sequence to other paths in the trellis. Observe furthermore that the surviving path's metric is now different from the number of errors encountered.

4.3.4 Soft-decision Viterbi Decoding

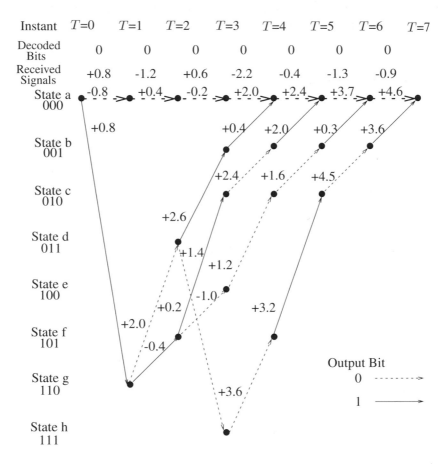

Figure 4.10: Soft-decision Viterbi decoding of the BCH(7,4,3) code.

So far, we have discussed hard-decision decoding. We now explore the techniques of soft-decision decoding. In this approach, the received signal at the output of the demodulator is sampled. The sampled values are then directly input to the Viterbi decoder.

Assuming that we are using *Binary Phase Shift Keying* (BPSK) at the transmitter, a logical 0 will be transmitted as -1.0 and a logical 1 is sent as $+1.0$. The transmitted sequence is $-1 -1 -1 -1 -1 -1 -1$, if we are transmitting a sequence of logical 0s. At the receiver, the soft outputs of the demodulator are $+0.8, -1.2, +0.6, -2.2, -0.4, -1.3, -0.9$, which correspond to the sequence of $1\ 0\ 1\ 0\ 0\ 0\ 0$, if we use hard-decision decoding, as in our last example.

The demodulator's soft outputs are used as a measure of confidence, and are shown in Figure 4.10. The first demodulator soft-output signal is $+0.8$, implying that the transmitted signal is likely to have been $+1$ and the confidence measure of the decision is 0.8. Considering the path $a \rightarrow g$, corresponding to a logical 1, the branch metric of the path is $+0.8$. However, path $a \rightarrow a$ does not tally with the received signal, and the branch metric of the path is therefore -0.8, accumulating a negative path metric, or a 'penalty' due to its dissimilarity. At second instant, the received signal is -1.2 which results in path metrics (accumulated confidence) of $+0.4$, -2.0, $+2.0$ and -0.4 for the paths $a \rightarrow a \rightarrow a$, $a \rightarrow a \rightarrow g$, $a \rightarrow g \rightarrow d$ and $a \rightarrow g \rightarrow f$, respectively.

Let us denote α_1 and α_2 as the paths $a \rightarrow a \rightarrow a \rightarrow a \rightarrow a$ and $a \rightarrow g \rightarrow d \rightarrow b \rightarrow a$. The total accumulated path metrics for paths α_1 and α_2 are $+2.0$ and $+0.4$, respectively. The Viterbi decoder selects the path associated with the larger path metric because of its stronger accumulated confidence. Hence, path α_1 is selected (instead of path α_2 which was selected in our previous hard-decision example). Hence, it was shown that soft-decision decoding performs better than hard-decision decoding.

4.3.5   Simulation Results

The following simulation results were obtained using simple Binary Phase Shift Keying (BPSK) over an Additive White Gaussian Noise (AWGN) channel.

4.3.5.1   The Berlekamp–Massey Algorithm

In this section, we characterise the performance of different BCH codes using the BM algorithm.

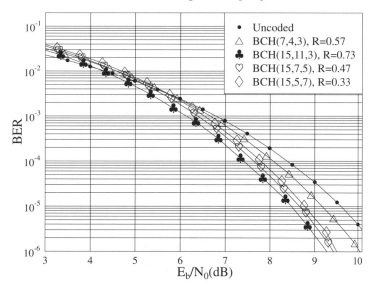

Figure 4.11: BER performance of the BCH codes with n=7 and n=15 over AWGN channels.

Figures 4.11, 4.12, 4.13 and 4.14 show the performance of the BCH codes having the same codeword length n. In Figure 4.14, we can see that the BCH(127,120,3) code, which has a coding rate of about 0.95, exhibits the worst performance at a bit error rate (BER) of 10^{-5}. The BCH(127,113,5) code, which has a coding rate of 0.89, gives a slightly better result than the BCH(127,120,3) scheme, with an improvement of 0.75 dB at a BER of 10^{-5}. As the coding rate

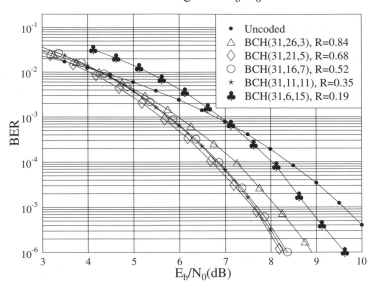

Figure 4.12: BER performance of the BCH codes with n=31 over AWGN channels.

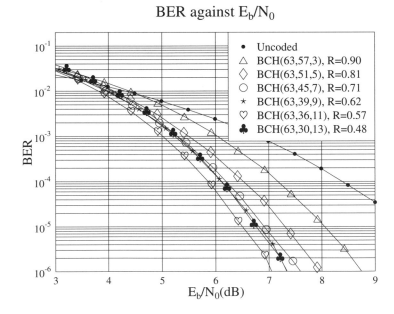

Figure 4.13: BER performance of the BCH codes with n=63 over AWGN channels.

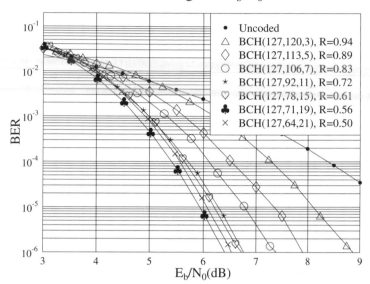

Figure 4.14: BER performance of the BCH codes with n=127 over AWGN channels.

decreases, the performance of the family codes from the BCH$(127, k, t)$ becomes better. However, this is not true when the coding rate decreases to about 0.5 or below since the amount of code redundancy becomes excessive, inevitably reducing the amount of energy per transmitted bit. For example, the BCH(127,71,19) code (code rate=0.56) performs better than the BCH(127,64,21) code (code rate=0.50). This applies to other families of the BCH code as well.

Figure 4.15 shows our performance comparison for different-rate BCH codes selected from each family. By contrast, Figure 4.16 provides the comparison of a set of near-half-rate codes. From these figures we surmise that the BCH-coded performance improves when n increases.

4.3.5.2 Hard-decision Viterbi Decoding

Figure 4.17 shows the performance of the BCH(31,21,5) and BCH(15,7,5) codes using two different decoding algorithms, namely the Hard-Decision Viterbi Algorithm (HD-VA) and that of the Hard-Decision Berlekamp–Massey algorithm (HD-BM). The performance of the algorithms appears to be fairly similar.

4.3.5.3 Soft-decision Viterbi Decoding

Figure 4.18 shows our performance comparison between the Soft-Decision Viterbi Algorithm (SD-VA) and Hard-Decision Berlekamp–Massey algorithm (HD-BM). As can be seen in the figure, there is an improvement of about 2 dB at a BER of 10^{-5}.

4.3.6 Conclusion on Block Coding

The performance of a range of BCH codes using the BM decoding algorithm has been investigated through simulations. The coding gain of the various BCH codes at a BER of 10^{-3} and a

BER against E_b/N_0

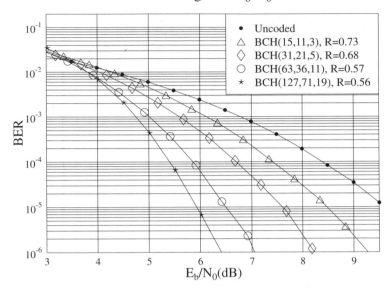

Figure 4.15: BER performance comparison of selected BCH codes over AWGN channels.

BER against E_b/N_0

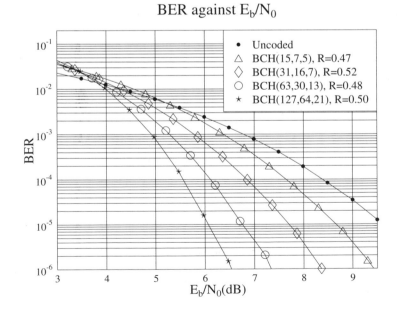

Figure 4.16: BER performance comparison of near-half-rate BCH codes over AWGN channels.

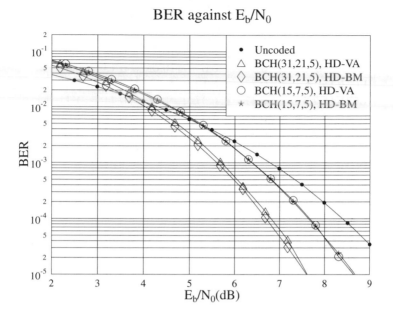

Figure 4.17: BER comparison between hard-decision Viterbi decoding and BM decoding for various BCH codes and decoding algorithms over AWGN channels.

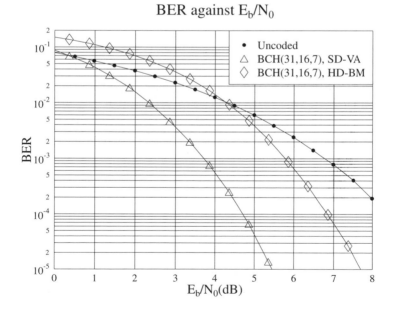

Figure 4.18: BER comparison between soft-decision Viterbi decoding and BM decoding over AWGN channels.

Code	Code Rate R	E_b/N_0 (dB) BER		Gain (dB) BER	
		10^{-3}	10^{-6}	10^{-3}	10^{-6}
Uncoded	1.00	6.78	10.53	0.00	0.00
BCH(7,4,3)	0.57	6.65	10.05	0.13	0.48
BCH(15,11,3)	0.73	6.12	9.27	0.66	1.26
BCH(15,7,5)	0.47	6.37	9.42	0.41	1.11
BCH(15,5,7)	0.33	6.52	9.50	0.26	1.03
BCH(31,26,3)	0.84	5.98	8.90	0.80	1.63
BCH(31,21,5)	0.68	5.61	8.23	1.17	2.30
BCH(31,16,7)	0.52	5.77	8.37	1.01	2.16
BCH(31,11,5)	0.35	5.75	8.29	1.03	2.24
BCH(31,6,7)	0.19	6.89	9.62	-0.11	0.91

Table 4.2: Coding gain of BCH codes with $n = 7, n = 15$ and $n = 31$ using the BM algorithm over AWGN channels.

Code	Code Rate R	E_b/N_0 (dB) BER		Gain (dB) BER	
		10^{-3}	10^{-6}	10^{-3}	10^{-6}
Uncoded	1.00	6.78	10.53	0.00	0.00
BCH(63,57,3)	0.90	6.03	8.75	0.75	1.78
BCH(63,51,5)	0.81	5.50	7.97	1.28	2.56
BCH(63,45,7)	0.71	5.29	7.60	1.49	2.93
BCH(63,39,9)	0.62	5.24	7.34	1.54	3.19
BCH(63,36,11)	0.57	5.02	7.05	1.76	3.48
BCH(63,30,13)	0.48	5.28	7.33	1.50	3.20
BCH(63,24,15)	0.38	5.78	7.77	1.00	2.76
BCH(63,18,21)	0.29	5.70	7.80	1.08	2.73
BCH(63,16,23)	0.25	5.80	7.83	0.98	2.70
BCH(63,10,27)	0.16	7.03	9.09	-0.25	1.44
BCH(63,7,31)	0.11	7.76	10.03	-0.98	0.50

Table 4.3: Coding gain of the BCH codes with $n = 63$ using the BM algorithm over AWGN channels.

BER of 10^{-6} over AWGN channels was tabulated in Tables 4.2, 4.3 and 4.4.

Figures 4.19 and 4.20 show the coding gain against the code rate for different BCH codes at a BER of 10^{-3} and a BER of 10^{-6}, respectively, suggesting the following conclusions. As k and the code rate $R = \frac{k}{n}$ increases, there is a maximum coding gain for each family (n is constant) of the BCH codes studied. Furthermore, the maximum coding gain is typically found when the code rate is between 0.5 and 0.6. For example, in Figure 4.20, the maximum coding gain of the BCH codes having a codeword length of $n = 127$ is 4.1 dB when the code rate is 0.56. For BCH codes with $n = 63$, the maximum coding gain is 3.5 dB when the code rate is 0.57.

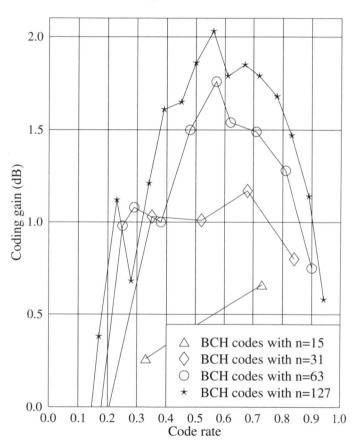

Figure 4.19: Coding gain against code rate for different BCH codes at a BER of 10^{-3} over AWGN channels using the codes summarised in Table 4.1.

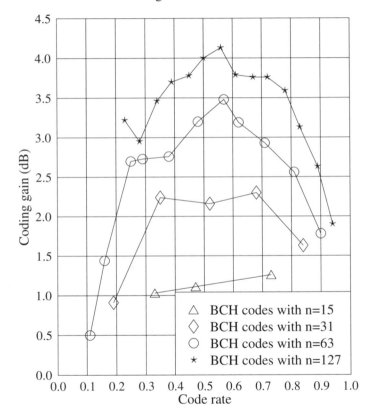

Figure 4.20: Coding gain against code rate for different BCH codes at a BER of 10^{-6} over AWGN channels using the codes summarised in Table 4.1.

Code	Code Rate R	E_b/N_0 (dB) BER		Gain (dB) BER	
		10^{-3}	10^{-6}	10^{-3}	10^{-6}
Uncoded	1.00	6.78	10.53	0.00	0.00
BCH(127,120,3)	0.94	6.20	8.63	0.58	1.90
BCH(127,113,5)	0.89	5.64	7.90	1.14	2.63
BCH(127,106,7)	0.83	5.31	7.40	1.47	3.13
BCH(127,99,9)	0.78	5.10	6.94	1.68	3.59
BCH(127,92.11)	0.72	4.99	6.77	1.79	3.76
BCH(127,85,13)	0.67	4.93	6.77	1.85	3.76
BCH(127,78,15)	0.61	4.99	6.74	1.79	3.79
BCH(127,71,19)	0.56	4.75	6.40	2.03	4.13
BCH(127,64,21)	0.50	4.92	6.53	1.86	4.00
BCH(127,57,23)	0.45	5.13	6.75	1.65	3.78
BCH(127,50,27)	0.39	5.17	6.83	1.61	3.70
BCH(127,43,29)	0.34	5.57	7.07	1.21	3.46
BCH(127,36,31)	0.28	6.10	7.58	0.68	2.95
BCH(127,29,43)	0.23	5.66	7.31	1.12	3.22
BCH(127,22,47)	0.17	6.40	8.08	0.38	2.45
BCH(127,15,55)	0.12	7.20	8.88	-0.42	3.25
BCH(127,8,63)	0.06	9.05	10.86	-2.27	-0.33

Table 4.4: Coding gain of the BCH codes with $n = 127$ using the BM algorithm over AWGN channels.

4.4 Soft input Algebraic Decoding

4.4.1 Introduction

In this section we investigate the benefits of using soft inputs in the context of the classic algebraic decoding. The decoding techniques, our simulation results and related conclusions will be outlined in this section. Since the discovery of BCH codes in 1960, numerous algorithms [17,29,31,32,34,115] have been suggested for their decoding. The BM algorithm [31–34] is widely recognised as an attractive decoding technique.

However, the BM algorithm assumes that the output of the demodulator is binary. This implies that the algorithm is incapable of directly exploiting the soft outputs provided by the demodulator at the receiver. In Section 4.3.5.3, we have shown that for the trellis decoding of BCH codes, there is an improvement of 2 dB if we use the soft-decision Viterbi algorithm rather than the hard-decision Viterbi algorithm.

In 1972, Chase [28] invented a class of decoding algorithms that utilise the soft outputs provided by the demodulator. At the receiver the demodulator provides the received signal value y_k, given that the corresponding data bit u_k was either 1 or 0, indicating two different features:

- Its polarity shows whether u_k is likely to be 1 (positive y_k) or 0 (negative y_k).

- Its magnitude $|y_k|$ indicates the confidence measure provided by the demodulator.

As mentioned earlier, the hard-decision-based BM algorithm only utilises the binary bit provided by the demodulator. The error correcting capability t of the BCH(n, k, d_{min}) code is related to the minimum Hamming distance d_{min} between the codewords. In general, the error correcting capability, t, of the BCH code is defined as the maximum number of guaranteed correctable errors per codeword, given by [96]:

$$t = \lfloor \frac{d_{min} - 1}{2} \rfloor \,, \tag{4.3}$$

where $\lfloor i \rfloor$ means the largest integer not exceeding i.

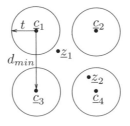

Figure 4.21: Stylised example of conventional algebraic decoding.

Figure 4.21 shows a stylised example of conventional algebraic decoding. There are four valid codewords $c_1, \ldots, c_4$ shown in Figure 4.21 and the minimum separation between them is d_{min}. Each codeword is surrounded by a decoding sphere of radius t. Let us assume that we have transmitted two identical BCH codewords, say $\underline{c}_2$, over the noisy channel. The associated received vectors of n binary bits, $\underline{z}_1$ and $\underline{z}_2$, are provided by the demodulator. As we can see, $\underline{z}_1$ is not within the decoding sphere of any valid codeword. Hence, the conventional algebraic decoder is incapable of correcting the errors in $\underline{z}_1$. On the other hand, the binary n-tuple $\underline{z}_2$ falls within the decoding sphere of codeword $\underline{c}_4$ and hence it is decoded to the valid codeword $\underline{c}_4$. However, if we additionally consider the soft-decision-based confidence measures $|\underline{y}_k|$ provided by the demodulator, the decoder might be able to correct more than t errors in the n-tuple $\underline{z}_1$. Moreover, the received n-tuple $\underline{z}_2$ might be more likely to be due to codeword $\underline{c}_2$ rather than $\underline{c}_4$. These problems are circumvented by the Chase algorithm of the forthcoming section.

4.4.2 Chase Algorithms

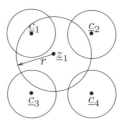

Figure 4.22: Stylised illustration of the Chase algorithm.

Figure 4.22 shows the geometric sketch of the decoding process aided by channel measurement information, which is elaborated on below. Accordingly, the received binary n-tuple $\underline{z}_1$ is perturbed with the aid of a set of test patterns TP, which is a binary sequence that contains 1s in the location of the bit positions that are to be tentatively inverted. By adding this test pattern, modulo 2, to the received binary sequence, a new binary sequence $\underline{z}_1'$ is obtained:

$$\underline{z}_1' = \underline{z}_1 \oplus TP. \tag{4.4}$$

As shown in Figure 4.22, r represents the maximum Hamming distance of the perturbed binary received sequence $\underline{z}_1'$ from the original binary received sequence $\underline{z}_1$. By using a number of test patterns, the perturbed received binary sequence $\underline{z}_1'$ may fall within the decoding sphere of a number of valid BCH codewords. If we increase r, the perturbed received sequence $\underline{z}_1'$ will fall within the decoding sphere of more valid BCH codewords.

If the perturbed received binary sequence $\underline{z}_1'$ falls within the decoding sphere of a valid BCH codeword c_1, by invoking algebraic decoding a new error pattern $\underline{e}'$ is obtained, which may be an all-zero or a non-zero tuple. The actual error pattern $\underline{e}$ associated with the binary received sequence $\underline{z}_1$ is given by:

$$\underline{e} = \underline{e}' \oplus TP, \tag{4.5}$$

which may or may not be different from the original test pattern TP, depending on whether the perturbed received binary sequence $\underline{z}_1'$ falls into the decoding sphere of a valid codeword. However, only those perturbed received binary sequences $\underline{z}_1'$ that fall into the decoding sphere of a valid codeword are considered.

A maximum likelihood decoder is capable of finding the codeword that satisfies:

$$\min_{m}\text{weight}(\underline{z} \oplus c_m), \tag{4.6}$$

where the range of m is over all possible codewords. Based on similar principles, Chase [28] defined a new channel decoder. However, for the sake of low complexity only a certain limited set of valid codewords is considered by Chase's technique, namely those surrounded by the decoding spheres that the perturbed received binary sequence $\underline{z}_1'$ may fall into. In this case, we are concerned with finding the error pattern $\underline{e}$ of minimum analogue weight, where the analogue weight of an error sequence $\underline{e}$ is defined as:

$$W(\underline{e}) = \sum_{i=1}^{n} e_i|y_i| . \tag{4.7}$$

The Chase algorithm can be summarised in the flowchart shown in Figure 4.23. Each time, the algorithm considers an n-tuple codeword of the BCH code, which is constituted by n number of the received bits $\underline{z}$ and their soft metrics y. The received bits $\underline{z}$ and their confidence values y are assembled, which is the first step shown in Figure 4.23. Then, a set of test patterns TP is generated. For each test pattern, a new sequence $\underline{z}'$ is obtained by modulo-2 addition of the particular test pattern TP and the received sequence $\underline{z}$. The conventional algebraic decoder is invoked to decode the new sequence $\underline{z}'$, as seen in Figure 4.21. If the conventional algebraic decoder found a non-zero error pattern $\underline{e}'$, we are able to find the actual error pattern $\underline{e}$, using Equation 4.5, associated with the received binary sequence $\underline{z}$. Using Equation 4.7, the analogue weight W of the actual error pattern $\underline{e}$ can be calculated. The generated test pattern TP will be stored in the memory, if the associated analogue weight W is found to be the lowest. The above procedure will be repeated for every test pattern generated. Upon completing the loop in Figure 4.23, the memory is checked. If there is an error pattern stored, the binary decoded sequence will be $\underline{z} \oplus \underline{e}$. Otherwise, the binary decoded sequence is the same as the received sequence $\underline{z}$.

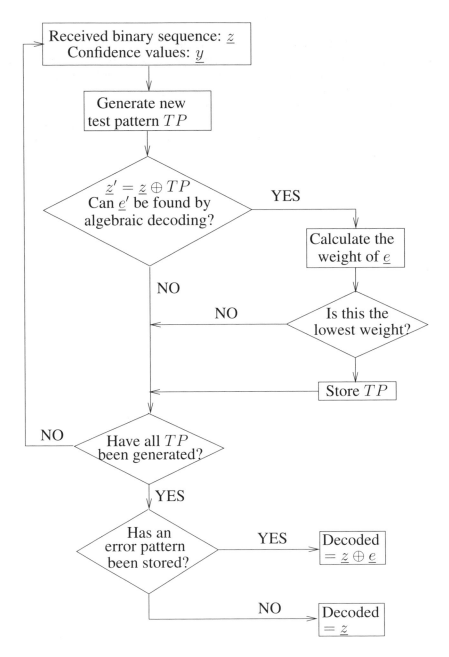

Figure 4.23: Flowchart of Chase algorithm.

The number of test patterns used can be varied according to the tolerable complexity, which also has an effect on the achievable performance. In the following sections, we present two variants of this algorithm, namely the Chase Algorithm 1 and Chase Algorithm 2. The nature of both algorithms depends essentially on the number of test patterns used.

4.4.2.1 Chase Algorithm 1

For this particular algorithm, typically a large set of test patterns TP is considered. In fact, the algorithm considers the entire set of possible test patterns within a sphere of radius $r = d_{min} - 1$ surrounding the received binary sequence $\underline{z}$. Thus, all possible test patterns of binary weight less than or equal to $d_{min} - 1$ are considered.

Let us illustrate the operation of the algorithm with the aid of an example, where the BCH(7,4,3) code is used, which has $d_{min} = 3$. Hence, all the test patterns having a binary weight less than or equal to $d_{min} - 1 = 2$ are generated, which is the second step in Figure 4.23. A fraction of the test patterns TP, which have a binary weight less than or equal to 1, are shown below:

$$
\begin{aligned}
TP_0 &= 0\,0\,0\,0\,0\,0\,0 \\
TP_1 &= 0\,0\,0\,0\,0\,0\,1 \\
TP_2 &= 0\,0\,0\,0\,0\,1\,0 \\
TP_3 &= 0\,0\,0\,0\,1\,0\,0 \\
TP_4 &= 0\,0\,0\,1\,0\,0\,0 \\
TP_5 &= 0\,0\,1\,0\,0\,0\,0 \\
TP_6 &= 0\,1\,0\,0\,0\,0\,0 \\
TP_7 &= 1\,0\,0\,0\,0\,0\,0\,.
\end{aligned}
\tag{4.8}
$$

Using the test patterns TP which have binary weights equal to 1, we are also able to generate the test patterns that have binary weights larger than 1.

Let us assume that BPSK is used and the transmitted sequence is $-1\,-1\,-1\,-1\,-1\,-1\,-1$. The soft demodulator outputs $\underline{y}$, the hard-decision decoded binary sequence $\underline{z}$ and the confidence measures $|\underline{y}|$ are shown in Table 4.5.

i	0	1	2	3	4	5	6
y_i	-0.9	-1.3	-0.4	-2.2	+0.6	-1.2	+0.8
z_i	0	0	0	0	1	0	1
δ_i	0.9	1.3	0.4	2.2	0.6	1.2	0.8

Table 4.5: Example of soft demodulator outputs, hard-decision decoded sequence and confidence measures.

Using for example the second test pattern TP_1 in Equation 4.8, the perturbed received sequence $\underline{z}'$ is:

$$
\begin{aligned}
\underline{z}' &= \underline{z} \oplus TP_1 \\
&= 0\,0\,0\,0\,1\,0\,1 \oplus 0\,0\,0\,0\,0\,0\,1 \\
&= 0\,0\,0\,0\,1\,0\,0\,.
\end{aligned}
\tag{4.9}
$$

i	0	1	2	3	4	5	6	7
TP_i	TP_0	TP_1	TP_2	TP_3	TP_4	TP_5	TP_6	TP_7
$W_{TP_i}(\underline{e})$	2.2	1.4	1.6	1.4	2.2	1.6	2.2	2.2

Table 4.6: Analogue weights associated with the test patterns in Equation 4.8, for the BCH(7,4,3) Chase decoding example.

Owing to this perturbation, we now have a sequence within a Hamming distance of one from a legitimate codeword, namely the $0\,0\,0\,0\,0\,0\,0$ sequence, and hence the perturbed received binary sequence $\underline{z}'$ is decoded by the algebraic decoder and the decoded sequence is $0\,0\,0\,0\,0\,0\,0$. Therefore, the associated error sequence $\underline{e}'$ is:

$$\underline{e}' = 0\,0\,0\,0\,1\,0\,0\,, \tag{4.10}$$

and the actual error sequence $\underline{e}$ is:

$$\begin{aligned} \underline{e} &= \underline{e}' \oplus TP_1 \\ &= 0\,0\,0\,0\,1\,0\,0 \oplus 0\,0\,0\,0\,0\,0\,1 \\ &= 0\,0\,0\,0\,1\,0\,1\,. \tag{4.11} \end{aligned}$$

As we will show below, this is the most likely error sequence $\underline{e}$, allowing us to correct two, rather than just one, error, which was a limitation of the hard-decision BM algorithm.

In order to quantify the probability of the possible error sequences, their analogue weight is determined next. The analog weight of the actual error sequence $\underline{e}$ is:

$$\begin{aligned} W_{TP_1}(\underline{e}) &= \sum_{i=1}^{n} \delta_i e_i \\ &= 0.9 \times 0 + 1.3 \times 0 + 0.4 \times 0 + 2.2 \times 0 + 0.6 \times 1 + 1.2 \times 0 + 0.8 \times 1 \\ &= 1.4. \tag{4.12} \end{aligned}$$

This process is repeated for other test patterns of binary weight less than or equal to $d_{min} - 1$ as shown in Equation 4.8. The analogue weights associated with each test pattern in Equation 4.8 are shown in Table 4.6. The lowest possible analogue weight is 1.4, taking into consideration all test patterns of binary weight less than or equal to $d_{min} - 1$. As we can see from Table 4.6, test patterns TP_1 and TP_3 have produced the lowest analogue weight, which is 1.4. It can be readily shown that the associated actual error pattern $\underline{e}$ of both test patterns TP_1 and TP_3 is the same. Therefore, the Chase-decoded sequence $\hat{\underline{z}}$ is:

$$\begin{aligned} \hat{\underline{z}} &= \underline{e}_1 \oplus \underline{z} \\ &= 1\,0\,1\,0\,0\,0\,0 \oplus 1\,0\,1\,0\,0\,0\,0 \\ &= 0\,0\,0\,0\,0\,0\,0\,. \tag{4.13} \end{aligned}$$

Computer simulations have shown that the performance of this algorithm is similar to that of the soft-decision Viterbi algorithm.

The BCH(7,4,3) code has a minimum free distance d_{min} of 3 and the number of possible test patterns is 29. If we employ the algorithm to decode the BCH(31,21,5) code, the minimum free distance d_{min} is now 5. Hence, we need to consider all the test patterns having a binary weight less than or equal to 4. In this case, the number of test patterns is over 36000. Generally, the number of test patterns increases exponentially with n and d_{min}.

i	0	1	2	3	4	5	6	7
TP_i	TP_0	TP_1	TP_2	TP_3	TP_4	TP_5	TP_6	TP_7
$W_{TP_i}(\underline{e})$	2.2	1.4	1.4	-	1.6	2.5	1.9	1.4

Table 4.7: Analogue weights, associated with the test patterns in Equation 4.14, for the BCH(7,4,3) example.

4.4.2.2 Chase Algorithm 2

For this variant of the Chase algorithm, a considerably smaller set of possible error patterns is used. Only the positions of the $\lfloor d_{min}/2 \rfloor$ lowest confidence measures are considered. The test patterns TP have any combination of 1s, which are located in the $\lfloor d_{min}/2 \rfloor$ positions of the lowest confidence values. Hence, there are only $2^{\lfloor d_{min}/2 \rfloor}$ possible test patterns, including the all-zero pattern.

For the BCH(7,4,3) code, the number of legitimate test positions is equal to one and the number of test patterns is two. Computer simulations have shown that the BER performance curve of this simplified algorithm is about a quarter of a dB worse than that of the soft-decision Viterbi algorithm. If the number of test positions were equal to three, i.e. there were eight test patterns, then the performance would be the same as that of the soft-decision Viterbi algorithm.

Using the same example as in Section 4.4.2.1, we have to search for the three test positions associated the lowest confidence measures. The eight test patterns are

$$
\begin{aligned}
TP_0 &= 0\,0\,0\,0\,0\,0\,0 \\
TP_1 &= 0\,0\,0\,0\,0\,0\,1 \\
TP_2 &= 0\,0\,0\,0\,1\,0\,0 \\
TP_3 &= 0\,0\,0\,0\,1\,0\,1 \\
TP_4 &= 0\,0\,1\,0\,0\,0\,0 \\
TP_5 &= 0\,0\,1\,0\,0\,0\,1 \\
TP_6 &= 0\,0\,1\,0\,1\,0\,0 \\
TP_7 &= 0\,0\,1\,0\,1\,0\,1,
\end{aligned}
\tag{4.14}
$$

while their associated analogue weights are summarised in Table 4.7. Notice that in the Chase Algorithm 2 the associated error sequence $\underline{e}'$ of test pattern TP_3 is not considered.

Computer simulations have shown that a better performance is achieved when the number of test positions increases, approaching that of the soft-decision Viterbi algorithm, as the number of test positions approaches d_{min}.

In closing, we note that unlike in the Chase Algorithm 1, the number of test patterns does not increase when n increases. However, the associated algorithmic complexity increases exponentially, if the number of test positions increases.

4.4.3 Simulation Results

Simulation results were obtained using simple BPSK over an AWGN channel. Figure 4.24 shows our performance comparison between the algorithms considered, namely algebraic decoding, the Chase Algorithm 1 and the soft-decision Viterbi algorithm using the BCH(15,11,3) code. It is shown in Figure 4.24 that the performance of the Chase Algorithm 1 and that of the soft-decision Viterbi algorithm is identical.

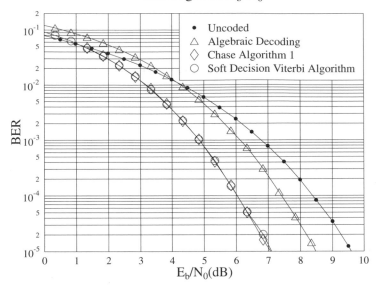

Figure 4.24: Performance comparison between algebraic decoding, the Chase Algorithm 1 and the soft-decision Viterbi algorithm using the BCH(15,11,3) code over AWGN channels.

Figure 4.25 portrays our performance comparison between algebraic decoding, the Chase Algorithm 2 and the soft-decision Viterbi algorithm using the more powerful BCH(31,21,5) code. For the Chase Algorithm 2, the performance of using $\lfloor d_{min}/2 \rfloor = 2$ test positions is 1.25 dB better than that of the algebraic decoding. As the number of test positions increases, the performance of the Chase Algorithm 2 improves. However, the relative improvement becomes smaller as the number of test positions increases. It is shown in Figure 4.25 that the performance lower bound, i.e. the best possible of the Chase Algorithm 2, is the same as that of the soft-decision Viterbi algorithm. The lower bound is achieved when the maximum number of test positions is equal to the minimum free distance d_{min} of the BCH code.

Often it is impractical to implement the Chase Algorithm 1, because the number of test patterns increases exponentially when n and d_{min} increase. However, a similar performance can be achieved using the Chase Algorithm 2 and the maximum number of test positions is equal to the minimum free distance d_{min} of the code. The Chase Algorithm 2 is less complex to implement than trellis decoding, if d_{min} is small. However, the number of test patterns increases exponentially when d_{min} increases.

4.5 Summary and Conclusions

This chapter served as a brief introduction to the trellis and Chase decoding of BCH codes, before we consider the more complex and novel turbo BCH codes in the next chapter. We commenced with the introduction of BCH codes and the associated definitions in Section 4.2. The detailed encoding process of systematic BCH codes was described in Section 4.2.1. We also showed in Section 4.2.2 that the BCH encoder could be implemented using its state diagram. The conventional Viterbi decoding algorithm was employed in Section 4.3.2 for decoding BCH codes.

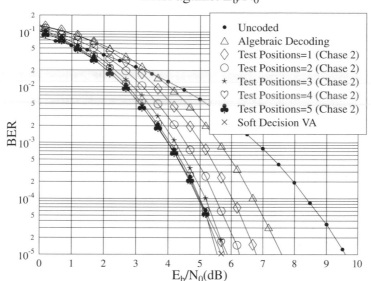

Figure 4.25: Performance comparison between algebraic decoding, the Chase Algorithm 2 (different number of test positions) and the soft-decision Viterbi algorithm using the BCH(31,21,5) code over AWGN channels.

Several examples of hard-decision and soft-decision Viterbi decoding were given in the section as well. In Section 4.3.5, we presented simulation results for both the Viterbi and Berlekamp–Massey decoding algorithms. It was found that the performance of employing the soft-decision Viterbi decoding algorithm is about 2 dB better than that of hard-decision decoding. The coding gains of various BCH codes having the same codeword length were evaluated and tabulated in Tables 4.2, 4.3 and 4.4. Then, the coding gain versus code rate was plotted in Figure 4.19 and 4.20 for different BCH codes having the same codeword length n. From both figures, we inferred that the maximum coding gain was achieved when the code rate was between 0.5 and 0.6.

Viterbi decoding of the BCH codes becomes prohibitively complex if $n - k$ increases, since the number of decoding states increases exponentially. The Chase algorithm, which offers a lower complexity associated with a slight performance degradation, was detailed in Section 4.4.2. Two variants of the Chase algorithm were described and the Chase Algorithm 2 was found to be more favourable owing to its lower complexity. A detailed example of the Chase decoding process was given in Section 4.4.2. Our simulation results were presented in Section 4.4.3. It was found that there is a lower and upper performance bound for the Chase algorithms. The upper bound performance is the same as that of the conventional algebraic decoding, whereas the lower bound performance is the same as that of the soft decision Viterbi algorithm. For the Chase Algorithm 2, the lower performance bound can be achieved if the maximum number of test positions is equal to the minimum free distance d_{min} of the code. However, the complexity of the Chase Algorithm 2 increases exponentially with the minimum free distance d_{min}.

Having acquired the necessary background for the understanding of turbo coding, in the next chapter we will highlight the operation and decoding algorithms of turbo convolutional codes.

Part II

Turbo Convolutional and Turbo Block Coding

Chapter 5

Turbo Convolutional Coding

J.P. Woodard, L. Hanzo[1]

5.1 Introduction

In Chapter 2 a rudimentary introduction to convolutional codes and their decoding using the Viterbi algorithm was provided. In this chapter we introduce the concept of turbo coding using convolutional codes as its constituent codes. Our discussions will become deeper, relying on basic familiarity with convolutional coding. In Chapter 6 we then further elaborate on the relationship between conventional convolutional codes and turbo convolutional codes and highlight the relationship between their trellis structures.

In insightful concept of turbo coding was proposed in 1993 in a seminal contribution by Berrou, Glavieux and Thitimajashima, who reported excellent coding gain results [12], approaching Shannonian predictions. The information sequence is encoded twice, with an interleaver between the two encoders serving to make the two encoded data sequences approximately statistically independent of each other. Often half-rate Recursive Systematic Convolutional (RSC) encoders are used, with each RSC encoder producing a systematic output which is equivalent to the original information sequence, as well as a stream of parity information. The two parity sequences can then be punctured before being transmitted along with the original information sequence to the decoder. This puncturing of the parity information allows a wide range of coding rates to be realised, and often half the parity information from each encoder is sent. Along with the original data sequence this results in an overall coding rate of 1/2.

At the decoder two RSC decoders are used. Special decoding algorithms must be used which accept soft inputs and give soft outputs for the decoded sequence. These soft inputs and outputs provide not only an indication of whether a particular bit was a 0 or a 1, but also a likelihood ratio which gives the probability that the bit has been correctly decoded. The turbo decoder operates iteratively. In the first iteration the first RSC decoder provides a soft output giving an estimation of the original data sequence based on the soft channel inputs alone. It also provides an *extrinsic* output. The extrinsic output for a given bit is based not on the channel input for that bit, but on the information for surrounding bits and the constraints imposed by the code being used. This

[1]This chapter is based on J. P. Woodard, L. Hanzo: Comparative Study of Turbo Decoding Techniques: An Overview; IEEE Transactions on Vehicular Technology, Nov. 2000, Vol. 49, No. 6, pp 2208-2234 ©IEEE.

extrinsic output from the first decoder is used by the second RSC decoder as *a-priori information*, and this information together with the channel inputs are used by the second RSC decoder to give its soft output and extrinsic information. In the second iteration the extrinsic information from the second decoder in the first iteration is used as the a-priori information for the first decoder, and using this a-priori information the decoder can hopefully decode more bits correctly than it did in the first iteration. This cycle continues, with at each iteration both RSC decoders producing a soft output and extrinsic information based on the channel inputs and a-priori information obtained from the extrinsic information provided by the previous decoder. After each iteration the Bit Error Rate (BER) in the decoded sequence drops, but the improvements obtained with each iteration fall as the number of iterations increases so that for complexity reasons usually only between 4 and 12 iterations are used.

In their pioneering proposal Berrou, Glavieux and Thitimajashima [12] invoked a modified version of the classic minimum BER Maximum *A-Posteriori* (MAP) algorithm due to Bahl *et al.* [11] in the above iterative structure for decoding the constituent codes. Since the conception of turbo codes a large amount of work has been carried out in the area, aiming for example to reduce the decoder complexity, as suggested by Robertson, Villebrun and Höher [52] as well as by Berrou *et al.* [122]. Le Goff, Glavieux and Berrou [55], Wachsmann and Huber [56] as well as Robertson and Wörz [57] suggested using the codes in conjunction with bandwidth-efficient modulation schemes. Further advances in understanding the excellent preformance of the codes are due, for example, to Benedetto and Montorsi [58, 59] as well as to Perez, Seghers and Costello [60]. A number of seminal contributors including Hagenauer, Offer and Papke [61], as well as Pyndiah [123], extended the turbo concept to parallel concatenated block codes. Jung and Nasshan [65] characterised the coded performance under the constraints of short transmission frame length, which is characteristic of speech systems. In collaboration with Blanz they also applied turbo codes to a CDMA system using joint detection and antenna diversity [67]. Barbulescu and Pietrobon [68] as well as a number of other authors addressed the equally important issues of interlever design. Because of space limitations here we have to curtail listing the range of further contributors to the field, without whose advances this treatise could not have been written. It is particulary important to note the tutorial paper authored by Sklar [69].

Here we embark on describing turbo codes in more detail. Specifically, in Section 5.2 we detail the encoder used, while in Section 5.3 the decoder is portrayed. Then in Section 5.4 we characterise the performance of various turbo codes over Gaussian channels using BPSK. Then in Section 5.5 we discuss the employment of turbo codes over Rayleigh channels, and characterise the system's speech performance when using the G.729 speech codec.

5.2 Turbo Encoder

The general structure used in turbo encoders is shown in Figure 5.1. Two component codes are used to code the same input bits, but an interleaver is placed between the encoders. Generally RSC codes are used as the component codes, but it is possible to achieve good performance using a structure like that seen in Figure 5.1 with the aid of other component codes, such as for example block codes, as advocated by Hagenauer and Offer and Papke [61] as well as by Pyndiah [123]. Furthermore, it is also possible to employ more than two component codes. However, in this chapter we concentrate entirely on the standard turbo encoder structure using two RSC codes. Turbo codes using block codes as component codes are described in Chapter 7.

The outputs from the two component codes are then punctured and multiplexed. Usually both component RSC codes are half rate, giving one parity bit and one systematic bit output for every input bit. Then to give an overall coding rate of one-half, half the output bits from the two

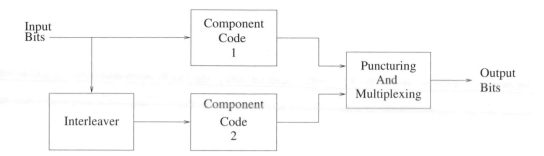

Figure 5.1: Turbo encoder schematic Berrou *et al.* ©IEEE, 1993 [12, 13].

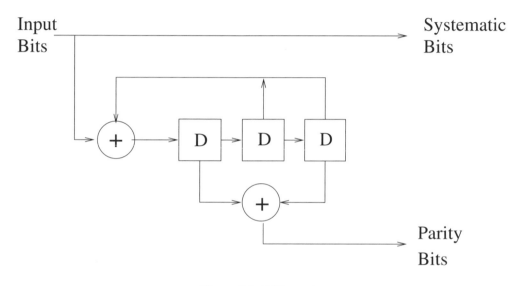

Figure 5.2: RSC encoder.

encoders must be punctured. The arrangement that is often favoured, and that we have used in our work, is to transmit all the systematic bits from the first RSC encoder, and half the parity bits from each encoder. Note that the systematic bits are rarely punctured, since this degrades the performance of the code more dramatically than puncturing the parity bits.

Figure 5.2 shows the $K=3$ RSC code we have used as the component codes in most of our simulations. This code has generator polynomials 7 (for the feedback polynomial) and 5.

Parallel concatenated codes had been well investigated before Berrou *et al.*'s breakthrough 1993 paper [12], but the dramatic improvement in performance that turbo codes gave arose because of the interleaver used between the encoders, and because recursive codes were used as the component codes. Recently theoretical papers have been published for example by Benedetto and Montorsi [58, 59] which endeavour to explain the remarkable performance of turbo codes. It appears that turbo codes can be thought of as having a performance gain proportional to the interleaver length used. However, the decoding complexity per bit does not depend on the interleaver length. Therefore extremely good performance can be achieved with reasonable complexity by using very long interleavers. However, for many important applications, such as speech transmission, extremely long frame lengths are not practical because of the delays they result in. Therefore in this chapter we have also investigated the use of turbo codes in conjunction with

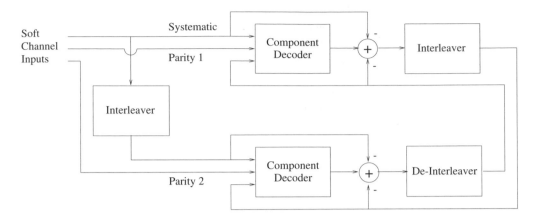

Figure 5.3: Turbo decoder schematic Berrou *et al.* ©IEEE, 1993 [12, 13].

short frame lengths of the order of 100 bits.

5.3   Turbo Decoder

5.3.1   Introduction

The general structure of an iterative turbo decoder is shown in Figure 5.3. Two component decoders are linked by interleavers in a structure similar to that of the encoder. As seen in the figure, each decoder takes three inputs–: the systematically encoded channel output bits, the parity bits transmitted from the associated component encoder, and the information from the other component decoder about the likely values of the bits concerned. This information from the other decoder is referred to as a-priori information. The component decoders have to exploit both the inputs from the channel and this a-priori information. They must also provide what are known as soft outputs for the decoded bits. This means that as well as providing the decoded output bit sequence, the component decoders must also give the associated probabilities for each bit that has been correctly decoded. The soft outputs are typically represented in terms of the so-called Log Likelihood Ratios (LLRs). The polarity of the LLR determines the sign of the bit, while its amplitude quantifies the probability of a correct decision. These LLRs are described in Section 5.3.2. Two suitable decoders are the Soft-Output Viterbi Algorithm (SOVA) proposed by Hagenauer and Höher [53] and Bahl's Maximum A-Posteriori (MAP) [11] algorithm, which are described in Sections 5.3.6 and 5.3.3, respectively.

   The decoder of Figure 5.3 operates iteratively, and in the first iteration the first component decoder takes channel output values only, and produces a soft output as its estimate of the data bits. The soft output from the first encoder is then used as additional information for the second decoder, which uses this information along with the channel outputs to calculate its estimate of the data bits. Now the second iteration can begin, and the first decoder decodes the channel outputs again, but now with additional information about the value of the input bits provided by the output of the second decoder in the first iteration. This additional information allows the first decoder to obtain a more accurate set of soft outputs, which are then used by the second decoder as a-priori information. This cycle is repeated, and with every iteration the BER of the decoded bits tends to fall. However, the improvement in performance obtained with increasing numbers of iterations decreases as the number of iterations increases. Hence, for complexity reasons, usually only about eight iterations are used.

Owing to the interleaving used at the encoder, care must be taken to properly interleave and de-interleave the LLRs which are used to represent the soft values of the bits, as seen in Figure 5.3. Furthermore, because of the iterative nature of the decoding, care must be taken not to reuse the same information more than once at each decoding step. For this reason the concept of so-called extrinsic and intrinsic information was used in their seminal paper by Berrou *et al.* describing iterative decoding of turbo codes [12]. These concepts and the reason for the subtraction circles shown in Figure 5.3 are described in Section 5.3.4.

Other, non-iterative, decoders have been proposed [124, 125] which give optimal decoding of turbo codes. However, the improvement in performance over iterative decoders was found to be only about 0.35 dB, and they are hugely complex. Therefore the iterative scheme shown in Figure 5.3 is commonly used. We now proceed with describing the concepts and algorithms used in the iterative decoding of turbo codes, commencing with the portrayal of LLRs.

5.3.2 Log Likelihood Ratios

The concept of LLRs was shown by Robertson [126] to simplify the passing of information from one component decoder to the other in the iterative decoding of turbo codes, and so is now widely used in the turbo coding literature. The LLR of a data bit u_k is denoted as $L(u_k)$ and is defined to be merely the log of the ratio of the probabilities of the bit taking its two possible values, i.e.:

$$L(u_k) \triangleq \ln \left(\frac{P(u_k = +1)}{P(u_k = -1)} \right). \tag{5.1}$$

Notice that the two possible values for the bit u_k are taken to be $+1$ and -1, rather than 1 and 0. This definition of the two values of a binary variable makes no conceptual difference, but it slightly simplifies the mathematics in the derivations which follow. Hence this convention is used throughout this chanpter. Figure 5.4 shows how $L(u_k)$ varies as the probability of $u_k = +1$ varies. It can be seen from this figure that the sign of the LLR $L(u_k)$ of a bit u_k will indicate whether the bit is more likely to be $+1$ or -1, and the magnitude of the LLR gives an indication of how likely it is that the sign of the LLR gives the correct value of u_k. When the LLR $L(u_k) \approx 0$, we have $P(u_k = +1) \approx P(u_k = -1) \approx 0.5$, and we cannot be certain about the value of u_k. Conversely, when $L(u_k) \gg 0$, we have $P(u_k = +1) \gg P(u_k = -1)$ and we can be almost certain that $u_k = +1$.

Given the LLR $L(u_k)$, it is possible to calculate the probability that $u_k = +1$ or $u_k = -1$ as follows. Remembering that $P(u_k = -1) = 1 - P(u_k = +1)$, and taking the exponent of both sides in Equation 5.1, we can write:

$$e^{L(u_k)} = \frac{P(u_k = +1)}{1 - P(u_k = +1)}, \tag{5.2}$$

so:

$$\begin{aligned} P(u_k = +1) &= \frac{e^{L(u_k)}}{1 + e^{L(u_k)}} \\ &= \frac{1}{1 + e^{-L(u_k)}}. \end{aligned} \tag{5.3}$$

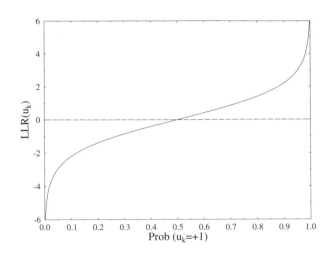

Figure 5.4: LLR $L(u_k)$ versus the probability of $u_k = +1$.

Similarly:

$$
\begin{aligned}
P(u_k = -1) &= \frac{1}{1 + e^{+L(u_k)}} \\
&= \frac{e^{-L(u_k)}}{1 + e^{-L(u_k)}},
\end{aligned}
\tag{5.4}
$$

and hence we can write:

$$
P(u_k = \pm 1) = \left(\frac{e^{-L(u_k)/2}}{1 + e^{-L(u_k)}} \right) \cdot e^{\pm L(u_k)/2}.
\tag{5.5}
$$

Notice that the bracketed term in this equation does not depend on whether we are interested in the probability that $u_k = +1$ or -1, and so it can be treated as a constant in certain applications, such as in Section 5.3.3 where we use this equation in the derivation of the MAP algorithm.

As well as the LLR $L(u_k)$ based on the unconditional probabilities $P(u_k = \pm 1)$, we are also interested in LLRs based on conditional probabilities. For example, in channel coding theory we are interested in the probability that $u_k = \pm 1$ based, or conditioned, on some received sequence $\underline{y}$, and hence we may use the conditional LLR $L(u_k|\underline{y})$, which is defined as:

$$
L(u_k|\underline{y}) \triangleq \ln \left(\frac{P(u_k = +1|\underline{y})}{P(u_k = -1|\underline{y})} \right).
\tag{5.6}
$$

The conditional probabilities $P(u_k = \pm 1|\underline{y})$ are known as the a-posteriori probabilities of the decoded bit u_k, and it is these a-posteriori probabilities that our soft-in soft-out decoders described in later sections attempt to find.

Apart from the conditional LLR $L(u_k|\underline{y})$ based on the a-posteriori probabilities $P(u_k = \pm 1|\underline{y})$, we will also use conditional LLRs based on the probability that the receiver's matched filter output would be y_k given that the corresponding transmitted bit x_k was either $+1$ or -1. This conditional LLR is written as $L(y_k|x_k)$ and is defined as:

$$L(y_k|x_k) \triangleq \ln \left(\frac{P(y_k|x_k = +1)}{P(y_k|x_k = -1)} \right). \tag{5.7}$$

Notice the conceptual difference between the definitions of $L(u_k|y)$ in Equation 5.6 and $L(y_k|x_k)$ in Equation 5.7, despite these two conditional LLRs being represented with very similar notation. This contrast in the definitions of conditional LLRs is somewhat confusing, but since these definitions are widely used in the turbo coding literature, we have introduced them here.

If we assume that the transmitted bit $x_k = \pm 1$ has been sent over a Gaussian or fading channel using BPSK modulation, then we can write for the probability of the matched filter output y_k that:

$$P(y_k|x_k = +1) = \frac{1}{\sigma\sqrt{2\pi}} \exp\left(-\frac{E_b}{2\sigma^2}(y_k - a)^2 \right), \tag{5.8}$$

where E_b is the transmitted energy per bit, σ^2 is the noise variance and a is the fading amplitude (we have $a = 1$ for non-fading AWGN channels). Similarly, we have:

$$P(y_k|x_k = -1) = \frac{1}{\sigma\sqrt{2\pi}} \exp\left(-\frac{E_b}{2\sigma^2}(y_k + a)^2 \right). \tag{5.9}$$

Therefore, when we use BPSK over a (possibly fading) Gaussian channel, we can rewrite Equation 5.7 as:

$$
\begin{aligned}
L(y_k|x_k) \quad &\triangleq \quad \ln\left(\frac{P(y_k|x_k = +1)}{P(y_k|x_k = -1)} \right) \\
&= \quad \ln\left(\frac{\exp\left(-\frac{E_b}{2\sigma^2}(y_k - a)^2\right)}{\exp\left(-\frac{E_b}{2\sigma^2}(y_k + a)^2\right)} \right) \\
&= \quad \left(-\frac{E_b}{2\sigma^2}(y_k - a)^2 \right) - \left(-\frac{E_b}{2\sigma^2}(y_k + a)^2 \right) \\
&= \quad \frac{E_b}{2\sigma^2} 4a \cdot y_k \\
&= \quad L_c y_k, \tag{5.10}
\end{aligned}
$$

where:

$$L_c = 4a\frac{E_b}{2\sigma^2} \tag{5.11}$$

is defined as the channel reliability value, and depends only on the signal-to-noise ratio (SNR) and fading amplitude of the channel. Hence, for BPSK over a (possibly fading) Gaussian channel, the conditional LLR $L(y_k|x_k)$, which is referred to as the soft output of the channel, is simply the matched filter output y_k multiplied by the channel reliability value L_c.

Having introduced LLRs, we now proceed to describe the operation of the MAP algorithm, which is one of the possible soft-in soft-out component decoders that can be used in an iterative turbo decoder.

5.3.3 The Maximum A-Posteriori Algorithm

5.3.3.1 Introduction and Mathematical Preliminaries

In 1974 an algorithm, known as the Maximum A-Posteriori (MAP) algorithm, was proposed by
Bahl, Cocke, Jelinek and Raviv for estimating the a-posteriori probabilities of the states and the
transitions of an observed Markov source, when subjected to memoryless noise. This algorithm
has also become known as the BCJR algorithm, named after its inventors. They showed how
the algorithm could be used for decoding both block and convolutional codes. When employed
for decoding convolutional codes, the algorithm is optimal in terms of minimising the decoded,
unlike the Viterbi algorithm [9], which minimises the probability of an incorrect path through the
trellis being selected by the decoder. Thus the Viterbi algorithm can be thought of as minimising
the number of *groups* of bits associated with these trellis paths, rather than the actual number
of bits, which are decoded incorrectly. Nevertheless, as stated by Bahl *et al.* in [11], in most
applications the performance of the two algorithms will be almost identical. However, the MAP
algorithm examines every possible path through the convolutional decoder trellis and therefore
initially seemed to be unfeasibly complex for application in most systems. Hence it was not
widely used before the discovery of turbo codes.

However, the MAP algorithm provides not only the estimated bit sequence, but also the probabilities for each bit that has been decoded correctly. This is essential for the iterative decoding
of turbo codes proposed by Berrou *et al.* [12], and so MAP decoding was used in this seminal
paper. Since then much work has been done to reduce the complexity of the MAP algorithm to
a reasonable level. In this section we describe the theory behind the MAP algorithm as used for
the soft-output decoding of the component convolutional codes of turbo codes. Throughout our
work it is assumed that binary codes are used.

We use Bayes' rule repeatedly throughout this section. This rule gives the joint probability
of *a and b*, $P(a \wedge b)$, in terms of the conditional probability of *a given b* as:

$$P(a \wedge b) = P(a|b) \cdot P(b). \tag{5.12}$$

A useful consequence of Bayes' rule is that:

$$P(\{a \wedge b\}|c) = P(a|\{b \wedge c\}) \cdot P(b|c), \tag{5.13}$$

which can be derived from Equation 5.12 by considering $x \equiv a \wedge b$ and $y \equiv b \wedge c$ as follows.
From Equation 5.12 we can write:

$$
\begin{aligned}
P(\{a \wedge b\}|c) &\equiv P(x|c) = \frac{P(x \wedge c)}{P(c)} \\
&= \frac{P(a \wedge b \wedge c)}{P(c)} \equiv \frac{P(a \wedge y)}{P(c)} \\
&= \frac{P(a|y) \cdot P(y)}{P(c)} \equiv P(a|\{b \wedge c\}) \cdot \frac{P(b \wedge c)}{P(c)} \\
&= P(a|\{b \wedge c\}) \cdot P(b|c). \tag{5.14}
\end{aligned}
$$

The MAP algorithm gives, for each decoded bit u_k, the probability that this bit was $+1$ or
-1, given the received symbol sequence $\underline{y}$. As explained in Section 5.3.2 this is equivalent to
finding the a-posteriori LLR $L(u_k|\underline{y})$, where:

$$L(u_k|\underline{y}) = \ln\left(\frac{P(u_k = +1|\underline{y})}{P(u_k = -1|\underline{y})}\right). \tag{5.15}$$

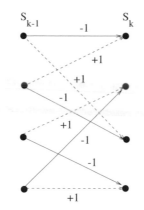

Figure 5.5: Possible transitions in $K=3$ RSC component code.

Bayes' rule allows us to rewrite this equation as:

$$L(u_k|\underline{y}) = \ln\left(\frac{P(u_k = +1 \wedge \underline{y})}{P(u_k = -1 \wedge \underline{y})}\right). \tag{5.16}$$

Let us now consider Figure 5.5 showing the transitions possible for the $K = 3$ RSC code shown in Figure 5.2, which we have used for the component codes in most of our work. For this $K = 3$ code there are four encoder states, and since we consider a binary code, in each encoder state two state transitions are possible, depending on the value of this bit. One of these transitions is associated with the input bit of -1 shown as a continuous line, while the other transition corresponds to the input bit of $+1$ shown as a broken line. It can be seen from Figure 5.5 that if the previous state S_{k-1} and the present state S_k are known, then the value of the input bit u_k, which caused the transition between these two states, will be known. Hence the probability that $u_k = +1$ is equal to the probability that the transition from the previous state S_{k-1} to the present state S_k is one of the set of four possible transitions that can occur when $u_k = +1$ (i.e. those transitions shown with broken lines). This set of transitions is mutually exclusive (i.e. only one of them could have occured at the encoder), and so the probability that any one of them occurs is equal to the sum of their individual probabilities. Hence we can rewrite Equation 5.16 as:

$$L(u_k|\underline{y}) = \ln\left(\frac{\displaystyle\sum_{\substack{(\grave{s},s)\Rightarrow \\ u_k=+1}} P(S_{k-1} = \grave{s} \wedge S_k = s \wedge \underline{y})}{\displaystyle\sum_{\substack{(\grave{s},s)\Rightarrow \\ u_k=-1}} P(S_{k-1} = \grave{s} \wedge S_k = s \wedge \underline{y})}\right), \tag{5.17}$$

where $(\grave{s}, s) \Rightarrow u_k = +1$ is the set of transitions from the previous state $S_{k-1} = \grave{s}$ to the present state $S_k = s$ that can occur if the input bit $u_k = +1$, and similarly for $(\grave{s}, s) \Rightarrow u_k = -1$. For brevity we shall write $P(S_{k-1} = \grave{s} \wedge S_k = s \wedge \underline{y})$ as $P(\grave{s} \wedge s \wedge \underline{y})$.

We now consider the individual probabilities $P(\grave{s} \wedge s \wedge \underline{y})$ from the numerator and denominator of Equation 5.17. The received sequence $\underline{y}$ can be split up into three sections: the received codeword associated with the present transition $\underline{y}_k$, the received sequence prior to the present transition $\underline{y}_{j<k}$ and the received sequence after the present transition $\underline{y}_{j>k}$. This split is shown in Figure 5.6, again for the example of our $K = 3$ RSC component code shown in Figure 5.2. We can thus write for the individual probabilities $P(\grave{s} \wedge s \wedge \underline{y})$:

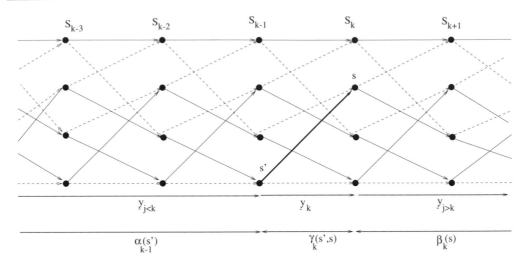

Figure 5.6: MAP decoder trellis for $K = 3$ RSC code.

$$P(\grave{s} \wedge s \wedge \underline{y}) = P(\grave{s} \wedge s \wedge \underline{y}_{j<k} \wedge \underline{y}_k \wedge \underline{y}_{j>k}). \tag{5.18}$$

Using Bayes' rule of $P(a \wedge b) = P(a|b)P(b)$ and the fact that if we assume that the channel is memoryless, then the future received sequence $\underline{y}_{j>k}$ will depend only on the present state s and not on the previous state $\grave{s}$ or the present and previous received channel sequences $\underline{y}_k$ and $\underline{y}_{j<k}$, we can write:

$$
\begin{aligned}
P(\grave{s} \wedge s \wedge \underline{y}) &= P(\underline{y}_{j>k} | \{\grave{s} \wedge s \wedge \underline{y}_{j<k} \wedge \underline{y}_k\}) \cdot P(\grave{s} \wedge s \wedge \underline{y}_{j<k} \wedge \underline{y}_k) \\
&= P(\underline{y}_{j>k} | s) \cdot P(\grave{s} \wedge s \wedge \underline{y}_{j<k} \wedge \underline{y}_k).
\end{aligned}
\tag{5.19}
$$

Again, using Bayes' rule and the assumption that the channel is memoryless, we can expand Equation 5.19 as follows:

$$
\begin{aligned}
P(\grave{s} \wedge s \wedge \underline{y}) &= P(\underline{y}_{j>k} | s) \cdot P(\grave{s} \wedge s \wedge \underline{y}_{j<k} \wedge \underline{y}_k) \\
&= P(\underline{y}_{j>k} | s) \cdot P(\{\underline{y}_k \wedge s\} | \{\grave{s} \wedge \underline{y}_{j<k}\}) \cdot P(\grave{s} \wedge \underline{y}_{j<k}) \\
&= P(\underline{y}_{j>k} | s) \cdot P(\{\underline{y}_k \wedge s\} | \grave{s}) \cdot P(\grave{s} \wedge \underline{y}_{j<k}) \\
&= \beta_k(s) \cdot \gamma_k(\grave{s}, s) \cdot \alpha_{k-1}(\grave{s}),
\end{aligned}
\tag{5.20}
$$

where:

$$\alpha_{k-1}(\grave{s}) = P(S_{k-1} = \grave{s} \wedge \underline{y}_{j<k}) \tag{5.21}$$

is the probability that the trellis is in state $\grave{s}$ at time $k - 1$ and the received channel sequence up to this point is $\underline{y}_{j<k}$, as visualised in Figure 5.6:

$$\beta_k(s) = P(\underline{y}_{j>k} | S_k = s) \tag{5.22}$$

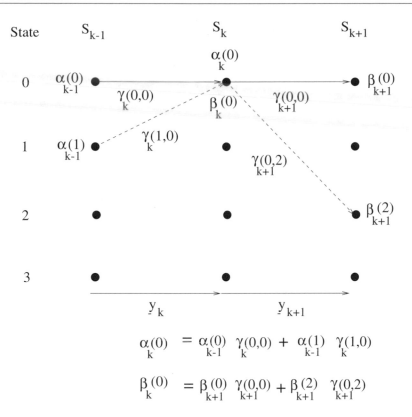

$$\alpha_k(0) = \alpha_{k-1}(0)\ \gamma_k(0,0) + \alpha_{k-1}(1)\ \gamma_k(1,0)$$

$$\beta_k(0) = \beta_{k+1}(0)\ \gamma_{k+1}(0,0) + \beta_{k+1}(2)\ \gamma_{k+1}(0,2)$$

Figure 5.7: Recursive calculation of $\alpha_k(0)$ and $\beta_k(0)$.

is the probability that given the trellis is in state s at time k the future received channel sequence will be $\underline{y}_{j>k}$, and lastly:

$$\gamma_k(\grave{s}, s) = P(\{\underline{y}_k \wedge S_k = s\}|S_{k-1} = \grave{s}) \tag{5.23}$$

is the probability that given the trellis was in state $\grave{s}$ at time $k - 1$, it moves to state s and the received channel sequence for this transition is $\underline{y}_k$.

Equation 5.20 shows that the probability $P(\grave{s} \wedge s \wedge \underline{y})$ that the encoder trellis took the transition from state $S_{k-1} = \grave{s}$ to state $S_k = s$ and the received sequence is $\underline{y}$, can be split into the product of three terms: $\alpha_{k-1}(\grave{s})$, $\gamma_k(\grave{s}, s)$ and $\beta_k(s)$. The meaning of these three probability terms is shown in Figure 5.6, for the transition $S_{k-1} = \grave{s}$ to $S_k = s$ shown by the bold line in this figure. The MAP algorithm finds $\alpha_k(s)$ and $\beta_k(s)$ for all states s throughout the trellis, i.e. for $k = 0, 1, \ldots, N - 1$, and $\gamma_k(\grave{s}, s)$ for all possible transitions from state $S_{k-1} = \grave{s}$ to state $S_k = s$, again for $k = 0, 1 \ldots, N - 1$. These values are then used to find the probabilities $P(S_{k-1} = \grave{s} \wedge S_k = s \wedge \underline{y})$ of Equation 5.20, which are then used in Equation 5.17 to give the LLRs $L(u_k|\underline{y})$ for each bit u_k. These operations are summarised in the flowchart of Figure 5.8 below. We now describe how the values $\alpha_k(s)$, $\beta_k(s)$ and $\gamma_k(\grave{s}, s)$ can be calculated.

5.3.3.2   Forward Recursive Calculation of the $\alpha_k(s)$ Values

Consider first $\alpha_k(s)$. From the definition of $\alpha_{k-1}(\grave{s})$ in Equation 5.21 we can write:

$$
\begin{aligned}
\alpha_k(s) &= P(S_k = s \wedge \underline{y}_{j<k+1}) \\
&= P(s \wedge \underline{y}_{j<k} \wedge \underline{y}_k) \\
&= \sum_{\text{all } \grave{s}} P(s \wedge \grave{s} \wedge \underline{y}_{j<k} \wedge \underline{y}_k),
\end{aligned}
\tag{5.24}
$$

where in the last line we split the probability $P(s \wedge \underline{y}_{y<k+1})$ into the sum of joint probabilities $P(s \wedge \grave{s} \wedge \underline{y}_{j<k+1})$ over all possible previous states $\grave{s}$. Using Bayes' rule and the assumption that the channel is memoryless again, we can proceed as follows:

$$
\begin{aligned}
\alpha_k(s) &= \sum_{\text{all } \grave{s}} P(s \wedge \grave{s} \wedge \underline{y}_{j<k} \wedge \underline{y}_k) \\
&= \sum_{\text{all } \grave{s}} P(\{s \wedge \underline{y}_k\}|\{\grave{s} \wedge \underline{y}_{j<k}\}) \cdot P(\grave{s} \wedge \underline{y}_{j<k}) \\
&= \sum_{\text{all } \grave{s}} P(\{s \wedge \underline{y}_k\}|\grave{s}) \cdot P(\grave{s} \wedge \underline{y}_{j<k}) \\
&= \sum_{\text{all } \grave{s}} \gamma_k(\grave{s}, s) \cdot \alpha_{k-1}(\grave{s}).
\end{aligned}
\tag{5.25}
$$

Thus, once the $\gamma_k(\grave{s}, s)$ values are known, the $\alpha_k(s)$ values can be calculated recursively. Assuming that the trellis has the initial state $S_0 = 0$, the initial conditions for this recursion are:

$$
\begin{aligned}
\alpha_0(S_0 = 0) &= 1 \\
\alpha_0(S_0 = s) &= 0 \quad \text{for all } s \neq 0.
\end{aligned}
\tag{5.26}
$$

Figure 5.7 shows an example of how one $\alpha_k(s)$ value, for $s = 0$, is calculated recursively using values of $\alpha_{k-1}(\grave{s})$ and $\gamma_k(\grave{s}, s)$ for our example $K = 3$ RSC code. Notice that, as we are considering a binary trellis, only two previous states, $S_{k=1} = 0$ and $S_{k-1} = 1$, have paths to the state $S_k = 0$. Therefore $\gamma_k(\grave{s}, s)$ will be non-zero only for $\grave{s} = 0$ or $\grave{s} = 1$ and hence the summation in Equation 5.25 is over only two terms.

5.3.3.3   Backward Recursive Calculation of the $\beta_k(s)$ Values

The values of $\beta_k(s)$ can similarly be calculated recursively as shown below. From the definition of $\beta_k(s)$ in Equation 5.22, we can write $\beta_{k-1}(\grave{s})$ as:

$$
\beta_{k-1}(\grave{s}) = P(\underline{y}_{j>k-1}|S_{k-1} = \grave{s}),
\tag{5.27}
$$

and again splitting a single probability into the sum of joint probabilities and using the derivation from Bayes' rule in Equation 5.13, as well as the assumption that the channel is memoryless, we have:

$$
\begin{aligned}
\beta_{k-1}(\grave{s}) &= P(\underline{y}_{j>k-1}|\grave{s}) \\
&= \sum_{\text{all } s} P(\{\underline{y}_{j>k-1} \wedge s\}|\grave{s}) \\
&= \sum_{\text{all } s} P(\{\underline{y}_k \wedge \underline{y}_{j>k} \wedge s\}|\grave{s}) \\
&= \sum_{\text{all } s} P(\underline{y}_{j>k}|\{\grave{s} \wedge s \wedge \underline{y}_k\}) \cdot P(\{\underline{y}_k \wedge s\}|\grave{s}) \\
&= \sum_{\text{all } s} P(\underline{y}_{j>k}|s) \cdot P(\{\underline{y}_k \wedge s\}|\grave{s}) \\
&= \sum_{\text{all } s} \beta_k(s) \cdot \gamma_k(\grave{s}, s). \quad (5.28)
\end{aligned}
$$

Thus, once the values $\gamma_k(\grave{s}, s)$ are known, a backward recursion can be used to calculate the values of $\beta_{k-1}(\grave{s})$ from the values of $\beta_k(s)$ using Equation 5.28. Figure 5.7 again shows an example of how the $\beta_k(0)$ value is calculated recursively using values of $\beta_{k+1}(s)$ and $\gamma_{k+1}(0, s)$ for our example $K = 3$ RSC code.

The initial conditions which should be used for $\beta_N(s)$ are not as clear as for $\alpha_0(s)$. From Equation 5.22 $\beta_k(s)$ is the probability that the future received sequence is $\underline{y}_{j>k}$, given that the present state is s. For the last stage in the trellis, however, i.e. when $k = N$, there is no future received sequence, and hence it is not clear what the initial values $\beta_N(s)$ should be set to. Berrou et al. [12] used the initial values $\beta_N(0) = 1$ and $\beta_N(s) = 0$ for all $s \neq 0$ for a trellis terminated in the all-zero state, and in [126] the initial conditions for an unterminated trellis were given by Robertson as $\beta_N(s) = \alpha_N(s)$ for all s. However, as pointed out by Breiling [127], if we consider $\beta_{N-1}(s)$ from Equation 5.22 we have:

$$
\begin{aligned}
\beta_{N-1}(\grave{s}) &= P(\underline{y}_N|\grave{s}) \\
&= \sum_{\text{all } s} P(\{\underline{y}_N \wedge s\}|\grave{s}) \\
&= \sum_{\text{all } s} \gamma_N(\grave{s}, s) \quad (5.29)
\end{aligned}
$$

and from the backward recursion for $\beta_{k-1}(\grave{s})$ in Equation 5.28 we have:

$$
\beta_{N-1}(\grave{s}) = \sum_{\text{all } s} \beta_N(s)\gamma_N(\grave{s}, s). \quad (5.30)
$$

For both Equation 5.29 and Equation 5.30 to be satisfied, we must have:

$$
\beta_N(s) = 1 \quad \text{for all } s. \quad (5.31)
$$

If the trellis is terminated, so that only the final state $S_N = 0$ is possible, this can be taken into account in the backward recursive calculation of the $\beta_k(s)$ values through the $\gamma_k(\grave{s}, s)$ values. In a terminated trellis for the last $K - 1$ transitions, where K is the constraint length of the convolutional code, for each state s only one transition (the one which takes the trellis towards the all-zero state) will be possible. Hence $\gamma_k(\grave{s}, s) = P(\{\underline{y}_k \wedge S_k = s\}|S_{k-1} = \grave{s})$ will be zero for all values of s except one, and with the initial values $\beta_N(s) = 1$ the correct values of

$\beta_{N-1}(s), \beta_{N-2}(s), \ldots, \beta_{N-K+1}$ will be calculated through Equation 5.28. Thus theory indicates that we should use $\beta_N(s) = 1$ for all s and account for the trellis termination by setting the values of $\gamma_k(\grave{s}, s)$ to zero for all transitions that are not possible owing to trellis termination. However, the same result can be achieved using β values of $\beta_N(0) = 1$ and $\beta_N(s) = 0$ for $s \neq 0$, as suggested by Berrou et $al.$ [12], and calculating $\gamma_k(\grave{s}, s)$ values in the same way as for all other transitions (i.e. directly from the channel inputs — see next section). This second method is simpler to implement and hence it is more commonly used in practice.

5.3.3.4 Calculation of the $\gamma_k(\grave{s}, s)$ Values

We now consider how the $\gamma_k(\grave{s}, s)$ values in Equation 5.20 can be calculated from the received channel sequence. Using the definition of $\gamma_k(\grave{s}, s)$ from Equation 5.23 and the derivation from Bayes' rule given in Equation 5.13 we have:

$$
\begin{aligned}
\gamma_k(\grave{s}, s) &= P(\{\underline{y}_k \wedge s\}|\grave{s}) \\
&= P(\underline{y}_k|\{\grave{s} \wedge s\}) \cdot P(s|\grave{s}) \\
&= P(\underline{y}_k|\{\grave{s} \wedge s\}) \cdot P(u_k),
\end{aligned} \tag{5.32}
$$

where u_k is the input bit necessary to cause the transition from state $S_{k-1} = \grave{s}$ to state $S_k = s$, and $P(u_k)$ is the a-priori probability of this bit. From Equation 5.5 this can be written as:

$$
\begin{aligned}
P(u_k) &= \left(\frac{e^{-L(u_k)/2}}{1 + e^{-L(u_k)}} \right) \cdot e^{(u_k L(u_k)/2)} \\
&= C_{L(u_k)}^{(1)} \cdot e^{(u_k L(u_k)/2)},
\end{aligned} \tag{5.33}
$$

where, as stated before:

$$
C_{L(u_k)}^{(1)} = \left(\frac{e^{-L(u_k)/2}}{1 + e^{-L(u_k)}} \right) \tag{5.34}
$$

depends only on the LLR $L(u_k)$ and not on whether u_k is $+1$ or -1.

The first term in the second and third lines of Equation 5.32, $P(\underline{y}_k|\{\grave{s} \wedge s\})$, is equivalent to $P(\underline{y}_k|\underline{x}_k)$, where $\underline{x}_k$ is the transmitted codeword associated with the transition from state $S_{k-1} = \grave{s}$ to state $S_k = s$. Again assuming the channel is memoryless we can write:

$$
P(\underline{y}_k|\{\grave{s} \wedge s\}) \equiv P(\underline{y}_k|\underline{x}_k) = \prod_{l=1}^{n} P(y_{kl}|x_{kl}), \tag{5.35}
$$

where x_{kl} and y_{kl} are the individual bits within the transmitted and received codewords $\underline{x}_k$ and $\underline{y}_k$, and n is the number of these bits in each codeword $\underline{y}_k$ or $\underline{x}_k$. Assuming that the transmitted bits x_{kl} have been transmitted over a Gaussian channel using BPSK, so that the transmitted symbols are either $+1$ or -1, we have for $P(y_{kl}|x_{kl})$:

$$
P(y_{kl}|x_{kl}) = \frac{1}{\sqrt{2\pi}\sigma} \exp\left(-\frac{E_b}{2\sigma^2}(y_{kl} - ax_{kl})^2 \right), \tag{5.36}
$$

where E_b is the transmitted energy per bit, σ^2 is the noise variance and a is the fading amplitude ($a=1$ for non-fading AWGN channels). Upon substituting Equation 5.36 in Equation 5.35 we have:

$$
\begin{aligned}
P(\underline{y}_k|\{\grave{s}\wedge s\}) &= \prod_{l=1}^{n}\frac{1}{\sqrt{2\pi}\sigma}\exp\left(-\frac{E_b}{2\sigma^2}(y_{kl}-ax_{kl})^2\right) \\
&= \frac{1}{(\sqrt{2\pi}\sigma)^n}\exp\left(-\frac{E_b}{2\sigma^2}\sum_{l=1}^{n}(y_{kl}-ax_{kl})^2\right) \\
&= \frac{1}{(\sqrt{2\pi}\sigma)^n}\exp\left(-\frac{E_b}{2\sigma^2}\sum_{l=1}^{n}(y_{kl}^2+a^2x_{kl}^2-2ax_{kl}y_{kl})\right) \\
&= C_{\underline{y}_k}^{(2)}\cdot C_{\underline{x}_k}^{(3)}\cdot\exp\left(\frac{E_b}{2\sigma^2}2a\sum_{l=1}^{n}y_{kl}x_{yl}\right),
\end{aligned}
\tag{5.37}
$$

where:

$$
C_{\underline{y}_k}^{(2)}=\frac{1}{(\sqrt{2\pi}\sigma)^n}\cdot\exp\left(-\frac{E_b}{2\sigma^2}\sum_{l=1}^{n}y_{kl}^2\right)
\tag{5.38}
$$

depends only on the channel SNR and on the magnitude of the received sequence $\underline{y}_k$, while:

$$
\begin{aligned}
C_{\underline{x}_k}^{(3)} &= \exp\left(-\frac{E_b}{2\sigma^2}a^2\sum_{l=1}^{n}x_{kl}^2\right) \\
&= \exp\left(-\frac{E_b}{2\sigma^2}a^2 n\right)
\end{aligned}
\tag{5.39}
$$

depends only on the channel SNR and on the fading amplitude. Hence we can write for $\gamma_k(\grave{s},s)$:

$$
\begin{aligned}
\gamma_k(\grave{s},s) &= P(u_k)\cdot P(\underline{y}_k|\{\grave{s}\wedge s\}) \\
&= C\cdot e^{(u_k L(u_k)/2)}\cdot\exp\left(\frac{E_b}{2\sigma^2}2a\sum_{l=1}^{n}y_{kl}x_{yl}\right) \\
&= C\cdot e^{(u_k L(u_k)/2)}\cdot\exp\left(\frac{L_c}{2}\sum_{l=1}^{n}y_{kl}x_{yl}\right),
\end{aligned}
\tag{5.40}
$$

where

$$
C=C_{L(u_k)}^{(1)}\cdot C_{\underline{y}_k}^{(2)}\cdot C_{\underline{x}_k}^{(3)}.
\tag{5.41}
$$

The term C does not depend on the sign of the bit u_k or the transmitted codeword $\underline{x}_k$ and so is constant over the summations in the numerator and denominator in Equation 5.17 and cancels out.

From Equations 5.17 and 5.20 we can write for the conditional LLR of u_k, given the received

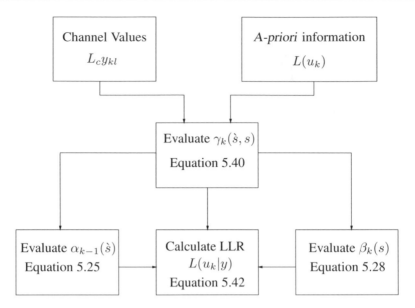

Figure 5.8: Summary of the key operations in the MAP algorithm.

sequence $\underline{y}_k$:

$$
\begin{aligned}
L(u_k|\underline{y}) &= \ln \left(\frac{\displaystyle\sum_{\substack{(\grave{s},s) \Rightarrow \\ u_k = +1}} P(S_{k-1} = \grave{s} \wedge S_k = s \wedge \underline{y})}{\displaystyle\sum_{\substack{(\grave{s},s) \Rightarrow \\ u_k = -1}} P(S_{k-1} = \grave{s} \wedge S_k = s \wedge \underline{y})} \right) \\[2ex]
&= \ln \left(\frac{\displaystyle\sum_{\substack{(\grave{s},s) \Rightarrow \\ u_k = +1}} \alpha_{k-1}(\grave{s}) \cdot \gamma_k(\grave{s},s) \cdot \beta_k(s)}{\displaystyle\sum_{\substack{(\grave{s},s) \Rightarrow \\ u_k = -1}} \alpha_{k-1}(\grave{s}) \cdot \gamma_k(\grave{s},s) \cdot \beta_k(s)} \right).
\end{aligned}
\tag{5.42}
$$

It is this conditional LLR $L(u_k|\underline{y})$ that the MAP decoder delivers.

5.3.3.5   Summary of the MAP Algorithm

From the description given above, we see that the MAP decoding of a received sequence $\underline{y}$ to give the a-posteriori LLR $L(u_k|\underline{y})$ can be carried out as follows. As the channel values y_{kl} are received, they and the a-priori LLRs $L(u_k)$ (which are provided in an iterative turbo decoder by the other component decoder — see Section 5.3.4) are used to calculate $\gamma_k(\grave{s},s)$ according to Equation 5.40. The constant C can be omitted from the calculation of $\gamma_k(\grave{s},s)$, as it will cancel out in the ratio in Equation 5.42. As the channel values y_{kl} are received, and the $\gamma_k(\grave{s},s)$ values are calculated, the forward recursion from Equation 5.25 can be used to calculate $\alpha_k(\grave{s},s)$. Once all the channel values have been received, and $\gamma_k(\grave{s},s)$ has been calculated for all $k = 1, 2, \ldots, N$, the backward recursion from Equation 5.28 can be used to calculate the $\beta_k(\grave{s},s)$ values. Finally, all the calculated values of $\alpha_k(\grave{s},s)$, $\beta_k(\grave{s},s)$ and $\gamma_k(\grave{s},s)$ are used in Equation 5.42 to calculate the values of $L(u_k|y)$. These operations are summarised in the flowchart of Figure 5.8. Care must be taken to avoid numerical underflow problems in the recursive calculation of $\alpha_k(\grave{s}, s)$ and $\beta_k(\grave{s}, s)$, but such problems can be avoided by careful normalisation of these values. Such normalisation cancels out in the ratio in Equation 5.42 and so causes no change in the LLRs produced by the algorithm.

The MAP algorithm is, in the form described in this section, extremely complex owing to the multiplications needed in Equations 5.25 and 5.28 for the recursive calculation of $\alpha_k(\grave{s}, s)$ and $\beta_k(\grave{s}, s)$, the multiplications and exponential operations required to calculate $\gamma_k(\grave{s}, s)$ using Equation 5.40, and the multiplication and natural logarithm operations required to calculate $L(u_k|y)$ using Equation 5.42. However, much work has been done to reduce this complexity, and the Log-MAP algorithm [52], which will be described in Section 5.3.5, gives the same performance as the MAP algorithm but with a much lower complexity and without the numerical problems described above. We will first describe the principles behind the iterative decoding of turbo codes, and how the MAP algorithm described in this section can be used in such a scheme, before detailing the Log-MAP algorithm.

5.3.4 Iterative Turbo Decoding Principles

5.3.4.1 Turbo Decoding Mathematical Preliminaries

In this section we explain the concepts of extrinsic and intrinsic information as used by Berrou *et al.* [12], and highlight how the MAP algorithm described in the previous section, and other soft-in soft-out decoders, can be used in the iterative decoding of turbo codes.

Consider first the expression for $\gamma_k(\grave{s}, s)$ in Equation 5.40, which is restated here for convenience:

$$\gamma_k(\grave{s}, s) = C \cdot e^{(u_k L(u_k)/2)} \cdot \exp\left(\frac{L_c}{2} \sum_{l=1}^{n} y_{kl} x_{yl}\right). \tag{5.43}$$

Since we are dealing with systematic codes, one of the n transmitted bits will be the systematic bit u_k. If we assume that this systematic bit is the first of the n transmitted bits, then we will have $x_{k1} = u_k$, and we can rewrite Equation 5.43 as:

$$
\begin{aligned}
\gamma_k(\grave{s}, s) &= C \cdot e^{(u_k L(u_k)/2)} \cdot \exp\left(\frac{L_c}{2} y_{ks} u_k\right) \cdot \exp\left(\frac{L_c}{2} \sum_{l=2}^{n} y_{kl} x_{yl}\right) \\
&= C \cdot e^{(u_k L(u_k)/2)} \cdot \exp\left(\frac{L_c}{2} y_{ks} u_k\right) \cdot \chi_k(\grave{s}, s),
\end{aligned}
\tag{5.44}
$$

where y_{ks} is the received version of the transmitted systematic bit $x_{k1} = u_k$ and:

$$\chi_k(\grave{s}, s) = \exp\left(\frac{L_c}{2} \sum_{l=2}^{n} y_{kl} x_{yl}\right). \tag{5.45}$$

Using Equation 5.44 and remembering that in the numerator we have $u_k = +1$ for all terms in the summation, whereas in the denominator we have $u_k = -1$, we can rewrite Equation 5.42 as:

$$
\begin{aligned}
L(u_k|\underline{y}) &= \ln\left(\frac{\displaystyle\sum_{\substack{(\grave{s},s)\Rightarrow\\u_k=+1}} \alpha_{k-1}(\grave{s})\cdot\gamma_k(\grave{s},s)\cdot\beta_k(s)}{\displaystyle\sum_{\substack{(\grave{s},s)\Rightarrow\\u_k=-1}} \alpha_{k-1}(\grave{s})\gamma_k(\grave{s},s)\cdot\beta_k(s)}\right)\\[2ex]
&= \ln\left(\frac{\displaystyle\sum_{\substack{(\grave{s},s)\Rightarrow\\u_k=+1}} \alpha_{k-1}(\grave{s})\cdot e^{+L(u_k)/2}\cdot e^{+L_c y_{ks}/2}\cdot\chi_k(\grave{s},s)\cdot\beta_k(s)}{\displaystyle\sum_{\substack{(\grave{s},s)\Rightarrow\\u_k=-1}} \alpha_{k-1}(\grave{s})\cdot e^{-L(u_k)/2}\cdot e^{-L_c y_{ks}/2}\cdot\chi_k(\grave{s},s)\cdot\beta_k(s)}\right)\\[2ex]
&= L(u_k) + L_c y_{ks} + \ln\left(\frac{\displaystyle\sum_{\substack{(\grave{s},s)\Rightarrow\\u_k=+1}} \alpha_{k-1}(\grave{s})\cdot\chi_k(\grave{s},s)\cdot\beta_k(s)}{\displaystyle\sum_{\substack{(\grave{s},s)\Rightarrow\\u_k=-1}} \alpha_{k-1}(\grave{s})\cdot\chi_k(\grave{s},s)\cdot\beta_k(s)}\right)\\[2ex]
&= L(u_k) + L_c y_{ks} + L_e(u_k), \tag{5.46}
\end{aligned}
$$

where:

$$
L_e(u_k) = \ln\left(\frac{\displaystyle\sum_{\substack{(\grave{s},s)\Rightarrow\\u_k=+1}} \alpha_{k-1}(\grave{s})\cdot\chi_k(\grave{s},s)\cdot\beta_k(s)}{\displaystyle\sum_{\substack{(\grave{s},s)\Rightarrow\\u_k=-1}} \alpha_{k-1}(\grave{s})\cdot\chi_k(\grave{s},s)\cdot\beta_k(s)}\right). \tag{5.47}
$$

Thus we can see that the a-posteriori LLR $L(u_k|\underline{y})$ calculated with the aid of the MAP algorithm can be viewed as comprising three additive soft-metric terms:- $L(u_k)$, $L_c y_{ks}$ and $L_e(u_k)$. The first soft-metric term is the a-priori LLR $L(u_k)$, which accrues from $P(u_k)$ in the expression for the branch transition probability $\gamma_k(\grave{s},s)$ in Equation 5.32. This probability should be generated by an independent source and is referred to as the a-priori probability of the kth information or systematic bit represented as $+1$ or -1, as illustrated in Figure 5.9. In most cases we will have no independent or a-priori knowledge of the likely value of the bit u_k at the decoder and hence the a-priori LLR $L(u_k)$ initially will be zero in the logarithmic domain, corresponding to an a-priori probability of $P(u_k) = 0.5$. However, in the case of an iterative turbo decoder, each component decoder is capable of providing the other decoder with an estimate of the a-priori LLR $L(u_k)$, as it will be described during our forthcoming discourse.

The second term $L_c y_{ks}$ in Equation 5.46 is the soft output of the channel representing the systematic bit u_k, which was directly transmitted across the channel and received as y_{ks}. In other words, this term corresponds to the systematic bits conveyed by the channel and to the extrinsic LLR values shown in Figure 5.9. When the channel SNR is high, the channel reliability value L_c of Equation 5.11 will be high and this systematic bit will have a large influence on the a-posteriori LLR $L(u_k|\underline{y})$. Conversely, when the channel SNR is low and hence L_c is low, the soft output of the channel for the received systematic bit y_{ks} will have less impact on the a-posteriori LLR delivered by the MAP algorithm.

The final term in Equation 5.46, $L_e(u_k)$, is derived using the constraints imposed by the code used, from the a-priori information sequence $L(u_n)$ and the received channel information

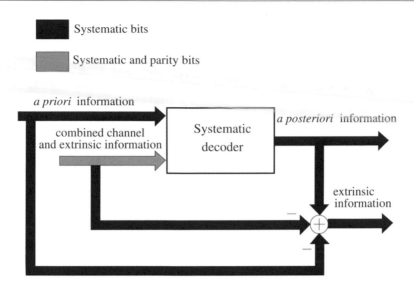

Figure 5.9: Schematic of a component decoder employed in a turbo decoder, showing the input information received and output information corresponding to the systematic and parity bits.

sequence $\underline{y}$, *excluding* the received systematic bit y_{ks} and the a-priori information $L(u_k)$ for the bit u_k. Hence it is referred to as the *extrinsic* LLR for the bit u_k. Figure 5.9 and Equation 5.46 demonstrate that the extrinsic soft-information related to a specific bit and generated by a MAP decoder can be obtained by subtracting the a-priori soft-information value $L(u_k)$ and the received systematic soft output of the channel representing the systematic bit u_k - namely $L_c y_{ks}$ - from the soft output $L(u_k|\underline{y})$ of the decoder. This is the reason for having the subtraction paths shown in Figure 5.3. The corresponding formulae similar to Equation 5.46 can be derived for the other component decoder, which can be used in iterative turbo decoding. At this stage we underline again that the the a-priori soft-information value $L(u_k)$ and the received systematic soft output of the channel representing the systematic bit u_k - namely $L_c y_{ks}$ - **are subtracted from the soft output $L(u_k|\underline{y})$ of the decoder only for the systematic information bit u_k, but not for the parity bit**, as shown explicitly in Figure 5.9 with the aid of the black lines. By contrast, when using turbo equalisation, which will be the subject of Part V of the book, it will be shown in the context of Figure 14.3 that the extrinsic information has to be generated also for the parity bits.

Notice that the expression describing the branch transition probabilities $\gamma_k(\grave{s}, s)$ in Equation 5.40 uses the a-priori soft-information $L(u_k)$ and all n bits, including the systematic bit y_{ks} of the received codeword $\underline{y}_k$. Observe that these branch transition probabilities are used in the recursive calculations of $\alpha_k(s)$ and $\beta_k(s)$ in Equations 5.25 and 5.28. Hence these terms appear in Equation 5.47 describing $L_e(u_k)$ and therefore it might seem that the received systematic soft-bit y_{ks} and the a-priori soft-information $L(u_k)$ of the bit u_k appear indirectly in the extrinsic soft-output $L_e(u_k)$. However, careful examination of Equation 5.47 shows that for the bit u_k we use the values of $\alpha_{k-1}(\grave{s})$ and $\beta_k(s)$. From Equations 5.25 and 5.28 derived for the recursive calculation of these values we observe that the branch transition probabilities $\gamma_n(\grave{s}, s)$ for $1 \leq n \leq k-1$ and $k+1 \leq n \leq N$ will be used for calculating the $\alpha_{k-1}(\grave{s})$ and $\beta_k(s)$ values. Notice, however, that the specific branch transition probability $\gamma_k(\grave{s}, s)$, which characterises the transition associated with the bit u_k, is not used. MOre explicitly, $L_e(u_k)$ uses the values of the branch transition probabilities $\gamma_n(\grave{s}, s)$ for all the branches *except* for the kth branch. Therefore,

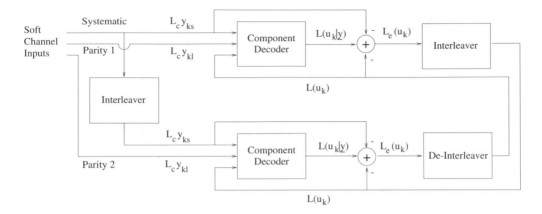

Figure 5.10: Turbo decoder schematic.

although it does depend on all the other a-priori information terms $L(u_n)$ and on the received systematic bits, the term $L_e(u_k)$ is actually independent of the a-priori information $L(u_k)$ and the received systematic bit y_{ks}. Hence it can be justifiably referred to as the extrinsic LLR of the bit u_k.

We summarise below what is meant by the terms a-priori, extrinsic and a-posteriori information, which we use throughout this treatise.

a priori The a-priori information related to a bit is information known before decoding commences, from a source other than the received sequence or the code constraints. It is also sometimes referred to as intrinsic information for contrasting it with the extrinsic information to be described next.

extrinsic The extrinsic information related to a bit u_k is the information provided by a decoder based on the received sequence and on the a-priori information, but *excluding* the received systematic bit y_{ks} and the a-priori information $L(u_k)$ related to the bit u_k. Typically the component decoder provides this information using the constraints imposed on the transmitted sequence by the code used. It processes the received bits and the a-priori information surrounding the systematic bit u_k, and uses this information and the code constraints for providing information about the value of the bit u_k.

a posteriori The a-posteriori information related to a bit is the information that the decoder generates by taking into account *all* available sources of information concerning u_k. It is the a-posteriori LLR, ie $L(u_k|\underline{y})$, that the MAP algorithm generates as its output.

5.3.4.2 Iterative Turbo Decoding

We now describe how the iterative decoding of turbo codes is carried out. Figure 5.3 from Section 5.3.1, showing the structure of an iterative turbo decoder, is repeated for convenience here as Figure 5.10. Figure 5.10 also shows the symbols we have used for the various inputs to and outputs from the component decoders.

Consider initially the first component decoder in the first iteration. This decoder receives the channel sequence $L_c\underline{y}^{(1)}$ containing the received versions of the transmitted systematic bits, $L_c y_{ks}$, and the parity bits, $L_c y_{kl}$, from the first encoder. Usually, to obtain a half-rate code, half of these parity bits will have been punctured at the transmitter, and so the turbo decoder must insert zeros in the soft channel output $L_c y_{kl}$ for these punctured bits. The first component decoder can then process the soft channel inputs and produce its estimate $L_{11}(u_k|\underline{y})$ of the conditional LLRs of the data bits u_k, $k = 1, 2 \ldots N$. In this notation the subscript 11 in $L_{11}(u_k|\underline{y})$ indicates that this is the a-posteriori LLR in the first iteration from the first component decoder. Note that in this first iteration the first component decoder will have no a-priori information about the bits, and hence $L(u_k)$ in Equation 5.40 giving $\gamma_k(\grave{s}, s)$ will be zero, corresponding to an a-priori probability of 0.5.

Next the second component decoder comes into operation. It receives the channel sequence $L_c \underline{y}^{(2)}$ containing the *interleaved* version of the received systematic bits, and the parity bits from the second encoder. Again, the turbo decoder will have to insert zeros into this sequence, if the parity bits generated by the encoder are punctured before transmission. However, now, in addition to the received channel sequence $L_c \underline{y}^{(2)}$, the decoder can use the conditional LLR $L_{11}(u_k|\underline{y})$ provided by the first component decoder to generate a-priori LLRs $L(u_k)$ to be used by the second component decoder. Ideally these a-priori LLRs $L(u_k)$ would be completely independent from all the other information used by the second component decoder. As can be seen in Figure 5.10 in iterative turbo decoders the extrinsic information $L_e(u_k)$ from the other component decoder is used as the a-priori LLRs, after being interleaved to arrange the decoded data bits $\underline{u}$ in the same order as they were encoded by the second encoder. Again, according to Equation 5.46, the reason for the subtraction paths shown in Figure 5.10 is that the a-posteriori LLRs from one decoder have the systematic soft channel inputs $L_c y_{ks}$ and the a-priori LLRs $L(u_k)$ (if any were available) subtracted to yield the extrinsic LLRs $L_e(u_k)$ which are then used as a-priori LLRs for the other component decoder. The second component decoder thus uses the received channel sequence $L_c \underline{y}^{(2)}$ and the a-priori LLRs $L(u_k)$ (derived by interleaving the extrinsic LLRs $L_e(u_k)$ of the first component decoder) to produce its a-posteriori LLRs $L_{12}(u_k|\underline{y})$. This is then the end of the first iteration.

For the second iteration the first component encoder again processes its received channel sequence $L_c \underline{y}^{(1)}$, but now it also has a-priori LLRs $L(u_k)$ provided by the extrinsic portion $L_e(u_k)$ of the a-posteriori LLRs $L_{12}(u_k|\underline{y})$ calculated by the second component encoder, and hence it can produce an improved a-posteriori LLR $L_{21}(u_k|\underline{y})$. The second iteration then continues with the second component decoder using the improved a-posteriori LLRs $L_{21}(u_k|\underline{y})$ from the first encoder to derive, through Equation 5.46, improved a-priori LLRs $L(u_k)$ which it uses in conjunction with its received channel sequence $L_c \underline{y}^{(2)}$ to calculate $L_{22}(u_k|\underline{y})$.

This iterative process continues, and with each iteration on average the BER of the decoded bits will fall. However, as will be seen in Figure 5.21, the improvement in performance for each additional iteration carried out falls as the number of iterations increases. Hence for complexity reasons usually only about eight iterations are carried out, as no significant improvement in performance is obtained with a higher number of iterations. This is the arrangement we have used in most of our simulations, i.e. the decoder carries out a fixed number of iterations. However, it is possible to use a variable number of iterations up to a maximum, with some termination criterion used to decide when it is deemed that further iterations will produce marginal gain. This allows the average number of iterations, and so the average complexity of the decoder, to be dramatically reduced [61] with only a small degradation in performance. Suitable termination criteria have been found to be the so-called cross-entropy of the outputs from the two component decoders [61], and the variance of the a-posteriori LLRs $L(u_k|\underline{y})$ of a component decoder [126].

Figure 5.11 shows how the a-posteriori LLRs $L(u_k|\underline{y})$ output from the component decoders in an iterative decoder vary with the number of iterations used. The output from the second component decoder is shown after one, two, four and eight iterations. The input sequence of

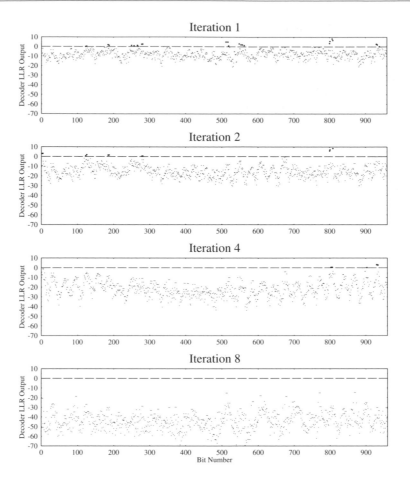

Figure 5.11: Soft outputs from the MAP decoder in an iterative turbo decoder for a transmitted stream of all -1.

the encoder consisted entirely of logical 0s, and consequently the negative a-posteriori LLR $L(u_k|\underline{y})$ values correspond to a correct hard decision, while the positive values to an incorrect hard decision. The input sequence was coded using a turbo encoder with two constraint length 3 recursive convolutional codes, and a block interleaver with 31 rows and 31 columns. This turbo encoder is used in the majority of our investigations and its performance is characterised in Section 5.4. The encoded bits were transmitted over an AWGN channel at a channel SNR of -1 dB, and then decoded using an iterative turbo decoder using the MAP algorithm. It can be seen that as the number of iterations used increases, the number of positive a-posteriori LLR $L(u_k|\underline{y})$ values, and hence the BER, decreases until after eight iterations there are no incorrectly decoded values. Furthermore, as the number of iterations increases, the decoders become more certain about the value of the bits and hence the magnitudes of the LLRs gradually become larger. The erroneous decisions in the figure appear in bursts, since deviating from the error-free trellis path typically inflicts several bit errors.

When the series of iterations halts, after either a fixed number of iterations or when a termination criterion is satisfied, the output from the turbo decoder is given by the de-interleaved a-posteriori LLRs of the second component decoder, $L_{i2}(u_k|\underline{y})$, where i is the number of iterations used. The sign of these a-posteriori LLRs gives the hard-decision output, i.e. whether the decoder believes that the transmitted data bit u_k was $+1$ or -1, and in some applications the magnitude of these LLRs, which gives the confidence the decoder has in its decision, may also be useful.

Ideally, for the iterative decoding of turbo codes, the a-priori information used by a component decoder should be completely independent from the channel outputs used by that decoder. However, in turbo decoders the extrinsic LLR $L_e(u_k)$ for the bit u_k, as explained above, uses all the available received parity bits and all the received systematic bits except the received value y_{ks} of the bit u_k. However, the same received systematic bits are also used by the other component decoder, which uses the interleaved or de-interleaved version of $L_e(u_k)$ as its a-priori LLRs. Hence the a-priori LLRs $L(u_k)$ are not truly independent from the channel outputs $\underline{y}$ used by the component decoders. However, owing to the fact that the component convolutional codes have a short memory, usually of only four bits or less, the extrinsic LLR $L_e(u_k)$ is only significantly affected by the received systematic bits relatively close to the bit u_k. When this extrinsic LLR $L_e(u_k)$ is used as the a-priori LLR $L(u_k)$ by the other component decoder, because of the interleaving used, the bit u_k and its neighbours will probably have been well separated. Hence the dependence of the a-priori LLRs $L(u_k)$ on the received systematic channel values $L_c y_{ks}$ which are also used by the other component decoder will have relatively little effect, and the iterative decoding provides good results.

Another justification for using the iterative arrangement described above is how well it has been found to work. In the limited experiments that have been carried out with optimal decoding of turbo codes [124, 125, 128] it has been found that optimal decoding performs only a fraction of a decibel (around 0.35–0.5 dB) better than iterative decoding with the MAP algorithm. Furthermore various turbo coding schemes have been found [63, 128] that approach the Shannonian limit, which gives the best performance theoretically available, to a similar fraction of a decibel. Therefore it seems that, for a variety of codes, the iterative decoding of turbo codes gives an almost optimal performance. Hence it is this iterative decoding structure, which is almost exclusively used with turbo codes, which we have used throughout our simulations.

Having described how the MAP algorithm can be used in the iterative decoding of turbo codes, we now proceed to describe other soft-in soft-out decoders, which are less complex and can be used instead of the MAP algorithm. We first describe two related algorithms, the Max-Log-MAP [50, 51, 129] and the Log-MAP [52], which are derived from the MAP algorithm, and then another, referred to as the Soft Output Viterbi Algorithm (SOVA) [53, 122], derived from the Viterbi algorithm.

5.3.5 Modifications of the MAP algorithm

5.3.5.1 Introduction

The MAP algorithm as described in Section 5.3.3 is much more complex than the Viterbi algorithm and with hard-decision outputs performs almost identically to it. Therefore for almost 20 years it was largely ignored. However, its application in turbo codes renewed interest in the algorithm, and it was realised that its complexity can be dramatically reduced without affecting its performance. The roots of employing the Max-Log-MAP algorithm based on the Jacobian logarithmic approximation to be outlined in Equation 5.56 go back to the pre-1993 era, preceding the publication of the turbo coding principles by Berrou et al. [12]. More specifically, to the authors' knowledge the Jacobian logarithmic approximation has been used in this context for the first time in 1989 in the master thesis of Erfanian at the University of Toronto [2]. and in the open literature in references [50, 51, 129]. The Max-Log-MAP technique simplified the MAP algorithm by transferring the recursions into the logarithmic domain and by invoking an approximation for the sake of dramatically reducing the associated implementational complexity. Because of this approximation the performance of the Max-Log-MAP algorithms is sub-optimal. However, Robertson et al. [52] in 1995 proposed the Log-MAP algorithm, which corrected the approximation used in the Max-Log-MAP algorithm and hence attained a performance virtually identical to that of the MAP algorithm at a fraction of its complexity. These two algorithms are described in more detail in this section.

5.3.5.2   Mathematical Description of the Max-Log-MAP Algorithm

The MAP algorithm calculates the a-posteriori LLRs $L(u_k|\underline{y})$ using Equation 5.42. To do this it requires the following values:

1) The $\alpha_{k-1}(\grave{s})$ values, which are calculated in a forward recursive manner using Equation 5.25,

2) the $\beta_k(s)$ values, which are calculated in a backward recursion using Equation 5.28, and

3) the branch transition probabilities $\gamma_k(\grave{s}, s)$, which are calculated using Equation 5.40.

The Max-Log-MAP algorithm simplifies this by transferring these equations into the log arithmetic domain and then using the approximation:

$$\ln \left(\sum_i e^{x_i} \right) \approx \max_i(x_i), \tag{5.48}$$

where $\max_i(x_i)$ means the maximum value of x_i. Then, with $A_k(s)$, $B_k(s)$ and $\Gamma_k(\grave{s}, s)$ defined as follows:

$$A_k(s) \triangleq \ln \left(\alpha_k(s) \right), \tag{5.49}$$

$$B_k(s) \triangleq \ln \left(\beta_k(s) \right), \tag{5.50}$$

and:

$$\Gamma_k(\grave{s}, s) \triangleq \ln \left(\gamma_k(\grave{s}, s) \right), \tag{5.51}$$

we can rewrite Equation 5.25 as:

$$\begin{aligned} A_k(s) &\triangleq \ln \left(\alpha_k(s) \right) \\ &= \ln \left(\sum_{\text{all } \grave{s}} \alpha_{k-1}(\grave{s}) \gamma_k(\grave{s}, s) \right) \\ &= \ln \left(\sum_{\text{all } \grave{s}} \exp \left[A_{k-1}(\grave{s}) + \Gamma_k(\grave{s}, s) \right] \right) \\ &\approx \max_{\grave{s}} \left(A_{k-1}(\grave{s}) + \Gamma_k(\grave{s}, s) \right). \end{aligned} \tag{5.52}$$

---

[2]Private communication by Erfanian and Pasupathy.

Equation 5.52 implies that for each path in Figure 5.6 from the previous stage in the trellis to the state $S_k = s$ at the present stage, the algorithm adds a branch metric term $\Gamma_k(\grave{s}, s)$ to the previous value $A_{k-1}(\grave{s})$ to find a new value $\tilde{A}_k(s)$ for that path. The new value of $A_k(s)$ according to Equation 5.52 is then the maximum of the $\tilde{A}_k(s)$ values of the various paths reaching the state $S_k = s$. This can be thought of as selecting one path as the 'survivor' and discarding any other paths reaching the state.

The value of $A_k(s)$ should give the natural logarithm of the probability that the trellis is in state $S_k = s$ at stage k, given that the received channel sequence up to this point has been $\underline{y}_{j \leq k}$. However, because of the approximation of Equation 5.48 used to derive Equation 5.52, only the maximum likelihood path through the state $S_k = s$ is considered when calculating this probability. Thus the value of A_k in the Max-Log-MAP algorithm actually gives the probability of the most likely path through the trellis to the state $S_k = s$, rather than the probability of *any* path through the trellis to state $S_k = s$. This approximation is one of the reasons for the sub-optimal performance of the Max-Log-MAP algorithm compared to the MAP algorithm.

We see from Equation 5.52 that in the Max-Log-MAP algorithm the forward recursion used to calculate $A_k(s)$ is exactly the same as the forward recursion in the Viterbi algorithm — for each pair of merging paths the survivor is found using two additions and one comparison. Notice that for binary trellises the summation, and maximisation, over all previous states $S_{k-1} = \grave{s}$ in Equation 5.52 will in fact be over only two states, because there will be only two previous states $S_{k-1} = \grave{s}$ with paths to the present state $S_k = s$. For all other values of $\grave{s}$ we will have $\gamma_k(\grave{s}, s) = 0$.

Similarly to Equation 5.52 for the forward recursion used to calculate the $A_k(s)$, we can rewrite Equation 5.28 as:

$$
\begin{aligned}
B_{k-1}(\grave{s}) &\triangleq \ln\left(\beta_{k-1}(\grave{s})\right) \\
&= \ln\left(\sum_{\text{all } s} \beta_k(s)\gamma_k(\grave{s}, s)\right) \\
&= \ln\left(\sum_{\text{all } s} \exp\left[B_k(s) + \Gamma_k(\grave{s}, s)\right]\right) \\
&\approx \max_{s}\left(B_k(s) + \Gamma_k(\grave{s}, s)\right),
\end{aligned}
\tag{5.53}
$$

and obtain the backward recursion used to calculate the $B_{k-1}(\grave{s})$ values. Again this is equivalent to the recursion used in the Viterbi algorithm — the value of $B_{k-1}(\grave{s})$ is found by adding, for every state $S_k = s$ having a path from $S_{k-1} = \grave{s}$ (two in a binary trellis), a branch metric $\Gamma_k(\grave{s}, s)$ to the value of $B_k(s)$ and selecting which path gives the highest $B_{k-1}(\grave{s})$ value.

Using Equation 5.40, the branch metrics $\Gamma_k(\grave{s}, s)$ in the recursive formulae of Equations 5.52 and 5.53 derived for $A_k(s)$ and $B_{k-1}(\grave{s})$, respectively, can be written as:

$$
\begin{aligned}
\Gamma_k(\grave{s}, s) &\triangleq \ln\left(\gamma_k(\grave{s}, s)\right) \\
&= \ln\left(C \cdot e^{(u_k L(u_k)/2)} \exp\left[\frac{E_b}{2\sigma^2}2a\sum_{l=1}^{n} y_{kl}x_{kl}\right]\right) \\
&= \ln\left(C \cdot e^{(u_k L(u_k)/2)} \exp\left[\frac{L_c}{2}\sum_{l=1}^{n} y_{kl}x_{kl}\right]\right) \\
&= \hat{C} + \frac{1}{2}u_k L(u_k) + \frac{L_c}{2}\sum_{l=1}^{n} y_{kl}x_{kl},
\end{aligned}
\tag{5.54}
$$

where $\hat{C} = \ln C$ does not depend on u_k or on the transmitted codeword $\underline{x}_k$ and so can be considered a constant and omitted. Hence the branch metric is equivalent to that used in the Viterbi algorithm, with the addition of the a-priori LLR term $u_k L(u_k)$. Furthermore the correlation term $\sum_{l=1}^{n} y_{kl} x_{kl}$ is weighted by the channel reliability value L_c of Equation 5.11.

Finally, from Equation 5.42, we can write for the a-posteriori LLRs $L(u_k|\underline{y})$ which the Max-Log-MAP algorithm calculates:

$$
\begin{aligned}
L(u_k|\underline{y}) &= \ln \left(\frac{\sum_{\substack{(\grave{s},s)\Rightarrow \\ u_k=+1}} \alpha_{k-1}(\grave{s}) \cdot \gamma_k(\grave{s},s) \cdot \beta_k(s)}{\sum_{\substack{(\grave{s},s)\Rightarrow \\ u_k=-1}} \alpha_{k-1}(\grave{s}) \cdot \gamma_k(\grave{s},s) \cdot \beta_k(s)} \right) \\[2ex]
&= \ln \left(\frac{\sum_{\substack{(\grave{s},s)\Rightarrow \\ u_k=+1}} \exp\left(A_{k-1}(\grave{s}) + \Gamma_k(\grave{s},s) + B_k(s)\right)}{\sum_{\substack{(\grave{s},s)\Rightarrow \\ u_k=-1}} \exp\left(A_{k-1}(\grave{s}) + \Gamma_k(\grave{s},s) + B_k(s)\right)} \right) \\[2ex]
&\approx \max_{\substack{(\grave{s},s)\Rightarrow \\ u_k=+1}} \left(A_{k-1}(\grave{s}) + \Gamma_k(\grave{s},s) + B_k(s)\right) \\
&\quad - \max_{\substack{(\grave{s},s)\Rightarrow \\ u_k=-1}} \left(A_{k-1}(\grave{s}) + \Gamma_k(\grave{s},s) + B_k(s)\right).
\end{aligned}
\tag{5.55}
$$

This means that in the Max-Log-MAP algorithm for each bit u_k the a-posteriori LLR $L(u_k|\underline{y})$ is calculated by considering every transition from the trellis stage S_{k-1} to the stage S_k. These transitions are grouped into those that might have occurred if $u_k = +1$, and those that might have occurred if $u_k = -1$. For both of these groups the transition giving the maximum value of $A_{k-1}(\grave{s}) + \Gamma(\grave{s},s) + B_k(s)$ is found, and the a-posteriori LLR is calculated based on only these two 'best' transitions. For a binary trellis there will be $2 \cdot 2^{K-1}$ transitions at each stage of the trellis, where K is the constraint length of the convolutional code. Therefore there will be 2^{K-1} transitions to consider in each of the maximisations in Equation 5.55.

The Max-Log-MAP algorithm can be summarised as follows. Forward and backward recursions, both similar to the forward recursion used in the Viterbi algorithm, are used to calculate $A_k(s)$ using Equation 5.52 and $B_k(s)$ using Equation 5.53. The branch metric $\Gamma_k(\grave{s},s)$ used is given by Equation 5.54, where the constant term $\hat{C}$ can be omitted. Once both the forward and backward recursions have been carried out, the a-posteriori LLRs can be calculated using Equation 5.55. Thus the complexity of the Max-Log-MAP algorithm is not hugely higher than that of the Viterbi algorithm — instead of one recursion two are carried out, the branch metric of Equation 5.54 has the additional a-priori term $u_k L(u_k)$ term added to it, and for each bit Equation 5.55 must be used to give the a-posteriori LLRs. This calculation of $L(u_k|\underline{y})$ from the $A_{k-1}(\grave{s})$, $B_k(s)$ and $\Gamma_k(\grave{s},s)$ values requires, for every bit 2 additions for each of the $2 \cdot 2^{K-1}$ transitions at each stage of the trellis, two maximisations and one subtraction. Viterbi states [130] that the complexity of the Log-MAP-Max algorithm is no greater than three times that of a Viterbi decoder. Unfortunately the storage requirements are much greater because of the need to store both the forward and backward recursively calculated metrics $A_k(s)$ and $B_k(s)$ before the $L(u_k|\underline{y})$ values can be calculated. However, Viterbi also states [130, 131] that it can be shown that by increasing the computational load slightly the associated memory requirements can be dramatically reduced.

5.3.5.3 Correcting the Approximation — the Log-MAP Algorithm

The Max-Log-MAP algorithm gives a slight degradation in performance compared to the MAP algorithm owing to the approximation of Equation 5.48. When used for the iterative decoding of turbo codes, this degradation was found by Robertson et al. [52] to result in a drop in performance of about 0.35 dB. However, the approximation of Equation 5.48 can be made exact by using the Jacobian logarithm:

$$
\begin{aligned}
\ln\left(e^{x_1} + e^{x_2}\right) &= \max(x_1, x_2) + \ln\left(1 + e^{-|x_1 - x_2|}\right) \\
&= \max(x_1, x_2) + f_c\left(|x_1 - x_2|\right) \\
&= g(x_1, x_2),
\end{aligned}
\tag{5.56}
$$

where $f_c(x)$ can be thought of as a correction term. This is then the basis of the Log-MAP algorithm proposed by Robertson et al. [52]. Similarly to the Max-Log-MAP algorithm, values for $A_k(s) \triangleq \ln(\alpha_k(s))$ and $B_k(s) \triangleq \ln(\beta_k(s))$ are calculated using a forward and a backward recursion. However, the maximisation in Equations 5.52 and 5.53 is complemented by the correction term in Equation 5.56. This means that the exact rather than approximate values of $A_k(s)$ and $B_k(s)$ are calculated. In binary trellises, as explained earlier, the maximisation will be over only two terms. Therefore we can correct the approximations in Equations 5.52 and 5.53 by merely adding the term $f_c(\delta)$, where δ is the magnitude of the difference between the metrics of the two merging paths. Similarly, the approximation in Equation 5.55 giving the a-posteriori LLRs $L(u_k|y)$ can be eliminated using the Jacobian logarithm. However, as explained earlier, there will be 2^{K-1} transitions to consider in each of the maximisations of Equation 5.55. Thus we must generalise Equation 5.48 in order to cope with more than two x_i terms. This is done by nesting the $g(x_1, x_2)$ operations as follows:

$$
\ln\left(\sum_{i=1}^{I} e^{x_i}\right) = g(x_I, g(x_{I-1}, \cdots, g(x_3, g(x_2, x_1)))\cdots).
\tag{5.57}
$$

The correction term $f_c(\delta)$ need not be computed for every value of δ, but instead can be stored in a look-up table. Robertson et al. [52] found that such a look-up table need contain only eight values for δ, ranging between 0 and 5. This means that the Log-MAP algorithm is only slightly more complex than the Max-Log-MAP algorithm, but it gives exactly the same performance as the MAP algorithm. Therefore it is a very attractive algorithm to use in the component decoders of an iterative turbo decoder.

Having described two techniques based on the MAP algorithm but with reduced complexity, we now describe an alternative soft-in soft-out decoder based on the Viterbi algorithm.

5.3.6 The Soft-output Viterbi Algorithm

5.3.6.1 Mathematical Description of the Soft-output Viterbi Algorithm

In this section we describe a variation of the Viterbi algorithm, referred to as the Soft-Output Viterbi Algorithm (SOVA) [53, 122]. This algorithm has two modifications over the classical Viterbi algorithm which allow it to be used as a component decoder for turbo codes. First the path metrics used are modified to take account of a-priori information when selecting the maximum likelihood path through the trellis. Second the algorithm is modified so that it provides a soft output in the form of the a-posteriori LLR $L(u_k|y)$ for each decoded bit.

The first modification is easily accomplished. Consider the state sequence $\underline{s}_k^s$ which gives the states along the surviving path at state $S_k = s$ at stage k in the trellis. The probability that this is the correct path through the trellis is given by:

$$p(\underline{s}_k^s | \underline{y}_{j \leq k}) = \frac{p(\underline{s}_k^s \wedge \underline{y}_{j \leq k})}{p(\underline{y}_{j \leq k})}. \tag{5.58}$$

As the probability of the received sequence $\underline{y}_{j \leq k}$ for transitions up to and including the kth transition is constant for all paths $\underline{s}_k$ through the trellis to stage k, the probability that the path $\underline{s}_k^s$ is the correct one is proportional to $p(\underline{s}_k^s \wedge \underline{y}_{j \leq k})$. Therefore our metric should be defined so that maximising the metric will maximise $p(\underline{s}_k^s \wedge \underline{y}_{j \leq k})$. The metric should also be easily computable in a recursive manner as we go from the $(k-1)$th stage in the trellis to the kth stage. If the path $\underline{s}_k^s$ at the kth stage has the path $\underline{s}_{k-1}^{\grave{s}}$ for its first $k-1$ transitions then, assuming a memoryless channel, we will have:

$$p(\underline{s}_k^s \wedge \underline{y}_{j \leq k}) = p(\underline{s}_{k-1}^{\grave{s}} \wedge \underline{y}_{j \leq k-1}) \cdot p(S_k = s \wedge \underline{y}_k | S_{k-1} = \grave{s}). \tag{5.59}$$

A suitable metric for the path $\underline{s}_k^s$ is therefore $M(\underline{s}_k^s)$, where

$$
\begin{aligned}
M(\underline{s}_k^s) &\triangleq \ln\left(p(\underline{s}_k^s \wedge \underline{y}_{j \leq k}) \right) \\
&= M(\underline{s}_{k-1}^{\grave{s}}) + \ln\left(p(S_k = s \wedge \underline{y}_k | S_{k-1} = \grave{s}) \right).
\end{aligned} \tag{5.60}
$$

Using Equation 5.23 we then have:

$$M(\underline{s}_k^s) = M(\underline{s}_{k-1}^{\grave{s}}) + \ln\left(\gamma_k(\grave{s}, s) \right), \tag{5.61}$$

where $\gamma_k(\grave{s}, s)$ is the branch transition probability for the path from $S_{k-1} = \grave{s}$ to $S_k = s$. From Equation 5.54 we can write:

$$\ln\left(\gamma_k(\grave{s}, s) \right) \triangleq \Gamma_k(\grave{s}, s) = \hat{C} + \frac{1}{2} u_k L(u_k) + \frac{L_c}{2} \sum_{l=1}^{n} y_{kl} x_{kl}, \tag{5.62}$$

and as the term $\hat{C}$ is constant, it can be omitted and we can rewrite Equation 5.61 as:

$$M(\underline{s}_k^s) = M(\underline{s}_{k-1}^{\grave{s}}) + \frac{1}{2} u_k L(u_k) + \frac{L_c}{2} \sum_{l=1}^{n} y_{kl} x_{kl}. \tag{5.63}$$

Hence our metric in the SOVA is updated as in the Viterbi algorithm, with the additional $u_k L(u_k)$ term included so that the a-priori information available is taken into account. Notice that this is equivalent to the forward recursion in Equation 5.52 used to calculate $A_k(s)$ in the Max-Log-MAP algorithm.

The possibility of modifying the metric used in the Viterbi algorithm to include a-priori information was mentioned by Forney [9] in his 1973 paper, although he proposed no application for such a modification. However, the requirement to use a-priori information in the soft-in soft-out component decoders of turbo decoders has provided an obvious application.

Let us now discuss the second modification of the algorithm required, i.e. to give soft outputs. In a binary trellis there will be two paths reaching state $S_k = s$ at stage k in the trellis. The modified Viterbi algorithm, which takes account of the a-priori information $u_k L(u_k)$, calculates

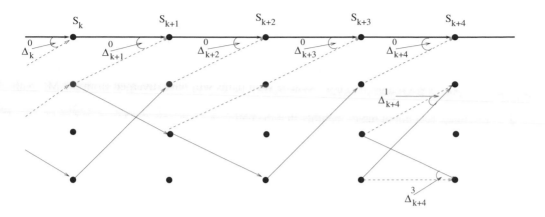

Figure 5.12: Simplified section of the trellis for our $K = 3$ RSC code with SOVA decoding.

the metric from Equation 5.63 for both merging paths, and discards the path with the lower metric. If the two paths $\underline{s}_k^s$ and $\hat{\underline{s}}_k^s$ reaching state $S_k = s$ have metrics $M(\underline{s}_k^s)$ and $M(\hat{\underline{s}}_k^s)$, and the path $\underline{s}_k^s$ is selected as the survivor because its metric is higher, then we can define the metric difference Δ_k^s as:

$$\Delta_k^s = M(\underline{s}_k^s) - M(\hat{\underline{s}}_k^s) \geq 0. \tag{5.64}$$

The probability that we have made the correct decision when we selected path $\underline{s}_k^s$ as the survivor and discarded path $\hat{\underline{s}}_k^s$ is then:

$$P(\text{correct decision at } S_k = s) = \frac{P(\underline{s}_k^s)}{P(\underline{s}_k^s) + P(\hat{\underline{s}}_k^s)}. \tag{5.65}$$

Upon taking into account our metric definition in Equation 5.60 we have:

$$\begin{aligned} P(\text{correct decision at } S_k = s) &= \frac{e^{M(\underline{s}_k^s)}}{e^{M(\underline{s}_k^s)} + e^{M(\hat{\underline{s}}_k^s)}} \\ &= \frac{e^{\Delta_k^s}}{1 + e^{\Delta_k^s}}, \end{aligned} \tag{5.66}$$

and the LLR that this is the correct decision is given by:

$$\begin{aligned} L(\text{correct decision at } S_k = s) &= \ln\left(\frac{P(\text{correct decision at } S_k = s)}{1 - P(\text{correct decision at } S_k = s)}\right) \\ &= \Delta_k^s. \end{aligned} \tag{5.67}$$

Figure 5.12 shows a simplified section of the trellis for our $K = 3$ RSC code, with the metric differences Δ_k^s marked at various points in the trellis.

When we reach the end of the trellis and have identified the Maximum Likelihood (ML) path through the trellis, we need to find the LLRs giving the reliability of the bit decisions along the ML path. Observations of the Viterbi algorithm have shown that all the surviving paths at a stage l in the trellis will normally have come from the same path at some point before l in the trellis. This point is taken to be at most δ transitions before l, where usually δ is set to be five times the constraint length of the convolutional code. Therefore the value of the bit u_k associated with the transition from state $S_{k-1} = \grave{s}$ to state $S_k = s$ on the ML path may have been different if, instead of the ML path, the Viterbi algorithm had selected one of the paths which merged with the ML path up to δ transitions later, i.e. up to the trellis stage $k + \sigma$. By the arguments above, if the algorithm had selected any of the paths which merged with the ML path after this point the value of u_k would not be affected, because such paths will have diverged from the ML path after the transition from $S_{k-1} = \grave{s}$ to $S_k = s$. Thus, when calculating the LLR of the bit u_k, the SOVA must take account of the probability that the paths merging with the ML path from stage k to stage $k + \delta$ in the trellis were incorrectly discarded. This is done by considering the values of the metric difference $\Delta_i^{s_i}$ for all states s_i along the ML path from trellis stage $i = k$ to $i = k + \delta$. It is shown by Hagenauer in [54] that this LLR can be approximated by:

$$L(u_k|\underline{y}) \approx u_k \min_{\substack{i=k\cdots k+\delta \\ u_k \neq u_k^i}} \Delta_i^{s_i}, \qquad (5.68)$$

where u_k is the value of the bit given by the ML path, and u_k^i is the value of this bit for the path which merged with the ML path and was discarded at trellis stage i. Thus the minimisation in Equation 5.68 is carried out only for those paths merging with the ML path which would have given a different value for the bit u_k if they had been selected as the survivor. The paths which merge with the ML path, but would have given the same value for u_k as the ML path, obviously do not affect the reliability of the decision of u_k.

For clarification of these operations refer again to Figure 5.12 showing a simplified section of the trellis for our $K = 3$ RSC code. In this figure, as before, continous lines represent transitions taken when the input bit is a -1, and broken lines represent transitions taken when the input bit is a $+1$. We assume that the all-zero path is identified as the ML path, and this path is shown as a bold line. Also shown are the paths which merge with this ML path. It can be seen from the figure that the ML path gives a value of -1 for u_k, but the paths merging with the ML path at trellis stages S_k, S_{k+1}, S_{k+3} and S_{k+4} all give a value of $+1$ for the bit u_k. Hence, if we assume for simplicity that $\sigma = 4$, from Equation 5.68 the LLR $L(u_k|\underline{y})$ will be given by -1 multiplied by the minimum of the metric differences Δ_k^0, Δ_{k+1}^0, Δ_{k+3}^0 and Δ_{k+4}^0.

5.3.6.2 Implementation of the SOVA

The SOVA can be implemented as follows. For each state at each stage in the trellis the metric $M(\underline{s}_k^s)$ is calculated for both of the two paths merging into the state using Equation 5.63. The path with the highest metric is selected as the survivor, and for this state at this stage in the trellis a pointer to the previous state along the surviving path is stored, just as in the classical Viterbi algorithm. However, in order to allow the reliability of the decoded bits to be calculated, the information used in Equation 5.68 to give $L(u_k|\underline{y})$ is also stored. Thus the difference Δ_k^s between the metrics of the surviving and the discarded paths is stored, together with a binary vector containing $\delta + 1$ bits, which indicate whether or not the discarded path would have given the same series of bits u_l for $l = k$ back to $l = k - \delta$ as the surviving path does. This series of bits is called the update sequence in [54], and as noted by Hagenauer it is given by the result of a modulo-2 addition (i.e. an exclusive-or operation) between the previous $\delta + 1$ decoded bits along the surviving and discarded paths. When the SOVA has identified the ML path, the stored update sequences and metric differences along this path are used in Equation 5.68 to calculate the values of $L(u_k|\underline{y})$.

The SOVA described in this section is the least complex of all the soft-in soft-out decoders discussed in this chapter. In [52] it is shown by Robertson *et al.* that the SOVA is about half as complex as the Max-Log-MAP algorithm. However, the SOVA is also the least accurate of

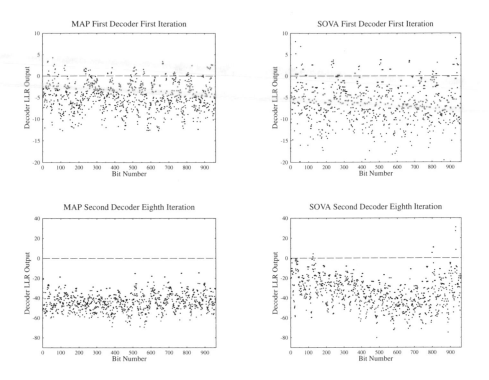

Figure 5.13: Soft outputs from the SOVA compared to the MAP algorithm for a transmitted stream of all
-1

the algorithms we have described in this chapter and, when used in an iterative turbo decoder, performs about 0.6 dB worse [52] than a decoder using the MAP algorithm. Figure 5.13 compares the LLRs output from the component decoders in an iterative turbo decoder using both the MAP and the SOVA algorithms. The same encoder, all -1 input sequence, and channel SNR as described for Figure 5.11 were used. It can be seen that the outputs of the SOVA are significantly more noisy than those from the MAP algorithm. For the second decoder at the eighth iteration the MAP algorithm gives LLRs which are all negative, and hence gives no bit errors. However, it can be seen from Figure 5.11 that, even after eight iterations, the SOVA still gives some positive LLRs, and hence will make several bit errors.

Let us now augment our understanding of iterative turbo decoding by considering a specific example.

5.3.7 Turbo Decoding Example

In this section we discuss an example of turbo decoding using the SOVA detailed in Section 5.3.6. This example serves to illustrate the details of the SOVA and the iterative decoding of turbo codes discussed in Section 5.3.4.

We consider a simple half-rate turbo code using the $K = 3$ RSC code, with generator polynomials expressed in octal form as 7 and 5, shown in Figure 5.2. Two such codes are combined, as shown in Figure 5.1, with a 3×3 block interleaver to give a simple turbo code. The parity bits from both the component codes are punctured, so that alternate parity bits from the first and the

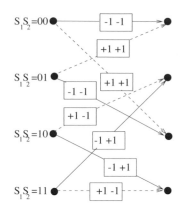

Figure 5.14: State transition diagram for our (2,1,3) RSC component codes.

second component encoder are transmitted. Thus the first, third, fifth, seventh and ninth parity bits from the first component encoder are transmitted, and the second, fourth, sixth and eighth parity bits from the second component encoder are transmitted. The first component encoder is terminated using two bits chosen to take this encoder back to the all-zero state. The transmitted sequence will therefore contain nine systematic and nine parity bits. Of the systematic bits seven will be the input bits, and two will be the bits chosen to terminate the first trellis. Of the nine parity bits five will come from the first encoder, and four from the second encoder.

The state transition diagram for the component RSC codes is shown in Figure 5.14. As in all our diagrams in this section, a continous line denotes a transition resulting from a -1 input bit, and a broken lines represents an input bit of $+1$. The figures within the boxes along the transition lines give the output bits associated with that transition:- the first bit is the systematic bit, which is the same as the input bit, and the second is the parity bit.

For the sake of simplicity we assume that an all -1 input sequence is used. Thus there will be seven input bits which are -1, and the encoder trellis will remain in the $S_1 S_2 = 00$ state. The two bits necessary to terminate the trellis will be -1 in this case and, as can be seen from Figure 5.14, the resulting parity bits will also be -1. Thus all 18 of the transmitted bits will be -1 for an all -1 input sequence. Assuming that BPSK modulation is used with the transmitted symbols being -1 or $+1$, the transmitted sequence will be a series of $18 - 1$s. The received channel output sequence for our example, together with the input and transmitted bits detailed above, is shown in Table 5.1. Notice that approximately half the parity bits from each component encoder are punctured — this is represented by a dash in Table 5.1. Also note that the received channel sequence values shown in Table 5.1 are the matched filter outputs, which were denoted by y_{kl} in previous sections. If hard-decision demodulation were used then negative values would be decoded as -1s, and positive values as $+1$s. It can be seen that from the 18 coded bits which were transmitted, all of which were -1, three would be decoded as $+1$ if hard-decision demodulation were used.

In order to illustrate the difference between iterative turbo decoding and the decoding of convolutional codes, we initially consider how the received sequence shown in Table 5.1 would be decoded by a convolutional decoder using the Viterbi algorithm. Imagine the half-rate $K = 3$ RSC code detailed above used as an ordinary convolutional code to encode an input sequence of seven -1s. If trellis termination was used then two -1s would be employed to terminate the trellis, and the transmitted sequence would consist of $18 - 1$s, just as for our turbo coding example. If the received sequence was as shown in Table 5.1, then the Viterbi algorithm decoding

Input Bit	Systematic Bit	Parity Bits Coder 1	Coder 2	Transmitted Sequence	Received Sequence
−1	−1	−1	–	−1,−1	−2.1, −0.1
−1	−1	–	−1	−1, −1	−1.4, −1.4
−1	−1	−1	–	−1, −1	−1.7, −0.5
−1	−1	–	−1	−1, −1	+0.9, +0.5
−1	−1	−1	–	−1, −1	+1.2, −1.7
−1	−1	–	−1	−1, −1	−1.1, −1.1
−1	−1	−1	–	−1, −1	−0.7, −0.8
-	−1	–	−1	−1, −1	−2.4, −1.9
-	−1	−1	–	−1, −1	−1.6, −0.9

Table 5.1: Input and transmitted bits for turbo decoding example.

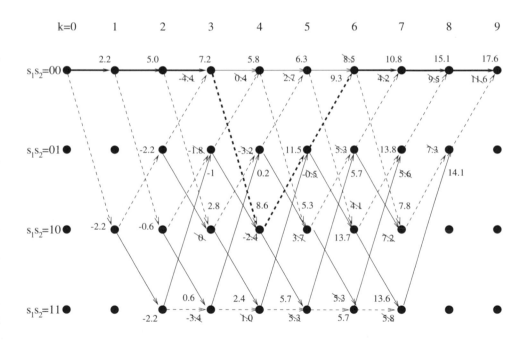

Received : (-2.1,-0.1) (-1.4,-1.4) (-1.7,-0.5) (+0.9,+0.5) (+1.2,-1.7) (-1.1,-1.1) (-0.7,-0.8) (-2.4,-1.9) (-1.6,-0.9)

Decoded : -1 -1 -1 +1 +1 +1 -1 - -

Figure 5.15: Trellis diagram for the Viterbi decoding of the received sequence shown in Table 5.1.

this sequence would have the trellis diagram shown in Figure 5.15. The metrics shown in this figure are given by the cross-correlation of the received and expected channel sequences for a given path, and the Viterbi algorithm maximises this metric to find the ML path, which is shown by the bold line in Figure 5.15. Notice that at each state in the trellis where two paths merge, the path with the lower metric is discarded and its metric is shown crossed out in the figure. As can be seen from Figure 5.15, the Viterbi algorithm makes an incorrect decision at stage $k = 6$ in the trellis and selects a path other than the all-zero path as the survivor. This results in three of the seven bits being decoded incorrectly as +1s.

Having seen how Viterbi decoding of a RSC code would fail and produce three errors given our received sequence, we now proceed to detail the operation of an iterative turbo decoder for the same channel sequence. Consider first the operation of the first component decoder in the first iteration. The component decoder uses the SOVA to decide upon not only the most likely input bits, but also the LLRs of these bits, as described in Section 5.3.6. We will describe here how the SOVA calculates these LLRs for the channel values given in Table 5.1.

The metric for the SOVA is given by Equation 5.63, which is repeated here for convenience:

$$M(\underline{s}_k^s) = M(\underline{s}_{k-1}^{\dot{s}}) + \frac{1}{2}u_k L(u_k) + \frac{L_c}{2}\sum_{l=1}^{Q} y_{kl}x_{kl}. \qquad (5.69)$$

Here $M(\underline{s}_{k-1}^{\dot{s}})$ is the metric for the surviving path through the state $S_{k-1} = \dot{s}$ at stage $k - 1$ in the trellis, u_k and x_{kl} are the input bit and the transmitted channel sequence associated with a given transition, y_{kl} is the received channel sequence for that transition and L_c is the channel reliability value, as defined in Equation 5.11. As initially we are considering the operation of the first decoder in the first iteration there is no a-priori information and hence we have $L(u_k) = 0$ for all k, which corresponds to an a-priori probability of 0.5. The received sequence given in Table 5.1 was derived from the transmitted channel sequence (which has $E_b = 1$) by adding AWGN with variance $\sigma = 1$. Hence, as the fading amplitude is $a = 1$, from Equation 5.11 we have for the channel reliability measure $L_c = 2$.

Figure 5.16 shows the trellis for this first component decoder in the first iteration. Owing to the puncturing of the parity bits used at the encoder, the second, fourth, sixth and eighth parity bits have been received as zeros. The a-priori and channel values shown in Figure 5.16 are given as $L(u_k)/2$ and $L_c y_{kl}/2$ so that the metric values, given by Equation 5.69, can be calculated by simple addition and subtraction of the values shown. As we have $L(u_k) = 0$ and $L_c = 2$, these metrics are again given by the cross-correlation of the expected and received channel sequences. Notice, however, that because of the puncturing used the metric values shown in Figure 5.16 are not the same as those in Figure 5.15.

To elaborate a little further, the metric calculation philosophy of the SOVA is also different from that of the Viterbi algorithm, since we may have more than one emerging branch at a given state in the trellis. For example, at trellis stage k=5, state $S_1 S_2$=01, both of the two paths emerging from this state survive. This is because these two paths lead to different states at the next stage in the trellis, where both of them win the metric comparison. More specifically, the path associated with the input bit 0 and hence the continuous line leads to state $S_1 S_2$=10 and has a total path metric of 10.7. This metric is compared to the metric of the other path leading to the state $S_1 S_2$=10, which is 4.3 for the path arriving from state $S_1 S_2$=00. Hence the path arriving from the state $S_1 S_2$=01 has the higher metric and therefore it is chosen as the winning path at the new state of $S_1 S_2$=10. Similarly, the other path emerging from the state of $S_1 S_2$=01 at stage k=5 survives. Specifically, this path associated with the input bit of 1 and hence the broken line arrives at state $S_1 S_2$=00 and has a total path metric of 8.5. This is compared to 6.5, namely to the metric of the other path arriving at state $S_1 S_2$=00 and hence this path is selected as the winner.

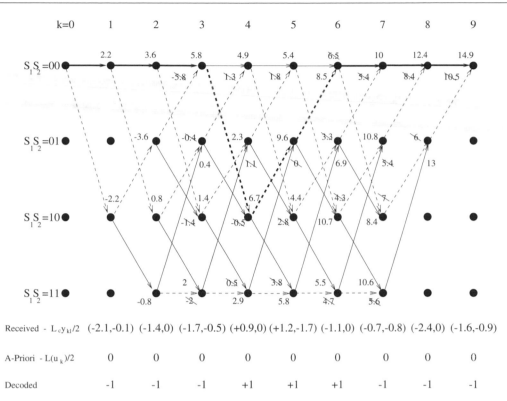

Figure 5.16: Trellis diagram for the SOVA decoding in the first iteration of the first decoder.

Again, owing to the puncturing employed the metric values shown in Figure 5.16 are not the same as those in the Viterbi decoding example of Figure 5.15. Despite this the ML path, shown by the bold line in Figure 5.16, is the same as the one that was chosen by the Viterbi algorithm shown in Figure 5.15, with three of the input bits being decoded as +1s rather than −1s.

We now discuss how, having determined the ML path, the SOVA finds the LLRs for the decoded bits. Figure 5.17 is a simplified version of the trellis from Figure 5.16, which shows only the ML path and the paths that merge with this ML path and are discarded. Also shown are the metric differences, denoted by Δ_k^s in Section 5.3.6, between the ML and the discarded paths. These metric differences, together with the previously defined update sequences that indicate for which of the bits the survivor and discarded paths would have given different values, are stored by the SOVA for each node at each stage in the trellis. When the ML path has been identified, the algorithm uses these stored values along the ML path to find the LLR for each decoded bit. Table 5.2 shows these stored values for our example trellis shown in Figures 5.16 and 5.17. The calculation of the decoded LLRs shown in this table is detailed at a later stage.

Notice in Table 5.2 that at trellis stages $k = 1$ and $k = 2$ there is no metric difference or update sequence stored because, as can be seen from Figures 5.16 and 5.17, there are no paths merging with the ML path at these stages. For all subsequent stages there is a merging path, and values of the metric differences and update sequences are stored. For the update sequence a 1 indicates that the ML and the discarded merging path would have given different values for a particular bit. At stage k in the trellis we have taken the Most Significant Bit (MSB), on the left-hand side, to represent u_k, the next bit to represent u_{k-1}, etc., until the Least Significant Bit (LSB), which represents u_1. For our RSC code any two paths merging at trellis stage k give

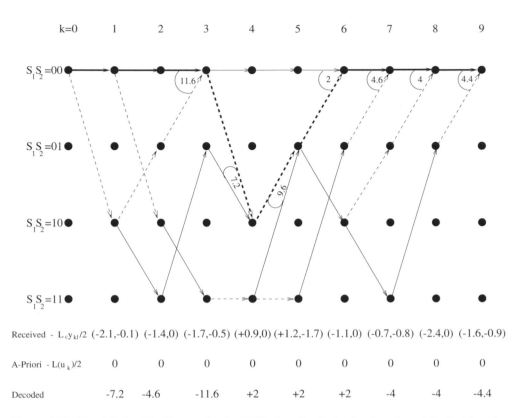

Figure 5.17: Simplified trellis diagram for the SOVA decoding in the first iteration of the first decoder.

Trellis Stage k	Decoded Bit u_k	Metric Difference Δ_k^s	Update Sequence	Decoded LLR
1	-1	$-$	$-$	-7.2
2	-1	$-$	$-$	-4.6
3	-1	11.6	111	-11.6
4	$+1$	7.2	1001	$+2$
5	$+1$	9.6	10010	$+2$
6	$+1$	2	111000	$+2$
7	-1	4.6	1100010	-4
8	-1	4	11100000	-4
9	-1	4.4	100100000	-4.4

Table 5.2: SOVA output for the first iteration of the first decoder.

different values for the bit u_k, and so the MSB in the update sequences in Table 5.2 is always 1. Notice furthermore that although in our example the update sequences are all of different lengths, this is only because of the very short frame length we have used. More generally, as explained in Section 5.3.6, all the stored update sequences will be $\delta + 1$ bits long, where δ is usually set to be five times the constraint length of the convolutional code (15 in our case). At this stage it is beneficial for the reader to verify the update sequences of Table 5.2 using Figure 5.17.

We now explain how the SOVA can use the stored update sequences and metric differences along the ML path to calculate the LLRs for the decoded bits. Equation 5.68, which is repeated here as Equation 5.70 for convenience, shows that the decoded a-posteriori LLR $L(u_k|\underline{y})$ for a bit u_k is given by the minimum metric difference of merging paths along the ML path:

$$L(u_k|\underline{y}) \approx u_k \min_{\substack{i=k\cdots k+\delta \\ u_k \neq u_k^i}} \Delta_i^{s_i}. \tag{5.70}$$

This minimum is taken only over the metric differences for stages $i = k, k+1, \ldots, k+\delta$ where the value u_k^i of the bit u_k given by the path merging with the ML path at stage i is different from the value given for this bit by the ML path. Whether or not the condition $u_k = u_k^i$ is met is determined using the stored update sequences. Denoting the update sequence stored at stage l along the ML path as $\underline{e}_l$, for each bit u_k the SOVA examines the MSB of $\underline{e}_k$, the second MSB of $\underline{e}_{k+1}$ etc. up to the $(\delta + 1)$th bit (which will be the LSB) of $\underline{e}_{k+\delta}$. For our example this examination of the update sequences is limited because of our short frame length, but the same principles are used. Taking the fourth bit u_4 as an example, to determine the decoded LLR $L(u_4|\underline{y})$ for this bit the algorithm examines the MSB of $\underline{e}_4$ in row 4 of Table 5.2, the second MSB of $\underline{e}_5$ in row 5, etc. up to the sixth MSB of $\underline{e}_9$ in row nine. It can be seen, from the corresponding rows in Table 5.2, that only the paths merging at stages $k = 4$ and $k = 6$ of the trellis give values different from the ML path for the bit u_4. Hence the decoded LLR $L(u_4|\underline{y})$ from the SOVA for this bit is calculated using Equation 5.70 as the value of the bit given by the ML path (+1) times the minimum of the metric differences stored at stages 4 and 6 of the trellis (7.2 and 2), yeilding $L(u_4|\underline{y}) = +2$. For the next bit, u_5, the MSB of $\underline{e}_5$ in row 5 of Table 5.2, the second MSB of $\underline{e}_6$ in row 6, etc. up to the fifth MSB of $\underline{e}_9$ in row 9 are examined, which indicate that a different value for this bit would have been given by the discarded paths at stages 5 and 6 of the trellis. Hence $L(u_5|\underline{y})$ also equals +2, as the metric difference, 2, at stage 6 in the trellis is less than the metric difference, 9.6, at stage 5. For the next bit, u_6, the update sequences indicate that all the merging paths from the sixth stage of the trellis to the end of the trellis would give values different to those given by the ML path. However, the minimum of all these metric differences is still 2, and so $L(u_6|\underline{y})$ also equals +2. This illustrates in the SOVA how one discarded path having a low metric difference can entirely determine the LLRs for all the bits, for which it gives a different value from the ML path, which is a consequence of taking the minimum in Equation 5.70. At stage 6 of the trellis, where the algorithm incorrectly chooses the non-zero path as the survivor, the metric difference between the chosen (incorrect) path and the discarded path is the lowest metric difference (2) encountered along the ML path. Hence the LLRs for the three incorrectly decoded bits, i.e. for u_4, u_5 and u_6, have the lowest magnitudes of any of the decoded bits.

The remaining decoded LLR values in Table 5.2 are computed following a similar procedure. However, also worth noting explicitly is the LLR for the bit u_3. Examination of the MSB of $\underline{e}_3$ in row 3 of Table 5.2, the second MSB of $\underline{e}_4$ in row 4, etc., up to the seventh MSB of $\underline{e}_9$, reveals that only the path merging with the ML path at the third stage of the trellis would give a different value for u_3. Hence the minimum for the LLR $L(u_3|\underline{y})$ in Equation 5.70 is over one term, and the magnitude of $L(u_3|\underline{y})$ is determined by the metric difference of the merging path at stage $k = 3$ of the trellis, which is 11.6.

Trellis Stage k	Decoder LLR Output $L(u_k|\underline{y})$	A-Priori Info. $L(u_k)$	Received Sys. Info. $L_c y_{ks}$	Extrinsic Information
1	-7.2	0	-4.2	-3
2	-4.6	0	-2.8	-1.8
3	-11.6	0	-3.4	-8.2
4	$+2$	0	$+1.8$	$+0.2$
5	$+2$	0	$+2.4$	-0.4
6	$+2$	0	-2.2	$+4.2$
7	-4	0	-1.4	-2.6
8	-4	0	-4.8	$+0.8$
9	-4.4	0	-3.2	-1.2

Table 5.3: Calculation of the extrinsic information from the first decoder in the first iteration using Equation 5.71.

We now move on to describe the operation of the second component decoder in the first iteration. This decoder uses the extrinsic information from the first decoder as a-priori information to assist its operation, and therefore should be able to provide a better estimate of the encoded sequence than the first decoder was. Equation 5.46 from Section 5.3.4 gives the extrinisic information from the MAP decoder as:

$$L_e(u_k) = L(u_k|\underline{y}) - L(u_k) - L_c y_{ks}. \qquad (5.71)$$

The same equation can be derived for all the soft-in soft-out decoders which are used as component decoders for turbo codes. This equation states that the extrinsic information $L_e(u_k)$ is given by the soft output $L(u_k|\underline{y})$ from the decoder with the a-priori information $L(u_k)$ (if any was available) and the received systematic channel information $L_c y_{ks}$ subtracted. Table 5.3 shows the extrinsic information calculated from Equation 5.71 from the first decoder, which is then interleaved by a 3×3 block interleaver and used as the a-priori information for the second component decoder. The second component decoder also uses the interleaved received systematic channel values, and the received parity bits from the second encoder which were not punctured (i.e. the second, fourth, sixth and eighth bits).

Figure 5.18 shows the trellis for the SOVA decoding of the second decoder in the first iteration. The extrinsic information values from Table 5.3 are shown after being interleaved and divided by two as $L(u_k)/2$. Also shown is the channel information $L_c y_{ks}/2$ used by this decoder. Notice that as the trellis is not terminated for the second component encoder, paths terminating in all four possible states of the trellis are considered at the decoder. However, the metric for the $S_1 S_2 = 00$ state is the maximum of the four final metrics, and hence this all-zero state is used as the final state of the trellis. Note furthermore that the metrics in Figure 5.18 are now calculated as the cross-correlation of the received and expected channel information, plus the a-priori information $u_k L(u_k)/2$.

The ML path chosen by the second component decoder is shown by a bold line in Figure 5.18, together with the LLR values output by the decoder. These are calculated, using update sequences and minumum metric differences, in the same way as was explained for the first decoder using Figure 5.17 and Table 5.2. It can be seen that the decoder makes an incorrect decision at stage $k = 5$ in the trellis and selects a path other than the all-zero path as the survivor. However, the incorrectly chosen path gives decoded bits of $+1$ for only two transitions, and hence only two, rather than three, decoding errors are made. Furthermore the difference in the metrics between the correct and the chosen path at trellis stage $k = 5$ is only 2.2, and so the magnitude of the

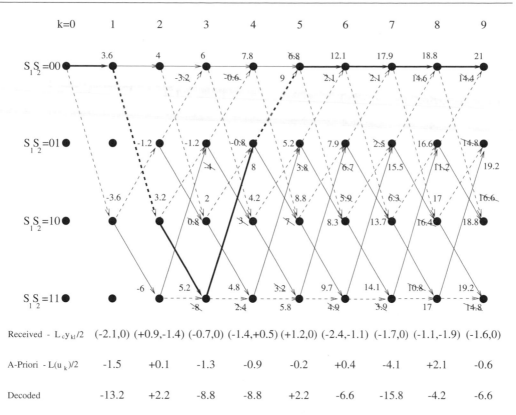

Figure 5.18: Trellis diagram for the SOVA decoding in the first iteration of the second decoder.

decoded LLRs $L(u_k|\underline{y})$ for the two incorrectly decoded bits, u_2 and u_5, is only 2.2. This is significantly lower than the magnitudes of the LLRs for the other bits, and indicates that the algorithm is less certain about these two bits being +1 than it is about the other bits being −1.

Having calculated the LLRs from the second component decoder, the turbo decoder has now completed one iteration. The soft-output LLR values from the second component decoder shown in the bottom line of Figure 5.18 could now be de-interleaved and used as the output from the turbo decoder. This de-interleaving would result in an output sequence which gave negative LLRs for all the decoded bits except u_4 and u_5, which would be incorrectly decoded as +1s as their LLRs are both +2.2. Thus, even after only one iteration, the turbo decoder has decoded the received sequence with one less error than the convolutional decoder did. However, generally better results are achieved with more iterations, and so we now progress to describe the operation of the turbo decoder in the second iteration.

In the second, and all subsequent, iterations the first component decoder is able to use the extrinsic information from the second decoder in the previous iteration as a-priori information. Table 5.4 shows the calculation of this extrinsic information using Equation 5.71 from the second decoder in the first iteration. It can be seen that it gives negative LLRs for all the bits except u_2 and u_5, and for these two bits the LLRs are close to zero. This extrinsic information is then de-interleaved and used as the a-priori information for the first decoder in the next (second) iteration. The trellis for this decoder is shown in Figure 5.19. It can be seen that this decoder uses the same channel information as it did in the first iteration. However, now, in contrast to Figure 5.16, it also has a-priori information, to assist it in finding the correct path through the trellis. The

Trellis Stage k	Decoder LLR Output $L(u_k\|\underline{y})$	A-Priori Info. $L(u_k)$	Received Sys. Info. $L_c y_{ks}$	Extrinsic Information
1	-13.2	-3	-4.2	-6
2	$+2.2$	$+0.2$	$+1.8$	$+0.2$
3	-8.8	-2.6	-1.4	-4.8
4	-8.8	-1.8	-2.8	-4.2
5	$+2.2$	-0.4	$+2.4$	$+0.2$
6	-6.6	$+0.8$	-4.8	-2.6
7	-15.8	-8.2	-3.4	-4.2
8	-4.2	$+4.2$	-2.2	-6.2
9	-6.6	-1.2	-3.2	-2.2

Table 5.4: Calculation of the extrinsic information from the second decoder in the first iteration using Equation 5.71.

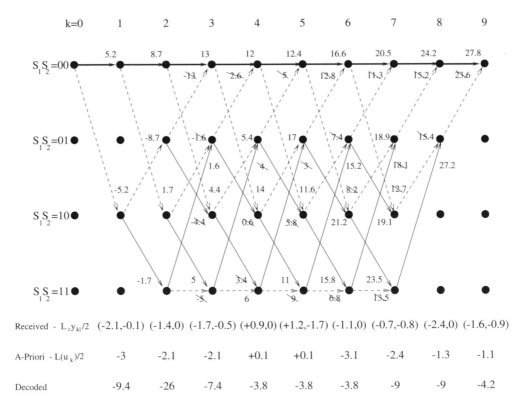

Figure 5.19: Trellis diagram for the SOVA decoding in the second iteration of the first decoder.

selected ML path is again shown by a bold line, and it can be seen that now the correct all-zero path is chosen. The second iteration is then completed by finding the extrinsic information from the first decoder, interleaving it and using it as a-priori information for the second decoder. It can be shown that this decoder will also now select the all-zero path as the ML path, and hence the output from the turbo decoder after the seond iteration will be the correct all -1 sequence. This concludes our example of the operation of an iterative turbo decoder using the SOVA.

5.3.8 Comparison of the Component Decoder Algorithms

In this section we have detailed the iterative structure and the component decoders used for decoding turbo codes. A numerical example illustrating this decoding was given in the previous section. We now conclude by summarising the operation of the algorithms which can be used as component decoders, highlighting the similarities and differences between these algorithms, and noting their relative complexities and performances.

The MAP algorithm is the optimal component decoder for turbo codes. It finds the probability of each bit u_k being a $+1$ or -1 by calculating the probability for each transition from state $S_{k-1} = \grave{s}$ to $S_k = s$ that could occur if the input bit was $+1$, and similarly for every transition that could occur if the input bit was -1. As these transitions are mutually exclusive, the probability of any one of them occurring is simply the sum of their individual probabilities, and hence the LLR for a bit u_k is given by the ratio of two sums of probabilities, as in Equation 5.17.

Owing to the Markov nature of the trellis and the assumption that the output from the trellis is observed in memoryless noise, the individual probabilities of the transitions in Equation 5.17 can be expressed as the product of three terms $\alpha_{k-1}(\grave{s})$, $\beta_k s$ and $\gamma_k(\grave{s}, s)$, as in Equation 5.20. By definition the $\alpha_{k-1}(\grave{s})$ term gives the probability that the trellis reaches state $S_{k-1} = \grave{s}$ and that the received sequence up to this point is $\underline{y}_{j<k}$. The state transition probability, $\gamma_k(\grave{s}, s)$, is defined as the probability that given the trellis is in state $S_{k-1} = \grave{s}$ it moves to state $S_k = s$ and the received sequence for that transition is $\underline{y}_k$. Finally, the $\beta_k(s)$ term gives the probability that given the trellis is in state $S_k = s$, the received sequence from this point to the end of the trellis is $\underline{y}_{j>k}$. The state transition probabilities $\gamma_k(\grave{s}, s)$ are calculated from the received and expected channel sequences, y_{kl} and x_{kl}, for a given transition and the a-priori LLR $L(u_k)$ of the bit associated with this transition, as seen in Equation 5.40 for an AWGN channel. The $\alpha_{k-1}(\grave{s})$ terms can then be calculated using a forward recursion through the trellis, as in Equation 5.25, and similarly the $\beta_k(s)$ terms are calculated using Equation 5.28 by a backward recursion through the trellis. The output LLR $L(u_k|\underline{y})$ from the MAP algorithm is then determined for each bit u_k by finding the probability of each transition from stage S_{k-1} to S_k in the trellis. These transitions are divided into two groups:- those that would have resulted if the bit u_k was a $+1$, and those that would have resulted if the bit u_k was a -1. The LLR $L(u_k|\underline{y})$ for the bit u_k is then given by the logarithm of the ratio of these probabilities.

The MAP algorithm is optimal for the decoding of turbo codes, but is extremely complex. Furthermore, because of the multiplications used in the recursive calculation of the $\alpha_{k-1}(\grave{s})$ and $\beta_k(s)$ terms, and the exponents used to calculate the $\gamma_k(\grave{s}, s)$ terms, it often suffers from numerical problems in practice. The Log-MAP algorithm is theoretically identical to the MAP algorithm, but transfers its operations to the log domain. Thus multiplications are replaced by additions, and hence the numerical problems of the MAP algorithm are circumvented, while the associated complexity is dramatically reduced.

The Max-Log-MAP algorithm further reduces the complexity of the Log-MAP algorithm using the maximisation approximation given in Equation 5.48. This has two effects on the operation of the algorithm compared to that of the Log-MAP algorithm. First, as can be seen by examining Equation 5.55, it means that only two transitions are considered when finding the LLR $L(u_k|\underline{y})$ for each bit u_k — the best transition from $S_{k-1} = \grave{s}$ to $S_k = s$ that would give $u_k = +1$ and the best that would give $u_k = -1$. Similarly in the recursive calculations of the $A_k(s) = \ln(\alpha_k(s))$ and $B_k(s) = \ln(\beta_k(s))$ terms of Equations 5.52 and 5.53 the approximation means that only one transition, the most likely one, is considered when calculating $A_k(s)$ from the $A_{k-1}(\grave{s})$ terms and $B_{k-1}(\grave{s})$ from the $B_k(s)$ terms. This means that although $A_{k-1}(\grave{s})$ should give the logarithm of the probability that the trellis reaches state $S_{k-1} = \grave{s}$ along *any* path from the initial state $S_0 = 0$, in fact it gives the logarithm of the probability of only the *most likely* path to state $S_{k-1} = \grave{s}$. Similarly $B_k(s) = \ln(\beta_k(s))$ should give the logarithm of the probability of the received sequence $\underline{y}_{j>k}$ given only that the trellis is in state $S_k = s$ at stage k. However, the maximisation in Equation 5.53 used in the recursive calculation of the $B_k(s)$ terms means that only the most likely path from state $S_k = s$ to the end of the trellis is considered, and not all paths.

Hence the Max-Log-MAP algorithm finds the LLR $L(u_k|\underline{y})$ for a given bit u_k by comparing the probability of the most likely path giving $u_k = +1$ to the probability of the most likely path giving $u_k = -1$. For the next bit, u_{k+1}, again the best path that would give $u_{k+1} = +1$ and the best path that would give $u_{k+1} = -1$ are compared. One of these 'best paths' will always be the ML path, and so will not change from one stage to the next, whereas the other may change. In contrast the MAP and the Log-MAP algorithms consider every path in the calculation of the LLR for each bit. All that changes from one stage to the next is the division of paths into those that give $u_k = +1$ and those that give $u_k = -1$. Thus the Max-Log-MAP algorithm gives a degraded performance compared to the MAP and Log-MAP algorithms.

In the SOVA the ML path is found by maximising the metric given in Equation 5.63. The recursion used to find this metric is identical to that used to find the $A_k(s)$ terms in Equation 5.52 in the Max-Log-MAP algorithm. Once the ML path has been found, the hard decision for a given bit u_k is determined by which transition the ML path took between trellis stages S_{k-1} and S_k. The LLR $L(u_k|\underline{y})$ for this bit is determined by examining the paths which merge with the ML path that would have given a different hard decision for the bit u_k. The LLR is taken to be the minimum metric difference for these merging paths which would have given a different hard decision for the bit u_k. Using the notation associated with the Max-Log-MAP algorithm, once a path merges with the ML path, it will have the same value of $B_k(s)$ as the ML path. Hence, as the metric in the SOVA is identical to the $A_k(s)$ values in the Max-Log-MAP algorithm, taking the difference between the metrics of the two merging paths in the SOVA is equivalent to taking the difference between two values of $(A_{k-1}(\grave{s}) + \Gamma_k(\grave{s}, s) + B_k(s))$ in the Max-Log-MAP algorithm, as in Equation 5.55. The only difference is that in the Max-Log-MAP algorithm one path will be the ML path, and the other will be the most likely path that gives a different hard decision for u_k. In the SOVA again one path will be the ML path, but the other may not be the most likely path that gives a different hard decision for u_k. Instead, it will be the most likely path that gives a different hard decision for u_k and survives to merge with the ML path. Other, more likely paths, which give a different hard decision for the bit u_k to the ML path, may have been discarded before they merge with the ML path. Thus the SOVA gives a degraded performance compared to the Max-Log-MAP algorithm. However, as pointed out in [52] by Robertson *et al.*, the SOVA and Max-Log-MAP algorithms will always give the same hard decisions, as in both algorithms these hard decisions are determined by the ML path, which is calculated using the same metric in both algorithms.

It was also noted in [52] that the outputs from the SOVA contain no bias when compared to those from the Max-Log-MAP algorithm, they are just more noisy. However, consideration of the arguments above makes it clear that when the most likely path that gives a different hard decision for u_k survives to merge with the ML path, the outputs from the SOVA and the Max-Log-MAP algorithms will be identical. Otherwise when this most likely path which gives a different hard decision for u_k does not survive to merge with the ML path, the merging path that is used to calculate the soft output from the SOVA will be less likely than the path which should have been used. Thus it will have a lower metric, and so the metric difference used in Equation 5.68 will be higher than it should be. Therefore, although the sign of the soft outputs from the SOVA will be identical to those from the Max-Log-MAP algorithm, their magnitudes will be either identical or

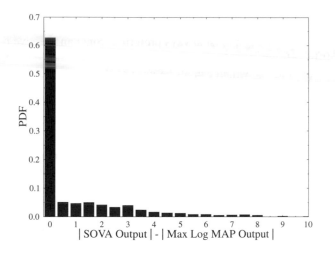

Figure 5.20: Probability density function of differences between absolute values of soft outputs from the SOVA and Max-Log-MAP algorithms.

higher. This can be seen in Figure 5.20, which shows the Probability Density Function (PDF) for the differences between the absolute values of the soft outputs from the SOVA and the Max-Log-MAP algorithms. Both decoders were used as the first decoder in the first iteration, and again the same encoder, all -1 input sequence, and channel SNR as described for Figure 5.11 were used. It can be seen that for more than half of the decoder outputs the two algorithms give identical values. It can also be seen that the absolute values given by the SOVA are never less than those given by the Max-Log-MAP algorithm.

A comparison of the complexities of the Log-MAP, the Max-Log-MAP and the SOVA algorithms is given in [52]. The relative complexity of the algorithms depends on the constraint length K of the convolutional codes used, but it is shown that the Max-Log-MAP algorithm is about twice as complex as the SOVA. The Log-MAP algorithm is about 50% more complex than the Max-Log-MAP algorithm owing to the look-up operation required for finding the correction factors $f_c(x)$. Viterbi noted furthermore [130, 131] that the Max-Log-MAP algorithm can be viewed as two generalised Viterbi decoders (one for the forward recursion, and one for the backward recursion) together with a generalised dual-maxima computation (for calculating the soft outputs using Equation 5.55). The complexity of the dual-maxima computation is lower than that of the Viterbi decoder, and hence Viterbi estimated the complexity of the Max-Log-MAP algorithm to be lower than three times that of the Viterbi algorithm. Our related calculations suggest for a $K = 3$ code that the Max-Log-MAP decoder is about 2.6 times as complex, and the Log-MAP decoder is about four times as complex, as the standard Viterbi algorithm.

The performance of the algorithms when used in the iterative decoding of turbo codes follows in the same order as their complexities, with the best performance given by the Log-MAP algorithm, then the Max-Log-MAP algorithm, and the worst performance exhibited by the SOVA. We will compare the performance of these three algorithms when decoding various turbo codes in Section 5.4.

5.3.9   Conclusions

In this section we have described the techniques used for the decoding of turbo codes. Although it is possible to optimally decode turbo codes in a single non-iterative step, for complexity reasons a non-optimum iterative decoder is almost always preferred. Such an iterative decoder employs two component soft-in soft-out decoders, and we have described the MAP, Log-MAP, Max-Log-MAP and SOVA algorithms, which can all be used as the component decoders. The MAP algorithm is optimal for this task, but it is extremely complex. The Log-MAP algorithm is a simplification of the MAP algorithm, and offers the same optimal performance with a reasonable complexity. The other two algorithms, the Max-Log-MAP and the SOVA, are both less complex again, but give a slightly degraded performance.

Having described the principles behind the encoding and decoding of turbo codes we now move on to present our simulation results demonstrating the excellent performance of turbo codes for various scenarios.

5.4   Turbo-coded BPSK Performance over Gaussian Channels

In the previous two sections we have discussed the structure of both the encoder and the decoder in a turbo codec. In this section we present simulation results for turbo codes using BPSK over AWGN channels. We show that there are many parameters, some of which are interlinked, which affect the performance of turbo codes. Some of these parameters are:

- The component decoding algorithm used.

- The number of decoding iterations used.

- The frame length or latency of the input data.

- The specific design of the interleaver used.

- The generator polynomials and constraint lengths of the component codes.

In this section we investigate how all of these parameters affect the performance of turbo codes. The standard parameters we have used in our simulations are shown in Table 5.5. All our results presented in this section consider turbo codes using BPSK modulation over an AWGN channel. The turbo encoder uses two component RS codes in parallel. Our standard RSC component codes are $K = 3$ codes with generator polynomials $G_0 = 7$ and $G_1 = 5$ in octal representation. These generator polynomials are optimum in terms of maximising the minimum free distance of the component codes [102]. The effects of varying these generator polynomials are examined in Section 5.4.5. The standard interleaver used between the two component RSC codes is a 1000-bit random interleaver with odd–even separation [68]. The effects of changing the length of the interleaver, and its structure, are examined in Sections 5.4.4 and 5.4.6. Unless otherwise stated, the results of this section are valid for half-rate codes, where half the parity bits generated by each of the two component RSC codes are punctured. However, for comparison, we also include some results for turbo codes where all the parity bits from both component encoders are transmitted, leading to a one-third-rate code. At the decoder two component, soft-in soft-out, decoders are used in parallel in the structure shown in Figure 5.3. In most of our simulations we use the Log-MAP decoder, but the effect of using other component decoders is investigated in Section 5.4.3. Usually eight iterations of the component decoders are used, but in the next section we consider the effect of the number of iterations.

Channel	Additive White Gaussian Noise (AWGN)
Modulation	Binary Phase Shift Keying (BPSK)
Component Encoders	Two identical recursive - convolutional codes
RSC Parameters	$n = 2, k = 1, K = 3$ $G_0 = 7, G_1 = 5$
Interleaver	1000-bit random interleaver with odd–even separation [68]
Puncturing Used	Half parity bits from each component encoder transmitted give half-rate code
Component Decoders	Log-MAP decoder
Iterations	8

Table 5.5: Standard turbo encoder and decoder parameters used.

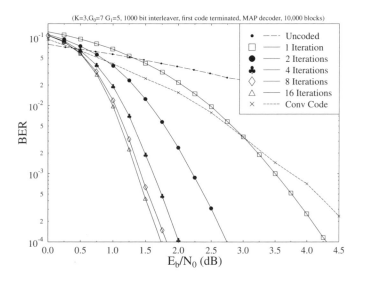

Figure 5.21: Turbo coding BER performance using different numbers of iterations of the MAP algorithm. Other parameters as in Table 5.5.

5.4.1 Effect of the Number of Iterations Used

Figure 5.21 shows the performance of a turbo decoder using the MAP algorithm versus the number of decoding iterations which were used. For comparison, the uncoded BER and the BER obtained using convolutional coding with a standard (2,1,3) non-recursive convolutional code are also shown. Like the component codes in the turbo encoder, the convolutional encoder uses the optimum octal generator polynomials of 7 and 5. It can be seen that the performance of the turbo code after one iteration is roughly similar to that of the convolutional code at low SNRs, but improves more rapidly than that of the convolutional coding as the SNR is increased. As the number of iterations used by the turbo decoder increases, the turbo decoder performs

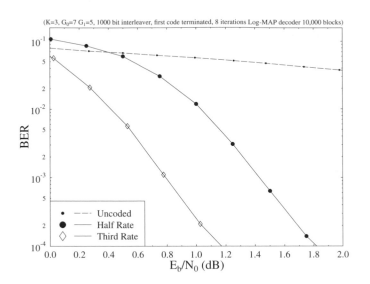

Figure 5.22: BER performance comparison between one-third- and half-rate turbo codes using parameters of Table 5.5.

significantly better. However, after eight iterations there is little improvement achieved by using further iterations. For example, it can be seen from Figure 5.21 that using 16 iterations rather than eight gives an improvement of only about 0.1 dB. Similar results are obtained when using the SOVA; again there is little improvement in the BER performance of the decoder from using more than eight iterations. Hence for complexity reasons usually only between about 4 and 12 iterations are used. Accordingly, unless otherwise stated, in our simulations we used eight iterations. In the next section we consider the effects of puncturing.

5.4.2 Effects of Puncturing

As described in Section 5.2, in a turbo encoder two or more component encoders are used for generating parity information from an input data sequence. In our work we have used two RSC component encoders, and this is the arrangement most commonly used for turbo codes having coding rates below two-thirds. Typically, in order to generate a half-rate code, half the parity bits from each component encoder are punctured. This was the arrangement used in the seminal paper by Berrou *et al.* on the concept of turbo codes [12]. However, it is of course possible to omit the puncturing and transmit all the parity information from both component encoders, which gives a one-third-rate code. The performance of such a code, compared to the corresponding half-rate code, is shown in Figure 5.22. In this figure the encoders use the same parameters as were described above for Figure 5.21. It can be seen that transmitting all the parity information gives a gain of about 0.6 dB, in terms of E_b/N_0, at a BER of 10^{-4}. This corresponds to a gain of about 2.4 dB in terms of channel SNR. Very similar gains are seen for turbo codes with different frame lengths. Let us now consider the performance of the various soft-in soft-out component decoding algorithms which were described in Section 5.3.

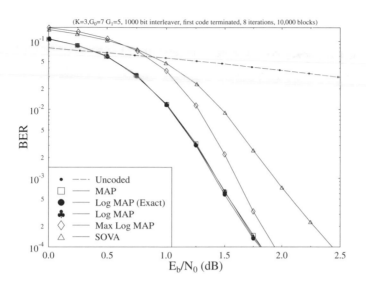

Figure 5.23: BER performance comparison between different component decoders for a random interleaver
with L=1000. Other parameters as in Table 5.5.

5.4.3 Effect of the Component Decoder Used

Figure 5.23 shows a comparison between turbo decoders using the different component decoders
described in Section 5.3 for a turbo encoder using the parameters described above. In this figure
the 'Log MAP (Exact)' curve refers to a decoder which calculates the correction term $f_c(x)$ in
Equation 5.56 of Section 5.3.5 exactly, i.e. using:

$$f_c(x) = \ln(1 + e^{-x}), \tag{5.72}$$

rather than using a look-up table as described in [52]. The Log-MAP curve refers to a decoder
which does use a look-up table with eight values of $f_c(x)$ stored, and hence introduces an approximation to the calculation of the LLRs. It can be seen that, as expected, the MAP and
the Log-MAP (exact) algorithms give identical performances. Furthermore, as Robertson *et al.*
found [52], the look-up procedure for the values of the $f_c(x)$ correction terms introduces no
degradation to the performance of the decoder.

It can also be seen from Figure 5.23 that the Max-Log-MAP and the SOVA both give a
degradation in performance compared to the MAP and Log-MAP algorithms. At a BER of 10^{-4}
this degradation is about 0.1 dB for the Max-Log-MAP algorithm, and about 0.6 dB for the
SOVA.

Figure 5.24 compares the Log-MAP, Max-Log-MAP and SOVA algorithms for a turbo decoder with a frame length of only 169 bits, rather than 1000 bits as was used for Figure 5.23. It
can be seen that although all three decoders give a worse BER performance than those shown in
Figure 5.23, the differences in the performances between the decoders are very similar to those
shown in Figure 5.23. Similarly, Figure 5.25 compares these three decoding algorithms for a one-
third-rate code, and again the degradations relative to a decoder using the Log-MAP algorithm
are about 0.1 dB for the Max-Log-MAP algorithm, and about 0.6 dB for the SOVA.

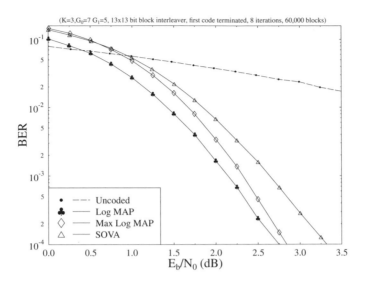

Figure 5.24: BER performance comparison between different component decoders for a L=169, 13×13, block interleaver. Other parameters as in Table 5.5.

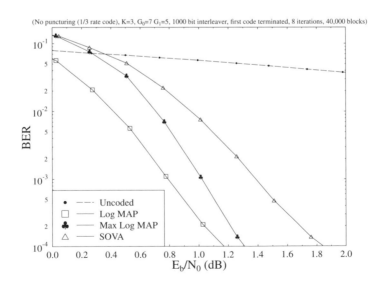

Figure 5.25: BER performance comparison between different component decoders for a random interleaver with L=1000 using a one-third rate code. Other parameters as in Table 5.5.

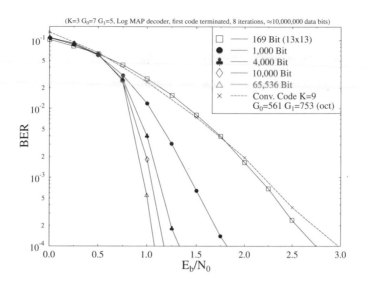

Figure 5.26: Effect of frame length on the BER performance of turbo coding. All interleavers except $L = 169$ block interleaver use random separated Interleavers [68]. Other Parameters as in Table 5.5.

5.4.4   Effect of the Frame Length of the Code

In the original paper on turbo coding by Berrou *et al.* [12], and in many of the subsequent papers, impressive results have been presented for coding with very large frame lengths. However, for many applications, such as for example speech transmission systems, the large delays inherent in using high frame lengths are unacceptable. Therefore an important area of turbo coding research is achieving as impressive results with short frame lengths as have been demonstrated for long frame length systems.

Figure 5.26 shows how dramatically the performance of turbo codes depends on the frame length L used in the encoder. The 169-bit code would be suitable for use in a speech transmission system at approximately 8 kbit/s with a 20 ms frame length [132], while the 1000-bit code would be suitable for video transmission. The larger frame length systems would be useful in data or non-real-time transmission systems. It can be seen from Figure 5.26 that the performance of turbo codes is very impressive for systems with long frame lengths. However, even for a short frame length system, using 169 bits per frame, it can be seen that turbo codes give good results, comparable to or better than a constraint length $K = 9$ convolutional code. The use of the $K = 9$ convolutional code as a bench-marker is justified below.

As noted in Section 5.3, a single decode with the Log-MAP decoder is about four times as complex as decoding the same code using a standard Viterbi decoder. The curves shown in Figure 5.26, and in most of our results, use two component decoders with eight iterations. Therefore the overall complexity of a turbo decoder is approximately $2 \times 8 \times 4 = 64$ times that of a Viterbi decoder for one of the component convolutional codes. This means that the complexity of our turbo decoder using eight iterations of two $K = 3$ component codes is approximately the same as the complexity of a Viterbi decoder for an ordinary $K = 9$ convolutional code. In

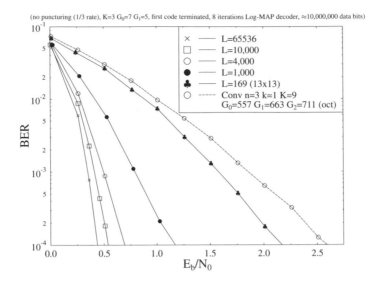

Figure 5.27: Effect of frame length on BER performance of one-third rate turbo coding. All interleavers
except $L = 169$ block interleaver use random interleavers. Other parameters as in Table 5.5.

order to provide a comparison between the performance of turbo codes and convolutional codes
for similar complexity decoders, we will compare our $K = 3$ turbo codes with an eight iteration
decoder to a $K = 9$ convolutional code.

Figure 5.26 shows the performance of such a convolutional code. A non-recursive (2,1,9)
convolutional code using the generator polynomials $G_0 = 561$ and $G_1 = 753$ in octal notation,
which maximise the free distance of the code [102], was used. These generator polynomials
provide the best performance in the AWGN channels we use in this section. A frame length of
169 bits is used, and the code is terminated. It can be seen that even for the short frame length
of 169 bits, turbo codes outperform similar complexity convolutional codes. As the frame length
is increased, the performance gain from using turbo codes, rather than high-constraint-length
convolutional codes, increases dramatically.

Figure 5.27 shows how the performance of a one-third rate turbo code varies with the frame
length of the code. Again, the performance of the turbo code is better the longer the frame length
of the code, but impressive results are still obtained with a frame length of only 169 bits. Again
the results for a $K = 9$ convolutional code are shown, this time using a third-rate $n = 3, k = 1$
code with the optimal generator polynomials of $G_0 = 557$, $G_1 = 663$ and $G_2 = 711$ [102]
in octal notation. Again it can be seen that the high-constraint-length convolutional code is
outperformed by turbo codes with frame lengths of 169 and higher.

Let us now consider the effect of using different RSC component codes.

5.4.5 The Component Codes

Both the constraint length and the generator polynomials used in the component codes of turbo
codes are important parameters. Often in turbo codes the generator polynomials which lead to

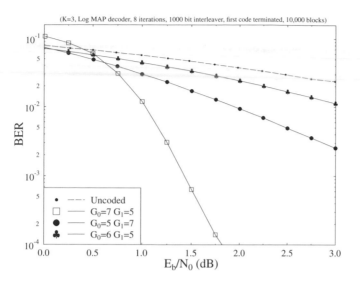

Figure 5.28: Effect of generator polynomials on BER performance of turbo coding. Other parameters as in Table 5.5.

the largest minimum free distance for ordinary convolutional codes are used, although when the effect of interleaving is considered these generator polynomials do not necessarily lead to the best minimum free distance for turbo codes. Figure 5.28 shows the huge difference in performance that can result from different generator polynomials being used in the component codes. The other parameters used in these simulations were the same as detailed above in Table 5.5.

Most of the results provided in this chapter were obtained using constraint length 3 component codes. For these codes we have used the optimum generator polynomials in terms of maximising the minimum free distance of the component convolutional codes, i.e. 7 and 5 in octal representation. These generator polynomials were also used for constraint length 3 turbo coding by Hagenauer *et al.* in [61] and Jung in [66]. It can be seen from Figure 5.28 that the order of these generator polynomials is important — the value 7 should be used for the feedback generator polynomial in Figure 5.1 (denoted in our work by G_0). If G_0 and G_1 are swapped round, the performance of a convolutional code (both regular and recursive systematic codes) would be unaffected, but for turbo codes this gives a significant degradation in performance.

The effect of increasing the constraint length of the component codes used in turbo codes is shown in Figure 5.29. For the constraint length 4 turbo code we again used the optimum minimum free distance generator polynomials for the component codes (15 and 17 in octal, 13 and 15 in decimal representations). The resulting turbo code gives an improvement of about 0.25 dB at a BER of 10^{-4} over the $K = 3$ curve.

For the constraint length 5 turbo code we used the octal generator polynomials 37 and 21 (31 and 17 in decimal), which were the polynomials used by Berrou *et al.* [12] in the original paper on turbo coding. We also tried using the octal generator polynomials 23 and 35 (19 and 29), which are again the optimum minimum free distance generator polynomials for the component codes, as suggested by Hagenauer *et al.* in [61]. We found that these generator polynomials

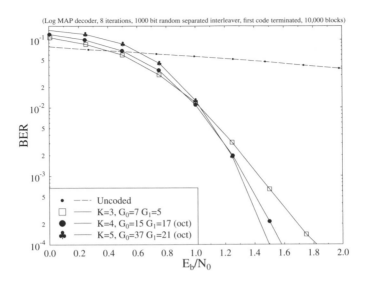

Figure 5.29: Effect of constraint length on the BER performance of turbo coding. Other parameters as in Table 5.5.

gave almost identical results to those used by Berrou *et al.* It can be seen from Figure 5.29 that increasing the constraint length of the turbo code does improve its performance, with the $K = 4$ code performing about 0.25 dB better than the $K = 3$ code at a BER of 10^{-4}, and the $K = 5$ code giving a further improvement of about 0.1 dB. However, these improvements are provided at the cost of approximately doubling or quadrupling the decoding complexity. Therefore, unless otherwise stated, we have used component codes with a constraint length of 3 in our work. Let us now focus on the effects of the interleaver used within the turbo encoder and decoder.

5.4.6   Effect of the Interleaver

It is well known that the interleaver used in turbo codes has a vital influence on the performance of the code. The interleaver design together with the generator polynomials used in the component codes, and the puncturing used at the encoder, have a dramatic affect on the free distance of the resultant turbo code. Several algorithms have been proposed, for example in references [126] and [65], that attempt to choose good interleavers based on maximising the minimum free distance of the code. However, this process is complex, and the resultant interleavers are not necessarily optimum. For example, in [133] random interleavers designed using the technique given in [65] are compared to a 12×16 dimensional block interleaver, and the 'optimised' interleavers are found to perform worse than the block interleaver.

In [68] a simple technique for designing good interleavers, which is referred to as 'odd–even separation' is proposed. With alternate puncturing of the parity bits from each of the component codes, which is the puncturing most often used, if an interleaver is designed so that the odd and even input bits are kept separate, then it can be shown that one (and only one) parity bit associated with each information bit will be left unpunctured. This is preferable to the more

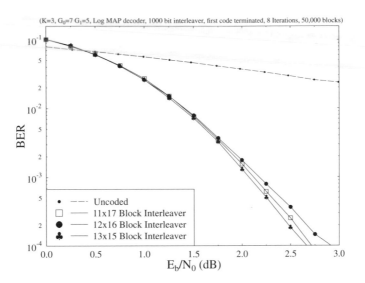

Figure 5.30: Effect of block interleaver choice for $L \approx 190$ frame length turbo codes. Other parameters as in Table 5.5.

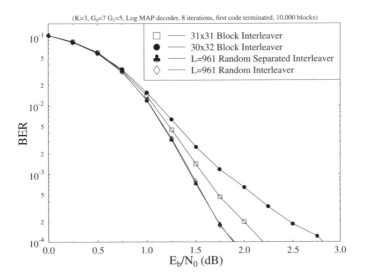

Figure 5.31: Effect of interleaver choice for $L \approx 961$ frame length turbo codes. Other parameters as in Table 5.5.

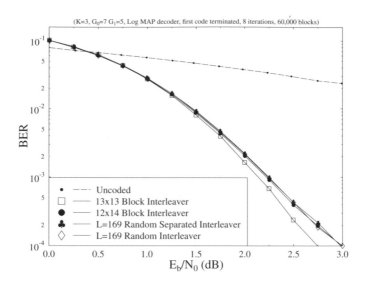

Figure 5.32: Effect of interleaver choice for $L \approx 169$ frame length turbo codes. Other parameters as in Table 5.5.

general situation, where some information bits will have their parity bits from both component codes transmitted, whereas others will have neither of their parity bits transmitted.

A convenient way of achieving odd–even separation in the interleaver is to use a block interleaver with an odd number of rows and columns [68]. The benefits of using an odd number of rows and columns with a block interleaver can be seen in Figure 5.30. This shows a comparison between turbo coders using several block interleavers with frame lengths of approximately 190 bits. The 12×16 dimensional block interleaver, proposed for short frame transmission systems in [133] and used by the same authors in other papers such as [66, 67, 134], clearly has a somewhat lower performance than the other block interleavers, which use an odd number of rows and columns. It is also interesting to note that of the two block interleavers with an odd number of rows and columns, the interleaver which is closer to being square (i.e. the 13×15 interleaver) performs better than the more rectangular 11×17 interleaver.

We also attempted using random interleavers of various frame lengths. The effect of the interleaver choice for a turbo coding system with a frame length of approximately 960 bits is shown in Figure 5.31. It can be seen from this figure that, as was the case with the codes with frame lengths around 192 bits shown in Figure 5.30, the block interleaver with an odd number of rows and columns (the 31×31 interleaver) performs significantly better than the interleaver with an even number of rows and columns (the 30×32 interleaver). However, both of these interleavers are outperformed by the two random interleavers. In the 'random separated' interleaver odd–even separation, as proposed by Barbulescu and Pietrobon [68], is used. This interleaver performs very slightly better than the other random interleaver, which does not use odd–even separation. However, the effect of odd–even separation is much less significant for the random interleavers than it is for the block interleavers.

Similar curves are shown in Figure 5.32 for turbo coding schemes with approximately 169

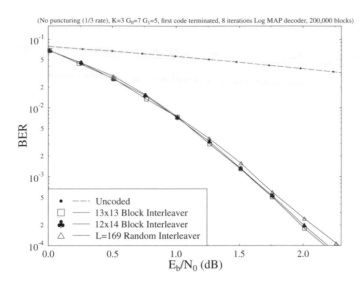

Figure 5.33: Effect of interleaver choice for third-rate $L \approx 169$ frame length turbo codes. Other parameters as in Table 5.5.

bits per frame. It can be seen again that the scheme using block interleaving with odd–even separation (i.e. the 13×13 interleaver) performs better than the scheme using block interleaving without odd–even separation (i.e. the 12×14 interleaver). However, for this short frame length system the two random interleavers perform worse than the best block interleaver. From our results it appears that although random interleavers give the best performance for turbo codes with long frame lengths, for short frame length systems the best performance is given using a block interleaver with an odd number of rows and columns.

When puncturing is not used, and we have a third-rate code, the benefit of using odd–even separation with block interleavers, i.e. using block interleavers with an odd number of rows and columns, disappears. This can be seen from Figure 5.33, which compares the performance of a turbo code with no puncturing using three different interleavers, all with a length of approximately 169 bits. As in the case of the half-rate turbo codes using puncturing in Figure 5.32, for a small frame length, such as 169 bits, the best performance is given by using a block rather than a random interleaver. However, it can be seen from Figure 5.33 that, unlike for half-rate codes, for turbo codes without puncturing there is little difference between the block interleavers with and without odd–even separation, i.e. between the 13×13 and 12×14 interleavers.

In [135] Herzberg suggests that a 'reverse block' interleaver, i.e. a block interleaver in which the output bits are read from the block in the reverse order relative to an ordinary block interleaver, gives an improved performance over ordinary block interleavers. He also suggests that for high SNRs, and hence for low BERs, reverse block interleavers having a short frame length give a better performance than random interleavers having a significantly higher frame length. However, as can be seen from Figure 5.34, which portrays the performance of ordinary and reverse block interleavers for various frame lengths, we found very little difference between the performances of block and reverse block interleavers. One difference between our results and

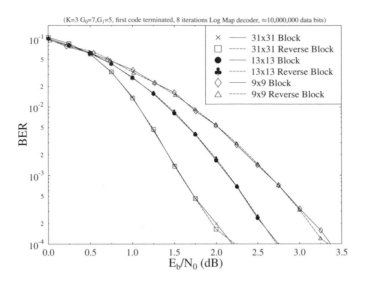

Figure 5.34: BER performance of block and reverse block interleavers. Other parameters as in Table 5.5.

those in [135] is that we have used punctured half-rate turbo codes, whereas Herzberg used turbo codes without puncturing. However, we found that even with third-rate turbo codes using no puncturing, and using 14×14 interleavers as Herzberg did, the performances of block and reverse block interleavers were almost identical. It appears in [135] that for turbo codes with long random interleavers, and with an ordinary block interleaver, Herzberg used the generator polynomials $G_0 = 5$ and $G_1 = 7$, whereas for the reverse block interleaver he used the generator polynomials $G_0 = 7$ and $G_1 = 5$. The generator polynomials $G_0 = 5$ and $G_1 = 7$ were used so that the performance of turbo codes with long random interleavers could be approximated using the Union bound and the error coefficients calculated by Benedetto and Montorsi in [59] for these generator polynomials. However, as was seen in Figure 5.28, these generator polynomials give a significantly worse performance than the generator polynomials $G_0 = 7$ and $G_1 = 5$ we have used for most of our simulations, and Herzberg used with his reverse block interleaver. Thus it appears that the reason Herzberg found such promising results for the reverse block interleaver was not because of this interleaver's superiority, but because of the inferiority of the generator polynomials he used with random and block interleavers.

Let us now focus our attention on the effect of the estimation of the channel reliability measure L_c.

5.4.7   Effect of Estimating the Channel Reliability Value L_c

In the previous section we highlighted how the component decoders of an iterative turbo decoder interacted using soft inputs, the channel inputs $L_c y_{kl}$ as well as the a-priori inputs $L(u_k)$, and provided the a-posteriori LLRs $L(u_k|y)$ as soft outputs. In the MAP and Log-MAP algorithms the channel inputs and a-priori information are used to calculate the transition probabilities $\gamma_k(\grave{s}, s)$ that are then used to recursively calculate the $\alpha_k(s)$ and $\beta_k(s)$ and finally the

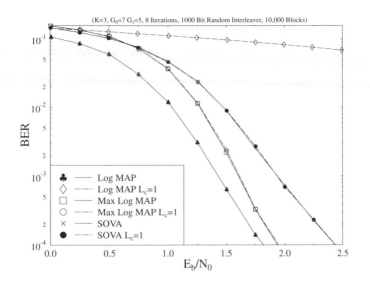

Figure 5.35: Effect of using incorrect channel reliability measures L_c on an iterative turbo decoder using various component decoders. Other parameters as in Table 5.5.

a-posteriori LLRs $L(u_k|\underline{y})$. Similarly, in the Max-Log-MAP and the SOVA algorithms the channel and a-priori information values are used to update metrics, which are then used to give the soft-output a-posteriori LLRs. In this section we investigate how important an accurate estimate of the channel reliability measure L_c is to the good performance of an iterative turbo decoder.

Figure 5.35 shows the performance of iterative turbo decoders using three different component decoders:- the Log-MAP, Max-Log-MAP and the SOVA algorithms. For each component decoder type the continuous line shows the performance of the codec when the channel reliability value L_c is calculated exactly using the known channel SNR. For all our previous results we assumed that the channel SNR, and hence the correct value of L_c, would be known at the decoder. The broken curves in Figure 5.35 show how the three component decoders perform when L_c is not known. For these curves the value of $L_c = 1$, which corresponds to a value of E_b/N_0 of -3 dB, was used at all channel SNRs. It can be seen from Figure 5.35 that for the SOVA and the Max-Log-MAP algorithms the turbo decoder performs equally well whether or not the correct value of L_c is known. However, for the Log-MAP algorithm the performance of the iterative turbo decoder is drastically affected by the value of L_c used.

The reason for these effects can be understood by considering the different operation of the three algorithms, as was described in Section 5.3. In the SOVA the channel values $L_c y_{kl}$ are used to recursively calculate the metrics M_n using Equation 5.63. The metric M_n for a state $S_k = s$ along a path is given by the metric M_{n-1} for the previous state along the path, added to an a-priori information term and to a cross-correlation term between the expected and the received channel values, x_{kl} and y_{kl}. The channel reliability measure L_c is used to scale this cross-correlation between the received and expected channel values. Then, once the ML path has been identified, the soft outputs from the algorithm are given by the minimum metric difference between the ML path and the paths merging with this ML path, as seen in Equation 5.68. When we use an incorrect value of L_c, effectively we are scaling the inputs to the component decoders by a factor. For instance, if we simply use $L_c = 1$, then we scale the channel input values by a factor of one over the correct value of L_c. In the SOVA this has the effect of scaling all the metrics M_n by the same factor, and as the a-posteriori LLRs from the algorithm are given by the difference between metrics for different paths, these output LLRs are also scaled by the same factor. Which path is chosen by the algorithm as the ML path will be unaffected by this scaling of the metrics, and so the hard decisions given by the algorithm will be unaffected by using the incorrect value of L_c.

Consider now the operation of the SOVA as a component decoder within an iterative turbo decoder. Assuming that no a-priori information about the values of the bits is available to the iterative turbo decoder, the first component decoder in the first iteration takes channel values only. The a-priori information values $L(u_k)$ are set equal to zero. If the correct value of the channel reliability measure for the channel SNR used is L_c, but an incorrect value of $\hat{L}_c$ is used instead, then effectively the channel input values will have been scaled by a factor X, where:

$$ X = \frac{\hat{L}_c}{L_c}. \tag{5.73} $$

The first component decoder will process these scaled channel values, and give soft-output LLRs $L(u_k|y)$. From our discussions above these soft outputs, and hence the extrinsic information $L_e(u_k)$ derived from them using Equation 5.71, will be equal to the correct soft outputs scaled by X. Next the second component decoder will take a-priori information, equal to the interleaved extrinsic information $L_e(u_k)$ from the first decoder and the channel inputs, and will use these values to calculate its soft outputs. Both the channel values $\hat{L}_c y_{kl}$ and the a-priori information $L(u_k)$ will have been scaled by X relative to their values if the correct L_c had been used, and so again all the metrics used in the SOVA will be scaled by X, and the soft outputs from this decoder will simply be the correct soft outputs scaled by the factor X. Hence we see that, because of the linearity in the SOVA, the effect of using an incorrect value of the channel reliability measure is that the output LLRs from the decoder are scaled by a constant factor. The relative importance of the two inputs to the decoder, i.e. the a-priori information and the channel information, will not change, since the LLRs for both these sources of information will be scaled by the same factor. The soft outputs from the final component decoder in the final iteration will have the same sign as those that would have been calculated using the correct value of L_c, and will merely have been scaled by X. Hence the hard outputs from an iterative turbo decoder using the SOVA are unaffected by the value of the channel reliability value L_c used, as can be seen from Figure 5.35.

The same linearity that is present in the SOVA is also found in the Max-Log-MAP algorithm. Instead of one metric two are calculated, but again only simple additions of the cross-correlation of the expected and received channel values are used. Hence, if an incorrect value of the channel reliability value is used, all the metrics are simply scaled by a factor X. As in the SOVA, the soft outputs are given by the differences in metrics between different paths, and so the same argument as was used above for the SOVA will also apply to the Max-Log-MAP algorithm:- if the input channel values are scaled by X, then the soft outputs of all the component decoders will also be scaled by X, and the final hard decisions given by the turbo decoder will be unaffected.

Let us now consider the Log-MAP algorithm. This is identical to the Max-Log-MAP algorithm, except for a correction factor $f_c(x) = \ln(1 + e^{-x})$ used in the calculation of the forward and backward metrics $A_k(s)$ and $B_k(s)$ and the soft-output LLRs. The function $f_c(x)$ is non-linear — it decreases asymptotically towards zero as x increases. Hence the linearity that is present in the Max-Log-MAP and SOVA algorithms is not present in the Log-MAP algorithm. We found that the effect of this non-linearity is two-fold if an incorrect value of the channel

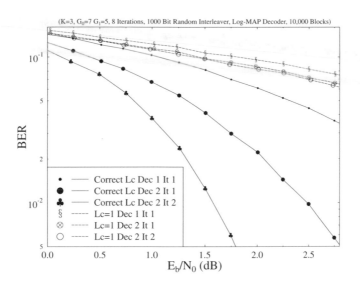

Figure 5.36: BER within an iterative turbo decoder using the Log-MAP decoder and the parameters of Table 5.5, with correct and incorrect channel reliability values L_c.

reliability value is used. First, even when only the channel values are used to calculate the soft outputs from the algorithm, the component decoder makes more hard-decision errors than if the correct value of L_c were used. This is the case for the first component decoder in the first iteration, where the a-priori information values $L(u_k)$ are assumed to be equal to zero. Figure 5.36 shows the performance of an iterative turbo decoder using the Log-MAP algorithm after the first component decoder in the first iteration, denoted by 'Dec 1 It 1' in the key, when both the correct value of the channel reliability measure L_c and an incorrect value of $L_c = 1$ are used. It can be seen from this figure that when the channel reliability value $L_c = 1$ is used the BER for the first component decoder in the first iteration is significantly increased.

As well as making more hard-decision errors, if an incorrect value of the channel reliability measure is used with the Log-MAP algorithm, then the extrinsic information derived from the soft-output values from the first component decoder has incorrect amplitudes. This means that the a-priori information that is used by the second decoder in the first iteration, and by both decoders in subsequent iterations, will have incorrect amplitudes relative to the soft channel inputs. In an iterative turbo decoder the feeding of a-priori information from one component decoder to the next allows a rapid decrease in the BER of the decoder as the number of iterations increases, as was seen in Figure 5.21. However, when the incorrect value of L_c is used in an iterative turbo decoder employing the Log-MAP algorithm, owing to the incorrect scaling of the a-priori information relative to the channel inputs, no such rapid fall in the BER with the number of iterations occurs. In fact the performance of the decoder is largely unaffected by the number of iterations used. This can be seen from Figure 5.36, which shows the BER from the second decoder after one and two iterations. A significant performance improvement can be observed between the first and second iterations when the correct value of L_c is used. By contrast, when the value of $L_c = 1$ is used, there is only a marginal improvement between the first and second

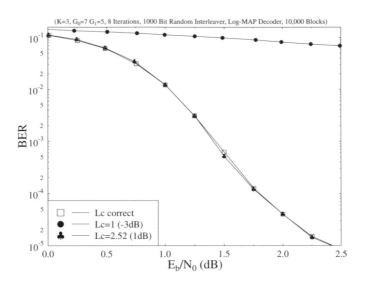

Figure 5.37: Turbo decoder performance using the Log-MAP algorithm and the parameters of Table 5.5 with a constant estimated channel reliability measure L_c.

iteration.

In reference [136] Summers and Wilson consider the degradation in the performance of an iterative turbo decoder using the MAP algorithm when the channel SNR is not correctly estimated. As explained in Section 5.3, the MAP and Log-MAP algorithms give identical outputs and hence our analysis of the Log-MAP algorithm also applies to the MAP algorithm. In [136] the authors propose a method for blind estimation of the channel SNR, using the ratio of the average squared received channel value to the square of the average of the magnitudes of the received channel values. For a one-third-rate code and a block length of 420 data bits (so 1260 coded bits are transmitted) it is shown that the SNR derived using this method rarely differs from the true SNR by more than 3 dB. It is also shown that using these estimated SNRs to derive a channel reliability measure gives a turbo decoder performance using the MAP algorithm almost identical to that given using channel reliability measures derived from the true SNR.

Our work presented here shows that if the Max-Log-MAP or SOVA algorithms are used as the component decoders, then no such SNR estimation is necessary for a turbo decoder. If the MAP, or equivalently the Log-MAP, algorithm is used then, as can be seen from Figure 5.35, if a very inaccurate value of L_c is used the turbo decoder performance will be drastically affected. However, we have found that the value of L_c used does not have to be very close to the true value for a good BER performance to be obtained. Figure 5.37 compares the performance of a turbo decoder using the Log-MAP algorithm with the correct value of L_c, to that of a scheme which uses a constant value of $L_c = 2.52$. This corresponds to a value of E_b/N_0 of 1 dB. It can be seen that using this estimated value of L_c gives a performance virtually identical to that given with the correct value of L_c for values of E_b/N_0 from 0 to 2.5 dB, or BERs from 10^{-1} to 10^{-5}. Hence, even when using the Log-MAP algorithm, only a rough estimate of L_c is needed.

Having investigated the performance of turbo codes when used with BPSK modulation over

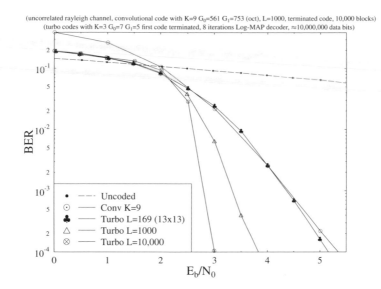

Figure 5.38: BER performance of turbo codes with different frame lengths L over perfectly interleaved Rayleigh fading channels. Other turbo codec parameters as in Table 5.5.

AWGN channels, in the next section we discuss the use of both convolutional and turbo codes with higher-order modulation schemes. This allows turbo codes to be used in systems which are both bandwidth and power efficient.

5.5 Turbo Coding Performance over Rayleigh Channels

5.5.1 Introduction

In the previous sections we have discussed the performance of turbo coding in conjunction with various modulation constellations over AWGN channels. We now move on to an investigation of using turbo coding over fading channels. In this work we have assumed Rayleigh fading, and that the receiver has exact estimates of the fading amplitude and phase inflicted by the channel. This assumption is justified, since several techniques, such as for example Pilot Symbol Assisted Modulation (PSAM) [137], are available for providing practical mechanisms of Channel State Information (CSI) recovery, achieving a performance close to that assuming perfect CSI recovery. In Section 5.5.2 we look at the performance of various turbo codes over Rayleigh fading channels which are perfectly interleaved. Then in Section 5.5.3 we consider the effects correlations experienced in real Rayleigh fading channels have, and evaluate the performance of turbo codes in such channels.

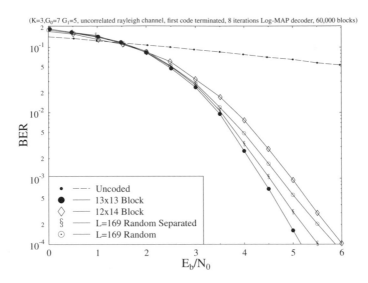

Figure 5.39: BER performance of turbo codes with different interleavers over perfectly interleaved Rayleigh fading channels. Other turbo codec parameters as in Table 5.5.

5.5.2 Performance over Perfectly Interleaved Narrowband Rayleigh Channels

Figure 5.38 shows the performance of three turbo codes with different frame lengths L over a perfectly interleaved Rayleigh fading channel using BPSK modulation. All the turbo codecs use two $K = 3$ RSC component codes with generator polynomials $G_0 = 7$ and $G_1 = 5$. At the decoder eight iterations of the Log-MAP decoder are used. Also shown in Figure 5.38 is the performance of a constraint length $K = 9$ convolutional code which, as explained earlier, has a decoder complexity which is similar to or slightly higher than that of the turbo decoder. It can be seen that the turbo codes with frame lengths of $L = 1000$ or $L = 10000$ give a significant increase in performance over the convolutional code. Even the turbo code with a short frame length of 169 bits outperforms the convolutional code for BERs below 10^{-3}.

Comparing the performance of the $L = 169$, $L = 1000$ and $L = 10,000$ turbo codes in Figure 5.38 to those in Figure 5.26 for the same codes over an AWGN channel, we see that the perfectly interleaved fading of the received channel values degrades the BER performance of the code by around 2 dB at a BER of 10^{-4}, with a larger degradation for the shorter frame length codes.

The short frame length $L = 169$ turbo codec in Figure 5.38 uses a 13×13 block interleaver. We found in Section 5.4.6 that when communicating over a Gaussian channel, for a half-rate turbo code having short frame lengths, a block interleaver with an odd number of rows and columns should be used. Figure 5.39 shows the effect of using different interleavers, all with a frame length of approximately 169 bits, on the BER performance of a turbo codec over a perfectly interleavered Rayleigh channel. It can be seen from this figure that again for the short frame length of 169 bits the best performance is given by a block interleaver with an odd number

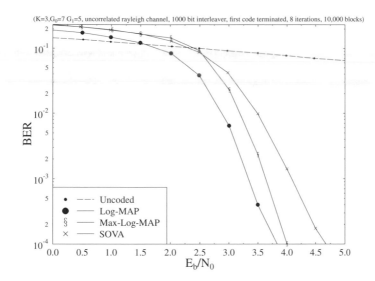

Figure 5.40: BER performance of turbo codes with $L = 1000$ using different component decoders over perfectly interleaved Rayleigh fading channels. Other turbo codec parameters as in Table 5.5.

of rows and columns. This block interleaver acheives odd–even separation [68] so that each data bit has one, and only one, of the two parity bits associated with it transmitted. The 12×14 block interleaver shown in Figure 5.39 does not give odd–even separation, and so even though its frame length is almost identical to that of the 13×13 block interleaver (168 rather than 169 bits) it performs almost 1 dB worse than the 13×13 block interleaver at a BER of 10^{-4}. The two random interleavers shown in Figure 5.39 also perform worse than the 13×13 block interleaver, although the random interleaver with odd–even separation does perform better than the non-separated interleaver.

Figure 5.40 shows how the choice of the component decoders used at the turbo decoder affects the performance of the codec over a perfectly interleavered Rayleigh channel. It can be seen that again the Log-MAP decoder gives the best performance, followed by the Max-Log-MAP decoder, with the SOVA decoder, the simplest of the three, giving the worst performance. It can also be seen that the differences in performances between the different decoders are slightly larger than they were over an AWGN channel — the Max-Log-MAP decoder performs about 0.2 dB worse than the Log-MAP decoder, and the SOVA decoder is about 0.8 dB worse than the Log-MAP decoder.

Figure 5.41 shows the effect of puncturing on a turbo code with frame length $L = 1000$ over the perfectly interleaved Rayleigh channel. In Figure 5.22 we saw that over the AWGN channel the third-rate code outperformed the half-rate code by about 0.6 dB in terms of E_b/N_0. We see from Figure 5.41 that again for the perfectly interleaved Rayleigh channel the difference in performance is bigger — about 1.5 dB in terms of E_b/N_0 or about 3.25 dB in terms of channel SNR.

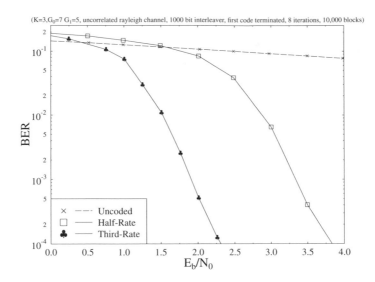

Figure 5.41: The BER performance comparison between one-third- and one-half-rate of turbo codes over perfectly interleaved Rayleigh fading channels. Other turbo codec parameters as in Table 5.5.

5.5.3   Performance over Correlated Narrowband Rayleigh Channels

Figure 5.42 shows the performance of a half-rate turbo coding system with $L = 1000$ over various Rayleigh fading channels. It can be seen that by far the best performance is achieved over the perfectly interleaved Rayleigh channel, where there is no correlation between successive fading values. The narrowband Rayleigh channel exhibits a normalised Doppler frequency of $f_d = 2.44 * 10^{-4}$, since we assumed a carrier frequency of 1.9 GHz, a symbol rate of 360 KBaud and a vehicular speed of 50 km/h. It can be seen that the turbo codes give a significant coding gain over the uncoded BER results even for this channel. We found that for Rayleigh fading channels exhibiting faster fading, i.e. a higher normalised Doppler frequency, the coding gain increased. Furthermore, it can be seen that interleaving the output bits of the turbo encoder before transmission over the Rayleigh fading channel improves the performance for the narrowband system by about 2.5 dB at a BER of 10^{-4}. This gain was acheived by merely interleaving over the 2000-bit length of the output block of the turbo encoder. Higher interleaving gains can be achieved at the cost of extra delay, by interleaving over longer periods. Near-perfect interleaving over a significantly longer period would give the performance indicated by the uncorrelated Rayleigh curve in Figure 5.42.

Also shown in Figure 5.42 is the performance of our turbo codec in the context of an Orthogonal Frequency Division Multiplexing (OFDM) system communicating over Rayleigh fading channels. The performance of turbo coded OFDM will be explored in more depth in various parts of the book, but suffice to say here that OFDM achieves a good turbo coded performance, which is close to that recorded, when communicating over the perfectly interleaved Rayleigh channel. Again, interleaving over the 2000 output bits of the turbo encoder substantially improves the achievable coded performance.

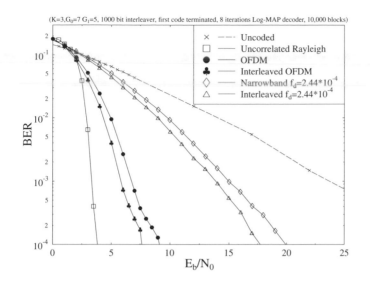

Figure 5.42: Performance of turbo coding over Rayleigh fading channels. Turbo codec parameters as in Table 5.5.

5.6 Summary and Conclusions

In this chapter we have characterised the performance of turbo coding schemes in conjunction with BPSK modulation, when communicating over both AWGN and Rayleigh channels. As expected, the turbo codes have been shown to perform significantly better than convolutional codes. We have demonstrated the effects of the various decoding algorithms, the constraint length and generator polynomials of the constituent codes, as well as the influence of the transmission frame length on the achievable performance. Furthermore, based on our detailed investigations we have demonstrated the importance of the choice of the interleaver in the context of turbo codes. More explicitly, we have reached the following conclusions regarding the choice of interleavers:

- When block interleavers are used in conjunction with half-rate codes, an odd number of rows and columns should be used.

- For long frame length systems random interleavers perform better than block interleavers, but for shorter frame length systems, such as those that might be used for speech transmission, block interleavers perform better.

Finally, in Section 5.5 we provided performance results obtained when using turbo codes in conjunction with BPSK and QPSK modulation for transmissions over Rayleigh fading channels.

Having explored the structure and decoding of convolutional constituent code-based turbo codes, in Chapter 6 we elaborate further on the relationship between conventional convolutional codes and turbo convolutional codes and highlight the relationship between their trellis structures.

The Super-trellis Structure of Convolutional Constituent Code-based Turbo Codes[1]

M. Breiling, L. Hanzo[2]

6.1 Introduction

Turbo convolutional codes were discussed in considerable depth in the previous chapter, and hence in this chapter familiarity with the basic concepts of turbo coding is assumed. In this chapter our aim is to illuminate the trellis structure of turbo codes and to provide further justification of their high error correction capability. This chapter will also allow us to appreciate the importance of employing the most appropriate interleaver design. The detailed discussions and proofs of this chapter are included for the benefit of researchers, but the associated concepts will become plausible also for readers who are new to turbo coding. Let us now commence with a brief introduction, in order to allow the reader to follow the subject without having to consult the previous introductory chapter.

As stated before, turbo codes were proposed by Berrou, *et al.* in 1993 [12]. At the time of writing, turbo codes constitute one of the most power-efficient family of binary channel codes, especially when aiming for medium Bit Error Ratios (BERs) [126]. They can also be viewed, as a class of soft-input decodable block codes of relatively large codeword lengths (typically longer than 1000 bits), where the codeword length is determined by the length of the turbo interleaver, as we will demonstrate. Turbo codes can be designed for various coding rates and their rate can be adjusted with the aid of puncturing. As seen in the previous chapter, the encoder of a turbo code consists of two so-called component encoders and a turbo interleaver. The first component

---

[1]This chapter presents an in-depth characterisation of the trellis structure of turbo codes, which is not required for following the forthcoming chapters. Nonetheless, advanced readers may find the contents of this chapter intriguing in terms of higlighting the relationship betwen the trellis structure of conventional convolutional codes and that of turbo codes. This may assist the readers in the design of efficient turbo interleavers.

[2]This chapter is based on M. Breiling, L. Hanzo: The Super-Trellis Structure of Turbo Codes, IEEE Tr. on Inf. Theory, Volume 46, Number 6, September 2000, pp 2212-2228, ©IEEE.

encoder encodes the information symbols directly, whereas the second component encoder encodes a permuted version of the same information symbols. The decoding of long turbo codes is typically accomplished using an iterative decoding algorithm, which is related to Gallager's classical probabilistic decoding algorithm [138]. In contrast to the latter algorithm, in a turbo decoder, there are two component decoders operating in unison and passing soft information to each other in a feedback loop. As argued in the previous chapter, the component decoders must be capable of accepting soft inputs (SIs) and generating soft outputs (SOs), hence they are often referred to as SISO decoders. State-of-the-art component decoders for carrying out this task are above all the Maximum A-Posteriori (MAP) algorithm [11] and the Soft-Output Viterbi Algorithm (SOVA) [61]. A vital part of a turbo encoder is its constituent interleaver. Reference [59] shows that if the average performance is evaluated for the range of all possible interleavers, the BER performance of the iterative decoder — although it is sub-optimal — converges to the performance of a Maximum Likelihood Sequence Estimation (MLSE) algorithm for medium and high Signal-to-Noise Ratios (SNRs). Further contributions in the previous literature focus on the weight distribution profile of turbo codes [60, 139]. Since the power efficiency of turbo codes deteriorates for low BERs, various methods have been proposed for improving the performance of turbo codes by designing the constituent interleaver appropriately [140, 141]. Conventional turbo codes rely on convolutional component codes, which is what we will restrict ourselves to in this chapter. However, there have been block component codes also proposed in the literature.

The outline of this chapter is as follows. *We commence our discussions by introducing a simple example, which assists us in highlighting the underlying concepts of non-iterative turbo decoding. During our later discourse we will generalise these introductory concepts and illuminate the super-trellis structure of turbo codes.* Specifically, Section 6.2.1 revisits the basic convolutional decoding philosophy introduced in Chapter 2 from a slightly different perspective. The approach employed can also be used for decoding turbo codes. By contrast, Section 6.3 concentrates on a simple non-iterative decoding technique developed from the conventional Viterbi decoder of Chapter 2 typically used for the decoding of convolutional codes, which can be applied for the decoding of turbo codes using simple interleavers having a low number of columns. Our discussions deepen in Section 6.3, introducing the model of the turbo encoder that is used in this chapter. Section 6.4 then provides a derivation of the super-trellis structure of turbo codes. Section 6.5 discusses the complexity of the super-trellises, while Section 6.6 gives a comparison between the performance of an optimum decoder for turbo codes and the conventional iterative decoder. Finally, Section 6.7 discusses the results obtained, which is followed by our conclusions in Section 6.8.

6.2   Non-iterative Turbo Decoding Example

6.2.1   Decoding Convolutional Codes

In this example we rely on the knowledge of the Viterbi algorithm of Chapter 2 and assume familiarity with the concept of iterative turbo decoding. In order to introduce our formalism, Figure 6.1 shows an example of a path through the trellis for a codeword c_i, where the quantities $\bar{c}_{i,j}$ along the path represent the symbol sequence within the codeword c_i that is associated with the trellis state transition j and the corresponding encoder input bit u_j. For finding the most likely transmitted codeword, we define the following path metrics (PMs):

$$\mathbf{M}_{\mathbf{c}_i, j \le k} \quad : = \quad \sum_{j=1}^{k} \| \bar{c}_{i,j} - \bar{r}_j \|^2 \hat{=} \text{forward PM} \tag{6.1}$$

$$\mathbf{M}_{\mathbf{c}_i, l < j} \quad : = \quad \sum_{j=l+1}^{N} \| \bar{c}_{i,j} - \bar{r}_j \|^2 \hat{=} \text{backward PM}. \tag{6.2}$$

These so-called PMs are constituted by the sum of consecutive branch metrics. Each branch-metric quantifies the similarity or dissimilarity between the received sequence $\bar{r}_j$ and the code-word $\bar{c}_{i,j}$ at instant j. When considering the trellis stage j, the two associated trellis paths depicted in Figure 6.1(a) will be referred to as the forward path and the backward path, respectively. The parameter N in Equation 6.2 is the length of the input dataword, in other words the total number of transitions in the trellis, while the $\bar{r}_j$ is the symbol sequence that has actually been received at stage j.

The terminology 'forward' and 'backward' paths were chosen, because their metrics can easily be calculated by a forward/backward recursion as follows:

$$\mathbf{M}_{\mathbf{c}_i, j \le k} \quad = \quad \mathbf{M}_{\mathbf{c}_i, j \le k-1} + \| \bar{c}_{i,k} - \bar{r}_k \|^2 \tag{6.3}$$

$$\mathbf{M}_{\mathbf{c}_i, l-1 < j} \quad = \quad \mathbf{M}_{\mathbf{c}_i, l < j} + \| \bar{c}_{i,l} - \bar{r}_l \|^2. \tag{6.4}$$

As we can see in Figure 6.1(a), any codeword $\mathbf{c}_i$ can now be broken up into a forward path ending at trellis state transition k and a backward path from this transition until the end of the trellis. Its total metric evaluated by the decoder consists therefore of two terms:

$$\mathbf{M}_{\mathbf{c}_i} = \mathbf{M}_{\mathbf{c}_i, j \le k} + \mathbf{M}_{\mathbf{c}_i, k < j}. \tag{6.5}$$

We make the following observation. If two codewords $\mathbf{c}_a$ and $\mathbf{c}_b$ differ only in terms of their forward paths with respect to the trellis stage k, while their backward paths are identical, then the codeword associated with the higher forward PM can be discarded, because its total metric $\mathbf{M}_{\mathbf{c}_i}$ is greater than that of the other one and it can thus never be the minimum metric path. Two partial paths are identical if they both commence and terminate in the same encoder state and are associated with the same data input bits along their way through the code trellis.

Moving on to the decoding process, this means that for any trellis stage k (i.e. the kth transition in the trellis), we have to look at each of the 2^{K-1} possible states in the trellis (where K is the constraint length of the encoder) and keep only the specific forward path with the minimum metric merging into this state. All other forward paths merging into the same state can be discarded. Then we can go on to the next decoding stage $k + 1$, extend the surviving forward paths of each of the 2^{K-1} states at stage k by one trellis transition, compute their metrics by a simple forward recursion, select for each state the forward path with the lowest metric, etc.

We shall now introduce a new type of path, which we refer to as the intermediate path. This is a trellis segment associated with an intermediate section of a codeword as can be seen in Figure 6.1(b) and its metric is accordingly given by:

$$\mathbf{M}_{\mathbf{c}_i, k < j \le l} \quad : = \quad \sum_{j=k+1}^{l} \| \bar{c}_{i,j} - \bar{r}_j \|^2. \tag{6.6}$$

With this notation, a codeword metric can be split up into three parts:

$$\mathbf{M}_{\mathbf{c}_i} = \mathbf{M}_{\mathbf{c}_i, j \le k} + \mathbf{M}_{\mathbf{c}_i, k < j \le l} + \mathbf{M}_{\mathbf{c}_i, l < j}. \tag{6.7}$$

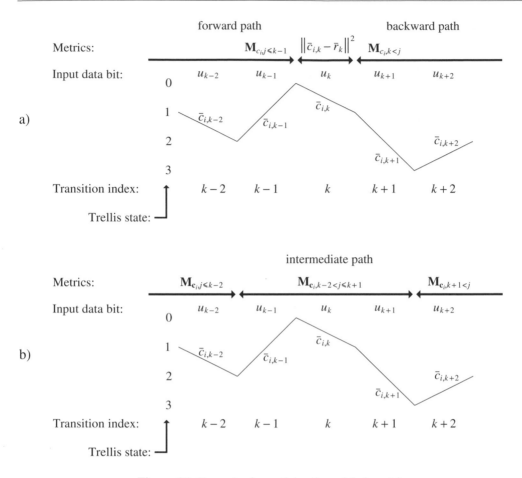

Figure 6.1: Examples for partial paths and their metrics.

We could therefore set up a dynamic programming approach as follows. If two codewords c_a and c_b differ only in an intermediate path, we may argue that the codeword with the higher intermediate PM $M_{c_i, k < j \leq l}$ can be discarded, before we go on to the next decoding stage. The reasoning for this is that their forward metric with respect to the kth symbol sequence and their backward metric with respect to the lth symbol sequence are identical, because their forward and backward paths are identical on these trellis segments. As discussed above for forward paths, the codeword with the higher intermediate metric has the higher total metric and can therefore be discarded. The surviving intermediate path could then be extended in either direction to explore the trellis in order to find the most likely path.

This shows that the Viterbi decoding process could also start in the middle of the trellis. But since 'identical apart from an intermediate path' now means that the forward paths have to merge into the same state at stage k and the backward paths have to start in the same state at stage l, we have to take into account every possible combination of states the intermediate path commences and terminates in. The number of states to take into account by this kind of algorithm is therefore squared in comparison to extending either forward or backward paths only, as in conventional Viterbi decoding.

Encoder

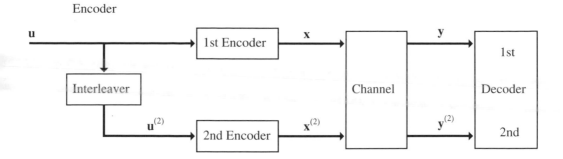

Figure 6.2: The turbo encoder/decoder structure.

6.2.2   Decoding Turbo Codes

Having explained the dynamic programming method for decoding convolutional codes, we are able to proceed to the more complex task of decoding turbo codes. We are going to highlight why conventional turbo Decoders use an iterative method and how we can define an optimum non-iterative decoder.

An important difference between conventional convolutional codes and turbo codes is that the decoding process of the latter is not sequential. The effect of changing a symbol in one part of the turbo-coded codeword will affect possible paths not only in this part, but also in distant parts of the codeword. In order to visualise this, the simplified encoder/decoder structure is displayed in Figure 6.2, where we use the following notation:

- $\mathbf{u} = (u_k)_{k=1..N}$ represents the original non-interleaved data bit sequence, which is used as input for the first encoder

- $\mathbf{u}^{(2)} = (u_k^{(2)})_{k=1..N}$ is the interleaved bit sequence, which is used as input for the second encoder

- $\mathbf{x} = \text{enc}(\mathbf{u})$ is the output sequence of the first encoder and

- $\mathbf{x}^{(2)} = \text{enc}(\mathbf{u}^{(2)})$ is the ouput of the second decoder, where $\text{enc}()$ denotes the encoding function

- $\mathbf{y}$ is the part of the received sequence belonging to $\mathbf{u}$, i.e. to the input sequence of the first decoder

- $\mathbf{y}^{(2)}$ is the part of the received sequence belonging to $\mathbf{u}^{(2)}$.

Figure 6.3 shows an example for the positions of the first seven bits of the input data sequence in $\mathbf{u}$ and $\mathbf{u}^{(2)}$, for the sake of illustration assuming a simple two-column interleaver algorithm, as becomes explicit from Figure 6.3.

Each codeword c_i is made up of its two parts x_i and $x_i^{(2)}$, and therefore we can refer to $\mathbf{x}$ and $\mathbf{x}^{(2)}$ as the partial codewords.

Figure 6.4 shows the stylised trellises corresponding to the interleaver of Figure 6.3 that are used to produce $\mathbf{x}$ and $\mathbf{x}^{(2)}$. Explicitly, we have to consider not one trellis but two, and hence we

$\mathbf{u}$ u_1 u_2 u_3 u_4 u_5 u_6 u_7

$\mathbf{u}^{(2)}$ $u_1^{(2)}$ $u_2^{(2)}$ $u_3^{(2)}$ $u_4^{(2)}$ $u_{n_1+1}^{(2)}$ $u_{n_1+2}^{(2)}$ $u_{n_1+3}^{(2)}$

               $= u_1$ $= u_3$ $= u_5$ $= u_7$ $= u_2$ $= u_4$ $= u_6$

Data bit no.:   1   2   3   4 $n_1 + 1$ $n_1 + 3$
                                                                  $n_1 + 2$

Figure 6.3: The original and the interleaved bit sequence.

have to introduce the following partial PMs:

$$\mathbf{M}_{\mathbf{x}_i, j \le k} \quad := \quad \sum_{j=1}^{k} \|\bar{x}_{i,j} - \bar{y}_j\|^2 \tag{6.8}$$

$$\mathbf{M}_{\mathbf{x}_i, l < j} \quad := \quad \sum_{j=l+1}^{N} \|\bar{x}_{i,j} - \bar{y}_j\|^2 \tag{6.9}$$

$$\mathbf{M}_{\mathbf{x}_i^{(2)}, m \le n} \quad := \quad \sum_{m=1}^{n} \left\|\bar{x}_{i,m}^{(2)} - \bar{y}_m^{(2)}\right\|^2 \tag{6.10}$$

$$\mathbf{M}_{\mathbf{x}_i^{(2)}, n < m \le o} \quad := \quad \sum_{m=n+1}^{o} \left\|\bar{x}_{i,m}^{(2)} - \bar{y}_m^{(2)}\right\|^2 \tag{6.11}$$

$$\mathbf{M}_{\mathbf{x}_i^{(2)}, o < m} \quad := \quad \sum_{m=o+1}^{N} \left\|\bar{x}_{\mathbf{x}_i,m}^{(2)} - \bar{y}_m^{(2)}\right\|^2, \tag{6.12}$$

where $\bar{x}_{i,j}$ is the symbol sequence belonging to the jth trellis stage in the partial codeword $\mathbf{x}_i$, and $\bar{y}_j$ is the segment of the received sequence belonging to the jth trellis stage. Equations 6.8 and 6.9 define the forward/backward metric for the partial codeword $\mathbf{x}_i$, while Equations 6.10 to 6.12 define the same metrics for $\mathbf{x}_i^{(2)}$. Their definition is analogous to Equations 6.1, 6.2 and 6.6 and they are visualised by the corresponding partial paths in Figure 6.1.

Since we are using the Euclidean distance for the metric $\mathbf{M}_{\mathbf{c}_i}$ of the complete turbo codeword $\mathbf{c}_i$, it is easy to show that the turbo-decoded metric is given by the sum of the constituent metrics:

$$\mathbf{M}_{\mathbf{c}_i} = \mathbf{M}_{\mathbf{x}_i} + \mathbf{M}_{\mathbf{x}_i^{(2)}}, \tag{6.13}$$

where $\mathbf{M}_{\mathbf{x}_i}$ and $\mathbf{M}_{\mathbf{x}_i^{(2)}}$ each can be broken up into three parts according to Equations 6.8 to 6.12 and Equation 6.7.

If we now attempt to use a dynamic programming approach for decoding $\mathbf{y}$ of Figure 6.2 and ignore $\mathbf{y}^{(2)}$, we start discarding forward paths in the upper trellis of Figure 6.4 while retaining the survivors. As shown in Section 6.2.1 in the context of decoding conventional convolutional codes, this way we are able to find the partial codeword $\mathbf{x}_i$ with the minimum metric $\mathbf{M}_{\mathbf{x}_i}$.

It would be very convenient if we were able to consider the decoding of the lower trellis having found the optimum sequence in the upper trellis. However, this is not possible, since

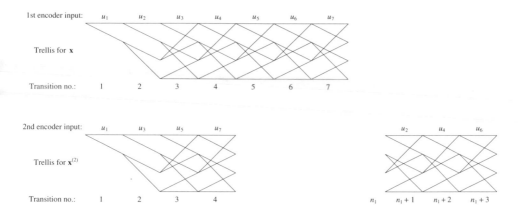

Figure 6.4: An example for the two encoder trellises.

having decided that $\mathbf{x}_i$ is the most probable partial codeword in the upper trellis, also the complete codeword $\mathbf{c}_i$ and the other partial codeword $\mathbf{x}_i^{(2)}$ are determined as there is a unique relationship between these three quantities, and hence there is only one possible path left in the lower trellis.

The optimal path in the upper trellis does not have to be associated with the most likely path exhibiting the lowest metric $\mathbf{M}_{\mathbf{x}^{(2)}}$ in the lower trellis. By minimising $\mathbf{M}_{\mathbf{x}_i}$, we do not necessarily minimise $\mathbf{M}_{\mathbf{c}_i}$, as other codewords $\mathbf{c}_j$ might have slightly greater metrics $\mathbf{M}_{\mathbf{x}_j}$, but much smaller metrics $\mathbf{M}_{\mathbf{x}_j^{(2)}}$, resulting in a smaller overall metric $\mathbf{M}_{\mathbf{c}_j}$. Owing to the random nature of the channel outputs $\mathbf{y}$ and $\mathbf{y}^{(2)}$, it would be easy to find such an example.

Following the above arguments, we conclude that decoding a parallel concatenated convolutional code cannot be achieved by serially decoding its constituent trellises with a standard dynamic programming approach.

Berrou *et al.* [12] proposed a solution to this problem by refraining from employing dynamic programming. Explicitly, instead of discarding potentially possible paths while identifying the most likely path, state-of-the-art techniques attempt to calculate the likelihood of each bit of the original dataword $\mathbf{u}$ of being 0 or 1 according to the first code trellis and the received sequence $\mathbf{y}$, and then pass this information on to the second decoder. The latter one uses this additional soft-decision information to recalculate the likelihood of the data sequence bits, but now according to the received sequence $\mathbf{y}^{(2)}$, and passes the new soft-decoded information back to the first decoder. Several of these iterations can be performed before the soft-decoded information is used to produce a hard-decision decoder output. This approach attempts to find iteratively the optimum dataword with the highest probability. The convergence speed varies and the computational power required to approach the optimum is fairly high. The performance of these decoders is close to the Shannonian limit.

This section presents a different approach. Instead of serially decoding each of the two trellises in turn, we decode both of them at the same time in a parallel fashion. As an introduction to this novel technique, let us consider the following example, assuming that we use a simple two-column block interleaver. Figure 6.3 shows the action of this interleaver with regards to $\mathbf{u}$ and $\mathbf{u}^{(2)}$ for the first seven bits. These seven bits are now encoded with the trellises as depicted in Figure 6.4. Let us now consider the operation of the decoder. In the first decoding stage, we consider the trellis paths of both trellises that are associated with the dataword bit u_1. In the upper trellis, there are only two possible path branches, because the upper trellis commences in the all-zero state. The left-hand-side section of the lower trellis starts also in the all-zero state,

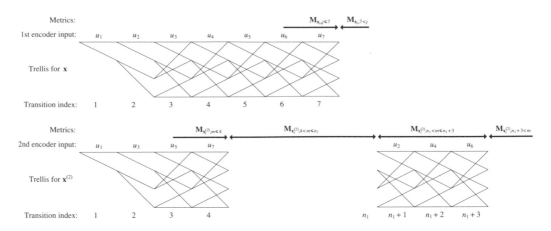

Figure 6.5: Calculation of the three-component PMs for the non-iterative turbo decoder.

and hence there are only two possible paths in this section as well.

We proceed to bit u_2. In the upper trellis of Figure 6.4, there are four possible paths now. In the lower trellis, bit u_2 is the input bit to the second encoder belonging to the $(n_1 + 1)$st trellis stage, since $u^{(2)}_{n_1+1} = u_2$. However, we do not know, as yet, which state the second encoder is in after the first n_1 transitions, hence the state at the start of the right section in the lower trellis is unkown. We must thus consider two paths emerging from all four possible states, resulting in eight possible paths associated with $u^{(2)}_{n_1+1}$.

Next we consider bit u_3. The number of possible paths in the upper trellis of Figure 6.4 increases to eight. In the lower trellis, bit $u^{(2)}_2 = u_3$, i.e. u_3 follows u_1. There are thus four possible paths in the bottom left section of Figure 6.4 now.

When we sequentially join the bits u_4, u_5 and u_6, the number of paths in the appropriate sections is doubled each time, corresponding to the logical 0 and 1 values of the bits. In the next decoding stage, i.e. after inputting bit u_7, we want to start discarding possible path combinations, which can be excluded from being a part of the optimal codeword. The reason for starting the decoding process only here will become obvious during the following explanation.

The bit combination $(u_1, .., u_7)$ can be considered as being the first seven bits of a dataword that generates a codeword c_i. There are of course many datawords starting with this bit combination and accordingly also many corresponding codewords.

Before we proceed, let us introduce a metric for the parts of the codeword (i.e. for the trellis stages) that are directly associated with the bits $u_1..u_7$. This three-component metric $\mathbf{M}_{c_i, j \leq 7}$ is the sum of the corresponding forward metric $\mathbf{M}_{x_i, j \leq 7}$ in the upper trellis in Figure 6.4, which is also depicted in Figure 6.5 as quantified by Equation 6.8, the forward metric $\mathbf{M}_{x^{(2)}_i, m \leq 4}$ for the left section in the lower trellis (see Equation 6.10) and the intermediate metric $\mathbf{M}_{x^{(2)}_i, n_1 \leq m \leq n_1+3}$ for the right section in the lower trellis (see Equation 6.11):

$$\mathbf{M}_{c_i, j \leq 7} := \mathbf{M}_{x_i, j \leq 7} + \mathbf{M}_{x^{(2)}_i, m \leq 4} + \mathbf{M}_{x^{(2)}_i, n_1 \leq m \leq n_1+3}. \tag{6.14}$$

We can now formulate the following algorithm:

Algorithm

If several of the trellis paths associated with the input bit combinations $(u_1, .., u_7)$ in the three considered sections of the two trellises exhibit the following properties:

1) their associated paths in the upper trellis terminate in the same state S_7 after the seventh transition AND

2) their associated paths in the left section of the lower trellis terminate in the same state $S_4^{(2)}$ after the fourth transition AND

3) their associated paths in the right section of the lower trellis commence in the same state $S_{n_1}^{(2)}$ after the n_1-st transition AND

4) their associated paths in the right section of the lower trellis terminate in the same state $S_{n_1+3}^{(2)}$ after the $(n_1 + 3)$rd transition,

then only the specific path with the lowest metric $M_{c_i,j \leq 7}$ must be kept as a survivor and all others can be discarded in the decoding process. Note that there are four potential path combinations associated with each of the $2^7 = 128$ possible bit combinations, since we do not know the decoder's state in the lower trellis after the n_1-st transition.

The reasoning follows exactly the rationale of Section 6.2.1. Explicitly, for any complete turbo codeword c_i, its complete metric can be split up as follows:

$$
\begin{aligned}
M_{c_i} &= M_{x_i} && + M_{x_i^{(2)}} \\
&= M_{x_i, j \leq 7} + M_{x_i, 7 < j} && + M_{x_i^{(2)}, m \leq 4} \\
& && + M_{x_i^{(2)}, 4 < m \leq n_1} \\
& && + M_{x_i^{(2)}, n_1 < m \leq n_1 + 3} \\
& && + M_{x_i^{(2)}, n_1 + 3 < m},
\end{aligned} \tag{6.15}
$$

where the various metric components become explicit in Figure 6.5. The second, fourth and sixth terms have not been encountered so far; they represent as yet unexplored sections of the trellises, namely the right-hand-side section of the upper trellis, the missing intermediate path and the missing right-hand-side section of the lower trellis in Figure 6.5 respectively, which have all been left blank. Upon rearranging Equation 6.15, we arrive at the following equation:

$$
\begin{aligned}
M_{c_i} &= M_{x_i, j \leq 7} && + M_{x_i^{(2)}, m \leq 4} + M_{x_i^{(2)}, n_1 < m \leq n_1 + 3} \\
&+ (M_{x_i, 7 < j} && + M_{x_i^{(2)}, 4 < m \leq n_1} + M_{x_i^{(2)}, n_1 + 3 < m}) \\
&= M_{c_i, j \leq 7} \\
&+ (M_{c_i, 7 < j}),
\end{aligned} \tag{6.16}
$$

where $M_{c_i, 7 < j}$ is the sum of the metrics of the still unexplored sections of the two trellises and hence cannot be evaluated as yet. Explicitly, $M_{c_i, 7 < j}$ is constituted by the backward PMs of both the upper and lower trellises as well as the metric of the missing central section in Figure 6.5. Suppose we have two codewords c_a and c_b, of which the associated paths

a) are different in the three considered sections of the two trellises, which constitute $M_{c_i, j \leq 7}$, but exhibit the four criteria 1)..4) listed above, and

b) are identical in all the three unexplored sections of the two trellises, which form part of $M_{c_i, 7 < j}$.

The assumption a) implies that $M_{c_a, j \leq 7} \neq M_{c_b, j \leq 7}$, whereas b) requires that $M_{c_a, 7 < j} = M_{c_b, 7 < j}$. We have thus $\min \{M_{c_a}; M_{c_b}\} = \min \{M_{c_a, j \leq 7}; M_{c_b, j \leq 7}\}$, such that the optimal codeword can never be the one with the higher metric, and this can therefore be discarded. We can repeat this procedure of selecting one of two possible codewords for any pair of codewords exhibiting the properties a) and b). Since the course of the paths does not depend on the bits $u_1..u_7$ outside the three considered sections constituting $\mathbf{M}_{\mathbf{c}_i, j \leq 7}$, we discard from the set of all the codewords sharing properties 1)..4) all those for which the metric $\mathbf{M}_{\mathbf{c}_i, j \leq 7}$ is not minimal.

When applying the above algorithm in order to identify the most likely path after the first seven bits, we have to evaluate the metrics of 512 possible paths within the considered sections, since there are $2^7 = 128$ different bit combinations and four possible starting states in the right-hand-side section of the lower trellis. We then have to identify 256 different survivors that differ in at least one of the properties 1)..4), since there are four legitimate states for each property, resulting in $4^4 = 256$ possible survivors. In other words, we can discard the less likely one of two paths sharing the same four properties, reducing the number of possible paths from 512 to 256.

In the following decoding stages, by concatenating a new bit we double the number of possible paths to 512, but since the same four properties still apply, the number of survivors remains 256. Clearly, the above algorithm constitutes a dynamic programming approach that restricts the number of paths to take into account to 256 at every decoding step.

The trellis states in the four open ends of the two trellises can be amalgamated into a *super-state* S_k^*. Our four properties 1)..4) are therefore uniquely associated with a single super-trellis state $S_7^* = s^*$, and our algorithm has to find the survivor for any possible super-state s^* at every decoding stage k. It can be readily shown that this dynamic programming approach always finds the optimum turbo codeword (in the sense of maximum likelihood).

6.2.2.1 Non-iterative Turbo Decoding Performance

Even though the non-iterative turbo decoder's performance is expected to be poor owing to the limited number of columns used by the turbo interleaver, it is interesting to evaluate the achievable performance. The employment of this defficient turbo interleaver is also expected to result in a poor performance for the iterative turbo decoder bench-marker to be used. We note, however, that when employing a more efficient 'square-shaped' odd–even turbo interleaver, a substantially better performance can be attained for the iterative turbo decoder. By contrast, the complexity of the non-iterative turbo decoder in conjunction with a 'square-shaped' interleaver would become excessive.

With the motivation of characterising the achievable performance we have evaluated the achievable BER of the proposed algorithm in comparison to the best iterative turbo decoding algorithm known, namely the MAP technique. We carried out simulations using a half-rate, memory-length two RSC code and a 3 columns×333 rows block interleaver over a Gaussian channel, the results of which are shown in Figure 6.6. The gap between the iterative MAP turbo decoder using 16 iterations and the non-iterative 'Flat' decoder is generally about 0.5 dB. In our example using a 4-state convolutional code and a block interleaver of width 2, we have shown that S_k^* can take on 256 different values, i.e. our super-trellis possesses 256 super-states, and 2.256 super-paths have to be treated in each decoding step. Our approach can be adapted for any interleaver and any convolutional component code, but it is clear that this complexity becomes prohibitive for more complex turbo codes, unless attractive sub-optimum detection techniques, such as the M-algorithm or T-algorithm, are invoked.

6.3   System Model and Terminology

Having introduced the concept of non-iterative turbo decoding, we will now generalise these

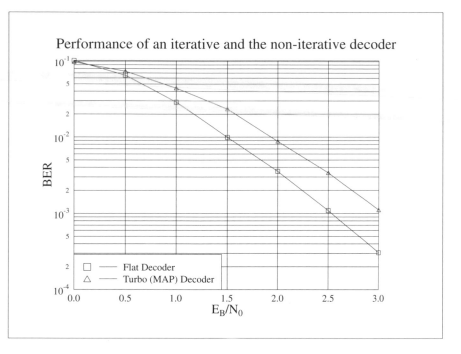

Figure 6.6: Performance comparison between a conventional MAP decoder us- ing 16 iterations and the proposed optimum turbo decoding scheme (memory length $M = 2$, 3×333 block interleaver, each point represents at least 1000 errors).

introductory concepts and illuminate the super-trellis structure of turbo codes in somewhat more depth. These generic concepts will be outlined using more generic notations in comparison to those used in the example of the previous section.

Figure 6.7 shows the model of a conventional turbo encoder. The encoder consists of three parallel branches connected to the encoder input, which generate the three constituent parts of the codeword $\mathbf{c} = (\mathbf{u}; \mathbf{c}^{(1)}; \mathbf{c}^{(2)})$. The vector of K binary encoder input symbols used for generating a codeword is referred to as the *information word* $\mathbf{u} = (u_1; \dots ; u_K)$. This information word $\mathbf{u}$ directly forms the systematic part of the codeword. The other two branches, referred to as the first and the second *component*, respectively, contain the scramblers $\mathrm{scr}^{(1)}$ and $\mathrm{scr}^{(2)}$, which are also often referred to as component encoders, processing the information symbols in order to generate the component codewords $\mathbf{c}^{(1)}$ and $\mathbf{c}^{(2)}$ constituting the parity part of the codeword $\mathbf{c}$.

Throughout the rest of this chapter we use the following notations, which are more generic notations than those used in the example of the previous section. Specifically, capitals denote random variables, whereas lower-case letters represent their specific manifestations. Variables with a superscript, such as $\bullet^{(1)}$ and $\bullet^{(2)}$, are associated with the first and second component, re- spectively. The first component scrambler $\mathrm{scr}^{(1)}$ is fed with the information symbols $\mathbf{u}^{(1)} = \mathbf{u}$ obeying their original ordering. The second component scrambler $\mathrm{scr}^{(2)}$ is fed with a *permuted* version $\mathbf{u}^{(2)}$ of the information word. The third branch therefore contains an interleaver, which is going to play a major role in our forthcoming elaborations. The vector containing the element- wise transpositions performed by the interleaver is denoted by $\boldsymbol{\pi} = (\pi_1; \dots ; \pi_K)$. Interleaving of $\mathbf{u}^{(1)} = \mathbf{u}$ is achieved by permuting the vector elements generating $\mathbf{u}^{(2)}$, such that $u_{\pi_i}^{(2)} = u_i^{(1)}$

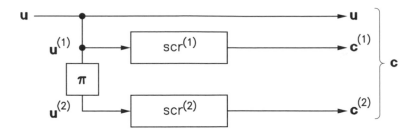

Figure 6.7: System model of the turbo encoder.

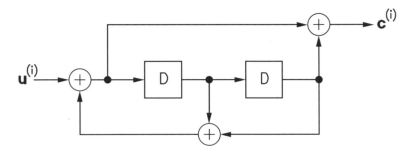

Figure 6.8: Example of a component scrambler or component encoder with feedback and feedforward polynomials expressed in octal form as 7 and 5, respectively

for $i = 1 \ldots K$. For the sake of simplicity, we assume identical scramblers in both components. As usual, the scrambler is a binary shift register with a feedback branch as exemplified in Figure 6.8.

This structure is also often referred to as a recursive systematic convolutional (RSC) code. Any such scrambler represents a Finite State Machine (FSM) $\mathbb{F}_s = (\mathcal{U}_s; \mathcal{C}_s; \mathcal{S}_s; \gamma_s)$, whose input value set is binary $\mathcal{U}_s \in \{0; 1\}$ and the corresponding output value set is also binary $\mathcal{C}_s \in \{0; 1\}$. We note here explicitly that the subscript s refers to the state machine of the component scrambler. For a scrambler of memory M, the state set $\mathcal{S}_s = \{0; \ldots; 2^M - 1\}$ consists of $\|\mathcal{S}_s\| = 2^M$ legitimate states, which correspond to the states of the scrambler shift register, i.e. to vectors of M binary elements. The mapping $\gamma_s : (\mathcal{S}_s \times \mathcal{U}_s) \to (\mathcal{S}_s \times \mathcal{C}_s)$ represents the state transitions, where for every pair (s, u) with $s \in \mathcal{S}_s$ (predecessor state) and $u \in \mathcal{U}_s$ (scrambler input) a pair (s', c) with $s' \in \mathcal{S}_s$ (successor state) and $c \in \mathcal{C}_s$ (scrambler output) is specified. A FSM can be graphically characterised by its state transition diagram or, if we also incorporate the elapse of time, by its trellis. Figure 6.9 displays the trellis of the scrambler of Figure 6.8. It consists of identical trellis segments, since the scrambler FSM is time-invariant. The labels along the trellis transitions, e.g. $1 \to 0$, denote the scrambler input and output symbols respectively, which are associated with the specific state transition.

Let us now consider a FSM $\mathbb{F}(t) = (\mathcal{U}(t); \mathcal{C}(t); \mathcal{S}(t); \gamma(t))$, which can be potentially time-variant, as indicated by its dependence on (t). Such a FSM can be described by a Markov model, where the state transition at a given discrete time instant $t + 1$, which is associated with the terminating state of S_{t+1} and output symbol of C_{t+1}, depends only on the starting state S_t and

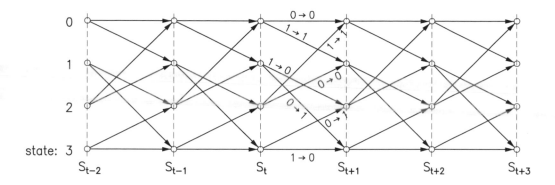

Figure 6.9: Trellis of the scrambler of Figure 6.8 with memory M=2.

on the input symbol U_{t+1}, provided that the FSM is time-invariant. More explicitly, the state transition of a FSM described by a Markov model does not depend on any of the previous states $S_{t'}$ or input symbols $U_{t'+1}$ associated with $t' < t$ at a given discrete time instant $t + 1$. If, however, the FSM is time-variant, the state transition at a given discrete time instant $t + 1$ may also depend on the time t, which explicitly manifests itself in a trellis structure that is different at different time instants t in the sequence of consecutive trellis stages. The issue of time-variant trellises will be augmented in more tangible terms at a later stage of our discourse. Observe that a transition and its associated input and output symbols have the same time index as the *terminating* state of the transition.

In our forthcoming elaborations, we will make use of the following three terms. The *pre-history* of a FSM with respect to the time instant t represents the state transitions and their associated FSM input that the FSM traversed through before and including the time instant t, i.e. the trellis section left of S_t in Figure 6.9. By contrast the *post-history* represents the transitions after and excluding time instant t, i.e. the trellis section right of S_t. The current state S_t represents therefore the *interface* between the pre-history of the FSM and its post-history in the following sense.

Let us assume first of all that a sequence of FSM states $\mathbf{S}_{t-} = (\dots ; S_{t-2}; S_{t-1}; S_t = s)$, $s \in \mathcal{S}_s$, corresponds to a valid sequence of state transitions constituting the pre-history with respect to time t, where the state $S_t = s$ was explicitly included for representing the *interface* state between the *pre-history* and *post-history*. Similarly, let us assume that the sequence $\mathbf{S}_{t+} = (S_t = s; S_{t+1}; S_{t+2}; \dots)$ corresponds to a valid sequence of state transitions constituting the post-history, where, again, the state $S_t = s$ was explicitly included for representing the *interface* state between the *pre-history* and *post-history*. Then we can combine the sequences, and $\mathbf{S} = (\dots ; S_{t-2}; S_{t-1}; S_t = s; S_{t+1}; S_{t+2}; \dots)$ represents a valid sequence of state transitions for the FSM. The only necessary and sufficient condition for combining the pre- and post-history is that the value s of the interface state S_t between the left and right trellis sections must be the same. If $\mathbf{S}_{t+} = (S_t = \tilde{s}; S_{t+1}; S_{t+2}; \dots)$ with $s \neq \tilde{s}$, then combining the pre- and post-history results in an invalid sequence, i.e. in an invalid history for the FSM. In other words, such a sequence of state transitions is illegitimate, since both $S_t = \tilde{s}$ and $S_t \neq \tilde{s}$ should hold, implying a contradiction. Note that the terms 'sequence of states' and 'sequence of state transitions' are used synonymously, since the transitions can directly be derived from the sequence of states.

Position	1	2	3	4	5	...	$K/2+1$	$K/2+2$	...	K
$\mathbf{U}^{(1)}$	U_1	U_2	U_3	U_4	U_5	...	$U_{K/2+1}$	$U_{K/2+2}$	...	K
$\mathbf{U}^{(2)}$	U_1	U_3	U_5	U_7	U_9	...	U_2	U_4	...	K

Table 6.1: Bit mapping scheme used in Example 1.

6.4 Introducing the Turbo Code Super-trellis

In this section we will develop a model of a FSM $\mathbb{F}_T(t) = (\mathcal{U}_T; \mathcal{C}_T; \mathcal{S}_T(t); \gamma_T(t))$ for a **turbo** encoder, where the subscript T refers to the state machine of the turbo encoder. The parameter t indicates that the turbo encoder FSM is time-variant, which manifests itself in a trellis structure that is different at different time instants t, exhibiting different legitimate transitions at different instants t, as we will show at a later stage. It is possible to model the turbo encoder's constituent elements seen in Figure 6.8 by their respective FSMs and combining these. For the two component scramblers of Figure 6.7 it is relatively straightforward to find the corresponding FSMs, which will be alluded to during our further discourse. For the systematic branch of the encoder in Figure 6.7 the corresponding FSM consists of a single state, where the output symbol equals the input symbol. The only device in the turbo encoder that poses difficulties is the interleaver, which introduces memory, since the complete vector $\mathbf{u}^{(1)}$ of input symbols must have been inserted into the interleaver before the vector $\mathbf{u}^{(2)}$ of output symbols can be read out. It is therefore cumbersome to model the interleaver by a FSM, which would be complex. Hence we will choose a different way of finding the FSM of the complete turbo encoder.

The trellis that is associated with the turbo encoder FSM will be referred to as the turbo code's *super-trellis* in order to distinguish it clearly from the component scrambler trellises. Several proposals have already been made to find a tree representation of a turbo code [142] and its super-trellis (e.g. [59], where it is referred to as the hyper-trellis). In contrast to [59], the structure presented in this chapter differs in that the complexity of the super-trellis is not only dependent on the interleaver length K but also on the interleaver structure. We will show that for simple interleavers the complexity of the super-trellis can be drastically reduced. In these cases, an optimum turbo decoder based on this super-trellis can be implemented and its performance can be directly compared to that of the sub-optimum iterative turbo decoder. Let us continue our discourse with an example.

6.4.1 Turbo Encoder Super-states

Example 1: Let us now illustrate the above concepts with the aid of an example, which provides a more detailed discussion in comparison to our elaborations provided in the context of Section 6.2. Explicitly, let us consider a turbo code incorporating the scrambler of Figure 6.8 and, again, a simple two-column interleaver of total interleaving length of K bits obeying the mapping seen in Table 6.1.

Let us now introduce the concept of super-trellis in the context of turbo codes. We assume that the turbo encoder's input symbols are given by $U_1 \ldots U_K$, where the ordering of the symbols is important and for the turbo encoder's set we have $\mathcal{U}_T \in \{0; 1\}$. The time-domain transition index of the super-trellis is denoted by t, which can be different from the corresponding transition index of the individual trellises associated with the component codes, as will be shown during our further discourse. Let us consider the positions of the first five input symbols, namely those of $U_1 \ldots U_5$ in both component trellises, which are illustrated in Figure 6.10. Entering U_5 in

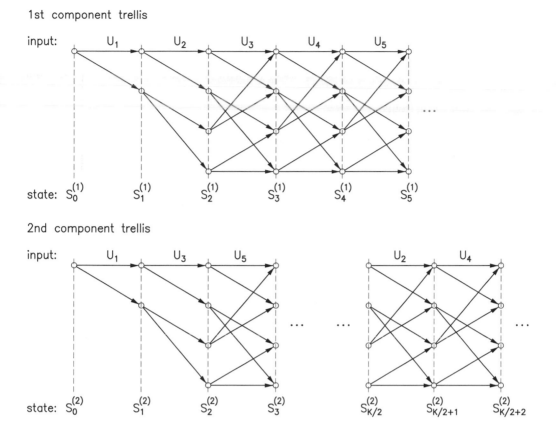

1st component trellis

2nd component trellis

Figure 6.10: Trellis transitions in the two component scrambler trellises up to time instant $t = 5$.

the turbo encoder's FSM corresponds to $t = 5$. Specifically, following the actions of the upper scrambler is straightforward, but that of the lower scrambler is somewhat more daunting to trace owing to the presence of the interleaver.

In the upper component trellis in Figure 6.10 the symbols $U_1^{(1)} \ldots U_5^{(1)}$ are in consecutive positions. The transitions $\mathbf{S}_{5-}^{(1)} = (S_0^{(1)} = 0; S_1^{(1)}; \ldots; S_5^{(1)})$ corresponding to these input symbols constitute therefore the pre-history of the first component trellis with respect to $t = 5$. (We note that the component trellises emanate at $t = 0$ from the zero state $S_0^{(1)} = 0$.) The post-history of the first component trellis *with respect to* (in the following abbreviated to *wrt*) $t = 5$ will be constructed from the input symbols $U_6^{(1)} \ldots U_K^{(1)}$ and from the corresponding state transitions $\mathbf{S}_{5+}^{(1)} = (S_5^{(1)}; S_6^{(1)}; \ldots)$. The interface between the pre- and post-history of the first component trellis *wrt* the time instant $t = 5$ is represented by $S_5^{(1)}$.

Owing to the permutation performed by the interleaver in Figure 6.7 in the second component trellis of Figure 6.10, the same five bits $U_1 \ldots U_5$ are transformed into the input symbols $U_1^{(2)}, U_2^{(2)}, U_3^{(2)}$ and $U_{K/2+1}^{(2)}, U_{K/2+2}^{(2)}$. Hence, the corresponding transitions are located in the second component trellis seen at the bottom of Figure 6.10 in two different sections. Consequently, the pre-history *wrt* $t = 5$, i.e. *wrt* bits $U_1 \ldots U_5$ which belong to the first five transitions of the turbo code super-trellis associated with the discrete time instants of $t = 0 \ldots 5$, corresponds to the transitions of the second component trellis in these two distinct trellis sections of Figure 6.10. The post-history of the second component trellis wrt $t = 5$ is constituted by the two remaining sections of the trellis, which are indicated by dots in Figure 6.10. The interface between the pre- and post-history, which is associated with $t = 5$, is formed in the second component trellis by the set $(S_3^{(2)}; S_{K/2}^{(2)}; S_{K/2+2}^{(2)})$, which we will refer to as the 'interface' states in the second component trellis.

These issues are further augmented below as follows. Both constituent encoders, namely $\mathrm{scr}^{(1)}$ and $\mathrm{scr}^{(2)}$, have the same set of input bit values, although these bits enter the constituent encoders in different order. For example, the 2. input bit of the upper scrambler is U_2, while that of the lower scrambler is U_3 for the two-column interleaver of Figure 6.7, as seen in Table 6.1. Owing to the two-column interleaver bit U_2 enters the lower scrambler of Figure 6.7 at instant $K/2 + 1$, again, as seen in Table 6.1. Hence for a number of further bits to be encoded we have at the inputs of the upper and lower scramblers the following string:

$$(U_1; U_2; U_3; U_4; U_5) = (U_1^{(2)}; U_{K/2+1}^{(2)}; U_2^{(2)}; U_{K/2+2}^{(2)}; U_3^{(2)})$$

where the corresponding state-transitions in the left and right sections of the second component trellis seen at the bottom of Figure 6.10 are represented by:

$$\mathbf{S}_{5-}^{(2)} = (\underbrace{S_0^{(2)} = 0; S_1^{(2)}; S_2^{(2)}; S_3^{(2)} = s_3^{(2)}}_{\text{left lower section in Figure 6.10}}; \underbrace{S_{K/2}^{(2)} = s_{K/2}^{(2)}; S_{K/2+1}^{(2)}; S_{K/2+2}^{(2)} = s_{K/2+2}^{(2)}}_{\text{right lower section in Figure 6.10}}),$$

which constitute the pre-history related to $t = 5$. Specifically, the first four state transitions correspond to the left-hand-side trellis section, and the remaining three to the right-hand-side trellis section of Figure 6.10. Following similar arguments, every valid post-history formulated as:

$$\mathbf{S}_{5+}^{(2)} = (\underbrace{S_3^{(2)} = s_3^{(2)}; S_4^{(2)}; \ldots; S_{K/2}^{(2)} = s_{K/2}^{(2)}}_{\text{left dotted section in Figure 6.10}}; \underbrace{S_{K/2+2}^{(2)} = s_{K/2+2}^{(2)}; S_{K/2+3}^{(2)}; \ldots}_{\text{right dotted section in Figure 6.10}})$$

can therefore be combined with $\mathbf{S}_{5-}^{(2)}$ and forms a valid sequence of state transitions for the second component trellis, if and only if the interface states $(S_3^{(2)}; S_{K/2}^{(2)}; S_{K/2+2}^{(2)})$ have identical values $(s_3^{(2)}; s_{K/2}^{(2)}; s_{K/2+2}^{(2)})$ in both $\mathbf{S}_{5-}^{(2)}$ and $\mathbf{S}_{5+}^{(2)}$.

So far we have elaborated on the details of the pre-history related to $t = 5$ in the first and second component trellis, as well as on their interfaces to the post-history. Hence, in the turbo encoder only these interface states are important for the encoding of the forthcoming input symbols $U_6; U_7; \ldots$. The complete memory of the turbo encoder at time instant $t = 5$, which is required for further encoding operations, can be bundled in the 4-tuple $(S_5^{(1)}; S_3^{(2)}; S_{K/2}^{(2)}; S_{K/2+2}^{(2)})$, which comprises all interfaces of both component trellises. It can be seen that every valid pre-history pair of state transitions $\mathbf{S}_{5-}^* = (\mathbf{S}_{5-}^{(1)}; \mathbf{S}_{5-}^{(2)})$ can be combined with every valid post-history pair $\mathbf{S}_{5+}^* = (\mathbf{S}_{5+}^{(1)}; \mathbf{S}_{5+}^{(2)})$, in order to form a valid pair of state transition sequences for the two component trellises, if and only if the interface states $(S_5^{(1)}; S_3^{(2)}; S_{K/2}^{(2)}; S_{K/2+2}^{(2)})$ in $\mathbf{S}_{5-}^*$ are identical to those in $\mathbf{S}_{5+}^*$, namely $(s_5^{(1)}; s_3^{(2)}; s_{K/2}^{(2)}; s_{K/2+2}^{(2)})$.

Hence, as a super-state of the example turbo encoder's FSM at time instant $t = 5$ we define the above 4-tuple, which is constituted by the states of the component scramblers of the FSMs as follows[3]:

$$S_5^* \overset{\text{def}}{=} (S_5^{(1)}; S_3^{(2)}; S_{K/2}^{(2)}; S_{K/2+2}^{(2)}).$$

---

[3]It is worth mentioning that at $t=1$ and $t = K - 1$ we have only two states.

6.4.2   Turbo Encoder Super-trellis

Let us now investigate the migration from $t = 5$ to $t = 6$, or equivalently, let us search for the super-state S_6^* under the assumption that the super-state S_5^* and the most recent turbo encoder input symbol U_6 are known. For the input symbols of the first component scrambler we have $U_6 = U_6^{(1)}$, such that the transition associated with U_6 in the first component trellis is directly adjacent to the transitions in the pre-history $\mathbf{S}_{5-}^{(1)}$ (upper trellis in Figure 6.10). Since the value $s_5^{(1)}$ of $S_5^{(1)}$ is known from S_5^*, the value u_6 of U_6 dictates the transition $(s_5^{(1)}; s_6^{(1)})$. The pre-history sequence of the state transitions of the first component code simply extends from $\mathbf{S}_{5-}^{(1)}$ to $\mathbf{S}_{6-}^{(1)} = (S_0^{(1)} = 0; \ldots; S_5^{(1)} = s_5^{(1)}; S_6^{(1)} = s_6^{(1)})$.

With the aid of Figure 6.10 we can see that for the input symbols of the second component scrambler the relation $U_6 = U_{K/2+3}^{(2)}$ holds. In Figure 6.10 the corresponding transition (index $K/2 + 3$ of the second component trellis) therefore emanates from $S_{K/2+2}^{(2)}$ in the right trellis section, which belongs to the pre-history. Since the value $s_{K/2+2}^{(2)}$ of $S_{K/2+2}^{(2)}$ is known from S_5^*, u_6 dictates the transition $(s_{K/2+2}^{(2)}; s_{K/2+3}^{(2)})$. Hence the pre-history of state transitions for the second component code becomes:

$$\mathbf{S}_{6-}^{(2)} = \underbrace{(S_0^{(2)} = 0; \ldots; S_3^{(2)} = s_3^2;}_{\text{first lower section in Figure 6.10}} \underbrace{S_{K/2}^{(2)}; \ldots; S_{K/2+2}^{(2)} = s_{K/2+2}^{(2)}; S_{K/2+3}^{(2)} = s_{K/2+3}^{(2)}).}_{\text{second lower section in Figure 6.10}}$$

From this we can easily infer the interfaces of the pair $\mathbf{S}_{6-}^* = (\mathbf{S}_{6-}^{(1)}; \mathbf{S}_{6-}^{(2)})$, which is formed by the two pre-histories regarding time instant $t = 6$. Hence the super-state of the turbo encoder at this time instant can be identified by exploiting the knowledge of $S_5^* = (s_5^{(1)}; s_3^{(2)}; s_{K/2}^{(2)}; s_{K/2+2}^{(2)})$ yielding:

$$S_6^* \stackrel{\text{def}}{=} (S_6^{(1)}; S_3^{(2)}; S_{K/2}^{(2)}; S_{K/2+3}^{(2)}) = (s_6^{(1)}; s_3^{(2)}; s_{K/2}^{(2)}; s_{K/2+3}^{(2)}),$$

where both the state transitions $(s_5^{(1)}; s_6^{(1)})$ and $(s_{K/2+2}^{(2)}; s_{K/2+3}^{(2)})$ are associated with the value u_6 of the input symbol U_6.

6.4.3   Generalised Definition of the Turbo Encoder Super-states

As an extension of our previous introductory elaborations in this section we will present a generalised definition of the super-state transitions in the super-trellis, which also defines implicitly the super-states associated with the super-trellis. For the sake of illustration the definition will be followed by the example of a simple super-trellis in Section 6.4.4.

First we have to introduce a set of indices $\mathcal{I}_{\text{pre}}^{(2)}(t)$, which describes the number of transitions belonging to the pre-history of the second component trellis wrt time instant t. If and only if an index obeys $i \in \mathcal{I}_{\text{pre}}^{(2)}(t)$, the transition with this index is part of the pre-history of the second component trellis at time t. It follows immediately that a state $S_j^{(2)}$ with index j is an interface state in the second component trellis at time t, if and only if one of the two adjacent transitions belongs to the pre-history and the other one does not, i.e. if and only if one of the following two cases is true: $\{ j \in \mathcal{I}_{\text{pre}}^{(2)}(t) \quad \wedge \quad (j+1) \notin \mathcal{I}_{\text{pre}}^{(2)}(t) \qquad \text{or} \qquad j \notin \mathcal{I}_{\text{pre}}^{(2)}(t) \quad \wedge \quad (j+1) \in \mathcal{I}_{\text{pre}}^{(2)}(t). \}$ For an interleaver π, which was shown in Figure 6.7, we can now define the super-states of the turbo encoder FSM as follows.

At time instant $t = 0$ both component trellises emanate from the zero state $S_0^{(1)} = S_0^{(2)} = 0$. Hence the following equation holds for the super-state:

$$S_0^{*} \overset{\text{def}}{=} (S_0^{(1)}; S_0^{(2)}) = (0; 0). \tag{6.17}$$

We initialise the index set by including only the (non-existent) transition index 0: $\mathcal{I}_{\text{pre}}^{(2)}(0) = \{0\}$.

Evolution from time instant t to $t + 1$: the previous super-state S_t^{*} is assumed to be known. For the first component code the relation $U_{t+1} = U_{t+1}^{(1)}$ holds, hence the input symbol $U_{t+1} = u_{t+1}$ specifies the transition $(S_t^{(1)}; S_{t+1}^{(1)}) = (s_t^{(1)}; s_{t+1}^{(1)})$ in the first component code. (In the first component trellis at time instant t the state with index t, $S_t^{(1)}$, is always the interface state contained in the super-state S_t^{*}, and hence the value $s_t^{(1)}$ is known from the vector s_t^{*}.) For the second component code we have $U_{t+1} = U_j^{(2)}$ with $j = \pi_{t+1}$. The index of the associated transition in the second component trellis for $t + 1$ is j, such that the corresponding transition becomes $(S_{j-1}^{(2)}; S_j^{(2)})$. This transition is not part of the pre-history wrt t and hence $j \notin \mathcal{I}_{\text{pre}}^{(2)}(t)$. Updating the set of indices means that $\mathcal{I}_{\text{pre}}^{(2)}(t+1) = \mathcal{I}_{\text{pre}}^{(2)}(t) \cup \{j\}$. In order to proceed from $(j - 1)$ to j, we have to distinguish between four different cases or scenarios, as far as the second component trellis is concerned, which will be detailed below and will also be augmented in the context of an example in Section 6.4.4 and Figure 6.12. In fact, the reader may find it beneficial to follow the philosophy of our forthcoming generic discussions on a case-by-case basis by referring to the specific example of Section 6.4.4 and Figure 6.12.

(1) In order to encounter our first scenario, the following condition must be met: $j - 1 \in \mathcal{I}_{\text{pre}}^{(2)}(t)$ and $j + 1 \notin \mathcal{I}_{\text{pre}}^{(2)}(t)$. The state with index $j - 1$, $S_{j-1}^{(2)}$, is thus one of the interface states in the vector, representing the super-state S_t^{*}, whereas the state with index j, $S_j^{(2)}$, is not an interface state at time instant t. This implies that the transition $(S_{j-1}^{(2)}; S_j^{(2)}) = (s_{j-1}^{(2)}; s_j^{(2)})$, which is due to $U_j^{(2)} = u_{t+1}$, is directly adjacent to the state transitions in the second component trellis already belonging to the pre-history in the second component code wrt the time instant t. From a graphical point of view the transition is located directly to the right of the transitions which have already been encountered in the trellis of Figure 6.10. The new super-state S_{t+1}^{*} of the turbo encoder FSM can be inferred from the old super-state S_t^{*}, by substituting the old interface in the first component trellis, namely $S_t^{(1)} = s_t^{(1)}$ – by the corresponding new one, namely by $S_{t+1}^{(1)} = s_{t+1}^{(1)}$, and also the old interface state $S_{j-1}^{(2)} = s_{j-1}^{(2)}$ by the corresponding new state $S_j^{(2)} = s_j^{(2)}$, where $(s_{j-1}^{(2)}; s_j^{(2)})$ is associated with u_{t+1}. We will refer to this transition from t to $t + 1$ as *right-extension*, since in the second component trellis a section is extended to the right in Figure 6.10. Again, these issues will be exemplified in Section 6.4.4 and Figure 6.12, where this scenario is encountered for transitions $t = 0 \rightarrow 1, t = 2 \rightarrow 3, t = 3 \rightarrow 4, t = 4 \rightarrow 5$ and $t = 7 \rightarrow 8$.

(2) The condition for encountering our second scenario is: $j - 1 \notin \mathcal{I}_{\text{pre}}^{(2)}(t)$ and $j + 1 \in \mathcal{I}_{\text{pre}}^{(2)}(t)$. The state with index j, $S_j^{(2)}$, then constitutes a component of the vector representing the super-state S_t^{*} ($S_j^{(2)}$ is hence an interface state in the second component trellis at time instant t), whereas the state with index $j - 1$ is not an interface state. Accordingly, the transition $(S_{j-1}^{(2)}; S_j^{(2)})$ associated with $U_j^{(2)}$ is directly adjacent to the state transitions which already belong to the pre-history wrt t. In order to obtain the new super-state S_{t+1}^{*} from the vector, which represents the old super-state S_t^{*}, one has to replace the old interface $S_t^{(1)}$ of the first component code by the new $S_{t+1}^{(1)}$ and also the old interface $S_j^{(2)}$ in the second component code by the corresponding new $S_{j-1}^{(2)}$. In analogy to scenario (1) we will refer to this case as *left-extension*.

(3) The third potential scenario encountered by the second trellis is when $j-1 \notin \mathcal{I}_{\mathrm{pre}}^{(2)}(t)$ and $j+1 \notin \mathcal{I}_{\mathrm{pre}}^{(2)}(t)$. This implies that neither the state associated with the index $j-1$, i.e. $S_{j-1}^{(2)}$, nor that with the index j, i.e. $S_j^{(2)}$, is contained in the vector corresponding to the super-state S_t^*. The transition $(S_{j-1}^{(2)}; S_j^{(2)})$, which is due to $U_j^{(2)}$, is not adjacent to any of the transitions which are already contained in the pre-history *wrt* to t. In the pre-history *wrt* $t+1$ this transition hence constitutes a separate section of the second component trellis, which consists of only one transition. This section possesses the two interface states $S_{j-1}^{(2)}$ and $S_j^{(2)}$. In order to obtain the super-state S_{t+1}^*, one has to substitute the interface state $S_j^{(1)}$ by $S_{t+1}^{(1)}$ in the super-state S_t^* and in addition, one has to extend the super-state vector by these two new interface states, namely by $S_{j-1}^{(2)}$ and $S_j^{(2)}$, in the second component trellis. Observe, however, that neither the value $s_{j-1}^{(2)}$ of $S_{j-1}^{(2)}$ nor the value $s_j^{(2)}$ of $S_j^{(2)}$ is specified by $S_t^* = s_t^*$, which represents the interface of the pre-history *wrt* t in the super-trellis.

The pair $(S_{j-1}^{(2)}; S_j^{(2)})$ can therefore assume all legitimate values $(s_{j-1}^{(2)}; s_j^{(2)})$ within the vector S_{t+1}^*, which represent valid state transitions and which correspond to the input symbol $U_j^{(2)} = u_{t+1}$. The consequence is that if the super-state S_t^* is known, several values are possible for the successor super-state S_{t+1}^* with equal probability. Specifically, the new interface state $S_{j-1}^{(2)}$, which is contained in the associated super-state vector, can assume all possible states $S_{j-1}^{(2)} \in \mathcal{S}_s$ and the other new interface state $S_j^{(2)}$ will be $s_j^{(2)}$, determined by the transition $(s_{j-1}^{(2)}; s_j^{(2)})$ due to the input symbol $U_j^{(2)} = u_{t+1}$. We will refer to this scenario as the *opening* of a new section in the second component trellis. These aspects will be revisited in Section 6.4.4 and Figure 6.12, where the transition $t = 1 \rightarrow 2$ corresponds to this specific scenario.

(4) The last possible scenario encountered by the second trellis is when $j-1 \in \mathcal{I}_{\mathrm{pre}}^{(2)}(t)$ and $j+1 \in \mathcal{I}_{\mathrm{pre}}^{(2)}(t)$. The state associated with the index $j-1$, i.e. $S_{j-1}^{(2)}$, as well as the one with the index j, i.e. $S_j^{(2)}$, are contained in the vector corresponding to super-state S_t^*. The specific states values $s_{j-1}^{(2)}$ and $s_j^{(2)}$ representing the interfaces $S_{j-1}^{(2)}$ and $S_j^{(2)}$ have already been determined by means of S_t^* representing the interface of the pre-history *wrt* t in the super-trellis. Although the transition $(S_{j-1}^{(2)}; S_j^{(2)})$, which is the transition at time instant $t+1$ in the second component trellis, is not contained in the pre-history *wrt* t, nonetheless this transition is determined by S_t^* due to its fixed start- and end-states constituted by $s_{j-1}^{(2)}$ and $s_j^{(2)}$, respectively. We will revisit these issues in the context of Section 6.4.4 and Figure 6.12, where the transition $t = 6 \rightarrow 7$ constitutes an example of scenario (4).

If this fixed transition $(s_{j-1}^{(2)}; s_j^{(2)})$ is legitimate and if it is associated with the value u_{t+1} of the most recent input symbol $U_j^{(2)}$, then two interface states are connected by means of this particular transition in the second component trellis. Hence, the gap between two disjoint trellis sections of the pre-history is closed, and therefore we refer to this scenario as the *fusion* of these two sections. The new super-state S_{t+1}^* evolves from the vector S_t^*

by removing the two interfaces $S_{j-1}^{(2)}$ and $S_j^{(2)}$ and by substituting $S_t^{(1)}$ by $S_{t+1}^{(1)}$.

However, it may happen that for the interface states $s_{j-1}^{(2)}$ and $s_j^{(2)}$, which are determined by $S_t^* = s_t^*$, the transition $(s_{j-1}^{(2)}; s_j^{(2)})$ is illegitimate, since this state transition is non-existent for the FSM of the component scramblers. In this case, assuming the value s_t^* for S_t^* constitutes a contradiction due to the illegitimate transition, this transition is therefore deemed invalid. Furthermore, it can occur that although the transition $(s_{j-1}^{(2)}; s_j^{(2)})$ is legitimate, it is not associated with the current specific value u_{t+1} of the input symbol for the component scrambler FSM. Hence for this reason pairing the super-state s_t^* and the input symbol u_{t+1} is impossible. Therefore in the super-trellis *no* transition emerges from the super-state $S_t^* = s_t^*$, which is associated with u_{t+1}. This may appear to be a contra-diction, since this super-state seems to have been reached owing to the sequence of input symbols $U_1 \ldots U_t$ and since in the super-trellis a path is supposed to exist for *all* possible sequences of input symbols $U_1 \ldots U_{t+1}$, regardless of the specific value of the input symbol U_{t+1}. However, the *fusion* of two trellis sections must always be preceded by the *opening* of a new section in the second component trellis, and as we infer from scenario (3) several super-states are reached with equal probability from a sequence of input symbols $U_1 \ldots U_t$. When for example the first *opening* is executed, a single sequence $U_1 \ldots U_t$ of input symbols leads to $\|\mathcal{S}_s\| = 2^M$ possible super-states (see scenario (3) above). Hence it is understandable that in the process of a section *fusion* in the super-trellis, a proportion of $(2^M - 1)/2^M$ of all (super-state/input-symbol) pairs $(s_t^*; u_{t+1})$ represent invalid tran-sitions. (For any interface state value $s_{j-1}^{(2)} \in \mathcal{S}_s$, which is determined by the super-state value s_t^*, and for any input symbol value $u_j^{(2)} \in \mathcal{U}_s$, there exists only *one* of 2^M possible interface state values $s_j^{(2)} \in \mathcal{S}_s$, such that the component scrambler transition $(s_{j-1}^{(2)}; s_j^{(2)})$ is associated with the input symbol $u_j^{(2)}$.) We will further augment this concept by means of an example in Section 6.4.4. Note that the trellis section in the first component trellis with its interface state $S_t^{(1)}$ is in all the above four scenarios extended to the right, i.e. the new interface state is $S_{t+1}^{(1)}$.

Let us now elaborate a little further. The output symbols of the turbo encoder FSM, which are associated with a super-state transition at time instant $t+1$, consist of the systematic code symbol U_{t+1} and the output symbols emitted during the most recent state transitions in both component scrambler trellises. These are the state transitions belonging in both of the component trellises to the input symbol $U_{t+1}^{(1)} = U_{t+1}$ and $U_j^{(2)} = U_{t+1}$, $j = \pi_{t+1}$, respectively, and which therefore generate the output symbols $C_{t+1}^{(1)}$ and $C_j^{(2)}$, respectively. The output of the turbo encoder FSM $\mathbb{F}_T$ at time instant $t + 1$ is therefore the vector:

$$\bar{C}_{t+1} = (U_{t+1}; C_{t+1}^{(1)}; C_{\pi_{t+1}}^{(2)}),$$

and hence the relation $\mathcal{C}_T = [\{0; 1\}]^3$ holds for the output set. This constitutes a turbo code of rate $1/3$. At higher coding rates, the scrambler outputs are punctured, which also has to be taken into account in the output vector $\bar{C}$ of the turbo encoder FSM. Let us now illustrate the above concepts with the aid of another example, which provides a more in-depth exposure in comparison to our elaborations presented in the context of Section 6.2.

6.4.4 Example of a Super-trellis

Example 2: Let us now examine the super-trellis of a simple turbo code, incorporating the

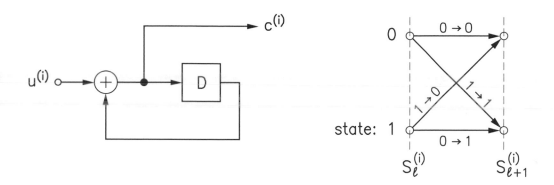

Figure 6.11: Simple component scrambler and its trellis segments.

Position	1	2	3	4	5	6	7	8
$\mathbf{U}^{(1)}$	U_1	U_2	U_3	U_4	U_5	U_6	U_7	U_8
$\mathbf{U}^{(2)}$	U_1	U_3	U_5	U_7	U_2	U_4	U_6	U_8

Table 6.2: Interleaving scheme for the component trellises of Example 2.

memory $M = 1$ scrambler of Figure 6.11 and a 4 row $\times$ 2 column rectangular interleaver resulting in $K = 8$, which is shown in Table 6.2. The output of the turbo encoder is not punctured. The segments of the super-trellis that is developed in the forthcoming paragraphs for time instants $t = 0 \ldots 8$ are displayed in Figure 6.12. The index $j = \pi_{t+1}$ denotes in the second component trellis the transition belonging to time $t + 1$, i.e. to the transition index $t + 1$ in the super-trellis. Let us now consider Figure 6.12, where at

$t = 0$: we initialise the super-state to $S_0^* = (S_0^{(1)}; S_0^{(2)}) = (0; 0)$ and the index set to $\mathcal{I}_{\text{pre}}^{(2)}(0) = 0$.

$0 \rightarrow 1$: We have $j = \pi_{t+1} = \pi_1 = 1$, since U_1 is at position 1 at the bottom of Table 6.2. As $j - 1 = 0 \in \mathcal{I}_{\text{pre}}^{(2)}(0)$ and $j + 1 = 2 \notin \mathcal{I}_{\text{pre}}^{(2)}(0)$, this corresponds to scenario (1) described in Section 6.4.3, i.e. the *right-extension* of an existing section in the second component trellis ($S_{j-1}^{(2)} = S_0^{(2)}$ is one of the interface states contained in S_0^*, while $S_j^{(2)}$ is not). Hence the new super-state is $S_1^* = (S_1^{(1)}; S_1^{(2)})$. The possible values for S_1^*, which depend on the value of U_1, are displayed in Table 6.3.

$1 \rightarrow 2$: We have $j = \pi_{t+1} = \pi_2 = 5$, since U_2 is at position 5 at the bottom of Table 6.2. Now we find that $j - 1 = 4 \notin \mathcal{I}_{\text{pre}}^{(2)}(1)$ and $j + 1 = 6 \notin \mathcal{I}_{\text{pre}}^{(2)}(1)$, thus neither $S_{j-1}^{(2)} = S_4^{(2)}$ nor $S_j^{(2)} = S_5^{(2)}$ is contained in S_1^* of Figure 6.12 at $t = 1$. This is therefore scenario (3), representing the *opening* of a new section in the second component trellis. The new super-state is $S_2^* = (S_2^{(1)}; S_1^{(2)}; S_4^{(2)}; S_5^{(2)})$. For every value of S_1^* and every value of U_2, there are $2^M = 2^1 = 2$ possible values for $S_4^{(2)}$, as seen in Table 6.3. At this stage careful further tracing of the super-trellis evolution with the aid of Figure 6.12 and Table 6.3 is helpful, in order to augment the associated operations.

$2 \rightarrow 3$: This is scenario (1) again, i.e. a *right-extension*. The new super-state is $S_3^* = (S_3^{(1)}; S_2^{(2)}; S_4^{(2)}; S_5^{(2)})$.

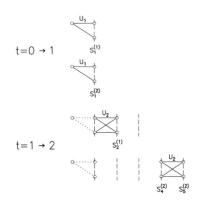

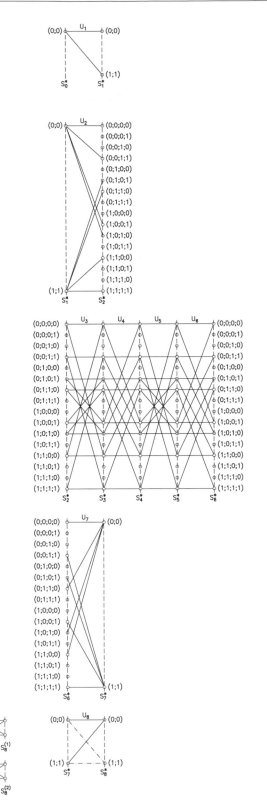

Figure 6.12: Segments of the super-trellis for the turbo encoder FSM of Example 2.

$3 \rightarrow 4$: Scenario (1), *right-extension*, $S_4^* = (S_4^{(1)}; S_2^{(2)}; S_4^{(2)}; S_6^{(2)})$.

$4 \rightarrow 5$: Scenario (1), *right-extension*, $S_5^* = (S_5^{(1)}; S_3^{(2)}; S_4^{(2)}; S_6^{(2)})$.

$5 \rightarrow 6$: Scenario (1), *right-extension*, $S_6^* = (S_6^{(1)}; S_3^{(2)}; S_4^{(2)}; S_7^{(2)})$.

$6 \rightarrow 7$: $j = \pi_{t+1} = \pi_7 = 4$, since U_7 is at position 7 at the bottom of Table 6.2. Since $j - 1 = 3 \in \mathcal{I}_{\text{pre}}^{(2)}(6)$ and $j + 1 = 5 \in \mathcal{I}_{\text{pre}}^{(2)}(6)$, i.e. both $S_3^{(2)}$ and $S_4^{(2)}$ are contained in S_6^*, this is scenario (4), corresponding to the *fusion* of two sections in the second component trellis of Figure 6.12. As inferred from the trellis of Figure 6.11, for $U_7 = 0$ the possible values of the state transition $(S_3^{(2)}; S_4^{(2)})$ are $(0; 0)$ and $(1; 1)$, for $U_7 = 1$, $(0; 1)$ and $(1; 0)$. Depending on the value of U_7, the super-state S_6^* containing other values for the pair $(S_3^{(2)}; S_4^{(2)})$ thus has no successor super-state. If there is a valid transition emerging from the super-state S_6^*, which is associated with u_7, the successor super-state is $S_7^* = (S_7^{(1)}; S_7^{(2)})$. Otherwise the transition is marked as 'invalid' or illegitimate. Observe that a portion of $(2^M - 1)/2^M = 1/2$ of all transitions is invalid (see scenario (4) above).

$7 \rightarrow 8$: Scenario (1), *right-extension* of the remaining section, $S_8^* = (S_8^{(1)}; S_8^{(2)})$.

t	S_t^*	U_{t+1}	S_{t+1}^*	$\bar{C}_{t+1}$
0	$(0; 0)$	0	$(0; 0)$	$(0; 0; 0)$
		1	$(1; 1)$	$(1; 1; 1)$
1	$(0; 0)$	0	$(0; 0; 0; 0)$	$(0; 0; 0)$
			$(0; 0; 1; 1)$	$(0; 0; 1)$
		1	$(1; 0; 0; 1)$	$(1; 1; 1)$
			$(1; 0; 1; 0)$	$(1; 1; 0)$
	$(1; 1)$	0	$(1; 1; 0; 0)$	$(0; 1; 0)$
			$(1; 1; 1; 1)$	$(0; 1; 1)$
		1	$(0; 1; 0; 1)$	$(1; 0; 1)$
			$(0; 1; 1; 0)$	$(1; 0; 0)$
2	$(0; 0; 0; 0)$	0	$(0; 0; 0; 0)$	$(0; 0; 0)$
		1	$(1; 1; 0; 0)$	$(1; 1; 1)$
	$(0; 0; 1; 1)$	0	$(0; 0; 1; 1)$	$(0; 0; 0)$
		1	$(1; 1; 1; 1)$	$(1; 1; 1)$
	$(1; 0; 0; 1)$	0	$(1; 0; 0; 1)$	$(0; 1; 0)$
		1	$(0; 1; 0; 1)$	$(1; 0; 1)$
	$(1; 0; 1; 0)$	0	$(1; 0; 1; 0)$	$(0; 1; 0)$
		1	$(0; 1; 1; 0)$	$(1; 0; 1)$
	$(1; 1; 0; 0)$	0	$(1; 1; 0; 0)$	$(0; 1; 1)$
		1	$(0; 0; 0; 0)$	$(1; 0; 0)$
	$(1; 1; 1; 1)$	0	$(1; 1; 1; 1)$	$(0; 1; 1)$
		1	$(0; 0; 1; 1)$	$(1; 0; 0)$
	$(0; 1; 0; 1)$	0	$(0; 1; 0; 1)$	$(0; 0; 1)$
		1	$(1; 0; 0; 1)$	$(1; 1; 0)$
	$(0; 1; 1; 0)$	0	$(0; 1; 1; 0)$	$(0; 0; 1)$
		1	$(1; 0; 1; 0)$	$(1; 1; 0)$
...				
		Continued on next page		

		Continued...		
t	S_t^*	U_{t+1}	S_{t+1}^*	$\bar{C}_{t+1}$
6	$(0;0;0;0)$	0	$(0;0)$	$(0;0;0)$
		1	invalid	
	$(1;1;0;0)$	0	invalid	
		1	$(0;0)$	$(1;0;0)$
	$(0;0;1;1)$	0	invalid	
		1	$(1;1)$	$(1;1;1)$
	$(1;1;1;1)$	0	$(1;1)$	$(0;1;1)$
		1	invalid	
	$(1;0;0;1)$	0	$(1;1)$	$(0;1;0)$
		1	invalid	
	$(0;1;0;1)$	0	invalid	
		1	$(1;1)$	$(1;1;0)$
	$(1;0;1;0)$	0	invalid	
		1	$(0;0)$	$(1;0;1)$
	$(0;1;1;0)$	0	$(0;0)$	$(0;0;1)$
		1	invalid	
7	$(0;0)$	0	$(0;0)$	$(0;0;0)$
		1	$(1;1)$	$(1;1;1)$
	$(1;1)$	0	$(1;1)$	$(0;1;1)$
		1	$(0;0)$	$(1;0;0)$

Table 6.3: State transitions and their respective output vectors for the turbo encoder super-trellis associated with the component codes of Figure 6.11.

We see from Table 6.3 that S_8^* can only assume the values $(0;0)$ and $(1;1)$, since the last bit is the same in both the original and the interleaved sequences. In other words, in our simple example at time instant $t = 8$ the memory of the interleaver in Table 6.2 has been exhausted. Hence both component scramblers have actually completed calculating the parity in the information word $\mathbf{u}$ and must hence be in the same state. For a random interleaver this may, however, not be the case.

If one or both component trellises are terminated by zeros, we impose the restriction of $S_K^{(1)} = 0$ and possibly $S_K^{(2)} = 0$. Only the super-states S_K^* satisfying these restrictions are valid, all others have to be discarded from the super-trellis. In the above example, if the component trellises were terminated, the only remaining legitimate value for S_8^* would be $(0;0)$. The invalid transitions in the last super-trellis segment of Figure 6.12 ($t = 7 \rightarrow 8$) are marked with dash–dotted lines.

From Table 6.3 and Figure 6.12, we can clearly see that the turbo code super-trellis is time-variant, i.e. the structure of the trellis sections depends on the time instant t, exhibiting different legitimate transitions for different t values. More explicitly, the trellis is different for $t = 2 \rightarrow 3$ and $t = 3 \rightarrow 4$, while it is identical for $t = 2 \rightarrow 3$ and for $t = 4 \rightarrow 5$. For the time instants $t = 2 \ldots 6$ we observe a periodicity in the super-trellis, manifesting itself in two different trellis sections, which are repeated alternately. This periodicity corresponds to the number of columns in the interleaver, which in our case was two. It is also easy to see that the constraint length of the super-trellis is three, while that of the component scramblers is two. Hence the memory introduced by the interleaver has increased the constraint length of the code. These issues will be augmented in more depth in Section 6.7. Following the above simple example it may now be worthwhile revisiting the generic super-trellis structure of Section 6.4.3 before proceeding further.

6.5 Complexity of the Turbo Code Super-trellis

With the goal of estimating the associated complexity of the turbo code super-trellis, we will assume that the turbo encoder considered comprises two indentical scramblers having a memory of M and an interleaver of length K.

6.5.1 Rectangular Interleavers

We will consider simple $\rho \times \chi$ rectangular interleavers, having ρ rows and χ columns. The data are written into the interleaver on a row-by-row basis and read out on a column-by-column basis. Upon using the previous definition of the time instant t (transition at time instant t in the super-trellis is due to the input symbol U_t of the turbo encoder FSM), the first component trellis is only extended to the right (scenario (1) in Section 6.4.3) upon increasing t. In the second component trellis for every $t = 2 \ldots \chi$ the *opening* (scenario (3) in Section 6.4.3) of a new section will occur, respectively. As a result of this, in the second component trellis a separate section exists for each of the χ columns of the interleaver. The section, which belongs to the first interleaver column (the leftmost section in the second component trellis), is associated with *one* interface state (at its right end), whereas the remaining $\chi - 1$ sections possess *two* interface states (left and right interface of the section, respectively). Therefore, together with the *single* interface state of the first component trellis the turbo code super-state is a vector consisting of 2χ interface states of the component trellises. Each of the interface states can assume 2^M legitimate values. Hence the following statement holds for the complexity of the turbo code super-trellis:

Bound for rectangular interleavers

A turbo code, which is associated with a $\rho \times \chi$ rectangular interleaver, can be described by a super-trellis having a maximum of $2^{M \cdot 2\chi}$ states at any super-trellis stage.

Hence for our example turbo code incorporating the 4×2 rectangular interleaver and a scrambler of memory $M = 1$, the super-trellis can have a maximum of $2^{(1 \cdot 2 \cdot 2)} = 16$ legitimate states. However, as seen in Figure 6.12, from the set of 16 possible states only eight are actually encountered. On the other hand, for a 2×4 rectangular interleaver, this upper bound is 256 super-states, although it can be readily shown that this turbo code has an equivalent super-trellis representation occupying only 8 of the 256 possible super-states (cf. Example 2 and simply swap the first and the second components).

In order to eliminate this ambiguity as regards the trellis complexity (one of the complexity upper bounds is associated with 16 states and the other one with 256 states), we can redefine the time t in the super-trellis. In contrast to the previous situation, we have to distinguish clearly between the input symbols $U_{\text{FSM},1} \ldots U_{\text{FSM},K}$ of an *abstract* or hypothetical turbo encoder FSM and the input symbols $U_1 \ldots U_K$ of the *real* turbo encoder. Let the input symbol $U_{\text{FSM},t}$ of the turbo encoder FSM at the time instant t be the input symbol U_l of the turbo encoder, so that the maximum number of super-states is minimised over all time instants t. This definition of t triggers a permutation of the input symbols U_l to $U_{\text{FSM},t}$. From this definition it follows for a $\rho \times \chi$-dimensional rectangular interleaver that the order of the input symbols $U_{\text{FSM},t}$ of the turbo encoder FSM.

- corresponds to the order of the input symbols of the turbo encoder for $\rho > \chi$, i.e. $U_{\text{FSM},t} = U_t$ (there is one section in the first component trellis and χ sections in the second component trellis),

- obeys the order of symbols at the output of the rectangular interleaver in the turbo encoder for $\rho < \chi$, i.e. $U_{\text{FSM},t} = U_{\phi_t}$, where $l = \phi_t$ is the inverse mapping of $t = \pi_l$ (there are ρ sections in the first component trellis and one section in the second component trellis, while the definition of the super-states is analogous to that in Section 6.4.3, where only the first and second component trellises have to be swapped),

- the trellis complexity for $\rho = \chi$ minimises for both of the above-mentioned orderings of the symbols.

To summarise of the above elaborations, when $\rho > \chi$, we can exchange the two decoders, as in Example 2 and the maximum number of super-trellis states can be formulated as $\leq 2^{2 \cdot M \cdot \min(\rho, \chi)}$.

6.5.2   Uniform Interleaver

Instead of considering a specific random interleaver, we will now derive an upper bound for the super-trellis complexity, averaged over all possible interleavers, a scenario which is referred to as the so-called uniform interleaver [59] of length K.

Without loss of generality, we define the time t such that the transition in the super-trellis at time instant t is associated with the input symbol U_t. This implies that the order of the input symbols of the abstract or hypothetical turbo encoder FSM is the same as for the real turbo encoder, i.e. $U_{\text{FSM},t} = U_t$. Hence, only one section exists in the first component trellis, which is extended to the right upon increasing t.

The specific input symbol of the second component scrambler, which belongs to the input symbol U_t and hence is input to the second scrambler simultaneously with U_t entering the first one, is $U_t^{(2)}$. Therefore the corresponding transition in the second component trellis has the index π_t. Let us consider a state $S_l^{(2)}$ with arbitrary index $l = 1 \ldots (K-1)$ in the second component trellis at time instant t. This state $S_l^{(2)}$ forms the interface at the right of a section in the second component trellis, which belongs to the pre-history wrt t, if and only if both of the following two conditions are satisfied:

(1) A section is situated directly to the left of $S_l^{(2)}$. This implies that a transition, which already belongs to the pre-history wrt t, is directly adjacent to the left of this state in the second component trellis. The input symbol $U_l^{(2)}$ which belongs to this transition must therefore have been already an input symbol of the turbo encoder FSM, hence there exists a value $m \in \{1 \ldots t\}$ for which $l = \pi_m$ must hold. For an arbitrary l and when averaged over all possible interleavers π, this condition is met with a probability of $p_1(t) = t/K$.

(2) no section exists directly to the right of $S_l^{(2)}$, or correspondingly that the transition with index $l+1$, which is directly adjacent at the right, does not belong to the pre-history of the super-trellis wrt t. Hence, the relation $l+1 \neq \pi_m$, $\forall m \in \{1 \ldots t\}$, must hold. Under the assumption that condition (1) is fulfilled, this condition will hold with a probability of $p_2(t) = (K-t)/(K-1)$.

For every state $S_l^{(2)}$, $l = 1 \ldots K-1$, the probability that it represents the right interface to a section in the second component trellis at time instant t is therefore $p_{12}(t) = p_1(t) \cdot p_2(t) = [t \cdot (K-t)]/[K \cdot (K-1)]$. Condition (2) is cancelled for state $S_K^{(2)}$, since the end of the trellis is located directly to the right of it. If the second component trellis is terminated, then we have $S_K^{(2)} = 0$, and hence $S_K^{(2)}$ does not appear as an interface state.

An arbitrary state therefore constitutes the right-hand-side interface of a section in the second component trellis with a probability of $p_3(t) = (K-1)/K \cdot p_{12}(t) + 1/K \cdot p_1(t) = t \cdot (K + 1 - t)/K^2$ at time instant t. Hence there are on average $n_r(t) = K \cdot p_3(t) = t \cdot (K + 1 - t)/K$ right-hand-side (RHS) interfaces in the second component trellis. This number is maximised for $t_m = (K+1)/2$, for which there are $n_r(t_m) = (K+1)^2/(4K)$ RHS interfaces on average in the second component trellis. Similarly, one can deduce that there are $n_l(t) = n_r(t)$ left-hand-side (LHS) interfaces on average in the second component trellis, when the edge effects at the start and end of the trellis are ignored.

Along with the single interface state $S_t^{(1)}$ in the first component trellis we obtain for the maximum total number of interfaces in both trellises under the assumption of a uniform interleaver, as defined above:

$$n_{\max} = 1 + \frac{(K+1)^2}{2K} \tag{6.18}$$

and therefore the following bound accrues:

Bound for the uniform interleaver
The following upper bound holds for the number of super-states in the minimal super-trellis of a turbo code, averaged over all possible interleavers of length K:

$$\mathrm{E}\left[\|\mathcal{S}_T(t)\|\right] \leq 2^{M \cdot n_{\max}} = 2^{M + M \cdot (K+1)^2/(2K)} \qquad \text{for } t = 1 \ldots K.$$

From Equation 6.18 for large values of K, i.e. for large interleavers, we have $n_{\max} \approx K/2$, for which a maximum of $2^{M \cdot K/2}$ super-states is expected in the super-trellis. For $M \leq 2$ this means that the trellis complexity of a turbo code with a random interleaver is on average only marginally lower than that of a random code [1] having 2^K codewords, which can be described by means of a trellis having a maximum of 2^K states.

6.6   Optimum Decoding of Turbo Codes

Having illuminated the super-trellis structure of turbo codes, we can now invoke this super-trellis, in order to optimally decode simple turbo codes. Let us commence by defining the task a decoder should carry out. Specifically, the goal of a decoder may be that of finding the codeword $\mathbf{c}_i$ with the highest probability of having been transmitted, upon reception of the sequence $\mathbf{y}$, which is formulated as identifying the index i:

$$i = \underset{l}{\operatorname{argmax}}\left(P(\mathbf{c}_l|\mathbf{y})\right), \tag{6.19}$$

where $\operatorname{argmax}()$ returns the index of the maximum and $P(\mathbf{c}_l|\mathbf{y})$ is the conditional probability that $\mathbf{c}_l$ is the transmitted codeword when $\mathbf{y}$ is received. These decoders are usually referred to as Maximum A-Posteriori Sequence Estimation (MAPSE) schemes, since they estimate a complete codeword or the associated information word. By contrast, the MAP algorithm [11] attempts to find the most probable values of all symbols contained in the information word or codeword.

It is straightforward to show that for memoryless AWGN channels, the decision rule of Eq. (6.19) for MAPSE decoding can be simplified to the following Maximum Likelihood Sequence Estimation (MLSE) type of decoding using Bayes' rule, if all codewords $\mathbf{c}_l$ appear with the same probability:

$$i = \underset{l}{\operatorname{argmin}}\left(\|\mathbf{c}_l - \mathbf{y}\|^2\right).$$

For MLSE decoding of a codeword transmitted over a memoryless AWGN channel, our decoding goal is equivalent to finding the valid codeword $\mathbf{c}_l$ that is closest to the received sequence $\mathbf{y}$ in terms of the Euclidean distance. We can thus introduce a Euclidean metric:

$$\mathbf{M}_{\mathbf{c}_l} = \|\mathbf{c}_l - \mathbf{y}\|^2 \qquad (6.20)$$

for each codeword, and having calculated the whole set of metrics, we opt for the codeword having the lowest metric, yielding:

$$i = \underset{l}{\operatorname{argmin}} \left(\mathbf{M}_{\mathbf{c}_l} \right). \qquad (6.21)$$

When the code used can be described by a trellis, i.e. when the code symbols are generated by a FSM, the task of finding the minimum metric can be accomplished by using a dynamic programming method, such as the Viterbi algorithm. This algorithm reduces the number of metrics to be calculated by introducing metrics associated with paths in the trellis, which can be recursively updated, if the path is extended by one transition. The maximum number of paths/metrics that has to be kept in memory is $2 \cdot \|\mathcal{S}\|$, if $\mathcal{S}$ is the set of states in the trellis and each state transition is associated with a binary input symbol. This can be achieved without ever discarding the optimum path in the sense of Equation (6.21).

6.6.1   Comparison to Iterative Decoding

Since we have found a super-trellis representation for turbo codes, we can now invoke the appropriately modified *non-iterative* Viterbi algorithm for the ML sequence decoding of turbo codes, as was also argued in the context of our introductory example in Section 6.2. In the Appendix of this chapter the optimality of this decoding approach is shown. Similarly, the MAP algorithm [11] proposed by Bahl *et al.* could be used for decoding the turbo code along its super-trellis, obtaining exact a-posteriori probabilities for the information and code symbols. Recall that the MAP algorithm was originally proposed by Bahl *et al.* for the minimum BER decoding of convolutional, rather than turbo, codes. Although the MAP algorithm slightly outperformed the MLSE-based Viterbi algorithm in the context of convolutional decoding, it was not widely used owing to its high complexity until the invention of turbo codes. For the non-iterative decoding of turbo codes here we invoked a MLSE decoder [128, 143, 144] since its complexity is considerably lower than that of the MAP decoder.

In contrast to the aforementioned proposals for MLSE decoding of turbo codes, when the super-trellis is used, its decoding imposes no restrictions on the information word length K of the code, only on the super-trellis complexity and therefore on the structure of the interleaver. Here we opted for rectangular interleavers having a low number of columns and compared the decoding BER results to those of an iterative 'turbo' decoder.

All turbo codes in the following examples contain two identical component scramblers. The first component scrambler was terminated at the end of the information word, whereas the second component trellis was left open at its right end. Puncturing was applied, in order to obtain a turbo code of rate $1/2$. The code symbols were transmitted over an AWGN channel using Binary Phase Shift Keying (BPSK). E_b/N_0 represents the energy per transmitted information bit E_b divided by the one-sided noise power spectral density N_0. The conventional, iterative 'turbo' decoder bench-marker used the MAP algorithm [11] for decoding the component codes.

First we consider a turbo code with component scramblers of memory $M = 1$ and a two-column interleaver, as used above for Example 2. We examine a turbo code with a 499×2 rectangular interleaver, i.e. the information word length is $K = 998$. Figure 6.13 shows the

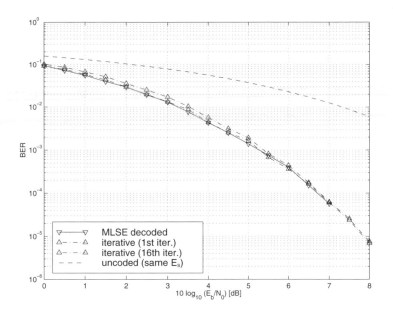

Figure 6.13: Turbo-coded BER performance for $M = 1$, 499×2 rectangular interleaver: comparison between optimum non-iterative and iterative decoding.

BER results of our simulations. We have used a Viterbi decoder for the 'MLSE decoded' super-trellis as well as the 'iterative' turbo decoder and plotted the BER results after the 1st and 16th iterations. We observe that for this example the code performance is quite low (also displayed is the BER for uncoded transmission with the same symbol energy $E_s = \frac{1}{2}E_b$ as for the coded transmission with rate $1/2$). There is virtually no difference between the MLSE decoder and the iterative one. Furthermore, we note that there is no BER improvement for more than one iteration.

All turbo codes used in the following simulations incorporate component scramblers of memory $M = 2$, which was chosen in order maintain a low complexity.

From Figure 6.14 we can see that the BER differences between the non-iterative MLSE and iterative decoding become clearer for three-column interleavers. We portrayed the simulation results for a 33×3 (i.e. $K = 99$) and a 333×3 ($K = 999$) interleaver, where the curves correspond to non-iterative MLSE decoding and iterative decoding. The 1st and 16th iterations are shown. Here, there is a gain in the iterative algorithm, when performing more than one iteration. However, the iterative decoder cannot attain the BER performance of the non-iterative MLSE decoder. The remaining SNR gap is about 0.5 dB at a BER of 10^{-3}. For lower BERs the associated curves seem to converge. The turbo codes have an identical number of 4096 super-states in conjunction with both interleavers, which explains why their coding performance is fairly similar. Since both interleavers have an identical number of columns, which determines the number of super-states, their respective number of rows hardly affects the coding performance, although at low BERs the curve for the 33×3 interleaver diverges from that of the 333×3 interleaver. This is due to the edge effect at the end of the super-trellis (or at the end of the two component trellises), which will not be discussed in more detail here. Figure 6.15 shows the performance of the iterative algorithm for the turbo code with the 333×3 interleaver. Specifically, the BER is shown after 1, 2, 4, 8 and 16 iterations, as is the BER curve for the non-iterative

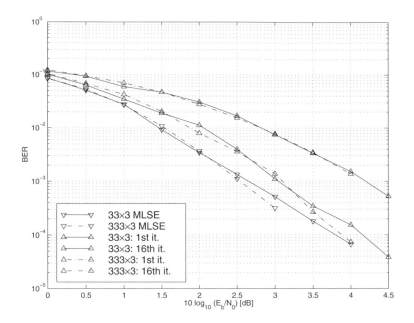

Figure 6.14: Turbo-coded BER performance for $M = 2$, 33×3 and 333×3 rectangular interleavers: comparison between optimum non-iterative and iterative decoding.

MLSE decoder. The performance improvements of the iterative decoder appear to saturate after a few iterations, exhibiting a performance gap with respect to the non-iterative MLSE decoder. In Figure 6.16 we see that a clear gap also exists between the Word Error Rate (WER) curves obtained for the non-iterative MLSE and the iterative decoder after 16 iterations using the 33×3 interleaver.

Figure 6.17 portrays the BER results for a turbo code incorporating a 39×5 rectangular interleaver, yielding $K = 195$. Observe that a gap of about 0.25 dB remains between the performance of the iterative and the non-iterative MLSE decoder. Compared to Figure 6.15, we note that the gap has narrowed upon increasing the number of columns in the interleaver from 3 to 5. Note, however, that only a low number of word errors were registered for each point in the BER curve of the optimum decoder. For example, only 1457 codewords were transmitted at an SNR of $10 \log_{10}(E_b/N_0) = 3$ (dB), of which 21 were in error after decoding. The reason for this low number of transmitted codewords is the complexity of the super-trellis for the turbo code, which possesses 2^{20} super-states. For the iterative decoder the associated complexity is considerably lower. The component scramblers have four states each. The iterative decoder exhibits a complexity per iteration, which is about four times that of MLSE decoding one of the component codes, since there is one forward and one backward recursion for both component codes [11, 12]. The optimum non-iterative decoder associated with the large super-trellis has therefore a complexity which corresponds to about $2^{20}/(4 \cdot 4) = 2^{16}$ iterations in the conventional 'turbo' decoder for the 39×5 interleaver.

An intermediate result is that the process of iterative decoding is sub-optimal. In practice the interleavers which have been investigated so far are not employed in practical turbo codes, since their performance is relatively low in comparison to other interleavers. For short information word lengths, such as $K \leq 200$, one normally employs rectangular interleavers [133], where

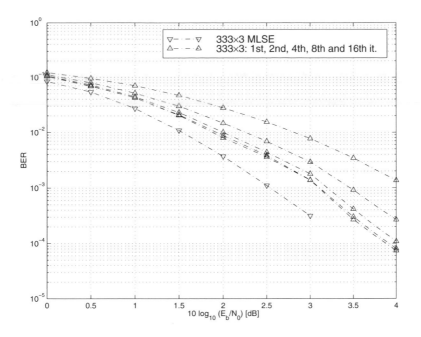

Figure 6.15: Turbo-coded BER performance for $M = 2$, 333×3 rectangular interleavers: convergence behaviour of the iterative algorithm compared to optimum non-iterative decoding.

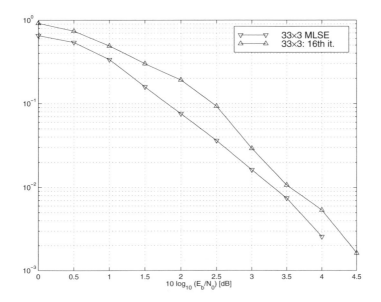

Figure 6.16: Turbo-coded Word Error Rate (WER) performance for $M = 2$, 33×3 rectangular interleaver: comparison between optimum non-iterative and iterative decoding.

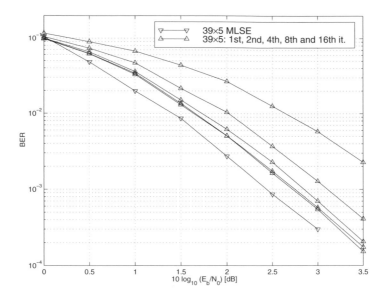

Figure 6.17: Turbo-coded BER performance for $M = 2$, 39×5 rectangular interleaver: comparison between optimum non-iterative and iterative decoding.

the number of rows approximately equals the number of columns. For high information word lengths K one normally employs random interleavers, in order to achieve a good coding performance. The super-trellises, which belong to these schemes are fairly complex, as was detailed in Section 6.5, and hence they cannot be used for non-iterative MLSE decoding in pratical codecs.

6.6.2 Comparison to Conventional Convolutional Codes

In this subsection we do not wish to assess the performance of the iterative decoding algorithm in comparison to the optimum one. Instead, we compare the performance of a turbo code to that of conventional convolutional codes, whose trellis exhibits the same complexity as that of the turbo code super-trellis.

First of all we consider a turbo code with component scramblers having a memory of $M = 1$ and a two-column interleaver, as in Example 2, but in this case with a 499×2 rectangular interleaver. As we have noted above in Example 2, the super-trellis is associated with a maximum of 16 super-states, of which only eight super-states are actually encountered in the super-trellis. Therefore in Figure 6.18 we compared the performance of the turbo code ('TC') having a 499×2 rectangular interleaver and invoking non-iterative MLSE decoding of the super-trellis with that of a convolutional code ('CC') having a memory of $M = 3$ (8 states, octal generator polynomials of 15_o and 17_o) and $M = 4$ (16 states, generator polynomials of 23_o and 35_o), respectively, which are also MLSE (Viterbi) decoded. We observe that both convolutional codes are more powerful than the turbo code of the same decoder complexity, if the turbo code is optimally decoded.

Furthermore, in Figure 6.19 we compared a turbo code having component scramblers of memory $M = 2$ and a 333×3 rectangular interleaver, which has a maximum of 4096 super-states, with a convolutional code of the same trellis complexity. Specifically, the convolutional code's memory was $M = 12$ and the octal generator polynomials were 10533_o and 17661_o, which are optimal according to [145]. Similarly to the previous case, for medium and high

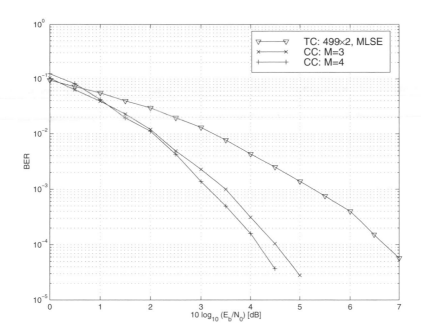

Figure 6.18: Turbo-coded BER performance with a maximum of 16 super-states, of which 8 are reached, compared to convolutional codes with 16 and 8 states.

SNRs the performance of this convolutional code is significantly better than that of the optimally decoded turbo code of the same trellis complexity.

In order to summarise, for both cases we found that a conventional convolutional code is far more powerful than a non-iterative MLSE-decoded turbo code of the same trellis complexity, since the achievable performance is degraded by the employment of a deficient turbo interleaver having a low number of columns.

6.7 Optimum Decoding of Turbo Codes: Discussion of the Results

As mentioned earlier, when inspecting the super-trellis of Figure 6.12 and Table 6.3, the trellis periodicity with period 2, which is based on the periodicity of the 4×2 rectangular interleavers employed, is apparent. In other words, we stated before that the structure of the trellis sections depends on the time instant t, exhibiting different legitimate transitions for different t values. More explicitly, the trellis was different for $t = 2 \rightarrow 3$ and $t = 3 \rightarrow 4$, while it was identical for $t = 2 \rightarrow 3$ and for $t = 4 \rightarrow 5$. We found that $\rho \times \chi$ rectangular interleavers exhibit a periodicity in the super-trellis (if edge effects are ignored), which has a period of $\min(\rho, \chi)$. By rearranging the interface states of the super-state vector one can eliminate the time-variant nature of the trellis and hence a time-invariant super-trellis can be generated, which is identical at all

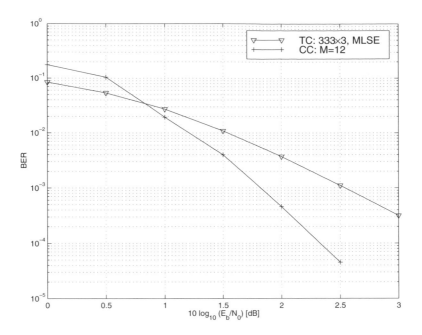

Figure 6.19: Turbo-coded BER performance with 4096 super-states compared to convolutional code with 4096 states.

values of t. In the context of Example 2, we could relabel the super-state vectors as follows:

$$\tilde{S}_2^* = (S_2^{(1)}; S_5^{(2)}; S_4^{(2)}; S_1^{(2)})$$
$$\tilde{S}_3^* = (S_3^{(1)}; S_2^{(2)}; S_4^{(2)}; S_5^{(2)})$$
$$\tilde{S}_4^* = (S_4^{(1)}; S_6^{(2)}; S_4^{(2)}; S_2^{(2)})$$
$$\tilde{S}_5^* = (S_5^{(1)}; S_3^{(2)}; S_4^{(2)}; S_6^{(2)})$$
$$\tilde{S}_6^* = (S_6^{(1)}; S_7^{(2)}; S_4^{(2)}; S_3^{(2)}),$$

in order to ensure that the two interface states that have replaced two previous interface states for generating super-state $\tilde{S}_{t+1}^*$ from $\tilde{S}_t^*$ (see Section 6.4.3, scenario (1)) are always placed as the first two vector elements, whereas the remaining interface states are the last two elements of the super-state vector. Table 6.4 shows the corresponding super-state transitions for these relabelled super-states. Note that relabelling of the super-states does not affect the output vector $\bar{C}_t$ associated with a (relabelled) super-state transition, which thus remains the same. Figure 6.20 shows a part of the super-trellis for Example 2, which has been rendered time-invariant by the above relabelling of the super-states. It is readily inferred that the constraint length for this example was 2 for the component scramblers, which increased to 3 for the super–trellis. The minimum free Hamming distance for this rate $1/3$ code is $\delta_{\text{free}} = 5$, if we take into consideration only paths in the time-invariant part of the super-trellis (i.e. ignoring edge effects at the super-trellis start and end). For comparison, the best conventional convolutional code with eight states (constraint length 4) and rate $1/3$ has $\delta_{\text{free}} = 10$ [145], which explains the convolutional code's superior BER performance.

    In general, the time-invariant super-trellis of a turbo code having a $\rho \times \chi$ rectangular inter-

$\tilde{S}_t^*$	U_{t+1}	$\tilde{S}_{t+1}^*$	$\bar{C}_{t+1}$
$(0;0;0;0)$	0	$(0;0;0;0)$	$(0;0;0)$
	1	$(1;1;0;0)$	$(1;1;1)$
$(0;0;1;1)$	0	$(0;1;1;0)$	$(0;0;1)$
	1	$(1;0;1;0)$	$(1;1;0)$
$(0;1;0;1)$	0	$(0;1;0;1)$	$(0;0;1)$
	1	$(1;0;0;1)$	$(1;1;0)$
$(0;1;1;0)$	0	$(0;0;1;1)$	$(0;0;0)$
	1	$(1;1;1;1)$	$(1;1;1)$
$(1;0;0;1)$	0	$(1;1;0;0)$	$(0;1;1)$
	1	$(0;0;0;0)$	$(1;0;0)$
$(1;0;1;0)$	0	$(1;0;1;0)$	$(0;1;0)$
	1	$(0;1;1;0)$	$(1;0;1)$
$(1;1;0;0)$	0	$(1;0;0;1)$	$(0;1;0)$
	1	$(0;1;0;1)$	$(1;0;1)$
$(1;1;1;1)$	0	$(1;1;1;1)$	$(0;1;1)$
	1	$(0;0;1;1)$	$(1;0;0)$

Table 6.4: State transitions and associated output after rearranging the interface states in the super-state vector for Example 2.

leaver, which has been obtained by relabelling the super-states, is reminiscent of the trellis of a conventional block-based or zero-terminated convolutional code. In general the constraint length of the turbo code super-trellis is $1 + m * M$ for component scramblers having a memory of M, where $m = \min(\rho, \chi)$, which is typically higher than the constraint length of a conventional convolutional code, if m is sufficiently high. The turbo code super-trellis exhibits $\leq 2^{M \cdot 2m}$ super-states. Unfortunately — as underlined by our simulation results — its BER performance is inferior to that of a good conventional convolutional code having the same number of states.

Although turbo codes which exhibit a complex super-trellis cannot be non-iterative MLSE decoded for complexity reasons, they can be subjected to iterative decoding. Indeed, the more complex the super-trellis, the closer the performance of the iterative decoding seems to be to that of the non-iterative MLSE decoding. The impact of the interleaver is that it results in a high number of super-states of the turbo code. Hence the statistical dependencies between the a-priori probabilities input to one of the component decoders, which stem from the extrinsic probabilities of the other component decoder, are reduced. This is also the reason that throughout our simulations the iterative algorithm was unable to reach the performance of the non-iterative MLSE decoding, while according to [59] this is possible on average across the range of all feasible interleavers, although nearly all of these interleavers result in a relatively high super-trellis complexity. Another example of convolutional codes having a high constraint length, which is decodable in an iterative fashion, is represented by the so-called Low-Density Parity Check Codes [146]. Conventional turbo codes have a high number of super-states in their super-trellis and hence they are superior to conventional convolutional codes having a low number of states. The advantage of turbo codes is that they are amenable to low-complexity iterative decoding, although their super-trellis structure is not optimal in terms of maximising the minimum Hamming distance $\delta_{\min}$. The low minimum Hamming distance of turbo codes is also reflected by the so-called *error floor* of the BER curve [59, 60]. In general the code performance is only determined at medium to high SNRs by $\delta_{\min}$ of the code. At low SNR, where turbo codes show a high power efficiency,

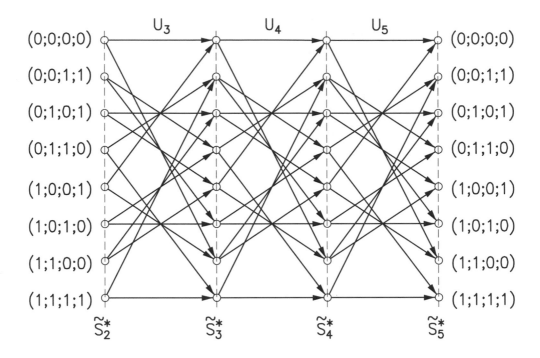

Figure 6.20: Three segments of the new time-invariant super-trellis of Example 2 obtained by relabelling the super-states.

the convenient shaping of the distance profile of turbo codes employing a random interleaver, which has been described in [139], is more important than achieving a high $\delta_{\min}$. The complex super-trellis of the turbo code does not endeavour to optimise $\delta_{\min}$, but instead it is responsible for the attractive shaping of the distance profile. Specifically, although there exist low-weight paths in the super-trellis, nonetheless, in comparison to other types of channel codes, only a low number of these paths exist.

If we consider for example a conventional non-recursive convolutional code with a free distance of δ_{free}, a single '1' in the information sequence at the input of the encoder results in a non-zero weight path, which is associated with a low output weight δ_1 with $\delta_{\text{free}} \leq \delta_1 \leq \lceil L/R \rceil$, where L is the constraint length, R is the rate of the code and $\lceil \bullet \rceil$ denotes the upwards rounding (ceiling) function. Accordingly, one can give $l \cdot \delta_1$ as the upper bound of the associated codeword weight for every information word of length K of a block-based or zero-terminated convolutional code, which contains l binary '1' symbols. In the case of a small l value there are at least $\binom{K}{l}$ codewords of a block-based or zero-terminated conventional convolutional code, which produce a low output weight of $w \leq l \cdot \delta_1$. For the class of recursive convolutional codes this statement holds as well, since for every recursive convolutional code, there is a non-recursive code that has the same set of codewords and only differs in the mapping from information bits to codewords. The same statement can be derived for turbo codes with a rectangular interleaver, which is due to the fact that a time-invariant super-trellis exists in this case, which exhibits a different trellis structure at different instants t. In a turbo code incorporating a random interleaver, however, there are much fewer codewords of a low weight [59, 60]. In contrast to rectangular interleavers, in conjunction with the random interleavers the number of low-weight paths does not increase with K, which is the reason for the interleaver gain [59, 60].

6.8 Summary and Conclusions

We have shown that turbo codes can be described by means of a super-trellis, if the trellises of the component scrambler and the interleaver structure are known. For rectangular interleavers one can model this super-trellis as time-invariant, so that it resembles the trellis of a conventional convolutional code. We have also argued that a 'good' conventional convolutional code of the same trellis complexity can be more powerful than a turbo code. On the basis of simulations we have found that iterative decoding is sub-optimal, at least for the investigated simple rectangular interleavers having a low number of columns. We have presented an upper bound for the super-trellis complexity of turbo codes based on rectangular interleavers and an upper bound for the super-trellis complexity averaged over all possible interleavers. The second bound gives rise to the supposition that the complexity of a turbo code having a random interleaver is of the same magnitude as that of a random code.

Having compared the trellis structures of conventional convolutional and convolutional constituent code-based turbo codes in this chapter, our discourse in Chapter 7 is focused on binary turbo codes based on BCH coding.

6.9 Appendix: Proof of Algorithmic Optimality

In order to show that using a Viterbi algorithm along the super-trellis is optimum in the sense of non-iterative MLSE decoding of turbo codes, we have to demonstrate that when paths are discarded in the super-trellis (in the following referred to as *super-paths*) during the course of the Viterbi algorithm, the specific super-path associated with the optimum codeword $\mathbf{c}_{\mathrm{opt}}$ in the sense of Equation 6.21 (for BPSK transmission over an AWGN channel) is never discarded. Let $\mathcal{M}_t$ be the set of codewords associated with the non-discarded super-paths at decoding stage t (corresponding to time instant t in the encoder), which implies at stage $t = 0$ that $\mathcal{M}_0 = \{\text{all valid codewords } \mathbf{c}_i\}$. Let us now invoke the method of induction below.

We claim that: $\mathbf{c}_{\mathrm{opt}} \in \mathcal{M}_t \ \forall t$

Proof:

Commence induction from: $\mathbf{c}_{\mathrm{opt}} \in \mathcal{M}_0$

Inductive assumption: Let us assume that for an arbitrary t the following is true: $\mathbf{c}_{\mathrm{opt}} \in \mathcal{M}_t$

Induction conclusion: When proceeding from the decoding stage t to the following stage $t + 1$ by extending the surviving super-paths and then discarding all but one path merging into any super-state, the super-path associated with the optimum codeword $\mathbf{c}_{\mathrm{opt}}$ is not discarded, i.e. $\mathbf{c}_{\mathrm{opt}} \in \mathcal{M}_{t+1}$, as shown below.

From the mapping γ_s of the component scrambler FSM $\mathbb{F}_s$ we see that the output symbol $c_k^{(i)}$ of the ith component scrambler at any transition index k is a function f_1 of the encoder state $s_{k-1}^{(i)}$ before this transition and the current input symbol $u_k^{(i)}$, which can be expressed as:

$$c_k^{(i)} = f_1(s_{k-1}^{(i)}; u_k^{(i)}). \tag{6.22}$$

Similarly to the output symbol $c_k^{(i)}$, the new encoder state $s_k^{(i)}$ is also a function of the previous state $s_{k-1}^{(i)}$ and the input symbol $u_k^{(i)}$, as shown below:

$$s_k^{(i)} = g_1(s_{k-1}^{(i)}; u_k^{(i)}). \tag{6.23}$$

A section $(c_{k+1}^{(i)}; \ldots ; c_l^{(i)})$ of a component codeword is therefore only dependent on the encoder state $s_k^{(i)}$ at encoding stage k and the input symbols $(u_{k+1}^{(i)}; \ldots ; u_l^{(i)})$:

$$(c_{k+1}^{(i)}; \ldots ; c_l^{(i)}) = \mathbf{f}_2(s_k^{(i)}; u_{k+1}^{(i)}; \ldots ; u_l^{(i)}). \tag{6.24}$$

We now consider decoding stage $t + 1$ of the Viterbi algorithm for the super-trellis. For the sake of simplicity, we omit the time index $t + 1$ in the variables. Let J be the total number of sections in the two component trellises belonging to the pre-history of the super-trellis as regards the current time $t+1$ entailing the sections in the component trellises, which have already been explored by the decoder at decoding stage $t + 1$. Let N be the number of trellis sections belonging to the post-history as regards $t + 1$, which entails the sections unexplored so far. Every turbo codeword $\mathbf{c}$ can be split into sub-vectors of two types, namely into J sub-vectors $\mathbf{c}_{\text{pre},j}$, $j = 1 \ldots J$, of code symbols, which are associated with the J component trellis sections belonging to the pre-history as regards $t + 1$, and in $\mathbf{c}_{\text{post},n} = (c_{k_n+1}^{(i_n)}; \ldots ; c_{l_n}^{(i_n)})$, $n = 1..N$, which are associated with the N sections belonging to the post-history.

Let $\mathbf{c}_{\text{pre}}$ be a vector that contains all code symbols belonging to the pre-history:

$$\mathbf{c}_{\text{pre}} = (\mathbf{c}_{\text{pre},1}; \ldots ; \mathbf{c}_{\text{pre},J}),$$

and let $\mathbf{c}_{\text{post}}$ represent all the code symbols belonging to the post-history at decoding stage $t+1$:

$$\mathbf{c}_{\text{post}} = (\mathbf{c}_{\text{post},1}; \ldots ; \mathbf{c}_{\text{post},N}).$$

Any codeword $\mathbf{c}$ is hence constituted by these two vectors. Explicitly, $\mathbf{c}$ is a specific permutation Π of a concatenation of these two vectors, where the actual nature of the permutation is irrelevant in this context, leading to the following formulation:

$$\mathbf{c} = \Pi \left(\mathbf{c}_{\text{pre}}; \mathbf{c}_{\text{post}} \right). \tag{6.25}$$

According to Equation 6.24, for the post-history $\mathbf{c}_{\text{post}}$ of any codeword $\mathbf{c}$, its N constituent parts $\mathbf{c}_{\text{post},n}$ depend only on the encoder state $s_{k_n}^{(i_n)}$ at the beginning of the specific section n, $n = 1 \ldots N$, in the corresponding component trellis and on the component scrambler input $(u_{k_n+1}^{(i_n)}; \ldots ; u_{l_n}^{(i_n)})$ associated with the transitions inside this section. Hence we obtain the following dependence:

$$\mathbf{c}_{\text{post}} = \mathbf{f}_3 \left(s_{k_1}^{(i_1)}; s_{k_2}^{(i_2)}; \ldots ; s_{k_N}^{(i_N)}; \mathbf{u}_{\text{post}} \right), \tag{6.26}$$

where

$$\mathbf{u}_{\text{post}} = \left(u_{k_1+1}^{(i_1)}; \ldots ; u_{l_1}^{(i_1)}; u_{k_2+1}^{(i_2)}; \ldots ; u_{l_N}^{(i_N)} \right)$$

represents the information symbols associated with the post-history of this codeword $\mathbf{c}$, and where $\mathbf{f}_3$ expresses a functional dependence which does not have to be further specified here. Furthermore, from Equation 6.23, we obtain a second dependence for the states at the end of the sections:

$$(s_{l_1}^{(i_1)}; s_{l_2}^{(i_2)}; \ldots ; s_{l_N}^{(i_N)}) = \mathbf{g}_2 \left(s_{k_1}^{(i_1)}; s_{k_2}^{(i_2)}; \ldots ; s_{k_N}^{(i_N)}; \mathbf{u}_{\text{post}} \right). \tag{6.27}$$

Again, $\mathbf{g}_2$ represents a functional dependence whose actual nature is irrelevant in this context.

The encoder states $s_{l_j}^{(i_j)}$, $j = 1 \ldots J$, and $s_{k_n}^{(i_n)}$, $n = 1 \ldots N$, constitute the super-state s^*, that the super-path associated with the codeword $\mathbf{c}$ is merging into at time instant $t + 1$, as discussed in Section 6.4.3. The codeword symbols belonging to the post-history can therefore be expressed by:

$$\mathbf{c}_{\mathrm{post}} = \mathbf{f}_4 \left(s^*; \mathbf{u}_{\mathrm{post}} \right), \tag{6.28}$$

and this vector $\mathbf{c}_{\mathrm{post}}$ exists if and only if $(s^*; \mathbf{u}_{\mathrm{post}})$ is a valid pair according to the condition of Equation 6.27.

When two super-paths merge into a super-state s^* at decoding stage $t + 1$, then the interface states in all the component trellis sections constituting the pre-history as regards to $t + 1$ are identical for the two super-paths, which is clear from the definition of the super-states in Section 6.4.3.

Suppose that the optimum codeword $\mathbf{c}_{\mathrm{opt}}$ is associated with the super-path $\overset{>}{p}_{\mathrm{opt}}$ (the $>$ identifies it as a path) merging into the super-state s^*_{opt} at decoding stage $t + 1$. Since we have imposed that $\mathbf{c}_{\mathrm{opt}} \in \mathcal{M}_t$, this super-path has not been discarded in earlier decoding stages and is present in the Viterbi decoder at the current stage $t + 1$ before the discarding process commences.

Let $\mathbf{c}_{\overset{>}{p}_{\mathrm{opt}}}$ denote the code symbols belonging to the super-path $\overset{>}{p}_{\mathrm{opt}}$ (i.e. the output of the turbo encoder FSM associated with this super-path of length $t + 1$). Thus, $\mathbf{c}_{\overset{>}{p}_{\mathrm{opt}}}$ is identical to $\mathbf{c}_{\mathrm{opt,pre}}$ (pre-history part of $\mathbf{c}_{\mathrm{opt}}$), and $\mathbf{c}_{\mathrm{opt,post}}$ represents the remaining code symbols of $\mathbf{c}_{\mathrm{opt}}$. We denote the information symbols associated with the post-history of $\mathbf{c}_{\mathrm{opt}}$ by $\mathbf{u}_{\mathrm{opt,post}}$. Then we see from Equation 6.28 that:

$$\mathbf{c}_{\mathrm{opt,post}} = \mathbf{f}_4 \left(s^*_{\mathrm{opt}}; \mathbf{u}_{\mathrm{opt,post}} \right), \tag{6.29}$$

and from Equation 6.25 we have:

$$\mathbf{c}_{\mathrm{opt}} = \Pi \left(\mathbf{c}_{\overset{>}{p}_{\mathrm{opt}}}; \mathbf{c}_{\mathrm{opt,post}} \right). \tag{6.30}$$

Let $\mathbf{c}_{\overset{>}{p}_2}$ represent the code symbols associated with another super-path $\overset{>}{p}_2$ merging into s^*_{opt} at time instant $t + 1$. Then for *every* valid $\overset{>}{p}_2$, we can construct a valid codeword:

$$\mathbf{c}_2 = \Pi \left(\mathbf{c}_{\overset{>}{p}_2}; \mathbf{c}_{\mathrm{opt,post}} \right), \tag{6.31}$$

where $\mathbf{c}_{\mathrm{opt,post}}$ is given in Equation 6.29.

If in the Viterbi algorithm, $\mathbf{M}_{\overset{>}{p}_{\mathrm{opt}}}$ is the Euclidean metric (cf. Equation 6.20) in the pre-history part (i.e. the metric of the code symbols $\mathbf{c}_{\overset{>}{p}_{\mathrm{opt}}}$ associated with super-path $\overset{>}{p}_{\mathrm{opt}}$) and $\mathbf{M}_{\mathrm{opt,post}}$ is the metric associated with the code symbols $\mathbf{c}_{\mathrm{opt,post}}$ in the post-history part of the codeword $\mathbf{c}_{\mathrm{opt}}$, then the Euclidean metric of $\mathbf{c}_{\mathrm{opt}}$ is given by:

$$\mathbf{M}_{\mathbf{c}_{\mathrm{opt}}} = \mathbf{M}_{\overset{>}{p}_{\mathrm{opt}}} + \mathbf{M}_{\mathrm{opt,post}}, \tag{6.32}$$

which is the minimum of all codeword metrics $\mathbf{M}_{\mathbf{c}_i}$.

For the metric $\mathbf{M}_{\mathbf{c}_2}$ of the codeword $\mathbf{c}_2$ this means that:

$$\mathbf{M}_{\mathbf{c}_{\mathrm{opt}}} = \mathbf{M}_{\overset{>}{p}_{\mathrm{opt}}} + \mathbf{M}_{\mathrm{opt,post}} \le \mathbf{M}_{\mathbf{c}_2} = \mathbf{M}_{\overset{>}{p}_2} + \mathbf{M}_{\mathrm{opt,post}} \tag{6.33}$$

and consequently for *all* valid $\overset{>}{p}_2$ merging into s^*_{opt}:

$$\mathbf{M}_{\overset{>}{p}_{\mathrm{opt}}} \le \mathbf{M}_{\overset{>}{p}_2}. \tag{6.34}$$

In the discarding process of decoding stage $t+1$, any super-path $\overrightarrow{p}_2$ leading to the super-state s^*_{opt} is discarded because of its higher metric $\mathbf{M}_{\overrightarrow{p}_2}$ and the optimum super-path $\overrightarrow{p}_{\text{opt}}$ leading to this super-state is retained, i.e. $\mathbf{c}_{\text{opt}} \in M_{t+1}$.

Chapter 7

Turbo BCH Coding

7.1 Introduction

Turbo coding [12] is a novel form of channel coding capable of achieving a performance near the Shannon limit [1]. As discussed in Chapter 5, typically so-called Recursive Systematic Convolutional (RSC) codes are used as their component codes. However, block codes can also be employed as their component codes and they have been shown for example by Hagenauer [61,63] to perform impressively even at near-unity coding rates. Block codes are typically more appropriate for near-unity coding rates, since lower-rate block-based turbo codes exhibit a high decoding complexity. Hence usually convolutional constituent code-based turbo codes are used for coding rates lower than 2/3 [61]. In this chapter, we will concentrate entirely on the standard turbo code structure using binary BCH codes as the component codes.

Block-coding-based turbo codes can be decoded either using algebraic decoding principles [123, 147, 148], as discussed in Chapter 3, or employing trellis-based decoding [61,63], which was the topic of Chapter 7. In this chapter we assume that the reader is familiar with the basic concepts discussed in Chapters 2–6 and hence only trellis-based turbo decoding will be discussed in this chapter.

In Chapter 5, we have provided a rudimentary introduction to the derivation of the log likelihood ratio, soft channel outputs and the Maximum A-Posteriori (MAP) algorithm [11] as well as the Max-Log-MAP and Log-MAP algorithms [50–52]. Here, we extend the derivation in the context of turbo BCH coding. Despite this, the Soft-Output Viterbi Algorithm (SOVA) is discussed in detail in Section 7.3.2. A simple example of turbo decoding is then given in Section 7.4. In Section 7.5, we propose a novel MAP algorithm for the family of extended BCH codes. The extended MAP algorithm is then also simplified for deriving the extended Max-Log-MAP and Log-MAP algorithms. Finally, various simulation results are presented in Section 7.6.

7.2 Turbo Encoder

The basic structure of the turbo BCH encoder is shown in Figure 7.1. Two BCH encoders, which we have been introduced in Section 4.2.1, are used and an interleaver is placed before the second BCH encoder in Figure 7.1. A number of interleaving techniques, such as block interleaving and random interleaving [68, 120], could potentially be employed for ensuring that the two BCH

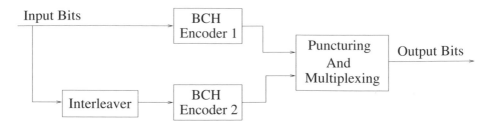

Figure 7.1: Turbo encoder schematic.

encoders are fed with near-uncorrelated bits. The significance of this will become clear during our further discourse.

Owing to its structure, the turbo encoder shown in Figure 7.1 is also often referred to as a parallel concatenated code [12, 61, 123]. Parallel concatenated codes constitute specific product codes. In general, product codes consist of two linear block codes C_1 and C_2 where C_1 and C_2 have parameters (n_1, k_1, d_{min1}) and (n_2, k_2, d_{min2}), respectively. Typically, $C_1 \equiv C_2$. As shown in Figure 7.2, product codes are obtained by placing the $k_1 \times k_2$ information data bits in an array of k_1 columns and k_2 rows. The k_1 columns and k_2 rows of the information data bits are encoded using C_1 and C_2, respectively. It is shown in [149] that the $(n_1 - k_1)$ last columns of Figure 7.2 are codewords of C_2, exactly as the $(n_2 - k_2)$ last rows are codewords of C_1 by construction. Furthermore, the parameters of the resulting product codes are given by $n = n_1 \times n_2$, $k = k_1 \times k_2$ and $d_{min} = d_{min1} \times d_{min2}$, while the code rate is given by $\frac{k_1}{n_1} \times \frac{k_2}{n_2}$. The structure of parallel concatenated codes is the same as that of product codes, except that the redundancy part arising from checking the parity of the parity part of both codes C_1 and C_2 is omitted. The major disadvantage of parallel concatenated codes is the loss in minimum free distance, which is only $d_{min1} + d_{min2} - 1$, compared to $d_{min1} \times d_{min2}$ in product codes.

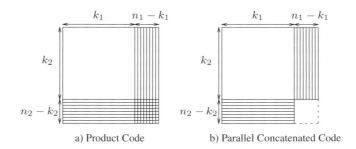

a) Product Code   b) Parallel Concatenated Code

Figure 7.2: Construction of product codes and parallel concatenated codes.

The output bits from the two BCH encoders of Figure 7.1 are then punctured and multiplexed. Table 4.1 shows a wide range of the BCH codes exhibiting a variety of coding rates. By appropriately designing the puncturer and the multiplexer, we are capable of achieving an overall coding rate which is identical to that of the original BCH codes. However, it is not a common practice to apply puncturing in turbo block codes [61, 123], since its code rate is high. Furthermore, their puncturing significantly degrades the achievable performance. Having considered the encoder structure, let us now concentrate on the decoder schematic in the next section.

7.3   Turbo Decoder

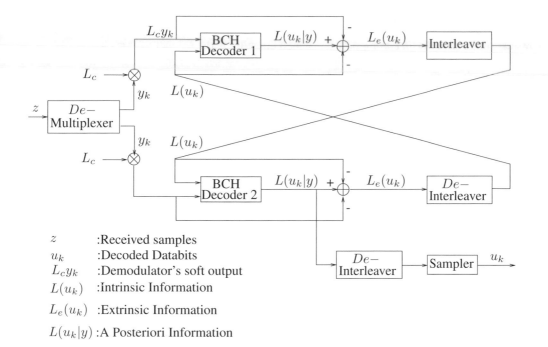

z :Received samples
u_k :Decoded Databits
$L_c y_k$:Demodulator's soft output
$L(u_k)$:Intrinsic Information
$L_e(u_k)$:Extrinsic Information
$L(u_k|y)$:A Posteriori Information

Figure 7.3: Turbo decoder schematic.

Figure 7.3 shows the general structure of a turbo BCH decoder. We summarise below the definition of the various quantities used in the figure.

Soft Channel Output, $L_c y$: The soft channel output is simply the matched filter-based demodulator's output, y_k, multiplied by the so-called channel reliability value, L_c. Both of these quantities were described in more depth in Section 5.3.2.

Intrinsic Information, $L(u_k)$: The intrinsic information concerning a data bit u_k is the information known before decoding starts. It can be achieved from any source other than the received channel output sequence or the code constraints and it is also known as *a-priori information*.

Extrinsic Information, $L_e(u_k)$: The extrinsic information related to a data bit u_k is the information carried by the bits surrounding u_k, which was imposed by the code constraints. In other words, as indicated by the terminology 'extrinsic', no information directly concerning the data bit itself is part of the extrinsic information.

A-Posteriori Information, $L(u_k|y)$: The a-posteriori information of a data bit u_k is given by the decoder after taking into account *all* the available sources of information about u_k, i.e. both the intrinsic and extrinsic sources.

Again in this chapter, the turbo decoder of Figure 7.3 is constituted by two BCH decoders. Since we opted for employing the trellis decoding method, two algorithms, namely the *Maximum A-Posteriori* (MAP) [11] and the *Soft-Output Viterbi Algorithm* (SOVA) [53,54] can be used.

The decoder uses the soft channel output $L_c y$ and the intrinsic information $L(u_k)$ to provide the a-posteriori information $L(u_k|y)$ at its output, as shown in Figure 7.3. The extrinsic information, $L_e(u_k)$ is given by subtracting the soft channel output $L_c y$ and the intrinsic information $L(u_k)$ from the a-posteriori information $L(u_k|y)$, which was justified in Section 5.3.3. After being interleaved or de-interleaved, as seen in the figure, the extrinsic information $L_e(u_k)$ becomes the intrinsic information $L(u_k)$ of the second decoder. Similarly, the extrinsic information gained by the second decoder is passed back to the first decoder as its intrinsic information. Basically, both decoders assist each other by exchanging their information related to the data bits and this results in the *iterative decoding* process, which constitutes the subject of this chapter. We note, however, that there is no intrinsic information for the first decoder in the first iteration, since the extrinsic information of the other decoder is unavailable at this stage.

7.3.1 Summary of the MAP Algorithm

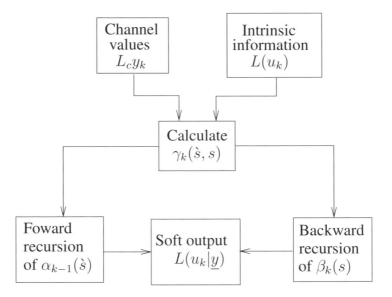

Figure 7.4: Summary of the key operations in the MAP algorithm, which constitutes the BCH decoder in Figure 7.3.

We have given a detailed derivation of the log likelihood ratio, soft channel outputs and the MAP algorithm in Sections 5.3.2 and 5.3.3. Therefore, in this section, we will provide only a summary of the MAP algorithm and explain the modification to the MAP algorithm in the context of the BCH codes.

Figure 7.4 shows a summary of the key operations in the MAP algorithm, which constitutes the BCH decoder in Figure 7.3. Initially, the algorithm relies on the demodulator's soft-output sequence y and on the intrinsic information $L(u_k)$, provided by the second component decoder shown in Figure 7.3. Both $\underline{y}$ and $L(u_k)$ are then used to calculate the $\gamma_k(\grave{s}, s)$ values according to Equation 5.40. After that, we use the $\gamma_k(\grave{s}, s)$ values in order to calculate the $\alpha_k(s)$ and $\beta_k(s)$ values by employing forward recursion of Equation 5.25 and backward recursion of Equation 5.28, respectively.

In Equation 5.20, we have shown that $P(\grave{s} \wedge s \wedge y)$ depends on the values $\alpha_k(s)$, $\gamma_k(\grave{s}, s)$ and $\beta_k(s)$, which we have already described. Using Equation 5.20, we can rewrite Equation 5.17 as follows:

$$L(u_k|\underline{y}) = \ln \left\{ \frac{\displaystyle\sum_{\substack{(\grave{s},s)\Rightarrow \\ u_k=+1}} P(S_{k-1}=\grave{s} \wedge S_k = s \wedge \underline{y})}{\displaystyle\sum_{\substack{(\grave{s},s)\Rightarrow \\ u_k=-1}} P(S_{k-1}=\grave{s} \wedge S_k = s \wedge \underline{y})} \right\}$$

$$= \ln \left\{ \frac{\displaystyle\sum_{\substack{(\grave{s},s)\Rightarrow \\ u_k=+1}} \alpha_{k-1}(\grave{s}) \cdot \gamma_k(\grave{s},s) \cdot \beta_k(s)}{\displaystyle\sum_{\substack{(\grave{s},s)\Rightarrow \\ u_k=-1}} \alpha_{k-1}(\grave{s}) \cdot \gamma_k(\grave{s},s) \cdot \beta_k(s)} \right\}. \tag{7.1}$$

Equation 7.1 is then used in the final stage of the decoding process to derive the conditional LLR of u_k, given the demodulator's soft-output sequence y and the intrinsic information $L(u_k)$.

Let us consider the expression of $\gamma_k(\grave{s},s)$ in Equation 5.40, which is restated and simplified here for convenience:

$$\gamma_k(\grave{s},s) = C. \exp\left\{ u_k \frac{L(u_k)}{2} \right\} \cdot \exp\left\{ \frac{L_c}{2} x_k y_k \right\}. \tag{7.2}$$

In this equation, we have omitted the summation in Equation 5.40 since there is only one data or parity bit for each trellis transition as shown in Section 4.2.2. On the other hand, there may be multiple data and parity bits for each trellis transition in convolutional codes. We have also shown in Section 4.2.2 that the systematic turbo BCH-coded transmitted sequence consists of the original data bit sequence, followed by the parity bit sequence. However, in Equation 7.1 we are interested in deriving the conditional LLRs of the original uncoded data bits u_k. Hence, the channel-coded transmitted bits x_k are replaced by u_k in Equation 7.2, yielding:

$$\gamma_k(\grave{s},s) = C. \exp\left\{ u_k . \frac{L(u_k)}{2} \right\} \cdot \exp\left\{ \frac{L_c}{2} u_k y_k \right\} \tag{7.3}$$

In Equation 7.3 we have seen that $\gamma_k(\grave{s},s)$ can be calculated from the demodulator's soft-output sample y_k and from the intrinsic information $L(u_k)$. Using the previously computed $\gamma_k(\grave{s},s)$ values, we can calculate the $\alpha_k(s)$ values, recursively using Equation 5.25. Figure 7.5

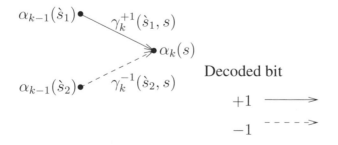

Figure 7.5: Forward recursion for the computation of $\alpha_k(s)$.

shows a simple example of the forward recursion calculation of $\alpha_k(s)$. For a binary trellis, for example in binary BCH codes, there are only two legitimate transitions from the previous state $S_{k-1}=\grave{s}$ to any present state $S_k = s$, as shown in Figure 7.5. Furthermore, one of these will be associated with an information input bit of $u_k = +1$, and the other with an input bit of $u_k = -1$. Hence, we can expand Equation 5.25 as:

$$\alpha_k(s) = \alpha_{k-1}(\grave{s}_1) \cdot \gamma_k^{+1}(\grave{s}_1, s) + \alpha_{k-1}(\grave{s}_2) \cdot \gamma_k^{-1}(\grave{s}_2, s) , \tag{7.4}$$

where $\gamma_k^{x_k}(\grave{s}, s)$ is the value of $\gamma_k(\grave{s}, s)$ for the transitions between states, when the transmitted bit is $x_k = -1$ or $+1$, as indicated in Figure 7.5.

In Section 4.2.2, we have shown that the trellis of the BCH code always commences from the all-zero state $S_0 = 0$. Hence, we have $P(S_0 = 0) = 1$ and $P(S_0 = s) = 0$ for all $s \neq 0$. The initial conditions for the forward recursion are:

$$\begin{aligned} \alpha_0(S_0 = 0) &= 1 \\ \alpha_0(S_0 = s) &= 0 \qquad \text{for all } s \neq 0 . \end{aligned} \tag{7.5}$$

Similarly to the forward recursion, once the $\gamma_k(\grave{s}, s)$ values are known, backward recursion can be used for calculating the values of $\beta_{k-1}(\grave{s})$ from the values of $\beta_k(s)$ using Equation 5.28. Figure 7.6 shows a simple example of the backward recursive calculation of $\beta_k(s)$. As for the for-

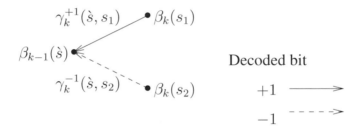

Figure 7.6: Backward recursion for the computation of $\beta_{k-1}(\grave{s})$.

ward recursion, there are only two transitions from the present states $S_k = s$ to the previous state $S_{k-1} = \grave{s}$, as shown in Figure 7.6. Hence similarly to Equation 7.4, we expand Equation 5.25 to:

$$\beta_{k-1}(\grave{s}) = \beta_k(s_1) \cdot \gamma_k^{+1}(\grave{s}, s_1) + \beta_k(s_2) \cdot \gamma_k^{-1}(\grave{s}, s_2) . \tag{7.6}$$

Again in Section 4.2.2, we have shown that the trellis of the BCH code always ends at the all-zero state $S_n = 0$. Hence, we have $P(S_n = 0) = 1$ and $P(S_n = s) = 0$ for all $s \neq 0$. The initial conditions for the backward recursion are:

$$\begin{aligned} \beta_n(S_n = 0) &= 1 \\ \beta_n(S_n = s) &= 0 \qquad \text{for all } s \neq 0. \end{aligned} \tag{7.7}$$

Using Equation 7.3 and remembering that in the numerator we have $u_k = +1$ for all terms in the summation, whereas in the denominator we have $u_k = -1$, we can rewrite Equation 7.1

yielding the a-posteriori information for our decision concerning the bit u_k as:

$$
\begin{aligned}
L(u_k|\underline{y}) &= \ln\left[\frac{\sum\limits_{\substack{(\grave{s},s)\Rightarrow\\u_k=+1}}\alpha_{k-1}(\grave{s})\cdot\gamma_k(\grave{s},s)\cdot\beta_k(s)}{\sum\limits_{\substack{(\grave{s},s)\Rightarrow\\u_k=-1}}\alpha_{k-1}(\grave{s})\cdot\gamma_k(\grave{s},s)\cdot\beta_k(s)}\right]\\
&= \ln\left[\frac{\sum\limits_{\substack{(\grave{s},s)\Rightarrow\\u_k=+1}}\alpha_{k-1}(\grave{s})\cdot e^{\{+\frac{L(u_k)}{2}\}}\cdot e^{\{+\frac{L_cy_k}{2}\}}\beta_k(s)}{\sum\limits_{\substack{(\grave{s},s)\Rightarrow\\u_k=-1}}\alpha_{k-1}(\grave{s})\cdot e^{\{-\frac{L(u_k)}{2}\}}\cdot e^{\{-\frac{L_cy_k}{2}\}}\beta_k(s)}\right]\\
&= L(u_k)+Lcy_k+\ln\left[\frac{\sum\limits_{\substack{(\grave{s},s)\Rightarrow\\u_k=+1}}\alpha_{k-1}(\grave{s})\cdot\beta_k(s)}{\sum\limits_{\substack{(\grave{s},s)\Rightarrow\\u_k=-1}}\alpha_{k-1}(\grave{s})\cdot\beta_k(s)}\right]\\
&= L(u_k)+Lcy_k+L_e(u_k),
\end{aligned}\tag{7.8}
$$

where:

$$
L_e(u_k)=\ln\left[\frac{\sum\limits_{\substack{(\grave{s},s)\Rightarrow\\u_k=+1}}\alpha_{k-1}(\grave{s}).\beta_k(s)}{\sum\limits_{\substack{(\grave{s},s)\Rightarrow\\u_k=-1}}\alpha_{k-1}(\grave{s}).\beta_k(s)}\right],\tag{7.9}
$$

and the following notations apply:

y_k	the demodulator's soft output for the transmitted data bit $x_k=u_k$
L_cy_k	soft channel output of y_k
$L(u_k)$	intrinsic or a-priori information
$L_e(u_k)$	extrinsic information
$L(u_k|\underline{y})$	a-posteriori information.

Now, we are ready to relate Equation 7.8 to Figure 7.7 which shows a single component decoder of Figure 7.3. In Figure 7.7, we can see that the decoder accepts two inputs, the intrinsic information $L(u_k)$ of the original uncoded information bits and the soft channel output $L_c\underline{y}$ constituted by the product of the demodulator's soft output $\underline{y}$ and the channel reliability value L_c of Equation 5.11. At its output, it produces the a-posteriori information $L(u_k|\underline{y})$ as shown in Figure 7.7. The decoder has to calculate the extrinsic information $L_e(u_k)$ imposed by the code constraints from the demodulator's soft-output sequence $\underline{y}$, but excluding the demodulator's soft-output sample y_k due to the transmitted data bit, x_k.

The MAP algorithm, in the form described in Section 5.3.3, is extremely complex owing to the floating point multiplication needed in Equations 5.25 and 5.28 for the recursive calculation of $\alpha_k(s)$ and $\beta_k(s)$, as well as because of the multiplications and exponential operations required to calculate $\gamma_k(\grave{s},s)$ using Equation 7.3. Further factors increasing the complexity are the multiplication and natural logarithm operations required to calculate $L(u_k|\underline{y})$ using Equation 7.8.

The Max-Log-MAP algorithm was initially proposed by Koch and Baier [50] and Erfanian *et al.* [51] for reducing the complexity of the MAP algorithm. This technique transfers the computation of the recursions into the logarithmic domain and invokes an approximation for dramatically reducing the complexity. Owing to the approximation, its performance is sub-optimal. However,

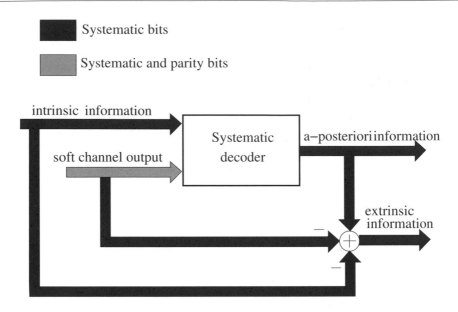

Figure 7.7: Schematic of a component decoder employed in a turbo decoder, showing the input information received and output informations corresponding to the systematic and parity bits.

Robertson *et al.* [52] later proposed the Log-MAP algorithm, which partially corrected the approximation made in the Max-Log-MAP algorithm. Hence the performance of the Log-MAP algorithm is similar to that of the MAP algorithm, but at a fraction of its complexity. The interested reader is referred to Section 5.3.5 for further discourse.

7.3.2 The Soft-output Viterbi Algorithm

Although the deduction of the mathematical formulation of the *Soft-Output Viterbi Algorithm* (SOVA) cast in the context of binary BCH codes is similar to that of convolutional codes, here we repeat the derivation of the corresponding formulae for the sake of providing a self-contained introduction to our SOVA decoding example in Sections 7.3.2.1 and 7.4. The SOVA was proposed by Hagenauer [53] in 1989. The SOVA can be implemented in the so-called register exchange mode [53] or in the trace-back mode [54]. The trace-back mode is used in this treatise.

The SOVA has two modifications with respect to the soft-decision Viterbi Algorithm (VA) that we have discussed in Section 4.3.4. First, the path metrics used are modified, in order to take account of the intrinsic information $L(u_k)$. Second, the algorithm is modified so that it provides the a-posteriori information $L(u_k|\underline{y})$ for each decoded data bit.

First of all, we revisit the fundamental idea of the Viterbi algorithm, where the state sequence $\underline{s}_k$ of the Maximum Likelihood (ML) path is decided on the basis of the demodulator's soft-output sequence $\underline{y}_{j \leq k}$. The state sequence $\underline{s}_k$ gives the states along the surviving path at state $S_k = s$ at instant k in the trellis. Notice that the state sequence $\underline{s}_k$ is chosen based purely on the demodulator's soft-output sequence $\underline{y}$ before and at instant k, $j \leq k$. Based on these arguments, we need to maximise the following equation:

$$P(\underline{s}_k|\underline{y}_{j\leq k}) = \frac{P(\underline{s}_k \wedge \underline{y}_{j\leq k})}{P(\underline{y}_{j\leq k})} , \qquad (7.10)$$

where Bayes' rule was invoked. Since the probability of encountering the received sequence $P(\underline{y}_{j\leq k})$ is constant for all the paths, we can equally maximise $P(\underline{s}_k \wedge \underline{y}_{j\leq k})$. Let us define the path metrics as:

$$M(\underline{s}_k) := \ln P(\underline{s}_k \wedge \underline{y}_{j\leq k}) . \qquad (7.11)$$

Assuming a memoryless channel, we will have:

$$P(\underline{s}_k \wedge \underline{y}_{j\leq k}) = P(\underline{s}_{k-1} \wedge \underline{y}_{j\leq k-1}) \cdot P(S_k = s \wedge y_k | S_{k-1} = \grave{s}) . \qquad (7.12)$$

Upon substituting Equation 7.12 into Equation 7.11, we arrive at:

$$\begin{aligned} M(\underline{s}_k) &= M(\underline{s}_{k-1}) + \ln\{P(S_k = s \wedge y_k | S_{k-1} = \grave{s})\} \\ &= M(\underline{s}_{k-1}) + \ln\{\gamma(\grave{s}, s)\} , \end{aligned} \qquad (7.13)$$

and $\gamma(\grave{s}, s)$ is the branch transition probability for the path from state $S_{k-1} = \grave{s}$ to $S_k = s$ which has been defined in Section 5.3.3. Using Equation 7.3, we can rewrite Equation 7.13 as:

$$\begin{aligned} \ln\{\gamma_k(\grave{s}, s)\} &= \ln C + \frac{1}{2}u_k L(u_k) + \frac{1}{2}u_k L_c y_k \\ &= C' + \frac{u_k}{2}\{L(u_k) + L_c y_k\} , \end{aligned} \qquad (7.14)$$

where $C' = \ln C$ is constant and it will cancel out, when we consider the path metric differences in the trellis. Hence, it can be omitted. Upon using Equation 7.14, we can rewrite Equation 7.13 as:

$$M(\underline{s}_k) = M(\underline{s}_{k-1}) + \frac{u_k}{2}\{L(u_k) + L_c y_k\} . \qquad (7.15)$$

In Equation 7.15, we have shown how the intrinsic information $L(u_k)$ is included in the path metric calculation. Hence, we have accomplished the first required modification of the conventional VA.

In Section 4.2.2 we have seen that a BCH codeword consists of data and parity bits. In turbo decoding, only the soft outputs of the original data bits are passed from one decoder to another. There is no intrinsic information $L(u_k)$ for the parity bits. Therefore, for the parity bits Equation 7.15 is further reduced to:

$$M(\underline{s}_k) = M(\underline{s}_{k-1}) + \frac{1}{2}u_k L_c y_k . \qquad (7.16)$$

Let us now discuss the above-mentioned second modification of the VA, namely the generation of the a-posteriori information $L(u_k|y)$, which allows us to compute the extrinsic information $L_e(u_k)$ from Equation 7.9. In the binary trellis there are two paths reaching state $S_k = s$ and one of the paths, namely the one having the smaller path metric, will be discarded. Let $\underline{s}_k$ and $\hat{\underline{s}}_k$ denote the ML path exhibiting the highest path metric, and the discarded path at instant k, having the lower path metric, respectively. Then we can define the path metric difference as:

$$\Delta_k = M(\underline{s}_k) - M(\hat{\underline{s}}_k) \geq 0 . \qquad (7.17)$$

The probability of the ML path $\underline{s}_k$ is:

$$
\begin{aligned}
P(\underline{s}_k) &= \frac{P(\underline{s}_k)}{P(\underline{s}_k) + P(\hat{\underline{s}}_k)} \\
&= \frac{e^{M(\underline{s}_k)}}{e^{M(\underline{s}_k)} + e^{M(\hat{\underline{s}}_k)}} \\
&= \frac{e^{\Delta_k}}{1 + e^{\Delta_k}} .
\end{aligned}
\tag{7.18}
$$

Therefore, the LLR of the decision concerning the ML path $\underline{s}_k$ is defined as follows:

$$
L\{P(\underline{s}_k)\} = \ln \frac{P(\underline{s}_k)}{1 - P(\underline{s}_k)} = \Delta_k .
\tag{7.19}
$$

We have shown previously how to derive the LLR of the ML path $\underline{s}_k$ at instant k. Now, we need

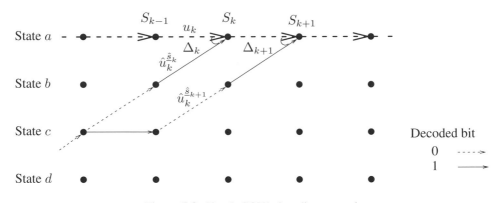

Figure 7.8: Simple SOVA decoding example.

to find the LLR of the decoded bits u_k associated with the survivor path $\underline{s}_k$. Figure 7.8 shows a simplified section of a trellis, where the ML path is the all-zero state sequence. We can see from the figure that path $c \rightarrow b \rightarrow a$, namely $\hat{\underline{s}}_k$, and path $c \rightarrow c \rightarrow b \rightarrow a$, namely $\hat{\underline{s}}_{k+1}$, are the discarded paths at instants k and $k + 1$, respectively. Let $\hat{u}_k^{\hat{\underline{s}}_k}$ and $\hat{u}_k^{\hat{\underline{s}}_{k+1}}$ be the decoded bits at instant k, which would have been the output associated with the discarded paths $\hat{\underline{s}}_k$ and $\hat{\underline{s}}_{k+1}$, had they not been discarded, respectively. As shown in Figure 7.8, u_k and $\hat{u}_k^{\hat{\underline{s}}_k}$ differ. Hence, the LLR of the actually decoded bit u_k is proportional to the LLR Δ_k of the survivor path. On the other hand, u_k and $\hat{u}_k^{\hat{\underline{s}}_{k+1}}$ are the same, namely '0'. Therefore, we would have made no mistake concerning the decoded bit u_k, irrespective of whether path $\underline{s}_k$ or $\hat{\underline{s}}_{k+1}$ was chosen as the most likely state sequence at trellis stage S_{k+1}. The LLR of the decoded bit u_k is then ∞. We define the LLR of u_k taking into account a discarded path $\hat{\underline{s}}_{k+i}$ as:

$$
L_{\hat{\underline{s}}_{k+i}}(u_k) := u_k . \begin{cases} \infty & \text{if } u_k = \hat{u}_k^{\hat{\underline{s}}_{k+i}} \\ \Delta_k & \text{if } u_k \neq \hat{u}_k^{\hat{\underline{s}}_{k+i}}, \end{cases}
\tag{7.20}
$$

where $i \geq 0$.

However, Equation 7.20 considers only one discarded path, while along the ML path, we would have a number of discarded paths. In Figure 7.8, at any node of the trellis there is a survivor and a discarded path. Upon tracing the trellis backwards across i trellis stages, we have to consider a total of $i + 1$ discarded paths in deriving the a-posteriori information $L(u_k|\underline{y})$. It was shown by Hagenauer [54] that the LLR of the decoded bit u_k can be approximated by:

$$L(u_k|\underline{y}) = u_k \cdot \min_{\substack{j=k\ldots n \\ u_k \neq \hat{u}_k^{\hat{s}j}}} \Delta_j \,. \tag{7.21}$$

In order to interpret Equation 7.21, let us consider the following arguments. The algorithm aims to explore whether any unreliable decisions have been encountered while traversing through the trellis, which may have resulted in a decision error. These potentially unreliable decisions are associated with discarding paths in the trellis which had a metric value similar to that of the survivor. These decisions are associated with a low difference Δ between the metrics of the survivor and that of the discarded path. This is, why the algorithm attempts to find the trellis node where the path metric difference was the lowest and, additionally, favouring the similar confidence discarded oath would have resulted in a different bit decision.

7.3.2.1 SOVA Decoding Example

In this section we will augment the concept of the SOVA decoding process using the same example as in Section 4.3.4. We employ BPSK modulation and hence a logical 0 will be transmitted as -1.0 and a logical 1 is sent as $+1.0$. Instead of producing the hard output of the decoded bits, we are going to calculate their soft-outputs. In order to simplify the calculations, we assume that $L_c = 1$ in Equation 5.11. Furthermore, initially there is no intrinsic information for the data bits u_k. Hence, the LLR values $L(u_k)$ are all reset to 0, implying a probability of 0.5.

Figure 7.9 shows our example of the SOVA decoding procedure using the BCH(7,4,3) code. The trellis diagram of the figure is the same as that in Figure 4.10 and so are the demodulator soft-output values as well as the accumulated path metrics. The Viterbi decoder proceeds in the usual way, finding the ML path $\underline{s}_k$, which is the all-zero sequence in this example, by calculating the path metrics. At the end of the trellis, i.e. at stage $T = 7$, the ML path is found and the trace-back procedure [54] begins, in order to find all discarded paths, which would have resulted in different bit decisions.

As seen in Figure 7.9, at instant $T = 7$, the path metric difference between the survivor path $\underline{s}_k$ and the discarded path $\hat{\underline{s}}_k$, according to Equation 7.17, is given by:

$$\begin{aligned} \Delta_7 &= M(\underline{s}_7) - M(\hat{\underline{s}}_7) \\ &= \frac{1}{2}(4.6 - 3.6) \\ &= 0.5 \,. \end{aligned}$$

In this SOVA decoder scenario the decoder starts tracing back the survivor path $\underline{s}_7$ and the discarded path $\hat{\underline{s}}_7$ which merge at $T = 7$. Hence their associated decoded bits are found, which are summarised in Table 7.1 for both paths. For each decoding instant, the exclusive-or (XOR) of the decoded data bits is taken, which will give '1' if and only if u_k and $\hat{u}_k^{\hat{s}_7}$ are different. If the decoded bits are different at an instant, then according to Equation 7.20 the associated LLR becomes $L_{\hat{\underline{s}}_7}(u_k) = u_k . \Delta_7$. On the other hand, if the decoded data bits are the same for an instant, then there is no ambiguity about the data bit given by the ML path. Therefore, the corresponding value of $L_{\hat{\underline{s}}_7}(u_k)$ is $u_k . \infty$.

We can see from Figure 7.9 that there are four discarded paths along the all-zero ML path, yielding the path metric of 4.6. Table 7.2 shows the calculated $|L_{\hat{\underline{s}}_k}(u_k)|$ of the data bits for

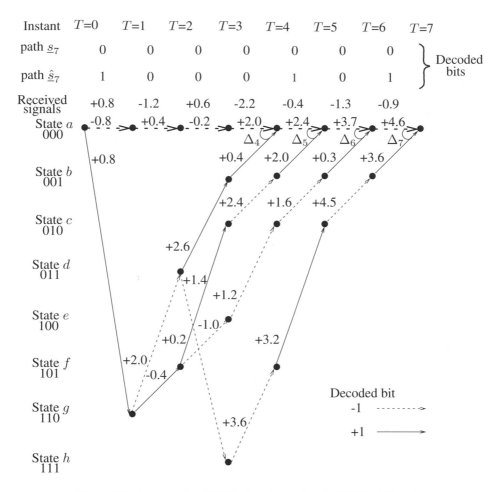

Figure 7.9: An example of SOVA decoding using the BCH(7,4,3) code.

	Instant T						
	0	1	2	3	4	5	6
u_T	0	0	0	0	0	0	0
$u_T^{\hat{\underline{s}}_7}$	1	0	0	0	1	0	1
u_T XOR $u_T^{\hat{\underline{s}}_7}$	1	0	0	0	1	0	1
$\|L_{\hat{\underline{s}}_7}(u_k)\|$	0.5	∞	∞	∞	0.5	∞	0.5

Table 7.1: Decoded bits for path $\underline{s}_k$ and $\hat{\underline{s}}_k$.

each discarded path in Figure 7.9. Specifically, the survivor path $\underline{s}_7$ and the discarded path $\hat{\underline{s}}_7$ are associated with the decoded bits 0, 0, 0, 0, 0, 0, 0 and 1, 0, 0, 0, 1, 0, 1, respectively. When considering the associated decoded bits u_k, they are different for $k = 0$, but identical for $k = 1, 2$ and 3. Hence in Table 7.2 we have $|L_{\hat{\underline{s}}_7}(u_k)| = 0.5, \infty, \infty$ and ∞ for $k = 0, 1, 2$ and 3, respectively.

Similarly, at $T = 6$ the survivor path results in the decoded bits 0, 0, 0, 0, 0, 0, 0, the discarded path in 1, 1, 0, 0, 0, 0, 1. Hence for $k = 0$ and 1 they are different, while for $k = 2$ and 3 the decoded bits are identical, yielding $|L_{\hat{\underline{s}}_6}(u_0)| = |L_{\hat{\underline{s}}_6}(u_1)| = 1.7$ and $|L_{\hat{\underline{s}}_6}(u_2)| = |L_{\hat{\underline{s}}_6}(u_3)| = 0$, as seen in Table 7.2. The remaining values in Table 7.2 can be derived similarly. Referring to Equation 7.21, we have to find the smallest Δ_k for which $u_k \neq \hat{u}_k$, i.e. the smallest $|L_{\hat{\underline{s}}_k}(u_k)|$ since this approximates the LLR of the survivor path. Hence, the minimum is taken along each column in Table 7.2 . According to Equation 7.21, the a-posteriori information $L(u_k|y)$ of bit u_k is then the polarity of the decoded data bit u_k times the minimum path metric difference Δ in Table 7.2. In Section 7.3.1, we have derived the relationship between

$L_{\hat{\underline{s}}_k}(u_k)$	Data bits u_k			
	u_0	u_1	u_2	u_3
$|L_{\hat{\underline{s}}_7}(u_k)|$	0.5	∞	∞	∞
$|L_{\hat{\underline{s}}_6}(u_k)|$	1.7	1.7	∞	∞
$|L_{\hat{\underline{s}}_5}(u_k)|$	0.2	0.2	0.2	∞
$|L_{\hat{\underline{s}}_4}(u_k)|$	0.8	∞	0.8	0.8
Minimum Δ	0.2	0.2	0.2	0.8
u_k	-1	-1	-1	-1
$L(u_k|y)$	-0.2	-0.2	-0.2	-0.8

Table 7.2: LLR for the decoded data bits.

the a-posteriori information $L(u_k|y)$ and the extrinsic information $L_e(u_k)$. In order to calculate the extrinsic information, we rewrite Equation 7.8 as:

$$L_e(u_k) = L(u_k|y) - L_c y_k - L(u_k). \qquad (7.22)$$

Applying Equation 7.22, we can find the extrinsic information $L_e(u_k)$ of the data bits by removing the soft channel output $L_c y_k$ and the intrinsic information $L(u_k)$ from the a-posteriori information $L(u_k|y)$. Specifically, $L(u_k|y)$ is given at the bottom of Table 7.2, $L_c y_k$ is the demodulator's soft-output seen as the 'received signal' in Figure 7.9, while $L(u_k) = 0$ in the first iteration, since all bit probabilities are 0.5. In summary, Table 7.3 shows the extrinsic information of the data bits in our example. The extrinsic information will then be passed to the second decoder in Figure 7.3.

7.4 Turbo Decoding Example

In this section, we discuss an example of turbo decoding using the SOVA detailed in Section 7.3.2. This example illustrates how iterative decoding assists in correcting multiple errors. Our elaborations are based on Sections 4.2, 4.3 and 7.3.2.

The component codes used in this example are two BCH(7,4,3) codes, which are combined, as shown in Figure 7.10, with a 2×2 block interleaver [120] for creating a simple turbo code. As

	Data bits u_k			
	u_0	u_1	u_2	u_3
$L(u_k|y)$	-0.2	-0.2	-0.2	-0.8
$L_c y_k$	$+0.8$	-1.2	$+0.6$	-2.2
$L(u_k)$	0.0	0.0	0.0	0.0
$L_e(u_k)$	-1.0	-1.4	-0.8	$+1.4$

Table 7.3: The a-posteriori information, soft channel outputs, intrinsic and extrinsic information of the data bits.

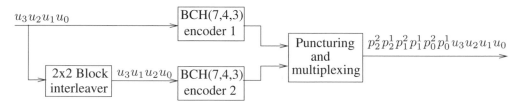

Figure 7.10: BCH(7,4,3) turbo encoder.

we can see from Figure 7.10, the parity bits p_k^1, generated by BCH encoder 1, and the parity bits p_k^2, of BCH encoder 2, are not punctured. Since the data bits of both component codes are the same, all the data bits of the second component code are punctured. The transmitted sequence therefore contains four data bits and six parity bits, resulting in a code rate of 0.4.

For the sake of simplicity, we assume that all data bits are binary zeros. Hence, both BCH(7,4,3) encoders generate an all-zero output sequence. Assuming that BPSK modulation is used, a logical 0 is transmitted as -1, whereas a logical 1 is sent as $+1$. The transmitted sequence of the turbo BCH encoder is hence a series of ten consecutive -1s.

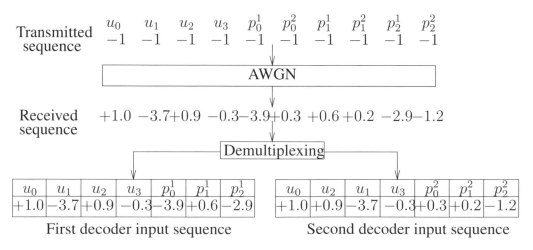

Figure 7.11: Demultiplexing process of the received sequence.

As we can see in Figure 7.11, the transmitted sequence, which has an energy per bit of

$E_b = 1$, is conveyed through an AWGN channel. Since there is no fading in an AWGN channel, the fading amplitude is $a = 1$ and the variance of the AWGN is $\sigma = \sqrt{2}$. Hence the channel reliability value L_c of Equation 5.11 is given by:

$$L_c = \frac{E_B}{2\sigma^2}.4a$$
$$= 1 . \tag{7.23}$$

The received sequence $\underline{y}$ has been corrupted by noise, which is passed to the demultiplexer that separates the received sequence to give the input sequence of the first and second decoder, as shown in Figure 7.11. In both decoders, the received samples due to the original data bits are the same, except that the sequence of the data bits of the second decoder was interleaved. This assists in increasing the error correction capability. This is because if the first decoder fails to correct the errors, the second decoder will use the rearranged received data bit sequence to correct the errors, and vice versa. In our following discourse we demonstrate how both decoders assist each other in correcting the errors.

The path metric of the SOVA is given by Equation 7.15, which is repeated here for convenience:

$$M(\underline{s}_k) = M(\underline{s}_{k-1}) + \frac{u_k}{2}\{L(u_k) + L_c y_k\} . \tag{7.24}$$

Here $M(\underline{s}_{k-1})$ is the path metric for the surviving path traversing through the state $S_{k-1} = \grave{s}$ at stage $k - 1$ in the trellis, u_k is the estimated transmitted bit associated with a given transition, y_k is the demodulator's soft-output sample for that transition and L_c is the associated channel reliability value, which was found to be unity in Equation 7.23. Hence, $L_c y_k = y_k$.

Initially, we consider the operation of the first decoder during its first iteration, where the associated trellis is shown in Figure 7.12. There is no a-priori information and hence we have $L(u_k) = 0$ for all k, which corresponds to the a-priori probability of 0.5, as shown in Figure 5.4. The received signal $L_c y_k$ constituted by the demodulator's soft output is the first decoder's input sequence in Figure 7.11. According to Equation 7.24, the transition metrics are derived by the summation of the intrinsic information $L(u_k)$ and the received signal $L_c y_k$ constituted by the demodulator's soft output, at each trellis transition. The values of the intrinsic information, the received signals and the resultant transition metrics are shown at the top of Figure 7.12 for every instant.

The SOVA proceeds in the same way as the VA in order to find the ML path (or survivor path) which is shown by the bold line in Figure 7.12. As we can see in the figure, the ML path of the first decoder in the first iteration is the state sequence of $a \rightarrow g \rightarrow d \rightarrow b \rightarrow a \rightarrow a \rightarrow a \rightarrow a$, which is not the all-zero state sequence. Similarly to our example in Section 7.3.2.1, there are four discarded paths along the ML path. The decoded bits of the ML path u_k and those that would have been produced by the four discarded paths $u_k^{\grave{s}_k}$ are shown at the bottom of Figure 7.12 for the convenience of the reader. In Section 7.3.2, we defined the path metric difference and here we restate this equation as follows:

$$\Delta_k = M(\underline{s}_k^s) - M(\hat{\underline{s}}_k^s) , \tag{7.25}$$

where $M(\underline{s}_k^s)$ and $M(\hat{\underline{s}}_k^s)$ are the path metrics of ML path $\underline{s}$ and the discarded path $\hat{\underline{s}}$, respectively. Since there are four discarded paths along the ML path, there will be four path metric differences Δ_k. The path metric difference Δ_k for instants 4 to 7 is shown at the right of the trellis in Figure 7.12. Specifically, at $T = 4$ in state a the discarded path is $a \rightarrow g \rightarrow d \rightarrow b \rightarrow a$ and the path metric difference is $\Delta_4 = (5.3 - 2.1)/2 = 1.6$. Similarly, at $T = 5$ in state a the discarded

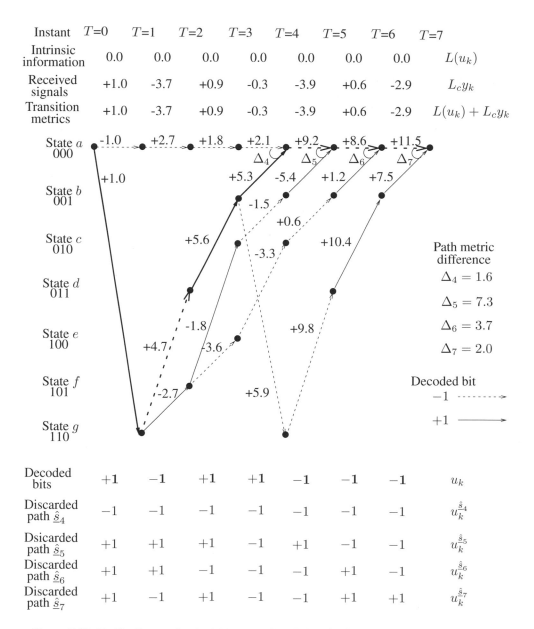

Figure 7.12: Trellis diagram for the SOVA decoding during the first iteration of the first decoder.

path is $a \to g \to f \to c \to b \to a$ and we have $\Delta_5 = \{9.2 - (-5.4)\}/2 = 7.3$. The remaining path metric difference can be derived similarly.

	$L_{\hat{\underline{s}}_k}(u_k)$				Min	u_k	$L(u_k|\underline{y})$	$L_c y_k$	$L(u_k)$	$L_e(u_k)$
k	$|L_{\hat{\underline{s}}_4}(u_k)|$	$|L_{\hat{\underline{s}}_5}(u_k)|$	$|L_{\hat{\underline{s}}_6}(u_k)|$	$|L_{\hat{\underline{s}}_7}(u_k)|$						
0	1.6	∞	∞	∞	1.6	$+1$	$+1.6$	$+1.0$	0.0	$+0.6$
1	∞	7.3	3.7	∞	3.7	-1	-3.7	-3.7	0.0	0.0
2	1.6	∞	3.7	∞	1.6	$+1$	$+1.6$	$+0.9$	0.0	$+0.7$
3	1.6	7.3	3.7	2.0	1.6	$+1$	$+1.6$	-0.3	0.0	$+1.9$

Table 7.4: SOVA output for the first iteration of the first decoder in terms of the a-posteriori information $L(u_k|\underline{y})$, soft channel outputs $L_c y_k$, intrinsic information $L(u_k)$ and extrinsic information $L_e(u_k)$.

The a-posteriori information $|L_{\hat{\underline{s}}_k}(u_k)|$ of the decoded bit u_k taking into account the discarded path $\hat{\underline{s}}_k$ is shown in Table 7.4, for all four discarded paths. We have shown in Equation 7.20 that $|L_{\hat{\underline{s}}_k}(u_k)| = \Delta_k$ if u_k and $u_k^{\hat{s}_k}$ differ, while $|L_{\hat{\underline{s}}_k}(u_k)| = \infty$, if the decoded bits associated with both the survivor path and the discarded path are the same. Again in Figure 7.12 the ML path $\underline{s}_4$ converges with the all-zero state sequence at instant $T = 4$ and the path metric difference is $\Delta_4 = 1.6$. Since the decoded bits of the ML path and the discarded path differ at instant $k = 0, 2$ and 3, the values of the a-posteriori information $|L_{\hat{\underline{s}}_4}(u_k)|$ for the decoded bits u_k, $k = 0, 2$ and 3, are equal to the path metric difference $\Delta_4 = 1.6$. By contrast, at $k = 1$ we have $|L_{\hat{\underline{s}}_4}(u_1)| = \infty$. This is shown in the second column of Table 7.4. Similarly, at $T = 5$ the survivor path is associated with the data bits 1, 0, 1, 1, and the discarded path with bits 1, 1, 1, 0. Hence at $k = 0$ and 2, we have $|L_{\hat{\underline{s}}_5}(u_k)| = \infty$, while at $k = 1$ and 3, $|L_{\hat{\underline{s}}_5}(u_k)| = 7.3$, as seen in the third column of Table 7.4. The associated a-posteriori information for other discarded paths is shown in the consecutive columns as well.

In order to derive the a-posteriori information $L(u_k|\underline{y})$ of the decoded bit u_k, we can approximate the LLR as follows:

$$L(u_k|\underline{y}) = u_k \cdot \min_{\substack{T=k\ldots n \\ u_k \neq \hat{u}_k^{\hat{s}_T}}} \Delta_T . \tag{7.26}$$

Equation 7.26 is the same as Equation 7.21 and restated here for convenience. In Table 7.4, at trellis stage $k = 3$, the values of $|L_{\hat{\underline{s}}_4}(u_3)|$, $|L_{\hat{\underline{s}}_5}(u_3)|$, $|L_{\hat{\underline{s}}_6}(u_3)|$ and $|L_{\hat{\underline{s}}_7}(u_3)|$ are 1.6, 7.3, 3.7 and 2.0 corresponding to the four discarded paths respectively and the minimum path metric difference in this set is 1.6. Using Equation 7.26, we can derive the a-posteriori information of u_4, yielding $L(u_4|\underline{y}) = +1.6$, since the decoded bit is $u_4 = +1$. Figure 7.3 shows that now we have to derive the extrinsic information $L_e(u_k)$, which is given by Equation 7.22 in Section 7.3.2.1 as:

$$L_e(u_k) = L(u_k|\underline{y}) - L(u_k) - L_c y_k . \tag{7.27}$$

This equation states that the extrinsic information $L_e(u_k)$ is given by subtracting the intrinsic information $L(u_k)$ and the received signal $L_c y_k$ from the a-posteriori information $L(u_k|\underline{y})$ of the decoder. The last column in Table 7.4 shows the extrinsic information calculated from Equation 7.27.

We now proceed to describe the operation of the second component decoder during its first iteration. The a-priori information $L(u_k)$ of the second decoder is then the extrinsic information produced by the first decoder, after interleaving by a 2×2 block interleaver which was shown in

Figure 7.3. It also uses the interleaved demodulator soft outputs $L_c y_k$ and the received parity bits produced by the second BCH encoder, i.e. the second decoder input sequence in Figure 7.11.

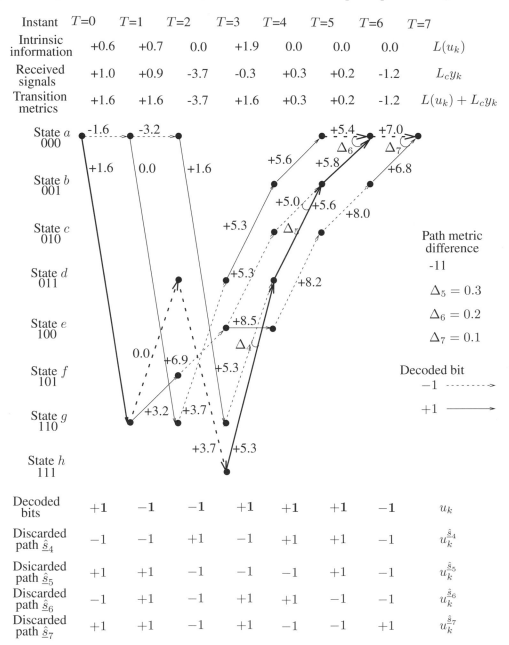

Figure 7.13: Trellis diagram for the SOVA decoding during the first iteration of the second decoder.

Figure 7.13 shows the SOVA decoding trellis of the second decoder during its first iteration. The extrinsic information values extracted from Table 7.4 — after interleaving — are shown as the a-priori information $L(u_k)$ at the top of the trellis. Also shown is the received signal

$L_c y_k$ constituted by the demodulator's soft output and the transition metrics $L(u_k) + L_c y_k$. The decoded bits of the ML path and those of the four other discarded paths are shown at the bottom of the trellis.

The ML path chosen by the second component decoder is shown by the bold line in Figure 7.13. The associated state sequence is $a \rightarrow g \rightarrow d \rightarrow h \rightarrow d \rightarrow b \rightarrow a \rightarrow a$ and, again, it is not the all-zero state sequence. Using Equation 7.25, we can calculate the path metric difference Δ_k of the ML path and the discarded paths, which are shown at the right of the trellis in Figure 7.13. Note that in Figure 7.12 the intrinsic information $L(u_k)$ was zero, while in Figure 7.13 we have valuable intrinsic information $L(u_k)$ provided by the first decoder. This substantially changes the associated transition metrics and path metric differences.

k	$|L_{\hat{s}_4}(u_k)|$	$|L_{\hat{s}_5}(u_k)|$	$|L_{\hat{s}_6}(u_k)|$	$|L_{\hat{s}_7}(u_k)|$	Min	u_k	$L(u_k|y)$	$L_c y_k$	$L(u_k)$	$L_e(u_k)$
			$L_{\hat{s}_k}(u_k)$							
0	6.9	∞	0.2	∞	0.2	+1	+0.2	+1.0	+0.6	−1.4
1	∞	0.3	0.2	0.1	0.1	−1	−0.1	+0.9	+0.7	−1.7
2	6.9	∞	∞	∞	6.9	−1	−6.9	−3.7	0.0	−3.2
3	6.9	0.3	∞	∞	0.3	+1	+0.3	−0.3	+1.9	−1.3

Table 7.5: SOVA output for the first iteration of the second decoder in terms of the a-posteriori information $L(u_k|y)$, soft channel outputs $L_c y_k$, intrinsic information $L(u_k)$ and extrinsic information $L_e(u_k)$.

Using the same steps as described in the context of the first decoder seen in Figure 7.12, we can calculate the a-posteriori information $L(u_k|y)$ and the extrinsic information $L_e(u_k)$, which are shown in Table 7.5. Table 7.5 has the same structure as Table 7.4. Considering Figure 7.5 in a little more depth, at $T = 4$ we observe in state d that the path metric difference is $\Delta_4 = 6.9$ between the survivor and the discarded paths. Their decoded bit sequences are 1, 0, 0, 1 and 0, 0, 1, 0, respectively. The only coincident bit decision is at $k = 1$, where we therefore have $|L_{\hat{s}_4}(u_1)| = \infty$ in Table 7.5. By contrast, for $k = 0, 2$ and 3 we have $|L_{\hat{s}_4}(u_0)| = |L_{\hat{s}_4}(u_2)| = |L_{\hat{s}_4}(u_3)| = 6.9$.

Similarly, at $T = 5$ and in state b the survivor and discarded paths have decoded patterns of 1, 0, 0, 1 and 1, 1, 0, 0, respectively, which are identical for $k = 0$ and 2. Hence the associated a-posteriori information in Table 7.5 at $k = 0$ and 2 is ∞, while at $k = 1$ and 3 it is $\Delta_5 = (5.6 - 5.0)/2 = 0.3$. The remaining a-posteriori information values seen in Table 7.5 accrue similarly. Their minima are found for each value of k, which are also listed in the sixth column of Table 7.5. According to Equation 7.26 the a-posteriori values $L(u_k|y)$ are given by the product of the decoded bits u_k, $k = 0, \ldots, 3$, and the above-mentioned minimum path metric difference. They are summarised in the eighth column of Table 7.5.

The ninth column of Table 7.5 repeats the soft demodulator outputs due to the received signal samples $+1.0$, $+0.9$, -3.7 and -0.3 extracted from Figure 7.13, while the tenth column summarises the intrinsic values of $+0.6$, $+0.7$, 0.0 and $+1.9$ provided by the other decoder. These values can be seen in both Figure 7.13 and in Table 7.5. Finally, the extrinsic values of the last column are derived from Equation 7.27.

Let us now compare the a-posteriori information $L(u_k|y)$ of the second decoder and the first decoder. In the first decoder characterised in Figure 7.12, we have three decoding errors in the data bit sequence and the number of errors is reduced to two in the second decoder of Figure 7.13. The soft outputs of the data bits u_0, u_1 and u_3, produced by the second decoder, which are the interleaved data bits u_0, u_2 and u_3 of the first decoder, are relatively low and are close to zero, which indicates a low confidence. Indeed, by comparing the eighth columns of Tables 7.4 and 7.5 we find that the confidence values on the whole have been reduced by invoking the second decoder. At first sight this confidence measure reduction might appear undesirable. However, we show that after the first decoding three of the bits were erroneous and hence the polarity of the associated LLRs was wrong. We expect the decoder to eventually change the polarity of these bits during the forthcoming iterations. This can only be achieved by first reducing the LLRs' magnitude and then changing their polarity during the successive iterations. Once the erroneous LLR polarity has been changed, it may become possible to increase the magnitude of the corresponding LLRs, which reflects their increased confidence. By contrast, data bit u_2, which is the interleaved data bit u_1 of the first decoder, shows a significant increase in terms of its reliability or confidence. In summary, we observe that the first decoder provides an estimation of the data bits and that the second decoder improves the reliability of the soft outputs.

In the second, and all subsequent, iterations the first component decoder is capable of using the extrinsic information provided by the second decoder in the previous iteration as intrinsic information. As a further step, Figure 7.14 shows the trellis diagram of SOVA decoding in the first decoder during the second iteration. It can be seen in the figure that this decoder uses the same channel information $L_c y_k$, as it did in the first iteration. In contrast to Figure 7.12, it has, however, intrinsic information provided by the second decoder during the first iteration, in order to assist in finding the correct path through the trellis. The selected ML path is again shown by the bold line in Figure 7.12 and it can be seen that now the correct all-zero path is chosen. Again, the a-posteriori information $L(u_k|\underline{y})$ and the extrinsic information $L_e(u_k)$ are calculated and summarised in Table 7.6.

k	$L_{\hat{s}_k}(u_k)$				Min	u_k	$L(u_k|\underline{y})$	$L_c y_k$	$L(u_k)$	$L_e(u_k)$
	$|L_{\hat{s}_4}(u_k)|$	$|L_{\hat{s}_5}(u_k)|$	$|L_{\hat{s}_6}(u_k)|$	$|L_{\hat{s}_7}(u_k)|$						
0	2.8	12.0	∞	3.5	2.8	-1	-2.8	$+1.0$	-1.4	-2.4
1	∞	12.0	∞	∞	12.0	-1	-12.0	-3.7	-3.2	-5.1
2	2.8	12.0	4.1	3.5	2.8	-1	-2.8	$+0.9$	-1.7	-2.0
3	2.8	∞	∞	∞	2.8	-1	-2.8	-0.3	-1.3	-1.2

Table 7.6: SOVA output for the second iteration of the first decoder in terms of the a-posteriori information $L(u_k|\underline{y})$, soft channel outputs $L_c y_k$, intrinsic information $L(u_k)$ and extrinsic information $L_e(u_k)$.

The second iteration is then completed by finding the extrinsic information generated by the first decoder, interleaving it and using it as intrinsic information for the second decoder. It can be shown that the second decoder will also select the all-zero path as the ML path, and hence the output of the turbo decoder after the second iteration will be the correct all -1 sequence. This concludes our example concerning the operation of iterative turbo decoding using the SOVA.

7.5 MAP Algorithm for Extended BCH codes

7.5.1 Introduction

This section presents a block turbo decoder using extended BCH codes as the component codes. The MAP algorithm is modified in order to incorporate the parity check bit in calculating the LLR of the decoded bits.

Conventional BCH codes are denoted by BCH(n, k, d_{min}), where n, k, d_{min} denote the codeword length, the number of information data bits and the minimum free distance, respectively. These codes are referred to as extended BCH codes [96], if a parity check bit is appended to the BCH codeword. This extends the primary BCH codeword by one bit, to $n + 1$, and it expands

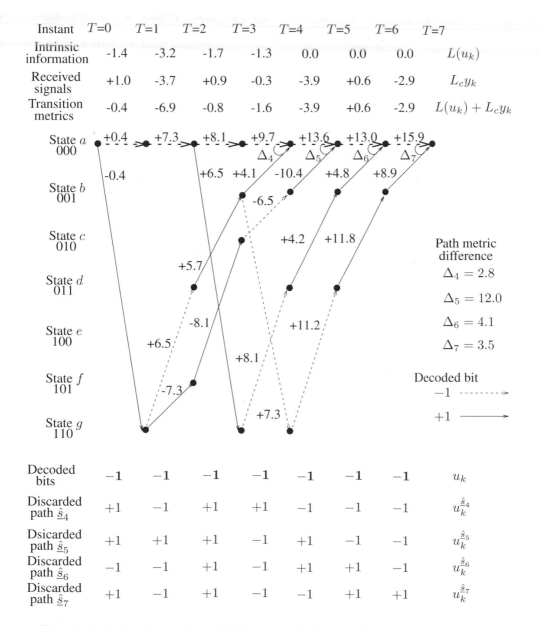

Instant	$T{=}0$	$T{=}1$	$T{=}2$	$T{=}3$	$T{=}4$	$T{=}5$	$T{=}6$	$T{=}7$
Intrinsic information	-1.4	-3.2	-1.7	-1.3	0.0	0.0	0.0	$L(u_k)$
Received signals	+1.0	-3.7	+0.9	-0.3	-3.9	+0.6	-2.9	$L_c y_k$
Transition metrics	-0.4	-6.9	-0.8	-1.6	-3.9	+0.6	-2.9	$L(u_k) + L_c y_k$

Path metric difference

$\Delta_4 = 2.8$

$\Delta_5 = 12.0$

$\Delta_6 = 4.1$

$\Delta_7 = 3.5$

Decoded bit

-1 - - - - - - - ->

$+1$ ———————>

Decoded bits	-1	-1	-1	-1	-1	-1	-1	u_k
Discarded path $\hat{\underline{s}}_4$	$+1$	-1	$+1$	$+1$	-1	-1	-1	$u_k^{\hat{s}_4}$
Dsicarded path $\hat{\underline{s}}_5$	$+1$	$+1$	$+1$	-1	$+1$	-1	-1	$u_k^{\hat{s}_5}$
Discarded path $\hat{\underline{s}}_6$	-1	-1	$+1$	-1	$+1$	$+1$	-1	$u_k^{\hat{s}_6}$
Discarded path $\hat{\underline{s}}_7$	$+1$	-1	$+1$	-1	-1	$+1$	$+1$	$u_k^{\hat{s}_7}$

Figure 7.14: Trellis diagram for the SOVA decoding in the second iteration of the first decoder.

its minimum free distance from d_{min} to $d_{min} + 1$. Hence the extended BCH code is denoted by BCH($n + 1, k, d_{min} + 1$). Figure 7.15 shows the weight distribution [19] of the BCH(31,26,3) and that of the extended BCH(32,26,4) codes. Notice that the extended BCH code has only even weight terms because of the effect of the parity check bit.

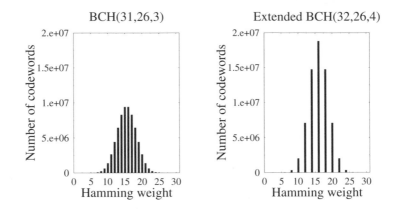

Figure 7.15: Weight distribution of the BCH(31,26,3) and BCH(32,26,4) codes.

7.5.2 Modified MAP Algorithm

7.5.2.1 The Forward and Backward Recursion

In order to incorporate the parity check bit of extended BCH codes in the MAP algorithm, we have to make modifications to Equations 7.4 as well as 7.6 and then to Equation 7.1. Let us define $\alpha_k^e(s)$ as the probability that the current trellis state is $S_k = s$ at time k, where the superscript indicates that the path arriving at this state gave an even number of transmitted bits, which were $+1$, given that the demodulator's soft output up to this point was $\underline{y}_{j \leq k}$. Similarly, $\alpha_k^o(s)$ is defined as the probability that the current trellis state is $S_k = s$ at time k, where the superscript indicates that the path leading to this state gave an odd number of transmitted bits, which were $+1$, given that the demodulator's soft output up to this point was $\underline{y}_{j \leq k}$. For each state in the trellis, we determine the probabilities $\alpha_k^e(s)$ and $\alpha_k^o(s)$, which are shown in Figure 7.16.

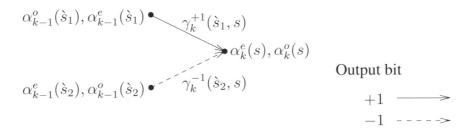

Figure 7.16: Forward recursion of $\alpha_k^e(s)$ and $\alpha_k^o(s)$. The corresponding conventional MAP forward recursion was shown in Figure 7.5.

Then, using Equation 7.4 and Figure 7.16, for a binary trellis we can derive the total probability of an even number of transmitted binary $+1$ bits as the sum of the probabilities of having
an even number of $+1$s at stage $k-1$ and encountering a -1 during the current transition, plus
the probability of an odd number of $+1$s at the previous stage and encountering a $+1$ during the
current transition, yielding:

$$\alpha_k^e(s) = \alpha_{k-1}^o(\grave{s}_1).\gamma_k^{+1}(\grave{s}_1,s) + \alpha_{k-1}^e(\grave{s}_2).\gamma_k^{-1}(\grave{s}_2,s) \;. \tag{7.28}$$

Explicitly, since transition $\gamma_k^{-1}(\grave{s}_2,s)$ from the previous state $\grave{s}_2$ to the present state s results in
the transmitted bit being -1, $\alpha_{k-1}^e(\grave{s}_2).\gamma_k^{-1}(\grave{s}_2,s)$ is the probability of the path to state s, which
gave an even number of transmitted bits that were $+1$. Conversely, $\alpha_{k-1}^o(\grave{s}_1)$ is the probability
of the path to state $\grave{s}_1$, which gave an odd number of transmitted $+1$ bits. Since transition
$\gamma_k^{+1}(\grave{s}_1,s)$ from the previous state $\grave{s}_1$ to the present state s results in the transmitted bit being $+1$,
$\alpha_{k-1}^o(\grave{s}_1).\gamma_k^{+1}(\grave{s}_1,s)$ is the probability of the path to state s, which also gave an even number of
transmitted bits that were $+1$.

Again, using Equation 7.4 and Figure 7.16, $\alpha_k^o(s)$ is derived for a binary trellis:

$$\alpha_k^o(s) = \alpha_{k-1}^e(\grave{s}_1).\gamma_k^{+1}(\grave{s}_1,s) + \alpha_{k-1}^o(\grave{s}_2).\gamma_k^{-1}(\grave{s}_2,s) \;. \tag{7.29}$$

The same modification is made to the backward recursion and hence we derive from Equation 7.6:

$$\beta_{k-1}^e(\grave{s}) = \beta_k^o(s_1).\gamma_k^{+1}(\grave{s},s_1) + \beta_k^e(s_2).\gamma_k^{-1}(\grave{s},s_2) \tag{7.30}$$

and:

$$\beta_{k-1}^o(\grave{s}) = \beta_k^e(s_1).\gamma_k^{+1}(\grave{s},s_1) + \beta_k^o(s_2).\gamma_k^{-1}(\grave{s},s_2) \;. \tag{7.31}$$

7.5.2.2 Transition Probability

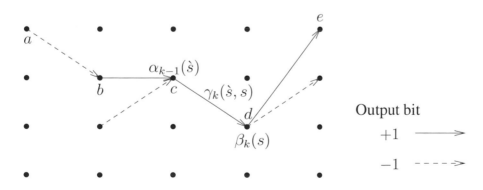

Figure 7.17: Probability of a transition in the trellis.

Figure 7.17 shows a simple trellis diagram which commences at state a and ends at state e.
The probability of the transition from state c to state d is formulated as:

$$P(c \to d) = \alpha_{k-1}(\grave{s}).\gamma_k(\grave{s},s).\beta_k(s) \;. \tag{7.32}$$

For extended BCH codes, however, the probability of a transition in the trellis no longer depends purely on $\alpha_{k-1}(\grave{s})$, $\gamma_k(\grave{s}, s)$ and $\beta_k(s)$. It also depends on whether the whole path gave an even or odd number of transmitted bits, which were $+1$. In Figure 7.17, path $a \to b \to c \to d \to e$ gives an odd number of transmitted bits, which were $+1$. Hence, the probability of the transition $c \to d$ is:

$$P(c \to d) = \alpha_{k-1}(\grave{s}).\gamma_k(\grave{s}, s).\beta_k(s).P(y_n|x_n = +1) \,, \qquad (7.33)$$

where x_n is the transmitted parity bit, while y_n is the corresponding demodulator soft output.

Similarly, if the path $a \to b \to c \to d \to e$ had resulted in an even number of transmitted bits, which were $+1$, the probability of the transition would be:

$$P(c \to d) = \alpha_{k-1}(\grave{s}).\gamma_k(\grave{s}, s).\beta_k(s).P(y_n|x_n = -1) \,. \qquad (7.34)$$

7.5.2.3   A-posteriori Information

For each trellis state, we now have $\alpha_k^e(s)$, $\alpha_k^o(s)$, $\beta_k^e(s)$ and $\beta_k^o(s)$ in contrast to $\alpha_k(s)$ and $\beta_k(s)$ in the generic MAP algorithm. In other words, we have separated the probabilities $\alpha_k(s)$ and $\beta_k(s)$ into two groups, respectively, depending on the nature of the trellis paths that reach state s. Additionally, we also have the probability $P(y_n|x_n)$ for the transmitted parity bit. Hence, we can derive the a-posteriori LLR $L(u_k|\underline{y})$ from Equation 7.1, as follows:

$$L(u_k|\underline{y}) =$$

$$\ln \frac{\displaystyle\sum_{\substack{(\grave{s},s) \Rightarrow \\ u_k = +1}} \gamma_k^{+1}(\grave{s}, s).[(\alpha^e.\beta^e + \alpha^o.\beta^o).P(y_n|x_n = +1) + (\alpha^e.\beta^o + \alpha^o.\beta^e).P(y_n|x_n = -1)]}{\displaystyle\sum_{\substack{(\grave{s},s) \Rightarrow \\ u_k = -1}} \gamma_k^{-1}(\grave{s}, s).[(\alpha^e.\beta^o + \alpha^o.\beta^e).P(y_n|x_n = +1) + (\alpha^e.\beta^e + \alpha^o.\beta^o).P(y_n|x_n = -1)]} \,,$$

$$(7.35)$$

where α^e, α^o, β^e and β^o are shorthand for $\alpha_{k-1}^e(\grave{s})$, $\alpha_{k-1}^o(\grave{s})$, $\beta_k^e(s)$ and $\beta_k^o(s)$, respectively.

By definition, in the numerator of Equation 7.35, the transition from the previous state $S_{k-1} = \grave{s}$ to the present state $S_k = s$ always results in a transmitted bit of $+1$. The path $\alpha_{k-1}^e(\grave{s}).\gamma_k^{+1}(\grave{s}, s).\beta_k^e(s)$ results in an odd number of transmitted bits that are $+1$, since both the paths to state $S_{k-1} = \grave{s}$, and from state $S_k = s$, gave an even number of transmitted bits which were $+1$, and the transition from state $S_{k-1} = \grave{s}$ to state $S_k = s$ results in a transmitted bit of $+1$. Similarly, the path $\alpha_{k-1}^o(\grave{s}).\gamma_k^{+1}(\grave{s}, s).\beta_k^o(s)$ also results in an odd number of transmitted bits that are $+1$. Therefore, the probabilities of both transitions are $\alpha^e.\gamma_k^{+1}(\grave{s}, s).\beta^e.P(y_n|x_n = +1)$ and $\alpha^o.\gamma_k^{+1}(\grave{s}, s).\beta^o.P(y_n|x_n = +1)$, respectively. Conversely, paths $\alpha_{k-1}^e(\grave{s}).\gamma_k^{+1}(\grave{s}, s).\beta_k^o(s)$ and $\alpha_{k-1}^o(\grave{s}).\gamma_k^{+1}(\grave{s}, s).\beta_k^e(s)$ would result in an even number of $+1$ transmitted bits. Hence, the probabilities of both transitions are $\alpha^e.\gamma_k^{+1}(\grave{s}, s).\beta^o.P(y_n|x_n = -1)$ and $\alpha^o.\gamma_k^{+1}(\grave{s}, s).\beta^e.P(y_n|x_n = -1)$, respectively. The same argument is applied to the denominator in Equation 7.35 in order to derive the probability of each transition.

7.5.3   Max-Log-MAP and Log-MAP Algorithms for Extended BCH codes

As mentioned in Section 5.3.5, conceptually the Max-Log-MAP and Log-MAP algorithms are the same, except that the approximation made in Equation 5.48 in the context of the Max-Log-MAP algorithm can be made exact by using the Jacobian logarithm [95]. Hence, in this section, we will only describe the Max-Log-MAP algorithm.

The proposed modified MAP algorithm calculates the a-posteriori LLRs $L(u_k|\underline{y})$ using Equation 7.35. Hence, it requires the following values:

1) The $\alpha^e_{k-1}(\grave{s})$ and $\alpha^o_{k-1}(\grave{s})$ values, which are calculated in a forward recursive manner using Equations 7.28 and 7.29, respectively.

2) The $\beta^e_k(s)$ and $\beta^o_k(s)$ values, which are calculated in a backward recursive manner using Equations 7.30 and 7.31, respectively.

3) The transition probabilities $\gamma_k(\grave{s}, s)$.

4) The probability that the number of transmitted bits which are $+1$ is even, $P(y_n|x_n = +1)$, or odd, $P(y_n|x_n = -1)$.

The Max-Log-MAP algorithm simplifies the preceding equations by transferring them into the logarithmic domain and then using the approximation [50, 51]:

$$\ln\left(\sum_i e^{x_i}\right) \approx \max_i(x_i) , \tag{7.36}$$

where $\max_i(x_i)$ means the maximum value of x_i.

Let us define $A^e_k(s)$, $A^o_k(s)$, $B^e_k(s)$, $B^o_k(s)$ and $\Gamma_k(\grave{s}, s)$ as follows:

$$A^e_k(s) \triangleq \ln[\alpha^e_k(s)] , \tag{7.37}$$

$$A^o_k(s) \triangleq \ln[\alpha^o_k(s)] , \tag{7.38}$$

$$B^e_k(s) \triangleq \ln[\beta^e_k(s)] , \tag{7.39}$$

$$B^o_k(s) \triangleq \ln[\beta^o_k(s)] , \tag{7.40}$$

$$\Gamma^{x_k}_k(\grave{s}, s) \triangleq \ln[\gamma^{x_k}_k(\grave{s}, s)] , \tag{7.41}$$

and

$$\epsilon^{x_n} \triangleq \ln[P(y_n|x_n)] . \tag{7.42}$$

Upon substituting Equation 7.28 into Equation 7.37 and using Equation 7.36, we arrive at:

$$
\begin{aligned}
A^e_k(s) &= \ln\left[\alpha^e_{k-1}(\grave{s}).\gamma^{-1}_k(\grave{s}, s) + \alpha^o_{k-1}(\grave{s}).\gamma^{+1}_k(\grave{s}, s)\right] \\
&= \ln\left[e^{\left\{A^e_{k-1}(\grave{s})+\Gamma^{-1}_k(\grave{s},s)\right\}} + e^{\left\{A^o_{k-1}(\grave{s})+\Gamma^{+1}_k(\grave{s},s)\right\}}\right] \\
&\approx \max\left[\left\{A^e_{k-1}(\grave{s}) + \Gamma^{-1}_k(\grave{s}, s)\right\}, \left\{A^o_{k-1}(\grave{s}) + \Gamma^{+1}_k(\grave{s}, s)\right\}\right] .
\end{aligned} \tag{7.43}
$$

Following the same approach and substituting Equation 7.29 into Equation 7.38, we get:

$$
\begin{aligned}
A^o_k(s) &= \ln\left[\alpha^o_{k-1}(\grave{s}).\gamma^{-1}_k(\grave{s}, s) + \alpha^e_{k-1}(\grave{s}).\gamma^{+1}_k(\grave{s}, s)\right] \\
&= \ln\left[e^{\left\{A^o_{k-1}(\grave{s})+\Gamma^{-1}_k(\grave{s},s)\right\}} + e^{\left\{A^e_{k-1}(\grave{s})+\Gamma^{+1}_k(\grave{s},s)\right\}}\right] \\
&\approx \max\left[\left\{A^o_{k-1}(\grave{s}) + \Gamma^{-1}_k(\grave{s}, s)\right\}, \left\{A^e_{k-1}(\grave{s}) + \Gamma^{+1}_k(\grave{s}, s)\right\}\right] ,
\end{aligned} \tag{7.44}
$$

Similarly to Equation 7.43 and 7.44, we can rewrite Equation 7.39 as:

$$
\begin{aligned}
B_k^e(s) &= \ln\left[\beta_k^e(s).\gamma_k^{-1}(\grave{s}, s) + \beta_k^o(s).\gamma_k^{+1}(\grave{s}, s)\right] \\
&= \ln\left[e^{\left\{B_k^e(s) + \Gamma_k^{-1}(\grave{s}, s)\right\}} + e^{\left\{B_k^o(s) + \Gamma_k^{+1}(\grave{s}, s)\right\}}\right] \\
&\approx \max\left[\left\{B_k^e(s) + \Gamma_k^{-1}(\grave{s}, s)\right\}, \left\{B_k^o(s) + \Gamma_k^{+1}(\grave{s}, s)\right\}\right] ,
\end{aligned}
\tag{7.45}
$$

and Equation 7.40 as:

$$
\begin{aligned}
B_k^o(s) &= \ln\left[\beta_k^o(s).\gamma_k^{-1}(\grave{s}, s) + \beta_k^e(s).\gamma_k^{+1}(\grave{s}, s)\right] \\
&= \ln\left[e^{\left\{B_k^o(s) + \Gamma_k^{-1}(\grave{s}, s)\right\}} + e^{\left\{B_k^e(s) + \Gamma_k^{+1}(\grave{s}, s)\right\}}\right] \\
&\approx \max\left[\left\{B_k^o(s) + \Gamma_k^{-1}(\grave{s}, s)\right\}, \left\{B_k^e(s) + \Gamma_k^{+1}(\grave{s}, s)\right\}\right] .
\end{aligned}
\tag{7.46}
$$

We quote and simplify the following equation from Equation 7.2:

$$
\Gamma_k^{x_k}(\grave{s}, s) = \frac{u_k}{2} L(u_k) + \frac{L_c}{2} y_k x_k ,
\tag{7.47}
$$

where $L(u_k)$ is the intrinsic information of the data bit u_k and L_c is the channel reliability value.

If we assume that the bit $x_n = \pm 1$ has been sent over a Gaussian or fading channel using BPSK modulation, then we can write for the conditional probability of the matched filter output y_n that:

$$
P(y_n|x_n = \pm 1) = \frac{1}{\sigma\sqrt{2\pi}} e^{-\frac{E_b}{2\sigma^2}(y_n \mp a)^2} ,
\tag{7.48}
$$

where E_b is the transmitted energy per bit, σ^2 is the noise variance and a is the fading amplitude. For non-fading AWGN channels, the value of a is unity, i.e. $a = 1$.

Using Equation 7.48, we can rewrite Equation 7.42 as:

$$
\begin{aligned}
e^{x_n = \pm 1} &= \ln\left[\frac{1}{\sigma\sqrt{2\pi}}\right] - \frac{E_b}{2\sigma^2}(y_n \mp a)^2 \\
&= C - \frac{E_b}{2\sigma^2}(y_n^2 \mp 2ay_n + a^2) \\
&= C - \frac{E_b}{2\sigma^2}(y_n^2 + a^2) \pm \frac{E_b}{2\sigma^2} 2ay_n \\
&= C - C' \pm \frac{L_c}{2} y_n ,
\end{aligned}
\tag{7.49}
$$

where $C = \ln\left[\frac{1}{\sigma\sqrt{2\pi}}\right]$, $C' = \frac{E_b}{2\sigma^2}(y_n^2 + a^2)$ and $L_c = \frac{2a}{\sigma^2}$. Both C and C' are independent of the transmitted parity check bit x_n and hence can be considered as constants, which are omitted, when comparing the various paths through the trellis.

Finally, using Equations 7.36 to 7.49, we can rewrite the a-posteriori LLRs $L(u_k|\underline{y})$ of the MAP algorithm in Equation 7.35 for the Max-Log-MAP algorithm, yielding:

$$
L(u_k|y) \approx \max_{\substack{(\grave{s}, s) \Rightarrow \\ u_k = +1}} [P(\grave{s}, s)] - \max_{\substack{(\grave{s}, s) \Rightarrow \\ u_k = -1}} [P(\grave{s}, s)] ,
\tag{7.50}
$$

where

$$
\begin{aligned}
P(\grave{s}, s) \quad = \quad & \max[\{\Gamma_k^{u_k}(\grave{s}, s) + A_{k-1}^e(\grave{s}) + B_k^e(s) + \epsilon^{x_n = u_k}\}, \\
& \{\Gamma_k^{u_k}(\grave{s}, s) + A_{k-1}^o(\grave{s}) + B_k^o(s) + \epsilon^{x_n = u_k}\}, \\
& \{\Gamma_k^{u_k}(\grave{s}, s) + A_{k-1}^e(\grave{s}) + B_k^o(s) + \epsilon^{x_n = -u_k}\}, \\
& \{\Gamma_k^{u_k}(\grave{s}, s) + A_{k-1}^o(\grave{s}) + B_k^e(s) + \epsilon^{x_n = -u_k}\}]\,.
\end{aligned}
\tag{7.51}
$$

7.6 Simulation Results

In this section, we are going to present our simulation results for turbo codes using simple BPSK over AWGN channels. We will show that some of the parameters are interlinked, which jointly affect the performance of turbo codes, depending on:

- The number of decoding iterations used.

- The decoding algorithm used.

- The component BCH code employed.

- The frame length of the input data.

- The design of the interleaver used.

All our investigations were conducted using the turbo encoder structure shown in Figure 7.1. The data bits of the upper encoder are transmitted through the channel, whereas the interleaved data bits of the lower encoder are punctured. However, the parity bits from both encoders are multiplexed alternately and none of them are punctured. The multiplexing and puncturing patterns are the same for all simulations, unless stated otherwise.

7.6.1 Number of Iterations Used

Figure 7.18 shows the performance of the rate $R = \frac{26}{36} = 0.72$ turbo BCH(31,26,3) code, using a 26×26-bit block interleaver and the Log-MAP algorithm, versus the number of decoding iterations used. Again, since all the parity bits of both the upper and lower encoder of Figure 7.1 are transmitted through the channel, this results in a coding rate of $R = 0.72$. As the number of iterations used by the decoder increases, the decoder performs significantly better. However, after four iterations there is only little improvement upon using further iterations.

In [148], Goalic and Pyndiah used the extended BCH(32,26,4) product code for real-time block turbo decoding. The authors invoked the soft-input and soft-decision output of algebraic decoding and achieved a coding gain of about 6 dB at a BER of 10^{-5} using four iterations. By contrast, the coding gain of our turbo BCH(31,26,3) code used in Figure 7.18 is about 5.5 dB at a BER of 10^{-5}. However, in [81, 148] the authors use 'factors optimised by simulation' for weighting the soft information and the associated code rate of $\frac{26}{36} = 0.72$ was also slightly lower. Moreover, this decoding method is only applicable to product code.

Figure 7.19 shows the performance of the rate $R = \frac{21}{41} = 0.51$ turbo BCH(31,21,5) code, using a 21×21-bit block interleaver and the Log-MAP algorithm, versus the number of decoding iterations. It can be seen that the coding gain between each iteration is higher than that of the BCH(31,26,3) component code. Specifically, the gain between the first and second iteration is

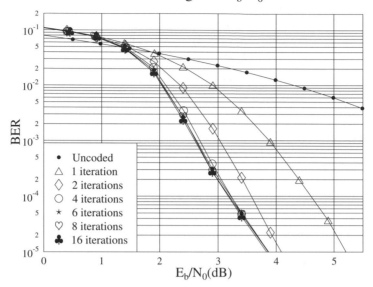

Figure 7.18: Performance comparison of different number of iterations using the rate $R = 0.72$ turbo BCH(31,26,3) code, in conjunction with a 26×26-bit block interleaver and the Log-MAP algorithm over AWGN channels.

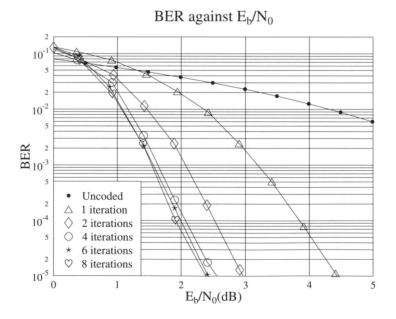

Figure 7.19: Performance comparison of different number of iterations using the rate $R = 0.51$ turbo BCH(31,21,5) code in conjunction with a 21×21-bit block interleaver and the Log-MAP algorithm over AWGN channels.

about 1.5 dB, whereas for the previous turbo BCH(31,26,3) code we had an improvement of about 1 dB only. When using two iterations, the coding gain is about 6.75 dB at a BER of 10^{-5}.

If we compare the complexity of both component codes, the rate $R = 0.72$ turbo BCH(31,26,3) code is significantly less complex than the rate $R = 0.51$ turbo BCH(31,21,3) scheme. Since the trellis decoding complexity of BCH codes is directly proportional to 2^{n-k}, the turbo BCH(31,21,5) code is about $\frac{2^{31-21}}{2^{31-26}} = 32$ times more complex, than the turbo BCH(31,26,3) scheme. Furthermore, the associated memory requirement is also 32 times higher. As an example, four iterations of the turbo BCH(31,26,3) code yield a BER of about 2×10^{-1} at an E_b/N_0 of 3 dB, which is about an order of magnitude better than the BER of the turbo BCH(31,21,5) scheme at one iteration, while its complexity is about eight times lower. At two iterations, however, the approximately 16 times more complex turbo BCH(31,21,5) code outperforms the BCH(31,26,3) turbo arrangement.

7.6.2   The Decoding Algorithm

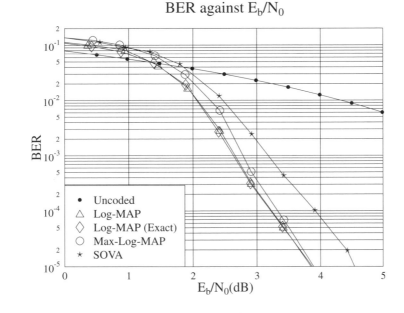

Figure 7.20: Performance comparison between different decoding algorithms for six iterations using the rate $R = 0.72$ turbo BCH(31,26,3) code in conjunction with a 26×26-bit block interleaver over AWGN channels.

Figure 7.20 shows a comparison between the different turbo BCH(31,26,3) component decoders described in Section 7.3, using a 26×26-bit block interleaver. In the figure, the performance of the MAP algorithm [11] is not shown, because it is identical to that of the Log-MAP algorithm [52]. This is justified, since the Log-MAP algorithm is a specific version of the MAP algorithm transformed into the logarithmic domain in order to simplify its operation and to reduce the numerical problems associated with the MAP algorithm, as described in Section 5.3.5.

The 'Log-MAP (Exact)' curve refers to a decoder which calculates the correction term $f_c(\delta)$

in Equation 5.56 of Section 5.3.5.3 exactly, i.e. using:

$$f_c(\delta) = \ln(1 + e^{-\delta}),\qquad (7.52)$$

rather than a look-up table, as described in [52]. The Log-MAP curve refers to a decoder which does use a look-up table with eight values of $f_c(\delta)$ stored, and hence introduces an approximation to the calculation of the LLRs. It can be seen in Figure 7.20 that the 'Log-MAP (Exact)' and 'Log-MAP' algorithms give identical performances. In [52], Robertson et al. found that the look-up table values of the $f_c(\delta)$ correction term in Equation 5.56 introduces no degradation to the performance of the decoder.

It can be seen from Figure 7.20 that the Max-Log-MAP and the SOVA both suffer some degradation in performance compared to the Log-MAP algorithm. At a BER of 10^{-4}, this degradation is about 0.1 dB for the Max-Log-MAP algorithm, and about 0.7 dB for the SOVA. However, at a BER of 10^{-5}, the Max-Log-MAP algorithm suffers only insignificant degradation.

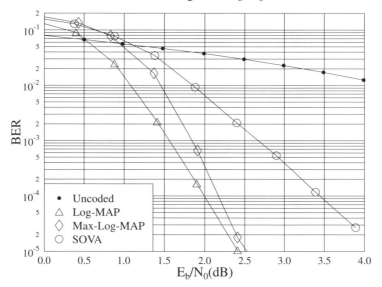

Figure 7.21: Performance comparison between different decoding algorithms for six iterations using the rate $R = 0.51$ turbo BCH(31,21,5) code in conjunction with a 21×21-bit block interleaver over AWGN channels.

Again, Figure 7.21 shows our performance comparison between different decoding algorithms. However, the component code used is the BCH(31,21,5) scheme employing a 21×21-bit block interleaver. The figure shows that the degradation incurred by using the Max-Log-MAP algorithm is approximately 0.2 dB at a BER of 10^{-4}, which is about the same as that for the BCH(31,26,3) component code in the previous case. However, the SOVA shows a significant E_b/N_0 performance degradation of 1.4 dB at a BER of 10^{-4}.

7.6.3   The Effect of Estimating the Channel Reliability Value L_c

In Section 7.3 we have highlighted how the component decoders use the soft inputs, the soft channel outputs $L_c y_k$ and the intrinsic information $L(u_k)$, in order to provide the a-posteriori information $L(u_k|y)$ as its soft outputs. In this section, we investigate how the imperfect estimation of the channel reliability value L_c affects the performance of the algorithms.

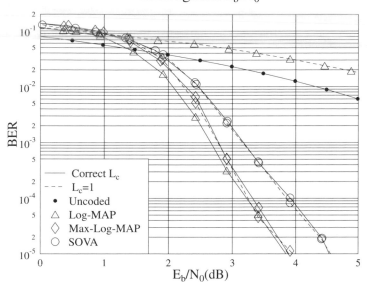

Figure 7.22: Effect of using incorrect channel reliability values L_c on the turbo BCH(31,26,3) code employing six iterations, the Log-MAP algorithm and a 26×26-bit block interleaver over AWGN channels.

Figure 7.22 shows the effect of using imperfect channel reliability values L_c for three different decoding algorithms, namely for the Log-MAP, Max-Log-MAP and SOVA algorithms. For each decoding algorithm, the continous line shows the performance of the algorithm when the channel reliability value L_c is calculated exactly using the known channel SNR. The broken curves in Figure 7.22 show how the three algorithms perform when the channel reliability L_c is not known. For these curves, the value $L_c = 1$, which according to Equation 5.11 corresponds to an E_b/N_0 value of -3 dB, was used at all channel SNRs. It can be seen from Figure 7.22 that the Max-Log-MAP algorithm and the SOVA perform equally well, whether or not the correct value of L_c is known. However, the performance of the Log-MAP algorithm is drastically affected by the incorrect L_c value used. We can see that the performance of the Log-MAP algorithm is even worse than that of the uncoded case, if $L_c = 1$.

The reason for these effects can be understood by considering the different operations described in Section 7.3. For the SOVA the soft channel output $L_c y_k$ is used recursively in order to calculate the path metric using Equation 7.15. In Equation 7.15, we can see that L_c is used to scale the demodulator's soft output y_k and this has an effect of scaling all the path metrics by the same factor. Since the soft-output LLRs generated by the algorithm are given by the path metric differences between the ML path and the discarded paths, the soft output LLRs are also scaled by the same factor. The same phenomenon was also observed for the Max-Log-MAP algorithm.

Let us now consider the Log-MAP algorithm. This is identical to the Max-Log-MAP algorithm, except for a correction factor of $f_c(\delta) = \ln(1+e^{-\delta})$ used in the calculation of the forward and backward recursion functions in Equations 5.25 and 5.28, respectively. The function $f_c(\delta)$ is a non-linear function which decreases asymptotically towards zero as δ increases. Since δ

depends directly on L_c value, the performance of the Log-MAP algorithm degrades if imperfect estimation of L_c is given.

7.6.4 The Effect of Puncturing

In their original turbo codec, Berrou and Glavieux [13] applied puncturing in order to obtain a half-rate code. An impressive performance was attained, even though half of the parity bits from both convolutional encoders were punctured. However, this might not be true if we use block codes as the component codes.

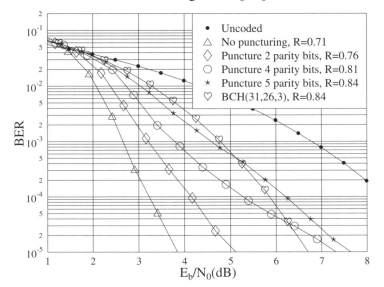

Figure 7.23: Performance comparison between different puncturing patterns of the rate $R = 0.72$ turbo BCH(31,26,3) code employing six iterations, the Log-MAP algorithm and a 26×26-bit block interleaver over AWGN channels.

Figure 7.23 shows our performance comparison between different puncturing patterns applied to the turbo BCH(31,26,3) code using a 26×26-bit block interleaver. We have seen in Section 7.2 that the turbo encoder consists of two BCH encoders. In Figure 7.23 two BCH(31,26,3) encoders are used for the turbo encoder and each of them generates five parity bits. Therefore, for every block of 26 data bits, the turbo encoder will produce ten parity bits. In our first example in Figure 7.23 no puncturing is applied, which results in ten transmitted parity bits and a code rate of $R = 0.72$. In our next example we show the performance of the code when two parity bits are randomly selected and punctured for every block of 26 data bits. In the following examples, we punctured more parity bits, in order to attain a higher coding rate.

We can see from Figure 7.23 that the performance of the turbo BCH(31,26,3) code decreases as the coding rates R increases. In the figure we also show the performance of the conventional BCH(31,26,3) using the soft-decision VA. It can be seen that if five parity bits are punctured in the turbo code, which results in the same coding rate as that of the conventional BCH(31,26,3) code, the performance of the turbo BCH(31,26,3) code is about 1 dB worse than that of the conventional BCH(31,26,3) code at a BER of 10^{-5}.

Generally, puncturing does not assist in improving the coding rate without sacrificing the performance of the turbo BCH code. Moreover, there are certain puncturing patterns which outperform others at a certain coding rate [150]. However, it is impossible to find the optimum puncturing patterns for long codes by exhaustive search. One possible solution is to study the distance profile of the code, since the performance of the code depends on the distance properties and we can investigate the effects of puncturing on the distance profile.

7.6.5 The Effect of the Interleaver Length of the Turbo Code

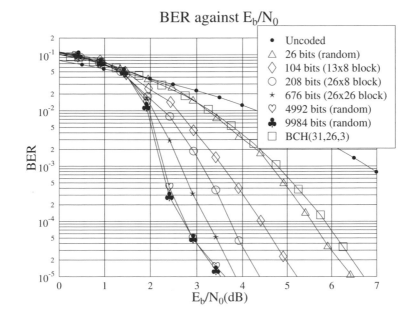

Figure 7.24: Performance comparison between different interleaver length for the rate $R = 0.72$ turbo BCH(31,26,3) code using six iterations, the Log-MAP algorithm and the conventional rate $R = 0.83$ BCH(31,26,3) code using the soft-decision VA over AWGN channels.

Many contributions [12, 13, 69] have shown impressive performances for large interleaver lengths. Although interleaver length is affordable in data transmission system (non-real-time systems) since a delay of say 10^4 bits is generally acceptable, for many other real-time applications, such as for example interactive speech and video transmission, the system often can only tolerate a delay of approximately 100 bits.

We show in Figure 7.24 how the interleaver length affects the performance of the rate $R = 0.72$ turbo BCH(31,26,3) code. The interleaver length of 26 and 104 bits, which uses a random and a 13×8-bit block interleaver, respectively, is suitable for the above-mentioned real-time systems. It can seen from the figure that the performance of the turbo BCH(31,26,3) code using a random interleaver having an interleaver depth of 26 bits is slightly better than the conventional BCH(31,26,3) code using the soft-decision VA at a BER of 10^{-3}. Notice also that the rate R of the turbo code is lower than that of the conventional BCH code. In terms of decoding complexity, the turbo code is more demanding than the conventional VA. As was shown in Section 5.3.5, both component decoders of the turbo code have to calculate the values of α, β and γ. This results in about three times higher complexity than decoding the same code using a standard

Viterbi decoder. The curves shown in Figure 7.24 were generated using two component decoders and six iterations. The overall complexity of the corresponding turbo decoder is approximately $2 \times 6 \times 3 = 36$ times higher than that of a Viterbi decoder. Therefore, the conventional BCH encoding and decoding method constitute a better choice, if the affordable delay of the system is low.

The performance of turbo codes increases as the interleaver length increases. However, as shown in Figure 7.24, the incremental coding gain becomes smaller as the interleaver length increases. It reaches its near-optimum performance when the interleaver length exceeds 5000 bits. Such a high interleaver length is only suitable for non-real-time systems.

7.6.6   The Effect of the Interleaver Design

It is well known that the interleaver design has a vital effect on the performance of turbo codes. The interleaver design together with the component codes used and the puncturing pattern play an important role in determining the minimum free distance d_{min} of turbo codes and in turn predetermine the performance of the code.

In the context of turbo BCH codes we face various problems in designing the interleaver. Let the BCH(n_1, k_1, d_{min1}) and BCH(n_2, k_2, d_{min2}) schemes be the two component codes of a turbo code. Since BCH codes are block codes, which encode k data bits each time, the length of the interleaver has to be multiple of k_1 and k_2. Therefore, the design of the interleaver is not as flexible as that of turbo convolutional codes, which can have more flexibility in terms of the interleaver length. In this section we consider how the interleaver design affects the performance of the code, while keeping the interleaver length, the component codes and the puncturing patterns the same.

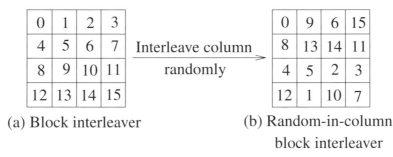

(a) Block interleaver   (b) Random-in-column
                                                block interleaver

Figure 7.25: Bit positions in the (a) block interleaver and (b) random-in-column block interleaver.

First, we define a new type of interleaver, which is referred to as the random-in-column block interleaver. The bit positions of the block interleaver and random-in-column block interleaver are shown in Figure 7.25. In the block interleaver the data bits are entered into the matrix on a row-by-row basis. After the matrix is full, the data bits are read out on a column-by-column basis. In order to design the random-in-column block interleaver, we rearrange randomly the data bits within each column of the block interleaver. In Figure 7.25, we can see that for each column the bit positions of the block and random-in-column block interleaver are the same, except for the arrangement of the bit positions.

The benefit of using the random-in-column block interleaver is shown in Figure 7.26 in terms of its improved BER performance. The figure shows a comparison of different interleaver designs for an interleaver length of 676 bits using the rate $R = 0.72$ turbo BCH(31,26,3) code. It can

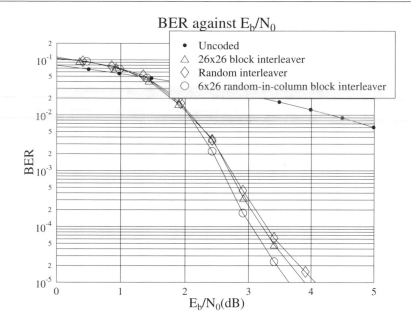

Figure 7.26: Performance comparison between different interleaver designs at an interleaver length of 676 bits using the rate $R = 0.72$ turbo BCH(31,26,3) code, six iterations and the Log-MAP decoding algorithm over AWGN channels.

be seen that at a BER of 10^{-5} the performance of the random-in-column block interleaver is about 0.2 dB better than that of the block interleaver. Figure 7.26 also shows that the random interleaver has the worst performance of all. However, this is not always the case when we use other interleaver lengths. In the following two figures, we show the effect of the interleaver design for both lower and higher interleaver lengths compared to the above-mentioned 676-bit interleaver length.

Figure 7.27 shows our performance comparison between different interleaver designs at an interleaver length of 104 bits using the rate $R = 0.72$ turbo BCH(31,26,3) code. Such short interleavers can be used for interactive speech transmission, for example. We can see from Figure 7.27 at a BER of 10^{-3}, which is often targeted by speech systems, that all the interleavers have essentially the same performance. The same trend is observed at low BERs for the 13×8-bit block interleaver, for the 104-bit random interleaver and for the 13×8-bit random-in-column block interleaver. However, at a BER of 10^{-5}, the performance of the 26×4-bit block interleaver is about 1 dB worse than that of the others.

Previously, we have seen how the interleaver design affects the performance of the code using small (104 bits) and medium (676 bits) interleaver sizes. Let us now characterise the performances of different interleaver designs at an interleaver length of 9984 bits, which is suitable for data transmission. Figure 7.28 shows our performance comparison between different interleaver designs using the turbo BCH(31,26,3) code. At a BER of 10^{-5}, the performance of all the interleavers is about the same. However, for higher BERs the random interleaver outperforms the others.

From the simulation results given above and from our simulation results using other component codes, we can draw some conclusions on the design of interleavers. Specifically, for a turbo BCH(n, k, d_{min}) code, the $k \times k$ random-in-column interleaver is preferred if the inter-

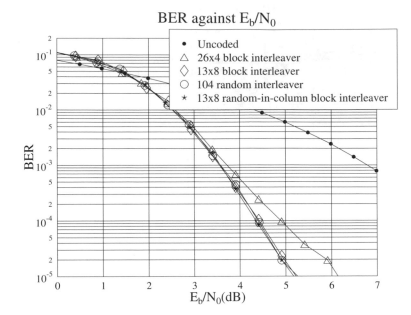

Figure 7.27: Performance comparison between different interleaver designs at an interleaver length of 104 bits using the rate $R = 0.72$ turbo BCH(31,26,3) code, six iterations and the Log-MAP decoding algorithm over AWGN channels.

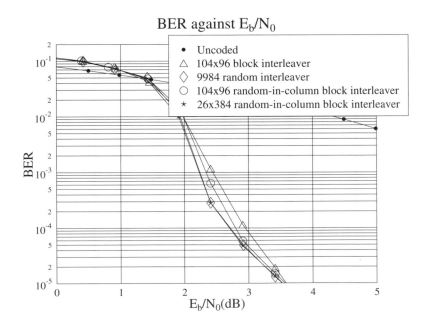

Figure 7.28: Performance comparison between different interleaver designs at an interleaver length of 9984 bits using the rate $R = 0.72$ turbo BCH(31,26,3) code, six iterations and the Log-MAP decoding algorithm over AWGN channels.

leaver length is $k \times k$. For a small interleaver length, below $k \times k$, the simple square block interleaver performs better. This implies that the number of columns and rows should not differ significantly. In a data transmission system we can have a high interleaver depth and it was found that the random interleaver is the best choice.

7.6.7 The Component Codes

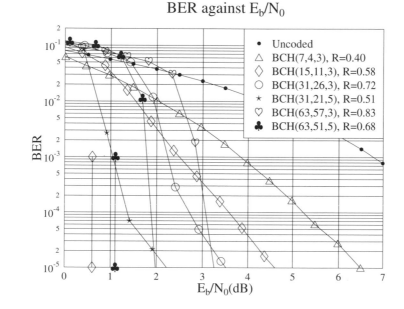

Figure 7.29: Performance comparison of different turbo BCH(n, k, d_{min}) codes employing six iterations, the Log-MAP decoding algorithm and an interleaver length of about 10000 bits over AWGN channels.

Figure 7.29 shows the performance of different turbo BCH(n, k, d_{min}) codes using six iterations of the Log-MAP algorithm. The interleaver length of each code is about 10000 bits and the coding rate R varies from 0.3 to 0.83.

In [61], Hagenauer *et al.* presented simulation results for various Hamming codes, which have the same minimum free distance as BCH codes. The MAP decoding algorithm was employed in the authors' simulations. After comparing the results of [61] and the simulation results of Figure 7.29, we conclude that the BER performance of turbo BCH and Hamming codes is similar.

In Figure 7.29, we have also included the Shannon capacity limit [1] of each BCH code, except for the turbo BCH(7,4,3) code. For the turbo BCH(15,11,3) code, which has a coding rate of $R = 0.58$, the associated performance is about 4 dB away in terms of E_b/N_0 from the Shannon capacity limit at a BER of 10^{-5}. As we increase the coding rate to 0.83 by using the BCH(63,57,3) scheme as component codes, the performance curve is within about 1.1 dB from the Shannon limit, again, when viewed at a BER of 10^{-5}. In Table 7.7 we tabulated the E_b/N_0 distances with respect to the Shannon capacity limit for the various turbo BCH codes studied.

A simple conclusion can be drawn from Table 7.7 at this point. As we increase the codeword length n, while keeping the minimum free distance d_{min} constant, the coding rate R increases.

Component code	Rate	Distance to Shannon limit
BCH(15,11,3)	0.58	4.0 dB
BCH(31,26,3)	0.72	2.1 dB
BCH(31,21,5)	0.51	2.0 dB
BCH(63,57,3)	0.83	1.1 dB
BCH(63,51,5)	0.68	0.8 dB

Table 7.7: Distance to the Shannon capacity limit for turbo BCH codes using various code rates R.

As the coding rate R increases, the discrepancy between the associated performance curve and the Shannon limit gets smaller. For example, we fix the minimum free distance to $d_{min} = 3$. Table 7.7 shows that the distance from the capacity limit is 4.0 dB when $n = 15$, which decreases to 1.1 dB as n increases to 63. However, in [63] Nickl *et al.* have performed several simulations over AWGN channels using different $(n = 2^N - 1, k = 2^N - 1 - N, 3)$ Hamming codes as component codes and a $k \times k$-bit block interleaver. They have shown that the performance improvement for very long Hamming codes $(N > 10)$ is marginal. Furthermore, they have also shown that the performance of the (1023,1013) Hamming code is only 0.27 dB away from the Shannon limit, which is the closest approximation to Shannon's limit reported so far for block codes.

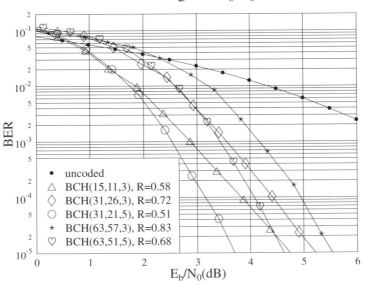

Figure 7.30: Performance comparison of different BCH(n, k, d_{min}) turbo codes employing six iterations, Log-MAP algorithm and an interleaver length of ≈ 100 over the AWGN channels.

Previously, we have studied the performance of different component codes using a high interleaver size of 10000 bits, which is suitable for non-real-time transmission systems. Let us now consider a real-time transmission system which requires a short delay. Hence the interleaver size was reduced to approximately 100 bits. Figure 7.30 shows our performance comparison for different turbo BCH(n, k, d_{min}) codes at an interleaver length of approximately 100 bits. In Figure 7.29, as we increase n while maintaining a minimum distance of $d_{min} = 3$, we observe that there is an increase in coding gain at a BER of 10^{-3}. However, if we limit the interleaver size to approximately 100 bits, the situation is reversed. As shown in Figure 7.30, the turbo code using the BCH(15,11,3) scheme as its component code achieves the highest coding gain at a BER of 10^{-3} compared to the BCH(31,26,3) and the BCH(63,57,3) codes. As n increases, while keeping d_{min} constant, k increases as well. Therefore, the number of BCH codewords that fit into an interleaver size of about 100 bits becomes smaller. Hence the number of BCH codewords per turbo-coded block reduces, and thus the correlation between the upper and lower codewords also increases. As the data bits become more dependent on each other, the turbo decoder is less likely to correct the errors that occurred in the turbo block.

In Figure 7.29, we observe that the BCH(63,51,5) component code outperforms the BCH(31,26,3) component code in the context of turbo coding, even though their coding rates are nearly the same, namely 0.68 and 0.72, respectively. This is because the BCH(63,51,5) code has a higher minimum free distance d_{min} and n is twice as high. However, for a small interleaver size of about 100, the BCH(31,26,3) code performs as well as the BCH(63,51,5) scheme at a BER of 10^{-3} and 0.5 dB worse at a BER of 10^{-5}. This simply implies that at a BER of 10^{-3} we can achieve a high coding gain by using the BCH(31,26,3) code, which has a higher coding rate and a lower complexity. Generally, if the interleaver size is small, it is better to choose BCH(n, k, d_{min}) turbo component codes which have small k.

7.6.8 BCH($31, k, d_{min}$) Family Members

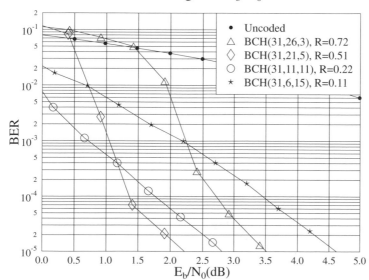

Figure 7.31: Performance comparison of turbo codes using BCH($31, k, d_{min}$) family members as component codes, employing six iterations, the Log-MAP decoding algorithm and an interleaver length of about 10000 bits over AWGN channels.

Both Figure 4.19 and Figure 4.20 in Section 4.3.6 demonstrate that for a certain family BCH code defined by a constant codeword length of n, a specific BCH code achieves the highest coding gain, which typically has a code rate in the range of $0.5 - 0.7$.

Here, we present similar results for different turbo BCH$(31, k, d_{min})$ code family members at an interleaver length of about 10000 bits in Figure 7.31. It is seen in the figure that the turbo code using the BCH(31,21,5) code as its component code has the highest coding gain. The coding rate R of the turbo code is about $\frac{1}{2}$.

7.6.9   Mixed Component Codes

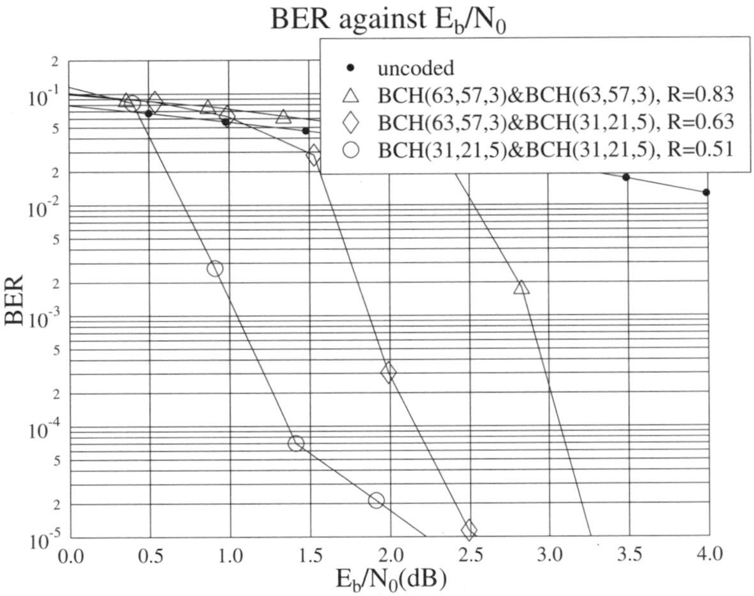

Figure 7.32: Performance comparison between turbo codes using both different and identical BCH component codes, employing six iterations, the Log-MAP decoding algorithm and an interleaver length of about 10000 bit over AWGN channels.

In the previous section, we have presented our simulation results using the same component code for both the upper and lower encoder shown in Figure 7.1. Let us now study the effect of different component codes on the performance of the turbo code.

Figure 7.32 shows our performance comparison between turbo codes using both different and identical component codes, at an interleaver length of about 10000 bits. The turbo coding scheme denoted by the diamond shapes in Figure 7.32 consisted of the BCH(63,67,3) scheme as the upper component code and the BCH(31,21,5) arrangement as the lower component code. This codec results in a coding rate of $R = 0.63$, which is between the coding rate of the first and the third turbo code shown in Figure 7.32.

In Figure 7.32, we have demonstrated that the performance of the rate $R = 0.63$ turbo code using mixed component codes is about 0.25 dB worse than that of the rate $R = 0.51$ turbo BCH(31,21,5) code. Hence, at the cost of slight degradation of the coding gain, we have increased the coding rate from 0.51 to 0.63. Furthermore, the complexity of the turbo code using mixed component codes is lower than that of the turbo BCH(31,21,5) code, but higher than that of the BCH(63,57,3) scheme.

7.6.10   Extended BCH Codes

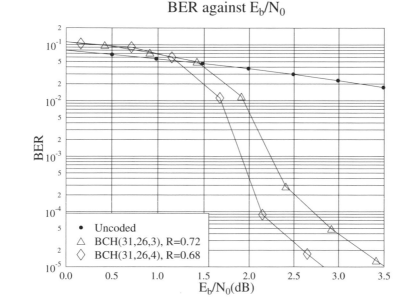

Figure 7.33: Performance comparison between turbo codes using the BCH(31,26,3) or the BCH(32,26,4) code as the component codes, employing six iterations, the Log-MAP decoding algorithm and a random interleaver with a depth of 9984 bits over AWGN channels.

In Section 7.5, we proposed the modified MAP and Log-MAP algorithms in order to incorporate the parity check bit of the extended BCH codes into calculating the LLR of the decoded bits. Figure 7.33 shows the performance of both the BCH(31,26,3) and the turbo BCH(32,26,4) codes. The Log-MAP algorithm was used and the number of iterations was six for both cases. The interleaver was a random interleaver using an interleaving depth of 9984 bits.

As explained in Section 7.5, a parity check bit was appended to the BCH(31,26,3) component code, which extends it to the BCH(32,26,4) code. Explicitly, the minimum free distance d_{min} was increased from three to four. In a turbo code the extra parity bit causes only a small degradation of the coding rate from 0.72 to 0.68. However, as shown in Figure 7.33, the performance of the turbo BCH(32,26,4) code is about 0.7 dB better than that of the turbo BCH(31,26,3) code.

Further simulation results were obtained for the BCH(32,21,6) and the BCH(31,21,5) coding based turbo codes. Unlike for the BCH(31,26,3) and BCH(32,26,4) codes, the performance of both codes remained about the same. This is probably because if we increase the minimum free distance d_{min} from three to four, this results in a 33% increase of d_{min}. By contrast, if we increase $d_{min} = 5$ to 6, the increase is only about 20%.

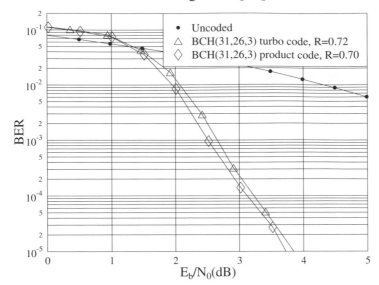

Figure 7.34: Performance comparison between the BCH(31,26,3) product code and the turbo BCH(31,26,3) code employing six iterations, the Log-MAP decoding algorithm and a 26×26-bit block interleaver over AWGN channels.

7.6.11 BCH Product Codes

In Section 7.2 we highlighted the differences between turbo codes and product codes. As shown in Figure 7.2, the structure of turbo codes is the same as that of product codes, except that in turbo codes the redundancy part arising from checking the parity of the parity part of both codes is neglected. Furthermore, the turbo code has a smaller minimum free distance than the product code, as argued in Section 7.2.

In order to exploit the parity of the parity bits of the product code, the Log-MAP algorithm was modified so that it provided the soft-output LLRs for the parity bits as well. Upon receiving the soft channel outputs of the product code, the decoder decodes the columns of the product code. The decoder provides the soft-output LLRs of both the data and the parity bits. Then, the decoder uses the soft channel outputs plus the intrinsic information of the data and parity bits, in order to decode the rows of the product code. The above process continues for a certain number of iterations. The same decoding techniques have also been proposed for product codes when using soft-in soft-out algebraic decoding [81, 123, 147, 148].

In Figure 7.34, we portray our performance comparison between the BCH(31,26,3) product code and the turbo BCH(31,26,3) code using a 26×26-bit block interleaver. It can be seen that there is no significant performance improvement when using the BCH(31,26,3) product code.

7.7 Summary and Conclusions

In this chapter our discussions revolved around turbo BCH codes, rather than the conventional turbo convolutional codes. The structure of the turbo encoder, which consists of two component codes, was discussed in Section 7.2. The difference between product codes and turbo codes was also highlighted in the section. Then the more complex structure of turbo decoders was presented in Section 7.3. The philosophy of iterative decoding was also detailed.

The MAP algorithm is the core of generating the soft information exchanged between the decoders in the turbo decoder. The detailed derivation of the algorithm was given in Section 5.3.3 but a brief summary of the algorithm was given in Section 7.3.1. We also explained the modification to the MAP algorithm in the context of the BCH codes. Briefly, the algorithm can be divided into three parts. It first calculates the transition probabilities of each trellis transition. Using the calculated transition probabilities, the algorithm performs forward and backward recursion, which are the remaining two parts of the algorithm. Owing to the high implementational complexity of the MAP algorithm, the less complex Max-Log-MAP and Log-MAP algorithms were, however, derived in a previous chapter in Section 5.3.5. The even less complex SOVA, which is a derivative of the VA, was presented in Section 7.3.2 and a decoding example was detailed in Section 7.3.2.1. In Section 7.4, a decoding example was given in the context of the turbo BCH(7,4,3) code. The SOVA decoding algorithm was employed in the example and we showed how iterative decoding improves the reliability of the soft outputs and hence corrects the channel errors that could not be removed by non-iterative decoding.

We modified the MAP algorithm in order to incorporate an additional parity check bit in extended BCH codes. The algorithm was modified such that it kept track of the probability of the paths, which gave an odd or even number of transmitted bits that were $+1$. This was vital in calculating the soft outputs, since the probability of the survivor path, which gives an odd or even number of $+1$ transmitted bits, is known, as detailed in Section 7.5. The reduced-complexity modified Max-Log-MAP and modified Log-MAP algorithms were also derived and presented in Section 7.5.3 for the class of extended BCH codes.

Finally, we presented our simulation results using BPSK over AWGN channels in Section 7.6. We first investigated the effect of iterative decoding on the performance of the turbo BCH codes. It was found that the performance of the BCH turbo code did not improve significantly after four iterations. Various decoding algorithms were compared in Section 7.6.2 and it was found that the Log-MAP and the Max-Log-MAP algorithms gave a similar performance. The SOVA decoding algorithm gives the worst performance, but it is the least complex algorithm. It was shown in Section 7.6.3 that imperfect estimation of the channel reliability value had no effect on the performance of the Max-Log-MAP and SOVA algorithms. However, the Log-MAP algorithm performs badly if the estimation of the channel reliability value is imperfect. The effect of puncturing was investigated in Section 7.6.4 and, as expected, it was found that puncturing degrades the performance of turbo BCH codes. Hence it was concluded that no puncturing should be applied to the parity bits of turbo BCH codes. In Section 7.6.5, we showed that, again, as anticipated, the performance of turbo BCH codes improves as the turbo interleaver length increases and a near-optimum performance was achieved when the interleaver length exceeded about 5000 bits. We proposed a novel interleaver design in Section 7.6.6. This interleaver design was referred to as the random-in-column block interleaver and it was shown to outperform other conventional block and random interleavers, if the interleaver had a dimension of $k \times k$, i.e. it was rectangular. Different turbo BCH codes which employ different BCH component codes were investigated in Section 7.6.7. It was found that turbo BCH codes perform impressively at near-unity code rates. The turbo BCH(63,51,5) code was found to perform within about 0.8 dB of the Shannon limit. In Section 7.6.9, two different BCH component codes were employed in the turbo encoder. The performance of this turbo BCH code was found to be between that of turbo BCH codes, which employ two identical BCH codes. The modified MAP algorithm was employed for decoding extended BCH codes and it was shown in Section 7.6.10 that the extended turbo BCH(32,26,4) code outperformed the turbo BCH(31,26,3) code by approximately 0.5 dB at a BER of 10^{-5}.

Finally, the performance of BCH product codes was found to be similar to that of turbo BCH codes in Section 7.6.11.

Now equipped with the understanding of a range of turbo BCH encoding and decoding techniques, we introduce in the next chapter a lesser known block coding technique, which has similar properties to those of the family of non-binary RS codes discussed in Chapter 3. These so-called residue number system codes are also maximum–minimum distance codes, just like RS codes, and they can also be turbo coded and decoded, as will be shown in the next chapter.

Chapter 8

Redundant Residue Number System Codes

T. H. Liew, L. L. Yang, L. Hanzo

8.1 Introduction

We have things of which we do not know the number,
If we count them by threes, the remainder is 2,
If we count them by fives, the remainder is 3,
If we count them by sevens, the remainder is 2,
How many things are there?
The answer is 23.

This verse [40] is quoted from a third-century book, *Suan-Ching*, by Sun Tzu. Motivated by the publication of the book [40], the study of the residue number system can begin. In Sun Tzu's historic work, he presents a formula for manipulating the remainders of an integer after division by 3, 5 and 7. Today, his contribution is referred to as the Chinese Remainder Theorem (CRT) [40, 41], which is one of the common rules of converting remainders, or residues, into integers.

The necessary background for following this chapter is that of understanding Chapters 3, 4 and 7.

The Residue Number System (RNS) [40–42] represents a further departure from two established and well-known number systems, the decimal and binary number systems. In many ways, the RNS is different from the decimal and binary number systems. Because of their basic differences, the RNS exhibits an unusual set of characteristics, which intrigued many researchers. The most interesting one is that the RNS provides the ability to add, subtract or multiply in parallel in the context of its moduli, regardless of the size of the numbers involved, without recourse to intermediate carry digits or internal processing delays [40, 41]. Additionally, there is a lack of ordered significance amongst the residues. Hence, without adding more logic circuits, it is possible in principle to produce a parallel computer.

However, a drawback of the RNS is the awkward nature of some operations, such as division [42, 151–153], magnitude comparison [41, 154–156], sign detection [41, 157, 158], scaling [42, 159, 160], additive and multiplicative overflow detection [41, 157, 160–164], etc., which are significantly much simpler in the decimal and binary number systems. Furthermore, significant complexity is involved in the conversion from the RNS to the decimal or binary number system [42, 165–168].

Owing to the above-mentioned disadvantages, until recently RNS arithmetic has not proved popular in general purpose computers. Instead, research interest has shifted to the fault tolerance characteristics of the RNS [41, 42, 42, 169–177], for applications in Digital Signal Processing (DSP) [157, 178–183], in modulation schemes [184], etc., for the correction of both computational errors and transmission errors. In digital signal processing, the carry-free and fault tolerance properties of the RNS renders it attractive for the implementation of digital filtering, that of Fast Fourier Transform (FFT) spectral analysis, correlation, matrix operations and image processing [157, 165, 178–182, 185–187]. A RNS-based M-ary modulation scheme has been proposed and analysed in [184, 188–193].

Using the RNS for the sake of maintaining fault tolerance during the processing of operands, as well as for error correction and detection, Szabo and Tanaka [41] are amongst the pioneering researchers who have derived a method for single error detection and correction. However, the error correction procedure proposed by Szabo and Tanaka is computationally inefficient and it is implementationally complex. Watson and Hastings [42] exploited the properties of the Redundant Residue Number System (RRNS) for detecting or correcting a single error and also for detecting multiple errors. The RRNS is based on a RNS, which has a number of redundant residues added to it. Watson's method for error correction needs a correction table, which may have a high memory requirement, thereby rendering the proposed technique impractical for the correction of more than a single error. Mandelbaum [194] showed how single error correction can be accomplished in a RRNS with the aid of less redundancy than that required in [42]. Later, Barsi and Maestrini [169] derived the necessary and sufficient conditions for the minimal amount of redundancy allowing the correction of an arbitrary single residue error. Watson's method was also used by Yau and Liu [170], but Watson's correction table was replaced by appropriate computations. Hence, the implementation proposed by Yau and Liu required less memory.

Recently, a coding theoretic approach to error control has been developed in [43, 44] in the context of the RRNS. The concepts of Hamming weight, minimum free distance, weight distribution, error detection capabilities and error correction capabilities were also introduced. A computationally efficient procedure was described in [43] for correcting a single error. In [44], the procedure was extended for correcting double errors and simultaneously correcting single and detecting multiple errors. In [195] a minimum distance RRNS decoding scheme was proposed, which is capable of correcting or detecting multiple residue errors. Recently, the RRNS codes have also been employed as component codes in turbo codes as well as in various adaptive transmission schemes [45, 196–198].

In this chapter, we commence our discourse by a rudimentary introduction to the RNS in Section 8.2. Some theorems associated with the RNS will be discussed in detail. Then, in Section 8.3 we present the coding theory of the RRNS. The options of implementing a channel codec using the RNS will be investigated in Section 8.7. Finally, our simulation results are given in Section 8.9.

8.2 Background

8.2.1 Conventional Number System

The decimal number system is the most widely used number system, in which any integer X can be represented by:

$$
\begin{aligned}
X &= a_{k-1}10^{k-1} + a_{k-2}10^{k-2} + \ldots + a_1 10 + a_0 \\
&= \sum_{j=0}^{k-1} a_j r^j \, ,
\end{aligned}
\tag{8.1}
$$

where r is the radix of the system, which is equal to 10 for the decimal number system. The decimal digits a_j are integers in the interval [0,9]. Thus, the integer $X = 17$ can be represented in the decimal number system as:

$$
17 = 1 \times 10^1 + 7 \times 10^0 \, .
\tag{8.2}
$$

Apparently, the decimal number system is of *unlimited range*, i.e. any integer can be expressed in the system regardless of its magnitude. Furthermore, this is a *unique representation*, since each integer has only one representation and this representation is *non-redundant*. A non-redundant system is defined as a system in which every combination of the digits a_j represents a number and there are no two different sets of digits a_j, which correspond to the same number.

In Equation 8.1, each digit a_j is multiplied by a constant, hence the decimal number system is a *weighted number system*. In this instance, the constant or weight is 10^j, which is a power of the radix 10. Since all the weights used in this system are powers of the same base or radix, the decimal number system is referred to as a *fixed radix system*, where the fixed radix is 10.

In modern computers the binary number system is used for representing the operands. Similarly to the decimal number system, the binary number system has an unlimited range, and a unique representation. It is a non-redundant, fixed radix number system. However, the radix is 2 and the coefficients a_j are either 0 or 1. As an example, the integer $X = 17$ is represented as:

$$
17 = 1 \times 2^4 + 0 \times 2^3 + 0 \times 2^2 + 0 \times 2^1 + 1 \times 2^0 \, .
\tag{8.3}
$$

A weighted number system exhibits numerous advantages:

- Magnitude comparison of two numbers can be readily carried out by comparing the most significant digits.

- The range of the number system is easily extended by adding more digits.

- Overflow can be detected by checking the carry of the most significant digit.

- Simple polarity detection.

- Multiplication and division by a power of the radix (e.g. 2 for a binary number system) can be accomplished by simple arithmetic shifts.

The attributes which lead to these advantages impose a limitation on the speed with which the arithmetic operations can be performed. During the arithmetic operations, the carry information must be passed from digits of lesser significance to those of higher significance. Thus, it is impossible to process all digits in parallel. Furthermore, any errors encountered during the process will be propagated through to the most significant digit.

8.2.2 Residue Number System

The RNS is defined in terms of a k-tuple of pairwise relatively prime positive integers, $m_1, m_2, \ldots, m_k$, where each individual member is referred to as a modulus. Hence the greatest common divisor of (m_i, m_j) is 1 for $i \neq j$. The product of the moduli represents the dynamic range, M, of the RNS which is formulated as:

$$M = \prod_{j=1}^{k} m_j . \qquad (8.4)$$

Any positive integer X in the range of $0 \leq X < M$ can be uniquely represented by a k-tuple residue sequence given by

$$X \longleftrightarrow (x_1, x_2, \ldots, x_k) , \qquad (8.5)$$

where the quantity x_j $(0 \leq x_j < m_j)$ is the least positive integer remainder of the division of X by m_j, which is expressed as the residue of X modulo m_j or $|X|_{m_j}$. The positive integer x_j is also referred to as the jth residue digit of X.

Let us consider a three-modulus RNS, having moduli of $m_1 = 2$, $m_2 = 3$ and $m_3 = 5$. The range M of the RNS is:

$$\begin{aligned} M &= m_1 \times m_2 \times m_3 \\ &= 2 \times 3 \times 5 \\ &= 30 , \end{aligned} \qquad (8.6)$$

which implies that the system can represent any integers in the interval [0, 29] uniquely. For example, an integer $X = 17$ can be represented by:

$$17 \longleftrightarrow (1, 2, 2) . \qquad (8.7)$$

If another integer of $X = 47$ is represented by the system, one would find that the integer 47 has the same residue representation as the integer 17, namely (1,2,2). The ambiguity of the residue representation is avoided, if only numbers from X to $X + 29$ are considered in this system, where X is an integer, which is normally 0. Table 8.1 shows the residue representation of integers -4 to $+31$. Clearly, we can see that the pattern of the residue representation repeats itself after 30 different patterns. For example, the integers 1 and 31 have the same residue representation.

For a *signed number system*, we can represent any integers from $-M/2 + 1$ to $M/2$. Again, we used the previous number system, which has the moduli of 2, 3 and 5. Hence, we can represent any integers from -14 to 15. Each number has an n-tuple representation where:

$$x_j = \begin{cases} |X|_{m_j} & X \geq 0 \\ |M - |X||_{m_j} & X < 0 . \end{cases} \qquad (8.8)$$

The signed RNS is often referred to as a symmetric system.

8.2.3 Mixed Radix Number System

In Section 8.2.1 we have seen that the conventional decimal and binary number systems are fixed radix number systems, since all the weights used in Equation 8.1 are powers of the same radices.

Inte-gers	Residues Moduli			Inte-gers	Residues Moduli			Inte-gers	Residues Moduli			Inte-gers	Residues Moduli		
	2	3	5		2	3	5		2	3	5		2	3	5
−4	0	2	1	+5	1	2	0	+14	0	2	4	+23	1	2	3
−3	1	0	2	+6	0	0	1	+15	1	0	0	+24	0	0	4
−2	0	1	3	+7	1	1	2	+16	0	1	1	+25	1	1	0
−1	1	2	4	+8	0	2	3	+17	1	2	2	+26	0	2	1
0	0	0	0	+9	1	0	4	+18	0	0	3	+27	1	0	2
+1	1	1	1	+10	0	1	0	+19	1	1	4	+28	0	1	3
+2	0	2	2	+11	1	2	1	+20	0	2	0	+29	1	2	4
+3	1	0	3	+12	0	0	2	+21	1	0	1	+30	0	0	0
+4	0	1	4	+13	1	1	3	+22	0	1	2	+31	1	1	1

Table 8.1: Residue representation of the integers −4 to +31 using the moduli 2, 3 and 5.

In a mixed radix system, the weights are products of a number of the radices. An integer X may be expressed in mixed radix form as [41]:

$$X = a_k \prod_{j=1}^{k-1} r_j + a_{k-1} \prod_{j=1}^{k-2} r_j + \ldots + a_3 r_1 r_2 + a_2 r_1 + a_1 \,, \tag{8.9}$$

where r_j are the radices and a_j, $0 \le a_j < r_j$, are the mixed radix digits. Explicitly, we can see that any positive integer in the interval $\left[0, \prod_{j=1}^{k} r_j - 1\right]$ may be represented by the system and that each number has an unique representation. Given a set of radices, $r_1, r_2, \ldots r_k$, the mixed radix representation of an integer X is given as follows:

$$X \longleftrightarrow (a_k, a_{k-1}, \ldots, a_2, a_1) \,. \tag{8.10}$$

Let us assume that we have a three-radix system, where $r_1 = 2$, $r_2 = 3$ and $r_3 = 5$, respectively. Hence, we can represent any integer in the interval $[0, 29]$, which are expressed as:

$$X = a_3 \times 6 + a_2 \times 2 + a_1 \,. \tag{8.11}$$

In Section 8.2.1 we have shown how the integer $X = 17$ is represented in the decimal and binary number systems with the aid of Equations 8.2 and 8.3, respectively. Applying the same principle, we can represent the integer $X = 17$ in the above-mentioned mixed radix number system as:

$$
\begin{aligned}
17 &= 2 \times 6 + 2 \times 2 + 1 \\
17 &\longleftrightarrow (2, 2, 1) \,.
\end{aligned}
\tag{8.12}
$$

Table 8.2 shows the mixed radix representation of the integers 0 to 29.

8.2.4   Residue Arithmetic Operations

Let us assume that we have two integers, namely X_1 and X_2. The corresponding RNS representations are $X_1 \longleftrightarrow (x_{11}, x_{12}, \ldots, x_{1k})$ and $X_2 \longleftrightarrow (x_{21}, x_{22}, \ldots, x_{2k})$, respectively. Then,

Number	a_3	a_2	a_1	Number	a_3	a_2	a_1	Number	a_3	a_2	a_1
0	0	0	0	10	1	2	0	20	3	1	0
1	0	0	1	11	1	2	1	21	3	1	1
2	0	1	0	12	2	0	0	22	3	2	0
3	0	1	1	13	2	0	1	23	3	2	1
4	0	2	0	14	2	1	0	24	4	0	0
5	0	2	1	15	2	1	1	25	4	0	1
6	1	0	0	16	2	2	0	26	4	1	0
7	1	0	1	17	2	2	1	27	4	1	1
8	1	1	0	18	3	0	0	28	4	2	0
9	1	1	1	19	3	0	1	29	4	2	1

Table 8.2: Mixed radix representation of the integers 0 to 29 for radices 2, 3, 5.

$(x_{11}, x_{12}, \ldots, x_{1k}) \odot (x_{21}, x_{22}, \ldots, x_{2k})$, where $\odot$ denotes addition, subtraction or multiplication, results in another unique k-tuple residue sequence, namely $X_3 \longleftrightarrow (x_{31}, x_{32}, \ldots, x_{3k})$, as long as the number X_3 is in the range $[0, M - 1]$. This is expressed as:

$$X_3 \;\; = \;\; X_1 \odot X_2$$
$$X_3 \;\; \longleftrightarrow \;\; \left(|x_{11} \odot x_{21}|_{m_1}, |x_{12} \odot x_{22}|_{m_2}, \ldots, |x_{1k} \odot x_{2k}|_{m_k} \right) . \tag{8.13}$$

Again, let us consider the previous example using moduli $m_1 = 2$, $m_2 = 3$ and $m_3 = 5$. The dynamic range M is 30. The three arithmetic operations, namely addition, subtraction and multiplication, are illustrated below as:

$$
\begin{array}{rclccc}
& 7 & \longleftrightarrow & (1, & 1, & 2) \\
+ & 4 & \longleftrightarrow & (0, & 1, & 4) \\
\hline
& 11 & \longleftrightarrow & (1 \bmod 2, & 2 \bmod 3, & 6 \bmod 5) \quad \equiv (1, 2, 1)
\end{array}
$$

$$
\begin{array}{rclccc}
& 7 & \longleftrightarrow & (1, & 1, & 2) \\
- & 4 & \longleftrightarrow & (0, & 1, & 4) \\
\hline
& 3 & \longleftrightarrow & (1 \bmod 2, & 0 \bmod 3, & -2 \bmod 5) \quad \equiv (1, 0, 3)
\end{array}
$$

$$
\begin{array}{rclccc}
& 7 & \longleftrightarrow & (1, & 1, & 2) \\
\times & 4 & \longleftrightarrow & (0, & 1, & 4) \\
\hline
& 28 & \longleftrightarrow & (0 \bmod 2, & 1 \bmod 3, & 8 \bmod 5) \quad \equiv (0, 1, 3) .
\end{array}
$$

It can be seen from the above example that the jth residue digit, namely x_{3j}, is uniquely and unambiguously defined in terms of $(x_{1j} \odot x_{2j})$ modulo m_j. That is, no carry information has to be communicated between the residue digits. Hence, the overhead of manipulating carry information in a weighted number system can be avoided. This results in high-speed parallel operations and it makes RNS attractive. In the event of an error occurring in a residue digit operation, the error is confined within the operation and it does not affect the result of other operations.

It should be noted that the previous examples satisfy the condition $0 \leq X_1 \odot X_2 < M$. If this condition is not satisfied, the correct results will not be obtained. For example:

$$
\begin{array}{rclccc}
& 7 & \longleftrightarrow & (1, & 1, & 2) \\
\times & 5 & \longleftrightarrow & (1, & 2, & 0) \\
\hline
& 35 & \longleftrightarrow\!\!\!/ & (1 \bmod 2, & 2 \bmod 3, & 0 \bmod 5) \quad \equiv (1, 2, 0) \longleftrightarrow 5 .
\end{array}
$$

Explicitly, the result represented by the residues is wrong, a condition which is termed as overflow. In the RNS, an overflow is difficult to detect, since the residue digits have the same significance. Other arithmetic operations, such as magnitude comparison, sign detection, division, etc., require more sophisticated procedures. Sometimes, we have to convert the residue digits to the decimal number system in order to accomplish the above-mentioned operations.

8.2.4.1 Multiplicative Inverse

In certain applications of the RNS the multiplicative inverse of an operand has to be determined. If $0 \leq L < m$ and $|X.L|_m = 1$, L is referred to as the *multiplicative inverse of X modulo m*. Table 8.3 shows the multiplicative inverse of X for moduli 2, 3 and 5. Observe that there exists no multiplicative inverse of $X = 2$ for modulus $m = 4$, since no L can be found, for which $|X.L|_m = 1$.

$m = 3$		$m = 4$		$m = 5$	
X	L	X	L	X	L
1	1	1	1	1	1
2	2	2	-	2	3
-	-	3	3	3	2
-	-	-	-	4	4

Table 8.3: Multiplicative inverse of various values of X modulo 3, 4, 5.

The multiplicative inverse L is useful, for example, for converting the division of a residue digit x_j by a number X to the multiplication of the residue digit x_j by the multiplicative inverse L of the number X modulo m. It is important to note, however, that this only applies if the remainder of the integer division $\frac{x_j}{X}$ is zero.

Let us now illustrate the above concepts further using an example. Let us assume that we have a modulus of $m = 5$ and the residue is $x_j = 4$. The result of $\frac{x_j}{X}$, $X = 2$, is to be calculated using the multiplicative inverse L of X modulo m. From Table 8.3, we know that the multiplicative inverse L of $X = 2$ modulo $m = 5$ is equal to 3. Therefore the required division can be carried out with the aid of a multiplication by the multiplicative inverse L as follows:

$$\frac{4}{2} = 2, \qquad \text{Remainder} = 0$$
$$\frac{4}{2} \equiv |4 \times 3|_5$$
$$= 2 . \tag{8.14}$$

As stated above, the division can only be replaced with the aid of multiplication by the multiplicative inverse, if the remainder of the integer division $\frac{x_j}{X}$ is zero. Hence in the following example we demonstrate that for a remainder of 1 the division cannot be replaced by the above-mentioned multiplication. Explicitly, if $X = 3$, the multiplicative inverse L of $X = 3$ modulo $m = 5$ is equal to 2:

$$\frac{4}{3} = 1, \qquad \text{Remainder} = 1$$
$$\frac{4}{3} \not\equiv |4 \times 2|_5$$
$$= 3 . \tag{8.15}$$

The concept of the multiplicative inverse is important when we have to convert the RNS representation of an integer to the decimal number system, using the Mixed Radix Conversion (MRC) method, which will be introduced in Section 8.2.5.2.

8.2.5   Residue to Decimal Conversion

There are two methods of residue to decimal conversion: the Chinese Remainder Theorem (CRT) [40, 41] and the MRC [41].

8.2.5.1   Chinese Remainder Theorem

The classic CRT uses the following expression for residue to decimal conversion:

$$
\begin{aligned}
X &= \left[M_1 \times |x_1 L_1|_{m_1} + M_2 \times |x_2 L_2|_{m_2} + \ldots + M_k \times |x_k L_k|_{m_k} \right] \bmod M \\
&= \left[\sum_{j=1}^{k} M_j |x_j L_j|_{m_j} \right] \bmod M \,,
\end{aligned}
\tag{8.16}
$$

where $M = \prod_{j=1}^{k} m_j$ and $M_j = \frac{M}{m_j}$, while L_j is the multiplicative inverse of M_j in the context of modulo m_j, which was formulated as $|(L_j M_j)|_{m_j} = 1$.

In order to show the validity of Equation 8.16, we take the modulo value m_i of both sides of the equation, yielding:

$$
\begin{aligned}
X \bmod m_i &= \left\{ \left[\sum_{j=1}^{k} M_j |x_j L_j|_{m_j} \right] \bmod M \right\} \bmod m_i \\
&= M_i |x_i L_i|_{m_i} \bmod m_i \,,
\end{aligned}
\tag{8.17}
$$

since $M_j \bmod m_i = 0$, except for $j = i$. Since we have shown previously that $(L_j M_j) \bmod m_j = 1$, we can reduce Equation 8.17 to:

$$
\begin{aligned}
X \bmod m_j &= x_j \bmod m_j \\
&= x_j \,,
\end{aligned}
\tag{8.18}
$$

which is simply the residue of the integer X upon division by m_j and hence it is valid by definition. Therefore, Equation 8.16 has been proven.

Let us now use an example in order to explain the operation of the CRT more clearly. A three-modulus RNS employing the moduli $m_1 = 3$, $m_2 = 4$ and $m_3 = 5$ is used. The residue representation of an unknown integer X is $(1, 3, 4)$. The corresponding values M_j and L_j are:

$$
\begin{array}{lll}
M = 3 \times 4 \times 5 & = 60 & \\
M_1 = 20 & M_2 = 15 & M_3 = 12 \\
|M_1|_{m_1} = 2 & |M_2|_{m_2} = 3 & |M_3|_{m_3} = 2 \\
L_1 = 2 & L_2 = 3 & L_3 = 3,
\end{array}
$$

where $(L_j \times |M_j|_{m_j})$ modulo $m_j = 1$. Using Equation 8.16, we arrive at:

$$
\begin{aligned}
X &= \left[M_1 |x_1 L_1|_{m_1} + M_2 |x_2 L_2|_{m_2} + M_3 |x_3 L_3|_{m_3} \right] \bmod M \\
&= \left[20 \times |1 \times 2|_3 + 15 \times |3 \times 3|_4 + 12 \times |4 \times 3|_5 \right] \bmod 60 \\
&= \left[40 + 15 + 24 \right] \bmod 60 \\
&= 19 \,.
\end{aligned}
\tag{8.19}
$$

In principle, this method is fairly simple and straightforward. However, the associated operations are not readily implementable if the dynamic range M is high. For instance, the summation of $M_j(x_j L_j)$, in a ten-modulus RNS having moduli ranging from 100 to 128, could easily exceed the typical dynamic range of 2^{64} of the computer used. We can solve this problem by modifying Equation 8.16 according to:

$$X = \left[\sum_{j=1}^{k} \left| M_j |x_j L_j|_{m_j} \right|_M \right] \bmod M \ . \tag{8.20}$$

However, this representation requires more computational power and the dynamic range M has to be less than that of the computer carrying out this conversion. By contrast, the MRC can be readily implemented, since it requires operations modulo m_j only.

8.2.5.2   Mixed Radix Conversion

In Section 8.2.3 we have highlighted the fundamental philosophy of the mixed radix number system and that of the multiplicative inverse. Let us now consider the conversion of the residues x_j in the RNS to the mixed radix digits a_j. Using the mixed radix digits a_j, the decimal number represented by the RNS can be readily calculated with the aid of Equation 8.9.

If a set of moduli $m_1, m_2, \ldots, m_k$ and a set of radices $r_1, r_2, \ldots, r_k$ are chosen so that $m_j = r_j$, the mixed radix system and the RNS are said to be associated. In this case both systems have the same dynamic range of values, that is $\left[0, \prod_{j=1}^{k} m_j - 1 \right]$. Below, we present the process of converting an operand from the RNS to the mixed radix number system.

Since $m_j = r_j$, we can rewrite Equation 8.9 as:

$$X = a_k \prod_{j=1}^{k-1} m_j + a_{k-1} \prod_{j=1}^{k-2} m_j + \ldots + a_3 m_1 m_2 + a_2 m_1 + a_1 \ , \tag{8.21}$$

where a_j are the mixed radix coefficients. Let us first take Equation 8.21 modulo m_1. As we can see in Equation 8.21, all the terms, except for the last one, are multiples of the modulus m_1. Hence, we have:

$$|X|_{m_1} = x_1 = a_1 \ . \tag{8.22}$$

Explicitly Equation 8.22 indicates that a_1 is the same as the first residue digit, namely x_1, of the RNS.

In order to obtain the coefficient a_2, we have to subtract a_1 from Equation 8.21:

$$X - a_1 = a_k \prod_{j=1}^{k-1} m_j + a_{k-1} \prod_{j=1}^{k-2} m_j + \ldots + a_3 m_1 m_2 + a_2 m_1 \ . \tag{8.23}$$

Equation 8.23 is now divisible by m_1, yielding:

$$\frac{X - a_1}{m_1} = a_n \prod_{j=2}^{k-1} m_j + a_{k-1} \prod_{j=2}^{k-2} m_j + \ldots + a_3 m_2 + a_2 \ . \tag{8.24}$$

Again, all the terms in Equation 8.24 are multiples of the modulus m_2. Hence we take Equation 8.24 modulo m_2 and we arrive at:

$$\left| \frac{X - a_1}{m_1} \right|_{m_2} = a_2 \ . \tag{8.25}$$

If we repeat the above procedures, namely subtracting, dividing and taking modulo, all the mixed radix digits may be obtained. Once all the mixed radix digits, a_j, have been found, the MRC is accomplished using Equation 8.9.

It is interesting to note that:

$$a_1 = |X|_{m_1}$$

$$a_2 = \left| \left[\frac{X}{m_1} \right] \right|_{m_2}$$

$$a_3 = \left| \left[\frac{X}{m_1 m_2} \right] \right|_{m_3}$$

$$.$$

$$.$$

$$a_k = \left| \left[\frac{X}{m_1 m_2 \ldots m_{k-1}} \right] \right|_{m_k} \ , \tag{8.26}$$

where $\left[\frac{X}{m_j} \right]$ is the integer value of the quotient $\frac{X}{m_j}$. More explicitly, let us augment the derivation of a_2 using Equation 8.21. Upon diving Equation 8.21 by m_1, we arrive at:

$$\frac{X}{m_1} = a_k \prod_{j=2}^{k-1} m_j + a_{k-1} \prod_{j=2}^{k-2} m_j + \ldots + a_3 m_2 + a_2 + \frac{a_1}{m_1} \ , \tag{8.27}$$

where $a_k \prod_{j=2}^{k-1} m_j + \ldots + a_2$ is the quotient and $\frac{a_1}{m_1}$ is the remainder. Observe that all the terms in the quotient, expect for the last one, are multiples of the modulus m_2. Hence, we have

$$a_2 = \left| \left[\frac{X}{m_1} \right] \right|_{m_2} \ . \tag{8.28}$$

Similar arguments can be applied when deriving the other mixed radix digits a_j.

In Section 8.2.5.1 we have shown, using an example, how to invoke the CRT for converting residue digits to a decimal number. Here, we use the same example, but we employ the MRC method of Equation 8.21. The RNS has the moduli of $m_1 = 3$, $m_2 = 4$ and $m_3 = 5$. The residue representation of the unknown integer X is $(1, 3, 4)$. We substitute m_j into Equation 8.21 and we get:

$$X = a_3(3 \times 4) + a_2(3) + a_1 \ , \tag{8.29}$$

which will be used to calculate the decimal representation of the unknown integer X. However, we have to determine the mixed radix coefficients a_j in Equation 8.29.

Applying Equation 8.22:

$$a_1 = x_1 = 1 \ . \tag{8.30}$$

We subtract $a_1 = 1$ from $X = 19$, yielding $19 - 1 = 18$, which is formulated in the residue domain as:

$$X \quad \overset{3,4,5}{\longleftrightarrow} \quad (1, \quad 3, \quad 4)$$
$$- \quad a_1 \quad \overset{3,4,5}{\longleftrightarrow} \quad (1, \quad 1, \quad 1)$$
$$\overline{X - a_1 \quad \overset{3,4,5}{\longleftrightarrow} \quad (0, \quad 2, \quad 3)} \quad ,$$

where $\overset{3,4,5}{\longleftrightarrow}$ denotes the residue representation of a number in the RNS using moduli 3, 4 and 5.

In Equation 8.23, we have shown that $X - a_1$ is divisible by m_1 and the remainder is equal to zero. Therefore, instead of employing the more cumbersome division operation of $\frac{X-a_1}{m_1}$ we can invoke the multiplication of $X - a_1$ with the multiplicative inverse L_j of m_1 modulo m_j. The required multiplicative inverse values are shown in Table 8.3, yielding:

$$X - a_1 \quad \overset{4,5}{\longleftrightarrow} \quad (2, \quad 3)$$
$$\times \quad L_j \quad \quad 3 \quad \quad 2$$
$$\overline{\frac{X-a_1}{m_1} \quad \overset{4,5}{\longleftrightarrow} \quad (|6|_4, \quad |6|_5)} \quad \longleftrightarrow \quad (2, \quad 1) \,.$$

As shown in Equation 8.25, $\left| \frac{X-a_1}{m_1} \right|_{m_2} = a_2$. Hence, a_2 is equal to the residue of $\frac{X-a_1}{m_1} = 6$ modulo 4, giving $a_2 = 2$.

We repeat the above process again and we subtract $a_2 = 2$ from $\frac{X-a_1}{m_1} = 6$, yielding 4, which is formulated in the residue domain as:

$$\frac{X-a_1}{m_1} \quad \overset{4,5}{\longleftrightarrow} \quad (2, \quad 1)$$
$$- \quad a_2 \quad \overset{4,5}{\longleftrightarrow} \quad (2, \quad 2)$$
$$\overline{\frac{X-a_1}{m_1} - a_2 \quad \overset{4,5}{\longleftrightarrow} \quad (|0|_4, \quad |-1|_5)} \quad \longleftrightarrow \quad (0, 4) \,.$$

Then, again, instead of carrying out the division we multiply with L_j, yielding:

$$\frac{X-a_1}{m_1} - a_2 \quad \overset{5}{\longleftrightarrow} \quad (4)$$
$$\times \quad L_j \quad \quad 4$$
$$\overline{\frac{\frac{X-a_1}{m_1} - a_2}{m_2} \quad \overset{5}{\longleftrightarrow} \quad (|16|_5)} \quad \longleftrightarrow \quad (1) \,,$$

and hence a_3 is found to be 1.

Finally, we substitute all mixed radix coefficients a_j, $j = 1, 2, 3$, into Equation 8.29, giving:

$$\begin{aligned} X &= 1 \times 12 + 2 \times 3 + 1 \\ &= 19 \,. \end{aligned} \tag{8.31}$$

Having exemplified the various conversion processes between the residue and decimal domains, let us now consider the construction of redundant RNSs, which are used in RNS-based error correction coding.

8.2.6 Redundant Residue Number System

As we have highlighted in Section 8.2.2, the RNS is defined by the choice of k moduli, namely $m_1, m_2, \dots, m_k$. The dynamic range M of the RNS is $\prod_{j=1}^{k} m_j$. In a Redundant Residue Number System (RRNS), extra moduli, namely $m_{k+1}, m_{k+2}, \dots, m_{k+n}$, are incorporated into the RNS. As in the RNS, the moduli, $m_1, m_2, \dots, m_k, m_{k+1}, \dots, m_n$, are chosen to be pairwise relatively prime positive integers and $m_{k+j} \geq \max\{m_1, m_2, \dots, m_k\}$. Moduli

$m_1, m_2, \ldots, m_k$ are considered to be non-redundant moduli and moduli $m_{k+1}, m_{k+2}, \ldots, m_n$ are the redundant moduli. The redundant moduli are not considered to increase the dynamic range, $M = \prod_{j=1}^{k} m_j$, even though the redundant residues related to an integer operand now become part of the residue representation of that integer. The interval $[0, M - 1]$ is referred to as the *legitimate range*, where $M = M_k = \prod_{j=1}^{k} m_j$, and the interval $[M_k, M_n - 1]$ is the *illegitimate range*, where $M_n = \prod_{j=1}^{n} m_j$.

In Section 8.2.5.1 the RNS having moduli $m_1 = 3$, $m_2 = 4$ and $m_3 = 5$ was used. The corresponding RNS representation of the integer $X = 19$ is $(1, 3, 4)$. If we incorporate a redundant modulus of $m_4 = 7$, the integer will be represented by the RRNS as:

$$X \longleftrightarrow (1, 3, 4, 5) . \tag{8.32}$$

In general, any k of the n residue digits could be used to calculate the decimal representation of the integer X, if and only if $m_{k+j} \geq \max\{m_1, m_2, \ldots, m_k\}$ [41]. The conversion method could be either the CRT or the MRC. Let us assume that a computer uses the above-mentioned RRNS as its number representation system. Because of a fault of a module in the computer, the residue x_2 associated with the modulus m_2 is no longer valid. However, we can still calculate the decimal representation of the integer X using the residues x_1, x_3 and x_4. These residues constitute the residue representation of a new RNS, which has m_1, m_3 and m_4 as its moduli. The dynamic range M of the new RNS is 105, which is more than the legitimate range of the RRNS. This implies that any number, within the legitimate range of the RRNS, is represented unambiguously by the new RNS. Furthermore, the new RNS and the RRNS have the same residues with respect to the moduli m_1, m_3 and m_4. Hence, we can calculate the original decimal number, even though we only have the residue digits x_1, x_3 and x_4. Employing the CRT, the corresponding values M_j and L_j of the new RNS are:

$$
\begin{array}{lll}
M = 3 \times 5 \times 7 & = 105 & \\
M_1 = 35 & M_3 = 21 & M_4 = 15 \\
|M_1|_{m_1} = 2 & |M_3|_{m_3} = 1 & |M_4|_{m_4} = 1 \\
L_1 = 2 & L_3 = 1 & L_4 = 1 ,
\end{array}
$$

where $(L_j \times |M_j|_{m_j})$ modulo $m_j = 1$. Using Equation 8.16, we arrive at:

$$
\begin{aligned}
X &= [35 \times |1 \times 2|_3 + 21 \times |4 \times 1|_5 + 15 \times |5 \times 1|_7] \bmod 105 \\
&= [70 + 84 + 75] \bmod 105 \\
&= 19 . \tag{8.33}
\end{aligned}
$$

The above simple example demonstrated an application of the RRNS. The RRNS is used extensively also in the field of error detection and correction. However, our detailed discussions on RRNS-based error correction coding are postponed to Section 8.3.

8.2.7   Base Extension

In the various applications of RNS arithmetic [41] it is often necessary to find the residue digits with respect to a new set of moduli, given the residue digits related to another set of moduli. In most cases the new set of moduli will be an extension of the original set; that is, one or more additional moduli are incorporated in the original set. This is the case, for example, when creating the redundant moduli from the non-redundant moduli of a RNS. Normally, the CRT is employed to find the decimal number represented by the original set of moduli. This decimal number is then used to calculate the residues of the new set of moduli. However, there is a simpler procedure known as *Base Extension* (BEX) [41, 175]. It is related to the MRC, with an additional final step, as highlighted below.

Consider a RNS consisting of moduli $m_1, m_2, \ldots, m_k$. The dynamic range of the RNS is $M = \prod_{j=1}^{k} m_j$. If another modulus, m_{k+1}, is incorporated into the RNS, the dynamic range will be extended and becomes $M = \prod_{j=1}^{k+1} m_j$. Therefore, we have to add another term to Equation 8.9, yielding the mixed radix representation of the integer X in the form of:

$$X = a_{k+1} \prod_{j=1}^{k} m_j + a_k \prod_{j=1}^{k-1} m_j + \ldots + a_3 m_1 m_2 + a_2 m_1 + a_1 . \tag{8.34}$$

Any integer X, represented by the original k moduli, will be in the interval $\left[0, \prod_{j=1}^{k} m_j - 1\right]$. If the integer X is represented by the extended moduli in the mixed radix form, as shown in Equation 8.34, a_{k+1} clearly will be equal to zero. In performing the MRC, the fact that $a_{k+1} = 0$ will be used to find $|X|_{m_{k+1}}$. The method is best illustrated with the aid of a numerical example.

In Section 8.2.6, we found the residue representation of the integer $X = 19$ in the RRNS. The process is straightforward and the residue representation of the integer $X = 19$ is $(1, 3, 4, 5)$, given that the moduli are $m_1 = 3$, $m_2 = 4$, $m_3 = 5$ and $m_4 = 7$. Let us assume that in this example we have no prior knowledge of the decimal representation of the integer X. The residue representation of the integer X is $(1, 3, 4)$, given that the moduli are $m_1 = 3$, $m_2 = 4$ and $m_3 = 5$. A redundant modulus, $m_4 = 7$, is appended to the RNS and hence we have to find $|X|_7$.

In the RRNS, the residue representation of X will be $(1, 3, 4, |X|_7)$. The process of determining $|X|_7$ is initiated by performing the MRC in Section 8.2.5.2 in the usual manner, but including the $|X|_7$ value in the operations. We commence by recalling that:

Moduli:	3	4	5	7	
Residue Representation		1	3	4	$|X|_7$	$a_1 = 1$
Subtract $a_1 = 1$		1	1	1	1	
		0	2	3	$|X|_7 + 6$	
Multiply by L_j, $|3.L_j|_{m_j} = 1$			3	2	5	
$\frac{X - a_1}{m_1}$			2	1	$5|X|_7 + 2$	$a_2 = 2$
Subtract $a_2 = 2$			2	2	2	
			0	4	$5|X|_7$	
Multiply by L_j, $|4.L_j|_{m_j} = 1$				4	2	
$\frac{\frac{X - a_1}{m_1} - a_2}{m_2}$				1	$3|X|_7$	$a_3 = 1$
Subtract $a_3 = 1$				1	1	
				0	$3|X|_7 + 6$	
Multiply by L_j, $|5.L_j|_{m_j} = 1$					3	
$\frac{\frac{\frac{X - a_1}{m_1} - a_2}{m_2} - a_3}{m_3}$					$2|X|_7 + 4 .$	

Since a_4 corresponds to a_{k+1} in Equation 8.34, a_4 is equal to zero. Therefore:

$$a_4 = |2|X|_7 + 4|_7 = 0 , \tag{8.35}$$

and

$$
\begin{aligned}
|2|X|_7|_7 &= |-4|_7 \\
&= 3 .
\end{aligned}
\tag{8.36}
$$

Multiplying by the multiplicative inverse of 2 modulo 7, $|L \times 2|_7 = 1 \Rightarrow L = 4$ yields:

$$|X|_7 = |3 \times 4|_7 = 5 . \tag{8.37}$$

Hence, the residue representation of integer X in the RRNS is $(1, 3, 4, 5)$.

The BEX operation is fundamental to several important arithmetic operations, such as scaling [40, 41], dynamic range extension, magnitude comparison, overflow detection and sign determination [41, 157].

8.3 Coding Theory of Redundant Residue Number Systems

As described earlier, RRNSs have been studied extensively for the protection of arithmetic and data transmission operations in general purpose computers [41–44, 169, 170, 194], in digital filters [178] and in modulation schemes [184]. In this section, we will discuss a coding theoretic approach to error control using the RRNS [43, 44]. The concepts of Hamming weight, minimum free distance, error detection and correction capabilities of RRNS-based codes are introduced. The necessary and sufficient conditions for the desired error control capability are derived from the minimum distance point of view. In a special case, we are capable of generating the Maximum Distance Separable RRNS (MDS-RRNS).

8.3.1 Minimum Free Distance of RRNS-based Codes

The minimum distance is a fundamental parameter associated with any error control code and it very much affects the performance of the code. As shown in Section 8.2.6, the integer number X, which is within the legitimate dynamic range M, can be represented in the RRNS by a set of residues or a valid codeword $\underline{x}$. The number of non-zero residues of $\underline{x}$ is the *Hamming weight*, weight($\underline{x}$), of the codeword. The *Hamming distance* between two codewords $\underline{x}_i$ and $\underline{x}_j$, $dist(\underline{x}_i, \underline{x}_j)$, is the number of residue positions in which $\underline{x}_i$ and $\underline{x}_j$ differ. Hence, we can define the minimum distance d_{min} of the RRNS-based codes as:

$$d_{min} = \min \left\{ dist(\underline{x}_i, \underline{x}_j) \right\} , \tag{8.38}$$

where $\underline{x}_i \neq \underline{x}_j$ and $\underline{x}_i$, $\underline{x}_j$ are legitimate codewords in the RRNS. However, if the number of possible codewords $\underline{x}$ in the RRNS is high, it may not be possible to compute d_{min} using a full-search based on Equation 8.38.

In [43], Krishna *et al.* derived the necessary and sufficient conditions imposed on the redundant moduli in order for a RRNS code to have a minimum distance equal to d_{min}. The minimum free distance of a RRNS code is d_{min} if and only if the product of the redundant moduli satisfies the following relation [43]:

$$\max \left\{ \prod_{i=1}^{d_{min}} m_{j_i} \right\} > M_{n-k} \geq \max \left\{ \prod_{i=1}^{d_{min}-1} m_{j_i} \right\} , \tag{8.39}$$

where $M_{n-k} = \prod_{j=k+1}^{n} m_j$ which is related to the $(n-k)$ number of redundant moduli, hence it is referred to as the product of the redundant moduli, and where m_{j_i} $1 \leq j_i \leq n$, is an arbitrary modulus of the RRNS code. In simple terms, a good code aims to attain the highest possible minimum distance, since this maximises the error correction capability of the code, whilst maximising the code rate. Maximising the code rate is achieved by maximising the useful information dynamic range M_k and hence minimising M_{n-k}, since the total dynamic range $M_n = M_k \cdot M_{n-k} = \prod_{j=1}^{n} m_j$ is constant.

Proof: We will show the validity of Equation 8.39 in two steps, namely the validity of the right-hand-side inequality first and then the validity of the left-hand-side inequality. Consider a codeword having a Hamming weight of α, which implies having non-zero residues in positions $j_1, j_2, \ldots, j_\alpha$ of the codeword, and zero residues elsewhere. As a result, X, which represents the set of residues describing the codewords having the above property, is a multiple of $m_j, j = 1, 2, \ldots, n; j \neq j_1, j_2, \ldots, j_\alpha$. Thus, we can write the integer X as:

$$X = X' \prod_{\substack{j=1 \\ j \neq j_1, j_2, \ldots, j_\alpha}}^{n} m_j , \tag{8.40}$$

where X' is an arbitrary integer satisfying:

$$0 < X' < m_{j_1} m_{j_2} \ldots m_{j_\alpha} . \tag{8.41}$$

For a RRNS code to have a minimum distance of d_{min}, the following conditions must be satisfied [43, 90]:

1) There are no valid codewords having a Hamming weight of $d_{min} - 1$ or less, except for the all-zero codeword;

2) There is at least one valid codeword having a Hamming weight of d_{min}.

The first condition implies that if the Hamming weight α of all illegitimate codewords $\underline{x}_i$ obeys $\alpha \leq d_{min} - 1$, then the corresponding integers obey $X_i \geq M_k = \prod_{j=1}^{k} m_j$, which implies that X_i is in the illegitimate range of the RRNS, because except for the all-zero codeword there are no legitimate codewords having a Hamming weight of $\alpha \leq d_{min} - 1$. On the other hand, the second condition implies that all the legitimate codewords $\underline{x}_j$ have to satisfy $\alpha > d_{min} - 1$. Then we have the corresponding integers $X_j < M_k$, which implies that X_j is in the legitimate range.

We now prove the first condition, which is satisfied trivially if and only if $X_i \geq M_k$ and the number of non-zero residues obeys $\alpha \leq d_{min} - 1$. Let us therefore show that even the smallest possible value of X_i is outside the legitimate range $[0, M_k - 1]$, i.e. it falls in the illegitimate range. The smallest possible value of X_i according to Equation 8.40 is obtained upon setting $X_i' = 1$, and including the smallest $(n - \alpha)$ number of moduli from the set of n moduli. Consequently, the moduli must satisfy [43]:

$$\min \left\{ \prod_{\substack{j=1 \\ j \neq j_1, j_2, \ldots, j_{d_{min}-1}}}^{n} m_{j_i} \right\} \geq M_k$$

$$\frac{M_k M_{n-k}}{\max \left\{ \prod_{i=1}^{d_{min}-1} m_{j_i} \right\}} \geq M_k$$

$$M_{n-k} \geq \max \left\{ \prod_{i=1}^{d_{min}-1} m_{j_i} \right\} . \tag{8.42}$$

The second condition implies that there is at least one codeword $\underline{x}_j$, represented by the integer $X_j < M_k$, that has a Hamming weight of $\alpha = d_{min}$. Again, we set $X_j' = 1$ and include the smallest $(n - \alpha)$ number of moduli from the set of n moduli in Equation 8.40. Hence, we have:

$$\min \left\{ \prod_{\substack{j=1 \\ j \neq j_1, j_2, \ldots, j_{d_{min}}}}^{n} m_{j_i} \right\} < M_k$$

$$\frac{M_k M_{n-k}}{\max \left\{ \prod_{i=1}^{d_{min}} m_{j_i} \right\}} < M_k$$

$$M_{n-k} < \max \left\{ \prod_{i=1}^{d_{min}} m_{j_i} \right\}, \tag{8.43}$$

which completes the proof. $\square$

Let us now consider an example of the RRNS using the following moduli [43, 169]:

$$(m_1, m_2, m_3, m_4, m_5, m_6) = (3, 7, 11, 13, 16, 17). \tag{8.44}$$

Here we show how various RRNS codes exhibiting different distance properties and error correction capabilities can be derived using these moduli. As mentioned before, our aim is to maximise the minimum distance and the useful information dynamic range M_k. This implies minimising M_{n-k}, since $M_n = M_{n-k} \cdot M_k = \prod_{j=1}^{n} m_j = C$, where C is a constant. Specifically, according to Equation 8.39, the minimum free distance of the RRNS becomes for example $d_{min} = 3$ if and only if the redundant moduli of the code satisfy:

$$\max \{ m_{j_1} m_{j_2} m_{j_3} \} > M_{n-k} \geq \max \{ m_{j_1} m_{j_2} \}$$
$$3536 > M_{n-k} \geq 272. \tag{8.45}$$

Therefore, we can have a RRNS code which has a minimum free distance of $d_{min} = 3$ by using the following set of moduli as the redundant moduli: $\{ m_5, m_6 : M_{n-k} = 272 \}$, $\{ m_1, m_2, m_4 : M_{n-k} = 273 \}$, $\{ m_1, m_2, m_5 : M_{n-k} = 336 \}$, and so on. One can also readily verify that if the redundant moduli of $\{ m_1, m_2, m_3, m_5 : M_{n-k} = 3696 \}$ are chosen, then the RRNS has a minimum free distance of $d_{min} = 4$.

It is plausible from Equation 8.39 and with the aid of the above example that for a given RRNS and for the minimum distance of d_{min}, the choice of the redundant moduli is not unique. However, we can find a set of redundant moduli, which requires the lowest possible M_{n-k} value for maintaining a given minimum distance d_{min}. The corresponding RRNS is termed the optimal RRNS. Note that minimising M_{n-k} means maximising the useful information dynamic range M_k, since $M_n = M_k \times M_{n-k}$ is a constant. This implies that the number of possible codewords $\underline{x}$ or the *dynamic range* of the code is maximised. In other words, the code rate is maximised, while maintaining a given minimum distance and error correction capability. Since the lowest illegitimate range is associated with the choice of $\{ m_5, m_6 : M_{n-k} = 272 \}$, this is our preferred option, since it ensures the highest useful information dynamic range M_k. Furthermore, it requires only two redundant moduli for achieving $d_{min} = 3$ and hence four information moduli can be used. The associated code can be denoted as a RRNS(6,4,3) code.

From Equation 8.42, the smallest value of M_{n-k} for a minimum distance of d_{min} is obtained by setting:

$$M_{n-k} = \max \left\{ \prod_{i=1}^{d_{min}-1} m_{j_i} \right\}, \tag{8.46}$$

where $1 \leq j_i \leq n$. Equation 8.46, which was derived from the right-hand side of Equation 8.39, shows that the left-hand side of the inequality in Equation 8.39 is satisfied trivially. It also shows that an optimal RRNS having a minimum distance of d_{min} uses the $(d_{min}-1)$ number of largest moduli from the set of n moduli as its redundant moduli. Therefore, we can write:

$$
\begin{aligned}
d_{min} - 1 &= n - k \\
n &= k + d_{min} - 1 .
\end{aligned}
\tag{8.47}
$$

Using a coding theoretic terminology, we will refer to a RRNS that satisfies Equation 8.47 as the maximum distance separable RRNS, which is a property introduced for characterising the family of the linear block codes in [90]. In our previous example, if moduli m_5 and m_6 are chosen as the redundant moduli, we obtain the maximum distance separable RRNS code having a minimum distance of 3. The useful information dynamic range of this RRNS is $[0, 3003]$. At this stage it is worth noting that RRNS codes exhibit strong similarities with the well-known class of Reed–Solomon (RS) code. They are both non-binary codes. RS codes convey a fixed number of bits per symbol, while RRNS codes may have a different number of bits per residue, as we will show in more depth during our further discourse. They also have similar distance properties.

8.3.2   Linearity of RRNS Codes

A RRNS(n, k) code is a block code, since it accepts k information symbols and generates n coded symbols. There are other block codes, such as the family of BCH and RS codes, which are linear block codes. The fundamental properties of linear block codes are [86]:

- All-zero vector is a valid codeword.

- The sum of two valid codewords is also a valid codeword.

Assume that there are two codewords $\underline{x}_1$ and $\underline{x}_2$ in a RRNS(n, k) code. The corresponding integers are X_1 and X_2, respectively, which are in the range of $[0, M_k - 1]$. Let $X_3 = X_1 + X_2$. If $X_3 < M_k$, then vector $\underline{x}_3$ is a valid codeword. However, if $M_k \leq X_3 < M_n$, then $\underline{x}_3$ is not a valid codeword. Clearly, this violates the definition of a linear code. Similarly, for a given scalar α such that $M_k \leq \alpha X_1 < M_n$, the vector $\alpha \underline{x}_1$ is not a valid codeword either. Hence, in [43] the RRNS codes are termed the *semi-linear* block codes. The term *semi-linearity* or *conditional linearity* implies that the property of linearity is satisfied under certain appropriate predefined conditions.

8.3.3   Error Detection and Correction in RRNS Codes

In this section we will relate the minimum free distance d_{min} of the RRNS code to the error detection and error correction capabilities of the code. In our forthcoming discourse, the triangular inequality will be used repeatedly. Consider three arbitrary residue vectors $\underline{A}$, $\underline{B}$ and $\underline{C}$ in the RRNS codes, where the corresponding integer values are X_A, X_B and X_C, respectively. The Hamming distance among the residue vectors or RRNS codewords satisfies the triangular inequality [43, 96]:

$$
dist(\underline{A}, \underline{B}) + dist(\underline{B}, \underline{C}) \geq dist(\underline{A}, \underline{C}) .
\tag{8.48}
$$

Proof: Let us assume that $X_C > X_A$. Therefore, recalling the definition of Hamming distance and weight from the beginning of Section 8.3.1, we have:

$$dist(\underline{A}, \underline{C}) = \text{weight}(\underline{C} - \underline{A}) , \qquad (8.49)$$

which indicates the number of residue positions where $\underline{A}$ and $\underline{C}$ differ.

There are three cases, depending on the value of X_B:

$$1.\, X_C > \quad X_B \quad > X_A$$
$$2.\, X_C > \quad X_A \quad > X_B$$
$$3.\, X_B > \quad X_C \quad > X_A.$$

For case 1, we can write:

$$
\begin{aligned}
dist(\underline{A}, \underline{C}) &= \text{weight}(\underline{C} - \underline{A}) \\
&= \text{weight}\,\{(\underline{C} - \underline{B}) + (\underline{B} - \underline{A})\} \\
&= \text{weight}(\underline{C} - \underline{B}) + \text{weight}(\underline{B} - \underline{A}) + \alpha_1 + 2\alpha_2 , \qquad (8.50)
\end{aligned}
$$

where α_1 is the number of residue positions, where the non-zero residue digits of $(\underline{C} - \underline{B})$ and $(\underline{B} - \underline{A})$ add to a non-zero number, and α_2 is the number of residue positions, where the non-zero residue digits of $(\underline{C} - \underline{B})$ and $(\underline{B} - \underline{A})$ add to zero. Since $\alpha_1 + 2\alpha_2 \geq 0$, we have:

$$\text{weight}(\underline{C} - \underline{B}) + \text{weight}(\underline{B} - \underline{A}) \geq \text{weight}(\underline{C} - \underline{A}) . \qquad (8.51)$$

For case 2, we can write:

$$
\begin{aligned}
dist(\underline{A}, \underline{C}) &= \text{weight}(\underline{C} - \underline{A}) \\
&= \text{weight}(\underline{C} - \underline{A} + \underline{B} + \underline{m} - \underline{B}) \\
&= \text{weight}\,\{(\underline{C} - \underline{B}) + \underline{m} - (\underline{A} - \underline{B})\} \\
&= \text{weight}(\underline{C} - \underline{B}) + \text{weight}\,\{\underline{m} - (\underline{A} - \underline{B})\} - \alpha_3 - 2\alpha_4 \\
&\leq \text{weight}(\underline{C} - \underline{B}) + \text{weight}\,\{\underline{m} - (\underline{A} - \underline{B})\} \\
&= \text{weight}(\underline{C} - \underline{B}) + \text{weight}(\underline{A} - \underline{B}) , \qquad (8.52)
\end{aligned}
$$

where $\underline{m}$ represents the vector of all moduli, α_3 is the number of residue positions in which the non-zero residue digits of $(\underline{C} - \underline{B})$ and weight $\{\underline{m} - (\underline{A} - \underline{B})\}$ add to a non-zero number, and α_4 is the number of residue positions where the non-zero residue digits $(\underline{C} - \underline{B})$ and weight $\{\underline{m} - (\underline{A} - \underline{B})\}$ add to zero. Finally, case 3 is similar to case 2 and hence we have shown the validity of the triangular inequality in Equation 8.48. $\square$

At the receiver of an RRNS-coded data transmission system, the demodulator provides the received residues $\underline{z}$ in response to a transmitted codeword $\underline{x}$, which can be modelled as:

$$\underline{z} = \underline{x} + \underline{e} \qquad (8.53)$$

where $\underline{e}$ is the error vector imposed by the channel, which may have an arbitrary number of non-zero components in the range between 0 and n. Each received residue can be written as follows:

$$z_i = (x_i + e_i)\bmod m_i , \qquad (8.54)$$

where $1 \leq i \leq n$ and e_i, $0 \leq e_i \leq m_i$, is the error magnitude of each residue. More explicitly, since the residue x_i may assume m_i different values, the error values are also m_i-ary. If $e_i = 0$ for all i, then the received residues are error free.

Let us denote the Hamming weight weight$(\underline{e})$ of the error vector $\underline{e}$ by α, which quantifies the number of non-zero e_i positions in the received residue vector $\underline{z}$. In a RRNS, there is no error vector $\underline{e}$, which has a weight of $0 < \alpha < d_{min}$, that can change codeword $\underline{x}$ into another valid codeword $\underline{\hat{x}}$. Since $dist(\underline{z}, \underline{x}) = \alpha$ and $0 < \alpha < d_{min}$, upon applying Equation 8.48, we can write:

$$\begin{aligned} dist(\underline{z}, \underline{\hat{x}}) + dist(\underline{z}, \underline{x}) &\geq dist(\underline{x}, \underline{\hat{x}}) \\ dist(\underline{z}, \underline{\hat{x}}) &\geq d_{min} - \alpha \\ &\geq 0 \,, \end{aligned}$$

(8.55)

for all possible codewords $\underline{\hat{x}}$ in RRNS and $\underline{\hat{x}} \neq \underline{x}$. Therefore, $\underline{z}$ cannot be a valid codeword. Let us assume that there is another error vector $\underline{e}$, which satisfies:

$$\underline{e} = \underline{\hat{x}} - \underline{x} \,,$$

(8.56)

where $\underline{\hat{x}}$ is an arbitrary codeword in the RRNS satisfying $\underline{\hat{x}} \neq \underline{x}$, and $dist(\underline{\hat{x}}, \underline{x}) = d_{min}$. Then, the received residue vector $\underline{z}$ is equal to codeword $\underline{\hat{x}}$. Therefore, there exist error vectors having weight$(\underline{e}) \geq d_{min}$ that are non-detectable. We can then characterise the error detection capability of a RRNS code as:

$$l = d_{min} - 1 \,,$$

(8.57)

which is the highest possible number of errors for which the residue vector $\underline{z}$ is not a valid codeword.

The error correction capability t of a RRNS code is defined as the highest possible number of errors that the decoder is capable of correcting. The error correction capability of an RRNS code is given by [43, 96]:

$$t = \left\lfloor \frac{d_{min} - 1}{2} \right\rfloor \,,$$

(8.58)

where $\lfloor i \rfloor$ means the largest integer not exceeding i. Here we note that l and t in Equations 8.57 and 8.58 are identical to the corresponding quantities in RS codes. Again, let $\underline{\hat{x}}$ be a codeword other than $\underline{x}$ in the RRNS code. The Hamming distance among $\underline{x}$, $\underline{\hat{x}}$ and $\underline{z}$ satisfies the triangular inequality:

$$dist(\underline{z}, \underline{x}) + dist(\underline{z}, \underline{\hat{x}}) \geq dist(\underline{x}, \underline{\hat{x}}) \,.$$

(8.59)

Since $dist(\underline{x}, \underline{\hat{x}}) \geq d_{min}$ and an error vector has a weight of $dist(\underline{z}, \underline{x}) = \alpha$, we can rewrite Equation 8.59 as:

$$dist(\underline{z}, \underline{\hat{x}}) > d_{min} - \alpha \,.$$

(8.60)

If the weight of the error vector is:

$$\alpha \leq \left\lfloor \frac{d_{min} - 1}{2} \right\rfloor \,,$$

(8.61)

then:

$$d_{min} - \alpha > \left\lfloor \frac{d_{min} - 1}{2} \right\rfloor \geq \alpha \, . \tag{8.62}$$

Therefore, we have shown that:

$$dist(\underline{z}, \underline{x}) < dist(\underline{z}, \hat{\underline{x}}) \, , \tag{8.63}$$

which implies that the received residue vector $\underline{z}$ is closer to $\underline{x}$ than to any other valid codeword $\hat{\underline{x}}$.

8.4 Multiple-error Correction Procedure

Based on the properties of modulus projection and the MRC, an algorithm was proposed in [169] for single residue error correction. It was then used in [178, 199, 200] for correcting single errors in digital filters and error checkers. Recently, this algorithm was extended [44] for detecting and correcting multiple residue errors.

In a RRNS having n moduli, we define the following quantity:

$$M^\alpha = \prod_{i=1}^{\alpha} m_{j_i} \, , \tag{8.64}$$

where $1 \leq j_i \leq n$ and $\alpha \leq n$. Hence, we can define the M^α-projection of integer X, denoted by X_{M^α}, as:

$$X_{M^\alpha} \equiv X \left(\mathrm{mod} \; \frac{M_n}{M^\alpha} \right) \, , \tag{8.65}$$

where X_{M^α} represents the reduced representation of X with the residues $x_{j_1}, x_{j_2}, \dots, x_{j_\alpha}$ deleted. If $\frac{M_n}{M^\alpha} \geq M_k$, it follows from Equation 8.65 that the M^α-projection of any legitimate number X in the RRNS is still the same legitimate number, that is $X_{M^\alpha} = X$.

For a RRNS(n, k) code which has a minimum free distance d_{min}, as described in Section 8.3.1, any valid codeword $\underline{x}_i$ represents a unique integer X_i which falls in the legitimate range of the code, $0 \leq X_i < M_k$. Conversely, any invalid codewords $\underline{x}_j$ are in the illegitimate range, $M_k \leq X_j < M_n$. It was also shown in Section 8.3.1 that any integer X_j differing from $X, 0 \leq X < M_k$, in at least one but no more than $d_{min} - 1$ residue digits is an illegitimate number. By using our previous arguments, we are able to detect if a set of residues $\underline{x}_j$ is erroneous, as long as the number of residue errors is less than or equal to $l = d_{min} - 1$.

Let us assume that no more than t number of errors occurred in a valid codeword $\underline{x}_i$, where t is the error correction capability of the RRNS code. The potentially corrupted vector, which may not be a valid codeword, can be represented as:

$$
\begin{aligned}
Z &\equiv X + E \; (\mathrm{mod} \; M_n) &\tag{8.66} \\
&\longleftrightarrow (z_1, z_2, \dots, z_n) \\
&= (x_1, x_2, \dots, x_n) + (0, \dots, 0, e_{j_1}, 0, \dots, 0, e_{j_2}, 0, \dots, 0, e_{j_t}, 0, \dots, 0) &\tag{8.67}
\end{aligned}
$$

where $X \longleftrightarrow (x_1, x_2, \dots, x_n)$ and $E \longleftrightarrow (0, \dots, 0, e_{j_1}, 0, \dots, 0, e_{j_2}, 0, \dots, 0, e_{j_t}, 0, \dots, 0)$. We have $E \equiv 0 \; (\mathrm{mod} \; m_j)$ for all $j \neq j_i, i = 1, 2, \dots, t$. Hence E is a multiple of all moduli except $m_{j_1}, m_{j_2}, \ldots, m_{j_t}$ and we can write it as:

$$
\begin{aligned}
E &= e \frac{M_n}{m_{j_1} m_{j_2} \ldots m_{j_t}} \\
&= e \frac{M_n}{M^t},
\end{aligned} \tag{8.68}
$$

where $0 < e < M^t$ and e cannot be a multiple of any of the moduli $m_{j_1}, m_{j_2}, \ldots, m_{j_t}$.

Let us now substitute Equation 8.68 into Equation 8.66 and take the M^t-projection of Z:

$$
\begin{aligned}
Z_{M^t} &\equiv Z \left(\bmod \frac{M_n}{M^t} \right) \\
&\equiv X + e \frac{M_n}{M^t} \left(\bmod \frac{M_n}{M^t} \right) \\
&\equiv X \left(\bmod \frac{M_n}{M^t} \right) \\
&= X_{M^t} = X < M_k .
\end{aligned} \tag{8.69}
$$

Therefore, we have shown that the valid codeword $\underline{x}_i$ can be recovered from the received vector $\underline{z}$, even though it has been corrupted by t errors. However, we have to show that for any other combination of t moduli, $m_{j_{1'}}, m_{j_{2'}}, \ldots, m_{j_{t'}}$, which is denoted by $M^{t'}$ and $M^{t'} \neq M^t$, the $M^{t'}$-projection of the integer Z results in an integer $Z_M^{t'}$, which represents an invalid codeword in the illegitimate range of $M_k \leq Z_M^{t'} < M_n$. Note that the total number of moduli combinations is equal to $^n C_t = \frac{n!}{t!(n-t)!}$. We define the $M^{t'}$-projection of Z, namely $Z_M^{t'}$, as its *illegitimate projection* and, conversely, Z_{M^t} as its *legitimate projection*. The illegitimate projection $Z_M^{t'}$ can be treated as a number originating from $X_M^{t'}$ which was projected with the aid of the moduli $m_{j_1}, m_{j_2}, \ldots, m_{j_t}$. We can then express $Z_M^{t'}$ as follows:

$$
\begin{aligned}
Z_M^{t'} &\equiv X_M^{t'} + e \frac{M_n}{M^t M^{t'}} \left(\bmod \frac{M_n}{M^t} \right) \\
&= X_M^{t'} + e \frac{M_n}{\prod_{i=1}^t m_{j_i} \prod_{i=1}^t m_{j_{i'}}} ,
\end{aligned} \tag{8.70}
$$

where $0 < e < \prod_{i=1}^t m_{j_{i'}}$. With the objective of quantifying the range, which an illegitimate projection of Z, namely $Z_M^{t'}$ results into, we find the minimum of $Z_M^{t'}$. Hence we choose $X_M^{t'} = 0$, $e = 1$ and we have:

$$
\begin{aligned}
\min \left(Z_M^{t'} \right) &= \frac{M_n}{\prod_{i=1}^t m_{j_i} \prod_{i=1}^t m_{j_{i'}}} \\
&= \frac{M_k M_{n-k}}{\max \left(\prod_{i=1}^t m_{j_i} \prod_{i=1}^t m_{j_{i'}} \right)} \\
&= \frac{M_k M_{n-k}}{M_{n-k}} \\
&= M_k .
\end{aligned} \tag{8.71}
$$

This shows that an illegitimate projection $Z_M^{t'}$ of integer Z is always larger than $M_k - 1$, i.e. $Z_M^{t'} \geq M_k$.

Let us now consider an example. We have a RRNS code based on the moduli $m_1 = 3$, $m_2 = 4$, $m_3 = 5$, $m_4 = 7$, $m_5 = 11$ and $m_6 = 13$, where m_3, m_4, m_5 and m_6 are the redundant moduli. Therefore, $n = 6$, $k = 2$ and:

$$
\begin{aligned}
M_k &= 3 \times 4 & M_n &= 3 \times 4 \times 5 \times 7 \times 11 \times 13 \\
&= 12 & &= 60060 .
\end{aligned}
\tag{8.72}
$$

The minimum free distance of the RRNS code is $d_{min} = 5$, since by applying Equation 8.39, we found:

$$
\begin{aligned}
4 \times 5 \times 7 \times 11 \times 13 &> M_{n-k} \geq 5 \times 7 \times 11 \times 13 \\
20020 &> M_{n-k} \geq 5005 .
\end{aligned}
\tag{8.73}
$$

The values of n, k and d_{min} satisfy Equation 8.47 and hence the designed RRNS code is a maximum distance separable RRNS code. The error correction capability is $t = \left\lfloor \frac{d_{min}-1}{2} \right\rfloor = 2$. Let $X = 10$, which is represented in the RRNS domain by:

$$
X \longleftrightarrow (1, 2, 0, 3, 10, 10) .
\tag{8.74}
$$

Since $0 \leq X < M_k = 12$, x is a valid codeword. Assume $E = 5720$ and with the aid of Equation 8.66, we can write:

$$
\begin{aligned}
Z &\equiv X + E \bmod (M_n) \\
&= 10 + 5720 \bmod (60060) \\
&= 5730 \\
(0, 2, 0, 4, 10, 10) &\longleftrightarrow \left| (1, 2, 0, 3, 10, 10) + (2, 0, 0, 1, 0, 0) \right|_{(m_1, m_2, m_3, m_4, m_5, m_6)} .
\end{aligned}
\tag{8.75}
$$

In order to correct the errors, their positions have to be found. Hence the integer Z is first calculated using the CRT of Section 8.2.5.1 and it is found that $Z \geq M_k$, which means that z is an invalid codeword. Errors have been detected and the error locations will have to be found in the next step. Since $n = 6$ and $t = 2$, we would have $^nC_t = \binom{6}{2} = 15$ different combinations of the residue error positions. Let us for example assume that the erroneous residue positions correspond to m_1 and m_2. The projection of Z using $M^{t'} = 3 \times 4 = 12$ is then:

$$
\begin{aligned}
Z_{M^{t'}} &\equiv Z \left(\bmod \frac{M_n}{M^{t'}} \right) \\
&= Z \left(\bmod \frac{60060}{12} \right) \\
&= 5730 \ (\bmod \ 5005) \\
&= 725 > M_k ,
\end{aligned}
\tag{8.76}
$$

which yields an illegitimate projection. Using all 15 different possible residue error moduli combinations and the steps described, we find the moduli projection of each combination. The results are shown in Table 8.4. It can be seen from the table that the moduli projection $M^{t'}$ of all moduli combinations is larger than M_k, except that corresponding to moduli m_1 and m_4. Hence, the residue positions 1 and 4 are declared to be in error. The moduli projection Z_{M^t} for moduli m_1 and m_4 is equal to 10, which is the same as integer X. The correct residues in positions 1

j_1,j_2	m_{j_1}	m_{j_2}	$M^{t'2}$	$\frac{M_n}{M^{t'2}}$	$Z_{Mt'}$	j_1,j_2	m_{j_1}	m_{j_2}	$M^{t'2}$	$\frac{M_n}{M^{t'2}}$	$Z_{Mt'}$
1,2	3	4	12	5005	725	2,6	4	13	52	1155	1110
1,3	3	5	15	4004	1726	3,4	5	7	35	1716	582
1,4	**3**	**7**	**21**	**2860**	**10**	3,5	5	11	55	1092	270
1,5	3	11	33	1820	270	3,6	5	13	65	924	186
1,6	3	13	39	1540	1110	4,5	7	11	77	780	270
2,3	4	5	20	3003	2727	4,6	7	13	91	660	450
2,4	4	7	28	2145	1440	5,6	11	13	143	420	270
2,5	4	11	44	1365	270	-,-	-	-	-	-	-

Table 8.4: Results of the 15 different moduli projections of integer $Z = 5730$.

and 4 are Z_{M^t} modulo m_1 and m_4, i.e. 10 mod 3 = 1 and 10 mod 4 = 2. Alternatively, we can apply the BEX algorithm in order to find the correct residues for both positions.

Above we have shown a simple example for correcting two residues errors. Let us now show using the same example that the same procedure can be used to correct one residue error in the RRNS. Again, let $X = 10 \longleftrightarrow (1,2,0,3,10,10)$ and assume $E = 40,040$. With the aid of Equation 8.66, we can write:

$$
\begin{aligned}
Z &\equiv 10 + 40040 \bmod (60060) \\
&= 40050
\end{aligned}
$$

$$(0,2,0,3,10,10) \longleftrightarrow (1,2,0,3,10,10) + (2,0,0,0,0,0). \tag{8.77}$$

During the first step of the procedure we calculate the integer Z, which is found to be in error. In order to locate the errors, 15 different moduli combinations are to be considered for finding the projection of Z. The corresponding results are shown in Table 8.5. From the table, we can see that there is more than one moduli combination which results in moduli projection $Z_{M^t} < M_k = 12$. Indeed, the moduli projections $Z_{M^t} < M_k = 12$ are all the same integers, namely 10. Actually, all moduli combinations which include modulus m_1 will produce a moduli projection of $Z_{M^t} = 10 < M_k = 12$. Hence, the procedure used to correct t residue errors can be applied to correct less than t errors as well.

j_1,j_2	m_{j_1}	m_{j_2}	$M^{t'2}$	$\frac{M_n}{M^{t'2}}$	$Z_{Mt'}$	j_1,j_2	m_{j_1}	m_{j_2}	$M^{t'2}$	$\frac{M_n}{M^{t'2}}$	$Z_{Mt'}$
1,2	**3**	**4**	**12**	**5005**	**10**	2,6	4	13	52	1155	780
1,3	**3**	**5**	**15**	**4004**	**10**	3,4	5	7	35	1716	582
1,4	**3**	**7**	**21**	**2860**	**10**	3,5	5	11	55	1092	738
1,5	**3**	**11**	**33**	**1820**	**10**	3,6	5	13	65	924	318
1,6	**3**	**13**	**39**	**1540**	**10**	4,5	7	11	77	780	270
2,3	4	5	20	3003	1011	4,6	7	13	91	660	450
2,4	4	7	28	2145	1440	5,6	11	13	143	420	150
2,5	4	11	44	1365	465	-,-	-	-	-	-	-

Table 8.5: Results of the 15 different moduli projections of integer $Z = 40050$.

In the previous two examples we have shown that the multiple errors correction procedure can be accomplished by using the CRT. However, if the number of moduli is increased in an effort to create larger and/or stronger RRNS codes, the range of M_k and M_n increases as well, which may lead to overflows. Therefore, special procedures have to be used for carrying out large integer operations, such as those involved in the CRT. This will impose extra complexity upon the decoding algorithm. In order to avoid large integer operations, the MRC was used by a number of authors [44, 169, 178] for correcting single or multiple errors in RRNS codes.

In Section 8.2.7, we explained that for any integer $X < M_k$ represented in the mixed radix form, as in Equation 8.34, the redundant mixed radix digits, $a_{k+j}, j = 1, 2, \ldots, n - k$, will be equal to zero. Hence, we can apply the MRC to a given residue representation $\underline{z}$ in order to generate the redundant mixed radix digits. By checking the redundant mixed radix digits, we are able to tell whether $Z < M_k$, i.e. whether $\underline{z}$ is a valid codeword. If $Z \geq M_k$, moduli projection can be used to locate the errors. In Equation 8.65, we defined the moduli projection of an integer. It can be shown that the M^t-projection of Z can also be represented as a reduced residue representation of Z with the residues $z_{j_1}, z_{j_2}, \ldots, z_{j_t}$ and their corresponding moduli deleted. Using the reduced residue representation, we are able to find the mixed radix representation of Z_{M^t}, as follows [41, 43, 44]:

$$Z_{M^t} = \sum_{\substack{i=1 \\ i \neq j_1, j_2, \ldots, j_t}}^{n} a_i \prod_{\substack{r=1 \\ r \neq j_1, j_2, \ldots, j_t}}^{i-1} m_r , \qquad (8.78)$$

which reflects the structure of the MRC definition in Equation 8.21 and where $\prod_{r=1}^{0} m_r \equiv 1$. The legitimate and illegitimate range of the reduced RRNS are $[0, M'_k - 1]$ and $[M'_k, \frac{M_n}{M^t} - 1]$, respectively. We can then specify the legitimate range of the reduced RRNS separately for two specific cases, where the dropped moduli are from the set of redundant and non-redundant moduli, i.e. $1 \leq j_r \leq n$, or where the dropped moduli are from the set of redundant moduli only, i.e. $k < j_r \leq n$. This is expressed explicitly as:

$$M'_k = \prod_{\substack{i=1 \\ i \neq j_1, j_2, \ldots, j_r}}^{k+r} m_i, \qquad 1 \leq j_r \leq n$$

$$M'_k = M_k = \prod_{i=1}^{k} m_i, \qquad k < j_r \leq n . \qquad (8.79)$$

We can see from the equation that $M'_k \geq M_k$ and $M'_k - 1$ is the highest integer that can be represented by Equation 8.78 with all the redundant mixed radix digits set to zero. Therefore, even if all the redundant mixed radix digits are zero, one can still argue that any legitimate projection Z_{M^t} can be larger than M_k since $M'_k \geq M_k$. However, we are able to show that any illegitimate projection of Z will result in $Z_{M^{t'}} \geq M'_k$. In order to find the minimum of $Z_M^{t'}$, we rewrite Equation 8.70 as:

$$Z_M^{t'} = X_M^{t'} + e \frac{M_k M_{n-k}}{m_{j_{1'}} \ldots m_{j_{r'}} \ldots m_{j_{t'}} \prod_{i=1}^{t} m_{j_i}}$$

$$= X_M^{t'} + e \frac{M_k}{m_{j_{1'}} \ldots m_{j_{r'}}} \frac{M_{n-k}}{m_{j_{r+1'}} \ldots m_{j_{t'}} \prod_{i=1}^{t} m_{j_i}} . \qquad (8.80)$$

Since $n - k \geq t + t' = 2t$, we have:

$$\frac{M_{n-k}}{m_{j_{r+1'}} \ldots m_{j_{t'}} \prod_{i=1}^{t} m_{j_i}} \geq m_{k+1} \ldots m_{k+r} . \qquad (8.81)$$

Following from Equation 8.81, we can then write:

$$Z_M^{t'} \geq X_M^{t'} + e\frac{M_k}{m_{j_1}, \ldots m_{j_r}}m_{k+1}\ldots m_{k+r}$$
$$= X_M^{t'} + eM_k' , \qquad (8.82)$$

where M_k' is the legitimate range of the reduced RRNS. Clearly, in Equation 8.82, any illegitimate projection of Z always results in $Z_M^{t'} > M_k'$. Hence, if not all the redundant mixed radix digits are equal to zero, the projection is illegitimate and vice versa. Once a legitimate projection is found, we can apply BEX to find the correct residues for the corresponding positions.

In order to augment our previous discussions let us explain in detail the multiple-error correction procedure employing MRC, using the above example. Previously, we have used moduli $m_1 = 3$, $m_2 = 4$, $m_3 = 5$, $m_4 = 7$, $m_5 = 11$ and $m_6 = 13$. The total number of moduli is $n = 6$ and the number of information moduli is $k = 2$. Therefore, we have $M_k = 12$ and $M_n = 60060$. It was shown in Equation 8.73 that the RRNS code has $d_{min} = 5$. Again, we have an integer message of $X = 10$, which has been corrupted by $E = 5520$ and became Z given by:

$$Z \equiv X + E \bmod (M_n)$$
$$= 5730$$
$$(0,2,0,4,10,10) \longleftrightarrow |(1,2,0,3,10,10) + (2,0,0,1,0,0)|_{(m_1,m_2,m_3,m_4,m_5,m_6)} .$$
$$(8.83)$$

We first assume that the erroneous residue positions correspond to m_1 and m_2. Then we find a reduced representation of Z with the residues z_1 and z_2 deleted. Since the moduli m_1 and m_2 are deleted, we have a new RRNS based on the moduli $m_1 = 5$, $m_2 = 7$, $m_3 = 11$ and $m_4 = 13$, where m_3 and m_4 are the redundant moduli and hence $M_k' = 5 \times 7 = 35$. Using the reduced residue representation, we can express the mixed radix representation of the reduced representation of Z with the residues z_1 and z_2 deleted, using the $M^{t'}$-projection of Z $Z_{M^{t'}}$ as follows:

$$Z_{M^{t'}} = a_4 m_1 m_2 m_3 + a_3 m_1 m_2 + a_2 m_1 + a_1$$
$$= 385a_4 + 35a_3 + 5a_2 + a_1 . \qquad (8.84)$$

Employing the procedures outlined in Section 8.2.5.2, we are able to calculate the values of a_i, $i = 1,2,3,4$, given the residues z_3, z_4, z_5 and z_6. However, for the reader's convenience, we can assume that we have $Z_{M^{t'}} = 725$ and hence rewrite Equation 8.84 as:

$$725 = 385a_4 + 35a_3 + 5a_2 + a_1 . \qquad (8.85)$$

Using Equation 8.85, we readily find the values $a_1 = 0$, $a_2 = 5$, $a_3 = 9$ and $a_4 = 1$. Since $a_3 \neq a_4 \neq 0$, we know that:

$$Z_{M^{t'}} > M_k'$$
$$725 > 35 . \qquad (8.86)$$

We can therefore conclude that $Z_{M^{t'}}$ is an illegitimate projection.

Let us now delete another set of moduli, namely m_1 and m_4, from the original set of moduli given by $m_1 = 3$, $m_2 = 4$, $m_3 = 5$, $m_4 = 7$, $m_5 = 11$ and $m_6 = 13$. Hence the original RRNS was reduced to a new set of RRNS based on the moduli $m_1 = 4$, $m_2 = 5$, $m_3 = 11$

and $m_4 = 13$, where m_3 and m_4 are the redundant moduli and the dynamic range is given by $M'_k = 4 \times 5 = 20$. Using the reduced residue representation, we can express the mixed radix representation of the reduced representation of Z with the residues z_1 and z_4 deleted, using the $M^{t'}$-projection of Z, $Z_{M^{t'}}$ as follows:

$$
\begin{aligned}
Z_{M^{t'}} &= a_4 m_1 m_2 m_3 + a_3 m_1 m_2 + a_2 m_1 + a_1 \\
&= 220 a_4 + 20 a_3 + 4 a_2 + a_1 .
\end{aligned}
\tag{8.87}
$$

Given the residues z_1, z_2, z_3 and z_4, we can apply the somewhat tedious procedures of Section 8.2.5.2 for calculating the values of a_i, $i = 1, 2, 3, 4$. Again, for the reader's convenience we assume that we have $Z_{M^{t'}} = 10$ and hence we arrive at:

$$
10 = 220 a_4 + 20 a_3 + 4 a_2 + a_1 .
\tag{8.88}
$$

Using Equation 8.88, we calculated $a_1 = 2$, $a_2 = 2$, $a_3 = 0$ and $a_4 = 0$. Since $a_3 = a_4 = 0$, $Z_{M^{t'}}$ is a legitimate projection. Once the legitimate projection is found, we can apply the BEX algorithm for finding the correct residues for the corresponding positions, namely for positions 1 and 4.

The flowchart of multiple-error correction procedures using the MRC method is shown in Figure 8.1. Initially, the corrupted residues $\underline{z}$ are received and MRC is applied. The redundant residues $a_{k+1}, \ldots , a_n$ are checked whether they are all zeros. If $a_{k+1} = a_{k+2} = \ldots = a_n = 0$, the received residues $\underline{z}$ are declared error free. If not all of them are zero, a set of t moduli is generated in an effort to find up to t error positions. MRC is applied to the reduced residue representation, where the set of t chosen moduli is deleted. The redundant mixed radix digits are checked to see whether all the redundant mixed radix digits are zero. If so, the error positions have been found and the errors can be corrected using the BEX. If either of the redundant mixed radix digits is non-zero, another set of t moduli is obtained and the MRC procedure is repeated again. The above process is repeated until all possible sets of t moduli combinations have been tested and hence the received residue vector is declared to have more than t errors.

8.5 RRNS Encoder

In the previous section we stated that a RRNS code is constituted by a set of residues with respect to a predefined set of moduli. Since the moduli and the residues can assume any positive integer value — representing an arbitrary number of binary bits — the RRNS code is a non-binary code, based on transmitting the residues conveying a number of bits. In this section, we propose two different mapping methods transforming the binary source bits to the non-binary RRNS code, which result in a so-called non-systematic and systematic RRNS code.

8.5.1 Non-systematic RRNS Code

Here we commence by summarising the non-systematic encoding process of Figure 8.2. The non-systematic encoder encodes k_b number of binary data bits per RRNS(n, k) codeword, where the integer 2^{k_b} must not be higher than the legitimate range M_k, hence:

$$
2^{k_b} \leq M_k .
\tag{8.89}
$$

In other words, the data bits are mapped to an integer X, which has to be in the range of $[0, 2^{k_b} - 1]$. Note that the full legitimate range M_k of the RRNS may not be actively exploited, since M_k

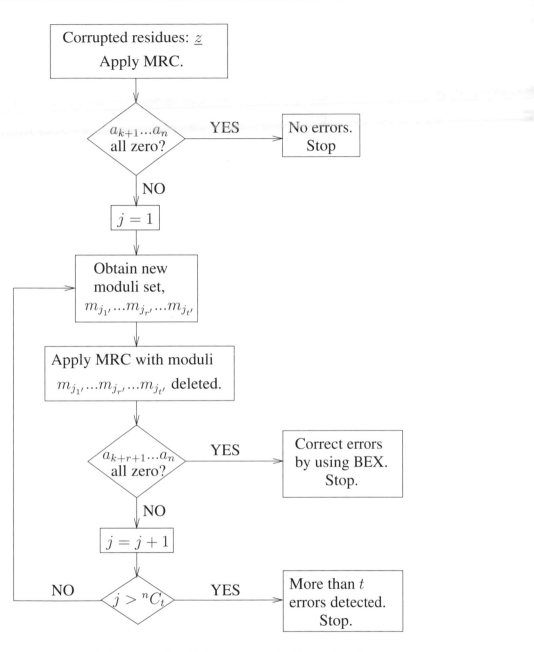

Figure 8.1: Flowchart of multiple-error correction RRNS decoding.

Figure 8.2: Non-systematic encoding procedures.

is typically not an integer power of 2. Considering now the mapping of the integer X to residues for transmission in Figure 8.2 and using the moduli in the RRNS, the residues x_j are simply obtained by invoking the conventional modulus operation:

$$x_j = |X|_{m_j} , \tag{8.90}$$

where $j = 1, 2, \ldots, n$. In order to represent an integer X in the RRNS seen in Figure 8.2, each residue x_j, non-redundant or redundant, has to be represented uniquely for transmission in terms of bits. Therefore, we have to ensure that:

$$2^{n_{bj}} \geq m_j , \tag{8.91}$$

where $j = 1, 2, \ldots, n$ and n_{bj} is the number of bits representing the residue x_j in Figure 8.2. The total number of coded bits per RRNS(n, k) codeword is therefore:

$$n_b = \sum_{j=1}^{n} n_{bj} . \tag{8.92}$$

The rate R of the code is then $\frac{k_b}{n_b}$. Since the residues $x_j, j = 1, \ldots, k$, do not directly represent the binary data bits, we refer to the above encoding process as non-systematic encoding.

Let us for example consider a RRNS based on the moduli $m_1 = 53$, $m_2 = 55$, $m_3 = 59$, $m_4 = 61$, $m_5 = 63$ and $m_6 = 64$, where m_5 and m_6 are the redundant moduli. We have $n = 6$, $k = 4$ and:

$$
\begin{aligned}
M_k &= 53 \times 55 \times 59 \times 61 \\
&= 10491085 ,
\end{aligned} \tag{8.93}
$$

where M_k is the legitimate dynamic range of the RRNS. In this case, we can represent an integer from 0 to 10491084 uniquely by the RRNS. Since $8388608 = 2^{23} < M_k < 2^{24} = 16777216$, $k_b = 23$ is the number of the binary data bits encoded by the RRNS code. Hence, the dynamic range of the RRNS is not fully utilised. Applying Equation 8.91, we calculate the number of coded bits by taking into account that each of the residues requires six bits for its unique representation, yielding:

$$
\begin{aligned}
n_b &= 6 + 6 + 6 + 6 + 6 + 6 \\
&= 36 .
\end{aligned} \tag{8.94}
$$

Therefore, the code rate is:

$$
\begin{aligned}
R &= \frac{k_b}{n_b} \\
&= \frac{23}{36} = 0.639
\end{aligned} \tag{8.95}
$$

8.5.2 Systematic RRNS Code

In contrast to the non-systematic encoder of Figure 8.2, Figure 8.3 characterises the systematic encoding process. Unlike the non-systematic encoder, which maps all the data bits to be transmitted to a single integer X, the systematic encoder divides the bit sequence to be encoded into

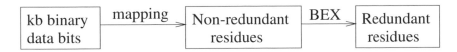

Figure 8.3: Systematic encoding procedures.

shorter groups of bits, each of which represents a non-redundant residue x_j. In contrast to the non-systematic mapping of Equation 8.89, in order to render this systematic mapping unique, the number of bits k_{b_j} mapped to residue x_j has to satisfy:

$$2^{k_{b_j}} < m_j . \tag{8.96}$$

The total number of data bits that the systematic encoder encodes into each RRNS codeword becomes:

$$k_b = \sum_{j=1}^{k} k_{b_j} . \tag{8.97}$$

Accordingly, as shown in Figure 8.3, the data bit sequences are mapped to the non-redundant residues directly. Then the so-called BEX algorithm of Section 8.2.7 can be invoked, in order to compute the redundant residues. Similarly to the non-systematic encoder, the number of bits needed to represent the redundant residues has to satisfy Equation 8.91 for $j = k + 1, ..., n$. The number of bits required for the unique representation of the non-redundant residues has been specified in Equation 8.96. Hence, we can write the number of coded bits n_{b_j} for each residue x_j as:

$$n_{b_j} = \begin{cases} 2^{k_{b_j}} < m_j & j = 1, 2, \dots, k \\ 2^{n_{b_j}} \geq m_j & j = k + 1, \dots, n. \end{cases} \tag{8.98}$$

The total number of coded bits can then be calculated using Equation 8.92.

We consider again the same moduli set, namely $m_1 = 53$, $m_2 = 55$, $m_3 = 59$, $m_4 = 61$, $m_5 = 63$ and $m_6 = 64$, as for our non-systematic coding example in Section 8.5.1. Applying Equations 8.96 and 8.97, the total number of data bits is:

$$k_b = 5 + 5 + 5 + 5 = 20 . \tag{8.99}$$

The number of coded bits, calculated using Equation 8.98, is then:

$$n_b = 5 + 5 + 5 + 5 + 6 + 6 = 32 . \tag{8.100}$$

The code rate is $R = \frac{k_b}{n_b} = \frac{20}{32} = 0.625$. If we compare the code rate of the non-systematic and systematic encoders, we can see that the code rate of the systematic encoder is lower than that of the non-systematic encoder. Furthermore, the dynamic range of the RRNS code is only $\left(2^5\right)^4 = 1048576$ as compared to the corresponding range of 8388608 for the non-systematic code. In order to increase the code rate and the dynamic range of the systematic RRNS encoder, we propose a more efficient mapping method, which is outlined in the next section.

8.5.2.1   Modified Systematic RRNS Code

As mentioned earlier, the legitimate dynamic range of a systematic encoder of Figure 8.3 is more limited than that of the non-systematic encoder seen in Figure 8.2. This also causes a reduction in the code rate compared to that of the non-systematic encoder. Here, we propose a modification to the mapping method used in the systematic encoder of the previous section. Instead of using Equation 8.96, the number of binary data bits mapped to each non-redundant residue is now increased by one, yielding:

$$2^{k_{b_j}} \geq m_j . \tag{8.101}$$

However, as a consequence of the new allocation of data bits, there may exist integers X, which are equal to or greater than the modulus, i.e. $X \geq m_j$. Hence, we define the new mapping method as follows:

$$x_j = \begin{cases} X & \text{if } X < m_j \\ 2^{k_{b_j}} - 1 - X & \text{if } X \geq m_j , \end{cases} \tag{8.102}$$

where k_{b_j} is the number of data bits mapped to the integer X for transmission in Figure 8.3. The mapping in the second line of Equation 8.102 has to be implemented on the basis of bitwise complement, if $X \geq m_j$. Although this implies that the mapping to the integers is ambiguous for some of the values, Equation 8.102 ensures the maximum Hamming distance separation of the ambiguous values. We will show in the following that this maximum Hamming distance separation allows the decoder to recognise the original integer messages.

We use the modulus $m_4 = 61$ as an example for further illustration. In the previous section, only five bits were assigned to this residue. Since $2^5 = 32, 0, 1, \ldots, 31$ are the possible residues. Hence, the previous bit assignment policy does not fully utilise the dynamic range of the modulus $m_4 = 61$, which is $0, 1, \ldots, 60$. Applying Equation 8.101, we now map six bits to the residue of modulus $m_4 = 61$ and $0, 1, \ldots, 63$ are the possible integers. However, integers $61, \ldots, 63$ are not valid residues. By applying Equation 8.102, we map the integers $61, 62, 63$ to the integers $2, 1, 0$, respectively. Figure 8.4 shows the mapping of integer 61 to 2 by implementing its bitwise complement.

Figure 8.4: An example of modified systematic mapping.

Using the same moduli set as in the previous two sections, and applying Equations 8.101 and 8.91, we arrive at $k_b = 6 + 6 + 6 + 6 = 24$ and $n_b = 6 + 6 + 6 + 6 + 6 + 6 = 36$. Consequently, the code rate is now $R = \frac{24}{36} = 0.667$, which is more than in the previous systematic and non-systematic encoding cases. Furthermore, the dynamic range of the systematic RRNS code was increased from 1048576 to $\left(2^6\right)^4 = 16777216$, which is more than $M_k = 10491085$.

8.6   RRNS Decoder

The structure of the systematic and non-systematic RRNS decoders is similar. Figure 8.5 shows the simplified block diagram of the RRNS decoder. The modulo of the received residues, $|x_j|_{m_j}$,

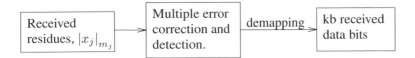

Figure 8.5: A block diagram of the RRNS decoder.

is taken, where $j = k + 1, \ldots, n$ for the systematic RRNS decoder and $j = 1, \ldots, n$ for the non-systematic RRNS decoder. Multiple residue error correction and detection is invoked, as described in Section 8.4, for correcting a maximum of t errors in the codeword. Then, the residues are de-mapped to the original k_b data bits according to the mapping method of the systematic and non-systematic RRNS encoders, respectively.

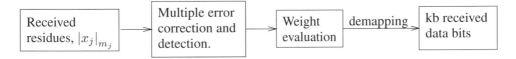

Figure 8.6: A block diagram of the modified systematic RRNS decoder.

For the modified systematic RRNS code of Section 8.5.2.1, the decoder structure needs an extra step, as shown in Figure 8.6. Owing to the potentially ambiguous mapping defined in Equation 8.102, a residue x_j may represent two integers, which exhibit maximum Hamming distance separation. Hence, after the multiple-error correction and detection procedures of Section 8.4, we have to determine which integer has to be used for extracting the k_{b_j} transmitted data bits. Soft-decision of the received bits y_{ji} can be used for calculating their decision metric W, which in turn determines the integer message transmitted. The soft decision metric W of the residue x_j is calculated as follows:

$$W = \sum_{i=1}^{k_{b_j}} \mathrm{asign}(x_{ji}) \times y_{ji} \, , \tag{8.103}$$

where:

$$\mathrm{asign}(x_{ji}) = \begin{cases} +1 & \text{if } x_{ji} = 1 \\ -1 & \text{if } x_{ji} = 0 \, . \end{cases} \tag{8.104}$$

Depending on the polarity of the soft-decision metric W, the transmitted integer is then inferred from:

$$X = \begin{cases} x_j & \text{if } W \geq 0.0 \\ 2^{k_{b_j}} - 1 - x_j & \text{if } W < 0.0 \, , \end{cases} \tag{8.105}$$

which is the reverse operation of Equation 8.102.

Again, we use the modulus $m_4 = 61$ as our example in augmenting the associated operations. After the multiple residue error correction and detection procedures, residue x_4 is found to be $x_4 = 2$. From Equation 8.105 and Figure 8.4 we know that there are two possible transmitted integers X associated with residue $x_4 = 2$. Therefore, soft decision of the received bits y_{ji} can be used for determining the transmitted integer by exploiting the maximum Hamming distance separation of the associated residues. Explicitly, let us assume that the received soft-decision bits are $+1.2, +1.5, -0.3, +0.8, -1.0, +0.2$. By applying Equation 8.103, we calculate the soft-decision metric as follows:

$$W = -1.2 - 1.5 + 0.3 - 0.8 - 1.0 - 0.2 = -4.4 < 0.0 . \qquad (8.106)$$

Therefore, with the aid of Equation 8.105 the transmitted integer is decided to be $X = 2^6 - 1 - 2 = 61$. By contrast, for example, the soft-decision values of $W \geq 0.0$ lead to a decision of $X = 2$.

8.7 Soft-input and Soft-output RRNS Decoder

If the output of the hard-decision-based demodulator is binary, the RRNS decoder is incapable of exploiting the potential advantages accruing from the soft outputs. In this section we contrive the soft decoding of RRNS codes by combining the classic Chase algorithm [28] of Section 4.4 with the hard-decision-based RRNS decoder. Hence only a brief account of the Chase algorithm will be given, since a detailed exposure to the Chase algorithm has been provided in Section 4.4. Recently, Pyndiah *et al.* [62,81,201,202] extended the Chase algorithm so that it became capable of providing soft outputs of the decoded bits. This modified algorithm was then invoked for RRNS codes [196] in order to provide the soft output of the decoded bits. Consequently, we contrived the Soft-Input Soft-Output (SISO) RRNS decoder. This advance allowed us to employ RRNS codes as component codes in turbo codes.

8.7.1 Soft-input RRNS Decoder

Let us consider the transmission of block-coded binary symbols $\{-1, +1\}$ using BPSK modulation over an AWGN channel. At the receiver, the demodulator provides the soft-decision values $\underline{y}$ for the RRNS decoder. A maximum likelihood decoder is capable of finding the codeword that satisfies:

$$\min_{m} \text{weight}(|\underline{y} - \underline{x}_j|^2) , \qquad (8.107)$$

where $x_{ji} \in \{-1, +1\}$ are the tentatively assumed transmitted binary bit representations of the RRNS-coded symbols and the range of j is over all possible legitimate RRNS codewords. The decision given by Equation 8.107 is optimum in the minimum BER sense, but the associated computational complexity increases exponentially with k and becomes prohibitive for block codes with $k > 6$. As a remedy, the reduced-complexity Chase algorithm [28] can be invoked for near maximum likelihood decoding of block codes. Again, the algorithm is sub-optimum, but it offers a significantly reduced complexity.

As already highlighted in Section 4.4 in the context of binary BCH codes, at the demodulator, the soft-decision outputs $\underline{y}$ are subjected to a tentative hard decision, yielding the binary sequence $\underline{z}$, and the associated soft-decision confidence values $|\underline{y}|$ are fed to the Chase algorithm. The tentative binary sequence $\underline{z}$ is perturbed with the aid of a set of test patterns TP, which are also binary sequences that contain binary 1s in the bit positions that are to be tentatively inverted. By adding this test pattern, modulo two, to the tentative binary sequence $\underline{z}$ a new sequence $\underline{z}'$ is obtained, where:

$$\underline{z}' = \underline{z} \oplus TP . \qquad (8.108)$$

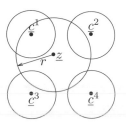

Figure 8.7: Simple coding-space illustration of the Chase algorithm.

Using the different test patterns, the perturbed received sequence $\underline{z}'$ falls within the decoding sphere of a number of different valid codewords, namely in that of $\underline{c}^1 \ldots \underline{c}^4$ for example in Figure 8.7. In the figure, r represents the maximum Hamming distance of the perturbed binary sequence $\underline{z}'$ from the original tentative binary sequence $\underline{z}$. Hence the value of r can be adjusted by varying the maximum Hamming weight of the TPs. If we increase r, the perturbed received sequence $\underline{z}'$ will fall within the decoding sphere of more valid RRNS codewords. In order to reduce the associated implementational complexity, typically only a small set of l bit positions associated with the least reliable confidence values $|y|$ is perturbed. The number of TPs for which tentative decoding is invoked is then equal to 2^l.

Specifically, if the perturbed received sequence $\underline{z}'$ falls within the decoding sphere of a valid codeword, a new error pattern $\underline{e}'$ is obtained with the aid of a tentative hard-decision RRNS decoding, which may be an all-zero or a non-zero tuple. Explicitly, the resultant error pattern is an all-zero tuple if the original hard-decision was erroneous, but the corruption by the TP in the low-reliability bit positions succeeded in bringing the hard-decision-based sequence $\underline{z}$ within the decoding sphere of the original transmitted codeword. By contrast, the resultant error pattern is a non-zero tuple, if the tentative corruption of the received codewords failed to move it into the decoding sphere of the original transmitted codeword. Instead, TP may have corrupted the received hard-decision-based codeword into the decoding sphere of another legitimate codeword. In this case the minimum number of decoding errors is d_{min}. The actual error pattern $\underline{e}$ associated with the received sequence $\underline{z}$ is given by:

$$\underline{e} = \underline{e}' \oplus TP, \tag{8.109}$$

which may or may not be different from the original test pattern TP, depending on whether or not the perturbed binary sequence $\underline{z}'$ falls into the decoding sphere of a valid codeword. However, only those perturbed binary sequences $\underline{z}'$ are tentatively RRNS decoded that fall into the decoding sphere of a valid codeword. Those $\underline{z}'$ binary sequences that do not fall within a legitimate decoding sphere cannot lead to a legitimate RRNS codeword and hence are discarded from our further procedures. More specifically, we derive the error pattern $\underline{e}'$ for all perturbed binary sequences $\underline{z}'$ within the decoding sphere of a valid RRNS codeword and find that particular one which is the most likely transmitted one, since it is the closest one to the received binary codeword $\underline{z}$. In this case, we are concerned with finding the error pattern $\underline{e}$ of minimum 'analogue weight', where the analogue weight of an error sequence $\underline{e}$ is defined as:

$$W(\underline{e}) = \sum_{i=1}^{n} e_i |y_i| . \tag{8.110}$$

The generated test pattern TP will be stored if the associated analogue weight W is found to be lower than the previously registered analogue weights associated with the other TPs. The above procedure will be repeated for the maximum number of TPs, which is tolerable in complexity terms. Upon completing this loop, the memory is checked in order to determine whether any error pattern has been stored, and, if so, the corrected decoded sequence will be $\underline{z} \oplus \underline{e}$. Otherwise, the decoded sequence is the same as the received sequence $\underline{z}$.

8.7.2 Soft-output RRNS Decoder

In the previous section, we have highlighted the philosophy of soft decoding RRNS codes. Following the philosophy of Section 7.3, which was cast in the context of turbo BCH codes, with the aim of contriving the turbo RRNS code, let us now determine here the Log Likelihood Ratio (LLR) of each decoded bit u_k, given that the demodulator's soft-output sequence is $\underline{y}$. This is equivalent to finding the LLR of:

$$L(u_k|\underline{y}) = \ln \frac{P(u_k = +1|\underline{y})}{P(u_k = -1|\underline{y})} , \qquad (8.111)$$

where k is a bit position in a RRNS codeword. Since the probability of $u_k = +1$ is equal to the sum of all the probabilities of all codewords $\underline{x}_i$, which have $u_k = +1$, we can rewrite the numerator of Equation 8.111 as follows:

$$P(u_k = +1|\underline{y}) = \sum_{\underline{x}_i \in \alpha^{+k}} P(\underline{x}_i|\underline{y}) , \qquad (8.112)$$

where α^{+k} is the set of codewords $\underline{x}_i$ such that $u_k = +1$. By applying Bayes' rule, which was detailed in Section 5.3.3, we can rewrite Equation 8.112 as:

$$P(u_k = +1|\underline{y}) = \sum_{\underline{x}_i \in \alpha^{+k}} \frac{P(\underline{y}|\underline{x}_i)P(\underline{x}_i)}{P(\underline{y})} . \qquad (8.113)$$

Similarly, the probability of $u_k = -1$ is equal to the sum of all the probabilities of all codewords $\underline{x}_i$ which have $u_k = -1$. Hence, the denominator of Equation 8.111 can be written as:

$$\begin{aligned} P(u_k = -1|\underline{y}) &= \sum_{\underline{x}_i \in \alpha^{-k}} P(\underline{x}_i|\underline{y}) \\ &= \sum_{\underline{x}_i \in \alpha^{-k}} \frac{P(\underline{y}|\underline{x}_i)P(\underline{x}_i)}{P(\underline{y})} , \end{aligned} \qquad (8.114)$$

where α^{-k} is the set of codewords $\underline{x}_i$ such that $u_k = -1$.

Substituting Equations 8.113 and 8.114 into Equation 8.111, and assuming that all codewords are equiprobable, we arrive at:

$$\begin{aligned} L(u_k|\underline{y}) &= \ln \frac{\sum_{\underline{x}_i \in \alpha^{+k}} \frac{P(\underline{y}|\underline{x}_i)P(\underline{x}_i)}{P(\underline{y})}}{\sum_{\underline{x}_i \in \alpha^{-k}} \frac{P(\underline{y}|\underline{x}_i)P(\underline{x}_i)}{P(\underline{y})}} \\ &= \ln \frac{\sum_{\underline{x}_i \in \alpha^{+k}} P(\underline{y}|\underline{x}_i)P(\underline{x}_i)}{\sum_{\underline{x}_i \in \alpha^{-k}} P(\underline{y}|\underline{x}_i)P(\underline{x}_i)} . \end{aligned} \qquad (8.115)$$

Let us assume that the transmitted bit x_k has been sent over a AWGN channel using BPSK modulation. Then, as we have seen in Section 5.3.2 in the context of turbo convolutional codes, the probability density function of the demodulator soft output y_k conditioned on the transmitted bit x_k can be expressed as:

$$P(\underline{y}|\underline{x}) = \left(\frac{1}{\sigma\sqrt{2\pi}}\right)^n \exp^{-\frac{E_B}{2\sigma^2}|\underline{y}-a\underline{x}|^2} \tag{8.116}$$

where:

n is the number of coded bits which is equal to n_b in RRNS code
σ^2 is the noise variance
E_B is the energy per bit
a is the fading amplitude (=1 for a non-fading AWGN channel).

Since the probability of a specific codeword, $P(\underline{x})$, is equal to the product of all probabilities of its constituent coded bits $x_j, j = 1, 2, \ldots, n$, we can then write:

$$\begin{aligned}
P(\underline{x}) &= P(x_1)P(x_2)\ldots P(x_n) \\
&= C\exp^{x_1\frac{L(x_1)}{2}} C\exp^{x_2\frac{L(x_2)}{2}} \ldots C\exp^{x_n\frac{L(x_n)}{2}} \\
&= C^n \exp^{\pm\underline{x}\frac{L(\underline{x})}{2}} ,
\end{aligned} \tag{8.117}$$

where $P(x_j) = C\exp^{x_j\frac{L(x_j)}{2}}$ and C is a constant, which will be cancelled out in Equation 8.115. The derivation of $P(x_j)$ can be found in Section 5.3.2, where the same problem was cast in the context of binary turbo convolutional codes.

Using Equations 8.116 and 8.117, we can rewrite Equation 8.115 as:

$$L(u_k|\underline{y}) = \ln \frac{\sum_{\underline{x}_i \in \alpha^{+k}} \exp^{-\frac{E_B}{2\sigma^2}|\underline{y}-a\underline{x}_i|^2} \exp^{\underline{x}_i\frac{L(\underline{x}_i)}{2}}}{\sum_{\underline{x}_i \in \alpha^{-k}} \exp^{-\frac{E_B}{2\sigma^2}|\underline{y}-a\underline{x}_i|^2} \exp^{\underline{x}_i\frac{L(\underline{x}_i)}{2}}} . \tag{8.118}$$

Let $\underline{x}^{+k} \in \alpha^{+k}$ and $\underline{x}^{-k} \in \alpha^{-k}$ be the codewords, which are at minimum Euclidean distance from the demodulator's soft-output sequence $\underline{y}$. Then, upon using the approximation [50,51]:

$$\ln\left(\sum_j \exp^{A_j}\right) \approx \max_j(A_j) , \tag{8.119}$$

where $\max_j(A_j)$ means the maximum value of A_j, and assuming that there were no transmission errors, i.e. $\underline{u} = \underline{x}$, then we have $L(\underline{x}) = L(\underline{u})$. Hence we can approximate Equation 8.118 as:

$$L(u_k|\underline{y}) \approx -\frac{E_B}{2\sigma^2}|\underline{y} - a\underline{x}^{+k}|^2 + \frac{1}{2}\underline{x}^{+k}L(\underline{u}) + \frac{E_B}{2\sigma^2}|\underline{y} - a\underline{x}^{-k}|^2 - \frac{1}{2}\underline{x}^{-k}L(\underline{u}) . \tag{8.120}$$

Since $|\underline{y} - a\underline{x}^{\pm k}|^2$ is the Euclidean distance between the demodulator's soft-output sequences $\underline{y}$ and the legitimate transmitted codewords $\underline{x}^{\pm k}$, we can write:

$$|\underline{y} - a\underline{x}^{\pm k}|^2 = \sum_{j=1}^{n} \left(y_j - ax_j^{\pm k}\right)^2 . \tag{8.121}$$

Upon substituting Equation 8.121 into Equation 8.120, we arrive at:

$$
\begin{aligned}
L(u_k|\underline{y}) &\approx \frac{E_b}{2\sigma^2} \left[\sum_{j=1}^{n} \left(y_j - ax_j^{-k}\right)^2 - \sum_{j=1}^{n} \left(y_j - x_j^{+k}\right)^2 \right] + \frac{L(\underline{u})}{2} \left(\underline{x}^{+k} - \underline{x}^{-k}\right) \\
&= \frac{E_b}{2\sigma^2} \left[\sum_{j=1}^{n} \left\{ y_j^2 - 2ay_j x_j^{-k} + \left(ax_j^{-k}\right)^2 \right\} - \sum_{j=1}^{n} \left\{ y_j^2 - 2ay_j x_j^{+k} + \left(ax_j^{+k}\right)^2 \right\} \right] \\
&\quad + \sum_{j=1}^{n} \frac{L(u_j)}{2} \left(x_j^{+k} - x_j^{-k}\right) .
\end{aligned}
\tag{8.122}
$$

Since $x^{\pm k} \in \{-1, +1\}$, $\left(x_j^{-k}\right)^2 = \left(x_j^{+k}\right)^2$ and $Lc = \frac{2E_b}{\sigma^2} a$, we can simplify Equation 8.122 to:

$$
\begin{aligned}
L(u_k|\underline{y}) &\approx \frac{E_b}{\sigma^2} a \left[\sum_{j=1}^{n} y_j x_j^{+k} - \sum_{j=1}^{n} y_j x_j^{-k} \right] + \sum_{j=1}^{n} \frac{L(u_j)}{2} \left(x_j^{+k} - x_j^{-k}\right) \\
&= \frac{L_c}{2} \sum_{j=1}^{n} y_j \left(x_j^{+k} - x_j^{-k}\right) + \frac{1}{2} \sum_{j=1}^{n} L(u_j) \left(x_j^{+k} - x_j^{-k}\right) \\
&= \frac{L_c}{2} y_k \left(x_k^{+k} - x_k^{-k}\right) + \frac{1}{2} L(u_k) \left(x_k^{+k} - x_k^{-k}\right) \\
&\quad + \frac{L_c}{2} \sum_{\substack{j=1 \\ i \neq k}}^{n} y_j \left(x_j^{+k} - x_j^{-k}\right) + \frac{1}{2} \sum_{\substack{j=1 \\ i \neq k}}^{n} L(u_j) \left(x_j^{+k} - x_j^{-k}\right) \\
&= L_c y_k + L(u_k) + \frac{1}{2} \sum_{\substack{j=1 \\ i \neq k}}^{n} [L_c y_j + L(u_j)] \left(x_j^{+k} - x_j^{-k}\right) \\
&= L_c y_k + L(u_k) + \sum_{\substack{j=1 \\ i \neq k}}^{n} e_j [L_c y_j + L(u_j)] ,
\end{aligned}
\tag{8.123}
$$

$$
\tag{8.124}
$$

where:

$$
e_j = \begin{cases} 0 & \text{if } x_j^{+k} = x_j^{-k} \\ 1 & \text{if } x_j^{+k} \neq x_j^{-k} . \end{cases}
\tag{8.125}
$$

In harmony with the corresponding turbo BCH coding formula of Equation 7.9, here we define the extrinsic information of bit u_k as:

$$
L_e(u_k) = \sum_{\substack{j=1 \\ i \neq k}}^{n} e_j [L_c y_j + L(u_j)] ,
\tag{8.126}
$$

which allows us to approximate the RRNS decoder's soft output as:

$$
L(u_k|\underline{y}) \approx L_c y_k + L(u_k) + L_e(u_k) ,
\tag{8.127}
$$

constituted by the sum of the soft channel output $L_c y_k$, the intrinsic information $L(u_k)$ and the extrinsic information $L_e(u_k)$.

8.7.3 Algorithm Implementation

Previously, we have shown in Equation 8.120 that in order to approximate the soft output $L(u_k|\underline{y})$, two codewords $\underline{x}^{+k}$ and $\underline{x}^{-k}$ which are nearest to $L_c\underline{y} + L(\underline{u})$ have to be found. Using the Chase algorithm described in Section 8.7.1, we can find a surviving codeword $\underline{x}$, which generates x_k on the basis of finding the codeword $\underline{x}$ having the lowest Euclidean distance from $L_c\underline{y} + L(\underline{u})$. The algorithm can be readily extended to finding another competing (or discarded) codeword $\hat{\underline{x}}$ which decodes to $\hat{x}_k \neq x_k$ and has the minimum Euclidean distance compared to all the other codewords, which decodes to $\hat{x}_k \neq x_k$. From Equation 8.123, we derive:

$$
\begin{aligned}
L(u_k|\underline{y}) &\approx \frac{1}{2}\left[\sum_{j=1}^{n} x_j^{+k}\{L_c y_j + L(u_j)\} - \sum_{j=1}^{n} x_j^{-k}\{L_c y_j + L(u_j)\}\right] \\
&= \frac{1}{2}\left[\sum_{j=1}^{n} x_j^{+k} y_j' - \sum_{j=1}^{n} x_j^{-k} y_j'\right] \\
&= \frac{1}{2}\left[\underline{x}^{+k}\underline{y}' - \underline{x}^{-k}\underline{y}'\right] ,
\end{aligned}
\tag{8.128}
$$

where $y_j' = L_c y_j + L(u_j)$. Given the surviving and discarded codewords and Equation 8.128, we approximate the soft output as:

$$
\begin{aligned}
L(x_k|\underline{y}) &\approx \frac{x_k}{4}\left[\left(\underline{y}'\right)^2 - 2\hat{\underline{x}}\underline{y}' + \hat{\underline{x}}^2 - \left(\underline{y}'\right)^2 + 2\underline{x}\underline{y}' - \underline{x}^2\right] \\
&= x_k\left[\frac{|\underline{y}' - \hat{\underline{x}}|^2 - |\underline{y}' - \underline{x}|^2}{4}\right] .
\end{aligned}
\tag{8.129}
$$

This expression can be interpreted physically as the difference between the Euclidean distances of the surviving codeword $\underline{x}$ and the discarded codeword $\hat{\underline{x}}$ from the sequence $\underline{y}$ constituted by $y_j' = L_c y_j + L(u_j)$. In order to find the transmitted codeword $\underline{x}$ with a high probability, we have to increase the radius of perturbation in Figure 8.7. Therefore, we increase the number of least reliable bit positions l considered in the Chase algorithm and also the number of test patterns TP. It is clear that the probability of finding the most likely codeword $\underline{x}$ increases with l. However, the complexity of the decoder increases exponentially with l and hence we must find a trade-off between complexity and performance. This also implies that in some cases, we shall be unable to find a discarded codeword $\hat{\underline{x}}$, which decodes to $\hat{x}_k \neq x_k$, given the l test positions. If, however, no such discarded codeword $\hat{\underline{x}}$ is found, we have to find another method of approximating the soft output. Pyndiah et al. [203] suggested that the soft output can be approximated as:

$$
L(u_k|\underline{y}) \approx y_k' + \beta \times L_c x_k ,
\tag{8.130}
$$

where $y_k' = L_c y_k + L(u_k)$ and β is a reliability factor which increases with the iteration index and that can be optimised by simulation. This rough approximation of the soft output is justified by the fact that if no discarded codewords $\hat{\underline{x}}$ were found by the Chase algorithm which decode to $\hat{x}_k \neq x_k$, then the discarded codewords $\hat{\underline{x}}$ which decode to $\hat{x}_k \neq x_k$ are probably far from $\underline{y}'$ in terms of the Euclidean distance. Since the discarded codewords $\hat{\underline{x}}$ are far from $\underline{y}'$, the probability that the decision u_k is correct is relatively high and hence the reliability of u_k, $L(u_k)$, is also high.

We note here that there is a distinct similarity between this algorithm and the Soft-Output Viterbi Algorithm (SOVA) proposed by Hagenauer [53,54], which was covered in Section 7.3.2.

According to Equation 7.15 in the SOVA, the surviving path $\underline{s}$ is decided on the basis of the demodulator's soft-output sequence $\underline{y}$ and the intrinsic information $L(\underline{u})$. The surviving path $\underline{y}$ determines the surviving codeword $\underline{x}$ in this case. Then, the soft output of the SOVA is given by Equation 7.21 in Section 7.3.2. Explicitly this soft output is proportional to the minimum path metric difference between the surviving path $\underline{s}$, which decodes to x_k, and a discarded path $\hat{\underline{s}}$, which decodes to $\hat{x}_k \neq x_k$. Observe that Equation 7.21 is similar to Equation 8.129. Specifically, Equation 8.129 identifies the codewords having the *minimum Euclidean distance difference* and evaluates the weight difference between the surviving codeword $\underline{x}$ and the discarded codeword $\hat{\underline{x}}$.

It was also proposed by Pyndiah [62] that a weighting factor α should be introduced in Equation 8.127, as follows:

$$L(u_k|\underline{y}) \approx L_c y_k + \alpha L(u_k) + L_e(u_k) . \tag{8.131}$$

The weighting factor α takes into account that the standard deviation of the demodulator's soft-output sequence $\underline{y}$ from its expected value and that of the intrinsic information $L(\underline{y})$ are different [12, 13, 62]. The standard deviation of the extrinsic information $L_e(u_k)$ is comparatively high in the first few decoding steps and decreases during future iterations as the associated reliability increases. This scaling factor α is also used to reduce the effect of the extrinsic information in the decoder during the first decoding steps, when the BER is relatively high. The value of α is small in the initial stages of decoding and it increases as the BER tends to zero.

The parameters α in Equation 8.131 and β in Equation 8.130 can be determined experimentally, in order to achieve an optimum performance. Both α and β were given in [62], which are reproduced in Table 8.6, where the decoding index j in Table 8.6 is the index of the decoding steps given by:

$$j = 2 \times \text{Number of iterations} + \text{Decoder index} . \tag{8.132}$$

	Decoding index j							
	1	2	3	4	5	6	7	8
$\alpha(j)$	0.0	0.2	0.3	0.5	0.7	0.9	1.0	1.0
$\beta(j)$	0.2	0.4	0.6	0.8	1.0	1.0	1.0	1.0

Table 8.6: The weighting factors α and reliability factors β versus the decoding index j.

8.8 Complexity

In Sections 8.4, 8.7.1 and 8.7.2, we have described the hard decision, soft decision and turbo decoding of RRNS codes. Let us now estimate the complexity of each decoding algorithm. It was shown in Section 8.7 that both the soft-decision decoding and turbo decoding of RRNS codes are dependent on their hard-decision decoding, which is the multiple-error correction procedure highlighted in Section 8.4. In Figure 8.1, we showed the flowchart of the associated multiple-error correction procedures, suggesting that for each new moduli set, the decoder has to perform MRC with t' number of moduli deleted. In Section 8.2.3, we have shown that in order to determine the mixed radix coefficients, one integer subtraction, one integer multiplication and one integer modulus operation have to be performed recursively. Let us denote the combined operation constituted by one integer subtraction, one integer multiplication and one integer modulus by δ. Since computer simulations have shown that MRC is the bottleneck of the decoding algorithm, the number of combined operations δ is used as the basis of our estimated complexity comparison in our forthcoming discussions.

Since the combined operation δ is invoked recursively in the MRC for determining the mixed radix coefficients, the algorithm can be summarised by the flowchart of Figure 8.8.

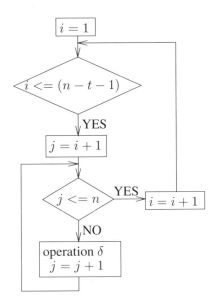

Figure 8.8: Flowchart of the MRC algorithm.

In Figure 8.8 n is the codeword length and t is the error correction capability of the RRNS code. Hence, we estimate the complexity of the MRC algorithm as:

$$comp(\text{MRC}) = \left[n(n-t-1) - \sum_{i=1}^{n-t-1} i \right] \times \delta . \qquad (8.133)$$

The estimated complexity of the hard-decision decoder is very much dependent on the number of moduli combinations to be employed in conjunction with the MRC algorithm in Equation 8.133. Let us here assume the worst-case scenario that t errors have been encountered. By contrast, for example in the case of 0 errors, no error correction has to be performed. In Section 8.4, we stated that the total number of moduli combinations is equal to ^nC_t. The decoder has to search through the ^nC_t number of moduli until a combination of moduli is found, which matches the position of t errors, as we also have seen in the context of our numerical example of Section 8.4. Assuming that t residue errors are randomly distributed among the n residues of a codeword, the probability of each moduli combination matching the position of t errors is equal. Statistically speaking, the number of moduli combinations to be tested for every codeword is equal to $\frac{^nC_t}{2}$. Since each moduli combination involves one MRC operation, we estimate the

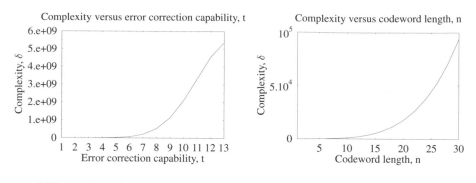

(a) Estimated complexity versus the error correction capability t. The codeword length was fixed
at $n = 26$.

(b) Estimated complexity versus the codeword
length n. The error correction capability was
fixed at 2.

Figure 8.9: Estimated complexity comparison of different hard-decision RRNS decoders under the assumption of encounting exactly t errors, where t is the maximum error correction capability of
the code.

complexity of the hard-decision decoder as:

$$comp(\text{Hard decision}) = \frac{^nC_t}{2}\left[n(n-t-1) - \sum_{i=1}^{n-t-1} i\right] \times \delta .\qquad (8.134)$$

Equation 8.134 suggests that the estimated complexity of the hard-decision decoder depends only
on the codeword length n and on the error correction capability t. Since ^nC_t increases exponentially as n or t increases, the value of the ^nC_t term is the dominant factor in Equation 8.134.

Figure 8.9 shows our estimated complexity comparisons for different hard-decision-based
RRNS decoders. The estimated complexity was computed using Equation 8.134. Figure 8.9(a)
shows the estimated complexity comparison of hard-decision RRNS decoders having a different error correction capability t. The codeword length n was fixed at $n = 26$. As expected,
the estimated complexity of the decoder increases exponentially with the error correction capability t. Figure 8.9(b) also shows our estimated complexity comparisons for the hard-decision
decoder using different codeword lengths n. The error correction capability t was fixed at $t = 2$.
Similarly, we can see that the estimated complexity of the decoder increases exponentially with
the codeword length n. However, the estimated complexity increase rate per codeword length
n appears to be slow compared to the estimated complexity increase rate per error correction
capability t. This is due to the dominance of the ^nC_t term in Equation 8.134, since ^nC_t increases
more rapidly with t than with n.

Let us now consider the estimated complexity of the soft-decision and turbo decoding based
RRNS decoder, which is directly proportional to the estimated complexity of the hard-decision
decoder. For a soft-decision decoder, we have seen in Section 8.7.1 that 2^l number of test patterns
are generated, where the tentatively corrupted l bit positions are associated with the least reliable
confidence values in the received sequence. For each test pattern, initially the hard-decision
decoder is invoked for decoding the perturbed binary codeword. Therefore, we can estimate the

complexity of the soft-decision coded RRNS decoder as:

$$comp(\text{Soft decision}) = 2^l \times comp(\text{Hard decision})$$

$$= 2^{l-1} \times {}^nC_t \left[n(n-t-1) - \sum_{i=1}^{n-t-1} i \right] \times \delta . \qquad (8.135)$$

Likewise, 2^l test patterns are generated for turbo RRNS decoding. Since the turbo RRNS decoding involves iterative decoding, the number of iterations has to be considered in calculating the decoding complexity. Furthermore, there are two component decoders for each iteration. Hence, the complexity of turbo RRNS decoding is estimated as:

$$comp(\text{Turbo decoding}) = 2 \times \text{Iteration no.} \times 2^l \times comp(\text{Hard decision})$$

$$= \text{Iteration no.} \times 2^l \times {}^nC_t \left[n(n-t-1) - \sum_{i=1}^{n-t-1} i \right] \times \delta . .$$

$$(8.136)$$

Let us consider an example for comparing the estimated complexity of hard-decision, soft-decision and turbo RRNS decoding of two RRNS(28,26) and RRNS(28,24) codes. They have an error correction capability of $t = 1$ and $t = 2$, respectively. Assuming that $l = 4$ and the number of iterations is four, we can apply Equations 8.134, 8.135 and 8.136 for finding the estimated complexity of the three decoding algorithms. The estimated complexity of each algorithm is shown in Table 8.7 for the RRNS(28,26) and RRNS(28,24) codes.

Code	t	Complexity, δ		
		Hard decision	Soft decision	Turbo decoding
RRNS(2826)	1	5278	84448	675584
RRNS(2824)	2	70875	1134000	9072000

Table 8.7: Estimated complexity of hard-decision, soft-decision and turbo decoding for the RRNS(28,26) and RRNS(28,24) codes. The number of turbo decoding iterations was four and $l = 4$ was used for both soft-decision-based and turbo decoding.

Using the values tabulated in Table 8.7, we plotted the estimated complexity bar chart of hard-decision, soft-decision and turbo decoding in Figure 8.10. We can see in the figure that the estimated complexity of soft-decision decoding is $2^4 = 16$ times higher than that of hard-decision decoding. For turbo decoding, the estimated complexity is about $2^4 \times 2 \times 4 = 128$ times higher than that of hard-decision decoding and it is $2 \times 4 = 8$ times higher than that of soft-decision decoding. Again, we can see that the error correction capability t plays an important role in determining the estimated complexity of the decoding algorithm. Consider turbo decoding as an example, where the estimated complexity of the turbo RRNS(28,24) code, which corrects $t = 2$ errors, is about 13.4 times higher than that of the turbo RRNS(28,26) code, where $t = 1$.

Finally, we conclude that error correction capability t is the major factor affecting the estimated complexity of all three decoding algorithms; namely that of hard decision, soft decision and turbo decoding. In order to obtain simulation results for $t > 1$, we attempted to simplify the considered operation δ. Explicitly, we replaced the operation with one integer subtraction and one two-dimensional table look-up operation. Explicitly, we invested memory for increasing the computational speed. Our computer simulations have shown that with the aid of this new technique, we were capable of reducing the computational complexity by about 70%.

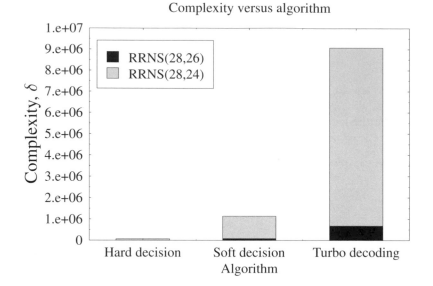

Figure 8.10: Estimated complexity comparison for hard-decision, soft-decision and turbo decoding of the RRNS(28,26) and RRNS(28,24) codes. The number of iterations was four and $l = 4$ was used for both soft-decision-based and turbo decoding.

8.9   Simulation Results

In this section, we present our simulation results for RRNS hard decision, soft decision and turbo decoding. All simulation results were obtained using BPSK over AWGN channels. We will demonstrate how the performance of RRNS codes is affected by the following parameters:

- The RRNS encoding algorithm — non-systematic, systematic and modified systematic.

- Error correction capability t and length n of the codeword;

- The decoding algorithm — hard decision, soft-decision and turbo decoding.

- The number of turbo decoding iterations used.

- The design of the turbo RRNS interleaver.

All the simulation results were generated using the encoder and decoder structures described in this chapter. The following parameters were the same for all simulations, unless otherwise stated:

- The RRNS codes used were maximum distance separable.

- The RRNS encoders used modified systematic bit-to-residue mapping.

- The number of bits k_{b_j} used to represent all residues in a codeword was the same.

- The weighting factors α and reliability factors β used were specified in Table 8.6.

- The number of turbo decoding iterations was four.

- No puncturing of the parity bits of the upper and lower encoders was used.

For a fixed number of Bits Per Symbol (BPS), k_{b_j}, used to represent all the residues in a codeword, the moduli set was chosen such that it maximised the code's legitimate dynamic range. Therefore, the moduli were chosen to be as large as possible, which are shown in Table 8.8.

BPS	max n	Moduli
4	4	$11, 13, 15, 16.$
5	8	$17, 19, 23, 25, 27, 29, 31, 32.$
6	10	$37, 41, 43, 47, 53, 55, 59, 61, 63, 64.$
7	18	$67, 71, 73, 79, 83, 89, 97, 101, 103, 107, 109, 113, 119, 121, 123, 125, 127, 128.$
8	28	$131, 137, 139, 149, 151, 157, 163, 167, 173, 179, 181, 191, 193, 197, 199, 211,$ $217, 223, 227, 229, 233, 239, 241, 247, 251, 253, 255, 256.$
9	50	$257, 263, 269, 271, 277, 281, 283, 293, 307, 311, 313, 317, 331, 337, 347, 349,$ $353, 359, 367, 373, 379, 383, 389, 397, 401, 409, 419, 421, 431, 433, 437, 439,$ $443, 449, 457, 461, 463, 467, 473, 479, 487, 491, 493, 499, 503, 505, 507, 509,$ $511, 512.$
10	82	$521, 523, 541, 547, 557, 563, 569, 571, 577, 587, 593, 599, 601, 607, 613, 617,$ $619, 631, 641, 643, 647, 653, 659, 661, 673, 677, 683, 691, 701, 709, 719, 727,$ $733, 739, 743, 751, 757, 761, 769, 773, 787, 797, 809, 811, 821, 823, 827, 829,$ $839, 853, 857, 859, 863, 877, 881, 883, 887, 907, 911, 919, 929, 937, 941, 947,$ $949, 953, 967, 971, 977, 983, 989, 991, 997, 1003, 1007, 1009, 1013, 1015,$ $1019, 1021, 1023, 1024.$
11	150	$1031, 1033, 1039, 1049, 1051, 1061, 1063, 1069, 1087, 1091, 1093, 1097, 1103,$ $1109, 1117, 1123, 1129, 1151, 1153, 1163, 1171, 1181, 1187, 1193, 1201, 1213,$ $1217, 1223, 1229, 1231, 1237, 1249, 1259, 1277, 1279, 1283, 1289, 1291, 1297,$ $1301, 1303, 1307, 1319, 1321, 1327, 1361, 1367, 1373, 1381, 1399, 1409, 1423,$ $1427, 1429, 1433, 1439, 1447, 1451, 1453, 1459, 1471, 1481, 1483, 1487, 1489,$ $1493, 1499, 1511, 1523, 1531, 1543, 1549, 1553, 1559, 1567, 1571, 1579, 1583,$ $1597, 1601, 1607, 1609, 1613, 1619, 1621, 1627, 1637, 1657, 1663, 1667, 1669,$ $1681, 1693, 1697, 1699, 1709, 1721, 1723, 1733, 1741, 1747, 1753, 1759, 1777,$ $1783, 1787, 1789, 1801, 1811, 1823, 1831, 1847, 1861, 1867, 1871, 1873, 1877,$ $1879, 1889, 1891, 1901, 1907, 1913, 1931, 1933, 1943, 1949, 1951, 1961, 1973,$ $1979, 1987, 1991, 1993, 1997, 1999, 2003, 2011, 2017, 2021, 2023, 2027, 2029,$ $2033, 2039, 2041, 2043, 2045, 2047, 2048.$

Table 8.8: List of the largest moduli for a given fixed number of BPS.

8.9.1   Hard-decision Decoding

8.9.1.1   Encoder Types

In Section 8.5, we have proposed two different methods for mapping the binary source bits to the non-binary RRNS code, which result in non-systematic and systematic RRNS codes, respectively. Furthermore, in Section 8.5.2.1 we modified the systematic RRNS mapping rule, in order to increase the associated code rate and to achieve the same number of BPS for every residue in

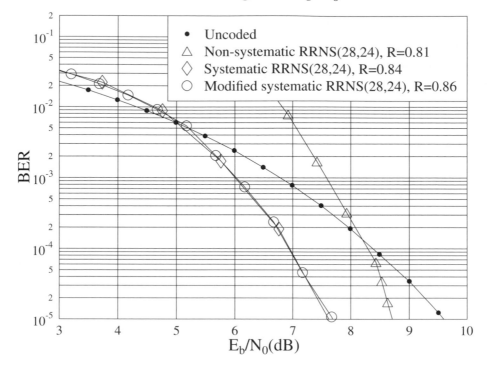

Figure 8.11: Performance comparison between the non-systematic, systematic and modified systematic 8
BPS RRNS(28,24) encoders over AWGN channels. The moduli set used is shown in Table 8.8.

a codeword. This has practical advantages in many applications and allows us to highlight the similarities with the well-known family of RS codes. In Figure 8.11, we show the performance of the non-systematic, systematic and modified systematic 8 BPS RRNS(28,24) code over AWGN channels. Owing to the associated different mapping methods used, the code rate of the different encoders varied. Since the number of BPS mapped to every residue in a codeword was the same for the modified systematic RRNS code, it exhibited the highest code rate.

In the figure, we can see that the BER performance of the non-systematic encoder is worse than that of the uncoded scenario for $E_b/N_0 < 8$ dB. However, there is a coding gain of about 0.8 dB at a BER of 10^{-5}. Both the systematic and the modified systematic RRNS(28,24) code exhibit a similar performance and they are about 1 dB better in E_b/N_0 terms than the non-systematic code. If there is a decoding failure, there will be about 50% bit decoding errors for the non-systematic decoder, which is the reason for its poor performance for $E_b/N_0 < 8$ dB. By contrast, owing to its systematic nature the systematic code simply outputs the bits associated with the k information moduli, if a decoder failure was detected. Hence, there will be significantly less than 50% bit decoding errors. Therefore, typically the systematic code is preferred to its non-systematic counterpart.

Throughout the rest of this chapter, the modified systematic RRNS code was used rather than the systematic RRNS code of Section 8.5.2. This is because the modified systematic RRNS code always has a higher code rate, while the BER performance of both codes is similar. Moreover, the

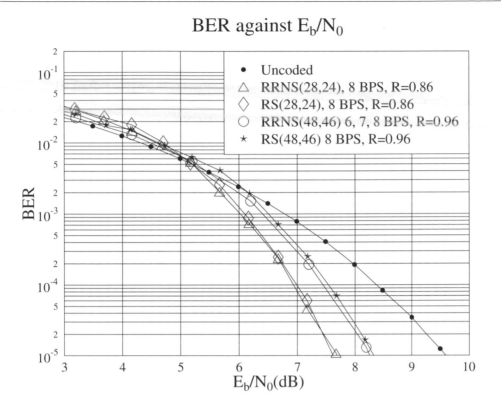

Figure 8.12: Performance comparison between modified systematic RRNS and systematic RS codes over
AWGN channels. The RRNS(28,24) code is compared to the RS(28,24) code constructed over
GF(256). Both codes transmit 8 BPS. Furthermore, the performance of the RRNS(48,46) code
transmitting 6, 7 and 8 BPS is compared to the 8 BPS RS(48,46) code.

modified systematic RRNS code has a fixed number of BPS, which is, again, directly comparable
to systematic RS codes.

8.9.1.2 Comparison of Redundant Residue Number System Codes and Reed-Solomon Codes

As argued earlier, RRNS codes which use the $d_{min} - 1$ number of largest moduli as their redundant moduli are maximum distance separable. Hence, they are similar to RS codes in terms
of their performance. Figure 8.12 compares the performance of the modified systematic RRNS
code and that of systematic RS codes over AWGN channels. Both the RRNS(28,24) and the
RS(28,24) codes constructed over GF(256) transmit 8 BPS. It can be seen from the figure that
the performance of both codes is similar. For the RRNS(48,46) code, we used moduli which
transmit 6, 7 and 8 BPS, which were compared to the 8 BPS RS(48,46) code. Again as expected,
the performance of both codes is similar.

There are a number of advantages when using RRNS codes instead of RS codes. For example,
if a code having a short codeword length n and high number of BPS is required, we have to
shorten the RS code. This implies that we have to incorporate dummy symbols at both the
encoder and decoder. Then the decoder has to decode the full-length padded RS codeword,

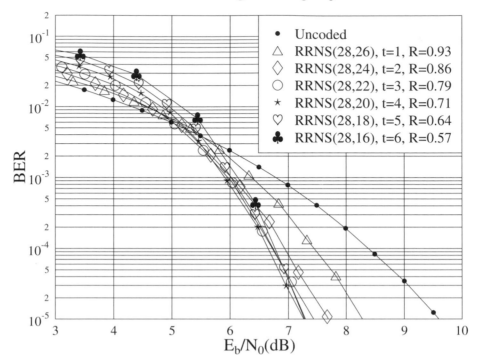

Figure 8.13: Performance comparison between different 8 BPS RRNS(28,k) codes having an error correction capability of t over AWGN channels.

which is wasteful in terms of decoding complexity. By contrast, these requirements can be readily met by using RRNS codes without introducing dummy symbols. This potentially reduces the decoding complexity.

In an Automatic Request (ARQ) system [204], a stronger RS code will be required for retransmission if a weaker RS code fails. This implies the transmission of another full codeword. However, for RRNS codes, a stronger code is obtained by introducing more redundant residues. If a RRNS code fails, stronger RRNS code can be obtained by transmitting more redundant residues without transmitting the whole codeword [198, 205].

However, as argued in Section 8.8, the estimated complexity of the RRNS decoder increases exponentially with the codeword length n and with the error correction capability t. Furthermore, it is extremely complex to implement a RRNS code with $n \approx 2^j$, where j is the number of BPS.

8.9.1.3 Comparison Between Different Error Correction Capabilities t

Figure 8.13 shows our performance comparison between different error correction capabilities t over AWGN channels. The codes used are 8 BPS RRNS(28,k) codes, which have a fixed codeword length. We can see in the figure that as t increases, the coding gain also increases. However, the performance improvements saturate when $t = 4$ and the best design trade-off is constituted by the RRNS(28,20) code. Since the complexity of the RRNS decoder is excessive for $t > 6$, no simulation results were provided. The associated theoretical results [206] suggest that the achievable coding gain decreases, once the code rate R is below 0.6, that is for $t > 4$.

Code	t	R	Complexity, δ	Coding gain (dB)
RRNS(28,26)	1	0.93	5278	1.3
RRNS(28,24)	2	0.86	70875	1.9
RRNS(28,22)	3	0.79	609336	2.2
RRNS(28,20)	4	0.71	3767400	2.3
RRNS(28,18)	5	0.64	17837820	2.2
RRNS(28,16)	6	0.57	67248090	2.2

Table 8.9: Coding gain and estimated complexity of 8 BPS RRNS(28,k) codes using hard-decision decoding for transmission over AWGN channels.

The coding gain of the RRNS(28,k) codes in Figure 8.13 was evaluated and is tabulated in Table 8.9. The estimated complexity of the decoder was calculated using Equation 8.134 and it is also shown in the table. Figure 8.14 shows the trend of the coding gain versus estimated complexity characteristic of the decoder. As we can see from the figure, increasing the estimated complexity does not necessarily increase the coding gain, and the best coding gain was achieved when the estimated complexity of the decoder was relatively low.

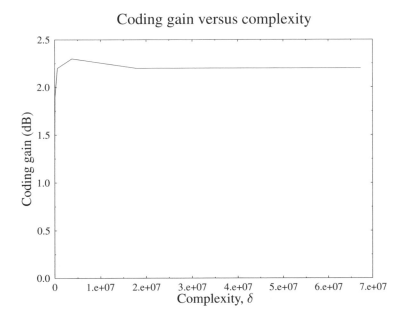

Figure 8.14: Coding gain versus estimated complexity for hard-decision decoding of various 8 BPS RRNS(28,k) codes for $k = 26, 24, 22, 20, 18$ and 16.

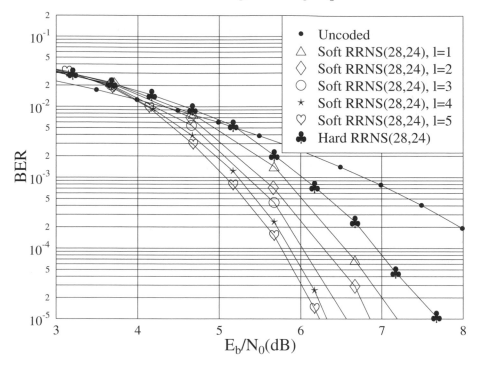

Figure 8.15: Performance of the soft-decision-assisted 8 BPS RRNS(28,24) decoder for a different number of test positions l over AWGN channels. The performance of the hard-decision decoder is also shown for comparison.

8.9.2   Soft-decision Decoding

8.9.2.1   Effect of the Number of Test Positions

In Section 8.7.1, we outlined how we can employ the Chase algorithm to exploit the soft outputs of the demodulator. In the decoder, only a limited number l bit positions associated with the least reliable confidence values were considered. Figure 8.15 shows the performance of the soft-decision-assisted 8 BPS RRNS(28,24) decoder for a different number of test positions l over AWGN channels. The performance of the hard-decision decoder is also shown in the figure for comparison. We can see from the figure that the performance of the soft-decision decoder improves as we increase the number of test positions l. However, the incremental improvement becomes more limited when $l > 4$. It is surmised that the best-case performance of the soft-decision RRNS decoder is similar to that of the soft decision Viterbi algorithm. The estimated complexity of the decoder increases exponentially, since the number of test patterns is equal to 2^l. In Figure 8.15, the performance of the soft-decision RRNS(28,24) decoder using $l = 4$ is about 1.5 dB better than that of the hard-decision-assisted RRNS(28,24) decoder at a BER of 10^{-5}.

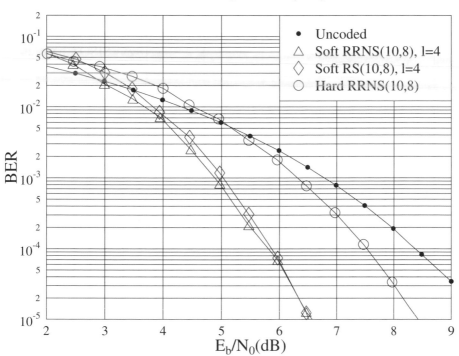

Figure 8.16: Performance comparison between the hard-decision-assisted 6 BPS RRNS(10,8) decoder and the soft-decision-assisted 6 BPS RRNS(10,8) decoder for $l = 4$ over AWGN channels. The performance of the RS(10,8) code constructed over GF(64) for $l = 4$ is also shown for comparison.

8.9.2.2   Soft-decision RRNS(10,8) Decoder

Figure 8.16 compares the performance of the hard-decision-assisted 6 BPS RRNS(10,8) decoder and that of the soft-decision-aided 6 BPS RRNS(10,8) decoder for $l = 4$ over AWGN channels. The soft-decision-aided decoder is about 2 dB better in terms of E_b/N_0 than that of the hard-decision-based decoder at a BER of 10^{-5}. We applied the Chase algorithm for decoding the RS(10,8) code constructed over GF(128), where for $l = 4$ we can see that the performance of the soft-decision-assisted RS(10,8) code constructed over GF(64) is similar to that of the soft-decision-aided RRNS(10,8) decoder.

8.9.3   Turbo RRNS Decoding

8.9.3.1   Algorithm Comparison

In Section 8.7.3, we stated that there is a similarity between the SOVA and the SISO Chase algorithm. In the SOVA, the surviving path is chosen and the soft outputs are derived on the basis of the minimum path metric differences between those specific surviving and discarded paths, which would lead to a different decoded bit. Similarly, the SISO Chase algorithm searches for the surviving codeword and then the associated soft output confidence measures are rendered propor-

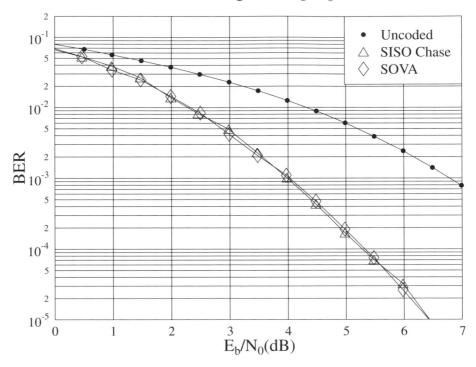

Figure 8.17: Performance comparison between the SOVA and the SISO Chase decoding algorithm using the turbo BCH(7,4) code over AWGN channels. For the SISO Chase algorithm, the weighting factors were set to $\alpha(j) = 1$ and the reliability factors to $\beta(j) = 1$ for every decoding index j and the number of test positions was $l = 5$. There were six decoding iterations and a 4×4-bit block interleaver was used.

tional to the minimum weight difference between the specific surviving and discarded path pair, which yield a different decoded result. Therefore, if the weighting factors of the Chase algorithm are set to $\alpha(j) = 1$ in Equation 8.131 and the reliability factors to $\beta(j) = 1$ in Equation 8.130 for all values of j, then all the valid codewords are considered by the Chase decoder during the decoding process and hence the operation of the SISO Chase algorithm becomes fairly similar to that of the SOVA.

In this section, we use the BCH(7,4) code as the component code of the turbo BCH(7,4) code. BCH codes were favoured, since a Viterbi decoder can also be invoked for their decoding and hence the Log-MAP, Max-Log-MAP and SOVA algorithms can be applied as bench-markers of the SISO Chase algorithm.

In Figure 8.17, we compare the performance of the SOVA and the SISO Chase decoding algorithm using the turbo BCH(7,4) code over AWGN channels. Since there are only $2^4 = 16$ valid codewords in the BCH(7,4) code, $2^{l=5} = 32$ test patterns are sufficient to perturb the received sequences to all the 16 valid codewords. The weighting factors $\alpha(j)$ and reliability factors $\beta(j)$ of Equations 8.131 and 8.130 are 1 for all decoding indices. Both algorithms have a total of six decoding iterations and a 4×4-bit block interleaver was used. From Figure 8.17, we can see that the performance of the algorithms is identical. Hence, we have shown that the

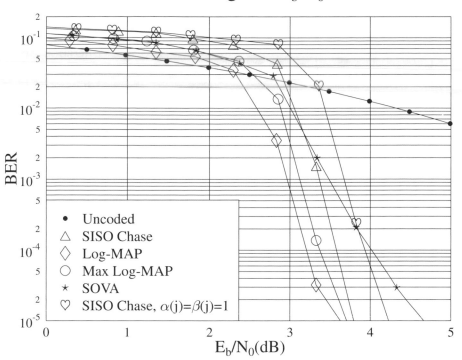

Figure 8.18: Performance comparison between different decoding algorithms using the turbo BCH(63,57) code over AWGN channels. There were six decoding iterations and a 57×57-bit block interleaver was used. For the SISO Chase algorithm, $\alpha(j)$ and $\beta(j)$ were specified in Table 8.6 and we had $l = 4$.

concept of the SISO Chase algorithm is similar to that of the SOVA.

Let us now compare the performance of the SISO Chase algorithm to that of other well-known algorithms, such as the Log-MAP, Max-Log-MAP and SOVA, in the context of binary turbo BCH codes. Figure 8.18 shows our performance comparison between the different decoding algorithms using the turbo BCH(63,57) code over AWGN channels. There were six decoding iterations and a 57×57-bit block interleaver was used. For the SISO Chase algorithm, $\alpha(j)$ and $\beta(j)$ were specified in Table 8.6 and we had $l = 4$. Since the Log-MAP decoding algorithm is optimum, its BER performance is the best in Figure 8.18. The Max-Log-MAP decoding algorithm suffers from a slight degradation of 0.1 dB at a BER of 10^{-5} compared to the Log-MAP decoding algorithm. With the optimum values of $\alpha(j)$ and $\beta(j)$ given by Pyndiah [62] as shown in Table 8.6, the SISO Chase algorithm exhibits a slight E_b/N_0 performance degradation of 0.2 and 0.1 dB at a BER of 10^{-5} compared to the Log-MAP and the Max-Log-MAP decoding algorithms, respectively. As compared to the SOVA, the SISO Chase algorithm performs better, having a 0.8 dB E_b/N_0 advantage at a BER of 10^{-5}. Also shown in Figure 8.18 is the performance of the SISO Chase algorithm in conjunction with $\alpha(j) = \beta(j) = 1$ for every decoding index. It can be seen that its E_b/N_0 performance is about 0.4 dB worse than that of the SISO Chase algorithm using the $\alpha(j)$ and $\beta(j)$ values shown in Table 8.6. Hence, we have shown that the BER performance can be improved by optimising the values of $\alpha(j)$ and $\beta(j)$.

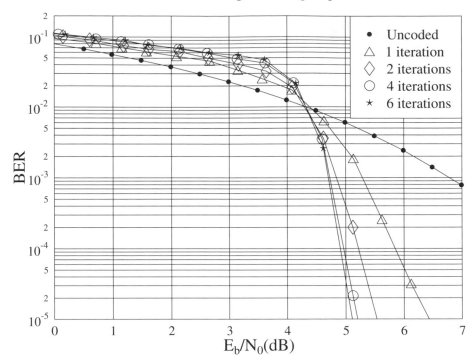

Figure 8.19: Performance comparison for different number of iterations using the rate $R = 0.87$ 8 BPS turbo RRNS(28,26) code, a 26×26 symbol block interleaver, and the $\alpha(j)$ and $\beta(j)$ values shown in Table 8.6, over AWGN channels.

8.9.3.2 Number of Iterations Used

Figure 8.19 shows the performance of a different number of iterations using the 8 BPS turbo RRNS(28,26) code, over AWGN channels. The interleaver used is a 26×26 symbol block interleaver, while the $\alpha(j)$ and $\beta(j)$ values used in our simulations are shown in Table 8.6. Since the parity symbols of both the upper and lower encoder are transmitted through the channel, the resultant code rate is 0.86. As the number of iterations performed by the turbo decoder increases, the performance of the turbo code improves. However, after four iterations the improvement becomes insignificant.

Figure 8.20 also shows our performance comparisons for a different number of iterations over AWGN channels. The turbo component codes are constituted by the 8 BPS RRNS(28,24) code and the code rate is 0.75. The values of $\alpha(j)$ and $\beta(j)$ used were shown in Table 8.6 and the turbo interleaver is a 24×24 symbol block interleaver. It can be seen from the figure that the performance improvement due to each extra iteration is higher compared to that of the turbo RRNS(28,26) code. For instance, at a BER of 10^{-5}, the E_b/N_0 performance of the turbo RRNS(28,24) code using two iterations is about 2 dB better than that of the first iteration. By contrast, for the turbo RRNS(28,26) code, the performance of the second iteration is only 1 dB better, in terms of E_b/N_0, than that of the first iteration. However, the performance of the turbo RRNS(28,24) code does not improve significantly after four iterations. Hence, the simulation

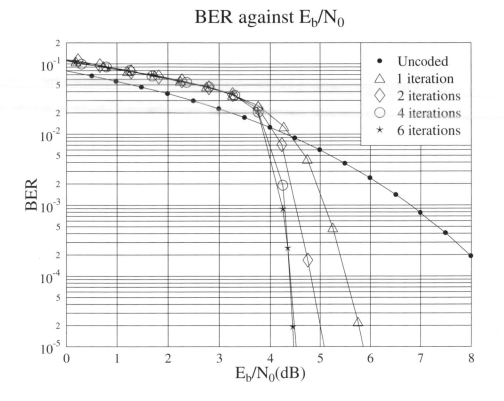

Figure 8.20: Performance comparison for different number of iterations using the rate $R = 0.75$ 8 BPS turbo RRNS(28,24) code, a 24×24 symbol block interleaver, and the $\alpha(j)$ and $\beta(j)$ values shown in Table 8.6, over AWGN channels.

results shown in the following sections will be based on four iterations.

8.9.3.3   Imperfect Estimation of the Channel Reliability Value L_c

In Equation 8.124 of Section 8.7.2, we have shown that we can approximate the soft output as:

$$L(u_k|\underline{y}) \approx L_c y_k + L(u_k) + \sum_{\substack{j=1 \\ i \neq k}}^{n} e_j \left[L_c y_j + L(u_j) \right] . \tag{8.137}$$

In the first iteration of the iterative decoding process, there is no intrinsic information $L(u_j) = L(u_k) = 0$ for the data bits. Therefore, we can simplify Equation 8.137 as follows:

$$L(u_k|\underline{y}) \approx L_c \left[y_k + \sum_{\substack{j=1 \\ i \neq k}}^{n} e_j y_j \right] . \tag{8.138}$$

Over AWGN channels the channel reliability value L_c becomes a constant in Equation 8.138. Therefore, we can omit the channel reliability value in deriving the soft outputs of the data bit

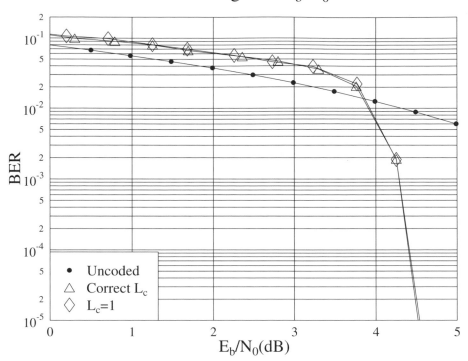

Figure 8.21: Effect of using incorrect channel reliability value L_c on the performance of the turbo RRNS(28,24) code employing four iterations, and a 24×24 symbol block interleaver. The values of $\alpha(j)$ and $\beta(j)$ were shown in Table 8.6.

u_k. Equation 8.137 is then simplified and rewritten as follows:

$$L_e(u_k) \approx y_k + L(u_k) + \sum_{\substack{j=1 \\ i \neq k}}^{n} e_j \left[y_j + L(u_j) \right] \ . \tag{8.139}$$

Figure 8.21 shows the associated performance, when using the correct L_c value and when setting $L_c = 1$ for the 8 BPS turbo RRNS(28,24) code over AWGN channels. The number of iterations was four and a 24×24 symbol block interleaver was used. The values of $\alpha(j)$ and $\beta(j)$ were shown in Table 8.6. The performance of the 8 BPS turbo RRNS(28,24) code using the correct values of L_c is shown in the figure for comparison purpose. It can be seen that the SISO Chase algorithm performs equally well whether or not the correct values of L_c are known. A similar observation is valid for the SOVA and the Max-Log-MAP algorithms, which were detailed in Section 7.6.3.

8.9.3.4   The Effect of the Turbo Interleaver

Conventionally, the turbo interleaver is used to disperse the parity information incorporated by the encoder into the transmitted sequence. In turbo RRNS codes the parity information is transmitted twice. Even if a number of bits are corrupted by channel errors, the second set of inter-

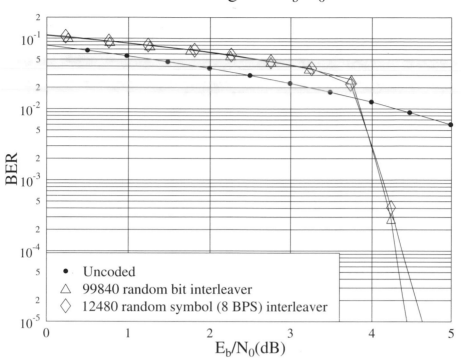

Figure 8.22: Performance comparison of the 8 BPS turbo RRNS(28,24) code for a 12480 random symbol (8 BPS) random interleaver and a 99840-bit random interleaver over AWGN channels. The number of iterations was four and the values of $\alpha(j)$ and $\beta(j)$ were shown in Table 8.6.

leaved parity information may assist the decoder in removing these errors, provided that these bits are not corrupted by the channel. For binary codes, such as the turbo BCH codes of Chapter 7, bit-based turbo interleavers were used. In this section we provide the performance results of a variety of interleavers.

Specifically, Figure 8.22 shows our performance comparison between a 12480 symbol (8 BPS) random interleaver and a 99840-bit random interleaver over AWGN channels. More explicitly, the interleaving depth of both interleavers is about 100000 bits. At a BER of 10^{-5} the performance of the bit interleaver is about 0.2 dB better, in terms of E_b/N_0, than that of the symbol interleaver. This result might suggest that using a bit interleaver is more beneficial than a symbol interleaver in terms of scrambling the input data bits and spreading the parity information. This assists in passing independent information between the two decoders, which is vital for decoding turbo codes iteratively.

It is well known that the design of the turbo interleaver affects the performance of the turbo code [12, 13, 69]. One of the most important factors is the size of the turbo interleaver. Impressive performance results have been obtained using large interleavers [12, 13, 69]. However, a large interleaver implies a long delay, which is an impediment in interactive speech and video transmissions. Hence below we will characterise the performance of low-latency interleavers.

We show in Figure 8.23 how the interleaver length affects the performance of the 8 BPS turbo

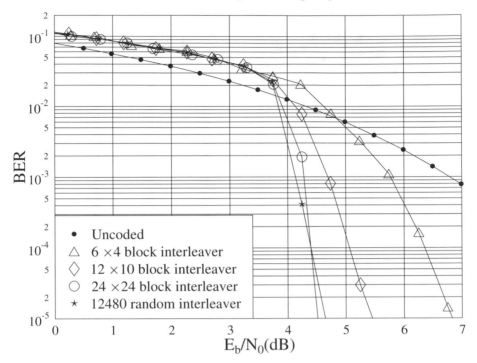

Figure 8.23: Performance comparison between different-length symbol interleavers for the 8 BPS
RRNS(28,24) code employing four iterations over AWGN channels. The values of $\alpha(j)$ and
$\beta(j)$ were shown in Table 8.6.

RRNS(28,24) code employing four iterations over AWGN channels. The smallest interleaver size
that can accommodate one RRNS(28,24) codeword is the 8 BPS 6×4 symbol block interleaver,
resulting in a delay of 192 bits. Owing to its small interleaver depth, the performance of this
scheme is moderate, achieving a BER of 10^{-5} at $E_b/N_0 = 6.8$ dB. If we increase the interleaver
size to a 12×10 8 BPS block symbol interleaver, the performance of the turbo code is about
1.5 dB better, in terms of the required E_b/N_0, than in the previous case at a BER of 10^{-5}. A
further increase of 0.8 dB E_b/N_0 improvement is observed at a BER of 10^{-5} when the interleaver
size is increased to using a 24×24 8 BPS block interleaver. However, beyond this interleaver size
there is no significant coding gain improvement when the interleaver size increases, for example,
to 12480 symbols or about 100000 bits.

8.9.3.5   The Effect of the Number of Bits Per Symbol

Figure 8.24 shows the performance of various 8, 9, 10 and 11 BPS turbo RRNS(28,24) codes employing four iterations over AWGN channels. The interleavers were implemented using random
symbol interleaving and their size was chosen such that the interleaver depth was approximately
100000 bits. Figure 8.24 shows that the BER performance degrades as we increase the number
of BPS. This is because over AWGN channels the errors occur randomly and independently.
Therefore, as the number of BPS increases, the probability of symbol errors increases.

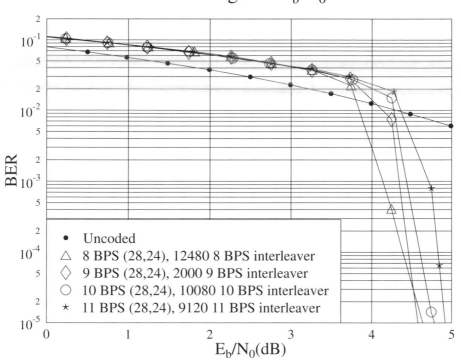

Figure 8.24: Performance comparison of different number of BPS turbo RRNS(28,24) codes employing four iterations over AWGN channels. All interleavers used were random and the values of $\alpha(j)$ and $\beta(j)$ were shown in Table 8.6.

8.9.3.6 Coding Gain versus Estimated Complexity

In Section 8.8, we have studied the estimated complexity of hard-decision, soft-decision and turbo decoding assisted RRNS codes. The estimated complexity of the decoding algorithms was also shown in Figure 8.10 for the RRNS(28,26) and RRNS(28,24) codes. Let us now investigate the amount of coding gain achievable by investing more complexity at the decoder.

Figure 8.25 shows our performance comparison for the hard-decision, soft-decision and turbo decoding assisted RRNS(28,24) codes over AWGN channels. Both the soft-decision and turbo decoding aided schemes had the same number of test positions, namely $l = 4$. For turbo decoding, the number of iterations was four and a 12480 8 BPS random interleaver was used. As shown in the figure, the BER performance of the RRNS(28,24) code improves as we invest more complexity. Specifically, with the aid of turbo decoding, the RRNS(28,24) code achieves a maximum coding of 5.0 dB at a BER of 10^{-5}. The estimated complexity versus coding gain performance of the three decoding algorithms is tabulated in Table 8.10.

Using the values seen in Table 8.10, we plotted the coding gain versus estimated complexity of this code in Figure 8.26. As we can see in the figure, the estimated complexity of hard-decision decoding is the lowest and hence the coding gain is only 1.9 dB. With a moderate increase of estimated complexity, a coding gain of 3.3 dB is achieved by soft-decision decoding. However, we have to invest a significantly higher complexity, in order to achieve a coding gain of 5 dB by

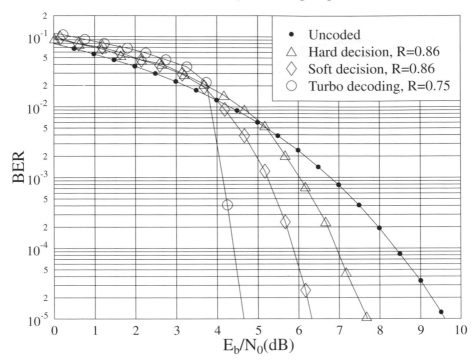

Figure 8.25: Performance comparison of hard-decision, soft-decision and turbo decoding of the RRNS(28,24) code over AWGN channels. For soft decision, the number of test positions was $l = 4$. Similarly, $l = 4$ was used for turbo decoding and the number of iterations was four. The turbo interleaver was a 12480 symbol 8 BPS random interleaver and the values of $\alpha(j)$ and $\beta(j)$ were shown in Table 8.6.

Algorithm	R	Complexity, δ	Coding gain (dB)
Hard decision	0.86	70875	1.9
Soft decision	0.86	1134000	3.3
Turbo decoding	0.75	9072000	5.0

Table 8.10: Estimated decoding complexity versus coding gain at a BER of 10^{-5} for hard-decision, soft-decision and turbo decoding assisted RRNS(28,24) codes over AWGN channels.

turbo decoding.

8.10   Summary and Conclusions

The ancient theory of the RNS was reviewed in Section 8.1. This was followed by a brief introduction in Section 8.2.1 to the conventional number systems, namely the decimal and the binary number systems. Then the RNS was introduced in more detail in Section 8.2.2. A simple exam-

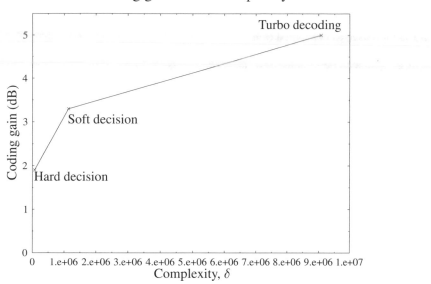

Figure 8.26: Coding gain versus estimated complexity for hard-decision, soft-decision and turbo decoding of RRNS(28,24) codes. The simulation results are obtained over the AWGN channels.

ple was given and the differences between the conventional number systems and the RNS were highlighted. In Section 8.2.3, we also discussed the differences between fixed radix number systems and mixed radix number systems. The conventional decimal number system is a fixed radix number system, whereas the RNS is a mixed radix system.

The various RNS-based arithmetic operations such as addition, subtraction and multiplication were demonstrated in Section 8.2.4 with the aid of simple examples. In Section 8.2.4.1 the multiplicative inverse was defined. Two known techniques were described in Section 8.2.5 for the conversion of operands from the RNS to the decimal number system, which are the CRT and the MRC. In Section 8.2.6, we showed that by incorporating extra moduli in the existing RNS, a redundant RNS is obtained. The BEX technique was suggested for finding the residue digits of an extended set of moduli, given the residue digits of the existing moduli.

In Section 8.3, we provided a brief overview of the coding theory of RRNS codes. Given a set of moduli, we showed that the minimum free distance d_{min} of the RRNS code was given by Equation 8.39. Furthermore we also showed that maximum distance separable RRNS codes are obtained if Equation 8.47 is satisfied. In Section 8.3.2, we highlighted why RRNS codes are semi-linear block codes. Then, we related the minimum free distance d_{min} of RRNS codes to their error detection and error correction capabilities. The procedure for multiple-error correction was summarised in Section 8.4. In Section 8.5, we demonstrated that different mapping methods result in systematic and non-systematic RRNS codes. Furthermore, we proposed a novel mapping method, which results in an increased code rate for systematic RRNS codes. The performance of the novel mapping method was found to be similar to that of the systematic RRNS codes. In Section 8.7, we modified the Chase algorithm so that it became capable of incorporating soft inputs and providing soft outputs. The estimated complexity of the hard-decision- and soft-decision-assisted decoder as well as that of iterative decoding was evaluated in Section 8.8. It was found that the estimated complexity of the algorithms depends on the codeword length n and on the error correction capability t. Moreover, the increase in estimated complexity is faster when t is increased as compared to n.

Finally in Section 8.9 we presented our simulation results for the proposed RRNS codes using BPSK over AWGN channels. It was found in Section 8.9.1.1 that systematic RRNS codes outperform their non-systematic counterparts. Modified systematic RRNS codes give a similar performance to systematic RRNS codes at a higher code rate. The performance of the RRNS codes was then compared to that of RS codes in Section 8.9.1.2. It was found that analogous codes of these code families give a similar performance. Fixing the codeword length n, the performance of the RRNS codes was evaluated by varying the error correction capability t. It was shown in Figure 8.14 that the increase in error correction capability t results in an increased estimated complexity, although no extra coding gain was obtained. In Section 8.9.2.1, a soft-decision-aided RRNS decoder was employed. It was found that the performance of the soft-decision-assisted decoder improves as the the number test position is increased. We also concluded that the best-case performance of the Chase algorithm is the same as that of the soft-decision Viterbi algorithm. The performance of the soft decoding of RS codes was found to be similar to that of RRNS codes. The performance of various decoding algorithms was compared in Figure 8.18 for the turbo BCH(63,57) code. It was shown in the figure that the SISO Chase algorithm, which offers a significantly reduced complexity, suffers only a small performance degradation. In Section 8.9.3.2, we concluded that the optimum number of decoding iterations is four. We then extended our research into the effects of the turbo interleaver in Section 8.9.3.4. Finally, we showed in Figure 8.26 that the coding gain can be increased by increasing the estimated complexity.

In conclusion, in this chapter, we have investigated a new class of codes referred to as RRNS codes. Their performance was found to be similar to that of the analogous RS codes, with some additional advantages over the well-known class of RS codes. For example, short non-binary block codes could be readily designed using RRNS codes without having to shorten long mother codes, which would be inevitable in the context of RS codes. Furthermore, in ARQ-assisted systems stronger RRNS codes are readily obtained for retransmission by transmitting extra redundant residues, without having to retransmit the whole codeword. In [198, 205, 207], we exploited the same principle in designing adaptive-rate RRNS codes for adaptive OFDM transmissions over mobile communication channels.

Part III

Coded Modulation: TCM, TTCM, BICM, BICM-ID

Chapter 9

Coded Modulation Theory and Performance

S. X. Ng, L. Hanzo

9.1 Introduction

In this chapter our elaborations are related to a set of combined error control techniques, where channel coding and modulation are carried out jointly. At the receiver it is possible to employ both iterative and non-iterative detection and, as expected, the latter family of iterative decoders typically achieves a better performance, although at the cost of an increased implementational complexity.

Since we are relying on convolutional coding principles, familiarity with the basic concepts of Chapters 2 and 5 is assumed.

The radio spectrum is a scarce resource. Therefore, one of the most important objectives in the design of digital cellular systems is the efficient exploitation of the available spectrum, in order to accommodate the ever-increasing traffic demands. Trellis-Coded Modulation (TCM) [208], which will be detailed in Section 9.2, was proposed originally for Gaussian channels, but it was further developed for applications in mobile communications [209, 210]. Turbo Trellis-Coded Modulation (TTCM) [211], which will be augmented in Section 9.4, is a more recent joint coding and modulation scheme that has a structure similar to that of the family of power-efficient binary turbo codes [12], but employs TCM schemes as component codes. TTCM [211] requires approximately 0.5 dB lower Signal-to-Noise Ratio (SNR) at a Bit Error Ratio (BER) of 10^{-4} than binary turbo codes when communicating using 8-level Phase Shift Keying (8PSK) over Additive White Gaussian Noise (AWGN) channels. TCM and TTCM invoked Set Partitioning (SP) based signal labelling, as will be discussed in the context of Figure 9.7 in order to achieve a higher Euclidean distance between the unprotected bits of the constellation, as we will show during our further discourse. It was shown in [208] that parallel trellis transitions can be associated with the unprotected information bits; as we will augment in Figure 9.2(b), this reduced the decoding complexity. Furthermore, in our TCM and TTCM oriented investigations random symbol interleavers, rather than bit interleavers, were utilised, since these schemes operate on the basis of symbol, rather than bit, decisions.

Another coded modulation scheme distinguishing itself by utilising bit-based interleaving in conjunction with Gray signal constellation labelling is referred to as Bit-Interleaved Coded Modulation (BICM) [212]. More explicitly, BICM combines conventional convolutional codes with several independent bit interleavers, in order to increase the achievable diversity order to the binary Hamming distance of a code for transmission over fading channels [212], as will be shown in Section 9.5.1. The number of parallel bit interleavers equals the number of coded bits in a symbol for the BICM scheme proposed in [212]. The performance of BICM is better than that of TCM over uncorrelated or perfectly interleaved narrowband Rayleigh fading channels, but worse than that of TCM in Gaussian channels owing to the reduced Euclidean distance of the bit-interleaved scheme [212], as will be demonstrated in Section 9.5.1. Recently iterative joint decoding and demodulation assisted BICM (BICM-ID) was proposed in an effort to further increase the achievable performance [213–218], which uses SP-based signal labelling. The approach of BICM-ID is to increase the Euclidean distance of BICM, as will be shown in Section 9.6, and hence to exploit the full advantage of bit interleaving with the aid of soft-decision feedback-based iterative decoding [219].

In this chapter, we are going to study the properties of the above-mentioned TCM, TTCM, BICM and BICM-ID schemes. Their performance will be evaluated in Section 9.7.

9.2 Trellis-Coded Modulation

The basic idea of TCM is that instead of sending a symbol formed by m information bits, for example two information bits for 4-level Phase Shift Keying (4PSK), we introduce a parity bit, while maintaining the same effective throughput of 2 bits/symbol by doubling the number of constellation points in the original constellation to eight, i.e. by extending it to 8PSK. As a consequence, the redundant bit can be absorbed by the expansion of the signal constellation, instead of accepting a 50% increase in the signalling rate, i.e. bandwidth. A positive coding gain is achieved when the detrimental effect of decreasing the Euclidean distance of the neighbouring phasors is outweighed by the coding gain of the convolutional coding incorporated.

Ungerböck has written an excellent tutorial paper [220], which fully describes TCM, and which this section is based upon. TCM schemes employ redundant non-binary modulation in combination with a finite state Forward Error Correction (FEC) encoder, which governs the selection of the coded signal sequences. Essentially the expansion of the original symbol set absorbs more bits per symbol than required by the data rate, and these extra bit(s) are used by a convolutional encoder which restricts the possible state transitions amongst consecutive phasors to certain legitimate constellations. In the receiver, the noisy signals are decoded by a trellis-based soft-decision maximum likelihood sequence decoder. This takes the incoming data stream and attempts to map it onto each of the legitimate phasor sequences allowed by the constraints imposed by the encoder. The best fitting symbol sequence having the minimum Euclidean distance from the received sequence is used as the most likely estimate of the transmitted sequence.

Simple four-state TCM schemes, where the four-state adjective refers to the number of possible states that the encoder can be in, are capable of improving the robustness of 8PSK-based TCM transmission against additive noise in terms of the required SNR by 3dB compared to conventional uncoded 4PSK modulation. With the aid of more complex TCM schemes the coding gain can reach 6 dB [220]. As opposed to traditional error correction schemes, these gains are obtained without bandwidth expansion, or without the reduction of the effective information rate. Again, this is because the FEC encoder's parity bits are absorbed by expanding the signal constel-

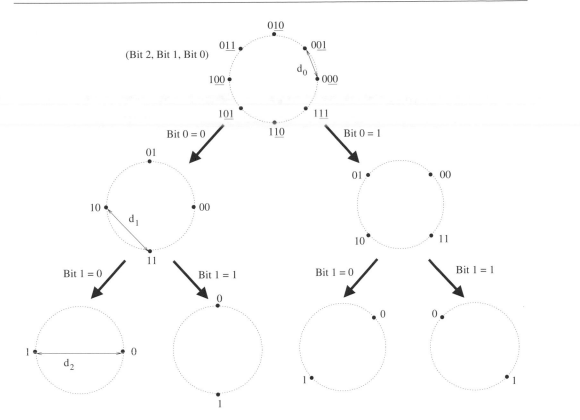

Figure 9.1: 8PSK set partitioning [208] ©IEEE, 1982, Ungerböck.

lation in order to transmit a higher number of bits per symbol. The term 'trellis' is used, because these schemes can be described by a state transition diagram similar to the trellis diagrams of binary convolutional codes [221]. The difference is that in the TCM scheme the trellis branches are labelled with redundant non-binary modulation phasors, rather than with binary code symbols.

9.2.1 TCM Principle

We now illustrate the principle of TCM using the example of a four-state trellis code for 8PSK modulation, since this relatively simple case assists us in understanding the principles involved.

The partitioned signal set proposed by Ungerböck [208, 220] is shown in Figure 9.1, where the binary phasor identifiers are now not Gray encoded. Observe in the figure that the Euclidean distance amongst constellation points is increased at every partitioning step. The underlined last two bits, namely bit 0 and bit 1, are used for identifying one of the four partitioned sets, while bit 2 finally pinpoints a specific phasor in each partitioned set.

The signal sets and state transition diagrams for (a) uncoded 4PSK modulation and (b) coded 8PSK modulation using four trellis states are given in Figure 9.2, while the corresponding four-state encoder-based modulator structure is shown in Figure 9.3. Observe that after differential encoding bit 2 is fed directly to the 8PSK signal mapper, whilst bit 1 is half-rate convolutionally encoded by a two-stage four-state linear circuit. The convolutional encoder adds the parity bit, bit 0, to the sequence, and again these two protected bits are used for identifying which constellation subset the bits will be assigned to, whilst the more widely spaced constellation points will

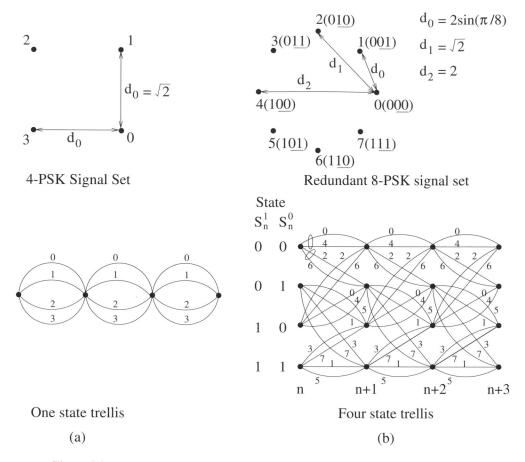

Figure 9.2: Constellation and trellis for 4- and 8PSK [220] ©IEEE, 1982, Ungerböck.

be selected according to the unprotected bit 2.

The trellis diagram for 4PSK is a trivial one-state trellis, which portrays uncoded 4PSK from the viewpoint of TCM. Every connected path through the trellis represents a legitimate signal sequence where no redundancy-related transition constraints apply. In both systems, starting from any state, four transitions can occur, as required for encoding two bits/symbol. The four parallel transitions in the state trellis diagram of Figure 9.2(a) do not restrict the sequence of 4PSK symbols that can be transmitted, since there is no channel coding and therefore all trellis paths are legitimate. Hence the optimum detector can only make nearest-phasor-based decisions for each individual symbol received. The smallest distance between the 4PSK phasors is $\sqrt{2}$, denoted as d_0, and this is termed the free distance of the uncoded 4PSK constellation. Each 4PSK symbol has two nearest neighbours at this distance. Each phasor is represented by a two-bit symbol and transitions from any state to any other state are legitimate.

The situation for 8PSK TCM is a little less simplistic. The trellis diagram of Figure 9.2(b) is constituted by four states according to the four possible states of the shift-register encoder of Figure 9.3, which we represent by the four vertically stacked bold nodes. Following the elapse of a symbol period a new two-bit input symbol arrives and the convolutional encoder's shift register is clocked. This event is characterised by a transition in the trellis from state S_n to state S_{n+1}, tracking one of the four possible paths corresponding to the four possible input symbols.

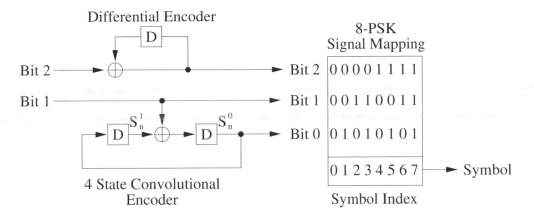

Figure 9.3: Encoder for the four-state 8PSK trellis [220] ©IEEE, 1982, Ungerböck.

In the four-state trellis of Figure 9.2(b) associated with the 8PSK TCM scheme, the trellis transitions occur in pairs and the states corresponding to the bold nodes are represented by the shift-register states S_n^0 and S_n^1 in Figure 9.3. Owing to the limitations imposed by the convolutional encoder of Figure 9.3 on the legitimate set of consecutive symbols only a limited set of state transitions associated with certain phasor sequence is possible. These limitations allow us to detect and to reject illegitimate symbol sequences, namely those which were not legitimately produced by the encoder, but rather produced by the error-prone channel. For example, when the shift register of Figure 9.3 is in state (0,0), only the transitions to the phasor points (0,2,4,6) are legitimate, whilst those to phasor points (1,3,5,7) are illegitimate. This is readily seen, because the linear encoder circuit of Figure 9.3 cannot produce a non-zero parity bit from the zero-valued input bits and hence the symbols (1,3,5,7) cannot be produced when the encoder is in the all-zero state. Observe in the 8PSK constellation of Figure 9.2(b) that the underlined bit 1 and bit 0 identify four twin-phasor subsets, where the phasors are opposite to each other in the constellation and hence have a high intra-subset separation. The unprotected bit 2 is then invoked for selecting the required phasor point within the subset. Since the redundant bit 0 constitutes also one of the shift-register state bits, namely S_n^0, from the initial states of $(S_n^1, S_n^0) = (0,0)$ or $(1,0)$ only the even-valued phasors (0,2,4,6) having $S_n^0 = 0$ can emerge, as also seen in Figure 9.2(b). Similarly, if we have $(S_n^1, S_n^0) = (0,1)$ or $(1,1)$ associated with $S_n^0 = 1$ then the branches emerging from these lower two states of the trellis in Figure 9.2(b) can only be associated with the odd-valued phasors of (1,3,5,7).

There are other possible codes, which result in for example four distinct transitions from each state to all possible successor states, but the one selected here proved to be the most effective [220]. Within the 8PSK constellation we have the following distances: $d_0 = 2\sin(\pi/8)$, $d_1 = \sqrt{2}$ and $d_2 = 2$. The 8PSK signals are assigned to the transitions in the four-state trellis in accordance with the following rules:

1) Parallel trellis transitions are associated with phasors having the maximum possible distance, namely (d_2), between them, which is characteristic of phasor points in the subsets (0,4), (1,5), (2,6) and (3,7). Since these parallel transitions belong to the same subset of Figure 9.2(b) and are controlled by the unprotected bit 2, symbols associated with them should be as far apart as possible.

2) All four-state transitions originating from, or merging into, any one of the states are labelled with phasors having a distance of *at least* $d_1 = \sqrt{2}$ between them. These are the

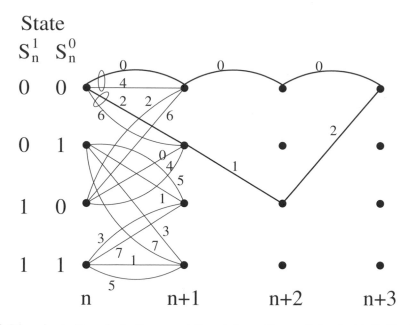

Figure 9.4: Diverging trellis paths for the computation of d_{free}. The parallel paths labelled by the symbols 0 and 4 are associated with the uncoded bits '0' and '1', respectively, as well as with the farthest phasors in the constellation of Figure 9.2(b).

phasors belonging to subsets (0,2,4,6) or (1,3,5,7).

3) All 8PSK signals are used in the trellis diagram with equal probability.

Observe that the assignment of bits to the 8PSK constellation of Figure 9.2(b) does not obey Gray coding and hence adjacent phasors can have arbitrary Hamming distances between them. The bit mapping and encoding process employed was rather designed for exploiting the high Euclidean distances between sets of points in the constellation. The underlined bit 1 and bit 0 of Figure 9.2(b) representing the convolutional codec's output are identical for all parallel branches of the trellis. For example, the branches labelled with phasors 0 and 4 between the identical consecutive states of (0,0) and (0,0) are associated with (bit 1)=0 and (bit 0)=0, while the uncoded bit 2 can be either '0' or '1', yielding the phasors 0 and 4, respectively. However, owing to appropriate code design this unprotected bit has the maximum protection distance, namely $d_2 = 2$, requiring the corruption of phasor 0 into phasor 4, in order to inflict a single bit error in the position of bit 2.

The effect of channel errors exhibits itself at the decoder by diverging from the trellis path encountered in the encoder. Let us consider the example of Figure 9.4, where the encoder generated the phasors 0-0-0 commencing from state (0,0), but owing to channel errors the decoder's trellis path was different from this, since the phasor sequence 2-1-2 was encountered. The so-called free distance of a TCM scheme can be computed as the lower one of two distances. Namely, the Euclidean distances between the phasors labelling the parallel branches in the trellis of Figure 9.2(b) associated with the uncoded bit(s), which is $d_2 = 2$ in our example, as well as the distances between trellis paths diverging and remerging after a number of consecutive trellis transitions, as seen in Figure 9.4 in the first and last of the four consecutive (0,0) states. The lower one of these two distances characterises the error resilience of the underlying TCM scheme, since the error event associated with it will be the one most frequently encountered owing to channel effects.

Specifically, if the received phasors are at a Euclidean distance higher than half of the code's free distance from the transmitted phasor, an erroneous decision will be made. It is essential to ensure that by using an appropriate code design the number of decoded bit errors is minimised in the most likely error events, and this is akin to the philosophy of using Gray coding in a non-trellis-coded constellation.

The Euclidean distance between the phasors of Figure 9.2(b) associated with the parallel branches is $d_2 = 2$ in our example. The distance between the diverging trellis paths of Figure 9.2(b) labelled by the phasor sequences of 0-0-0 and 2-1-2 following the states $\{(0,0),(0,0),(0,0),(0,0)\}$ and $\{(0,0),(0,1),(1,0),(0,0)\}$ respectively, portrayed in Figure 9.4, is inferred from Figure 9.2(b) as d_1-d_0-d_1. By inspecting all the remerging paths of the trellis in Figure 9.2(b) we infer that this diverging path has the shortest accumulated Free Euclidean Distance (FED) that can be found, since all other diverging paths have higher accumulated FED from the error-free 0-0-0 path. Furthermore, this is the only path having the minimum free distance of $\sqrt{d_1^2 + d_0^2 + d_1^2}$. More specifically, the free distance of this TCM sequence is given by:

$$d_{free} = min\{d_2; \sqrt{d_1^2 + d_0^2 + d_1^2}\} \tag{9.1}$$

$$= min\{2; \sqrt{2 + (2.\sin\frac{\pi}{8})^2 + 2}\}. \tag{9.2}$$

Explicitly, since the term under the square root in Equation 9.2 is higher than $d_2 = 2$, the free distance of this TCM scheme is given ultimately by the Euclidean distance between the parallel trellis branches associated with the uncoded bit 2, i.e.:

$$d_{free} = 2. \tag{9.3}$$

The free distance of the uncoded 4PSK constellation of Figure 9.2(a) was $d_0 = \sqrt{2}$ and hence the employment of TCM has increased the minimum distance between the constellation points by a factor of $g = \frac{d_{free}^2}{d_0^2} = \frac{2^2}{(\sqrt{2})^2} = 2$, which corresponds to 3 dB. There is only one nearest-neighbour phasor at $d_{free} = 2$, corresponding to the π-rotated phasor in Figure 9.2(b). Consequently the phasor arrangement can be rotated by π, whilst retaining all of its properties, but other rotations are not admissible.

The number of erroneous decoded bits induced by the diverging path 2-1-2 is seen from the phasor constellation of Figure 9.2(b) to be 1-1-1, yielding a total of three bit errors. The more likely event of a bit 2 error, which is associated with a Euclidean distance of $d_2 = 2$, yields only a single bit error.

Soft-decision-based decoding can be accomplished in two steps. The first step is known as subset decoding, where within each phasor subset assigned to parallel transitions, i.e. to the uncoded bit(s), the phasor closest to the received channel output in terms of Euclidean distance is determined. Having resolved which of the parallel paths was more likely to have been encountered by the encoder, we can remove the parallel transitions, hence arriving at a conventional trellis. In the second step the Viterbi algorithm is used for finding the most likely signal path through the trellis with the minimum sum of squared Euclidean distances from the sequence of noisy channel outputs received. Only the signals already selected by the subset decoding are considered. For a description of the Viterbi algorithm the reader is referred to references [32, 121].

9.2.2 Optimum TCM Codes

Ungerböck's TCM encoder is a specific convolutional encoder selected from the family of Recursive Systematic Convolutional (RSC) codes [208], which attaches one parity bit to each informa-

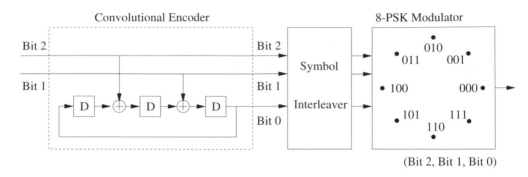

Figure 9.5: Ungerböck's RSC encoder and modulator forming the TCM encoder. The SP-based mapping of bits to the constellation points was highlighted in Figure 9.1.

tion symbol. Only $\tilde{m}$ out of m information bits are RSC encoded and hence only $2^{\tilde{m}}$ branches will diverge from and merge into each trellis state. When not all information bits are RSC encoded, i.e. $\tilde{m} < m$, $2^{m-\tilde{m}}$ parallel transitions are associated with each of the $2^{\tilde{m}}$ branches. Therefore a total of $2^{\tilde{m}} \times 2^{m-\tilde{m}} = 2^m$ transitions occur at each trellis stage. The memory length K of a code defines the number of shift-register stages in the encoder. Figure 9.5 shows the TCM encoder using an eight-state Ungerböck code [208], which has a high FED for the sake of attaining a high performance over AWGN channels. It is a systematic encoder, which attaches an extra parity bit to the original 2-bit information word. The resulting 3-bit codewords generated by the 2-bit input binary sequence are then interleaved by a symbol interleaver in order to disperse the bursty symbol errors induced by the fading channel. Then, these 3-bit codewords are modulated onto one of the $2^3 = 8$ possible constellation points of an 8PSK modulator.

The connections between the information bits and the modulo-2 adders, as shown in Figure 9.5, are given by the generator polynomials. The coefficients of these polynomials are defined as:

$$H^j(D) := h^j_K.D^K + h^j_{K-1}.D^{K-1} + \ldots + h^j_1.D + h^j_0, \tag{9.4}$$

where D represents the delay due to one register stage. The coefficient h^j_i takes the value of '1', if there is a connection at a specific encoder stage or '0', if there is no connection. The polynomial $H^0(D)$ is the feedback generator polynomial and $H^j(D)$ for $j \leq 1$ is the generator polynomial associated with the jth information bit. Hence, the generator polynomial of the encoder in Figure 9.5 can be described in binary format as:

$$
\begin{aligned}
H^0(D) &= 1001 \\
H^1(D) &= 0010 \\
H^2(D) &= 0100,
\end{aligned}
$$

or equivalently in octal format as:

$$
\begin{aligned}
\mathbf{H(D)} &= \begin{bmatrix} H^0(D) & H^1(D) & H^2(D) \end{bmatrix} \\
&= \begin{bmatrix} 11 & 02 & 04 \end{bmatrix}. \tag{9.5}
\end{aligned}
$$

Ungerböck suggested [208] that all feedback polynomials should have coefficients $h^0_K = h^0_0 = 1$. This guarantees the realisability of the encoders shown in Figures 9.3 and 9.5. Furthermore, all generator polynomials should also have coefficients $h^j_K = h^j_0 = 0$ for $j > 0$. This ensures that

Code	State	$\tilde{\mathsf{m}}$	$H^0(D)$	$H^1(D)$	$H^2(D)$
4QAM	8	1	13	06	-
4QAM	64	1	117	26	-
8PSK	8	2	11	02	04
8PSK	32	2	45	16	34
8PSK	64	2	103	30	66
8PSK	128	2	277	54	122
8PSK	256	2	435	72	130
16QAM	64	2	101	16	64

Table 9.1: Ungerböck's TCM codes [208, 220, 222, 223].

at time n the input bits of the TCM encoder have no influence on the parity bit to be generated, nor on the input of the first binary storage element in the encoder. Therefore, whenever two paths diverge from or merge into a common state in the trellis, the parity bit must be the same for these transitions, whereas the other bits differ in at least one bit [208]. Phasors associated with diverging and merging transitions therefore have at least a distance of d_1 between them, as can be seen from Figure 9.2(b). Table 9.1 summarises the generator polynomials of some TCM codes, which were obtained with the aid of an exhaustive computer search conducted by Ungerböck [220], where $\tilde{\mathsf{m}}$ ($\leq \mathsf{m}$) indicates the number of information bits to be encoded, out of the m information bits in a symbol.

9.2.3   TCM Code Design for Fading Channels

It was shown in Section 9.2.1 that the design of TCM for transmission over AWGN channels is motivated by the maximisation of the FED, d_{free}. By contrast, the design of TCM concerned for transmission over fading channels is motivated by minimising the length of the shortest error event path and the product of the branch distances along that particular path [209].

The average bit error probability of TCM using M-ary PSK (MPSK) [208] for transmission over Rician channels at high SNRs is given by [209]:

$$P_b \cong \frac{1}{B} C \left(\frac{(1 + \bar{K}) e^{-\bar{K}}}{E_s/N_0} \right)^L ; E_s/N_0 \gg \bar{K} \tag{9.6}$$

where C is a constant that depends on the weight distribution of the code, which quantifies the number of trellises associated with all possible Hamming distances measured with respect to the all-zero path [49]. The variable B in Equation 9.6 is the number of binary input bits of the TCM encoder during each transmission interval, while $\bar{K}$ is the Rician fading parameter [49] and E_s/N_0 is the channel's symbol energy to noise spectral density ratio. Furthermore, L is the 'length' of the shortest error event path which is expressed in terms of the number of trellis stages encountered before remerging with the all-zero path. It is clear from Equation 9.6 that P_b varies inversely proportionally with $(E_s/N_0)^L$ and this ratio can be increased by increasing the code's diversity [209], which was defined in [224] as the 'length' L of the shortest error event path or the Effective Code Length (ECL). More specifically, in [224], the authors pointed out that the shortest error event paths are not necessarily associated with the minimum accumulated FED error events. For example, let the all-zero path be the correct path. Then the code characterised

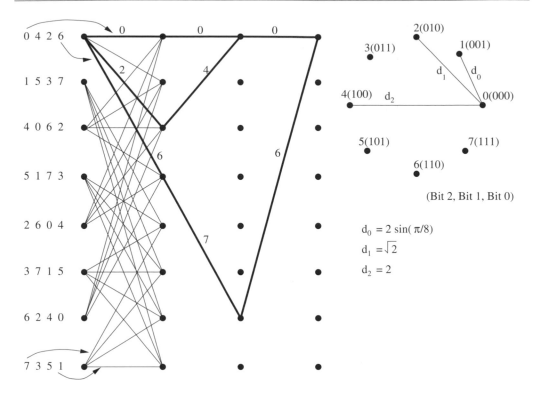

Figure 9.6: Ungerböck's eight-state 8PSK code.

by the trellis seen in Figure 9.6 exhibits a minimum squared FED of:

$$d_{free}^2 = d_1^2 + d_0^2 + d_1^2$$
$$= 4.585, \tag{9.7}$$

from the 0-0-0 path associated with the transmission of three consecutive 0 symbols from the path labelled with the transmitted symbols of 6-7-6. However, this is not the shortest error event path, since its length is $L = 3$, which is longer than the path labelled with transmitted symbols of 2-4, which has a length of $L = 2$ and a FED of $d_{free}^2 = d_1^2 + d_0^2 + d_1^2 = 6$. Hence, the 'length' of the shortest error event path is $L = 2$ for this code, which, again, has a squared Euclidean distance of $d_1^2 + d_2^2 = 6$. In summary, the number of bit errors associated with the above $L = 3$ and $L = 2$ shortest error event paths is seven and two, respectively, clearly favouring the $L = 2$ path, which had a higher accumulated FED of 6 than that of the 4.585 FED of the $L = 3$ path. Hence, it is worth noting that if the code was designed based on the minimum FED, it may not minimise the number of bit errors. Hence, as an alternative design approach, in Section 9.5 we will study BICM, which relies on the shortest error event path L or the bit-based Hamming distance of the code and hence minimises the BER.

The design of coded modulation schemes is affected by a variety of factors. A high squared FED is desired for AWGN channels, while a high ECL and a high minimum product distance are desired for fading channels [209]. In general, a code's diversity or ECL is quantified in terms of the shortest error event path L, which may be increased for example by simple repetition coding, although at the cost of reducing the effective data rate proportionally. Alternatively, space-time-coded multiple transmitter/receiver structures can be used, which increase the scheme's cost and

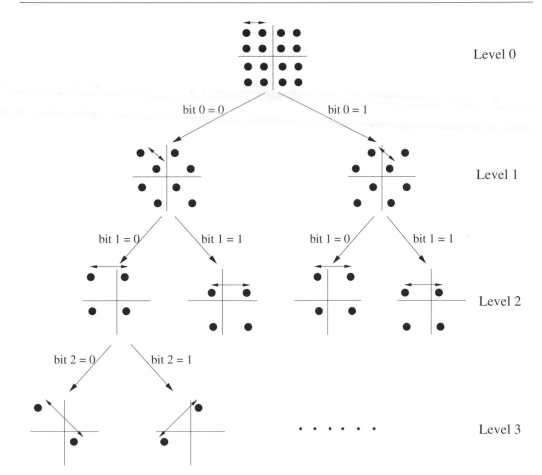

Figure 9.7: Set partitioning of a 16QAM signal constellation. The minimum Euclidean distance at a partition level is denoted by the line between the signal points [208] ©IEEE, 1982, Ungerböck.

complexity. Finally, simple interleaving can be invoked, which induces latency. In our approach, symbol-based interleaving is employed in order to increase the code's diversity.

9.2.4 Set Partitioning

As we have seen in Figure 9.4, if higher-order modulation schemes, such as 16-level Quadrature Amplitude Modulation (16QAM) or 64QAM, are used, parallel transitions may appear in the trellis diagram of the TCM scheme, when not all information bits are convolutional channel encoded or when the number of states in the convolutional encoder has to be kept low for complexity reasons. As noted before, in order to avoid encountering high error probabilities, the parallel transitions should be assigned to constellation points exhibiting a high Euclidean distance. Ungerböck solved this problem by introducing the set partitioning technique. Specifically, the signal set is split into a number of subsets, such that the minimum Euclidean distance of the signal points in the new subset is increased at every partitioning step.

In order to elaborate a little further, Figure 9.7 illustrates the set partitioning of 16QAM. Here we used the $R = \frac{3}{4}$-rate code of Table 9.1. This is a relatively high-rate code, which would not be sufficiently powerful if we employed it for protecting all three original information bits. Moreover, if we protect for example two out of the three information bits, we can use a more potent $\frac{2}{3}$-rate code for the protection of the more vulnerable two information bits and leave the most error-resilient bit of the 4-bit constellation unprotected. This is justifiable, since we can observe in Figure 9.7 that the minimum Euclidean distance of the constellation points increases from Level 0 to Level 3 of the constellation partitioning tree. This indicates that the bits labelling or identifying the specific partitions have to be protected by the RSC code, since they label phasors that have a low Euclidean distance. By contrast, the intra-set distance at Level 3 is the highest, suggesting a low probability of corruption. Hence the corresponding bit, bit 3, can be left unprotected. The partitioning in Figure 9.7 can be continued, until there is only one phasor or constellation point left in each subset. The intra-subset distance increases as we traverse down the partition tree. The first partition level, *Level* 0, is labelled by the parity bit, and the next two levels by the coded bits. Finally, the uncoded bit labels the lowest level, *Level* 3, in the constellation, which has the largest minimum Euclidean distance.

Conventional TCM schemes are typically decoded/demodulated with the aid of the appropriately modified Viterbi Algorithm (VA) [225]. Furthermore, the VA is a maximum likelihood sequence estimation algorithm, which does not guarantee that the Symbol Error Ratio (SER) is minimised, although it achieves a performance near the minimum SER. By contrast, the symbol-based MAP algorithm [211] guarantees the minimum SER, albeit at the cost of a significantly increased complexity. Hence the symbol-based MAP algorithm has been used for the decoding of TCM sequences. We will, however, in Section 9.4, also consider Turbo TCM (TTCM), where instead of the VA-based sequence estimation, symbol-by-symbol-based soft information has to be exchanged between the TCM decoders of the TTCM scheme. Hence in the next section we will present the symbol-based MAP algorithm.

9.3 The Symbol-based MAP Algorithm

In this section, the non-binary or symbol-based MAP decoding algorithm will be presented. The non-binary MAP algorithm was proposed in [211], while the binary MAP algorithm was first presented in [11] and it has been described in detail in Section 5.3.3. In our forthcoming discourse we use $P(e)$ to denote the probability of the event e, and, given a received symbol sequence $\underline{y}$ of length N, the received channel output symbol y_k associated with the present transition, $\underline{y}_{j<k}$, is constituted by the symbol sequence received prior to the present transition, as seen in Figure 9.9 below. Similarly, the symbol sequence $\underline{y}_{j>k}$ received after the present transition is shown in Figure 9.9. We note here in closing that while in Section 5.3.3 $\underline{y}_k$ associated with the present transition was a codeword of length n, in the context of the symbol-based MAP algorithm of this section y_k denotes a channel output sample representing a specific received symbol.

9.3.1 Problem Description

The problem that the MAP algorithm has to solve is presented in Figure 9.8. An information source produces a sequence of N information symbols u_k, $k = 1, 2, \ldots, N$. Each information symbol can assume M different values, i.e. $u_k \in \{0, 1, \ldots, M-1\}$, where M is typically a power of two, so that each information symbol carries $\mathsf{m} = \log_2 M$ information bits. We assume here that the symbols are to be transmitted over an AWGN channel. To this end, the m-bit symbols are first fed into an encoder for generating a sequence of N channel symbols $x_k \in X$, where X denotes the set of complex values belonging to some phasor constellations such as an increased-order QAM or PSK constellation, having $\bar{M}$ possible values carrying $\bar{\mathsf{m}} = \log_2 \bar{M}$ bits. Again, the channel symbols are transmitted over an AWGN channel and the received symbols are:

$$y_k = x_k + n_k, \tag{9.8}$$

where n_k represents the complex AWGN samples. The received symbols are fed to the decoder, which has the task of producing an estimate $\hat{u}_k$ of the 2^m-ary information sequence, based on the $2^{\bar{\mathsf{m}}}$-ary received sequence, where $\bar{\mathsf{m}} > \mathsf{m}$. If the goal of the decoder is that of minimising the number of symbol errors, where a symbol error occurs when $u_k \neq \hat{u}_k$, then the best decoder is the MAP decoder [11]. This decoder computes the A-Posteriori Probability (APP) $A_{k,m}$ for every 2^m-ary information symbol u_k that the information symbol value was m given the received sequence, i.e. computes $A_{k,m} = p(u_k = m | \underline{y})$, for $m = 0, 1, \ldots, M-1$, $k = 1, 2, \ldots, N$. Then it decides that the information symbol was the one having the highest probability, i.e. $\hat{u}_k = m$ if $A_{k,m} \geq A_{k,i}$ for $i = 0 \ldots M-1$. In order to realise a MAP decoder one has to devise a suitable algorithm for computing the APP.

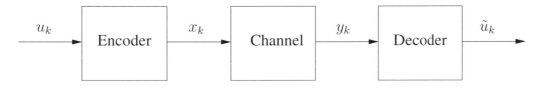

Figure 9.8: The transmission system.

In order to compute the APP, we must specify how the encoder operates. We consider a trellis encoder. The operation of a trellis encoder can be described by its trellis. The trellis seen in Figure 9.9 is constituted by $(N+1) \cdot S$ nodes arranged in $(N+1)$ columns of S nodes. There are M branches emerging from each node, which arrive at nodes in the immediately following column. The trellis structure repeats itself identically between each pair of columns.

It is possible to identify a set of paths originating from the nodes in the first column and terminating in a node of the last column. Each path will comprise exactly N branches. When employing a trellis encoder, the input sequence unambiguously determines a single path in the trellis. This path is identified by labelling the M branches emerging from each node by the M possible values of the original information symbols, although only the labelling of the first branch at $m = 0$ and the last branch at $m = M-1$ IS shown in Figure 9.9 owing to space limitations. Then, commencing from a specified node in the first column, we use the first input symbol, u_1, to decide which branch is to be chosen. If $u_1 = m$, we choose the branch labelled with m, and move to the corresponding node in the second column that this branch leads to. In this node we use the second information symbol, u_2, for selecting a further branch and so on. In this way the information sequence identifies a path in the trellis. In order to complete the encoding operation, we have to produce the symbols to be transmitted over the channel, namely $x_1, x_2, \ldots, x_N$ from the information symbols $u_1, u_2, \ldots, u_N$. To this end we add a second label to each branch, which is the corresponding phasor constellation point that is transmitted when the branch is encountered.

In a trellis it is convenient to attach a time index to each column, from 0 to N, and to number the nodes in each column from 0 to $S-1$. This allows us to introduce the concept of trellis states at time k. Specifically, during the encoding process, we say that the trellis is in state s at time k, and write $S_k = s$, if the path determined by the information sequence crosses the sth node of the kth column. There is a branch leading from state $S_{k-1} = \grave{s}$ to state $S_k = s$, which

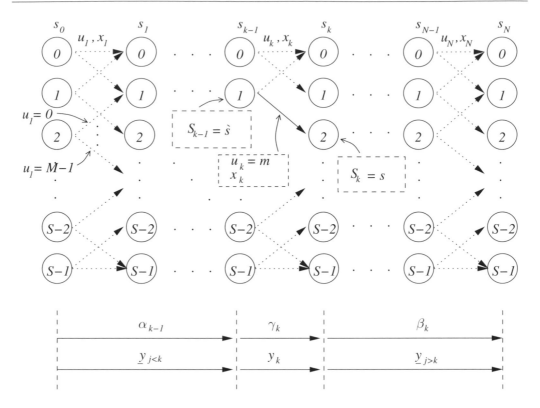

Figure 9.9: The non-binary trellis and its labelling, where there are M branches emerging from each node. (The binary trellis for the binary MAP algorithms is illustrated in Figure 5.6, where there are only two possible branches emerging from each node.)

is encountered if the input symbol is $u_k = m$, and the corresponding transmitted symbol is x_k. The aim of the MAP decoding algorithm is to find the path in the trellis that is associated with the most likely transmitted symbols, i.e. that of minimising the SER. By contrast, the VA-based detection of TCM signals aims to identify the most likely transmitted symbol sequence, which does not automatically guarantee attaining the minimum SER.

9.3.2   Detailed Description of the Symbol-based MAP Algorithm

Having described the problem to be solved by the MAP decoder and the encoder structure, we now seek an algorithm capable of computing the APP, i.e. $A_{k,m} = P(u_k = m|\underline{y})$. The easiest way of computing these probabilities is by determining the sum of a different set of probabilities, namely $P(u_k = m \wedge S_{k-1} = \grave{s} \wedge S_k = s|\underline{y})$, where, again, $\underline{y}$ denotes the received symbol sequence. This is because we can devise a recursive way of computing the second set of probabilities as we traverse through the trellis from state to state, which reduces the detection complexity. Thus we write:

$$A_{k,m} = P(u_k = m|\underline{y}) = \sum_{\text{all } \grave{s}, s} P(u_k = m \wedge S_{k-1} = \grave{s} \wedge S_k = s|\underline{y}), \qquad (9.9)$$

where the summation implies adding all probabilities associated with the nodes $\grave{s}$ and s labelled by $u_k = m$ and the problem is now that of computing $P(u_k = m \wedge S_{k-1} = \grave{s} \wedge S_k = s|\underline{y})$. As a preliminary consideration we note that this probability is zero, if the specific branch of the trellis emerging from state $\grave{s}$ and merging into state s is not labelled with the input symbol m. Hence, we can eliminate the corresponding terms of the summation. Thus, we can rewrite Equation 9.9 as:

$$A_{k,m} = \sum_{\substack{(\grave{s},s) \Rightarrow \\ u_k = m}} P(S_{k-1} = \grave{s} \wedge S_k = s | \underline{y}), \tag{9.10}$$

where $(\grave{s}, s) \Rightarrow u_k = m$ indicates the specific set of transitions emerging from the previous state $S_{k-1} = \grave{s}$ to the present state $S_k = s$ that can be encountered when the input symbol is $u_k = m$. If the transitions $(\grave{s}, s) \Rightarrow u_k = m$ exist, then we can compute the probabilities $P(S_{k-1} = \grave{s} \wedge S_k = s | \underline{y})$, using Bayes' rule, as:

$$P(\grave{s} \wedge s | \underline{y}) = \frac{1}{P(\underline{y})} \cdot P(\grave{s} \wedge s \wedge \underline{y}). \tag{9.11}$$

Using Equations 5.18 to 5.20 of Section 5.3.3.1, we can rewrite Equation 9.11 as:

$$P(\grave{s} \wedge s | \underline{y}) = \frac{1}{P(\underline{y})} \cdot \beta_k(s) \cdot \gamma_k(\grave{s}, s) \cdot \alpha_{k-1}(\grave{s}), \tag{9.12}$$

where:

$$\begin{aligned} \alpha_{k-1}(\grave{s}) &= P(S_{k-1} = \grave{s} \wedge \underline{y}_{j<k}) \\ \beta_k(s) &= P(\underline{y}_{j>k} | S_k = s) \\ \gamma_k(\grave{s}, s) &= P(\{y_k \wedge S_k = s\} | S_{k-1} = \grave{s}) \end{aligned} \tag{9.13}$$

similar to those in Equations 5.21, 5.22 and 5.23 for the binary MAP algorithm, except for the slight differences that will be discussed during our discourse below. More specifically, in Sections 5.3.3.2 and 5.3.3.3 we have shown how the $\alpha_{k-1}(\grave{s})$ values and the $\beta_k(s)$ values can be efficiently computed using the $\gamma_k(\grave{s}, s)$ values. However, the computation of the $\gamma_k(\grave{s}, s)$ values in the symbol-based MAP is different from that of the binary MAP. In our forthcoming discourse we study the $\gamma_k(\grave{s}, s)$ values and further simplify Equation 9.10.

Upon substituting Equation 9.12 into Equation 9.10 we have:

$$A_{k,m} = C_k^1 \cdot \sum_{\substack{(\grave{s},s) \Rightarrow \\ u_k = m}} \beta_k(s) \cdot \gamma_k(\grave{s}, s) \cdot \alpha_{k-1}(\grave{s}), \tag{9.14}$$

where $C_k^1 = \frac{1}{p(\underline{y})}$ is a common normalisation factor. The first simplification is to note that we do not necessarily need the exact $A_{k,m}$ values, only their ratios. In fact, for a fixed time instant k, the vector $A_{k,m}$, which is a vector of probabilities, has to sum to unity. Thus, by normalising the sum in Equation 9.10 to unity, we can compute the exact value of $A_{k,m}$ from $\bar{A}_{k,m}$ with the aid of:

$$A_{k,m} = C_k \cdot \bar{A}_{k,m}. \tag{9.15}$$

For this reason we can omit the common normalisation factor of C_k^1 in Equation 9.14.

Let us now consider the term $\gamma_k(\grave{s}, s)$ of Equation 9.14, which can be rewritten using Bayes' rule as:

$$\begin{aligned} \gamma_k(\grave{s}, s) &= P(\{y_k \wedge s\} | \grave{s}) \\ &= P(y_k | \{\grave{s} \wedge s\}) \cdot P(s | \grave{s}) \\ &= P(y_k | \{\grave{s} \wedge s\}) \cdot P(m), \end{aligned} \tag{9.16}$$

where $u_k = m$ is the input symbol necessary to cause the transition from state $S_{k-1} = \grave{s}$ to state $S_k = s$, and $P(m)$ is the a-priori probability of this symbol. Let us now study the multiplicative terms at the right of Equation 9.16, where $P(y_k|\{\grave{s} \wedge s\})$ is the probability that we receive y_k when the branch emerges from state $S_{k-1} = \grave{s}$ of Figure 9.9 to state $S_k = s$. When this branch is encountered, the symbol transmitted is x_k, as seen in Figure 9.9. Thus, the probability of receiving the sample y_k, given that the previous state was $S_{k-1} = \grave{s}$ and the current state is $S_k = s$, can be written as:

$$P(y_k|\{\grave{s} \wedge s\}) = p(y_k|x_k). \tag{9.17}$$

By remembering that $y_k = x_k + n_k$, where n_k is the complex AWGN, we can compute Equation 9.17 as [226]:

$$
\begin{aligned}
P(y_k|\{\grave{s} \wedge s\}) &= \frac{1}{2\pi\sigma^2} \cdot e^{-\frac{|y_k - x_k|^2}{2\sigma^2}} \\
&= C_k^2 \cdot \eta_k(\grave{s}, s), \tag{9.18}
\end{aligned}
$$

where $\sigma^2 = N_0/2$ is the noise's variance, N_0 is the noise's Power Spectral Density (PSD), $C_k^2 = \frac{1}{2\pi\sigma^2}$ and $\eta_k(\grave{s}, s) = e^{-\frac{|y_k - x_k|^2}{2\sigma^2}}$. In verbal terms, Equation 9.18 indicates that the probability expressed in Equation 9.17 is a function of the distance between the received noisy sample y_k and the transmitted noiseless sample x_k. Observe in Equation 9.18 that we can drop the multiplicative factor of $C_k^2 = \frac{1}{2\pi\sigma^2}$, since it constitutes another scaling factor. As to the second multiplicative term on the right-hand side of Equation 9.16, note that $p(u_k = m|S_{k-1} = \grave{s}) = p(u_k = m)$, since the original information to be transmitted is independent of the previous trellis state. The probabilities:

$$\Pi_{k,m} = p(u_k = m) \tag{9.19}$$

are the a-priori probabilities of the information symbols. Typically the information symbols are independent and equiprobable, hence $\Pi_{k,m} = 1/M$. However, if we have some prior knowledge about the transmitted symbols, this can be used as their a-priori probability. As we will see, a turbo decoder will have some a-priori knowledge about the transmitted symbols after the first iteration. We now rewrite Equation 9.16 using Equations 9.18 and 9.19 as:

$$\gamma_k(\grave{s}, s) = C_k^2 \cdot \Pi_{k,m} \cdot \eta_k(\grave{s}, s) . \eta_k(j, m). \tag{9.20}$$

Then, by substituting Equation 9.20 into Equation 9.14 and and exchanging the order of summations we can portray the APPs in their final form, yielding:

$$A_{k,m} = C_k \cdot \Pi_{k,m} \cdot \sum_{\substack{(\grave{s}, s) \Rightarrow \\ u_k = m}} \beta_k(s) \cdot \alpha_{k-1}(\grave{s}) \cdot \eta_k(\grave{s}, s), \tag{9.21}$$

where $C_k = C_k^1 \cdot C_k^2$ is a common scaling factor at instant k that can be dropped for the sake of simplicity without any ambiguity. Therefore, we only have to consider $\bar{A}_{k,m}$ of Equation 9.15, yielding:

$$\bar{A}_{k,m} = \Pi_{k,m} \cdot \sum_{\substack{(\grave{s}, s) \Rightarrow \\ u_k = m}} \beta_k(s) \cdot \alpha_{k-1}(\grave{s}) \cdot \eta_k(\grave{s}, s). \tag{9.22}$$

The symbol-based MAP algorithm can be evaluated also in the logarithmic domain (log-domain) for the sake of reducing the computational complexity and for mitigating the numerical stability

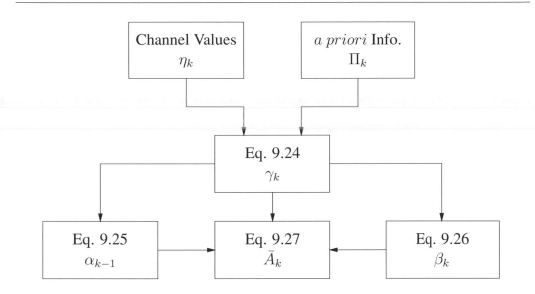

Figure 9.10: Summary of the symbol-based MAP algorithm operations. (The summary of the LLR-based binary MAP algorithm operations is shown in Figure 5.8.)

problems associated with the MAP algorithm [11], when processing small numbers representing the associated probabilities. The modifications of the symbol-based MAP algorithm required for transforming it to the log-domain are similar to that of the binary MAP algorithm, which was discussed in Section 5.3.5.

9.3.3   Symbol-based MAP Algorithm Summary

Let us now summarise the operations of the symbol-based MAP algorithm using Figure 9.10. We assume that the a-priori probabilities $\Pi_{k,m}$ in Equation 9.19 were known. These are either all equal to $1/M$ or constituted by additional external information. The first step is to compute the set of probabilities $\eta_k(\grave{s}, s)$ from Equation 9.18 as:

$$\eta_k(\grave{s}, s) = e^{-\frac{|y_k - x_k|^2}{2\sigma^2}}. \tag{9.23}$$

From these and the a-priori probabilities, the $\gamma_k(\grave{s}, s)$ values are computed according to Equation 9.20 as:

$$\gamma_k(\grave{s}, s) = \Pi_{k,m} \cdot \eta_k(\grave{s}, s). \tag{9.24}$$

The above values are then used to recursively compute the values $\alpha_{k-1}(\grave{s})$ employing Equation 5.25 as:

$$\alpha_k(s) = \sum_{\text{all } \grave{s}} \gamma_k(\grave{s}, s) \cdot \alpha_{k-1}(\grave{s}), \tag{9.25}$$

and the values $\beta_k(s)$ using Equation 5.28 as:

$$\beta_{k-1}(\grave{s}) = \sum_{\text{all } s} \beta_k(s) \cdot \gamma_k(\grave{s}, s). \tag{9.26}$$

Finally, the APP can be obtained using Equation 9.22:

$$\bar{A}_{k,m} \;=\; \Pi_{k,m} \cdot \sum_{\substack{(\grave{s},s)\Rightarrow \\ u_k=m}} \beta_k(s) \cdot \alpha_{k-1}(\grave{s}) \cdot \eta_k(\grave{s},s). \qquad (9.27)$$

When considering the implementation of the MAP algorithm, one can opt for computing and storing the $\eta_k(\grave{s},s)$ values, and use these values together with the a-priori probabilities for determining the values $\gamma_k(\grave{s},s)$ during decoding. In order to compute the probabilities $\eta_k(\grave{s},s)$ it is convenient to separately evaluate the exponential function of Equation 9.23 for every k and for every possible value of the transmitted symbol. As described in Section 9.3.1, a sequence of N information symbols was produced by the information source and each information symbol can assume M possible values, while the number of encoder states is S. There are $\bar{M} = 2 \cdot M$ possible transmitted symbols, since the size of the original signal constellation was doubled by the trellis encoder. Thus $N \cdot 2 \cdot M$ evaluations of the exponential function of Equation 9.23 are needed. Using the online computation of the $\gamma_k(\grave{s},s)$ values, two multiplications are required for computing one additive term in each of Equations 9.25 and 9.26, and there are $N \cdot S$ terms to be computed, each requiring M terms to be summed. Hence $2 \cdot N \cdot M \cdot S$ multiplications and $N \cdot M \cdot S$ additions are required for computing the forward recursion α or the backward recursion β. Approximately three multiplications are required for computing each additive term in Equation 9.27, and there are $N \cdot M$ terms to be computed, each requiring S terms to be summed. Hence, the total implementational complexity entails $7 \cdot N \cdot M \cdot S$ multiplications, $3 \cdot N \cdot M \cdot S$ summations and $N \cdot 2 \cdot M$ exponential function evaluations, which is directly proportional to the length N of the transmitted sequence, to the number of code states S and to the number of different values M assumed by the input symbols.

The computational complexity can be reduced by implementing the algorithm in the log-domain, where the evaluation of the exponential function in Equation 9.23 is avoided. The multiplications and additions in Equations 9.24 to 9.27 are replaced by additions and Jacobian comparisons, respectively. Hence the total implementational complexity imposed is $7 \cdot N \cdot M \cdot S$ additions and $3 \cdot N \cdot M \cdot S$ Jacobian comparisons.

When implementing the MAP decoder presented here it is necessary to control the dynamic range of the likelihood terms computed in Equations 9.25 to 9.27. This is because these values tend to become lower and lower owing to the multiplication of small values. The dynamic range can be controlled by normalising the sum of the $\alpha_k(s)$ and the $\beta_k(s)$ values to unity at every particular k symbol. The resulting symbol values will not be affected, since the normalisation only affects the scaling factors C_k in Equation 9.15. However, this problem can be avoided when the MAP algorithm is implemented in the log-domain.

To conclude, let us note that the MAP decoder presented here is suitable for the decoding of finite-length, preferably short, sequences. When long sequences are transmitted, the employment of this decoder is impractical, since the associated memory requirements increase linearly with the sequence length. In this case the MAP decoder has to be modified. A MAP decoder designed for long sequences was first presented in [227]. An efficient implementation, derived by adapting the algorithm of [11], was proposed by Piazzo in [228]. Having described the symbol-based MAP algorithm, let us now consider Turbo TCM (TTCM) and the way it invokes the MAP procedure.

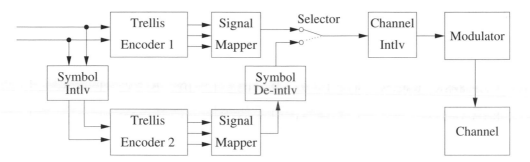

Figure 9.11: Schematic of the TTCM encoder. The selector enables the transmission of the information bits only once and selects alternative parity bits from the constituent encoders seen at the top and bottom [211] ©IEEE, 1998, Robertson and Wörz.

9.4   Turbo Trellis-coded Modulation

9.4.1   TTCM Encoder

It is worth describing the signal set dimensionality ($\bar{D}$) [229, 230] before we proceed. For a specific $2\bar{D}$ code, we have one $2\bar{D}$ symbol per codeword. For a general multidimensional code having a dimensionality of $D = 2 \cdot n$ where $n > 0$ is an integer, one $D\bar{D}$ codeword is comprised of n $2\bar{D}$ sub-codewords. The basic concept of the multidimensional signal mapping [229] is to assign more than one $2\bar{D}$ symbol to one codeword, in order to increase the spectral efficiency, which is defined as the number of information bits transmitted per channel symbol. For instance, a $2\bar{D}$ 8PSK TCM code seen in Table 9.2 maps $n = \frac{D}{2} = 1$ three-bit $2\bar{D}$ symbol to one $2\bar{D}$ codeword, where the number of information bits per $2\bar{D}$ codeword is $\mathsf{m} = 2$ yielding a spectral efficiency of $\mathsf{m}/n = 2$ information bits per symbol. However, a $4\bar{D}$ 8PSK TCM code seen in Table 9.2 maps $n = \frac{D}{2} = 2$ three-bit $2\bar{D}$ symbols to one six-bit $4\bar{D}$ codeword using the mapping rule of [229], where the number of information bits per $4\bar{D}$ codeword is $\mathsf{m} = 5$, yielding a spectral efficiency of $\mathsf{m}/n = 2.5$ information bits per symbol. However, during our further discourse we only consider $2\bar{D}$ signal sets.

Employing TTCM [211] avoids the obvious disadvantage of rate loss that one would incur when applying the principle of parallel concatenation to TCM without invoking puncturing. Specifically, this is achieved by puncturing the parity information in a particular manner, so that all information bits are sent only once, and the parity bits are provided alternatively by the two component TCM encoders. The TTCM encoder is shown in Figure 9.11, which comprises two identical TCM encoders linked by a symbol interleaver.

Let the memory of the interleaver be N symbols. The number of modulated symbols per block is $N.n$, where $n = \frac{D}{2}$ is an integer and D is the number of dimensions of the signal set. The number of information bits transmitted per block is $N.\mathsf{m}$, where m is the number of information bits per symbol. The encoder is clocked at a rate of $n.T$, where T is the symbol duration of each transmitted $2^{(\mathsf{m}+1)/n}$-ary $2\bar{D}$ symbol. At each step, m information bits are input to the TTCM encoder and n symbols each constituted by $\mathsf{m} + 1$ bits are transmitted, yielding a coding rate of $\frac{\mathsf{m}}{\mathsf{m}+1}$.

Each component TCM encoder consists of an Ungerböck encoder and a signal mapper. The first TCM encoder operates on the original input bit sequence, while the second TCM encoder manipulates the interleaved version of the input bit sequence. The signal mapper translates the codewords into complex symbols using the SP-based labelling method of Section 9.2.4. A complex symbol represents the amplitude and phase information passed to the modulator in the

Code	$\tilde{m}$	$H^0(D)$	$H^1(D)$	$H^2(D)$	$H^3(D)$	$d^2_{free}/\triangle^2_0$
$2\bar{D}$, 8PSK, 4 states	2	07	02	04	-	
$2\bar{D}$, 8PSK, 8 states	2	11	02	04	-	3
$4\bar{D}$, 8PSK, 8 states	2	11	06	04	-	3
$2\bar{D}$, 8PSK, 16 states	2	23	02	10	-	3
$4\bar{D}$, 8PSK, 16 states	2	23	14	06	-	3
$2\bar{D}$, 16QAM, 8 states	3	11	02	04	10	2
$2\bar{D}$, 16QAM, 16 states	3	21	02	04	10	3
$2\bar{D}$, 64QAM, 8 states	2	11	04	02	-	3
$2\bar{D}$, 64QAM, 16 states	2	21	04	10	-	4

Table 9.2: 'Punctured' TCM codes exhibiting the best minimum distance for 8PSK, 16QAM and 64QAM, where octal format is used for specifying the generator polynomials [211] ©IEEE, 1998, Robertson and Wörz. The notation $\bar{D}$ denotes the dimensionality of the code while $\triangle^2_0$ denotes the squared Euclidean distance of the signal set itself and d^2_{free} denotes the squared FED of the TCM code.

system seen in Figure 9.11. The complex output symbols of the signal mapper at the bottom of Figure 9.11 are symbol de-interleaved according to the inverse operation of the interleaver. Again, the interleaver and de-interleaver are symbol interleavers [231]. Owing to invoking the de-interleaver of Figure 9.11 at the output of the component encoder seen at the bottom, the TTCM codewords of both component encoders have identical information bits before the selector. Hence, the selector that alternatively selects the symbols of the upper and lower component encoders is effectively a puncturer that punctures the parity bits of the output symbols.

The output of the selector is then forwarded to the channel interleaver, which is, again, another symbol interleaver. The task of the channel interleaver is to effectively disperse the bursty symbol errors experienced during transmission over fading channels. This increases the diversity order of the code [209, 224]. Finally, the output symbols are modulated and transmitted through the channel.

Table 9.2 shows the generator polynomials of some component TCM codes that can be employed in the TTCM scheme. These generator polynomials were obtained by Robertson and Wörz [211] using an exhaustive computer search of all polynomials and finding the one that maximises the minimal Euclidean distance, taking also into account the alternative selection of parity bits for the TTCM scheme. In Table 9.2, $\tilde{m}$ denotes the number of information bits to be encoded out of the total m information bits in a symbol, $\triangle^2_0$ denotes the squared Euclidean distance of the signal set itself, i.e. after TCM signal expansion, and d^2_{free} denotes the squared FED of the TCM constituent codes, as defined in Section 9.2.1. Since $d^2_{free}/\triangle^2_0 > 0$, the 'punctured' TCM codes constructed in Table 9.2 exhibit a positive coding gain in comparison to the uncoded but expanded signal set, although not necessarily in comparison to the uncoded and unexpanded original signal set. Nonetheless, the design target is to provide a coding gain also in comparison to the uncoded and unexpanded original signal set at least for the targeted operational SNR range of the system.

Considering the 8PSK example, where $\triangle^2_0 = d^2_{8PSK}$, we have $d^2_{free}/d^2_{8PSK} = 3$, but when we compare the 'punctured' 8PSK TCM codes with the original uncoded QPSK signal set we have $d^2_{free}/d^2_{QPSK} = d^2_{free}/2 = 0.878$ [211], which implies a negative coding gain. However, when the iterative decoding scheme of TTCM is invoked, we can attain a significant positive coding gain, as we will demonstrate in Section 9.7.

9.4.2   TTCM Decoder

Recall that in Figure 5.9 of Section 5.3.4 the concept of $a-priori$, $a-posteriori$ and $extrinsic$ information was introduced. This illustration is repeated here in Figure 9.12(a) for the sake of convenient comparison with its non-binary counterpart seen in Figure 9.12(b). The associated concept is portrayed in more detail in Figure 9.13, which will be detailed during our further discourse.

The TTCM decoder structure of Figure 9.13(b) is similar to that of binary turbo codes shown in Figure 9.13(a), except that there is a difference in the nature of the information passed from one decoder to the other and in the treatment of the very first decoding step. Specifically, each decoder alternately processes its corresponding encoder's channel-impaired output symbol, and then the other encoder's channel-impaired output symbol.

In a binary turbo coding scheme the component encoders' output can be split into three additive parts for each information bit u_k at step k, when operating in the logarithmic or LLR domain [52] as shown in Figure 9.13(a), which are:

1) the systematic component (S/s), i.e. the corresponding received systematic value for bit u_k;

2) the $a-priori$ or intrinsic component (A/a), i.e. the information provided by the other component decoder for bit u_k; and

3) the extrinsic information component related to bit u_k (E/e), which depends not on bit u_k itself but on the surrounding bits.

These components are impaired by independent noise and fading effects. In turbo codes, only the extrinsic component should be passed on to the other component decoder, so that the intrinsic information directly related to a bit is not reused in the other component decoder [12]. This measure is necessary in turbo codes for avoiding the prevention of achieving iterative gains, due to the dependence of the constituent decoders' information on each other.

However, in a symbol-based non-binary TTCM scheme the m systematic information and the parity bit are transmitted together in the same non-binary symbol. Hence, the systematic component of the non-binary symbol, namely the original information bits, cannot be separated from the extrinsic component, since the noise and/or fading that affects the parity component also affects the systematic component. Therefore, in this scenario the symbol-based information can be split into only two components:

1) the a-priori component of a non-binary symbol (A/a), which is provided by the other component decoder, and

2) the inseparable extrinsic as well as systematic component of a non-binary symbol $([E\&S]/[e\&s])$, as can be seen from Figure 9.13(b).

Each decoder passes only the latter information to the next component decoder while the a-priori information is removed at each component decoder's output, as seen in Figure 9.13(b), where, again, the extrinsic and systematic components are inseparable.

As described in Section 9.4.1, the number of modulated symbols per block is $N \cdot n$, with $n = \frac{D}{2}$, where D is the number of dimensions of the signal set. Hence for a $2\bar{D}$ signal set we have $n = 1$ and the number of modulated symbols per block is N. Therefore the symbol interleaver of length N will interleave a block of N complex symbols. Let us consider $2\bar{D}$ modulation having a coding rate of $\frac{m}{m+1}$ for the following example.

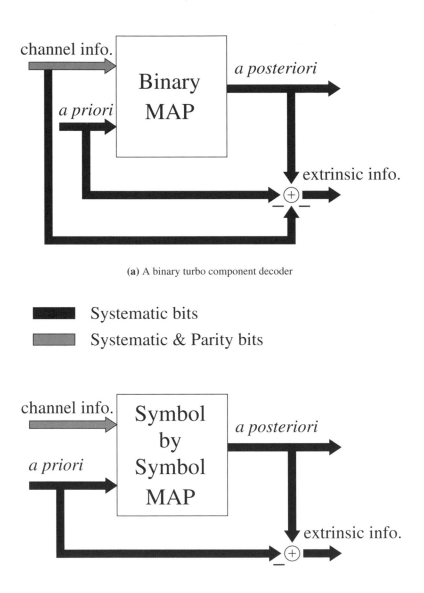

(a) A binary turbo component decoder

(b) A non-binary TTCM component decoder

Figure 9.12: Schematic of the component decoders for binary turbo codes and non-binary TTCM.

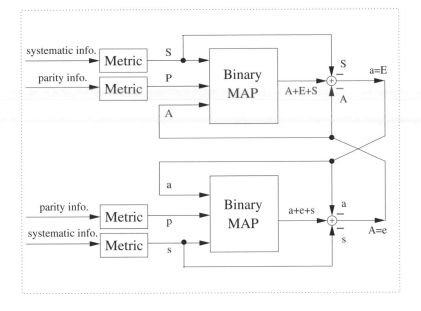

(a) Binary Turbo Decoder at step k

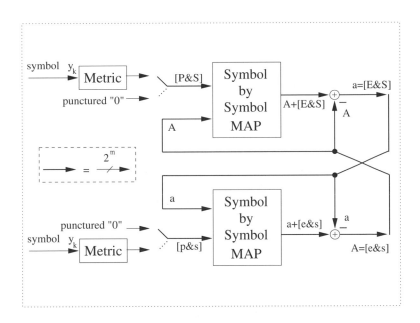

(b) TTCM Decoder at step k

Figure 9.13: Schematic of the decoders for binary turbo codes and TTCM. Note that the labels and arrows apply only to one specific information bit for the binary turbo decoder, or a group of m information bits for the TTCM decoder [211] ©IEEE, 1998, Robertson and Wörz. The interleavers/de-interleavers are not shown and the notations P, S, A and E denote the parity information, systematic information, $a - priori$ probabilities and $extrinsic$ probabilities, respectively. Upper (lower) case letters represent the probabilities of the upper (lower) component decoder.

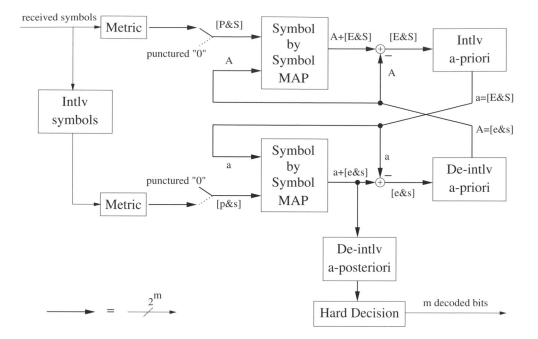

Figure 9.14: Schematic of the TTCM decoder. P, S, A and E denote the parity information, systematic information, $a-priori$ probabilities and $extrinsic$ probabilities, respectively. Upper (lower) case letters represent the probabilities of the upper (lower) component decoder.

The received symbols are input to the 'Metric' block of Figure 9.14, in order to generate a set of $\bar{M} = 2^{m+1}$ symbol probabilities for quantifying the likelihood that a certain symbol of the $\bar{M}$-ary constellation was transmitted. The selector switches seen at the input of the 'Symbol by Symbol MAP' decoder select the current symbol's reliability metric, which is produced at the output of the 'Metric' block, if the current symbol was not punctured by the corresponding encoder. Otherwise puncturing will be applied where the probabilities of the various legitimate symbols at index k are set to 1 or to 0 in the log-domain. The upper (lower) case letters denote the set of probabilities of the upper (lower) component decoder, as shown in the figure. The 'Metric' block provides the decoder with the inseparable parity and systematic ($[P\&S]$ or $[p\&s]$) information, and the second input to the decoder is the a-priori (A or a) information provided by the other component decoder. The MAP decoder then provides the a-posteriori ($A + [E\&S]$ or $a + [e\&s]$) information at its output. Then A ($or\ a$) is subtracted from the a-posteriori information, so that the same information is not used more than once in the other component decoder, since otherwise the component decoders' corresponding information would become dependent on each other, which would preclude the achievement of iteration gains. The resulting $[E\&S\ or\ e\&s]$ information is symbol interleaved (or de-interleaved) in order to present the a ($or\ A$) input for the other component decoder in the required order. This decoding process will continue iteratively, in order to offer an improved version of the set of symbol reliabilities for the other component decoder. One iteration comprises the decoding of the received symbols by both the component decoders once. Finally, the a-posteriori information of the lower component decoder will be de-interleaved in order to extract m decoded information bits per symbol. Hard decision implies selecting the specific symbol which exhibits the maximum a-posteriori probability associated with the m-bit information symbol out of the 2^m probability values. Having described the operation of the symbol-based TTCM technique, which does not protect all transmitted bits of the

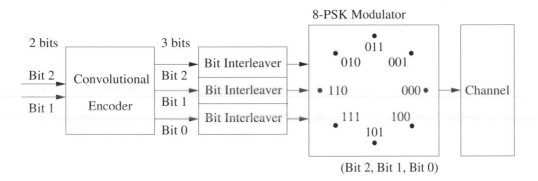

Figure 9.15: BICM encoder schematic employing independent bit interleavers and and protecting all transmitted bits. Instead of the SP-based labelling of TCM in Figure 9.1 here Gray labelling is employed [212] ©IEEE, 1992, Zehavi.

symbols, let us now consider bit-interleaved coded modulation as a design alternative.

9.5 Bit-interleaved Coded Modulation

Bit-interleaved Coded Modulation (BICM) was proposed by Zehavi [212] with the aim of increasing the diversity order of Ungerböck's TCM schemes which was quantified in Section 9.2.3. Again, the diversity order of a code is defined as the 'length' of the shortest error event path expressed in terms of the number of trellis stages encountered, before remerging with the all-zero path [224] or, equivalently, defined as the minimum Hamming distance of the code [232] where the diversity order of TCM using a symbol-based interleaver is the minimum number of different symbols between the erroneous path and the correct path along the shortest error event path. Hence, in a TCM scenario having parallel transitions, as shown in Figure 9.4, the code's diversity order is one, since the shortest error event path consists of one branch. This implies that parallel transitions should be avoided in TCM codes at all was possible, and if there were no parallel branches, any increase in diversity would be obtained by increasing the constraint length of the code. Unfortunately no TCM codes exist where the parallel transitions associated with the unprotected bits are avoided. In order to circumvent this problem, Zehavi's idea [212] was to render the code's diversity equal to the smallest number of different bits, rather than to that of the different channel symbols, by employing bit-based interleaving, as will be highlighted below.

9.5.1 BICM Principle

The BICM encoder is shown in Figure 9.15. In comparison to the TCM encoder of Figure 9.5, the differences are that BICM uses independent bit interleavers for all the bits of a symbol and non-systematic convolutional codes, rather than a single symbol-based interleaver and systematic RSC codes protecting some of the bits. The number of bit interleavers equals the number of bits assigned to the non-binary codeword. The purpose of bit interleaving is:

- to disperse the bursty errors induced by the correlated fading and to maximise the diversity order of the system;

- to render the bits associated with a given transmitted symbol uncorrelated or independent of each other.

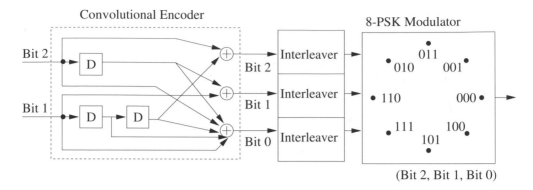

Figure 9.16: Paaske's non-systematic convolutional encoder, bit-based interleavers and modulator forming the BICM encoder [96, 212], where none of the bits are unprotected and instead of the SP-based labelling as seen in Figure 9.1 here Gray labelling is employed.

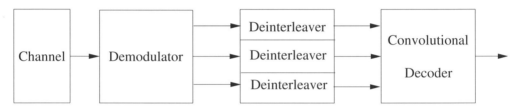

Figure 9.17: BICM decoder [212].

The interleaved bits are then grouped into non-binary symbols, where Gray-coded labelling is used for the sake of optimising the performance of the BICM scheme. The BICM encoder uses Paaske's non-systematic convolutional code proposed on p. 331 of [96], which exhibits the highest possible free Hamming distance, hence attaining optimum performance over Rayleigh fading channels. Figure 9.16 shows Paaske's non-systematic eight-state code of rate-2/3, exhibiting a free bit-based Hamming distance of four. The BICM decoder implements the inverse process, as shown in Figure 9.17. In the demodulator module six bit metrics associated with the three bit positions, each having binary values of 0 and 1, are generated from each channel symbol. These bit metrics are de-interleaved by three independent bit de-interleavers, in order to form the estimated codewords. Then the convolutional decoder of Figure 9.17 is invoked for decoding these codewords, generating the best possible estimate of the original information bit sequence.

From Equation 9.6 we know that the average bit error probability of a coded modulation scheme using MPSK over Rayleigh fading channels at high SNRs is inversely proportional to $(E_s/N_0)^L$, where E_s/N_0 is the channel's symbol energy to noise spectral density ratio and L is the minimum Hamming distance or the code's diversity order. When bit-based interleavers are employed in BICM instead of the symbol-based interleaver employed in TCM, the minimum Hamming distance of BICM is quantified in terms of the number of different bits between the erroneous path in the shortest error event and the correct path. Since in BICM the bit-based minimum Hamming distance is maximised, BICM will give a lower bit error probability in Rayleigh fading channels than that of TCM maximising the FED. Again, the design of BICM is aimed at providing maximum minimum Hamming distance, rather than providing maximum FED, as in TCM schemes. Moreover, we note that attaining a maximum FED is desired for transmission over Gaussian channels, as shown in Section 9.2.1. Hence, the performance of BICM is not as good as that of TCM in AWGN channels. The reduced FED of BICM is due to the 'ran-

Rate	K	$g^{(1)}$	$g^{(2)}$	$g^{(3)}$	$g^{(4)}$	d_{free}
1/2	3	15	17	-	-	5
	6	133	171	-	-	10
2/3	3	4	2	6	-	4
		1	4	7	-	
	4	7	1	4	-	5
		2	5	7	-	
	6	64	30	64	-	7
		30	64	74	-	
3/4	3	4	4	4	4	4
		0	6	2	4	
		0	2	5	5	
	5	6	2	2	6	5
		1	6	0	7	
		0	2	5	5	
	6	6	1	0	7	6
		3	4	1	6	
		2	3	7	4	

Table 9.3: Paaske's non-systematic convolutional codes, p. 331 of [96], where K denotes the code memory and d_{free} denotes the free Hamming distance. Octal format is used for representing the generator polynomial coefficients.

Rate	K	$g^{(1)}$	$g^{(2)}$	puncturing matrix	d_{free}
5/6	3	15	17	1 0 0 1 0	3
				0 1 1 1 1	
	6	133	171	1 1 1 1 1	3
				1 0 0 0 0	

Table 9.4: Rate-Compatible Punctured Convolutional (RCPC) codes [233,234], where K denotes the code memory and d_{free} denotes the free Hamming distance. Octal format is used for representing the generator polynomial coefficients.

dom' modulation imposed by the 'random' bit interleavers [212], where the $\bar{m}$-bit BICM, coded symbol is randomised by the $\bar{m}$ number of bit interleavers. Again, m denotes the number of information bits, while $\bar{m}$ denotes the total number of bits in a $2^{\bar{m}}$-ary modulated symbol.

Table 9.3 summarises the parameters of a range of Paaske's non-systematic codes utilised in BICM. For a rate-k/n code there are k generator polynomials, each having n coefficients. For example, $\mathbf{g_i} = (g^0, g^1, \ldots, g^n)$, $i \leq k$, is the generator polynomial associated with generating the ith information bit. The generator matrix of the encoder seen in Figure 9.16 is:

$$\mathbf{G(D)} = \begin{bmatrix} 1 & D & 1+D \\ D^2 & 1 & 1+D+D^2 \end{bmatrix},$$ (9.28)

while the equivalent polynomial expressed in octal form is given by:

$$\mathbf{g_1} = \begin{bmatrix} 4 & 2 & 6 \end{bmatrix} \quad \mathbf{g_2} = \begin{bmatrix} 1 & 4 & 7 \end{bmatrix}.$$ (9.29)

Observe in Table 9.3 that Paaske generated codes of rate-1/2, 2/3 and 3/4, but not 5/6. In order to study rate-5/6 BICM/64QAM, we created the required punctured code from the rate-1/2 code of Table 9.3. Table 9.4 summarises the parameters of the Rate-Compatible Punctured

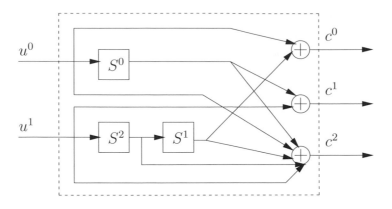

Figure 9.18: Paaske's non-systematic convolutional encoder [96].

State	Information Word $u = (u^1,\ u^0)$			
$S = (S^2, S^1, S^0)$	$00 = 0$	$01 = 1$	$10 = 2$	$11 = 3$
$000 = 0$	$000 = 0$	$101 = 5$	$110 = 6$	$011 = 3$
$001 = 1$	$110 = 6$	$011 = 3$	$000 = 0$	$101 = 5$
$010 = 2$	$101 = 5$	$000 = 0$	$011 = 3$	$110 = 6$
$011 = 3$	$011 = 3$	$110 = 6$	$101 = 5$	$000 = 0$
$100 = 4$	$100 = 4$	$001 = 1$	$010 = 2$	$111 = 7$
$101 = 5$	$010 = 2$	$111 = 7$	$100 = 4$	$001 = 1$
$110 = 6$	$001 = 1$	$100 = 4$	$111 = 7$	$010 = 2$
$111 = 7$	$111 = 7$	$010 = 2$	$001 = 1$	$100 = 4$
Codeword $c = (c^2,\ c^1, c^0)$				
$000 = 0$	$000 = 0$	$001 = 1$	$100 = 4$	$101 = 5$
$001 = 1$	$000 = 0$	$001 = 1$	$100 = 4$	$101 = 5$
$010 = 2$	$000 = 0$	$001 = 1$	$100 = 4$	$101 = 5$
$011 = 3$	$000 = 0$	$001 = 1$	$100 = 4$	$101 = 5$
$100 = 4$	$010 = 2$	$011 = 3$	$110 = 6$	$111 = 7$
$101 = 5$	$010 = 2$	$011 = 3$	$110 = 6$	$111 = 7$
$110 = 6$	$010 = 2$	$011 = 3$	$110 = 6$	$111 = 7$
$111 = 7$	$010 = 2$	$011 = 3$	$110 = 6$	$111 = 7$
Next State $S = (S^2,\ S^1,\ S^0)$				

Table 9.5: The codeword generation and state transition table of the non-systematic convolutional encoder of Figure 9.18. The state transition diagram is seen in Figure 9.19.

Convolutional (RCPC) codes that can be used in rate=5/6 BICM/64QAM schemes. Specifically, rate-1/2 codes were punctured according to the puncturing matrix of Table 9.4 in order to obtain the rate-5/6 codes, following the approach of [233, 234]. Let us now consider the operation of BICM with the aid of an example.

9.5.2   BICM Coding Example

Considering Paaske's eight-state convolutional code [96] in Figure 9.18 as an example, the BICM encoding process is illustrated here. The corresponding generator polynomial is shown in Equation 9.29. A two-bit information word, namely $u = (u^1, \ u^0)$, is encoded in each cycle in order to form a three-bit codeword, $c = (c^2, \ c^1, \ c^0)$. The encoder has three shift registers, namely S^0, S^1 and S^2, as shown in the figure. The three-bit binary contents of these registers represent eight states, as follows:

$$S = (S^2, \ S^1, \ S^0) \in \{000, \ 001, \ \ldots, \ 111\} = \{0, \ 1, \ \ldots, \ 7\}. \tag{9.30}$$

The input sequence, u, generates a new state S and a new codeword c at each encoding cycle. Table 9.5 illustrates the codewords generated and the associated state transitions. The encoding process can also be represented with the aid of the trellis diagram of Figure 9.19. Specifically, the top part of Table 9.19 contains the codewords $c = (c^2, \ c^1, c^0)$ as a function of the encoder state $S = (S^2, S^1, S^0)$ as well as that of the information word $u = (u^1, \ u^0)$, while the bottom section contains the next states, again as a function of S and u. For example, if the input is $u = (u^1, \ u^0) = (1, 1) = 3$ when the shift register is in state $S = (S^2, S^1, S^0) = (1, 1, 0) = 6$, the shift register will change its state to state $S = (S^2, S^1, S^0) = (1, 1, 1) = 7$ and $c = (c^2, \ c^1, c^0) = (0, 1, 0) = 2$ will be the generated codeword. Hence, if the input binary sequence is $\{01\ 10\ 01\ 00\ 10\ 10 \rightarrow\}$ with the rightmost being the first input bit, the corresponding information words are $\{1\ 2\ 1\ 0\ 2\ 2 \rightarrow\}$. Before any decoding takes place, the shift register is initialised to zero. Therefore, as seen at the right of Figure 9.19, when the first information word of $u_1 = 2$ arrives, the state changes from $S^{-1} = 0$ to $S = 4$, generating the first codeword $c_1 = 6$ as seen in the bottom and top sections of Table 9.5, respectively. Then the second information word of $u_2 = 2$ changes the state from $S^{-1} = 4$ to $S = 6$, generating the second codeword of $c_2 = 2$. The process continues in a similiar manner according to the transition table, namely Table 9.5. The codewords generated as seen at the right of Figure 9.19 are $\{4\ 0\ 0\ 1\ 2\ 6 \rightarrow\}$, and the state transitions are $\{2 \leftarrow 4 \leftarrow 1 \leftarrow 2 \leftarrow 6 \leftarrow 4 \leftarrow 0\}$. Then the bits constituting the codeword sequence are interleaved by the three bit interleavers of Figure 9.16, before they are assigned to the corresponding 8PSK constellation points.

9.6 Bit-Interleaved Coded Modulation Using Iterative Decoding

BICM using Iterative Decoding (BICM-ID) was proposed by Li [213, 214] for further improving the FED of Zehavi's BICM scheme, although BICM already improved the diversity order of Ungerböck's TCM scheme. This FED improvement can be achieved with the aid of combining SP-based constellation labelling, as in TCM, and by invoking soft-decision feedback from the decoder's output to the demodulator's input, in order to exchange soft-decision-based information between them. As we will see below, this is advantageous, since upon each iteration the channel decoder improves the reliability of the soft information passed to the demodulator.

9.6.1 Labelling Method

Let us now consider the mapping of the interleaved bits to the phasor constellation in this section. Figure 9.20 shows the process of subset partitioning for each of the three bit positions for both Gray labelling and in the context of SP labelling. The shaded regions shown inside the circle correspond to the subset $\chi(i, 1)$ defined in Equation 9.36, and the unshaded regions to $\chi(i, 0)$, $i = 0, 1, 2$, where i indicates the bit position in the three-bit BICM/8PSK symbol. These are also the decision regions for each bit, if hard-decision-based BICM demodulation is used for detecting each bit individually. The two labelling methods seen in Figure 9.20 have the same

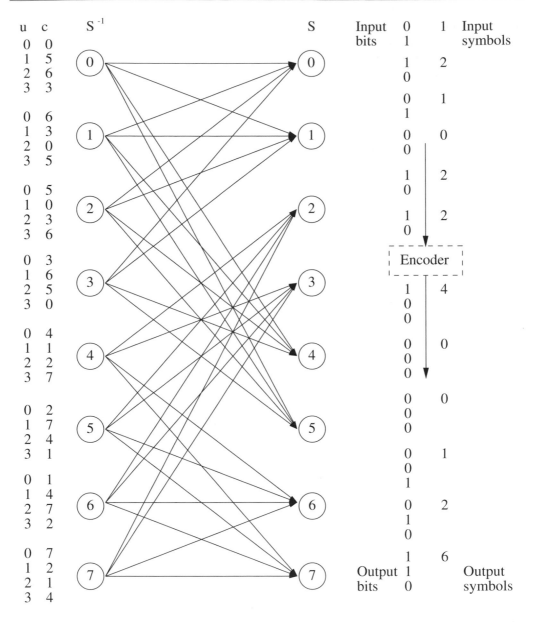

Figure 9.19: Trellis diagram for Paaske's eight-state convolutional code, where u indicates the information word, c indicates the codeword, S^{-1} indicates the previous state and S indicates the current state. As an example, the encoding of the input bit sequence of $\{011001001010 \rightarrow\}$ is shown at the right. The encoder schematic is portrayed in Figure 9.18, while the state transitions are summarised in Table 9.5.

intersubset distances, although a different number of nearest neighbours. For example, $\chi(0, 1)$, which denotes the region where bit 0 equals to 1, is divided into two regions in the context of Gray labelling, as can be seen in Figure 9.20(a). By contrast, in the context of SP labelling seen in Figure 9.20(b), $\chi(0, 1)$ is divided into four regions. Clearly, Gray labelling has a lower number of nearest neighbours compared to SP-based labelling. The higher the number of nearest

(Bit 2, Bit 1, Bit 0)

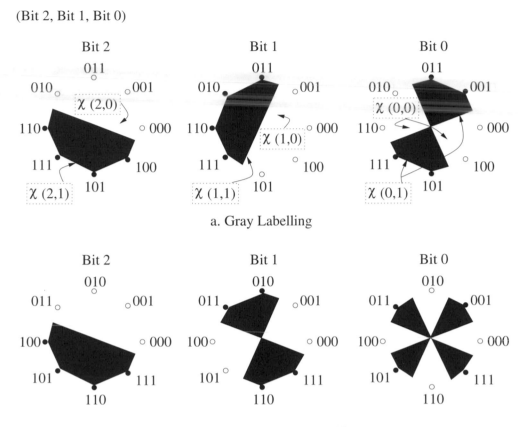

a. Gray Labelling

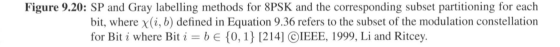

b. Set Partitioning Based Labelling

Figure 9.20: SP and Gray labelling methods for 8PSK and the corresponding subset partitioning for each bit, where $\chi(i, b)$ defined in Equation 9.36 refers to the subset of the modulation constellation for Bit i where Bit $i = b \in \{0, 1\}$ [214] ©IEEE, 1999, Li and Ritcey.

neighbours, the higher the chances for a bit to be decoded into the wrong region. Hence, Gray labelling is a more appropriate mapping during the first decoding iteration, and hence it was adopted by the non-iterative BICM scheme of Figure 9.17.

During the second decoding iteration in BICM-ID, given the feedback information representing the original uncoded information bits in Figure 9.16, namely Bit 1 and Bit 2, the constellation associated with Bit 0 is confined to a pair of constellation points, as shown at the right of Figure 9.21. Therefore, as far as Bit 0 is concerned, the 8PSK phasor constellation is translated into four binary constellations, where one of the four possible specific BPSK constellations is selected by the feedback Bit 1 and Bit 2. The same is true for the constellations associated with both Bit 1 and Bit 2, given the feedback information of the corresponding other two bits.

In order to optimise the second-pass decoding performance of BICM-ID, one must maximise the minimum Euclidean distance between any two points of all the $2^{m-1} = 4$ possible phasor pairs at the left (Bit 2), centre (Bit 1) and the right (Bit 0) of Figure 9.21. Clearly, SP-based labelling serves this aim better, when compared to Gray labelling, since the corresponding minimum Euclidean distance of SP-based labelling is higher than that of Gray labelling for both Bit 1 and Bit 2, as illustrated at the left and the centre of Figure 9.21. Although the first-pass

(Bit 2, Bit 1, Bit 0)

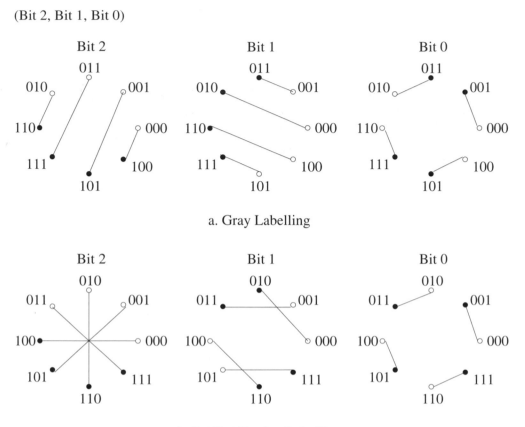

a. Gray Labelling

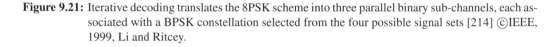

b. Set Partitioning Labelling

Figure 9.21: Iterative decoding translates the 8PSK scheme into three parallel binary sub-channels, each associated with a BPSK constellation selected from the four possible signal sets [214] ©IEEE, 1999, Li and Ritcey.

performance is important, in order to prevent error precipitation due to erroneous feedback bits, the error propagation is effectively controlled by the soft feedback of the decoder. Therefore, BICM-ID assisted by soft decision feedback uses SP labelling.

Specifically, the desired high Euclidean distance for Bit 2 in Figure 9.21(b) is only attainable when Bit 1 and Bit 0 are correctly decoded and fed back to the SP-based demodulator. If the values to be fed back are not correctly decoded, the desired high Euclidean distance will not be achieved and error propagation will occur. On the other hand, an optimum convolutional code having a high binary Hamming distance is capable of providing a high reliability for the decoded bits. Therefore, an optimum convolutional code using appropriate signal labelling is capable of 'indirectly' translating the high binary Hamming distance between coded bits into a high Euclidean distance between the phasor pairs portrayed in Figure 9.21. In short, BICM-ID converts a 2^m-ary signalling scheme to $\bar{m}$ independent parallel binary schemes by the employment of $\bar{m}$ number of independent bit interleavers and involves an iterative decoding method. This simultaneously facilitates attaining a high diversity order with the advent of the bit interleavers, as well as achieving a high FED with the aid of the iterative decoding and SP-based labelling. Hence, BICM-ID effectively combines powerful binary codes with bandwidth-efficient modulation.

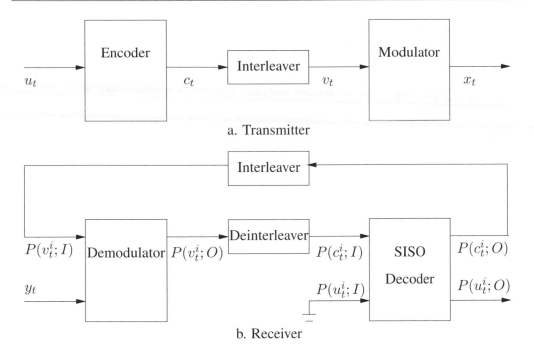

Figure 9.22: The transmitter and receiver modules of the BICM-ID scheme using soft-decision feedback [213] ©IEEE, 1998, Li.

9.6.2   Interleaver Design

The interleaver design is important as regards the performance of BICM-ID. In [215], Li introduced certain constraints on the design of the interleaver, in order to maximise the minimum Euclidean distance between the two points in the $2^{\bar{m}-1}$ possible specific BPSK constellations. However, we advocate a more simple approach, where the $\bar{m}$ number of interleavers used for the $2^{\bar{m}}$-ary modulation scheme are generated randomly and separately, without any interactions between them. The resultant minimum Euclidean distance is less than that of the scheme proposed in [215], but the error bursts inflicted by correlated fading are expected to be randomised effectively by the independent bit interleavers. This was expected to give a better performance over fading channels at the cost of a slight performance degradation over AWGN channels, when compared to Li's scheme [215]. However, as we will demonstrate in the context of our simulation results in Section 9.7.2.2, our independent random interleaver design and Li's design perform similarly.

Having described the labelling method and the interleaver design in the context of BICM-ID, let us now consider the operation of BICM-ID with the aid of an example.

9.6.3   BICM-ID Coding Example

The BICM-ID scheme using soft-decision feedback is shown in Figure 9.22. The interleavers used are all bit-based, as in the BICM scheme of Figure 9.16, although for the sake of simplicity here only one interleaver is shown. A Soft-Input Soft-Output (SISO) [235] decoder is used in the receiver module and the decoder's output is fed back to the input of the demodulator. The SISO decoder of the BICM-ID scheme is actually a MAP decoder that computes the a-posteriori probabilities for the non-systematic channel-coded bits and the original information bits.

For an (n, k) binary convolutional code the encoder's input symbol at time t is denoted by $u_t = [u_t^0, u_t^1, \ldots, u_t^{k-1}]$ and the coded output symbol by $c_t = [c_t^0, c_t^1, \ldots, c_t^{n-1}]$, where u_t^i or c_t^i is the ith bit in a symbol as defined in the context of Table 9.5 and Figure 9.19. The coded bits are interleaved by $\bar{m}$ independent bit interleavers, then $\bar{m}$ interleaved bits are grouped together in order to form a channel symbol $v_t = [v_t^0, v_t^1, \ldots, v^{\bar{m}-1}]$ as seen in Figure 9.22(a), for transmission using $2^{\bar{m}}$-ary modulation. Let us consider 8PSK modulation, i.e. $\bar{m} = 3$ as an example.

A signal labelling method μ maps the symbol v_t to a complex phasor according to $x_t = \mu(v_t)$, $x_t \in \chi$, where the 8PSK signal set is defined as $\chi = \{\sqrt{E_s}\, e^{j2n\pi/8},\ n = 0, \ldots, 7\}$ and E_s is the energy per transmitted symbol. In conjunction with a rate-$2/3$ code, the energy per information bit is $E_b = E_s/2$. For transmission over Rayleigh fading channels using coherent detection, the received discrete time signal is:

$$y_t = \rho_t x_t + n_t, \tag{9.31}$$

where ρ_t is the Rayleigh-distributed fading amplitude [49] having an expectation value of $E(\rho_t^2) = 1$, while n_t is the complex AWGN exhibiting a variance of $\sigma^2 = N_0/2$ where N_0 is the noise's PSD. For AWGN channels we have $\rho_t = 1$ and the Probability Density Function (PDF) of the non-faded but noise-contaminated received signal is expressed as [226]:

$$P(y_t | x_t, \rho_t) = \frac{1}{2\pi\sigma^2}\, e^{-\frac{1}{2}\left(\frac{n_t}{\sigma}\right)^2}, \tag{9.32}$$

where $\sigma^2 = N_0/2$ and the constant multiplicative factor of $\frac{1}{2\pi\sigma^2}$ does not influence the shape of the distribution and hence can be ignored when calculating the branch transition metric η, as described in Section 9.3.3. For AWGN channels, the conditional PDF of the received signal can be written as:

$$P(y_t | x_t) = e^{-\frac{|y_t - x_t|^2}{2\sigma^2}}. \tag{9.33}$$

Considering AWGN channels, the demodulator of Figure 9.22(b) takes y_t as its input for computing the confidence metrics of the bits using the maximum APP criterion [219]:

$$P(v_t^i = b | y_t) = \sum_{x_t \in \chi(i,b)} P(x_t | y_t), \tag{9.34}$$

where $i \in \{0, 1, 2\}$, $b \in \{0, 1\}$ and $x_t = \mu(v_t)$. Furthermore, the signal after the demodulator of Figure 9.22 is described by the demapping of the bits $[\nabla^0(x_t), \nabla^1(x_t), \nabla^2(x_t)]$ where $\nabla^i(x_t) \in \{0, 1\}$ is the value of the ith bit of the three-bit label assigned to x_t. With the aid of Bayes' rule in Equation 5.12 we obtain:

$$P(v_t^i = b | y_t) = \sum_{x_t \in \chi(i,b)} P(y_t | x_t) P(x_t), \tag{9.35}$$

where the subset $\chi(i, b)$ is described as:

$$\chi(i,b) = \{\mu([\nabla^0(x_t), \nabla^1(x_t), \nabla^2(x_t)]) \mid \nabla^j(x_t) \in \{0,1\}, j \neq i\}, \tag{9.36}$$

which contains all the phasors for which $\nabla^i(x_t) = b$ holds. For 8PSK, where $\bar{m} = 3$, the size of each such subset is $2^{\bar{m}-1} = 4$ as portrayed in Figure 9.20. This implies that only the a-priori probabilities of $\bar{m} - 1 = 2$ bits out of the total of $\bar{m} = 3$ bits per channel symbol have to be considered, in order to compute the bit metric of a particular bit.

Now using the notation of Benedetto *et al.* [235], the a-priori probabilities of an original uncoded information bit at time index t and bit index i, namely u_t^i being 0 and 1, are denoted by $P(u_t^i = 0; I)$ and $P(u_t^i = 1; I)$ respectively, while I refers to the a-prIori probabilities of the bit. This notation is simplified to $P(u_t^i; I)$, when no confusion arises, as shown in Figure 9.22. Similarly, $P(c_t^i; I)$ denotes the a-priori probabilities of a legitimate coded bit at time index t and position index i. Finally, $P(u_t^i; O)$ and $P(c_t^i; O)$ denote the extrinsic a-pOsteriori information of the original information bits and coded bits, respectively.

The a-priori probability $P(x_t)$ in Equation 9.35 is unavailable during the first-pass decoding, hence an equal likelihood is assumed for all the $2^{\overline{m}}$ legitimate symbols. This renders the extrinsic a-posteriori bit probabilities, $P(v_t^i = b; O)$, equal to $P(v_t^i = b|y_t)$, when ignoring the common constant factors. Then, the SISO decoder of Figure 9.22(b) is used for generating the extrinsic a-posteriori bit probabilities $P(u_t^i; O)$ of the information bits, as well as the extrinsic a-posteriori bit probabilities $P(c_t^i; O)$ of the coded bits, from the de-interleaved probabilities $P(v_t^i = b; O)$, as seen in Figure 9.22(b). Since $P(u_t^i; I)$ is unavailable, it is not used in the entire decoding process.

During the second iteration $P(c_t^i; O)$ is interleaved and fed back to the input of the demodulator in the correct order in the form of $P(v_t^i; I)$, as seen in Figure 9.22(b). Assuming that the probabilities $P(v_t^0; I)$, $P(v_t^1; I)$ and $P(v_t^2; I)$ are independent by the employment of three independent bit interleavers, we have for each $x_t \in \chi$:

$$
\begin{aligned}
P(x_t) &= P(\mu([\nabla^0(x_t), \nabla^1(x_t), \nabla^2(x_t)])) \\
&= \prod_{j=0}^{2} P(v_t^j = \nabla^j(x_t); I),
\end{aligned}
\tag{9.37}
$$

where $\nabla^j(x_t) \in \{0, 1\}$ is the value of the jth bit of the three-bit label for x_t. Now that we have the a-priori probability $P(x_t)$ of the transmitted symbol x_t, the extrinsic a-posteriori bit probabilities for the second decoding iteration can be computed using Equations 9.35 and 9.37, yielding:

$$
\begin{aligned}
P(v_t^i = b; O) &= \frac{P(v_t^i = b|y_t)}{P(v_t^i = b; I)} \\
&= \sum_{x_t \in \chi(i,b)} \left(P(y_t|x_t) \prod_{j \neq i} P(v_t^j = \nabla^j(x_t); I) \right) \\
& \quad i \in \{0, 1, 2\}, \quad b \in \{0, 1\}.
\end{aligned}
\tag{9.38}
$$

As seen from Equation 9.38, in order to recalculate the metric for a bit we only need the a-priori probabilities of the other two bits in the same channel symbol. After interleaving in the feedback loop of Figure 9.22, the regenerated bit metrics are tentatively soft demodulated again and the process of passing information between the demodulator and decoder is continued. The final decoded output is the hard-decision-based *extrinsic* bit probability $P(u_t^i; O)$.

So far in Sections 9.2–9.6 we have studied the conceptual differences between four coded modulation schemes in terms of their coding structure, signal labelling philosophy, interleaver type and decoding philosophy. The symbol-based non-binary MAP algorithm was also highlighted, when operating in the log-domain. In the next section we will proceed to study the performance of TCM, BICM, TTCM and BICM-ID when communicating over both narrowband and wideband channels.

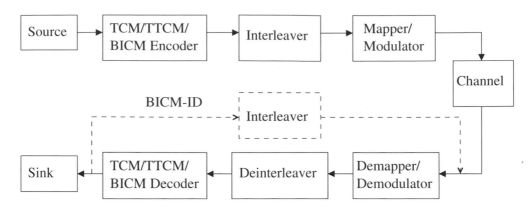

Figure 9.23: System overview of different coded modulation schemes.

9.7   Coded Modulation Performance

9.7.1   Introduction

Having described the principles of TCM, BICM, TTCM and BICM-ID in Sections 9.2–9.6, in this section their performance will be evaluated for transmission over both narrowband and wideband fading channels. Specifically, in Section 9.7.2 we will evaluate the performance of these coded modulation schemes for transmissions over narrowband channels, while in Section 9.7.3 we will consider their performance in the context of wideband channels.

Owing to the Inter-Symbol Interference (ISI) inflicted by wideband channels, the employment of equalisers is essential in assisting the operation of the coded modulation schemes considered. Hence a Decision Feedback Equaliser (DFE) is introduced in Section 9.7.3.2, while Section 9.7.3.3 will evaluate the performance of a DFE-aided wideband burst-by-burst adaptive coded modulation system. Another approach to overcoming the ISI in wideband channels is the employment of a multi-carrier Orthogonal Frequency Division Multiplexing (OFDM) system. Hence, OFDM is studied in Section 9.7.3.4, while Section 9.7.3.5 evaluates the performance of an OFDM-assisted coded modulation scheme.

9.7.2   Coded Modulation in Narrowband Channels

In this section, a comparative study of TCM, TTCM, BICM and BICM-ID schemes over both Gaussian and uncorrelated narrowband Rayleigh fading channels is presented in the context of eight-level Phase Shift Keying (PSK), 16-level Quadrature Amplitude Modulation (QAM) and 64QAM. We comparatively study the associated decoding complexity, the effects of the encoding block length and the achievable bandwidth efficiency. It will be shown that TTCM constitutes the best compromise scheme, followed by BICM-ID.

9.7.2.1   System Overview

The schematic of the coded modulation schemes under consideration is shown in Figure 9.23. The source generates random information bits, which are encoded by one of the TCM, TTCM or BICM encoders. The coded sequence is then appropriately interleaved and used for modulating the waveforms according to the symbol mapping rules. For a narrowband Rayleigh fading channel in conjunction with coherent detection, the relationship between the transmitted discrete

Rate	State	$\tilde{\mathsf{m}}$	H^0	H^1	H^2	H^3
2/3	8	2	11	02	04	-
(8PSK)	64 *	2	103	30	66	-
3/4	8	3	11	02	04	10
(16QAM)	64 *	3	101	16	64	-
5/6	8	2	11	02	04	-
(64QAM)	64 *	2	101	16	64	-

Table 9.6: 'Punctured' TCM codes with best minimum distance for PSK and QAM, ©Robertson and Wörz [211]. '*' indicates Ungerböck's TCM codes [208]. Two-dimensional ($2\bar{D}$) modulation is utilised. Octal format is used for representing the generator polynomials H^i and $\tilde{\mathsf{m}}$ denotes the number of coded information bits out of the total m information bits in a modulated symbol.

time signal x_t and the received discrete time signal y_t is given by:

$$y_t = \rho_t x_t + n_t, \tag{9.39}$$

where ρ_t is the Rayleigh-distributed fading amplitude having an expected value of $\mathrm{E}(\rho_t^2) = 1$, while n_t is the complex AWGN having a variance of $\sigma^2 = N_0/2$ where N_0 is the noise's PSD. For AWGN channels we have $\rho_t = 1$. The receiver consists of a coherent demodulator followed by a de-interleaver and one of the TCM, TTCM or BICM decoders. TTCM schemes consist of two component TCM encoders and two parallel decoders. In BICM-ID schemes the decoder output is appropriately interleaved and fed back to the demodulator input, as shown in Figure 9.23.

The log-domain branch metric required for the maximum likelihood decoding of TCM and TTCM over fading channels is given by the squared Euclidean distance between the faded transmitted symbol x_t and the noisy received symbol y_t, which is formulated as:

$$\pi_t = |y_t - \rho_t x_t|^2. \tag{9.40}$$

By contrast, the corresponding branch metric for BICM and BICM-ID is formed by summing the de-interleaved bit metrics λ of each coded bit v_t^i which quantifies the reliability of the corresponding symbol, yielding:

$$\pi_t = \sum_{i=0}^{\mathsf{m}} \lambda(v_t^i = b), \tag{9.41}$$

where i is the bit position of the coded bit in a constellation symbol, m is the number of information bits per symbol and $b \in (0,1)$. The number of coded bits per symbol is $(\mathsf{m}+1)$, since the coded modulation schemes add one parity bit to the m information bits by doubling the original constellation size, in order to maintain the same spectral efficiency of m bits/s/Hz. The BICM bit metrics $\tilde{\lambda}$ before the de-interleaver are defined as [214]:

$$\tilde{\lambda}(v_t^i = b) = \sum_{x \in \chi(i,b)} |y_t - \rho_t x|^2, \tag{9.42}$$

where $\chi(i,b)$ is the signal set, for which the bit i of the symbol has a binary value b.

To elaborate a little further, the coded modulation schemes that we comparatively studied are Ungerböck's TCM [208], Robertson's TTCM [211], Zehavi's BICM [212] and Li's BICM-ID [219]. Table 9.6 shows the generator polynomials of both the TCM and TTCM codes in octal

Rate	State	g^1	g^2	g^3	g^4	d_{free}
2/3	8	4	2	6	-	4
(8PSK)	(K=3)	1	4	7	-	
	16	7	1	4	-	5
	(K=4)	2	5	7	-	
	64	15	6	15	-	7
	(K=6)	6	15	17	-	
3/4	8	4	4	4	4	4
(16QAM)	(K=3)	0	6	2	4	
		0	2	5	5	
	32	6	2	2	6	5
	(K=5)	1	6	0	7	
		0	2	5	5	

Rate	State	g^1	g^2	Puncturing	d_{free}
5/6	8	15	17	1 0 0 1 0	3
(64QAM)	(K=3)			0 1 1 1 1	
	64	133	171	1 1 1 1 1	3
	(K=6)			1 0 0 0 0	

Table 9.7: Top table shows the generator polynomials of Paaske's code, p. 331 of [96]. Bottom table shows those of the rate-compatible puncture convolutional codes [234]. K is the code memory and d_{free} is the free Hamming distance. Octal format is used for the polynomial coefficients g^i, while '1' and '0' in the puncturing matrix indicate the position of the unpunctured and punctured coded bits, respectively.

format. These are RSC codes that add one parity bit to the information bits. Hence, the coding rate of a 2^{m+1}-ary PSK or QAM signal is $R = \frac{m}{m+1}$. The number of decoding states associated with a code of memory K is 2^K. When the number of protected/coded information bits $\tilde{m}$ is less than the total number of original information bits m, there are $(m - \tilde{m})$ uncoded information bits and $2^{m-\tilde{m}}$ parallel transitions in the trellis of the code. Parallel transitions assist in reducing the decoding complexity and the memory required, since the dimensionality of the corresponding trellis is smaller than that of a trellis having no parallel branches.

Table 9.7 shows the generator polynomials for the BICM and BICM-ID codes in octal format. These codes are non-systematic convolutional codes having a maximum free Hamming distance. Again, only one extra bit is added to the information bits. Hence, the achievable coding rate and the bandwidth efficiency are similar to that of TCM and TTCM for the 2^{m+1}-ary modulation schemes used. In order to reduce the required decoding memory, the BICM and BICM-ID schemes based on 64QAM were obtained by puncturing the rate-1/2 codes following the approach of [234], since for a non-punctured rate-5/6 code there are $2^{(m=5)} = 32$ branches emerging from each trellis state for a block length of $\bar{L}$, whereas for the punctured rate-1/2 code, there are only $2^{(m=1)} = 2$ branches emerging from each trellis state for a block length of $m\bar{L} = 5\bar{L}$. Therefore the required decoding memory is reduced by a factor of $\frac{2^m \cdot \bar{L}}{2^1 \cdot m \cdot L} = 3.2$.

Soft-decision trellis decoding utilising the Log-Maximum A-Posteriori (Log-MAP) algorithm [52] was invoked for the decoding of the coded modulation schemes. As discussed in Section 9.3.3, the Log-MAP algorithm is a numerically stable version of the MAP algorithm operating in the log-domain, in order to reduce its complexity and to mitigate the numerical problems associated with the MAP algorithm [11].

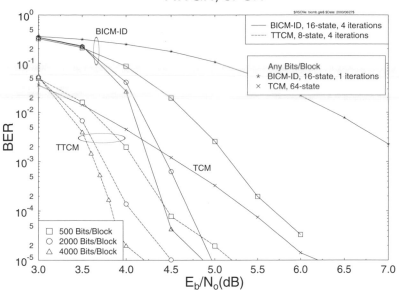

Figure 9.24: Effects of block length on the TCM, TTCM and BICM-ID performance in the context of an 8PSK scheme for transmissions over AWGN channels.

9.7.2.2   Simulation Results and Discussions

In this section we study the performance of TCM, TTCM, BICM and BICM-ID using computer simulations. The complexity of the coded modulation schemes is compared in terms of the number of decoding states, and the number of decoding iterations. For a TCM or BICM code of memory K, the corresponding complexity is proportional to the number of decoding states namely to $S = 2^K$. Since TTCM schemes invoke two component TCM codes, a TTCM code with t iterations and using an S-state component code exhibits a complexity proportional to $2.t.S$ or $t.2^{K+1}$. As for BICM-ID schemes, only one decoder is used but the demodulator is invoked in each decoding iteration. However, the complexity of the demodulator is assumed to be insignificant compared to that of the channel decoder. Hence, a BICM-ID code with t iterations using an S-state code exhibits a complexity proportional to $t.S$ or $t.2^K$.

9.7.2.2.1   Coded Modulation Performance over AWGN Channels
It is important to note that in terms of the total number of trellis states the decoding complexity of 64-state TCM and 8-state TTCM using two TCM decoders in conjunction with four iterations can be considered similar. The same comments are valid also for 16-state BICM-ID using four iterations or for 8-state BICM-ID using eight iterations. In our forthcoming discourse we will always endeavour to compare schemes of similar decoding complexity, unless otherwise stated. Figure 9.24 illustrates the effects of interleaving block length on the TCM, TTCM and BICM-ID performance in an 8PSK scheme over AWGN channels. It is clear from the figure that a high interleaving block length is desired for the iterative TTCM and BICM-ID schemes. The block length does not affect the BICM-ID performance during the first pass, since it constitutes a BICM scheme using SP-based phasor labelling. However, if we consider four iterations, the performance improves,

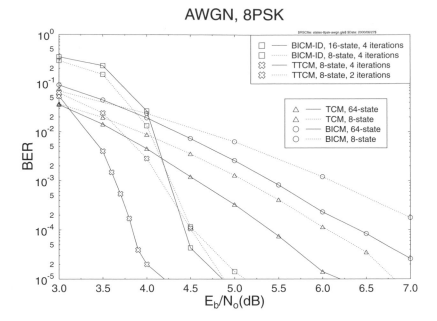

Figure 9.25: Effects of decoding complexity on the TCM, TTCM, BICM and BICM-ID schemes' performance in the context of an 8PSK scheme for transmissions over AWGN channels using a
block length of 4000 information bits (2000 symbols).

converging faster to the Error-Free-Feedback (EFF) bound[1] [214] for larger block lengths. At a
BER of 10^{-4} a 500-bit block length was about 1 dB inferior in terms of the required SNR to the
2000-bit block length in the context of the BICM-ID scheme. A slight further SNR improvement
was obtained for the 4000-bit block length. In other words, the advantage of BICM-ID over TCM
for transmissions over AWGN channels is more significant for larger block lengths. The 8-state
TTCM performance also improves, when using four iterations, as the block length is increased
and, on the whole, TTCM is the best performer in this scenario.

Figure 9.25 shows the effects of the decoding complexity on the TCM, TTCM, BICM and
BICM-ID schemes' performance in the context of an 8PSK scheme for transmissions over
AWGN channels using a block length of 4000 information bits (2000 symbols). Again, the
64-state TCM, 64-state BICM, 8-state TTCM using four iterations and 16-state BICM-ID along
with four iterations exhibit a similar decoding complexity. At a BER of 10^{-4}, TTCM requires
about 0.6 dB lower SNR than BICM-ID, 1.6 dB less energy than TCM and 2.5 dB lower SNR
than BICM. When the decoding complexity is reduced such that 8-state codes are used in the
TCM, BICM and BICM-ID schemes, their corresponding performance becomes worse than that
of the 64-state codes, as shown in Figure 9.25. In order to be able to compare the associated
performance with that of 8-state BICM-ID using four iterations, 8-state TTCM along with two
iterations is employed. Observe that due to the insufficient number of iterations, TTCM exhibits
only marginal advantage over BICM-ID.

Figure 9.26 shows the performance of TCM, TTCM and BICM-ID invoking 16QAM for
transmissions over AWGN channels using a block length of 6000 information bits (2000 symbols). Upon comparing 64-state TCM with 32-state BICM-ID using two iterations, we observed

---

[1]The EFF bound is defined as the BER upper bound performance achieved for the idealised situation, when the
decoded values fed back to the demodulator in Figure 9.23 are error free.

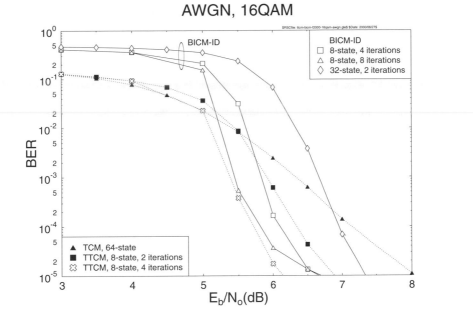

Figure 9.26: Performance comparison of TCM, TTCM and BICM-ID employing 16QAM for transmissions over AWGN channels using a block length of 6000 information bits (2000 symbols).

that BICM-ID outperforms TCM for E_b/N_0 values in excess of 6.8 dB. However, 8-state BICM-ID using an increased number of iterations, such as four or eight, outperforms the similar complexity 32-state BICM-ID scheme employing two iterations as well as 64-state TCM. An approximately 1.2 dB E_b/N_0 gain was obtained at a BER of 10^{-4} for 8-state BICM-ID using eight iterations over 64-state TCM at a similar decoding complexity. Comparing 8-state TTCM using two iterations and 8-state BICM-ID employing four iterations reveals that BICM-ID performs better for the E_b/N_0 range of 5.7 dB to 7 dB. When the number of iterations is increased to four for TTCM and to eight for BICM-ID, TTCM exhibits a better performance, as seen in Figure 9.26.

Owing to the associated SP, the intra-subset distance of TCM and TTCM increases as we traverse down the partition tree of Figure 9.7, for example. It was shown in [211] that we only need to encode $\tilde{m} = 2$ out of $m = 5$ information bits in the 64QAM/TTCM to attain target BERs around 10^{-5} in AWGN channels. Hence in this scenario there are $2^{m-\tilde{m}} = 8$ parallel transitions due to the $m - \tilde{m} = 3$ uncoded information bits in the trellis of 64QAM/TTCM. Figure 9.27 illustrates the performance of TCM, TTCM, BICM and BICM-ID using 64QAM over AWGN channels. When using a block length of 10000 information bits (2000 symbols), 8-state TTCM invoking four iterations is the best candidate, followed by the similar complexity 8-state BICM-ID scheme employing eight iterations. Again, TCM performs better than BICM in AWGN channels. When a block length of 1250 information bits (250 symbols) was used, both TTCM and BICM-ID experienced a performance degradation. It is also seen in Figure 9.27 that BICM-ID performs close to TTCM, when a longer block length is used.

9.7.2.2.2 Performance over Uncorrelated Narrowband Rayleigh Fading Channels The uncorrelated Rayleigh fading channels implied using an infinite-length interleaver over narrowband Rayleigh fading channels. Figure 9.28 shows the performance of 64-state TCM, 64-state

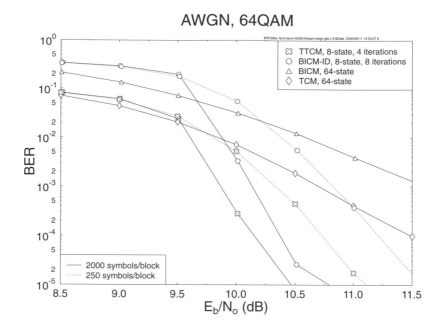

Figure 9.27: Performance comparison of TCM, TTCM, BICM and BICM-ID using 64QAM over AWGN channels.

BICM, 8-state TTCM using four iterations and 16-state BICM-ID employing four iterations in the context of an 8PSK scheme communicating over uncorrelated narrowband Rayleigh fading channels using a block length of 4000 information bits (2000 symbols). These four coded modulation schemes have a similar complexity. As can be seen from Figure 9.28, TTCM performs best, followed by BICM-ID, BICM and TCM. At a BER of 10^{-4}, TTCM performs about 0.7 dB better in terms of the required E_b/N_0 value than BICM-ID, 2.3 dB better than BICM and 4.5 dB better than TCM. The error floor of TTCM [211] was lower than the associated EFF bound of BICM-ID. However, the BERs of TTCM and BICM-ID were identical at $E_b/N_0 = 7$ dB.

Figure 9.29 compares the performance of TCM, TTCM and BICM-ID invoking 16QAM for communicating over uncorrelated narrowband Rayleigh fading channels using a block length of 6000 information bits (2000 symbols). Observe that 32-state BICM-ID using two iterations outperforms 64-state TCM for E_b/N_0 in excess of 9.6 dB. At the same complexity, 8-state BICM-ID invoking eight iterations outperforms 64-state TCM beyond $E_b/N_0 = 8.2$ dB. Similarly to 8PSK, the coding gain of BICM-ID over TCM in the context of 16QAM is more significant over narrowband Rayleigh fading channels compared to AWGN channels. Near E_b/N_0 of 11 dB the 8-state BICM-ID scheme approaches the EFF bound, hence 32-state BICM-ID using two iterations exhibits a better performance due to its lower EFF bound. Observe also that 8-state BICM-ID using four iterations outperforms 8-state TTCM employing two iterations in the range of $E_b/N_0 = 8.5$ dB to 12.1 dB. Increasing the number of iterations only marginally improves the performance of BICM-ID, but results in a significant gain for TTCM. The performance of 8-state TTCM using four iterations is better than that of 8-state BICM-ID along with eight iterations for E_b/N_0 values in excess of 9.6 dB.

Figure 9.30 illustrates the performance of TCM, TTCM, BICM and BICM-ID when invoking 64QAM for communicating over uncorrelated narrowband Rayleigh fading channels. Using a block length of 10000 information bits (2000 symbols), 64-state BICM performs better than

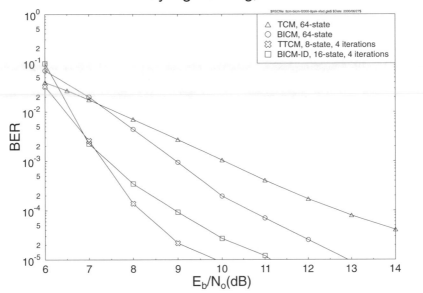

Figure 9.28: Performance comparison of TCM, TTCM, BICM and BICM-ID for 8PSK transmissions over uncorrelated Rayleigh fading channels using a block length of 4000 information bits (2000 symbols).

64-state TCM for E_b/N_0 values in excess of 15 dB. BICM-ID exhibits a lower error floor than TTCM in this scenario, since BICM-ID protects all the five information bits, while TTCM protects only two information bits of the six-bit 64QAM symbol. The three unprotected information bits of TCM and TTCM render these schemes less robust to the bursty error effects of the uncorrelated fading channel. If we use a TCM or TTCM code generator that encodes all the five information bits, a better performance is expected. Reducing the block length from 2000 symbols to 250 symbols resulted in a small performance degradation for TTCM, but yielded a significant degradation for BICM-ID.

9.7.2.2.3   Coding Gain versus Complexity and Interleaver Block Length   In this section, we will investigate the coding gain (G) of the coded modulation schemes utilising an 8PSK scheme versus the Decoding Complexity (DC) and the Interleaver Block Length (IL) at a BER of 10^{-4}. The coding gain G is measured by comparing to the uncoded 4PSK scheme, which exhibits a BER of 10^{-4} at $E_b/N_0 = 8.35$ dB and $E_b/N_0 = 35$ dB for transmissions over AWGN channels and uncorrelated narrowband Rayleigh fading channels, respectively. Again, the DC is measured using the associated number of decoding states and the notations S and t represent the number of decoding states and the number of decoding iterations, respectively. Hence, the relative complexity of TCM, BICM, TTCM and BICM-ID is given by $S, S, 2 \times t \times S$ and $t \times S$, respectively. The IL is measured in terms of the number of information bits in the interleaver.

Figure 9.31 portrays the coding gain G versus DC plot of the coded modulation schemes for 8PSK transmissions over (a) AWGN channels and (b) uncorrelated narrowband Rayleigh fading channels, using an IL of 4000 information bits (2000 symbols). At a DC as low as 8, the non-iterative TCM scheme exhibits the highest coding gain G for transmissions over AWGN

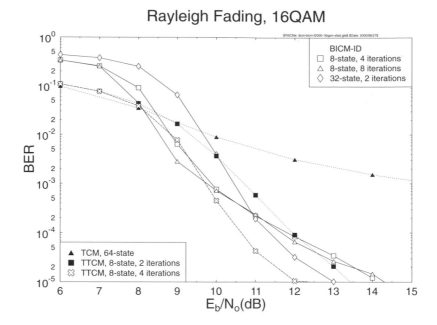

Figure 9.29: Performance comparison of TCM, TTCM and BICM-ID for 16QAM transmissions over uncorrelated narrowband Rayleigh fading channels transmitting 2000 symbols/block (6000 information bits/block).

channels, as seen in Figure 9.31(a). By contrast, the BICM scheme exhibits the highest coding gain G for transmissions over uncorrelated narrowband Rayleigh fading channels, as seen in Figure 9.31(b). However, for a DC higher than 16, the iterative TTCM and BICM-ID schemes exhibit higher coding gains than their non-iterative counterparts for transmission over both channels.

For the iterative schemes different combinations of t and S may yield different performances at the same DC. For example, the coding gain G of BICM-ID in conjunction with $t \cdot S = 8 \times 8$ is better than that of $t \cdot S = 4 \times 16$ at DC=64 for transmissions over AWGN channels, as seen in Figure 9.31(a), since BICM-ID invoking a constituent code associated with $S = 16$ has not reached its optimum performance at iteration $t = 4$. However, the coding gain G of BICM-ID in conjunction with $t \cdot S = 4 \times 16$ is better than that of $t \cdot S = 8 \times 8$ at DC=64, when communicating over uncorrelated narrowband Rayleigh fading channels, as seen in Figure 9.31(b). This is because BICM-ID invoking a constituent code associated with $S = 8$ has reached its EFF bound at iteration $t = 4$, while BICM-ID invoking a constituent code associated with $S = 16$ has not reached its EFF bound, because the EFF bound for code associated with $S = 16$ is lower than that of a code associated with $S = 18$. In general, the coding gain G of TTCM is the highest for DC values in excess of 32 for transmissions over both channels.

Figure 9.32 portrays the coding gain G versus IL plot of the coded modulation schemes for 8PSK transmissions over (a) AWGN channels and (b) uncorrelated narrowband Rayleigh fading channels in conjunction with a DC of 64 both with and without code termination. We can observe in Figure 9.32(a) that IL affects the performance of the schemes using no code termination, since the shorter the IL, the higher the probability for the decoding trellis to end at a wrong state. For transmissions over AWGN channels and upon using code-terminated schemes, only the performance of the BICM-ID scheme is affected by the IL, since the performance of

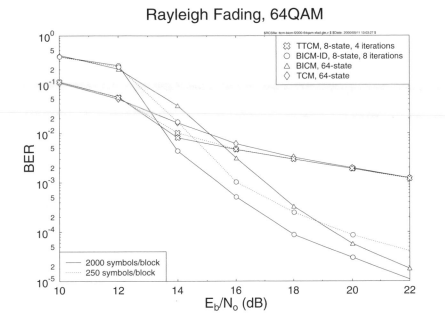

Figure 9.30: Performance comparison of TCM, TTCM, BICM and BICM-ID using 64QAM for transmissions over uncorrelated narrowband Rayleigh fading channels.

the scheme communicating over AWGN channels depends on the FED, while the high FED of BICM-ID depends on the reliability of the feedback values. Therefore, when the IL is short, BICM-ID suffers from a performance degradation. However, the other schemes are not affected by the IL when communicating over AWGN channels, as seen in Figure 9.32(a), since there are no bursty channel errors to be dispersed by the interleaver and hence there is no advantage in utilising a long IL. To elaborate a little further, as seen in Figure 9.32 for transmissions over uncorrelated narrowband Rayleigh fading channels using code-terminated schemes, the IL does not significantly affect the performance of the schemes, since the error events are uncorrelated in the uncorrelated Rayleigh fading scenario. These results constitute the upper bound performance achievable when an infinitely long interleaver is utilised for rendering the error events uncorrelated.

Figure 9.33 portrays the coding gain G versus IL plot of the coded modulation schemes for 8PSK transmissions over correlated narrowband Rayleigh fading channels, in conjunction with a decoding complexity of 64, when applying code termination. The normalised Doppler frequency of the channel is 3.25×10^{-5}, which corresponds to a Baud rate of 2.6 MBaud, a carrier frequency of 1.9 GHz and a vehicular speed of 30 mph. This is a slow fading channel and hence the fading envelope is highly correlated. It is demosntrated by Figure 9.33 that the coding gain G of all coded modulation schemes improves as the IL increases. This is because the MAP decoder is unable to perform at its best when the channel errors occur in bursts. However, the performance improves when the error bursts are dispersed by the employment of a long interleaver. In general, TTCM is the best performer for a variety of IL values. However, BICM-ID is the worst performer for an IL of 4000 bits, while performing similarly to TTCM for long IL values.

On one hand, TCM performs better than BICM for short IL values, which follows the performance trends observed for transmissions over AWGN channels, as shown in Figure 9.32(a). This is because slowly fading channels are highly correlated and hence they behave as near-Gaussian

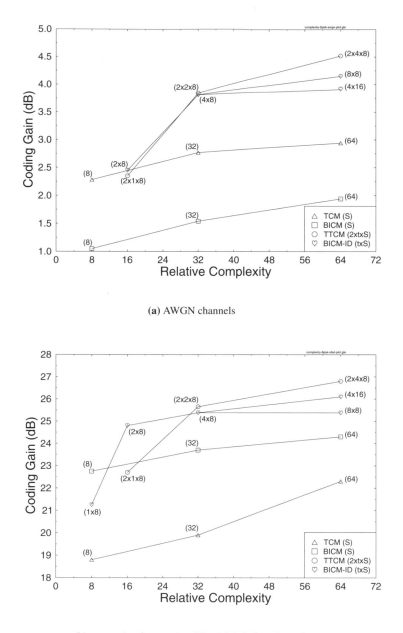

(a) AWGN channels

(b) uncorrelated narrowband Rayleigh fading channels

Figure 9.31: Coding gain at a BER of 10^{-4} over the uncoded 4PSK scheme, against the decoding complexity of TCM, TTCM, BICM and BICM-ID for 8PSK transmissions over (a) AWGN channels and (b) uncorrelated narrowband Rayleigh fading channels, using an interleaver block length of 4000 information bits (2000 symbols). The notations S and t represent the number of decoding states and the number of decoding iterations, respectively.

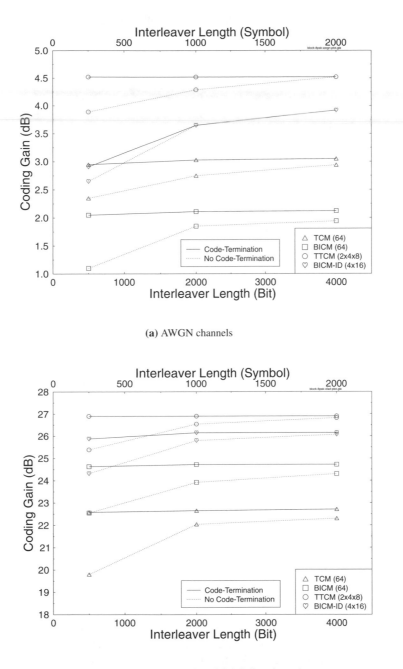

(a) AWGN channels

(b) uncorrelated narrowband Rayleigh fading channels

Figure 9.32: Coding gain at a BER of 10^{-4} over the uncoded 4PSK scheme, against the IL of TCM, TTCM, BICM and BICM-ID for 8PSK transmissions over (a) AWGN channels and (b) uncorrelated narrowband Rayleigh fading channels, invoking a DC of 64 applying code termination or no code termination.

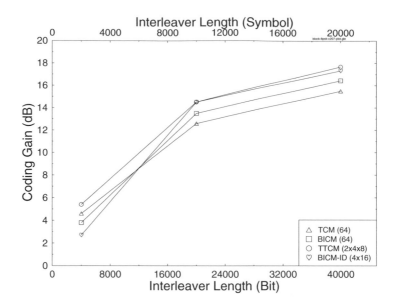

Figure 9.33: Coding gain at a BER of 10^{-4} over the uncoded 4PSK scheme, against the IL of TCM, TTCM, BICM and BICM-ID for 8PSK transmissions over correlated narrowband Rayleigh fading channels, invoking a DC of 64 applying code termination. The normalised Doppler frequency of the channel is 3.25×10^{-5}, which corresponds to a Baud rate of 2.6 MBaud, a carrier frequency of 1.9 GHz and a vehicular speed of 30 mph.

channels, where TCM is at its best, since TCM was designed for Gaussian channels. By contrast, although BICM was designed for fading channels, when the channel-induced error bursts are inadequately dispersed owing to the employment of a short IL, the performance of BICM suffers. In other words, when communicating over slowly fading channels, extremely long interleavers may be necessary for over-bridging the associated long fades and for facilitating the dispersion of bursty transmission errors, which is a prerequisite for the efficient operation of channel codecs.

On the other hand, BICM performs better than TCM for long IL values, which is reminiscent of the performance trends observed when communicating over uncorrelated Rayleigh fading channels, as evidenced by Figure 9.32(b). This is justified, since the correlation of the fading channel is broken when a long IL is employed for dispersing the error bursts.

9.7.2.3 Conclusion

In conclusion, at a given complexity TCM performs better than BICM in AWGN channels, but worse in uncorrelated narrowband Rayleigh fading channels. However, BICM-ID using soft-decision feedback outperforms TCM and BICM for transmissions over both AWGN and uncorrelated narrowband Rayleigh fading channels at the same DC. TTCM has shown superior performance over the other coded modulation schemes studied, but exhibited a higher error floor for the 64QAM scheme due to the presence of uncoded information bits for transmissions over uncorrelated narrowband Rayleigh fading channels. Comparing the coding gain against the DC, the iterative decoding schemes of TTCM and BICM-ID are capable of providing a high coding gain even in conjunction with a constituent code exhibiting a short memory length, although only at the cost of a sufficiently high number of decoding iterations, which may imply a relatively high decoding complexity. Comparing the achievable coding gain against the IL, TTCM is the best

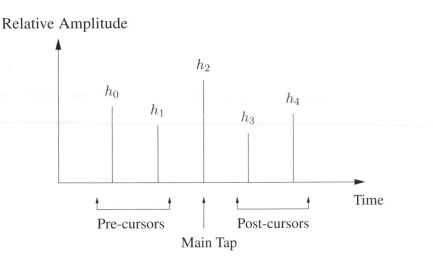

Figure 9.34: Channel Impulse Response (CIR) having pre-cursors, the main tap and post-cursors.

performer for a variety of ILs, while the performance of BICM-ID is highly dependent on the IL for transmissions over both AWGN and Rayleigh fading channels.

9.7.3 Coded Modulation in Wideband Channels

In this section we will consider the performance of the various coded modulation schemes in the context of practical dispersive channels.

9.7.3.1 Inter-symbol Interference

The mobile radio channels [49] can be typically characterised by band-limited linear filters. If the modulation bandwidth exceeds the coherence bandwidth of the radio channel, Inter-Symbol Interference (ISI) occurs and the modulation pulses are spread or dispersed in the time domain. The ISI, inflicted by band-limited frequency, selective time-dispersive channels, distorts the transmitted signals. At the receiver, the linearly distorted signal has to be equalised in order to recover the information.

The linearly distorted instantaneous signal received over the dispersive channel can be visualised as the superposition of the channel's response due to several information symbols in the past and in the future. Figure 9.34 shows the Channel's Impulse Response (CIR) exhibiting three distinct parts. The main tap h_2 possesses the highest relative amplitude. The taps before the main tap, namely h_0 and h_1, are referred to as pre-cursors, whereas those following the main tap, namely h_3 and h_4, are referred to as post-cursors.

The energy of the wanted signal is received mainly over the path described by the main channel tap. However, some of the received energy is contributed by the convolution of the pre-cursors with future symbols and the convolution of the post-cursor with past symbols, which are termed pre-cursor ISI and post-cursor ISI, respectively. Thus the received signal is constituted by the superposition of the wanted signal, pre-cursor ISI and post-cursor ISI.

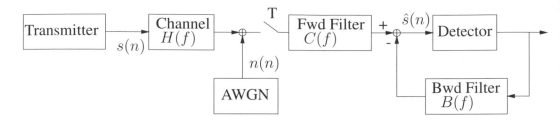

Figure 9.35: Schematic of the transmission system portraying the oriented feedforward (Fwd) and backward (Bwd) filter of the DFE, where $C(f)$ and $B(f)$ are the corresponding frequency-domain transfer functions, respectively.

9.7.3.2 Decision Feedback Equaliser

Channel equalisers that are utilised for compensating the effects of ISI can be classified structurally as linear equalisers or DFEs. They can be distinguished also on the basis of the criterion used for optimising their coefficients. When applying the Minimum Mean Square Error (MMSE) criterion, the equaliser is optimised such that the mean squared error between the distorted signal and the actual transmitted signal is minimised. For time varying dispersive channels, a range of adaptive algorithms can be invoked for updating the equaliser coefficients and for tracking the channel variations [207].

9.7.3.2.1 Decision Feedback Equaliser Principle   The simple Zero Forcing Equaliser (ZFE) [236] forces all the impulse response contributions of the concatenated system constituted by the channel and the equaliser to zero at the signalling instants nT for $n \neq 0$, where T is the signalling interval duration. The ZFE provides gain in the frequency domain at frequencies where the channel's transfer function experiences attenuation and vice versa. However, both the signal and the noise are enhanced simultaneously and hence the ZFE is ineffective owing to the associated noise enhancement effects. Furthermore, no finite-gain ZFE can be designed for channels that exhibit spectral nulls [207, 237].

Linear MMSE equalisers [236] are designed for mitigating both the pre-cursor ISI and the post-cursor ISI, as defined in Section 9.7.3.1. The MMSE equaliser is more intelligent than the ZFE, since it jointly minimises the effects of both the ISI and noise. Although the linear MMSE equaliser approaches the same performance as the ZFE at high SNRs, an MMSE solution does exists for all channels, including those that exhibit spectral nulls.

The idea behind the DFE [207, 236, 237] is that once an information symbol has been detected and decided upon, the ISI that these detected symbols inflicted on future symbols can be estimated and the corresponding ISI can be cancelled before the detection of subsequent symbols.

The DFE employs a feedforward filter and a backward-oriented filter for combating the effects of dispersive channels. Figure 9.35 shows the general block diagram of the transmission system employing a DFE. The forward-oriented filter partially eliminates the ISI introduced by the dispersive channel. The feedback filter, in the absence of decision errors, is fed with the error-free transmitted signal in order to further reduce the ISI.

The feedback filter, denoted as the Bwd Filter in Figure 9.35, receives the detected symbol. Its output is then subtracted from the estimates generated by the forward filter, denoted as the Fwd Filter, in order to produce the detector's input. Since the feedback filter uses the ISI-free signal as its input, the feedback loop mitigates the ISI without introducing enhanced noise into

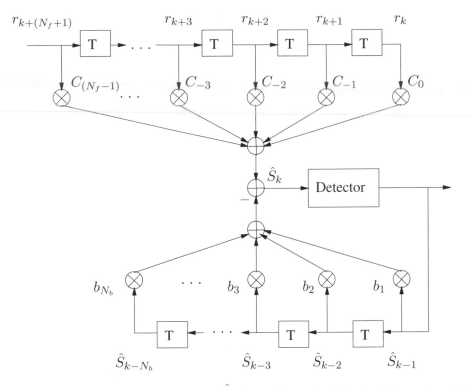

Figure 9.36: Structure of the DFE where r_k and $\hat{S}_k$ denote the received signal and detected symbol, respectively, while C_m, b_q represent the coefficient taps of the forward- and backward-oriented filters, respectively.

the system. The drawback of the DFE is that when wrong decisions are fed back into the feedback loop, error propagation is inflicted and the BER performance of the equaliser is degraded.

The detailed DFE structure is shown in Figure 9.36. The feedforward filter is constituted by the coefficients or taps labelled as $C_0 \rightarrow C_{N_f-1}$, where N_f is the number of taps in the feedforward filter, as shown in the figure. The causal feedback filter is constituted by N_b feedback taps, denoted as $b_1 \rightarrow b_{N_b}$. Note that the feedforward filter contains only the present input signal r_k, and future input signals $r_{k+1} \ldots r_{k+(N_f+1)}$, which implies that no latency is inflicted. Therefore, the feedforward filter eliminates only the pre-cursor ISI, but not the post-cursor ISI. By contrast, the feedback filter mitigates the ISI caused by the past data symbols, i.e. post-cursor ISI. Since the feedforward filter only eliminates the pre-cursor ISI, the noise enhancement effects are less problematic in DFEs compared to the linear MMSE equaliser.

Here, the MMSE criterion [236] is used for deriving the optimum coefficients of the feedforward section of the DFE. The Mean Square Error (MSE) between the transmitted signal, s_k, and its estimate, $\hat{s}_k$, at the equaliser's output is formulated as:

$$MSE = E[|s_k - \hat{s}|^2], \tag{9.43}$$

where $E[|s_k - \hat{s}|^2]$ denotes the expected value of $|s_k - \hat{s}|^2$. In order to minimise the MSE, the orthogonality principal [238] is applied, stating that the residual error of the equaliser, $e_k = s_k - \hat{s}_k$, is orthogonal to the input signal of the equaliser, r_k, when the equaliser taps are optimal, yielding:

$$E[e_k r_{k+l}^*] = 0, \tag{9.44}$$

where the superscript * denotes conjugation. Following Cheung's approach [207, 239], the optimum coefficient of the feedforward section can be derived from the following set of N_f equations:

$$\sum_{m=0}^{N_f-1} C_m \left[\sum_{v=0}^{l} h_v^* h_{v+m-l} \sigma_S^2 + N_0 \delta_{m-l} \right] = h_l^* \sigma_S^2, \ l = 0 \dots N_f - 1, \tag{9.45}$$

where σ_s^2 and $N_0/2$ are the signal and noise variance, respectively, while h^* denotes the complex conjugate of the CIR and δ is the delta function. By solving these N_f simultaneous equations, the equaliser coefficients, C_m, can be obtained. For the feedback filter the following set of N_b equations were used, in order to derive the optimum feedback coefficient, b_q [239]:

$$b_q = \sum_{m=0}^{N_f-1} C_m h_{m+q}, \ q = 1 \dots N_b. \tag{9.46}$$

9.7.3.2.2 Equalizer Signal To Noise Ratio Loss   The equaliser's performance can be measured in terms of the equaliser's SNR loss, BER performance and MSE [237]. Here the SNR loss is considered, since this parameter will be used in next section.

The SNR loss of the equaliser was defined by Cheung [239] as:

$$SNR_{loss} = SNR_{input} - SNR_{output}, \tag{9.47}$$

where SNR_{input} is the SNR measured at the equaliser's input, given by:

$$SNR_{input} = \frac{\sigma_s^2}{2\sigma_N^2}, \tag{9.48}$$

with σ_s^2 being the average received signal power, assuming wide sense stationary conditions, and σ_N^2 is the variance of the AWGN.

The equaliser's output contains the wanted signal, the effective Gaussian noise, the residual ISI and the ISI caused by the past data symbols. In order to simplify the calculation of SNR_{output}, we assume that the SNR is high and hence we consider the low-BER range, where effectively correct bits are fed back to the DFE's feedback filter. Thus the post-cursor ISI is completely eliminated from the equaliser's output. Hence SNR_{output} is given by [239]:

$$SNR_{output} = \frac{\text{Wanted Signal Power}}{\text{Residual ISI Power} + \text{Effective Noise Power}}, \tag{9.49}$$

where the residual ISI is assumed to be an extra noise source that possessed a Gaussian distribution. Therefore, we have:

$$\text{Wanted Signal Power} = E \left[|s_k \sum_{m=0}^{N_f-1} C_m h_m|^2 \right], \tag{9.50}$$

$$\text{Effective Noise Power} = N_0 \sum_{i=0}^{N_f-1} |C_i|^2, \tag{9.51}$$

and:

$$\text{Residual ISI Power} = \sum_{q=-(N_f-1)}^{-1} E\left[|f_q s_{k-q}|^2\right], \tag{9.52}$$

where $f_q = \sum_{m=0}^{N_f-1} C_m h_{m+q}$ and the remaining notations are accrued from Figure 9.36. By substituting Equations 9.50, 9.51 and 9.52 into 9.49, the SNR_{output} can be written as [237]:

$$SNR_{output} = \frac{E\left[|s_k \sum_{m=0}^{N_f-1} C_m h_m|^2\right]}{\sum_{q=-(N_f-1)}^{-1} E\left[|f_q s_{k-q}|^2\right] + N_0 \sum_{i=0}^{N_f-1} |C_i|^2}. \tag{9.53}$$

Following this rudimentary introduction to channel equalisation, we focus our attention on quantifying the performance of various wideband coded modulation schemes, referring the reader to [207] for an in-depth discourse on channel equalisation.

9.7.3.3 Decision Feedback Equalizer Aided Adaptive Coded Modulation

In this section, DFE-aided wideband Burst-by-Burst (BbB) adaptive TCM, TTCM, BICM and BICM-ID schemes are proposed and characterised in performance terms, when communicating over the COST207 Typical Urban (TU) wideband fading channel. These schemes are evaluated using a practical near-instantaneous modem mode switching regime. **System I** represents schemes without channel interleaving, while **System II** invokes channel interleaving over four transmission bursts. **System I** exhibited a factor four delay in lower overall modem mode signalling, and hence it was capable of more prompt modem mode reconfiguration. By contrast, **System II** was less agile in terms of modem mode reconfiguration, but benefited from a longer interleaver delay. We will show in Section 9.7.3.3.4 that a substantially improved Bit Per Symbol (BPS) and BER performance was achieved by **System II** in comparison to **System I**. We will also show that BbB adaptive TTCM was found to perform better than the BbB adaptive TCM in **System II** at a similar DC, when aiming for a target BER of below 0.01%.

9.7.3.3.1 Introduction
In general fixed-mode transceivers fail to adequately counteract the time varying nature of the mobile radio channel and hence typically result in bursts of transmission errors. By contrast, in BbB adaptive schemes [240] a higher-order modulation mode is employed when the instantaneous estimated channel quality is high in order to increase the number of BPS transmitted and, conversely, a more robust lower-order modulation mode is employed when the instantaneous channel quality is low, in order to improve the mean BER performance. Uncoded adaptive schemes [240–245] and coded adaptive schemes [246–248] have been investigated for transmissions over narrowband fading channels. Finally, a turbo-coded wideband adaptive scheme assisted by a DFE was investigated in [249].

In our practical approach the local transmitter is informed about the channel quality estimate generated by the remote receiver upon receiving the transmission burst of the remote transmitter. In other words, the modem mode required by the remote receiver for maintaining its target integrity is superimposed on the transmission burst emitted by the remote transmitter. Hence a delay of one transmission burst duration is incurred. In the literature, adaptive coding designed for time varying channels using outdated fading estimates has been investigated for example in [250].

Over wideband fading channels the DFE employed will eliminate most of the ISI. Consequently, the MSE at the output of the DFE can be calculated and used as the metric invoked for switching the modulation modes [245]. This ensures that the performance is optimised by employing equalization and BbB adaptive TCM/TTCM jointly, in order to combat both the signal power fluctuations and the time variant ISI of the wideband channel.

In Section 9.7.3.3.2, the system's schematic is outlined. In Section 9.7.3.3.3, the performance of various fixed-mode TCM and TTCM schemes is evaluated, while Section 9.7.3.3.4 contains the detailed characterisation of the BbB adaptive TCM/TTCM schemes in the context of the non-interleaved **System I** and interleaved **System II**. In Section 9.7.3.3.5 we compare the performance of the proposed schemes with that of other adaptive coded modulation schemes, such as BICM and BICM-ID. Finally, we will conclude with our findings in Section 9.7.3.3.6.

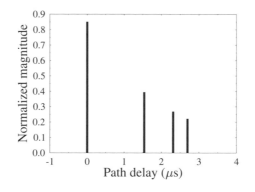

Figure 9.37: The impulse response of a COST207 Typical Urban (TU) channel [251].

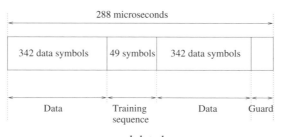

non-spread data burst

Figure 9.38: Transmission burst structure of the FMA1 non-spread data as specified in the FRAMES proposal [252].

9.7.3.3.2 System Overview The multi-path channel model is characterised by its discretised symbol-spaced COST207 Typical Urban (TU) CIR [251], as shown in Figure 9.37. Each path is faded independently according to a Rayleigh distribution and the corresponding normalised Doppler frequency is 3.25×10^{-5}, the system's Baud rate is 2.6 MBaud, the carrier frequency is 1.9 GHz and the vehicular speed is 30 mph. The DFE incorporated 35 feed forward taps and 7 feedback taps and the transmission burst structure used is shown in Figure 9.38. When considering a Time Division Multiple Access (TDMA)/Time Division Duplex (TDD) system providing 16 slots per 4.615 ms TDMA frame, the transmission burst duration is 288 μs, as specified in the Pan-European FRAMES proposal [252].

The following assumptions are stipulated. First, we assume that the equaliser is capable of estimating the CIR perfectly with the aid of the equaliser training sequence of Figure 9.38. Second, the CIR is time-invariant for the duration of a transmission burst, but varies from burst to burst according to the Doppler frequency, which corresponds to assuming that the CIR is slowly varying. We refer to this scenario as encountering burst-invariant fading. The error propagation of the DFE will degrade the estimated performance, but the effect of error propagation is left for further study. At the receiver, the CIR is estimated, and is then used for calculating the DFE coefficients [207]. Subsequently, the DFE is used for equalising the ISI-corrupted received signal. In addition, both the CIR estimate and the DFE feedforward coefficients are utilised for computing the SNR at the output of the DFE. More specifically, by assuming that the residual ISI is near-Gaussian distributed and that the probability of decision feedback errors is negligible, the SNR at the output of the DFE, γ_{dfe}, is calculated as [239]:

$$\gamma_{dfe} = \frac{\text{Wanted Signal Power}}{\text{Residual ISI Power} + \text{Effective Noise Power}}.$$

$$= \frac{E\left[\left|s_k \sum_{m=0}^{N_f} C_m h_m\right|^2\right]}{\sum_{q=-(N_f-1)}^{-1} E\left[\left|\sum_{m=0}^{N_f-1} C_m h_{m+q} s_{k-q}\right|^2\right] + N_0 \sum_{m=0}^{N_f} |C_m|^2}, \quad (9.54)$$

where C_m and h_m denote the DFE's feedforward coefficients and the CIR, respectively. The transmitted signal is represented by s_k and N_0 denotes the noise spectral density. Finally, the number of DFE feedforward coefficients is denoted by N_f.

The equaliser's SNR, γ_{dfe}, in Equation 9.54, is then compared against a set of adaptive modem mode switching thresholds f_n, and subsequently the appropriate modulation mode is selected [245, 253]. The modem mode required by the remote receiver for maintaining its target integrity is then fed back to the local transmitter. The modulation modes that are utilised in this scheme are 4QAM, 8PSK, 16QAM and 64QAM [207].

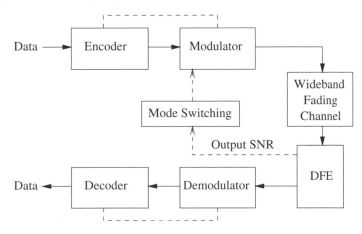

Figure 9.39: System I employing no channel interleaver. The equaliser's output SNR is used for selecting a suitable modulation mode, which is fed back to the transmitter on a burst-by-burst basis.

The simplified block diagram of the BbB adaptive TCM/TTCM **System I** is shown in Figure 9.39, where no channel interleaving is used. Transmitter A extracts the modulation mode required by receiver B from the reverse-link transmission burst in order to adjust the adaptive TCM/TTCM mode suitable for the currently experienced instantaneous channel quality. This incurs one TDMA/TDD frame delay between estimating the actual channel condition at receiver B and the selected modulation mode of transmitter A. Better channel quality prediction can be achieved using the techniques proposed in [254]. We invoke four encoders, each adding one parity bit to each information symbol, yielding the coding rate of $1/2$ in conjunction with the TCM/TTCM mode of 4QAM, $2/3$ for 8PSK, $3/4$ for 16QAM and $5/6$ for 64QAM.

The design of TCM schemes contrived for fading channels relies on the time and space diversity provided by the associated channel coder [209,224]. Diversity may be achieved by repetition coding, which reduces the effective data rate, spacedtime- coded multiple transmitter/receiver structures [79], which increases cost and complexity, or by simple interleaving, which induces latency. In [255] adaptive TCM schemes were designed for narrowband fading channels utilising repetition-based transmissions during deep fades along with ideal channel interleavers and assuming zero delay for the feedback of the channel quality information.

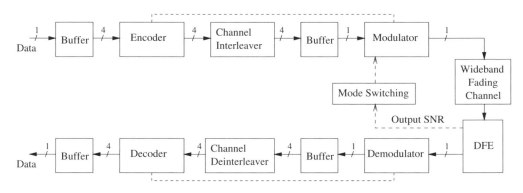

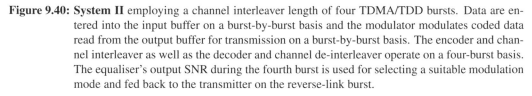

Figure 9.40: System II employing a channel interleaver length of four TDMA/TDD bursts. Data are entered into the input buffer on a burst-by-burst basis and the modulator modulates coded data read from the output buffer for transmission on a burst-by-burst basis. The encoder and channel interleaver as well as the decoder and channel de-interleaver operate on a four-burst basis. The equaliser's output SNR during the fourth burst is used for selecting a suitable modulation mode and fed back to the transmitter on the reverse-link burst.

Figure 9.40 shows the block diagram of **System II**, where symbol-based channel interleaving over four transmission bursts is utilised, in order to disperse the bursty symbol errors. Hence, the coded modulation module assembles four bursts using an identical modulation mode, so that they can be interleaved using the symbol-by-symbol random channel interleaver without the need of adding dummy bits. Then, these four-burst TCM/TTCM packets are transmitted to the receiver. Once the receiver has received the fourth burst, the equaliser's output SNR for this most recent burst is used for choosing a suitable modulation mode. The selected modulation mode is fed back to the transmitter on the reverse-link burst. Upon receiving the modulation mode required by receiver B (after one TDMA frame delay), the coded modulation module assembles four bursts of data from the input buffer for coding and interleaving, which are then stored in the output buffer ready for the next four bursts' transmission. Thus the first transmission burst exhibits one TDMA/TDD frame delay and the fourth transmission burst exhibits four frame delay which is the worst-case scenario.

Soft-decision trellis decoding utilising the Log-MAP algorithm [52] of Section 5.3.5 was invoked for TCM/TTCM decoding. The Log-MAP algorithm is a numerically stable version of the MAP algorithm operating in the log-domain, in order to reduce its complexity and to mitigate the numerical problems associated with the MAP algorithm [11]. As stated in Section 9.2, the TCM scheme invokes Ungerböck's codes [208], while the TTCM scheme invokes Robertson's codes [211]. A component TCM code memory of 3 was used for the TTCM scheme. The number of turbo iterations for TTCM was fixed to four and hence it exhibited a similar DC to the TCM code memory of 6. In the next section we present simulation results for our fixed-mode transmissions.

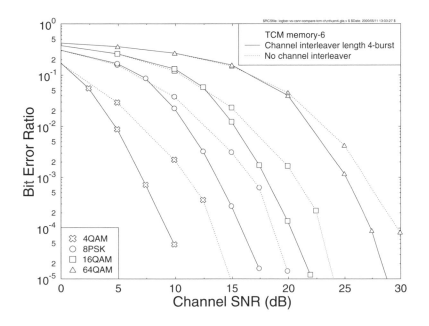

Figure 9.41: TCM performance of each individual modulation mode over the Rayleigh fading COST207 TU channel of Figure 9.37. A TCM code memory of 6 was used, since it had a similar decoding complexity to TTCM in conjunction with four iterations using a component TCM code memory of 3.

9.7.3.3.3 Fixed-mode Performance Before characterising the proposed wideband BbB adaptive scheme, the BER performance of the fixed modem modes of 4QAM, 8PSK, 16QAM and 64QAM are studied both with and without channel interleavers. These results are shown in Figure 9.41 for TCM, and in Figure 9.42 for TTCM. The random TTCM symbol-interleaver memory was set to 684 symbols, corresponding to the number of data symbols in the transmission burst structure of Figure 9.38, where the resultant number of bits was the number of data bits per symbol $(BPS) \times 684$. A channel interleaver of 4×684 symbols was utilised, where the number of bits was $(BPS+1) \times 4 \times 684$ bits, since one parity bit was added to each TCM/TTCM symbol.

As expected, in Figures 9.41 and 9.42 the BER performance of the channel-interleaved scenario was superior compared to that without channel interleaver, although at the cost of an associated higher transmission delay. The SNR gain difference between the channel-interleaved and non-interleaved scenarios was about 5 dB in the TTCM/4QAM mode, but this difference reduced for higher-order modulation modes. Again, this gain was obtained at the cost of a four-burst

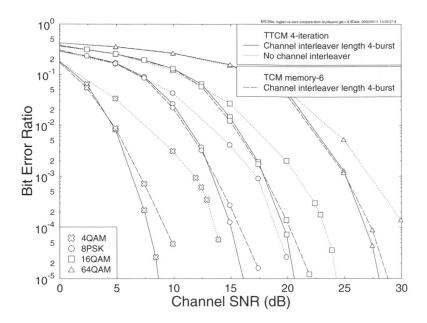

Figure 9.42: TTCM performance of each individual modulation mode over the Rayleigh fading COST207
TU channel of Figure 9.37. A component TCM code memory of 3 was used and the number
of turbo iterations was four. The performance of the TCM code with memory 6 utilising a
channel interleaver was also plotted for comparison.

channel interleaving delay. This SNR gain difference shows the importance of time diversity in
coded modulation schemes.

TTCM has been shown to be more efficient than TCM for transmissions over AWGN channels and narrowband fading channels [211, 256]. Here, we illustrate the advantage of TTCM in
comparison to TCM over the dispersive or wideband Gaussian CIR of Figure 9.37, as seen in
Figure 9.43. In conclusion, TTCM is superior to TCM in a variety of channels.

Let us now compare the performance of the BbB adaptive TCM/TTCM **system I** and **II**.

9.7.3.3.4   System I and System II Performance   The modem mode switching mechanism of
the adaptive schemes is characterised by a set of switching thresholds, the corresponding random
TTCM symbol interleavers and the component codes, as follows:

$$
\text{Modulation Mode} = \begin{cases} 4QAM, I_0 = 684, R_0 = 1/2 & \text{if } \gamma_{DFE} \leq f_1 \\ 8PSK, I_1 = 1368, R_1 = 2/3 & \text{if } f_1 < \gamma_{DFE} \leq f_2 \\ 16QAM, I_2 = 2052, R_2 = 3/4 & \text{if } f_2 < \gamma_{DFE} \leq f_3 \\ 64QAM, I_3 = 3420, R_3 = 5/6 & \text{if } \gamma_{DFE} > f_3, \end{cases} \tag{9.55}
$$

where $f_n, n = 1 \ldots 3$, are the equaliser's output SNR thresholds, while I_n represents the random
TTCM symbol interleaver size in terms of the number of bits, which is not used for the TCM
schemes. The switching thresholds f_n were chosen experimentally, in order to maintain a BER
of below 0.01%, and these thresholds are listed in Table 9.8.

Let us consider the adaptive TTCM scheme in order to investigate the performance of **System I** and **System II**. The performance of the non-interleaved **System I** was found to be identical

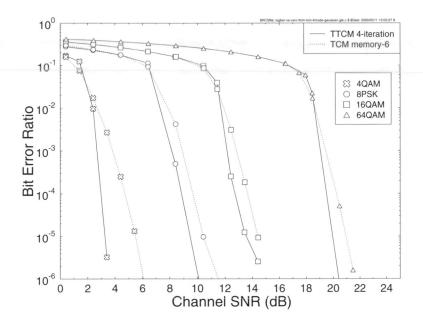

Figure 9.43: TTCM and TCM performance of each individual modulation mode for transmissions over the unfaded COST207 TU channel of Figure 9.37. The TTCM scheme used component TCM codes of memory 3 and the number of turbo iterations was four. The performance of the TCM scheme in conjunction with memory 6 was plotted for comparison with the similar-complexity TTCM scheme.

BER < 0.01 %		Switching Thresholds		
Adaptive System Type		f_1	f_2	f_3
TCM, Memory 3	System I	19.56	23.91	30.52
	System II	17.17	21.91	29.61
TCM, Memory 6	System I	19.56	23.88	30.07
	System II	17.14	21.45	29.52
TTCM, 4 iterations	System I	19.69	23.45	30.29
	System II	16.66	21.40	28.47
BICM, Memory 3	System I	19.94	24.06	31.39
BICM-ID, 8 iterations	System II	16.74	21.45	28.97

Table 9.8: The switching thresholds were set experimentally for transmissions over the COST207 TU channel of Figure 9.37, in order to achieve a target BER of below 0.01%. **System I** does not utilise a channel interleaver, while **System II** uses a channel interleaver length of four TDMA/TDD bursts.

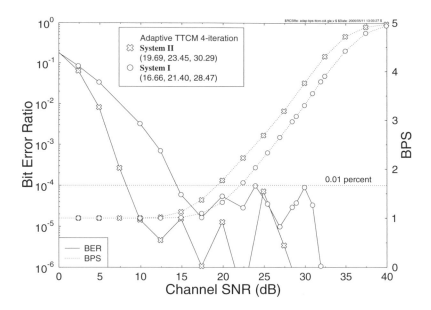

Figure 9.44: BER and BPS performance of adaptive TTCM for transmissions over the COST207 TU channel of Figure 9.37, using four turbo iterations in **System I** (without channel interleaver) and in **System II** (with a channel interleaver length of four bursts) for a target BER of less than 0.01%. The legends indicate the associated switching thresholds expressed in dB, as seen in the brackets.

to that of the same scheme employing interleaving over one transmission burst, which was described in the context of Figure 9.38. This is because in the context of the burst-invariant fading scenario the channel behaves like a dispersive Gaussian channel, encountering a specific fading envelope and phase trajectory across a transmission burst. The CIR is then faded at the end or at the commencement of each transmission burst. Hence the employment of a channel interleaver having a memory of one transmission burst would not influence the distribution of the channel errors experienced by the decoder. The BER and BPS performances of both adaptive TTCM systems using four iterations are shown in Figure 9.44, where we observed that the throughput of **System II** was superior to that of **System I**. Furthermore, the overall BER of **System II** was lower than that of **System I**. In order to investigate the switching dynamics of both systems, the mode switching together with the equaliser's output SNR was plotted versus time at an average channel SNR of 25 dB in Figures 9.45 and 9.46. Observe in Table 9.8 that the switching thresholds f_n of **System II** are lower than those of **System I**, since the fixed-mode-based results of **System II** in Figure 9.42 were better. Hence higher-order modulation modes were chosen more frequently than in **System I**, giving a better BPS throughput. From Figures 9.45 and 9.46, it is clear that **System I** was more flexible in terms of mode switching, while **System II** benefited from higher diversity gains due to the four-burst channel interleaver. This diversity gain compensated for the loss of switching flexibility, ultimately providing a better performance in terms of BER and BPS, as seen in Figure 9.44.

In our next endeavour, the adaptive TCM and TTCM schemes of **System I** and **System II** are compared. Figure 9.47 shows the BER and BPS performance of **System I** for adaptive TTCM using four iterations, adaptive TCM of memory 3 (which was the component code of our TTCM scheme) and adaptive TCM of memory 6 (which had a similar decoding complexity to

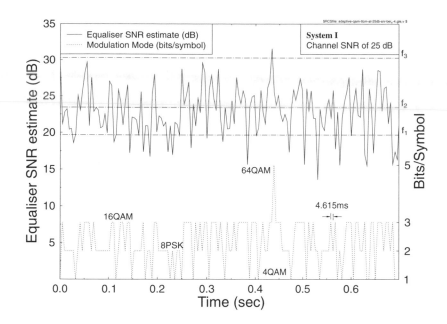

Figure 9.45: Channel SNR estimate and BPS versus time plot for adaptive TTCM for transmissions over the COST207 TU channel of Figure 9.37, using four turbo iterations in **System I** at an average channel SNR of 25 dB, where the modulation mode switching is based upon the equaliser's output SNR, which is compared to the switching thresholds f_n defined in Table 9.8. The duration of one TDMA/TDD frame is 4.615 ms. The TTCM mode can be switched after one frame duration.

our TTCM scheme). As can be seen from the fixed-mode results of Figures 9.41 and 9.42 in the previous section, TCM and TTCM performed similarly in terms of their BER, when no channel interleaver was used for the slow fading wideband COST207 TU channel of Figure 9.37. Hence, they exhibited a similar performance in the context of the adaptive schemes of System I, as shown in Figure 9.47. Even the TCM scheme of memory 3 associated with a lower complexity could give a similar BER and BPS performance. This shows that the equaliser plays a dominant role in **System I**, where the coded modulation schemes could not benefit from sufficient diversity due to the lack of interleaving.

When the channel interleaver is introduced in **System II**, the bursty symbol errors are dispersed. Figure 9.48 illustrates the BER and BPS performance of **System II** for adaptive TTCM using four iterations, adaptive TCM of memory 3 and adaptive TCM of memory 6. The performance of all these schemes improved in the context of **System II**, as compared to the corresponding schemes in **System I**. The TCM scheme of memory 6 had a lower BER than TCM of memory 3, and also exhibited a small BPS improvement. As expected, TTCM had the lowest BER and also the highest BPS throughput compared to the other coded modulation schemes.

In summary, we have observed BER and BPS gains for the channel-interleaved adaptive coded schemes of **System II** in comparison to the schemes without channel interleaver as in **System I**. Adaptive TTCM exhibited a superior performance in comparison to adaptive TCM in the context of **System II**.

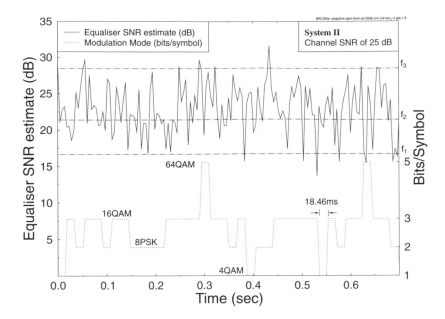

Figure 9.46: Channel SNR estimate and BPS versus time plot for adaptive TTCM for transmissions over the COST207 TU channel of Figure 9.37, using four turbo iterations in **System II** at an average channel SNR of 25 dB, where the modulation mode switching is based upon the equaliser's output SNR which is compared to the switching thresholds f_n defined in Table 9.8. The duration of one TDMA/TDD frame is 4.615 ms. The TTCM mode is maintained for four frame durations, i.e. for 18.46 ms.

9.7.3.3.5   Overall Performance   Figure 9.49 shows the fixed modem modes' performance for TCM, TTCM, BICM and BICM-ID in the context of **System II**. For the sake of a fair comparison of the DC, we used a TCM code memory of 6, TTCM code memory of 3 in conjunction with four turbo iterations, BICM code memory of 6 and a BICM-ID code memory of 3 in conjunction with eight decoding iterations. However, BICM-ID had a slightly higher DC, since the demodulator was invoked in each BICM-ID iteration, whereas in the BICM, TCM and TTCM schemes the demodulator was only visited once in each decoding process. As illustrated in the figure, the BICM scheme performed marginally better than the TCM scheme at a BER below 0.01%, except in the 64QAM mode. Hence, adaptive BICM is also expected to be better than adaptive TCM in the context of **System II**, when a target BER of less than 0.01% is desired. This is because when the channel interleaver depth is sufficiently high, the diversity gain of the BICM's bit interleaver is higher than that of the TCM's symbol interleaver [212, 232].

Figure 9.50 compares the adaptive BICM and TCM schemes in the context of **System I**, i.e. without channel interleaving, although the BICM scheme invoked an internal bit interleaver of one burst memory. As can be seen from the figure, adaptive TCM exhibited a better BPS throughput and BER performance than BICM, due to employing an insufficiently high channel interleaving depth for the BICM scheme, for transmissions over our slow fading wideband channels.

As observed in Figure 9.49, we noticed that BICM-ID had the worst performance at low SNRs in each modulation mode compared to other coded modulation schemes. However, it exhibited a steep slope and therefore at high SNRs it approached the performance of the TTCM

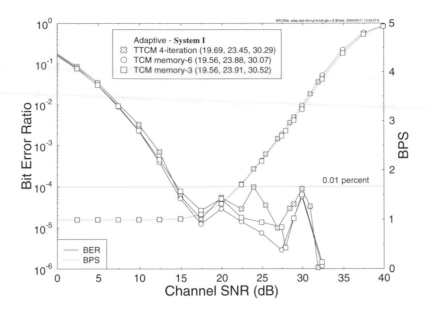

Figure 9.47: BER and BPS performance of adaptive TCM and TTCM without channel interleaving in **System I**, for transmissions over the Rayleigh fading COST207 TU channel of Figure 9.37. The switching mechanism was characterised by Equation 9.55. The switching thresholds were set experimentally, in order to achieve a BER of below 0.01%, as shown in Table 9.8.

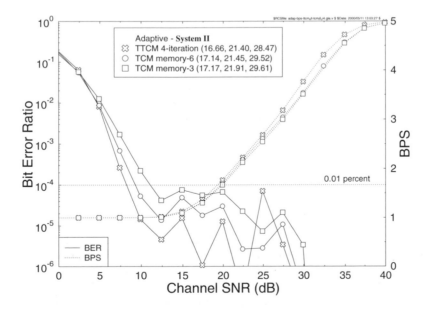

Figure 9.48: BER and BPS performance of adaptive TCM and TTCM using a channel interleaver length of four bursts in **System II**, for transmissions over the Rayleigh fading COST207 TU channel of Figure 9.37. The switching mechanism was characterised by Equation 9.55. The switching thresholds were set experimentally, in order to achieve a BER of below 0.01%, as shown in Table 9.8.

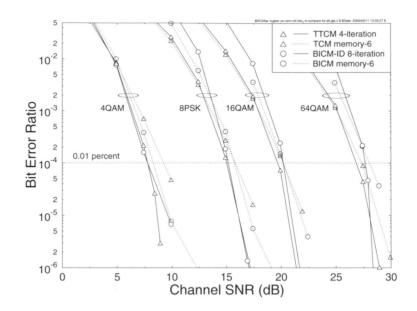

Figure 9.49: BER performance of the fixed modem modes of 4QAM, 8PSK, 16QAM and 64QAM utilising TCM, TTCM, BICM and BICM-ID schemes in the context of **System II** for transmissions over the COST207 TU channel of Figure 9.37. For the sake of maintaining a similar DC, we used a TCM code memory of 6, TTCM code memory of 3 in conjunction with four turbo iterations, BICM code memory of 6 and a BICM-ID code memory of 3 in conjunction with eight decoding iterations. However, BICM-ID had a slightly higher complexity than the other systems, since the demodulator module was invoked eight times as compared to only once for its counterparts during each decoding process.

scheme. The adaptive BICM-ID and TTCM schemes employed in the context of **System II** were compared in Figure 9.50. The adaptive TTCM scheme exhibited a better BPS throughput than adaptive BICM-ID, since TTCM had a better performance in fixed modem modes at a BER of 0.01%. However, adaptive BICM-ID exhibited a lower BER performance than adaptive TTCM owing to the high steepness of the BER curve of BICM-ID in its fixed modem modes.

9.7.3.3.6 Conclusions In this section, BbB adaptive TCM, TTCM, BICM and BICM-ID were proposed for transmissions over wideband fading channels both with and without channel interleaving and they were characterised in performance terms when communicating over the COST207 TU fading channel. When observing the associated BPS curves, adaptive TTCM exhibited up to 2.5 dB SNR gain for a channel interleaver length of four bursts in comparison to the non-interleaved scenario, as evidenced in Figure 9.44. Upon comparing the BPS curves, adaptive TTCM also exhibited up to 0.7 dB SNR gain compared to adaptive TCM of the same complexity in the context of **System II** for a target BER of less than 0.01%, as shown in Figure 9.48. Lastly, adaptive TCM performed better than the adaptive BICM bench-marker in the context of **System I**, and the adaptive BICM-ID scheme performed marginally worse than adaptive TTCM in the context of **System II** as discussed in Section 9.7.3.3.5.

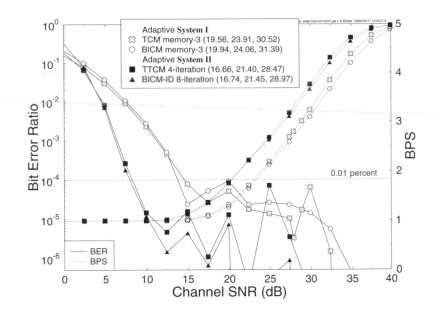

Figure 9.50: BER and BPS performance of the adaptive TCM/BICM **System I**, using memory 3 codes and that of the adaptive TTCM/BICM-ID **System II**, for transmissions over the Rayleigh fading COST207 TU channel of Figure 9.37. The switching mechanism was characterised by Equation 9.55. The switching thresholds were set experimentally, in order to achieve a BER of below 0.01%, as shown in Table 9.8.

9.7.3.4 Orthogonal Frequency Division Multiplexing

The employment of equalisers for removing the ISI has been discussed in Section 9.7.3.2. By contrast, as an attractive design alternative here Discrete Fourier Transform (DFT)-based Orthogonal Frequency Division Multiplexing (OFDM) [207] will be utilised for removing the ISI.

9.7.3.4.1 Orthogonal Frequency Division Multiplexing Principle
In this section we briefly introduce Frequency Division Multiplexing (FDM), also referred to as Orthogonal Multiplexing (OMPX), as a means of dealing with the problems of frequency-selective fading encountered when transmitting over a high-rate wideband radio channel. The fundamental principle of orthogonal multiplexing originates from Chang [257], and over the years a number of researchers have investigated this technique [258, 259]. Despite its conceptual elegance, until recently its employment has been mostly limited to military applications because of the associated implementational difficulties. However, it has recently been adopted as the new European Digital Audio Broadcasting (DAB) standard; it is also a strong candidate for Digital Terrestrial Television Broadcast (DTTB) and for a range of other high-rate applications, such as 155 Mbit/s wireless Asynchronous Transfer Mode (ATM) local area networks. These wide-ranging applications underline its significance as an alternative technique to conventional channel equalisation in order to combat signal dispersion [260–262].

In the FDM scheme of Figure 9.51 the serial data stream of a traffic channel is passed through a serial-to-parallel converter, which splits the data into a number of parallel sub-channels. The data in each sub-channel are applied to a modulator, such that for N channels there are N modulators whose carrier frequencies are $f_0, f_1, \ldots, f_{N-1}$. The centre frequency difference between

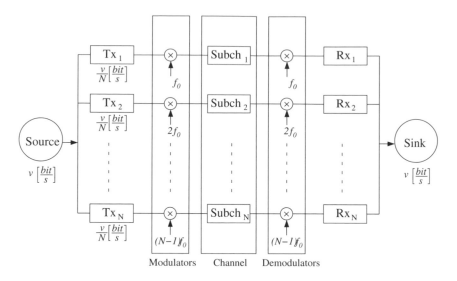

Figure 9.51: Simplified block diagram of the orthogonal parallel modem.

adjacent channels is Δf and the overall bandwidth W of the N modulated carriers is $N\Delta f$.

These N modulated carriers are then combined to give a FDM signal. We may view the serial-to-parallel converter as applying every Mth symbol to a modulator. This has the effect of interleaving the symbols entered into each modulator, hence symbols $S_0, S_N, S_{2N}, \ldots$ are applied to the modulator whose carrier frequency is f_1. At the receiver the received FDM signal is demultiplexed into N frequency bands, and the N modulated signals are demodulated. The baseband signals are then recombined using a parallel-to-serial converter.

The main advantage of the above FDM concept is that because the symbol period has been increased, the channel's delay spread is a significantly shorter fraction of a symbol period than in the serial system, potentially rendering the system less sensitive to ISI than the conventional N times higher-rate serial system. In other words, in the low-rate sub-channels the signal is no longer subject to frequency-selective fading, hence no channel equalisation is necessary.

A disadvantage of the FDM approach shown in Figure 9.51 is its increased complexity in comparison to the conventional system caused by employing N modulators and filters at the transmitter and N demodulators and filters at the receiver. It can be shown that this complexity can be reduced by employing the Discrete Fourier Transform (DFT), typically implemented with the aid of the Fast Fourier Transform (FFT) [207]. The sub-channel modems can use almost any modulation scheme, and 4- or 16-level QAM is an attractive choice in many situations.

The FFT-based QAM/FDM modem's schematic is portrayed in Figure 9.52. The bits provided by the source are serial/parallel converted in order to form the n-level Gray-coded symbols, N of which are collected in TX buffer 1, while the contents of TX buffer 2 are being transformed by the Inverse Fast Fourier Transform (IFFT) in order to form the time-domain modulated signal. The Digital-to-Analogue (D/A) converted, low-pass filtered modulated signal is then transmitted via the channel and its received samples are collected in RX buffer 1, while the contents of RX buffer 2 are being transformed to derive the demodulated signal. The twin buffers are alternately filled with data to allow for the finite FFT-based demodulation time. Before the data are Gray coded and passed to the data sink, they can be equalised by a low-complexity method, if there are some dispersions within the narrow sub-bands. For a deeper tutorial exposure the interested reader is referred to reference [207].

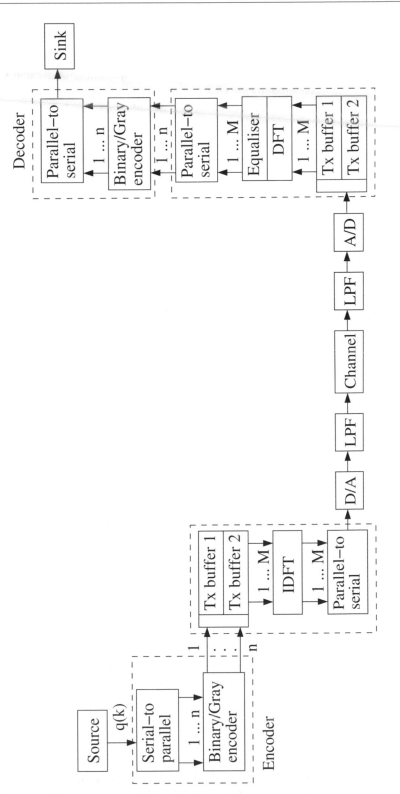

Figure 9.52: FFT-based OFDM modem schematic ©Webb, Hanzo, 1994, [207]

9.7.3.5 Orthogonal Frequency Division Multiplexing Aided Coded Modulation

9.7.3.5.1 Introduction The coded modulation schemes of Sections 9.2–9.6 are here integrated with OFDM by mapping coded symbols to the OFDM modulator at the transmitter and channel decoding the symbols output by the OFDM demodulator at the receiver.

When the channel is frequency selective and OFDM modulation is used, the received symbol is given by the product of the transmitted frequency-domain OFDM symbol and the frequency-domain transfer function of the channel. This direct relationship facilitates the employment of simple frequency-domain channel equalisation techniques. Essentially, if an estimate of the complex frequency-domain transfer function H_k is available at the receiver, channel equalisation can be performed by dividing each received value by the corresponding frequency-domain channel transfer function estimate. The channel's frequency-domain transfer function can be estimated with the aid of known frequency-domain pilot subcarriers inserted into the transmitted signal's spectrum [207]. These known pilots effectively sample the channel's frequency-domain transfer function according to the Nyquist frequency. These frequency-domain samples then allow us to recover the channel's transfer function between the frequency-domain pilots with the aid of interpolation. In addition to this simple form of channel equalisation, another advantage of the OFDM-based modulation is that it turns a channel exhibiting memory into a memoryless one, where the memory is the influence of the past transmitted symbols on the value of the present symbol.

9.7.3.5.2 System Overview The encoder produces a block of N_i channel symbols to be transmitted. These symbols are transmitted by the OFDM modulator. As the OFDM modulator transmits N_u modulated symbols per OFDM symbol, if $N_u = N_i$ then the whole block of N_u modulated symbols can be transmitted in a single OFDM symbol. We refer to this case as the single-symbol mapping based scenario. By contrast, if $N_u < N_i$, then more than one OFDM symbol is required for the transmission of the channel-coded block. We refer to this case as the multiple-symbol mapping based scenario. Both the single- and the multiple-symbol scenarios are interesting. The single-symbol scenario is more appealing from an implementation point of view, as it is significantly more simple. However, it is well known that the performance of a turbo-coded scheme improves upon increasing the IL. Since the number of subcarriers in an OFDM system is limited by several factors, such as for example the oscillator's frequency stability, by using the single-symbol solution we would be limited in terms of the TTCM block length as well. Thus the multiple-symbol solution also has to be considered in order to fully exploit the advantages of the TTCM scheme. Since the single-symbol based scheme is conceptually more simple, we will consider this scenario first. Its extension to the multiple-symbol scenario is straightforward.

In the single carrier system discussed in Section 9.7.3.2 the received signal is given by $y_k = x_k * h_k + n_k$, where $*$ denotes the convolution of the transmitted sequence x_k with the channel's impulse response h_k. An equaliser is used for removing the ISI before channel decoding, giving $\tilde{y}_k = \tilde{x}_k + \tilde{n}_k$. The associated branch metrics can be computed as:

$$P(\tilde{y}_k|\tilde{x}_k) = e^{-\frac{|\tilde{n}|^2}{2\sigma^2}}$$
$$= e^{-\frac{|\tilde{y}_k - \tilde{x}_k|^2}{\sigma^2}}. \tag{9.56}$$

However, in a multi-carrier OFDM system, the received signal is given by $Y_k = X_k \cdot H_k + N_k$, which facilitates joint channel equalisation and channel decoding by computing the branch

OFDM Parameters	
Total number of subcarriers, N	2048 (2K mode)
Number of effective subcarriers, N_u	1705
OFDM symbol duration T_s	224 μs
Guard interval	$T_s/4 = 56\mu$s
Total symbol duration (inc. guard interval)	280 μs
Consecutive subcarrier spacing $1/T_s$	4464 Hz
DVB channel spacing	7.61 MHz
QPSK and QAM symbol period	7/64 μs
Baud rate	9.14 MBaud

Table 9.9: Parameters of the OFDM module [264].

metrics as:

$$P(y_k|x_k) = e^{-\frac{|Y_k - H_k \cdot X_k|^2}{2\sigma^2}}, \tag{9.57}$$

where H_k is the channel's frequency-domain transfer function at the centre frequency of the kth subcarrier. Hence, as long as the the channel transfer function estimation is of sufficiently high quality, simple frequency-domain equalisation could be invoked during the decoding process. If iterative channel decoding is invoked, the channel transfer function estimation is expected to improve during the consecutive iterative steps, in a fashion known in the context of turbo equalisation [105]. Indeed, a performance as high as that in conjunction with perfect channel estimation can be attained [263].

Let us now consider the effect of the channel interleaver. When the channel is frequency selective, it exhibits frequency-domain nulls, which may obliterate several OFDM subcarriers. Thus the quality of several consecutive received OFDM symbols will be low. If the quality is inferior, the channel decoder is unable to correctly estimate the transmitted symbols. When, however, a channel interleaver is present, the received symbols are shuffled before channel decoding and hence these clusters of corrupted subcarriers are disperse. Thus, after the channel interleaver we expect to have only isolated low-quality subcarriers surrounded by unimpaired ones. In this case the decoder is more likely to be able to recover the symbol transmitted on the corrupted subcarrier, using the redundancy added by the channel coding process, conveyed by the surrounding unimpaired subcarriers.

Finally, we consider the multiple-symbol scenario. The system requires only a minor modification. When more than one OFDM symbol is required for transmitting the block of channel-coded symbols, the OFDM demodulator has to store both the demodulated symbols and the channel transfer function estimates in order to form a whole channel-coded block. This block is then fed to the channel interleaver and then to the channel decoder, together with the channel transfer function estimate. Exactly the same MAP decoder as in the single-symbol case can be used for performing the joint channel equalisation and channel decoding. Similarly, the function of the channel interleaver is the same as in the single-symbol scenario.

9.7.3.5.3 Simulation Parameters   The Digital Video Broadcasting (DVB) standard's OFDM scheme [207] was used for this study. The parameters of the OFDM DVB system are presented in Table 9.9. Since the OFDM modem has 2048 subcarriers, the subcarrier signalling rate is effectively 2000 times lower than the maximum DVB transmission rate of 20 Mbit/s, corresponding to about 10 kbit/s. At this sub-channel rate, the individual sub-channels can be considered nearly frequency-flat.

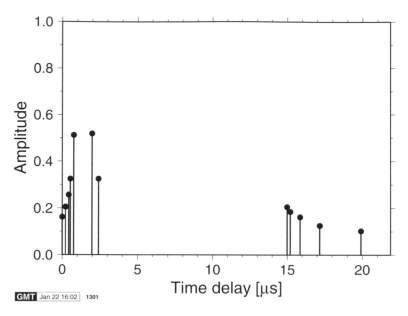

Figure 9.53: COST207 Hilly Terrain (HT) type of impulse response [265].

The channel model employed is the 12-path COST207 [265] Hilly Terrain (HT) type of impulse response, exhibiting a maximum relative path delay of 19.9 μs. The unfaded impulse response is depicted in Figure 9.53. The carrier frequency is 500 MHz and the sampling rate is 7/64 μs. Each of the channel paths was faded independently, obeying a Rayleigh fading distribution, according to a normalised Doppler frequency of 10^{-5} [49]. This corresponds to a worst-case vehicular velocity of about 200 km/h.

9.7.3.5.4 Simulation Results and Discussions In this section, the system performance of the OFDM-based coded modulation schemes is evaluated using QPSK, 8PSK and 16QAM. The coding rate of the coded modulation schemes changes according to the modem mode used. Hence the effective throughput for QPSK is 1 bit/subcarrier, for 8PSK it is 2 bits/subcarrier and finally for 16QAM it is 3 bits/subcarrier.

Figure 9.54 shows the performance of the integrated systems in conjunction with a channel interleaver of 5000 symbols using a QPSK modem. We can see from the figure that BICM performs about 3.8 dB better than TCM at a BER of 10^{-4}. When the iterative schemes of TTCM and BICM-ID are used, a better performance is achieved. When using eight iterations, TTCM and BICM-ID reach their optimum performance. Clearly, in this scenario TTCM is superior to BICM-ID.

Figure 9.55 illustrates the performance of the systems considered for an effective throughput of 2 bits/subcarrier using a channel interleaver of 5000 symbols. TCM, BICM and BICM-ID exhibit a similar complexity; however, TTCM using eight iterations is somewhat more complex. Specifically, in order to be able to compare the associated complexities we assumed that the number of decoder trellis states determined the associated complexity. For example, a memory-length $K = 3$ TTCM scheme had $2^K = 8$ trellises per decoding iteration and there are two decoders. Hence after four iterations we encounter a total of $8 \cdot 2 \cdot 4 = 64$ trellis states. Hence

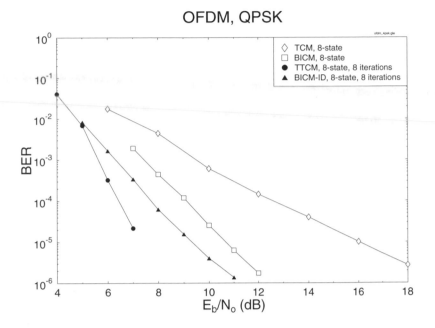

Figure 9.54: Comparison of TCM, TTCM, BICM and BICM-ID using QPSK-OFDM modem for transmissions over the Rayleigh fading COST207 HT channel of Figure 9.53 at a normalised Doppler frequency of 10^{-5}. The OFDM parameters are listed in Table 9.9, while the coded modulation parameters were summarised in Section 9.7.2.1.

a 64-state, $K = 6$ TCM scheme has a similar complexity to a $K = 4$ TTCM arrangement using four iterations. For the sake of fair comparisons, TTCM employing four iterations should be used, although the performance difference between four and eight iterations is only marginal. Again, TTCM is the best scheme. At a BER of 10^{-4}, TTCM performs about 1.2 dB better than BICM-ID, 2.2 dB better than BICM and 3.6 dB better than TCM.

Figure 9.56 compares the performance of the systems for a throughput of 3 bits/subcarrier using a channel interleaver of 5000 symbols. Again, the systems exhibited a similar complexity, except for TTCM. There is one uncoded information bit in the 4-bit 16QAM symbol of the rate-3/4 TCM code having 64 states, as can be seen from its generator polynomial in Table 9.6. For this reason, this TCM scheme is only potent at lower SNRs, while exhibiting modest performance improvements in the higher SNR region, as demonstrated by Figure 9.56. The rest of the schemes do not have uncoded information bits in their 16QAM symbols. BICM-ID outperforms BICM for E_b/N_0 values in excess of about 1.2 dB, while it is inferior in comparison to TTCM by about 1 dB at a BER of 10^{-4}. However, TTCM using 8-state TCM component codes exhibits an error floor at a BER around 10^{-5}.

9.7.3.5.5 Conclusions In this section OFDM was studied and integrated with the coded modulation schemes of Sections 9.2–9.6. The performance of OFDM-assisted TCM, TTCM, BICM and BICM-ID was investigated for transmissions over the dispersive COST207 HT Rayleigh fading channel of Figure 9.53 using QPSK, 8PSK and 16QAM modulation modes. TTCM was found to be the best compromise scheme, followed by BICM-ID, BICM and TCM.

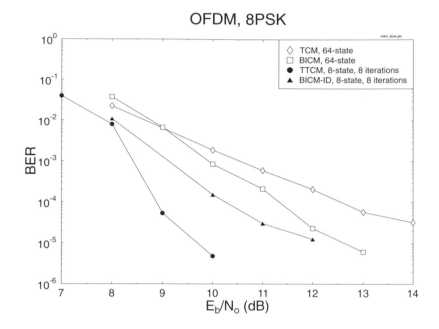

Figure 9.55: Comparison of TCM, TTCM, BICM and BICM-ID using 8PSK-OFDM modem for transmissions over the Rayleigh fading COST207 HT channel of Figure 9.53 at a normalised Doppler frequency of 10^{-5}. The OFDM parameters are listed in Table 9.9 while the coded modulation parameters were summarised in Section 9.7.2.1.

9.8 Summary and Conclusions

In Sections 9.2–9.6 we have studied the conceptual differences between four coded modulation schemes in terms of their coding structure, signal labelling philosophy, interleaver type and decoding philosophy. The symbol-based non-binary MAP algorithm was also highlighted, when operating in the logarithmic domain.

Furthermore, in Section 9.7 the performance of the above-mentioned four coded modulation schemes was evaluated for transmissions over narrowband channels in Section 9.7.2 and over wideband channels in Section 9.7.3. Over the dispersive Rayleigh fading channels a DFE was utilised for supporting the operation of the coded modulation schemes, as investigated in Section 9.7.3.3.

TCM was found to perform better than BICM in AWGN channels owing to its higher Euclidean distance, while BICM faired better in fading channels, since it was designed for fading channels with a higher grade of time diversity in mind. BICM-ID gave better performance both in AWGN and fading channels compared to TCM and BICM, although it exhibited a higher complexity due to employing several decoding iterations, while TTCM struck the best balance between performance and complexity.

The performance of the BbB DFE-assisted adaptive coded modulation scheme was investigated in Section 9.7.3.3, and attained an improved performance in comparison to the fixed-mode-based coded modulation schemes in terms of both its BER and BPS performance.

OFDM was also invoked for assisting the operation of coded modulation schemes in highly dispersive propagation environments, as investigated in Section 9.7.3.5 in a multi-carrier transmission scenario. Again, TTCM/OFDM struck the most attractive trade-off in terms of its perfor-

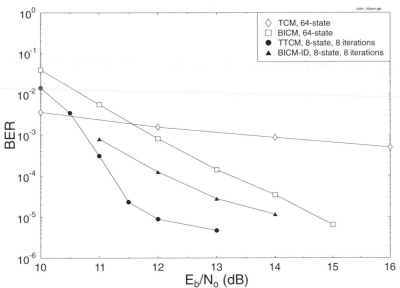

Figure 9.56: Comparison of TCM, TTCM, BICM and BICM-ID using 16QAM-OFDM modem for transmissions over the Rayleigh fading COST207 HT channel of Figure 9.53 at a normalised Doppler frequency of 10^{-5}. The OFDM parameters are listed in Table 9.9 while the coded modulation parameters were summarised in Section 9.7.2.1.

mance versus complexity balance in these investigations, closely followed by BICM-ID/OFDM.

Having introduced a host of channel coding techniques in the previous chapters, in the next two chapters we will concentrate on the family of transmit-diversity-aided space-time coding arrangements, which are very powerful in terms of mitigating the effects of fading when communicating over wireless channels.

Part IV

Space-Time Block and Space-Time Trellis Coding

Chapter 10

Space-time Block Codes

10.1 Introduction[1]

In this chapter, as well as in the forthcoming one, we will concentrate our attention mainly on systems designed for transmissions over wireless links, where the channel errors tend to occur in bursts, rather than to be randomly distributed. Hence here we assume a basic background in the area of mobile communication channel properties. We also assume that the reader is well versed in channel coding in general, since a range of previously studied channel coding schemes will be combined with various space-time coding arrangements in these two chapters.

The-third generation (3G) mobile communications standards [266] are expected to provide a wide range of bearer services, ranging from voice to high-rate data services, supporting rates of at least 144 kbit/s in vehicular, 384 kbit/s in outdoor-to-indoor and 2 Mbit/s in indoor as well as picocellular applications [266].

In an effort to support such high rates, the bit/symbol capacity of band-limited wireless channels can be increased by employing multiple antennas [267]. The classic approach is to use multiple antennas at the receiver and invoke Maximum Ratio Combining (MRC) [268–270] of the received signals for improving the performance. However, applying receiver diversity at the Mobile Stations (MSs) increases their complexity. Hence receiver diversity techniques typically have been applied at the Base Stations (BSs). The BSs provide services for many MSs and hence upgrading the BSs is economically viable. However, the drawback of this scheme is that it only provides diversity gain for the BSs' receivers.

In the past, different transmit diversity techniques have been introduced, in order to provide diversity gain for MSs by upgrading the BSs. These transmit diversity techniques can be classified into three main categories: schemes using information feedback [271, 272], arrangements invoking feedforward or training information [273–275] and blind schemes [276, 277]. Recently, Tarokh *et al.* proposed space-time trellis coding [70, 80, 278–281] by jointly designing the channel coding, modulation, transmit diversity and the optional receiver diversity scheme. The performance criteria for designing space-time trellis codes were derived in [70], under the assumption that the channel is fading slowly and that the fading is frequency non-selective. These advances were then also extended to fast fading channels. The encoding and decoding complex-

---

[1]This chapter is based on T. H. Liew and L. Hanzo: Space-time Block Codes and Concatenated Channel Codes: A Historical Perspective and Comparative Study, Proc. of the IEEE, February 2001.

ity of these *space-time trellis codes* is comparable to that of conventional trellis codes [46–48] often employed in practice over non-dispersive Gaussian channels.

Space-time trellis codes [70, 80, 278–281] perform extremely well at the cost of relatively high complexity. In addressing the issue of decoding complexity, Alamouti [71] discovered a remarkable scheme for transmissions using two transmit antennas. A simple decoding algorithm was also introduced by Alamouti [71], which can be generalised to an arbitrary number of receiver antennas. This scheme is significantly less complex than space-time trellis coding using two transmitter antennas, although there is a loss in performance [72]. Despite the associated performance penalty, Alamouti's scheme is appealing in terms of its simplicity and performance. This proposal motivated Tarokh *et al.* [72, 73] to generalise Alamouti's scheme to an arbitrary number of transmitter antennas, leading to the concept of *space-time block codes*.

Intrigued by the decoding simplicity of the space-time block codes proposed in [71–73], in this chapter we commence our discourse by detailing their encoding and decoding processes. Subsequently, we investigate the performance of the space-time block codes over perfectly interleaved, non-dispersive Rayleigh fading channels. A system which consists of space-time block codes and different channel coders will be proposed. Finally, the performance and estimated complexity of the different systems will be compared and tabulated.

Following a rudimentary introduction to space-time block codes in Section 10.3 and to channel-coded space-time codes in Section 10.4, the associated estimated complexity issues and memory requirements are addressed in Section 10.4.3. The bulk of this chapter is constituted by the performance study of various space-time and channel-coded transceivers in Section 10.5. Our aim is first to identify a space-time code, channel code combination constituting a good engineering trade-off in terms of its effective throughput, BER performance and estimated complexity in Section 10.5.1. Specifically, the issue of bit-to-symbol mapping is addressed in the context of convolution codes and convolutional coding as well as Bose–Chaudhuri–Hocquenghem (BCH) coding-based turbo codes in conjunction with an attractive unity-rate space-time code and multilevel modulation in Section 10.5.2. These schemes are also benchmarked against a range of powerful Trellis-Coded Modulation (TCM) and Turbo Trellis-Coded Modulation (TTCM) schemes. Our conclusions concerning the merits of the various schemes are presented in Section 10.5.4 in the context of their coding gain versus estimated complexity.

10.2   Background

In this section, we present a brief overview of space-time block codes by considering the classical Maximum Ratio Combining (MRC) technique [71, 102, 282]. The introduction of this classical technique is important, since at a later stage it will assist us in highlighting the philosophy of space-time block codes.

10.2.1   Maximum Ratio Combining

In conventional transmission systems we have a single transmitter which transmits information to a single receiver. In Rayleigh fading channels the transmitted symbols experience severe magnitude fluctuation and phase rotation. In order to mitigate this problem, we can employ several receivers that receive replicas of the same transmitted symbol through independent fading paths. Even if a particular path is severely faded, we may still be able to recover a reliable estimate of the transmitted symbols through other propagation paths. However, at the station we have to combine the received symbols of the different propagation paths, which involves additional complexity. Again, the classical method often used in practice is referred to as the MRC technique [71, 102, 282].

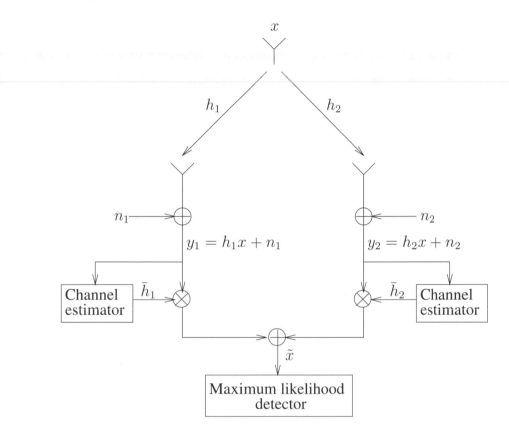

Figure 10.1: Baseband representation of the MRC technique using two receivers.

Figure 10.1 shows the baseband representation of the classical MRC technique in conjunction with two receivers. At a particular instant, a symbol x is transmitted. As we can see from the figure, the transmitted symbol x propagates through two different channels, namely h_1 and h_2. For simplicity, all channels are assumed to be constituted by a single propagation path and can be modelled as complex multiplicative distortion, which consists of a magnitude and phase response given as follows:

$$
\begin{aligned}
h_1 &= |h_1|e^{j\theta_1} \\
h_2 &= |h_2|e^{j\theta_2} \,,
\end{aligned}
$$
(10.1)
(10.2)

where $|h_1|$, $|h_2|$ are the fading magnitudes and θ_1, θ_2 are the phase values. Noise is added by each receiver, as shown in Figure 10.1. Hence, the resulting received baseband signals are:

$$
\begin{aligned}
y_1 &= h_1 x + n_1 \\
y_2 &= h_2 x + n_2 \,,
\end{aligned}
$$
(10.3)
(10.4)

where n_1 and n_2 are complex noise samples. In matrix form this can be written as follows:

$$\begin{pmatrix} y_1 \\ y_2 \end{pmatrix} = x \begin{pmatrix} h_1 \\ h_2 \end{pmatrix} + \begin{pmatrix} n_1 \\ n_2 \end{pmatrix} . \tag{10.5}$$

Assuming that we have perfect channel information, i.e. a perfect channel estimator, the received signals y_1 and y_2 can be multiplied by the conjugate of the complex channel transfer functions $\bar{h}_1$ and $\bar{h}_2$, respectively, in order to remove the channel's effects. Then the corresponding signals are combined at the input of the maximum likelihood detector of Figure 10.1 as follows:

$$\begin{aligned} \tilde{x} &= \bar{h}_1 y_1 + \bar{h}_2 y_2 \\ &= \bar{h}_1 h_1 x + \bar{h}_1 n_1 + \bar{h}_2 h_2 x + \bar{h}_2 n_2 \\ &= \left(|h_1|^2 + |h_2|^2 \right) x + \bar{h}_1 n_1 + \bar{h}_2 n_2 . \end{aligned} \tag{10.6}$$

The combined signal $\tilde{x}$ is then passed to the maximum likelihood detector, as shown in Figure 10.1. Based on the Euclidean distances between the combined signal $\tilde{x}$ and all possible transmitted symbols, the most likely transmitted symbol is determined by the maximum likelihood detector. The simplified decision rule is based on choosing x_i if and only if:

$$dist(\tilde{x}, x_i) \leq dist(\tilde{x}, x_j) , \quad \forall i \neq j , \tag{10.7}$$

where $dist(A, B)$ is the Euclidean distance between signals A and B, and the index j spans all possible transmitted signals. From Equation 10.7 we can see that maximum likelihood transmitted symbol is the one having the minimum Euclidean distance from the combined signal $\tilde{x}$.

10.3 Space-time Block Codes

In the previous section we have briefly introduced the classic MRC technique. In this section we will present the basic principles of space-time block codes. In analogy to the MRC matrix formula of Equation 10.5, a space-time block code describing the relationship between the original transmitted signal x and the signal replicas artificially created at the transmitter for transmission over various diversity channels is defined by an $n \times p$ dimensional transmission matrix. The entries of the matrix are constituted by linear combinations of the $k$ary input symbols $x_1, x_2, \ldots, x_k$ and their conjugates. The $k$ary input symbols x_i, $i = 1 \ldots k$, are used to represent the information bearing binary bits to be transmitted over the transmit diversity channels. In a signal constellation having 2^b constellation points, b number of binary bits are used to represent a symbol x_i. Hence, a block of $k \times b$ binary bits is entered into the space-time block encoder at a time and it is therefore referred to as a space-time block code. The number of transmitter antennas is p and n represents the number of time slots used to transmit k input symbols. Hence, a general form of the transmission matrix of a space-time block code is as follows:

$$\begin{pmatrix} g_{11} & g_{21} & \cdots & g_{p1} \\ g_{12} & g_{22} & \cdots & g_{p2} \\ . & . & . & . \\ . & . & . & . \\ g_{1n} & g_{2n} & \cdots & g_{pn} \end{pmatrix} , \tag{10.8}$$

where the entries g_{ij} represent linear combinations of the symbols $x_1, x_2, \ldots, x_k$ and their conjugates. More specifically, the entries g_{ij}, where $i = 1 \ldots p$, are transmitted simultaneously from transmit antennas $1, \ldots, p$ in each time slot $j = 1, \ldots, n$. For example, in time slot $j = 2$, signals $g_{12}, g_{22}, \dots, g_{p2}$ are transmitted simultaneously from transmit antennas $Tx\ 1, Tx\ 2, \dots, Tx\ p$. We can see in the transmission matrix defined in Equation 10.8 that encoding is carried out in both space and time; hence the term space-time coding.

The $n \times p$ transmission matrix in Equation 10.8, which defines the space-time block code, is based on a complex generalised orthogonal design, as defined in [71–73]. Since there are k symbols transmitted over n time slots, the code rate of the space-time block code is given by:

$$R = k/n . \tag{10.9}$$

At the receiving end, we can have an arbitrary number of q receivers. It was shown in [71] that the associated diversity order is $p \times q$. A combining technique [71, 73] similar to MRC can be applied at the receiving end, which may be generalised to q number of receivers. At the current state of the art the associated diversity channels are often assumed to be flat fading channels. A possible approach to satisfying this condition for high-rate transmissions over frequency-selective channels is to split the high-rate bit-stream into a high number of low-rate streams transmitted over flat fading sub-channels This can be achieved with the aid of Orthogonal Frequency Division Multiplexing (OFDM) [207]. It is also typically assumed that the complex fading envelope is constant over n consecutive time slots.

10.3.1 A Twin-transmitter-based Space-time Block Code

As mentioned above, the simplest form of space-time block codes was proposed by Alamouti in [71], which is a simple twin-transmitter-based scheme associated with $p = 2$. The transmission matrix is defined as follows:

$$\mathbf{G}_2 = \begin{pmatrix} x_1 & x_2 \\ -\bar{x}_2 & \bar{x}_1 \end{pmatrix} . \tag{10.10}$$

We can see in the transmission matrix $\mathbf{G}_2$ that there are $p = 2$ (number of columns in the matrix $\mathbf{G}_2$) transmitters, $k = 2$ possible input symbols, namely x_1, x_2, and the code spans over $n = 2$ (number of rows in the matrix $\mathbf{G}_2$) time slots. Since $k = 2$ and $n = 2$, the code rate given by Equation 10.9 is unity. The associated encoding and transmission process is shown in Table 10.1. At any given time instant T, two signals are simultaneously transmitted from the antennas $Tx\ 1$

Time	antenna	
slot, T	$Tx\ 1$	$Tx\ 2$
1	x_1	x_2
2	$-\bar{x}_2$	$\bar{x}_1$

Table 10.1: The encoding and transmission process for the $\mathbf{G}_2$ space-time block code of Equation 10.10.

and $Tx\ 2$. For example, in the first time slot associated with $T = 1$, signal x_1 is transmitted from antenna $Tx\ 1$ and *simultaneously* signal x_2 is transmitted from antenna $Tx\ 2$. In the next time slot corresponding to $T = 2$, signals $-\bar{x}_2$ and $\bar{x}_1$ (the conjugates of symbols x_1 and x_2) are simultaneously transmitted from antennas $Tx\ 1$ and $Tx\ 2$, respectively.

10.3.1.1 The Space-time Code $\mathbf{G}_2$ Using One Receiver

Lets us now consider an example of encoding and decoding the $\mathbf{G}_2$ space-time block code of Equation 10.10 using one receiver. This example can be readily extended to an arbitrary number

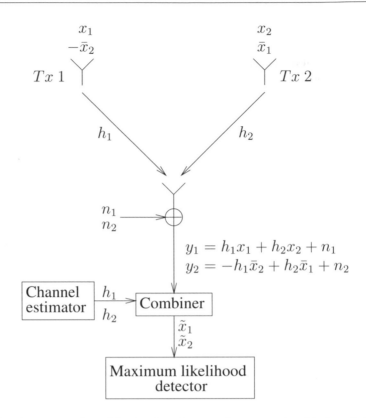

Figure 10.2: Baseband representation of the simple twin-transmitter space-time block code $\mathbf{G}_2$ of Equation 10.10 using one receiver. ©IEEE, Alamouti [71], 1998

of receivers. In Figure 10.2 we show the baseband representation of a simple two-transmitter space-time block code, namely that of the $\mathbf{G}_2$ code seen in Equation 10.10 using one receiver. We can see from the figure that there are two transmitters, namely Tx 1 as well as Tx 2, and they transmit two signals simultaneously. As mentioned earlier, the complex fading envelope is assumed to be constant across the corresponding two consecutive time slots. Therefore, we can write

$$
\begin{aligned}
h_1 &= h_1(T=1) = h_1(T=2) & (10.11) \\
h_2 &= h_2(T=1) = h_2(T=2) \,. & (10.12)
\end{aligned}
$$

Independent noise samples are added at the receiver in each time slot and hence the received signals can be expressed with the aid of Equation 10.10 as:

$$
\begin{aligned}
y_1 &= h_1 x_1 + h_2 x_2 + n_1 & (10.13) \\
y_2 &= -h_1 \bar{x}_2 + h_2 \bar{x}_1 + n_2 \,, & (10.14)
\end{aligned}
$$

where y_1 is the first received signal and y_2 is the second. Notice that the received signal y_1 consists of the transmitted signals x_1 and x_2, and y_2 of their conjugates. In order to determine the transmitted symbols, we have to extract the signals x_1 and x_2 from the received signals y_1 and y_2. Therefore, both signals y_1 and y_2 are passed to the combiner, as shown in Figure 10.2.

In the combiner — aided by the channel estimator, which provides perfect estimation of the diversity channels in this example — simple signal processing is performed in order to separate the signals x_1 and x_2. Specifically, in order to extract the signal x_1, we combine the received signals y_1 and y_2 as follows:

$$
\begin{aligned}
\tilde{x}_1 &= \bar{h}_1 y_1 + h_2 \bar{y}_2 \\
&= \bar{h}_1 h_1 x_1 + \bar{h}_1 h_2 x_2 + \bar{h}_1 n_1 - h_2 \bar{h}_1 x_2 + h_2 \bar{h}_2 \bar{x}_1 + h_2 \bar{n}_2 \\
&= \left(|h_1|^2 + |h_2|^2 \right) x_1 + \bar{h}_1 n_1 + h_2 \bar{n}_2 .
\end{aligned}
\tag{10.15}
$$

Similarly, for signal x_2 we generate:

$$
\begin{aligned}
\tilde{x}_2 &= \bar{h}_2 y_1 - h_1 \bar{y}_2 \\
&= \bar{h}_2 h_1 x_1 + \bar{h}_2 h_2 x_2 + \bar{h}_2 n_1 + h_1 \bar{h}_1 x_2 - h_1 \bar{h}_2 x_1 - h_1 \bar{n}_2 \\
&= \left(|h_1|^2 + |h_2|^2 \right) x_2 + \bar{h}_2 n_1 - h_1 \bar{n}_2 .
\end{aligned}
\tag{10.16}
$$

Clearly, from Equations 10.15 and 10.16 we can see that we have separated the signals x_1 and x_2 by simple multiplications and additions. From the orthogonality of the space-time block code G_2 in Equation 10.10 [72], the unwanted signal x_2 is cancelled out in Equation 10.15 and, vice versa, signal x_1 is removed from Equation 10.16. Both signals $\tilde{x}_1$ and $\tilde{x}_2$ are then passed to the maximum likelihood detector of Figure 10.2, which applies Equation 10.7 to determine the most likely transmitted symbols.

From Equations 10.15 and 10.16 we can derive a simple rule of thumb for manipulating the received signal in order to extract a symbol x_i. For each received signal y_j, we would have a linear combination of the transmitted signals x_i convolved with the corresponding Channel Impulse Response (CIR) h_i. The non-dispersive CIR is assumed to be constituted by a single CIR tap corresponding to a complex multiplicative factor. The conjugate of the CIR $\bar{h}_i$ should be multiplied with the received signal y_j, if x_i is in the expression of the received signal y_j. However, if the conjugate of x_i, namely $\bar{x}_i$, is present in the expression, we should then multiply the CIR h_i with the conjugate of the received signal y_j, namely $\bar{y}_j$. The product should then be added to or subtracted from the rest, depending on the sign of the term in the expression of the received signal y_j.

10.3.1.2   The Space-time Code G_2 Using Two Receivers

In Section 10.3.1.1 we have shown an example of the encoding and decoding process for the G_2 space-time block code of Equation 10.10 using one receiver. However, this example can be readily extended to an arbitrary number of receivers. The encoding and transmission sequence will be identical to the case of a single receiver. For illustration, we discuss the specific case of two transmitters and two receivers, as shown in Figure 10.3. We will show, however, that the generalisation to q receivers is straightforward. In Figure 10.3, the subscript i in the notation h_{ij}, n_{ij} and y_{ij} represents the receiver index. By contrast, the subscript j denotes the transmitter index in the CIR h_{ij}, but it denotes the time slot T in n_{ij} and y_{ij}. Therefore, at the first receiver $Rx\,1$ we have:

$$
y_{11} = h_{11} x_1 + h_{12} x_2 + n_{11}
\tag{10.17}
$$

$$
y_{12} = -h_{11} \bar{x}_2 + h_{12} \bar{x}_1 + n_{12} ,
\tag{10.18}
$$

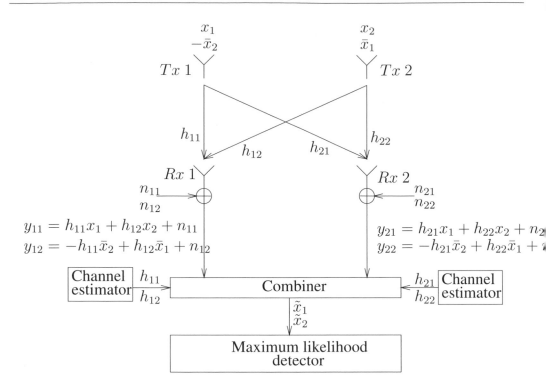

Figure 10.3: Baseband representation of the simple twin-transmitter space-time block code $\mathbf{G}_2$ of Equation 10.10 using two receivers. ©IEEE, Alamouti [71], 1998

while at receiver $Rx\ 2$ we have:

$$y_{21} = h_{21}x_1 + h_{22}x_2 + n_{21} \tag{10.19}$$

$$y_{22} = -h_{21}\bar{x}_2 + h_{22}\bar{x}_1 + n_{22}\ . \tag{10.20}$$

We can, however, generalise these equations to:

$$y_{i1} = h_{i1}x_1 + h_{i2}x_2 + n_{i1} \tag{10.21}$$

$$y_{i2} = -h_{i1}\bar{x}_2 + h_{i2}\bar{x}_1 + n_{i2}\ , \tag{10.22}$$

where $i = 1, \ldots, q$ and q is the number of receivers, which is equal to two in this example. At the combiner of Figure 10.3, the received signals are combined to extract the transmitted signals x_1 and x_2 from the received signals y_{11}, y_{12}, y_{21} and y_{22}, as follows:

$$\tilde{x}_1 = \bar{h}_{11}y_{11} + h_{12}\bar{y}_{12} + \bar{h}_{21}y_{21} + h_{22}\bar{y}_{22} \tag{10.23}$$

$$\tilde{x}_2 = \bar{h}_{12}y_{11} - h_{11}\bar{y}_{12} + \bar{h}_{22}y_{21} - h_{21}\bar{y}_{22}\ . \tag{10.24}$$

Again, we can generalise the above expressions to q receivers, yielding:

$$\tilde{x}_1 = \sum_{i=1}^{q}\left(\bar{h}_{i1}y_{i1} + h_{i2}\bar{y}_{i2}\right) \tag{10.25}$$

$$\tilde{x}_2 = \sum_{i=1}^{q}\left(\bar{h}_{i2}y_{i1} - h_{i1}\bar{y}_{i2}\right)\ . \tag{10.26}$$

Finally, we can simplify Equations 10.23 and 10.24 to:

$$\tilde{x}_1 = \left(|h_{11}|^2 + |h_{12}|^2 + |h_{21}|^2 + |h_{22}|^2 \right) x_1 + \bar{h}_{11}n_{11} + h_{12}\bar{n}_{12} + \bar{h}_{21}n_{21} + h_{22}\bar{n}_{22}$$

$$(10.27)$$

$$\tilde{x}_2 = \left(|h_{11}|^2 + |h_{12}|^2 + |h_{21}|^2 + |h_{22}|^2 \right) x_2 + \bar{h}_{12}n_{11} - h_{11}\bar{n}_{12} + \bar{h}_{22}n_{21} - h_{21}\bar{n}_{22} .$$

$$(10.28)$$

In the generalised form of q receivers we have:

$$\tilde{x}_1 = \sum_{i=1}^{q} \left[\left(|h_{i1}|^2 + |h_{i2}|^2 \right) x_1 + \bar{h}_{i1}n_{i1} + h_{i2}\bar{n}_{i2} \right] \qquad (10.29)$$

$$\tilde{x}_2 = \sum_{i=1}^{q} \left[\left(|h_{i1}|^2 + |h_{i2}|^2 \right) x_2 + \bar{h}_{i2}n_{i1} - h_{i1}\bar{n}_{i2} \right] . \qquad (10.30)$$

Signals $\tilde{x}_1$ and $\tilde{x}_2$ are finally derived and passed to the maximum likelihood detector seen in Figure 10.3. Again, Equation 10.7 is applied to determine the maximum likelihood transmitted symbols.

We observe in Equation 10.29 that signal x_1 is multiplied by a term related to the fading amplitudes, namely $|h_{i1}|^2 + |h_{i2}|^2$. Hence, in order to acquire a high reliability signal $\tilde{x}_1$, the amplitudes of the CIRs must be high. If the number of receivers is equal to one, i.e. $q = 1$, then Equation 10.29 is simplified to Equation 10.15. In Equation 10.15, we can see that there are two fading amplitude terms, i.e. two independent paths associated with transmitting the symbol x_1. Therefore, if either of the paths is in a deep fade, the other path still may provide a high reliability for the transmitted signal x_1. This explains why the performance of a system having two transmitters and one receiver is better than that of the system employing one transmitter and one receiver. On the other hand, in the conventional single-transmitter, single-receiver system there is only a single propagation path, which may be severely attenuated by a deep fade. To elaborate further, if the number of receivers is increased to $q = 2$, Equation 10.27 accrues from Equation 10.29. We can see in Equation 10.27 that there are now twice as many propagation paths as in Equation 10.15. This increases the probability of providing a high reliability for the signal $\tilde{x}_1$.

10.3.2 Other Space-time Block Codes

In Section 10.3.1 we have detailed Alamouti's simple two-transmitter space-time block code, namely the $\mathbf{G}_2$ code of Equation 10.10. This code is significantly less complex than the space-time trellis codes of [70, 80, 278–281] using two transmit antennas. However, again, there is a performance loss compared to the space-time trellis codes of [70, 80, 278–281]. Despite its performance loss, Alamouti's scheme [71] is appealing in terms of its simplicity. This motivated Tarokh *et al.* [72] to search for similar schemes using more than two transmit antennas. In [72] the theory of orthogonal code design was invoked, in order to construct space-time block codes having more than two transmitters. The half-rate space-time block code employing three transmitters was defined as [72]:

$$\mathbf{G}_3 = \begin{pmatrix} x_1 & x_2 & x_3 \\ -x_2 & x_1 & -x_4 \\ -x_3 & x_4 & x_1 \\ -x_4 & -x_3 & x_2 \\ \bar{x}_1 & \bar{x}_2 & \bar{x}_3 \\ -\bar{x}_2 & \bar{x}_1 & -\bar{x}_4 \\ -\bar{x}_3 & \bar{x}_4 & \bar{x}_1 \\ -\bar{x}_4 & -\bar{x}_3 & \bar{x}_2 \end{pmatrix}, \tag{10.31}$$

and the four-transmitter half-rate space-time block code was specified as [72]:

$$\mathbf{G}_4 = \begin{pmatrix} x_1 & x_2 & x_3 & x_4 \\ -x_2 & x_1 & -x_4 & x_3 \\ -x_3 & x_4 & x_1 & -x_2 \\ -x_4 & -x_3 & x_2 & x_1 \\ \bar{x}_1 & \bar{x}_2 & \bar{x}_3 & \bar{x}_4 \\ -\bar{x}_2 & \bar{x}_1 & -\bar{x}_4 & \bar{x}_3 \\ -\bar{x}_3 & \bar{x}_4 & \bar{x}_1 & -\bar{x}_2 \\ -\bar{x}_4 & -\bar{x}_3 & \bar{x}_2 & \bar{x}_1 \end{pmatrix}. \tag{10.32}$$

By employing the space-time block codes $\mathbf{G}_3$ and $\mathbf{G}_4$, we can see that the bandwidth efficiency has been reduced by a factor of two compared to the space-time block code $\mathbf{G}_2$. Besides, the number of transmission slots across which the channels are required to have a constant fading envelope is eight, namely four times higher than that of the space-time code $\mathbf{G}_2$.

In order to increase the associated bandwidth efficiency, Tarokh *et al.* constructed the rate-3/4 so-called generalised complex orthogonal sporadic codes [72, 73]. The corresponding rate-3/4 three-transmitter space-time block code is given by [72]:

$$\mathbf{H}_3 = \begin{pmatrix} x_1 & x_2 & \frac{x_3}{\sqrt{2}} \\ -\bar{x}_2 & \bar{x}_1 & \frac{x_3}{\sqrt{2}} \\ \frac{\bar{x}_3}{\sqrt{2}} & \frac{\bar{x}_3}{\sqrt{2}} & \frac{(-x_1-\bar{x}_1+x_2-\bar{x}_2)}{2} \\ \frac{\bar{x}_3}{\sqrt{2}} & -\frac{\bar{x}_3}{\sqrt{2}} & \frac{(x_2+\bar{x}_2+x_1-\bar{x}_1)}{2} \end{pmatrix}, \tag{10.33}$$

while the rate-3/4 four-transmitter space-time block code is defined as [72]:

$$\mathbf{H}_4 = \begin{pmatrix} x_1 & x_2 & \frac{x_3}{\sqrt{2}} & \frac{x_3}{\sqrt{2}} \\ -\bar{x}_2 & \bar{x}_1 & \frac{x_3}{\sqrt{2}} & -\frac{x_3}{\sqrt{2}} \\ \frac{\bar{x}_3}{\sqrt{2}} & \frac{\bar{x}_3}{\sqrt{2}} & \frac{(-x_1-\bar{x}_1+x_2-\bar{x}_2)}{2} & \frac{(-x_2-\bar{x}_2+x_1-\bar{x}_1)}{2} \\ \frac{\bar{x}_3}{\sqrt{2}} & -\frac{\bar{x}_3}{\sqrt{2}} & \frac{(x_2+\bar{x}_2+x_1-\bar{x}_1)}{2} & \frac{(-x_1-\bar{x}_1-x_2+\bar{x}_2)}{2} \end{pmatrix}. \tag{10.34}$$

In Table 10.2 we summarise the parameters associated with all space-time block codes proposed by Alamouti [71] and Tarokh, *et al.* [72,73]. The decoding algorithms and the corresponding performance results of the space-time block codes were given in [73].

10.3.3 MAP Decoding of Space-time Block Codes

Recently Bauch [283] derived a simple symbol-by-symbol Maximum-A-Posteriori (MAP) decoding rule for space-time block codes. The soft outputs provided by the space-time MAP decoder can be used as the input to channel decoders, such as for example turbo codes, which may be concatenated for further improving the system's performance.

Space-time code	Rate	No. of transmitters, p	No. of input symbols, k	Code span, n
$\mathbf{G}_2$	1	2	2	2
$\mathbf{G}_3$	1/2	3	4	8
$\mathbf{G}_4$	1/2	4	4	8
$\mathbf{H}_3$	3/4	3	3	4
$\mathbf{H}_4$	3/4	4	3	4

Table 10.2: Table of different space-time block codes.

By using Bayes' rule, the a-posteriori probability of the transmitted k-ary symbols $x_1, \ldots, x_k$ given the received signals $y_{11}, \ldots, y_{1n}, y_{21}, \ldots, y_{qn}$ can be expressed as [283]:

$$P(x_1, \ldots, x_k | y_{11}, \ldots, y_{qn}) = P(y_{11}, \ldots, y_{qn} | x_1, \ldots, x_k) \cdot P(x_1, \ldots, x_k), \qquad (10.35)$$

where $P(x_1, \ldots, x_k)$ is the associated a-priori information of the transmitted symbols which can be obtained from other independent sources, for example from channel decoders. Furthermore, according to Bauch [283], over non-dispersive Rayleigh fading channels we have:

$$P(y_{11}, \ldots, y_{qn} | x_1, \ldots, x_k) = \frac{1}{(\sigma \sqrt{2\pi})^{qn}} \exp \left\{ -\frac{1}{2\sigma^2} \sum_{l=1}^{q} \sum_{i=1}^{n} \left| y_{li} - \sum_{j=1}^{p} h_{lj} g_{ji} \right|^2 \right\},$$

$$(10.36)$$

where σ is the noise variance and g_{ij} are the entries of the transmission matrix in Equation 10.8. We can, however, simplify Equation 10.35 and obtain the expression for the a-posteriori probability of each transmitted symbol x_i as [283]:

$$P(x_i | y_{11}, \ldots, y_{qn}) = P(y_{11}, \ldots, y_{qn} | x_i) \cdot P(x_i), \qquad (10.37)$$

where $i = 1, \ldots, k$.

Let us now consider as an example the simplest possible space-time code, namely $\mathbf{G}_2$ associated with $k = 2$, $n = 2$ and $p = 2$. Assuming that there is no a-priori information, i.e. that $P(x_1, \ldots, x_k) = C$ where C is a constant, we obtain the a-posteriori information of the transmitted k-ary symbols from Equations 10.35 and 10.36 as [283]:

$$P(x_1, \ldots, x_k | y_{11}, \ldots, y_{qn})$$

$$= C \cdot \frac{1}{(\sigma \sqrt{2\pi})^{qn}} \exp \left\{ -\frac{1}{2\sigma^2} \sum_{l=1}^{q} \left[\left| y_{l1} - \sum_{j=1}^{p} h_{lj} g_{j1} \right|^2 + \left| y_{l2} - \sum_{j=1}^{p} h_{lj} g_{j2} \right|^2 \right] \right\}$$

$$= C' \cdot \exp \left\{ -\frac{1}{2\sigma^2} \sum_{l=1}^{q} \left[|y_{l1} - h_{l1} g_{11} - h_{l2} g_{21}|^2 + |y_{l2} - h_{l1} g_{12} - h_{l2} g_{22}|^2 \right] \right\}$$

$$= C' \cdot \exp \left\{ -\frac{1}{2\sigma^2} \sum_{l=1}^{q} \left[|y_{l1} - h_{l1} x_1 - h_{l2} x_2|^2 + |y_{l2} + h_{l1} \bar{x}_2 - h_{l2} \bar{x}_1|^2 \right] \right\}, \quad (10.38)$$

where $C' = C \cdot \frac{1}{(\sigma \sqrt{2\pi})^{qn}}$. In order to obtain the expression of the a-posteriori probability for symbol x_1, x_2-related terms can be eliminated in Equation 10.38 because of the orthogonality of the code, arriving at:

$$
\begin{aligned}
& P\left(x_1 | y_{11}, \ldots, y_{q2}\right) \\
&= C' \cdot \exp\left\{-\frac{1}{2\sigma^2} \sum_{l=1}^{q}\left[\left|y_{l1} - h_{l1}x_1\right|^2 + \left|y_{l2} - h_{l2}\bar{x}_1\right|^2\right]\right\} \\
&= C'' \cdot \exp\left\{-\frac{1}{2\sigma^2} \sum_{l=1}^{q}\left[-h_{l1}x_1\bar{y}_{l1} - \bar{h}_{l1}\bar{x}_1 y_{l1} - h_{l2}\bar{x}_1\bar{y}_{l2} - \bar{h}_{l2}x_1 y_{l2} + |x_1|^2 \sum_{i=1}^{2}|h_{li}|^2\right]\right\},
\end{aligned}
$$
(10.39)

where $|y_{l1}|^2$ and $|y_{l2}|^2$ are constants which do not depend on x_1 and hence incorporated into C''. Following a few further manipulations, we can simplify Equation 10.39 to:

$$
\begin{aligned}
& P\left(x_1 | y_{11}, \ldots, y_{q2}\right) \\
&= C \cdot \exp\left\{-\frac{1}{2\sigma^2}\left[\left|\left[\sum_{l=1}^{q}\left(\bar{h}_{l1}y_{l1} + h_{l2}\bar{y}_{l2}\right)\right] - x_1\right|^2 + \left(-1 + \sum_{l=1}^{q}\sum_{i=1}^{2}|h_{li}|^2\right)|x_1|\right]\right\}.
\end{aligned}
$$
(10.40)

Similarly, we can eliminate the x_1-related terms in Equation 10.38 and simplify it to:

$$
\begin{aligned}
& P\left(x_2 | y_{11}, \ldots, y_{q2}\right) \\
&= C \cdot \exp\left\{-\frac{1}{2\sigma^2}\left[\left|\left[\sum_{l=1}^{q}\left(\bar{h}_{l2}y_{l1} - h_{l1}\bar{y}_{l2}\right)\right] - x_2\right|^2 + \left(-1 + \sum_{l=1}^{q}\sum_{i=1}^{2}|h_{li}|^2\right)|x_2|\right]\right\}.
\end{aligned}
$$
(10.41)

It can be seen that Equations 10.40 and 10.41 resemble the equations given in [73] for the maximum likelihood decoding of the space-time code $\mathbf{G}_2$. Besides considering the space-time code $\mathbf{G}_2$, the maximum likelihood decoding algorithms were also given for the space-time codes $\mathbf{G}_3$, $\mathbf{G}_4$, $\mathbf{H}_3$ and $\mathbf{H}_4$ in [73]. It can be shown that Bauch's MAP algorithms [283] applicable to the space-time codes $\mathbf{G}_3$, $\mathbf{G}_4$, $\mathbf{H}_3$ and $\mathbf{H}_4$ also resemble the maximum likelihood algorithms given in [73].

10.4 Channel-coded Space-time Block Codes

In Section 10.3, we have given a detailed illustration of the concept of space-time block codes. The MAP decoding rules were also applied to space-time block codes in Section 10.3.3. This enables the space-time decoder to provide soft outputs, which in turn can be used by the concatenated channel decoders. Hence, in this section we concatenate space-time block codes with Convolutional Codes (CCs) [3, 284, 285], Turbo Convolutional (TC) codes [12, 13], Turbo BCH (TBCH) codes [61], Trellis-Coded Modulation (TCM) [46, 47] and Turbo-Trellis Coded Modulation (TTCM) [57]. The performance and estimated complexity of each scheme will be studied and compared. We will also address the issue of mapping channel coded bits of the TC and TBCH schemes to different protection classes in multilevel modulation [207].

Convolutional codes (CCs) were first suggested by Elias [3] in 1955. The so-called Viterbi Algorithm (VA) was proposed by Viterbi [8, 9] in 1967 for the maximum likelihood decoding of convolutional codes. As an alternative decoder, the more complex MAP algorithm was proposed by Bahl *et al.* [11], which provided the optimum Bit Error Rate (BER) performance, although this was not significantly better than that of the Viterbi algorithm. In the early 1970s, convolutional codes were used in deep-space and satellite communications. They were then also adopted by the Global System of Mobile communications (GSM) [49] for the pan-European digital cellular mobile radio system.

In 1993, Berrou *et al.* [12, 13] proposed a novel channel code, referred to as a turbo code. As detailed in Chapter 7, the turbo encoder consists of two component encoders. Generally, convolutional codes are used as the component encoders, and the corresponding turbo codes are termed here as a TC code. However, BCH [49, 86] codes can also be employed as their component codes, resulting in the so-called turbo BCH codes (TBCH). They have been shown for example by Hagenauer *et al.* [61, 63] to perform impressively at near-unity coding rates, although at a higher decoding complexity than that of the corresponding-rate TC codes.

In 1987, Ungerboeck [46, 47] invented Trellis Coded Modulation (TCM) by combining the design of channel coding and modulation. TCM optimises the Euclidean distance between code-words and hence maximises the coding gain. In [57], Robertson and Wörz applied the basic idea of turbo codes [12, 13] to TCM by retaining the important properties and advantages of both structures. In the resultant TTCM scheme, two Ungerboeck codes [46, 47] are employed in combination with TCM as component codes in an overall structure similar to that of turbo codes.

10.4.1 System Overview

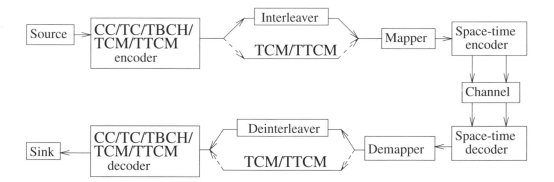

Figure 10.4: System overview of space-time block codes and different channel coding schemes.

The schematic of the proposed concatenated space-time block codes and the different channel coding schemes is shown in Figure 10.4. As mentioned above, the investigated channel coding schemes are CCs, TC codes, TBCH codes, TCM and TTCM. The information source at the transmitter of Figure 10.4 generates random data bits. The information bits are then encoded by each of the above five different channel coding schemes. However, only the output binary bits of the CC, TBCH and TC coding schemes are channel interleaved, as seen in Figure 10.4. The role of the interleaver will be detailed in Section 10.5.2.

The output bits of the TCM and TTCM schemes are passed directly to the mapper in Figure 10.4, which employs two different mapping techniques. Gray mapping [55, 86, 286] is used for the CC, TBCH and TC schemes, whereas set partitioning [46–48, 57] is utilised for the TCM and TTCM schemes. Different modulation schemes are employed, namely Binary Phase Shift

Keying (BPSK), Quadrature Phase Shift Keying (QPSK), 8-level Phase Shift Keying (8PSK), 16-level Quadrature Amplitude Modulation (16QAM) and 64-level Quadrature Amplitude Modulation (64QAM) [207].

Following the mapper, the channel-coded symbols are passed to the space-time block encoder, as shown in Figure 10.4. Below, we will investigate the performance of all the previously mentioned space-time block codes, namely that of the G_2, G_3, G_4, H_3 and H_4 codes proposed in [71–73]. The corresponding transmission matrices are given in Equations 10.10, 10.31, 10.32, 10.33 and 10.34, respectively. The coding rate and number of transmitters of the associated space-time block codes are shown in Table 10.2. The channels are uncorrelated or — synonymously — perfectly interleaved narrowband or non-dispersive Rayleigh fading channels. This assumption does not contradict requiring a constant channel magnitude and phase over p (number of rows in the transmission matrix) consecutive symbols, since upon applying a sufficiently high channel interleaving depth the channels' fading envelope can become indeed near-uncorrelated. We assumed that the narrowband fading amplitudes received from each transmitter antenna were mutually uncorrelated Rayleigh-distributed processes. The average signal power received from each transmitter antenna was the same. Furthermore, we assumed that the receiver had a perfect estimate of the channels' fading amplitudes. In practice, the channels' fading amplitude can be estimated for example with the aid of pilot symbols [207].

At the receiver, the number of receiver antennas constitutes a design parameter, which was fixed to one, unless specified otherwise. The space-time block decoders then apply the MAP or Log-MAP decoding algorithm of Section 10.3.3 for the decoding of the signals received from the different antennas. Owing to its implementational simplicity, the Log-MAP decoding algorithm is preferred in the proposed system. The soft outputs associated with the received bits or symbols are passed through the channel de-interleaver or directly to the TCM/TTCM decoder, respectively, as seen in Figure 10.4. The channel-de-interleaved soft outputs of the received bits are then passed to the CC, TC or TBCH decoders. The Viterbi algorithm [8,9] is applied in the CC and TCM decoder. By contrast, all turbo decoder schemes apply the Log-MAP [13,57,61] decoding algorithm. The decoded bits are finally passed to the information sink for calculation of the BER, as shown in Figure 10.4.

10.4.2   Channel Codec Parameters

In Figure 10.4, we have given an overview of the proposed system. As we can see in the figure, there are different channel encoders to be considered, namely the CC, TC, TBCH, TCM and TTCM schemes. In this section, we present the parameters of all the channel codecs to be used in our investigations.

Table 10.3 shows the parameters of each channel encoder proposed in the system. We commence with the most well-known channel code, namely the convolutional code. A convolutional code is described by three parameters n, k and K and it is denoted as $CC(n,k,K)$. At each instant, a $CC(n,k,K)$ encoder accepts k input bits and outputs n coded bits. The constraint length of the code is K and the number of encoder states is equal to 2^{K-1}. The channel coded rate is given by:

$$R = \frac{k}{n} . \tag{10.42}$$

However, different code rates can be obtained by suitable puncturing [233] and we will elaborate on this issue later in the section. The first entry of Table 10.3 is the $CC(2,1,5)$ code, which was adopted by the Groupe Speciale Mobile (GSM) committee in 1982 [49,287]. Then in 1996,

Code	Octal generator polynomial	No. of states	Decoding algorithm	No. of iterations
Convolutional Code (CC)				
CC(2,1,5)	23,33	16	VA	−
CC(2,1,7)	171,133	64	VA	−
CC(2,1,9)	561,753	256	VA	−
Turbo Convolutional Code (TC)				
TC(2,1,3)	7,5	4	Log-Map	8
TC(2,1,4)	13,15	8	Log-Map	8
TC(2,1,5)	23,35	16	Log-Map	8
Turbo BCH Code (TBCH)				
TBCH(31,26)	45	32	Log-Map	8
TBCH(32,26)	45	64	Log-Map	8
TBCH(31,21)	3551	1024	Log-Map	8
TBCH(63,57)	103	64	Log-Map	8
TBCH(127,120)	211	128	Log-Map	8
Trellis-Coded Modulation (TCM)				
8PSK-TCM	103,30,66	64	VA	−
16QAM-TCM	101,16,64	64	VA	−
Turbo Trellis-Coded Modulation (TTCM)				
8PSK-TTCM	11,2,4	8	Log-Map	8
16QAM-TTCM	23,2,4,10	16	Log-Map	8

Table 10.3: Parameters of the different channel encoders used in Figure 10.4.

a more powerful convolutional code, the CC(2, 1, 7) arrangement, was employed by the Digital Video Broadcasting (DVB) [39] standard for television, sound and data services. Recently, the Universal Mobile Telecommunication System (UMTS) proposed the use of the CC(2, 1, 9) scheme, which is also shown in Table 10.3. The implementation of this scheme is about 16 times more complex than that of the CC(2, 1, 5) scheme adopted by GSM some 15 years ago. This clearly shows that the advances of integrated circuit technology have substantially contributed towards the performance improvement of mobile communication systems.

As mentioned earlier, a turbo encoder consists of two component encoders. Generally, two identical Recursive Systematic Convolutional (RSC) codes are used. Berrou *et al.* [12, 13] used two constraint length $K = 3$, RSC codes, each having four trellis states. We denote a turbo convolutional code as TC(n, k, K) where n, k and K have their usual interpretations, as in CC. In [12, 13], the MAP algorithm [11] was employed for iterative decoding. However, in our systems the Log-MAP decoding algorithm [52] was utilised. The Log-MAP algorithm is a more attractive version of the MAP algorithm, since it operates in the logarithmic domain, in order to reduce the computational complexity and to mitigate the numerical problems associated with the MAP algorithm [52]. The number of turbo iterations was set to eight, since this yielded a performance close to the optimum performance associated with an infinite number of iterations. In our investigations we will consider the TC(2, 1, 3) code, as proposed in [12, 13]. However, the more complex TC(2, 1, 4) code [288] was proposed by UMTS to be employed in the third-generation (3G) mobile communication systems [49, 266, 289]. The TC(2, 1, 5) code is also interesting,

since it is expected to provide further significant coding gains over that of the TC$(2, 1, 3)$ and
TC$(2, 1, 4)$ codes.

BCH codes [49] are used as the component codes in the TBCH codes of Table 10.3. Again,
TBCH codes have been shown for example by Hagenauer *et al.* [61,63] to perform impressively
at near-unity coding rates, although at high complexity. Hence in our study the BCH component
codes BCH$(31, 26)$, BCH$(31, 21)$, BCH$(63, 57)$ and BCH$(127, 120)$ are employed, as shown in
Table 10.3. Finally, we also investigate TCM and TTCM. Both of them are employed in 8PSK
and 16QAM modulation modes. This results in 8PSK-TCM, 16QAM-TCM and 8PSK-TTCM,
16QAM-TTCM, respectively.

In Table 10.3, we have given the encoding and decoding parameters of the different channel
encoders employed. However, as mentioned earlier, we can design codes of variable code rates R
by employing suitable puncturing patterns. By combining puncturing with different modulation
modes, we could design a system having a range of various throughputs, expressed in terms of
the number of Bits Per Symbol (BPS), as shown in Tables 10.4 and 10.5. Some of the parameters
in Tables 10.4 and 10.5 are discussed in depth during our further discourse, but significantly more
information can be gleaned concerning these systems by carefully studying both tables.

Code	Code rate R	Puncturing pattern	Modula-tion mode	BPS	Random interleaver depth
Convolutional Code (CC)					
CC(2,1,5)	0.50	1,1	QPSK	1.00	20000
CC(2,1,7)	0.50	1,1	QPSK	1.00	20000
	0.75	101, 110	64QAM	4.50	13320
	0.83	10101, 11010	64QAM	5.00	12000
CC(2,1,9)	0.50	1,1	QPSK	1.00	20000
			16QAM	2.00	20000
			64QAM	3.00	20004
Trellis-Coded Modulation (TCM)					
8PSK-TCM	0.67	1,1	8PSK	2.00	–
16QAM-TCM	0.75	1,1	16QAM	3.00	–

Table 10.4: Simulation parameters associated with the CC and TCM channel encoders in Figure 10.4.

In Table 10.4 we summarise the simulation parameters of the CC and TCM schemes employed. Since there are two coded bits ($n = 2$) for each data bit ($k = 1$), we have two possible
puncturing patterns, as shown in the table. A binary 1 means that the coded bit is transmitted,
whereas a binary 0 implies that the coded bit is punctured. Accordingly, the puncturing pattern
$(1, 1)$ simply implies that no puncturing is applied and hence results in a half-rate CC. However, for example in the DVB standard [39] different puncturing patterns were proposed for the
CC$(2, 1, 7)$ code, which result in different coding rates. These are also shown in Table 10.4.

In Table 10.5 we show the simulation parameters of three different turbo schemes, namely
that of the TC, TBCH and TTCM arrangements. Again, different code rates can be designed using suitable puncturing patterns, where the puncturing patterns seen in Table 10.5 consist of two
parts. Specifically, the associated different puncturing patterns represent the puncturing patterns
of the parity bits emanating from the first and the second encoder, respectively. These patterns
are different from the puncturing patterns seen in Table 10.4. For the TC$(2, 1, 3)$ scheme different puncturing patterns are employed for the various code rates R. The puncturing patterns were

Code	Code rate R	Puncturing pattern	Modulation mode	BPS	Random turbo interleaver depth	Random (separation) interleaver depth
Turbo Convolutional Code (TC)						
TC(2,1,3)	0.50	10, 01	16QAM	2.00	10000	20000
TC(2,1,4)	0.33	1, 1	64QAM	2.00	10000	30000
	0.50	10, 01	16QAM	2.00	10,000	20000
			64QAM	3.00	10002	20004
	0.67	1000, 0001	64QAM	4.00	10000	15000
	0.75	100000, 000001	16QAM	3.00	9990	13320
			64QAM	4.50	9990	13320
	0.83	1000000000, 0000000001	64QAM	5.00	10000	12000
	0.90	10000000000000000, 00000000000000001	64QAM	5.40	10044	11160
TC(2,1,5)	0.50	10, 01	16QAM	2.00	10000	20000
Turbo BCH Code (TBCH)						
TBCH(31,26)	0.72	1,1	16QAM	2.89	9984	13824
			64QAM	4.33	9984	13824
TBCH(32,26)	0.68	1,1	8PSK	2.05	9984	14592
TBCH(31,21)	0.51	1,1	16QAM	2.04	9996	19516
TBCH(63,57)	0.83	1,1	64QAM	4.96	10032	12144
TBCH(127,120)	0.90	1,1	64QAM	5.37	10080	11256
Turbo Trellis Coded Modulation (TTCM)						
8PSK-TTCM	2/3	10, 01	8PSK	2.00	10000	–
16QAM-TTCM	3/4	10, 01	16QAM	3.00	13332	–

Table 10.5: Simulation parameters associated with the TC, TBCH and TTCM channel encoders in Figure 10.4.

optimised experimentally by simulations, in order to attain the best possible BER performance. The design procedure for punctured turbo codes was proposed by Acikel and Ryan [64] in the context of BPSK and QPSK.

10.4.3 Complexity Issues and Memory Requirements

In this section we address the complexity issues and memory requirements of the proposed system. We will mainly focus on the relative estimated complexity and memory requirements of the proposed channel decoders rather than attempting to determine their exact complexity. Therefore, in order to simplify our comparative study, several assumptions are made. In our simplified approach the estimated complexity of the whole system is deemed to depend only on that of the channel decoders. In other words, the complexity associated with the modulator, demodulator, space-time encoder and decoder as well as channel encoders is assumed to be insignificant compared to the complexity of channel decoders.

Since the estimated complexity of the channel decoders depends directly on the number of trellis transitions, the number of trellis transitions per information data bit will be used as the basis of our comparison. Several channel encoders schemes in Table 10.3 are composed of convolutional codes. For the binary CC(2, 1, K) code, two trellis transitions diverge from each of the 2^{K-1} states. Hence, we can approximate the complexity of a CC(2, 1, K) code as:

$$
\begin{aligned}
comp\left\{\text{CC(2,1,K)}\right\} &= 2 \times 2^{K-1} \\
&= 2^K .
\end{aligned} \tag{10.43}
$$

The number of trellis transitions in the Log-MAP decoding algorithm is assumed to be three times higher than that of the conventional Viterbi algorithm, since the Log-MAP algorithm has to perform forward as well as backward recursion and soft-output calculations, which results in traversing through the trellis three times. The reader is referred to Section 7.3.1 for further details of the algorithm. For TC codes we apply the Log-MAP decoding algorithm for iterative decoding, assisted by the two component decoders. Upon taking into account the number of turbo decoding iterations as well, the complexity of TC decoding is then approximated by:

$$
\begin{aligned}
comp\left\{TC(2, 1, K)\right\} &= 3 \times 2 \times 2^{K-1} \times 2 \times \text{No. of Iterations} \\
&= 3 \times 2^{K+1} \times \text{No. of Iterations} .
\end{aligned} \tag{10.44}
$$

In TCM we construct a non-binary decoding trellis [48]. The TCM schemes of Table 10.3 have 2^{BPS-1} trellis branches diverging from each trellis state, where BPS is the number of transmitted bits per modulation symbol. However, for each trellis transition we would have $BPS - 1$ transmitted information data bits, since the TCM encoder typically adds one parity bit per non-binary symbol. Therefore, we can estimate the complexity of the proposed TCM schemes as:

$$
comp\left\{\text{TCM}\right\} = 2^{BPS-1} \times \frac{\text{No. of States}}{BPS - 1} . \tag{10.45}
$$

Similarly to TC, TTCM consists of two TCM codes and the Log-MAP decoding algorithm [57] is employed for iterative decoding. The associated TTCM complexity is then estimated as:

$$
\begin{aligned}
comp\left\{\text{TTCM}\right\} &= 3 \times 2^{BPS-1} \times \frac{\text{No. of States}}{BPS - 1} \times 2 \times \text{No. of Iterations} \\
&= \frac{3 \times 2^{BPS} \times \text{No. of States} \times \text{No of Iterations}}{BPS - 1} .
\end{aligned} \tag{10.46}
$$

For TBCH(n, k) codes the estimated complexity calculation is not as straightforward as in the previous cases. Its component codes are BCH(n, k) codes and the decoding trellis can be divided into three sections [18]. Assuming that $k > n - k$, for every decoding instant j the number of trellis states is given as [18]:

$$
\text{No. of States}_j = \begin{cases} 2^j & j = 0, 1, \dots, n - k - 1 \\ 2^{n-k} & j = n - k, n - k + 1, \dots, k \\ 2^{n-j} & j = k + 1, k + 2, \dots, n . \end{cases} \tag{10.47}
$$

It can be readily shown that:

$$
2^{n-k} - 1 = \sum_{j=0}^{n-k-1} 2^j \tag{10.48}
$$

$$
= \sum_{j=k+1}^{n} 2^{n-j} . \tag{10.49}
$$

Upon using the approximation $\sum_{j=0}^{n-k-1} 2^j = \sum_{j=k+1}^{n} 2^{n-j} = 2^{n-k} - 1 \approx 2^{n-k}$, we can write the number of decoding trellis states per information data bit as:

$$\text{No. of States} = \frac{2 \times 2^{n-k} + \{k - (n-k)\} \times 2^{n-k}}{k}$$

$$= \frac{(2k - n + 2) \times 2^{n-k}}{k} . \tag{10.50}$$

Having derived the number of decoding trellis states per information data bit, we can approximate the complexity of TBCH codes as:

$$comp\left\{\text{TBCH}(n,k)\right\} = 3 \times 2 \times \frac{(2k - n + 2) \times 2^{n-k}}{k} \times 2 \times \text{No. of Iterations}$$

$$= \frac{3 \times (2k - n + 2) \times 2^{n-k+2} \times \text{No. of Iterations}}{k} . \tag{10.51}$$

Having approximated the complexity of each channel decoder, we will now derive their approximate memory requirements. Typically, the memory requirement of a channel decoder depends directly on the number of trellis states in the entire coded block. Therefore in this section the number of trellis states per coded block serves as the basis of a relative memory requirement comparison between the channel decoders studied. For a binary CC, observation of the VA has shown that typically all surviving paths of the current trellis state emerge from trellis states not 'older' than approximately five times the constraint length, K. Therefore at any decoding instant, only a section of $5 \times K$ trellis transitions has to be stored. We can then approximate the associated memory requirement as:

$$mem\left\{CC(2, 1, K)\right\} = 2^{K-1} \times 5 \times K . \tag{10.52}$$

Again, as highlighted in Section 7.3.1, the Log-MAP algorithm requires the storage of γ, α and β values. Hence for the same number of decoding trellis states, the Log-MAP algorithm would require about three times more memory than the classic VA. Consequently, we can estimate the memory requirement of the TC code as:

$$mem\left\{TC(2, 1, K)\right\} = 3 \times 2^{K-1} \times \text{Block Length} . \tag{10.53}$$

Similarly to CCs, we can approximate the memory requirements of TCM as:

$$mem\left\{TCM\right\} = \text{No. of States} \times \text{Block Length} . \tag{10.54}$$

Following similar arguments, the memory requirements of TTCM employing the Log-MAP algorithm can be approximated as:

$$mem\left\{TTCM\right\} = 3 \times \text{No. of States} \times \text{Block Length} . \tag{10.55}$$

The estimation of the memory requirements of TBCH codes is again different from that of the other channel codes considered. Specifically, their memory requirement does not directly depend on the number of decoding trellis states in a coded TBCH block. Instead, it depends on the number of decoding trellis states in the constituent BCH codewords. From Equation 10.50, we can estimate the associated memory requirements as:

$$mem\left\{\text{TBCH}(n,k)\right\} = 3 \times (2k - n + 2) \times 2^{n-k} . \tag{10.56}$$

Applying Equations 10.43 to 10.56, we can summarise the estimated complexity and memory requirements of the channel decoders characterised in Table 10.3. Explicitly, assuming that there are 10000 information data bits per coded block, the associated estimated complexity and memory requirements are then given in Table 10.6. Note that the block length of TCM and TTCM is expressed in terms of the number of symbols per coded block, since these schemes are symbol-oriented rather than bit-oriented.

Code	No. of states	No. of states per data bit	Iteration No.	Block length	Complexity	Memory requirement
Convolutional Code (CC)						
CC(2,1,5)	16	16	–	10000	32	400
CC(2,1,7)	64	64	–	10000	128	2240
CC(2,1,9)	256	256	–	10000	512	11520
Turbo Convolutional Code (TC)						
TC(2,1,3)	4	4	8	10000	384	120000
TC(2,1,4)	8	8	8	10000	768	240000
TC(2,1,5)	16	16	8	10000	1536	480000
Turbo BCH Code (TBCH)						
TBCH(31,26)	32	28	8	31	2718	2208
TBCH(32,26)	64	54	8	32	5199	4224
TBCH(31,21)	1024	634	8	31	60855	39936
TBCH(63,57)	64	60	8	63	5713	10176
TBCH(127,120)	128	123	8	127	11776	44160
Trellis Coded Modulation (TCM)						
8PSK-TCM	64	32	–	5000	128	320000
16QAM-TCM	64	21	–	3333	171	213312
Turbo Trellis Coded Modulation (TTCM)						
8PSK-TTCM	8	4	8	5000	768	120000
16QAM-TTCM	16	5	8	3333	2048	159984

Table 10.6: Complexity and memory requirements of the different channel decoders as characterised in Table 10.3.

10.5   Performance Results

In this section, unless otherwise stated, all simulation results are obtained over uncorrelated or — synonymously — perfectly interleaved narrowband or non-dispersive Rayleigh fading channels. As stated before, this does not contradict requiring a constant channel magnitude and phase over n consecutive time slots in Equation 10.8, since upon applying a sufficiently high interleaving depth the channel's fading envelope can indeed be uncorrelated. Our assumptions were that:

1) The fading amplitudes were constant across n consecutive transmission slots of the space-time block codes' transmission matrix.

2) The average signal power received from each transmitter antenna was the same.

3) The receiver had a perfect knowledge of the channels' fading amplitudes.

We note that the above assumptions are unrealistic, yielding the best-case performance, but nonetheless facilitating the performance comparison of the various techniques under identical circumstances.

In the following sections, we compare the performance of various combinations of space-time block codes and channel codes. As mentioned earlier, various code rates can be used for both the space-time block codes and for the associated channel codes. The different modulation schemes employed result in various effective throughput. Hence, for a fair comparison, all different systems are compared on the basis of the same effective BPS throughput given by:

$$\text{BPS} = R_{st} \times R_{cc} \times \text{Modulation Throughput} , \qquad (10.57)$$

where R_{st} and R_{cc} are the code rates of the space-time block code and the channel code, respectively.

10.5.1 Performance Comparison of Various Space-time Block Codes Without Channel Codecs

In this section, the performance of various space-time block codes without channel codes is investigated and compared. All the investigated space-time block codes, namely the $\mathbf{G}_2$, $\mathbf{G}_3$, $\mathbf{G}_4$, $\mathbf{H}_3$ and $\mathbf{H}_4$ codes [71–73], have their corresponding transmission matrices given in Equations 10.10, 10.31, 10.32, 10.33 and 10.34, respectively. The encoding parameters are summarised in Table 10.2.

10.5.1.1 Maximum Ratio Combining and the Space-time Code $\mathbf{G}_2$

Figure 10.5 shows the performance of MRC and the space-time code $\mathbf{G}_2$ using BPSK over uncorrelated Rayleigh fading channels. It is assumed that the total power received from both transmit antennas in the space-time-coded system using $\mathbf{G}_2$ of Equation 10.10 is the same as the transmit power of the single transmit antenna MRC-assisted system. It can be seen in Figure 10.5 that the performance of the space-time code $\mathbf{G}_2$ is about 3 dB worse than that of the MRC technique using two receivers, even though both systems have the same diversity order of two. The 3 dB penalty is incurred, because the transmit power of each antenna in the $\mathbf{G}_2$ space-time-coded arrangement is only half of the transmit power in the MRC-assisted system. It is shown in Figure 10.5, however, that at a BER of 10^{-5} a diversity gain of 20 dB is achieved by the space-time code $\mathbf{G}_2$. If we increase the diversity order to four by using two receivers, the space-time code $\mathbf{G}_2$ achieves a diversity gain of 32 dB. However, there is still a 3 dB performance penalty as compared to the conventional MRC technique using four receivers. The advantage of the space-time-coded scheme is nonetheless that the increased complexity of the space-time-coded transmitter is more affordable at the BS than at the MS, where the MRC receiver would have to be located.

10.5.1.2 Performance of 1 BPS Schemes

Figures 10.6 and 10.7 compare the performance of the space-time codes $\mathbf{G}_2$, $\mathbf{G}_3$ and $\mathbf{G}_4$ having an effective throughput of 1 BPS over uncorrelated Rayleigh fading channels using one and two receivers, respectively. BPSK modulation was employed in conjunction with the space-time code $\mathbf{G}_2$. As shown in Table 10.2, the space-time codes $\mathbf{G}_3$ and $\mathbf{G}_4$ are half-rate codes. Therefore, QPSK modulation was used in the context of $\mathbf{G}_3$ and $\mathbf{G}_4$ in order to retain a throughput of 1 BPS.

BER against E_b/N_0

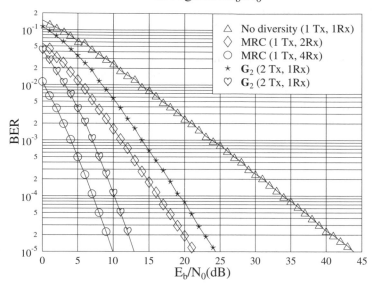

Figure 10.5: Performance comparison of the MRC technique and space-time code $\mathbf{G}_2$ using **BPSK** over uncorrelated Rayleigh fading channels.

BER against E_b/N_0

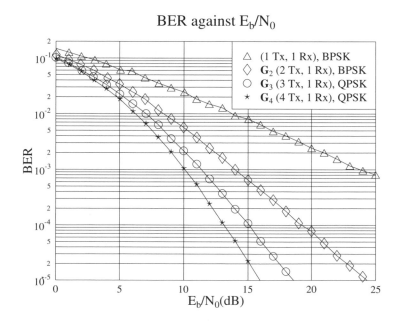

Figure 10.6: Performance comparison of the space-time codes $\mathbf{G}_2$, $\mathbf{G}_3$ and $\mathbf{G}_4$ of Table 10.2 at an effective throughput of **1 BPS** using **one receiver** over uncorrelated Rayleigh fading channels.

It can be seen in Figure 10.6 that at a BER of 10^{-5} the space-time codes $\mathbf{G}_3$ and $\mathbf{G}_4$ give about 2.5 and 7.5 dB gain over the $\mathbf{G}_2$ code, respectively. If the number of receivers is increased to two, as shown in Figure 10.7, the associated E_b/N_0 gain reduces to about 1 and 3.5 dB, respectively. The reason is that over the perfectly interleaved flat fading channel encountered much of the attainable diversity gain is already achieved using the $\mathbf{G}_2$ code and two receivers. The associated gains of the various schemes at a BER of 10^{-5} are summarised in Table 10.7 below.

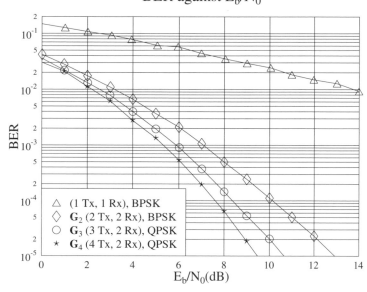

Figure 10.7: Performance comparison of the space-time codes $\mathbf{G}_2$, $\mathbf{G}_3$ and $\mathbf{G}_4$ of Table 10.2 at an effective throughput of **1 BPS** using **two receivers** over uncorrelated Rayleigh fading channels.

10.5.1.3 Performance of 2 BPS Schemes

In Figure 10.8 we compare the performance of the space-time codes $\mathbf{G}_2, \mathbf{G}_3, \mathbf{G}_4, \mathbf{H}_3$ and $\mathbf{H}_4$ proposed in [71–73] using the encoding parameters summarised in Table 10.2. The performance results were obtained over uncorrelated Rayleigh fading channels using one receiver and the effective throughput of the system is about 2 BPS. For the $\mathbf{G}_2$ code QPSK modulation was used, while the $\mathbf{G}_3$ and $\mathbf{G}_4$ codes employ 16QAM conveying 4 BPS. Hence the effective throughput is 2 BPS, since $\mathbf{G}_3$ and $\mathbf{G}_4$ are half-rate codes. Since the code rate of the $\mathbf{H}_3$ and $\mathbf{H}_4$ codes is $\frac{3}{4}$, 8PSK modulation was employed in this context, resulting in a throughput of $3 \times 3/4 = 2.25$ BPS, which is approximately 2 BPS. We can see in Figure 10.8 that at high BERs or low E_b/N_0 values the $\mathbf{G}_2$ code slightly outperforms the others. However, the situation is reversed when the system is operated at low BER or high E_b/N_0 values. At a BER of 10^{-5} the code $\mathbf{G}_4$ only gives a diversity gain of 5 dB over the $\mathbf{G}_2$ code. This is a 2.5 dB loss compared to the 7.5 dB gain achieved by the system transmitting at an effective throughput of 1 BPS in the previous section. This is because the more vulnerable 16QAM scheme was used for the space-time code $\mathbf{G}_4$. Since the 16QAM signal constellation is more densely packed compared to QPSK, it is more prone to errors. Moreover, the space-time code $\mathbf{G}_4$ has no error correction capability to correct the extra errors induced by employing a more vulnerable, higher-order modulation

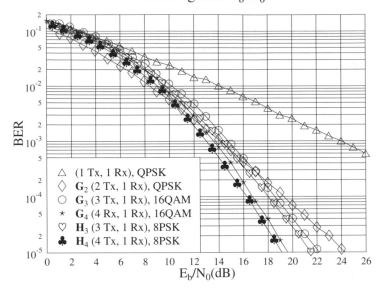

Figure 10.8: Performance comparison of the space-time codes $\mathbf{G}_2$, $\mathbf{G}_3$, $\mathbf{G}_4$, $\mathbf{H}_3$ and $\mathbf{H}_4$ at an effective throughput of approximately **2 BPS** using **one receiver** over uncorrelated Rayleigh fading channels. The associated parameters of the space-time codes are summarised in Table 10.2.

scheme. Hence, this results in a poorer performance. If the throughput of the system is increased by employing an even higher-order modulation scheme, the space-time code $\mathbf{G}_4$ will suffer even higher performance degradations, as will be shown in the next section. Since the space-time code $\mathbf{G}_3$ of Table 10.2 is also a half-rate code, similarly to the $\mathbf{G}_4$ code, it suffers from the same drawbacks.

In Figure 10.8, we also show the performance of the rate-$\frac{3}{4}$ space-time codes $\mathbf{H}_3$ and $\mathbf{H}_4$ of Table 10.2. Both the $\mathbf{H}_4$ and $\mathbf{G}_4$ codes have the same diversity order of four in conjunction with one receiver. However, at a BER of 10^{-5} the performance of the $\mathbf{H}_4$ code is about 0.5 dB better than that of the $\mathbf{G}_4$ code. This is again due to the higher-order modulation employed in conjunction with the half-rate code $\mathbf{G}_4$, in order to maintain the same throughput. As alluded to earlier, the higher-order modulation schemes are more susceptible to errors and hence the performance of the system in conjunction with the $\mathbf{G}_3$ or $\mathbf{G}_4$ code of Table 10.2 is worse than that of the $\mathbf{H}_3$ or $\mathbf{H}_4$ code having the same diversity orders, respectively. The associated gains of the various schemes at a BER of 10^{-5} are summarised in Table 10.7 below.

10.5.1.4   Performance of 3 BPS Schemes

Figures 10.9 and 10.10 show our performance comparisons for the space-time codes $\mathbf{G}_2$, $\mathbf{G}_3$, $\mathbf{G}_4$, $\mathbf{H}_3$ and $\mathbf{H}_4$ of Table 10.2 at an effective throughput of 3 BPS over uncorrelated Rayleigh fading channels using one and two receivers, respectively. When using the $\mathbf{G}_2$ code we employed 8PSK modulation. Since $\mathbf{G}_3$ and $\mathbf{G}_4$ are half-rate codes, 64QAM was employed, in order to obtain an effective throughput of 3 BPS. By contrast, for the $\mathbf{H}_3$ and $\mathbf{H}_4$ codes, which have a code rate of $\frac{3}{4}$, 16QAM was used in order to ensure the same throughput of $4 \times 3/4 = 3$ BPS.

In Figure 10.9 we can see that at a BER of 10^{-5} the diversity gain of the $\mathbf{G}_4$ code over the

BER against E_b/N_0

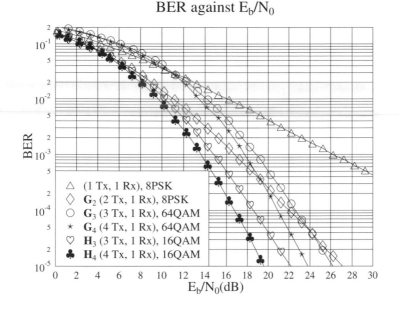

Figure 10.9: Performance comparison of the space-time codes G_2, G_3, G_4, H_3 and H_4 of Table 10.2 at an effective throughput of **3 BPS** using **one receiver** over uncorrelated Rayleigh fading channels.

G_2 code is further reduced to about 3 dB. There is only a marginal diversity gain for the G_3 code over the G_2 code. As alluded to in the previous section, 64QAM in conjunction with the space-time code G_3 or G_4 has a densely packed signal constellation and hence this scheme is prone to errors. At the higher BER of 10^{-2} the G_2 code outperforms the G_3 and G_4 codes by approximately 3 and 4 dB, respectively.

Owing to the associated higher-order modulation scheme employed, we can see in Figure 10.9 that at a BER of 10^{-5} the H_3 and H_4 codes of Table 10.2 outperform both the G_3 and the G_4 codes. Specifically, we can see that the H_3 code attains about 2 dB gain over the G_4 code, even though it has a lower diversity order.

If we increase the number of receivers to two, a scenario characterised in Figure 10.10, the performance degradation of the space-time codes G_3 and G_4 is even more pronounced. At a BER of 10^{-5} the performance gain of the H_4 code over the G_4 code is approximately 4 dB compared to the 0.5 dB gain, when the system's effective throughput is only 2 BPS, as was shown in Figure 10.8 of the previous section.

Having studied Figures 10.6 to 10.10, we may conclude two important points. First, the space-time codes G_3 and G_4 of Table 10.2 suffer from having a code rate of a half, since this significantly reduces the effective throughput of the system. In order to maintain the same throughput as the unity-rate G_2 code, higher-order modulation schemes, such as for example 64QAM, have to be employed. This results in a preponderance of channel errors, since the constellation points of the higher-order modulation schemes are more densely packed. By their lack of error correcting capability, the G_3 and G_4 codes suffer performance losses compared to the G_2 code. Second, if the number of receivers is increased to two, the performance gain of the G_3, G_4, H_3 or H_4 codes over the G_2 code becomes lower. The reason behind this phenomenon is that much of the attainable diversity gain was already achieved using the G_2 code and two receivers. The

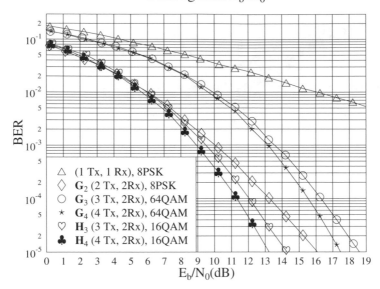

Figure 10.10: Performance comparison of the space-time codes G_2, G_3, G_4, H_3 and H_4 of Table 10.2 at an effective throughput of **3 BPS** using **two receivers** over uncorrelated Rayleigh fading channels.

associated gains of the various schemes at a BER of 10^{-5} are summarised in Table 10.7.

		One receiver			Two receivers		
Code	Rate	1 BPS	2 BPS	3 BPS	1 BPS	2 BPS	3 BPS
G_2	1	19.5	19.6	19.1	30.9	30.9	30.1
G_3	1/2	25.2	21.8	20.0	33.2	29.6	27.6
G_4	1/2	27.9	24.3	22.4	34.3	30.7	28.8
H_3	3/4	–	22.4	24.0	–	30.1	31.9
H_4	3/4	–	24.8	22.6	–	31.2	33.0

Table 10.7: Coding gain of the space-time block codes of Table 10.2 over uncorrelated Rayleigh fading channels.

10.5.1.5 Channel-coded Space-time Block Codes

In the previous sections, we have shown that without channel coding the performance of the unity-rate space-time G_2 code is inferior to the lower-rate space-time codes, namely to that of the G_3, G_4, H_3 and H_4 schemes. Since the space-time code G_2 has a unity code rate, half-rate turbo codes can be employed for improving the performance of the system. In Figure 10.11, we compare the performance of the half-rate TC(2,1,4) code concatenated with the space-time code G_2 and with the space-time block codes G_4 and H_4. Both the space-time codes G_4 and H_4 have a diversity gain of four and a code rate of $\frac{1}{2}$ and $\frac{3}{4}$, respectively. The associated parameters are shown in Tables 10.2, 10.3 and 10.5. Suitable modulation schemes were chosen so that all sys-

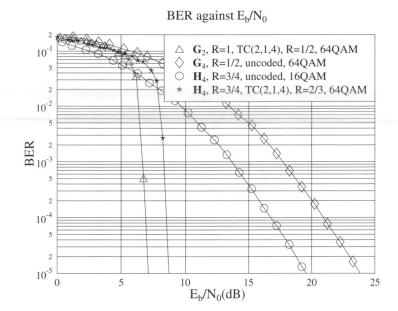

Figure 10.11: Performance comparison of the half-rate TC(2,1,4) code concatenated with the space-time code G_2 and the space-time block codes G_4 and H_4. The associated parameters are shown in Tables 10.2, 10.3 and 10.5. All simulation results were obtained at an effective throughput of **3 BPS** over uncorrelated Rayleigh fading channels.

tems had the same throughput of 3 BPS. All simulation results were obtained over uncorrelated Rayleigh fading channels.

From Figure 10.11, we can see that a huge performance improvement is achieved by concatenating the space-time code G_2 with the half-rate code TC(2,1,4). At a BER of 10^{-5} this concatenated scheme attains a coding gain of 16 dB and 13 dB compared to the space-time codes G_4 and H_4, respectively. This clearly shows that it is better to invest the parity bits associated with the code-rate reduction in the concatenated turbo code, rather than in non-unity-rate space-time block codes. In Figure 10.11 we also show the performance of the space-time code H_4 concatenated with the punctured two-thirds-rate code TC(2,1,4). The figure shows that the TC(2,1,4) code improves the performance of the system tremendously, attaining a coding gain of 11 dB compared to the non-turbo-coded space-time code H_4, at a BER of 10^{-5}. However, its performance is still inferior to that of the half-rate TC(2,1,4)-coded space-time code G_2.

In conclusion, in Figure 10.11 we have seen that the reduction in coding rate is best assigned to turbo channel codes, rather than to space-time codes. Therefore, in all our forthcoming simulations, all channel codecs of Table 10.3 are concatenated with the unity-rate space-time code G_2, instead of the non-unity-rate space-time codes G_3, G_4, H_3 and H_4 of Table 10.2.

10.5.2   Mapping Binary Channel Codes to Multilevel Modulation

As mentioned earlier, in our investigations different modulation schemes are employed in conjunction with the binary channel codecs CC, TC and TBCH. Specifically, the modulation schemes used are BPSK, QPSK, 8PSK, 16QAM and 64QAM. Gray mapping [86, 102, 207] is employed to map the bits to the QPSK, 8PSK, 16QAM and 64QAM symbols. In higher-order modulation schemes, such as 8PSK, 16QAM and 64QAM, we have several transmitted bits per constellation point. However, the different bit positions of the constellation points have different noise-protection distances [207]. More explicitly, the protection distance is the Euclidean distance from one constellation point to another, which results in the corruption of a particular bit. A larger noise-protection distance results in a higher integrity of the bit and vice versa. Therefore, for the different bit positions in the symbol we have different protection for the transmitted bits within the phasor constellation of the non-binary modulation schemes. It can be readily shown that in 8PSK and 16QAM we have two protection classes, namely class I and II [86, 207], where the class I transmitted bits are more protected. Similarly, in 64QAM we have three protection classes, namely I, II and III [207], where the transmitted bits in class I are most protected, followed by class II and class III.

In our system the parity bits are generated by binary channel encoders, such as the CC, TC and TBCH schemes for protecting the binary data bits. However, it is not intuitive whether the integrity of the data or parity bits is more important in yielding a better overall BER performance. For example, if the parity bits are more important, it is better to allocate the parity bits to the better protection classes in higher-order modulation schemes and vice versa. Therefore, in this section, we will investigate the performance of different channel codes along with different bit mapping schemes. The effect of the bit interleaver seen in Figure 10.4 is studied in conjunction with binary channel codes as well.

10.5.2.1 Turbo Convolutional Codes: Data and Parity Bit Mapping

We commence here by studying half-rate TC codes, which are characterised in Table 10.3. An equal number of parity and data bits are generated by the half-rate TC codes and they are then mapped to the protection classes of the 16QAM scheme considered. Again, in the Gray-mapping-assisted 16QAM constellation there are two protection classes [207], class I and II, depending on the bit position. Explicitly, there are four BPS in the 16QAM constellation and two of the bit positions are more protected than the remaining two bits.

In Figure 10.12 we compare the performance of various parity and data bit mapping schemes for the (a) TC(2,1,3), (b) TC(2,1,4) and (c) TC(2,1,5) codes. The curve marked by triangles represents the performance of the TC codes, when allocating the parity bits to the higher-integrity protection class I and the data bits to the lower-integrity protection class II. On the other hand, the performance curve marked by diamonds indicates the allocation of data bits to protection class I, while the parity bits are assigned to protection class II.

In Figure 10.12(a), we can see that at low E_b/N_0 values the performance of the TC(2,1,3) code, when allocating the parity bits to protection class I, is worse than upon allocating the data bits to protection class I. However, for E_b/N_0 values in excess of about 4 dB, the situation is reversed. At a BER of 10^{-5}, there is a performance gain of about 1 dB when using the TC(2,1,3) arrangement with the parity bits allocated to protection class I. We surmise that by protecting the parity bits better, we render the TC(2,1,3) code more powerful. It is common that stronger channel codes perform worse than weaker codes at low E_b/N_0 values, but outperform their less powerful counterparts for higher E_b/N_0 values.

This is further justified in Figure 10.13. Here, we show the performance of hard-decision algebraic decoding of the BCH(7,4), BCH(63,36) and BCH(127,71) codes using BPSK over AWGN channels. All BCH codes characterised in the figure have approximately the same code rate, which is $R = 0.57$. From the figure we can see that at a BER of 10^{-3} the performance of the BCH codes improves with an increasing codeword length n. However, at a high BER or low E_b/N_0 value we can see that the performance of the BCH(7,4) code is better than that of the BCH(63,36) and BCH(127,71) codes, which are stronger channel codes. This is because

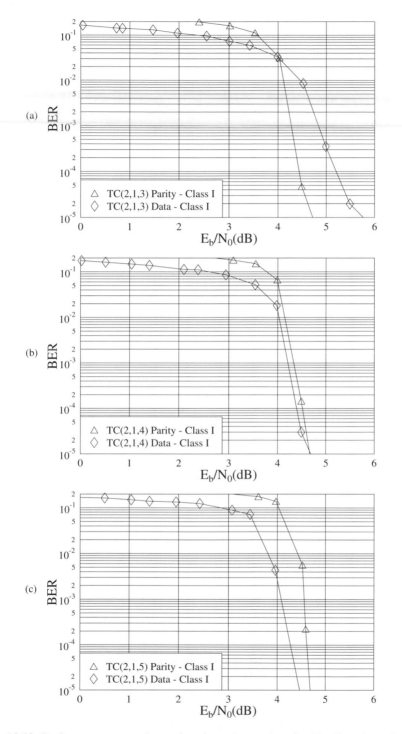

Figure 10.12: Performance comparison of various data and parity bit allocation schemes for the (a) TC(2,1,3), (b) TC(2,1,4) and (c) TC(2,1,5) codes, where the parameters are shown in Table 10.3. All simulation results were obtained upon employing the space-time code $\mathbf{G}_2$ using one receiver and 16QAM over uncorrelated Rayleigh fading channels at an effective throughput of 2 BPS.

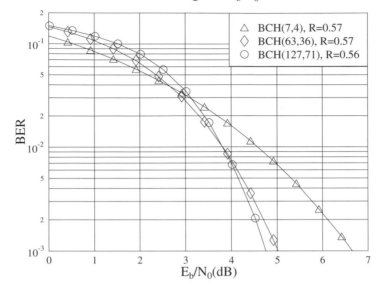

Figure 10.13: Performance comparison of hard-decision algebraic decoding of different BCH codes having approximately the same code rate of $R = 0.57$, using BPSK over AWGN channels.

stronger codes have many codewords having a large free distance. At low SNRs we have bad channel conditions and hence the channel might corrupt even those codewords having a large free distance. Once they are corrupted, they produce many erroneous information bits, a phenomenon which results in a poorer BER performance.

In Figure 10.12(b) we showed the performance of the TC(2,1,4) code using the same data and parity bit allocation as in Figure 10.12(a). The figure clearly shows that the TC(2,1,4) scheme exhibits a better performance for E_b/N_0 values below about 4.7 dB, if the data bits are more strongly protected than the parity bits. It is also seen from the figure that the situation is reversed for E_b/N_0 values above this point. This phenomenon is different from the behaviour of the TC(2,1,3) scheme, since the crossing point of both curves occurs at a significantly lower BER. The same situation can be observed for the BCH codes characterised in Figure 10.13, where we can see that the performance curve of the BCH(127,71) code crosses the performance curve of the BCH(63,36) scheme at $E_b/N_0 \approx 4$ dB. This value is lower than the crossing point of the performance curves of the BCH(63,36) and BCH(7,4) codes. Hence the trend is that the crossing point of stronger codes is shifted to right of the figure. Hence the crossing point of the performance curves of stronger codes will occur at lower BERs and shifted to the right on the E_b/N_0 scale. From the above argument we can speculate also in the context of TC codes that since the TC(2,1,4) scheme is a stronger code than the TC(2,1,3) arrangement, the crossing point of the associated performance curves for TC(2,1,4) is at a lower BER than that of the TC(2,1,3) code and appears to be shifted to right of the E_b/N_0 scale.

Let us now consider the same performance curves in the context of the significantly stronger TC(2,1,5) code in Figure 10.12(c). The figure clearly shows that better performance is yielded in the observed range when the data bits are more strongly protected. Unlike in Figure 10.12(a) and 10.12(b), there is no visible crossing point in Figure 10.12(c). However, judging from the gradient of both curves, if we were to extrapolate the curves in Figure 10.12(c), they might cross at a BER of 10^{-6}. The issue of data and parity bit mapping to multilevel modulation schemes was also addressed by Goff *et al.* [55, 286]. However, the authors only investigated the performance of the TC(2,1,5) code and stated that better performance is achieved by protecting the data bits more strongly. Additionally, we note here that the situation was reversed for the TC(2,1,3) code, where better performance was achieved by protecting the parity bits more strongly.

Hence, from the three diagrams of Figure 10.12 we can draw the following conclusions for the mapping of the data and parity bits to the different protection classes of the modulated symbol. For weaker half-rate turbo codes, such as the TC(2,1,3) arrangement, it is better to protect the parity bits more strongly. On the other hand, for stronger half-rate turbo codes, such as the TC(2,1,4) and TC(2,1,5) schemes, better performance is achieved by protecting the data bits more strongly. From our simulation results, we found that the same scenario also applies to turbo codes having code rates lower or higher than halfrates, as shown in Table 10.5. Based on these facts, we continue our investigations into the effect of interleavers, in an effort to achieve an improved performance.

10.5.2.2   Turbo Convolutional Codes: Interleaver Effects

In Figure 10.4 we have seen that a bit-based interleaver is employed for the CC, TC and TBCH codes. Since our performance results are obtained over uncorrelated Rayleigh fading channels, the purpose of the bit-based interleaver is to disperse bursts of channel errors within a modulated symbol, when it experiences a deep fade. This is vital for TC codes, because according to the turbo code structure proposed by Berrou *et al.* in [12, 13], at the output of the turbo encoder, a data bit is followed by the parity bits generated for its protection against errors. Therefore in multilevel modulation schemes a particular modulated symbol could consist of the data bit and its corresponding parity bits generated for its protection. If the symbol experiences a deep fade, the demodulator would provide low-reliability values for both the data bit and the associated parity bits. In conjunction with low-reliability information the turbo decoder may fail to correct errors induced by the channel. However, we can separate both the data bit and the parity bits generated for its protection into different modulation symbols. By doing so, there is a better chance that the demodulator can provide high-reliability parity bits, which are represented by another modulation symbol, even if the data bit experienced a deep fade and vice versa. This will assist the turbo decoder in correcting errors.

More explicitly, the random interleaver shown in Figure 10.4 has two different effects on the binary channel codes, namely:

1) It separates the data bit and the parity bits generated for its protection into different modulated symbols.

2) It randomly maps the data and parity bits into different protection classes in multilevel modulation schemes.

The first effect of the random interleaver will improve the performance of the binary channel codecs. By contrast, the second effect might have a negative impact on the performance of the channel codecs, because the data and parity bits are randomly mapped to the different protection classes, rather than assigning the more vulnerable bits consistently to the higher-integrity protection class.

In order to eliminate the potentially detrimental second effect of the random interleaver, we propose to invoke a so-called random separation-based interleaver. Explicitly, Figure 10.14 shows an example of the random separation based interleaving employed. The objective of such interleaving is to randomly interleave the bits within the same protection class of the multilevel

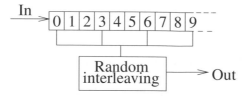

Figure 10.14: Random separation-based interleaving.

modulated symbols. If 8PSK modulation is used, 3 BPS are transmitted. Hence, for every three-bit spaced position, the bits will be randomly interleaved. For example, in Figure 10.14 we randomly interleaved the bit positions $0, 3, 6, 9, \ldots$ Similarly, bit positions $1, 4, 7, \ldots$ and $2, 5, 8, \ldots$ will be randomly interleaved as well.

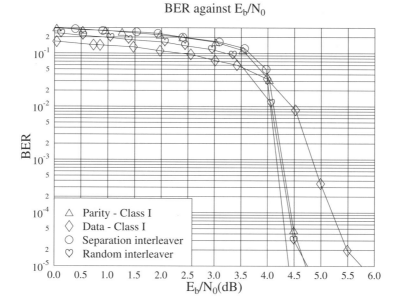

Figure 10.15: Performance comparison between different bit-to-symbol mapping methods for the **TC(2,1,3)** code in conjunction with the space-time code $\mathbf{G}_2$ using one receiver and **16QAM** over uncorrelated Rayleigh fading channels at an effective throughput of **2 BPS**. The encoding parameters of the TC(2,1,3) code are shown in Table 10.3.

In Figure 10.15 we investigated the effects of both a random interleaver and those of a random separation-based interleaver on the performance of the TC(2,1,3) code. The encoding parameters of the TC(2,1,3) code are shown in Table 10.3. The simulation results were obtained in conjunction with the space-time code $\mathbf{G}_2$ using one receiver and 16QAM over uncorrelated Rayleigh fading channels. The performance curves marked by the triangles and diamonds were obtained by protecting the parity bits and data bits more strongly, respectively. Recall that the same performance curves were also shown in Figure 10.12(a).

As mentioned earlier, the random interleaver has two different effects on the performance of binary channel codes. It randomly maps the data and parity bits into different protection classes which might have a negative impact on the performance of the channel codecs. Additionally, it may separate the data bits and parity bits generated for their protection into different modulated symbols, which on the other hand might improve the performance. In Figure 10.15 the random interleaver-based performance curve is marked by the hearts, which is similar to that of the TC(2,1,3)-coded scheme protecting the parity bits more strongly. This suggest that the above-mentioned positive effect of the random interleaver is more pronounced than the negative effect in the context of the TC(2,1,3)-coded scheme. On the other hand, based on the evidence of Figure 10.12(a) the random separation-based interleaver was ultimately applied in conjunction with the allocation of the parity bits, rather than the data bits, into protection class I. The interleaver randomly interleaved the coded bits within the same protection class of a block of transmitted symbols. Therefore, the parity bits remained more protected compared to the data bits and yet they have been randomly interleaved within the set of parity. In Figure 10.15 the performance of the random separation-based interleaver is marked by circles, and is about 0.5 dB better than that of the TC(2,1,3)-coded scheme with the parity bits allocated to protection class I.

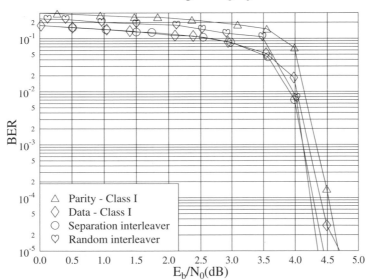

Figure 10.16: Performance comparison between different bit-to-symbol mapping methods for the **TC(2,1,4)** code in conjunction with the space-time code G_2 using one receiver and **16QAM** over uncorrelated Rayleigh fading channels at an effective throughput of **2 BPS**. The encoding parameters of the TC(2,1,4) code are shown in Table 10.3.

Similarly to Figure 10.15, in Figures 10.16 and 10.17 we show the performance of the TC(2,1,4) and TC(2,1,5) codes, respectively, using different bit-to-symbol mapping methods. All simulation results were obtained in conjunction with the space-time code G_2 using one receiver and 16QAM over uncorrelated Rayleigh fading channels. The encoding parameters of the TC(2,1,4) and TC(2,1,5) codes are shown in Table 10.3. Unlike in Figure 10.15, the random separation-based interleaver was applied in conjunction with the allocation of the data bits, rather than the parity bits, to protection class I. It can be seen from Figures 10.16 and 10.17 that the performance of the random interleaver and random separation-based interleaver is similar. This again suggests that the above-mentioned positive effect yielded by the random interleaver is more pronounced than its detrimental effect in the context of both the TC(2,1,4) and TC(2,1,5) schemes.

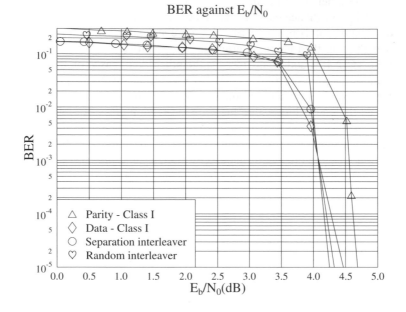

Figure 10.17: Performance comparison between different bit-to-symbol mapping methods for the **TC(2,1,5)** code in conjunction with the space-time code G_2 using one receiver and **16QAM** over uncorrelated Rayleigh fading channels at an effective throughput of **2 BPS**. The encoding parameters of the TC(2,1,5) code are shown in Table 10.3.

In conclusion, our simulation results presented in this section demonstrated that at a BER of 10^{-5} the half-rate turbo codes using a random separation-based interleaver attain the best performance, albeit for certain schemes only by a small margin. Therefore in our forthcoming performance comparisons we will be employing the random separation-based interleaver in conjunction with the various TC codes.

10.5.2.3 Turbo BCH Codes

Figure 10.18 characterises the performance of the TBCH(32,26) code in conjunction with different bit-to-symbol mapping to the two protection classes of 8PSK. All simulation results were obtained with the aid of the space-time code G_2 using one receiver and 8PSK over uncorrelated Rayleigh fading channels. Again, the encoding parameters of the TBCH(32,26) code are shown in Tables 10.3 and 10.5. The TBCH(32,26) code was chosen for our investigations because the parity bits of the constituent encoders were not punctured and hence this resulted in a code rate of $R \approx \frac{2}{3}$. Roughly speaking, for every two data bits, there is one parity bit. Similarly to 16QAM, in the Gray-mapping-assisted 8PSK constellation there are also two protection classes, depending on the bit position in the three-bit symbols. From the three bits of the 8PSK constellation two of the bit positions are more protected than the remaining bit. In Figure 10.18, we portray the performance of the TBCH(32,26) scheme for four different bit-to-symbol mapping methods. First, one data bit and one parity bit were mapped to the two better protected 8PSK bit positions. The corresponding BER curve was marked by the triangles in Figure 10.18. According to the

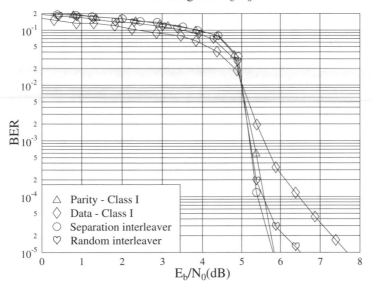

Figure 10.18: Performance comparison between different bit-to-symbol mapping methods for the TBCH(32,26) code in conjunction with the space-time code G_2 using one receiver and **8PSK** over uncorrelated Rayleigh fading channels at an effective throughput of **2 BPS**. The encoding parameters of the TBCH(32,26) code are shown in Tables 10.3 and 10.5.

second method, the data bits were mapped to the two better protected bit positions of the 8PSK symbol. This scenario was marked by the diamonds in Figure 10.18. As we can see from the figure, the first mapping method yields a substantial E_b/N_0 gain of 1.5 dB at a BER of 10^{-5} over the second method. By applying the random separation-based interleaver of Figure 10.14, while still better protecting one of the data bits and the parity bit than the remaining data bits, we disperse the bursty bit errors associated with a transmitted symbol over several BCH codewords of the TBCH code. As shown in Figure 10.18, the performance curve marked by the circles shows a slight improvement compared to the above-mentioned first method, although the difference is marginal. Finally, we show the performance of applying random interleaving, which randomly distributes the data and parity bits between the two 8PSK protection classes. It can be seen that the associated performance is worse than that of the first bit-to-symbol mapping method.

In Figure 10.18, we have shown that it is better to protect the parity bits more strongly for the TBCH(32,26) code and a slight further improvement can be achieved by applying a random separation-based interleaver. More simulation results were obtained in conjunction with the other TBCH codes shown in Tables 10.3 and 10.5 with the aid of the space-time code G_2 and 64QAM over uncorrelated Rayleigh fading channels. From the simulation results we have found that all TBCH codes shown in Tables 10.3 and 10.5 perform better if the parity bits are more protected. In general, a slight further improvement can be obtained for TBCH codes when a random separation-based interleaver is applied. A possible explanation is that the component encoders of the TBCH codes are BCH encoders, where a block of parity bits is generated by a block of data bits. Hence, every parity bit has an influence on the whole codeword. Moreover, we used high-rate TBCH codes and hence there are more data bits compared to the parity bits. Hence, in our forthcoming TBCH comparisons, we will use bit-to-symbol mappers protecting the parity bits better.

10.5.2.4   Convolutional Codes

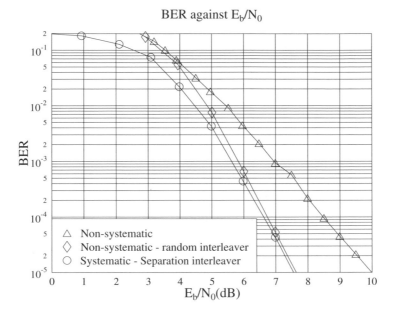

Figure 10.19: Performance comparison between the **systematic and non-systematic half-rate CC(2,1,9)** code in conjunction with the space-time code G_2 and **16QAM** over uncorrelated Rayleigh fading channels at a throughput of **2 BPS**. The encoding parameters of the CC(2,1,9) code are shown in Tables 10.3 and 10.4.

Let us now investigate the space-time code G_2 in conjunction with the half-rate CC(2,1,9) code proposed for UMTS. The CC(2,1,9) code is a non-systematic non-recursive CC, where the original information bits cannot be explicitly recognised in the encoded sequence. Its associated performance curve is shown in Figure 10.19 marked by the triangles. A random interleaver was then applied, in order to disperse the bursty channel errors, and the associated performance curve is marked by the diamonds in Figure 10.19. At a BER of 10^{-5} there is a performance gain of 2.5 dB if the random interleaver is applied. As a further scheme we invoked a systematic CC(2,1,9) code, which was obtained using a recursive CC [49, 102]. Hence, in this scenario we have explicitly separable data bits and parity bits. In Figure 10.19 the performance curve marked by the circles is obtained by mapping the data bits of the systematic CC(2,1,9) code to protection class I of the associated 16QAM scheme in conjunction with the random separation-based interleaver of Figure 10.14. From the figure we can see that there is only a marginal performance improvement over the non-systematic CC(2,1,9) code using the random interleaver.

10.5.3   Performance Comparison of Various Channel Codecs Using the G_2 Space-time Code and Multilevel Modulation

In this section we compare the G_2 space-time-coded performance of all channel codecs summarised in Table 10.3. In order to avoid having an excessive number of curves in one figure, only one channel codec will be characterised from each group of the CC, TC, TBCH, TCM and TTCM schemes. The choice of the channel codec considered depends on its performance, complexity and code rate. Unless otherwise stated, all channel codecs are concatenated with the space-time code G_2 using one receiver. All comparisons are carried out on the basis of the same BPS throughput over uncorrelated Rayleigh fading channels. Let us now briefly discuss in the forthcoming sections how each channel codec is selected from the codec families considered.

10.5.3.1 Comparison of Turbo Convolutional Codes

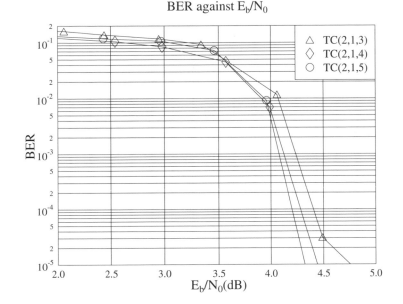

Figure 10.20: Performance comparison between the half-rate codes TC(2,1,3), TC(2,1,4) and TC(2,1,5), where the encoding parameters are shown in Tables 10.3 and 10.5. All simulation results were obtained with the aid of the space-time code G_2 using **16QAM** over uncorrelated Rayleigh fading channels and the throughput was **2 BPS**.

In Figure 10.20 we compare the performance of the half-rate turbo codes TC(2,1,3), TC(2,1,4) and TC(2,1,5), where the encoding parameters are shown in Tables 10.3 and 10.5. The simulation results were obtained with the aid of the space-time code G_2 using 16QAM over uncorrelated Rayleigh fading channels. The three performance curves in the figure are the best performance curves chosen from Figures 10.17, 10.16 and 10.15 for the half-rate codes TC(2,1,5), TC(2,1,4) and TC(2,1,3), respectively. It can be seen from the figure that the performance of the turbo codes improves when we increase the constraint length of the component codes from 3 to 5. However, this performance gain is obtained at the cost of a higher decoding complexity. At a BER of 10^{-5} the TC(2,1,4) code has an E_b/N_0 improvement of approximately 0.25 dB over the TC(2,1,3) scheme at a penalty of twice the complexity. However, at the cost of the same complexity increment over that of the TC(2,1,4) arrangement the TC(2,1,5) scheme only achieves a marginal performance gain of 0.1 dB at a BER of 10^{-5}. Therefore, in our following investigations only the TC(2,1,4) scheme will be characterised as it exhibits a significant coding gain at a moderate complexity. Furthermore, the TC(2,1,4) code has been adopted by the

3G UTRA mobile communication system [49].

10.5.3.2 Comparison of Different-rate TC(2,1,4) Codes

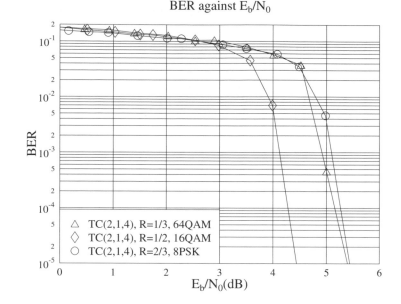

Figure 10.21: Performance of the TC(2,1,4) code using coding rates of $\frac{1}{3}$, $\frac{1}{2}$ and $\frac{2}{3}$, where the associated encoding parameters are shown in Tables 10.3 and 10.5. All simulation results were obtained with the aid of the space-time code $\mathbf{G}_2$ at an effective throughput of **2 BPS** over uncorrelated Rayleigh fading channels.

In their seminal paper on turbo coding [12, 13], Berrou *et al.* applied alternate puncturing of the parity bits. This results in half-rate turbo codes. However, additionally a range of different puncturing patterns can be applied, which results in different code rates [64]. In Figure 10.21 we portray the performance of the punctured TC(2,1,4) code having coding rates of $\frac{1}{3}$, $\frac{1}{2}$ and $\frac{2}{3}$. The associated coding parameters are shown in Tables 10.3 and 10.5. Suitable multilevel modulation schemes are chosen so that all systems have the same effective throughput of 2 BPS. Explicitly, 64QAM, 16QAM and 8PSK are used. All simulation results were obtained with the aid of the space-time code $\mathbf{G}_2$ over uncorrelated Rayleigh fading channels. As expected, from Figure 10.21 we can clearly see that the best performance is achieved by the half-rate TC(2,1,4) scheme. At a BER of 10^{-5} the half-rate TC(2,1,4) code achieved a performance gain of approximately 1 dB over the third-rate and the two-thirds-rate TC(2,1,4) codes. Even though the third-rate TC(2,1,4) code has a higher amount of redundancy than the half-rate TC(2,1,4) scheme, its performance is worse than that of the half-rate TC(2,1,4) arrangement. We speculate that this is because the constellation points in 64QAM are more densely packed than those of 16QAM. Therefore, they are more prone to errors and hence the extra coding power of the thirds-rate TC(2,1,4) code is insufficient to correct the extra errors. This results in a poorer performance. On the other hand, there are less errors induced by 8PSK, but the two-third-rate TC(2,1,4) code is a weak code due to the puncturing of the parity bits. Again, this results in an inferior performance.

In Figure 10.22 we show the performance of the TC(2,1,4) code at coding rates of $\frac{1}{2}$ and $\frac{3}{4}$.

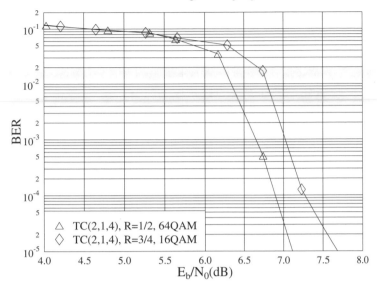

BER against E_b/N_0

Figure 10.22: Performance of the punctured TC(2,1,4) code at coding rates of $\frac{1}{2}$ and $\frac{3}{4}$, where the associated parameters are shown in Tables 10.3 and 10.5. All simulation results were obtained with the aid of the space-time code $\mathbf{G}_2$ at an effective throughput of **3 BPS** over uncorrelated Rayleigh fading channels.

The associated coding parameters were shown in Tables 10.3 and 10.5. Again, suitable modulation schemes were chosen so that both systems have the same effective throughput, namely 3 BPS. All simulation results were obtained with the aid of the space-time code $\mathbf{G}_2$ over uncorrelated Rayleigh fading channels. As compared to Figure 10.21, the throughput of the systems in Figure 10.22 has been increased from 2 BPS to 3 BPS. In order to maintain a high BPS throughput, 64QAM was employed in conjunction with the half-rate TC(2,1,4) code. We can see from the figure that the performance gain of the half-rate TC(2,1,4) code over the three-quarters-rate TC(2,1,4) code has been reduced to only 0.5 dB, as compared to 1 dB over the two-thirds-rate TC(2,1,4) code characterised in Figure 10.21. Moreover, the three-quarters-rate TC(2,1,4) code is weaker than the two-thirds-rate TC(2,1,4) code, since fewer parity bits are transmitted over the channel. Based on the fact that the performance gain of the half-rate TC(2,1,4) code has been reduced, we surmise that high-rate turbo codes will outperform the half-rate TC(2,1,4) code if the throughput of the system is increased to 4 BPS or even further.

From Figures 10.21 and 10.22 we can see that the best performance is achieved by the half-rate TC(2,1,4) code for an effective throughput of 2 and 3 BPS. However, we are also interested in the system's performance at higher effective BPS throughputs. Hence, during our later discourse in Section 10.5.3.6 the performance of high-rate TC and TBCH codes will be studied for throughput values in excess of 5 BPS.

10.5.3.3 Convolutional Codes

In Figure 10.23 we compare the performance of the $\mathbf{G}_2$ space-time-coded non-recursive half-rate CC(2,1,5), CC(2,1,7) and CC(2,1,9) codes. These schemes were standardised in the GSM

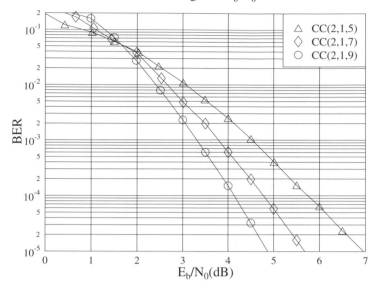

Figure 10.23: Performance comparison between the non-recursive half-rate CC(2,1,5), CC(2,1,7) and
CC(2,1,9) codes, where the coding parameters are shown in Tables 10.3 and 10.4. All simulation results were obtained with the aid of the space-time code G_2 using **QPSK** over
uncorrelated Rayleigh fading channels. The effective throughput is **1BPS**.

[49, 287], DVB [39] and the 3G UTRA systems [49, 266, 289], respectively. The associated
coding parameters were shown in Tables 10.3 and 10.4. All simulation results were obtained with
the aid of the space-time code G_2 using QPSK over uncorrelated Rayleigh fading channels. We
can see from the figure that at a BER of 10^{-5} the performance of the non-recursive CCs improves
by approximately 1 dB if the complexity is increased by a factor of $2^2 = 4$. However, the extra
performance gain attainable becomes smaller as the affordable complexity further increases. In
our forthcoming channel code comparisons, only the CC(2,1,9) code will be used, since it has
the best performance amongst the above three schemes and it has a comparable complexity to
that of the TC codes studied. Moreover, the CC(2,1,9) code is also proposed for the 3G UTRA
mobile communication system [49].

10.5.3.4 G_2-coded Channel Codec Comparison: Throughput of 2 BPS

Having narrowed down the choice of the G_2 space-time-coded CCs and the turbo codes, we
are now ready to compare the performance of the different proposed channel codecs belonging
to different codec families. Our comparison is carried out on the basis of the same throughput
and all channel codecs are concatenated with the space-time code G_2, when transmitting over
uncorrelated Rayleigh fading channels. Figure 10.24 shows the performance of our channel
codecs selected from the CC, TC, TBCH, TCM and TTCM families on the basis of the same
throughput of 2 BPS, regardless of their coding rates. The associated coding parameters are
shown in Tables 10.3, 10.4 and 10.5. The throughput is 2 BPS.

From Figure 10.24 we can see that the half-rate TC(2,1,4) code outperforms the other channel
codecs. At a BER of 10^{-5} the TC(2,1,4) code achieves a gain of approximately 0.5 dB over the

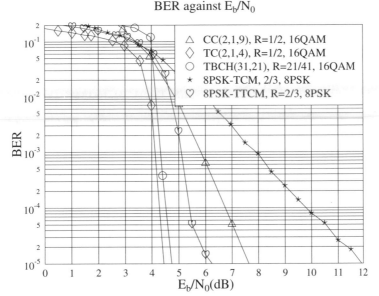

Figure 10.24: Performance comparison between different CC, TC, TBCH, TCM and TTCM schemes where the coding parameters are shown in Tables 10.3, 10.4 and 10.5. All simulation results were obtained with the aid of the space-time code G_2 at a throughput of **2 BPS** over uncorrelated Rayleigh fading channels.

TBCH(31,21) scheme at a much lower complexity. At the same BER, the TC(2,1,4) code also outperforms 8PSK-TTCM by approximately 1.5 dB. The poor performance of TTCM might be partially due to using generator polynomials, which are optimal for AWGN channels [57]. However, to date only limited research has been carried out on finding optimum generator polynomials for TTCM over fading channels [290].

In Figure 10.24 we also characterise the performance of the CC(2,1,9) and 8PSK-TCM schemes. The figure clearly demonstrates that the invention of turbo codes invoked in our TC, TBCH and TTCM G_2-coded schemes resulted in substantial improvements over the conventional G_2-coded channel codecs, such as the CC and TCM schemes considered. At a BER of 10^{-5}, the TC(2,1,4) code outperforms the CC(2,1,9) and 8PSK-TCM arrangements by approximately 3.0 dB and 7.5 dB, respectively.

10.5.3.5 G_2-coded Channel Codec Comparison: Throughput of 3 BPS

In Figure 10.25 we portray the performance of various channel codecs belonging to the CC, TC, TBCH, TCM and TTCM codec families on the basis of a constant throughput of 3 BPS, regardless of their coding rates. The associated coding parameters are shown in Tables 10.3, 10.4 and 10.5. The simulation results were obtained with the aid of the space-time code G_2 over uncorrelated Rayleigh fading channels.

From Figure 10.25 we can infer a few interesting points. As mentioned earlier, the half-rate TC(2,1,4) code suffers from the effects of puncturing as we increase the throughput of the system. In order to maintain a throughput of 3 BPS, 64QAM has to be employed in the systems using the half-rate TC(2,1,4) code. The rather vulnerable 64QAM modulation scheme

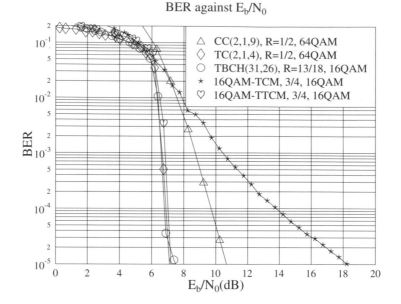

Figure 10.25: Performance comparison between different CC, TC, TBCH, TCM and TTCM schemes where the coding parameters are shown in Tables 10.3, 10.4 and 10.5. All simulation results were obtained with the aid of the space-time code G_2 at an effective throughput of **3 BPS** over uncorrelated Rayleigh fading channels.

appears to over-stretch the coding power of the half-rate TC(2,1,4) code attempting to saturate the available channel capacity. At a BER of 10^{-5} there is no obvious performance gain over the TBCH(31,26)/16QAM and 16QAM-TTCM schemes. Hence, we have reasons to postulate that if the throughput of the system is increased beyond 3 BPS, high-rate turbo codes should be employed for improving the performance, rather than invoking a higher-throughput modulation scheme.

10.5.3.6   Comparison of G_2-coded High-rate TC and TBCH Codes

In the previous section we have shown that at the BER of 10^{-5}, the required E_b/N_0 is increased by about 2.5 dB for the half-rate turbo code TC(2,1,4), as the throughput of the system is increased from 2 BPS to 3 BPS. A range of schemes having a throughput in excess of 5 BPS is characterised in Figure 10.26. Specifically, the figure shows the performance of high-rate TC and TBCH codes concatenated with the space-time code G_2 employing 64QAM over uncorrelated Rayleigh fading channels. The parameters of the TC and TBCH codes used are shown in Tables 10.3 and 10.5. The performance of half-rate turbo codes along with such a high throughput is not shown, because a modulation scheme having at least 1024 constellation points would be needed, which is practically infeasible over non-stationary wireless channels. Moreover, the turbo codes often would be overloaded with the plethora of errors induced by the densely packed constellation points.

In Figure 10.26 we can clearly see that there is not much difference in performance terms between the high-rate TC(2,1,4) and TBCH codes employed, although the TBCH codes exhibit marginal gains. This gain is achieved at a cost of high decoding complexity, as evidenced by Ta-

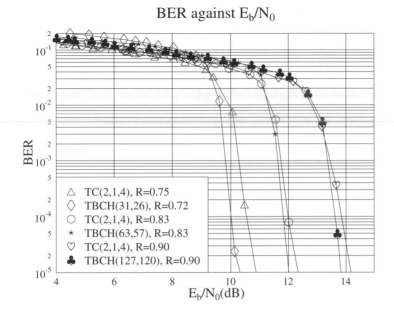

Figure 10.26: Performance comparison between high-rate TC and TBCH codes concatenated with the space-time code $\mathbf{G}_2$ employing 64QAM over uncorrelated Rayleigh fading channels. The parameters of the TC and TBCH codes were shown in Tables 10.3 and 10.5.

ble 10.6. The slight performance improvement of the TBCH(31,26) code over the three-quarters rate TC(2,1,4) scheme is probably due to its slightly lower code rate of $R = 0.72$, compared to the rate of $R = 0.75$ associated with the TC(2,1,4) code. It is important to note that all BCH component codes used in the TBCH codes have a minimum distance d_{min} of 3. We speculate that the performance of the TBCH codes might improve if d_{min} is increased to 5. However, owing to the associated complexity we will refrain from employing $d_{min} = 5$ BCH component codes in the TBCH schemes studied.

10.5.3.7 Comparison of High-rate TC and Convolutional Codes

In Figure 10.27, we compare the performance of the high-rate punctured TC(2,1,4) and CC(2,1,7) codes concatenated with the space-time code $\mathbf{G}_2$ employing 64QAM over uncorrelated Rayleigh fading channels. The puncturing patterns employed for the CC(2,1,7) scheme were proposed in the DVB standard [39]. The parameters of the TC(2,1,4) and CC(2,1,7) codes are shown in Tables 10.3, 10.4 and 10.5. From the figure we can see that both high-rate TC(2,1,4) codes outperform their equivalent rate CC(2,1,7) counterparts by about 2 dB at a BER of 10^{-5}, whilst maintaining a similar estimated decoding complexity, as was evidenced by Table 10.6. This fact indicates that at a given tolerable complexity, better BER performance can be attained by an iterative turbo decoder. These findings motivated the investigations of our next section, where the performance of the various schemes was studied in the context of the achievable coding gain versus the estimated decoding complexity.

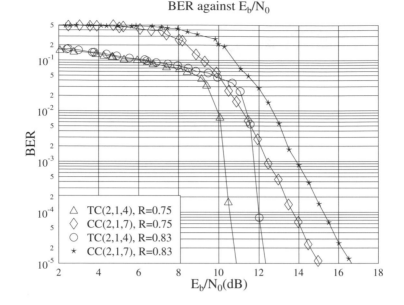

Figure 10.27: Performance comparison between high-rate TC and CC codes concatenated with the space-time code G_2 employing 64QAM over uncorrelated Rayleigh fading channels. The parameters of the TC and CC codes were shown in Tables 10.3, 10.4 and 10.5.

10.5.4   Coding Gain versus Complexity

In Section 10.4.3 we have estimated the various channel decoders' complexity based on a few simplifying assumptions. All the complexities estimated in our forthcoming discourse were calculated based on Equations 10.43 to 10.51. Again, our performance comparison of the channel codes was made on the basis of the coding gain defined as the E_b/N_0 difference, expressed in decibels, at a BER of 10^{-5} between the various channel-coded and uncoded systems having the same throughput, while using the space-time code G_2.

10.5.4.1   Complexity Comparison of Turbo Convolutional Codes

Figure 10.28 shows (a) the coding gain versus the number of iterations and (b) the coding gain versus estimated complexity for the TC(2,1,3), TC(2,1,4) and TC(2,1,5) codes, where the coding parameters used are shown in Tables 10.3, 10.5 and 10.6. All simulation results were obtained upon employing the space-time code G_2 using one receiver and 64QAM over uncorrelated Rayleigh fading channels at an effective throughput of 3 BPS. We can see from Figure 10.28(a) that there is a huge performance improvement of approximately 3-4 dB between the first and second turbo decoding iteration. However, the further coding gain improvements become smaller, as the number of iterations increases. It can be seen from the figure that the performance of turbo codes does not significantly improve after eight iterations, as indicated by the rather flat coding gain curve. Figure 10.28(a) also shows that as we increase the constraint length K of the turbo codes from 3 to 5, the associated performance improves.

In Figure 10.28(b) the coding gains of the various turbo codes using different number of iterations were compared on the basis of their estimated complexity. This was necessary, since we

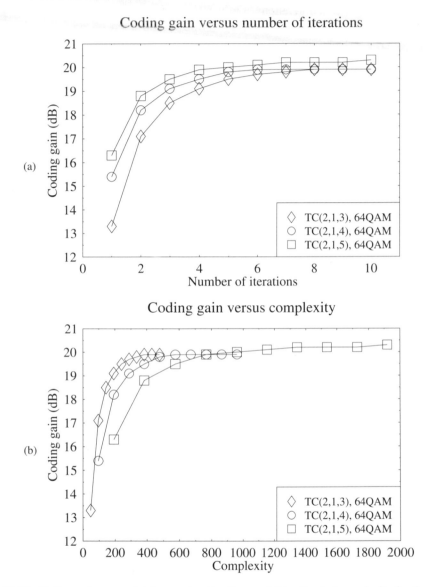

Figure 10.28: Coding gain versus (a) the number iterations and (b) estimated complexity for the TC(2,1,3), TC(2,1,4) and TC(2,1,5) codes, where the coding parameters are shown in Tables 10.3, 10.5 and 10.6. All simulation results were obtained upon employing the space-time code G_2 using one receiver and **64QAM** over uncorrelated Rayleigh fading channels at an effective throughput of **3 BPS**.

have seen in Section 10.4.3 that the estimated complexity of turbo codes depends exponentially on the constraint length K, but only linearly on the number of iterations. From Figure 10.28(b), we can see that the estimated complexity of the TC(2,1,5) code ranges from approximately 200 to 2000, when using one to ten iterations. On the other hand, the estimated complexity of the TC(2,1,3) scheme ranges only from approximately 50 to 500 upon invoking one to ten iterations. This clearly shows that the estimated complexity of the turbo codes is dominated by the constraint length K. Figure 10.28(b) also shows that the coding gain curve of the TC(2,1,3) code saturates faster, which is demonstrated by the steep increase in coding gain as the estimated complexity increases. For achieving the same coding gain of 19 dB, we can see that the TC(2,1,3) scheme requires the lowest estimated complexity. We would require two to three times higher computational power for the TC(2,1,5) code to achieve the above-mentioned coding gain of 19 dB.

10.5.4.2   Complexity Comparison of Channel Codes

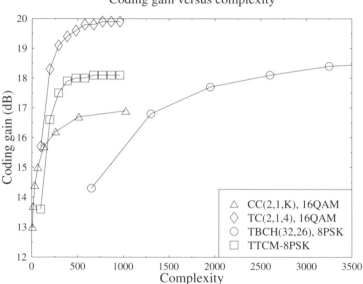

Figure 10.29: Coding gain versus estimated complexity for the CC(2,1,K), TC(2,1,4), TBCH(32,26) and TTCM-8PSK schemes where the parameters are shown in Tables 10.3, 10.4, 10.5 and 10.6. All simulation results were obtained upon employing space-time code G_2 using one receiver over uncorrelated Rayleigh fading channels at an effective throughput of **2 BPS**.

In the previous section we have compared the coding gain versus estimated complexity of the G_2-coded turbo schemes TC(2,1,3), TC(2,1,4) and TC(2,1,5). Here we compare the TC(2,1,4) arrangement that faired best amongst them to the CC(2,1,9) code and to the TBCH(32,26)/8PSK as well as to the TTCM-8PSK arrangements, representing the other codec families studied. Specifically, Figure 10.29 shows the coding gain versus estimated complexity for the CC(2,1,K), TC(2,1,4), TBCH(32,26) and TTCM-8PSK schemes, where the associated parameters are shown in Tables 10.3, 10.4, 10.5 and 10.6. All simulation results were obtained upon employing the space-time code G_2 using one receiver over uncorrelated Rayleigh fading channels at an effective throughput of 2 BPS. For the turbo schemes TC(2,1,4), TBCH(32,26) and TTCM-8PSK the increased estimated complexity is achieved by increasing the number of iterations from one to ten. However, CCs are decoded non-iteratively. Therefore in Figure 10.29 we vary the constraint length K of the CCs from three to ten, which results in increased estimated complexity. The generator polynomials of the CC(2,1,K) codec, where $K = 3 \ldots 10$, are given in [102] and they define the corresponding maximum minimum free distance of the codes. From Figure 10.29 we can see that there is a steep increase in the coding gain achieved by the TC(2,1,4) code as the estimated complexity is increased. Moreover, the TC(2,1,4) scheme asymptotically achieves a maximum coding gain of approximately 20 dB. At a low estimated complexity of approximately 200, the TC(2,1,4) code attains a coding gain of approximately 18 dB, which exceeds that of the other channel codes studied. The TBCH(32,26) arrangement is the least attractive one, since a huge estimated complexity is incurred when aiming for a high coding gain.

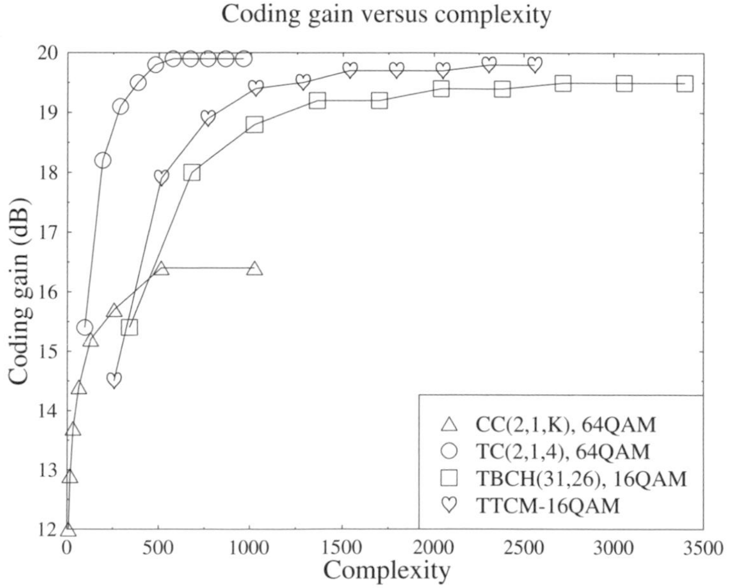

Coding gain versus complexity

Figure 10.30: Coding gain versus estimated complexity for the CC(2,1,K), TC(2,1,4), TBCH(31,26) and TTCM-16QAM schemes where the coding parameters are shown in Tables 10.3, 10.4, 10.5 and 10.6. All simulation results were obtained upon employing space-time code G_2 using one receiver over uncorrelated Rayleigh fading channels at an effective throughput of **3 BPS**.

In contrast to the 2 BPS schemes of Figure 10.29, Figure 10.30 shows the corresponding coding gain versus estimated complexity curves for the CC(2,1,K), TC(2,1,4), TBCH(31,26) and TTCM-16QAM 3 BPS arrangements, where the coding parameters are shown in Tables 10.3, 10.4, 10.5 and 10.6. Again, all simulation results were obtained upon employing the space-time code G_2 using one receiver over uncorrelated Rayleigh fading channels at an effective throughput of 3 BPS. As before, the increased estimated complexity of the turbo schemes is incurred by increasing the number of iterations from one to ten. For the CCs the constraint length K is varied from three to ten. Similarly to Figure 10.29, the TC(2,1,4) scheme achieves a considerable coding gain at a relatively low estimated complexity. For example, in order to achieve a coding gain of 18 dB, the TTCM and TBCH(31,26) arrangements would require an approximately three and four times higher computational power compared to the TC(2,1,4) code.

From Figures 10.29 and 10.30 we can clearly see that turbo codes are the most attractive ones of all the channel codes studied in conjunction with the space-time code $\mathbf{G}_2$, offering an impressive coding gain at a moderate estimated decoding complexity.

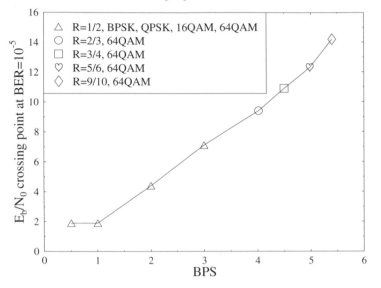

Figure 10.31: The E_b/N_0 value required for maintaining BER= 10^{-5} versus the effective throughput BPS for the space-time block code $\mathbf{G}_2$ concatenated with the TC(2,1,4) code where the coding parameters are shown in Tables 10.3, 10.5 and 10.6. All simulation results were obtained upon employing space-time code $\mathbf{G}_2$ using **one receiver** over uncorrelated Rayleigh fading channels.

In Figure 10.31, we show the E_b/N_0 value required for maintaining a BER of 10^{-5} versus the effective throughput BPS for the space-time block code $\mathbf{G}_2$ concatenated with the TC(2,1,4) code where the coding parameters are shown in Tables 10.3, 10.5 and 10.6. All simulation results were obtained upon employing space-time code $\mathbf{G}_2$ using **one receiver** over uncorrelated Rayleigh fading channels. Half-rate TC(2,1,4) code was employed for BPS = 3. Then TC(2,1,4) code with various rates was employed with 64QAM in order to achieve increasing effective throughput BPS. It can be seen from the figure that the E_b/N_0 value required for maintaining a BERof 10^{-5} increases linearly as the effective throughput BPS increases.

10.6   Summary and Conclusions

The state of the art of transmission schemes based on multiple transmitters and receivers was reviewed in Section 10.1. This was followed by a rudimentary introduction to the MRC [71] technique, using a simple example in Section 10.2.1. Space-time block codes were introduced in Section 10.3 employing the unity-rate space-time code $\mathbf{G}_2$. In Sections 10.3.1.1 and 10.3.1.2 two examples of employing the space-time code $\mathbf{G}_2$ were provided using one and two receivers, respectively. The transmission matrix of a range of different-rate space-time codes, namely that of the codes $\mathbf{G}_3$, $\mathbf{G}_4$, $\mathbf{H}_3$ and $\mathbf{H}_4$ of Table 10.2, was also given. Additionally, a brief description of the MAP decoding algorithm [283] was provided in Section 10.3.3 in the context of space-time block codes.

In Section 10.4 we proposed a system which consists of the concatenation of the above-mentioned space-time block codes and a range of different channel codes. The channel coding schemes investigated were convolutional codes, turbo convolutional codes, turbo BCH codes, trellis-coded modulation and turbo trellis-coded modulation. The estimated complexity and memory requirements of the channel decoders were summarised in Section 10.4.3.

Finally, we presented our simulation results in Section 10.5, which were divided into four categories. In Section 10.5.1, we first compared the performance results of the space-time codes G_2, G_3, G_4, H_3 and H_4 without using channel codecs. It was found that as we increased the effective throughput of the system, the performance of the half-rate space-time codes G_3 and G_4 degraded in comparison to that of the unity-rate space-time code G_2. This was because in order to maintain the same effective throughput, higher modulation schemes had to be employed in conjunction with the half-rate space-time codes G_3 and G_4, which were more prone to errors and hence degraded the performance of the system. On the other hand, for the sake of maintaining the same diversity gain and same effective throughput, we found that the performance of the space-time codes H_3 and H_4 was better than that of the space-time codes G_3 and G_4, respectively. Since the space-time code G_2 has a code rate of unity, we were able to concatenate it with half-rate TC codes, while maintaining the same effective throughput as the half-rate space-time code using no channel coding. Hence for the same effective throughput, the unity-rate G_2 space-time-coded and half-rate channel-coded schemes provided substantial performance improvement over the three-quarters rate space-time code H_4 and half-rate space-time code G_4, which were unable to benefit from channel coding. We concluded that the reduction in coding rate was best invested in turbo channel codes, rather than space-time block codes. Therefore, all channel codes studied were concatenated with the unity-rate space-time code G_2 only.

In the second category of our investigations in Section 10.5.1.5 we studied the effect of the binary channel codes' data and parity bits mapped into different protection classes of multilevel modulation schemes. It was found that TC codes having different constraint lengths K require different mapping methods, as evidenced by Figure 10.12. By contrast, in the TBCH codes studied mapping of the parity bits to the higher-integrity protection class of a multilevel modulation scheme yielded a better performance. The so-called random separation-based interleaver was proposed, in order to improve the performance of the system.

The third set of results compared the performance of all proposed channel codes in conjunction with the space-time code G_2. In order to avoid confusion, we only selected one channel code from each group of channel codes in Table 10.3. Specifically, only half-rate TC codes were studied, as they gave better coding gain performance compared to other TC codes having lower and higher rates. It was then found that the performance of the half-rate TC codes was better than that of the CC, TBCH, TCM and TTCM codes. Then, we compared the performance of high-rate TC codes with high-rate TBCH codes in conjunction with 64QAM. It was found that the TBCH codes provided a slight performance improvement over high-rate TC codes, but at the cost of high complexity. Finally, the chapter was concluded by comparing the G_2 space-time-coded channel codes upon taking their estimated complexity into consideration. In Figures 10.29 and 10.30, we can clearly see that the half-rate TC codes give the best coding gain at a moderate estimated complexity.

Following our discussions on space-time block codes in this chapter, which have been contrived for communications over non-dispersive wireless channels, in the next chapter space-time trellis codes are discussed.

Chapter 11

Space-Time Trellis Codes

11.1 Introduction[1]

In the previous chapter, we have detailed the encoding and decoding processes of *space-time block codes* [71–73]. Various proposed space-time block codes [72, 73] have been discussed and their performance was investigated, when communicating over perfectly interleaved, non-dispersive Rayleigh fading channels. A range of systems consisting of space-time block codes and different channel codecs were proposed. The performance versus estimated complexity trade-off of the different systems was investigated and compared.

While the understanding the basic space-time trellis coding principles does not require deep prior knowledge in the field of channel coding, an appreciation of the concepts of trellis-based decoding techniques, which have been extensively discussed in Chapters 2, 4 and 9, is necessary. Furthermore, as in the previous chapter, we will combine space-time trellis coding with the various channel codecs previously highlighted in the book.

In an effort to provide as comprehensive a technology road-map as possible and to identify the most promising schemes in the light of their performance versus estimated complexity, in this chapter we shall explore the family of *space-time trellis codes* [70, 80, 278–281] which were proposed by Tarokh *et al*. Space-time trellis codes incorporate jointly designed channel coding, modulation, transmit diversity and optional receiver diversity. The performance criteria for designing space-time trellis codes were outlined in [70], under the assumption that the channel is fading slowly and that the fading is frequency non-selective. It was shown in [70] that the system's performance is determined by matrices constructed from pairs of distinct code sequences. Both the *diversity gain* and *coding gain* of the codes are determined by the minimum rank and the minimum determinant [70, 291] of the matrices, respectively. The results were then also extended to fast fading channels. The space-time trellis codes proposed in [70] provide the best trade-off between data rate, diversity advantage and trellis complexity.

The performance of both space-time trellis and block codes over narrowband Rayleigh fading channels was investigated by numerous researchers [70, 71, 73, 79, 80]. The investigation of space-time codes was then also extended to the class of practical wideband fading channels. The effect of multiple paths on the performance of space-time trellis codes was studied in [281]

---

[1] This chapter is based on T. H. Liew and L. Hanzo: Space-Time Trellis Coding and Space-Time Block Coding versus Adaptive Modulation: A Comparative Study over Wideband Channels, submitted to IEEE Tr. on Vehicular Technology, 2001

for transmission over slowly varying Rayleigh fading channels. It was shown in [281] that the presence of multiple paths does not decrease the diversity order guaranteed by the design criteria used to construct the space-time trellis codes. The evidence provided in [281] was then also extended to rapidly fading dispersive and non-dispersive channels. As a further performance improvement, turbo equalisation was employed in [74] in order to mitigate the effects of dispersive channels. However space-time-coded turbo equalisation involved an enormous complexity. In addressing the complexity issues, Bauch and Al-Dhahir [75] derived finite-length Multi-Input Multi-Output (MIMO) channel filters and used them as prefilters for turbo equalisers. These prefilters significantly reduce the number of turbo equaliser states and hence mitigate the decoding complexity. As an alternative solution, the effect of Inter Symbol Interference (ISI) could be eliminated by employing Orthogonal Frequency Division Multiplexing (OFDM) [207]. A system using space-time trellis-coded OFDM is attractive, since the associated decoding complexity is relatively modest. The attractive performance versus complexity trade-offs associated with this system resulted in a recent surge of related research interests [76–79]. In [76, 78, 79], non-binary Reed–Solomon (RS) codes were employed in the context of space-time trellis-coded OFDM systems for improving the achievable performance.

Similarly, the performance of space-time block codes was also investigated over frequency-selective Rayleigh fading channels. In [292], a multiple-input multiple-output equaliser was utilised for equalising the dispersive multipath channels. Furthermore, the advantages of OFDM were also exploited in space-time block-coded systems [79, 293, 294].

We commence our discussion with a detailed description of the encoding and decoding processes of the space-time trellis codes in Section 11.2. The state diagrams of a range of other space-time trellis codes are also given in Section 11.2.2. In Section 11.3, a specific system was proposed, which enables the comparison of space-time trellis codes and space-time block codes over wideband channels. Our simulation results are then given in Section 11.4. We continue our investigations by proposing space-time-coded adaptive modulation-based OFDM in Section 11.5. Finally, we conclude the chapter in Section 11.6.

11.2   Space-time Trellis Codes

In this section, we will detail the encoding and decoding processes of space-time trellis codes. Space-time trellis codes are defined by the number of transmitters p, by the associated state diagram and by the modulation scheme employed. For ease of explanation, as an example we shall use the simplest 4-state, 4-level Phase Shift Keying (4PSK) space-time trellis code, which has $p = 2$ two transmit antennas.

11.2.1   The 4-state, 4PSK Space-time Trellis Encoder

At any time instant k, the 4-state 4PSK space-time trellis encoder transmits symbols $x_{k,1}$ and $x_{k,2}$ over the transmit antennas $Tx\,1$ and $Tx\,2$, respectively. The output symbols at time instant k are given by [70]:

$$x_{k,1} = 0.d_{k,1} + 0.d_{k,2} + 1.d_{k-1,1} + 2.d_{k-1,2} \qquad (11.1)$$

$$x_{k,2} = 1.d_{k,1} + 2.d_{k,2} + 0.d_{k-1,1} + 0.d_{k-1,2} \qquad (11.2)$$

where $d_{k,i}$ represents the current input bits, and $d_{k-1,i}$ the previous input bits, and $i = 1, 2$. More explicitly, we can represent Equation 11.2 with the aid of a shift register, as shown in Figure 11.1, where $\oplus$ represents modulo-4 addition. Let us explain the operation of the shift-register encoder

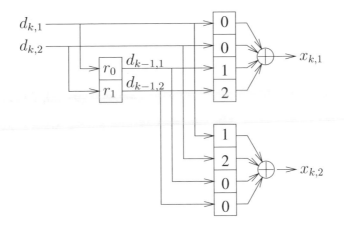

Figure 11.1: The 4-state, 4PSK space-time trellis encoder.

for the random input data bits 01111000. The shift-register stages r_0 and r_1 must be reset to zero before the encoding of a transmission frame starts. They represent the state of the encoder. The operational steps are summarised in Table 11.1. Again, given the register stages $d_{k-1,1}$ and $d_{k-1,2}$ as well as the input bits $d_{k,1}$ and $d_{k,2}$, the output symbols seen in the table are determined according to Equation 11.2 or Figure 11.1. Note that the last two binary data bits in Table 11.1 are

Input queue	Instant k	Input bits $(d_{k,1}; d_{k,2})$	Shift register $(d_{k-1,1}; d_{k-1,2})$	State S_k	Transmitted symbols $(x_{k,1}; x_{k,2})$
00011110	0	- -	0 0	0	- -
000111	1	0 1(2)	0 0	0	0 2
0001	2	1 1(3)	0 1	2	2 3
00	3	1 0(1)	1 1	3	3 1
	4	0 0(2)	1 0	1	1 0
	5	- -	0 0	0	- -

Table 11.1: Operation of the space-time encoder of Figure 11.1.

intentionally set to zero in order to force the 4-state 4PSK trellis encoder back to the zero state, which is common practice at the end of a transmission frame. Therefore, the transmit antenna Tx 1 will transmit symbols $0, 2, 3, 1$. By contrast, symbols $2, 3, 1, 0$ are then transmitted by the antenna Tx 2.

According to the shift-register encoder shown in Figure 11.1, we can find all the legitimate subsequent states, which result in transmitting the various symbols $x_{k,1}$ and $x_{k,2}$, depending on a particular state of the shift register. This enables us to construct the state diagram for the encoder. The 4PSK constellation points are seen in Figure 11.2, while the corresponding state diagram of the 4-state 4PSK space-time trellis code [70] is shown in Figure 11.3. In Figure 11.3, we can see that for each current state there are four possible trellis transitions to the states $0, 1, 2$ and 3, which correspond to the legitimate input symbols of $0(d_{k,1} = 0, d_{k,2} = 0), 1(d_{k,1} = 1, d_{k,2} = 0), 2(d_{k,1} = 0, d_{k,2} = 1)$ and $3(d_{k,1} = 1, d_{k,2} = 1)$, respectively. Correspondingly, there are four sets of possible transmitted symbols associated with the four trellis transitions, shown at right of the state diagram. Each trellis transition is associated with two transmitted symbols,

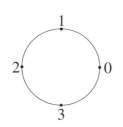

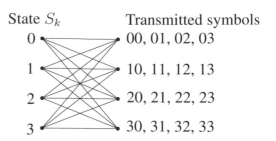

Figure 11.2: The 4PSK constellation points.   Figure 11.3: The 4-state, 4PSK space-time trellis code.

namely with x_1 and x_2, which are transmitted by the antennas Tx 1 and Tx 2, respectively. In Figure 11.4, we have highlighted the trellis transitions from state zero, $S_k = 0$, to various states. The associated input symbols and the transmitted symbols of each trellis transitions are shown on top of each trellis transition. If the input symbol is 0, then the symbol $x_1 = 0$ will be

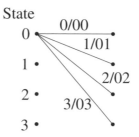

Figure 11.4: The trellis transitions from state $S_k = 0$ to various states.

sent by the transmit antenna Tx 1, and symbol $x_2 = 0$ by the transmit antenna Tx 2 as seen in Figure 11.4 or Figure 11.3. The next state remains $S_{k+1} = 0$. However, if the input symbol is 2 associated with $d_{k,1} = 0$, $d_{k,2} = 1$ in Table 11.1, then the trellis traverses from state $S_k = 0$ to state $S_{k+1} = 2$ and the symbols $x_1 = 0$ and $x_2 = 2$ are transmitted over the antennas Tx 1 and Tx 2, respectively. Again, the encoder is required to be in the zero state both at the beginning and at the end of the encoding process.

11.2.1.1   The 4-state, 4PSK Space-time Trellis Decoder

In Figure 11.5 we show the baseband representation of the 4-state, 4PSK space-time trellis code using two receivers. At any transmission instant, we have symbols x_1 and x_2 transmitted by the antennas Tx 1 and Tx 2, respectively. At the receivers Rx 1 and Rx 2, we would have:

$$y_1 = h_{11}x_1 + h_{12}x_2 + n_1 \qquad (11.3)$$
$$y_2 = h_{21}x_1 + h_{22}x_2 + n_2 , \qquad (11.4)$$

where h_{11}, h_{12}, h_{21} and h_{22} represent the corresponding complex time-domain channel transfer factors. Aided by the channel estimator, the Viterbi-Algorithm-based maximum likelihood sequence estimator [70] first finds the branch metric associated with every transition in the decoding trellis diagram, which is identical to the state diagram shown in Figure 11.3. For each trellis transition, we have two estimated transmit symbols, namely $\tilde{x}_1$ and $\tilde{x}_2$, for which the

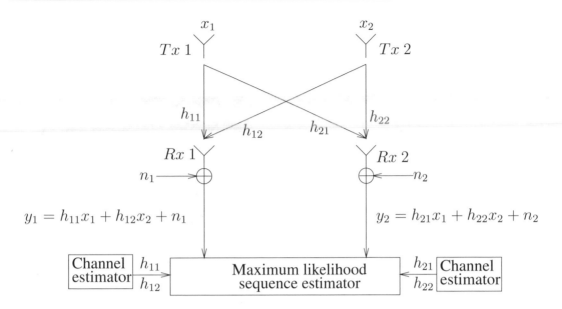

Figure 11.5: Baseband representation of the 4-state, 4PSK space-time trellis code using two receivers.

branch metric BM is given by:

$$
\begin{aligned}
BM &= |y_1 - h_{11}\tilde{x}_1 - h_{12}\tilde{x}_2 + y_2 - h_{21}\tilde{x}_1 - h_{22}\tilde{x}_2|^2 \\
&= \sum_{i=1}^{2} |y_i - h_{i1}\tilde{x}_1 - h_{i2}\tilde{x}_2|^2 \\
&= \sum_{i=1}^{2} \left| y_i - \sum_{j=1}^{2} h_{ij}\tilde{x}_j \right|^2 .
\end{aligned}
\tag{11.5}
$$

We can, however, generalise Equation 11.5 to p transmitters and q receivers, as follows:

$$
BM = \sum_{i=1}^{p} \left| y_i - \sum_{j=1}^{q} h_{ij}\tilde{x}_j \right|^2 .
\tag{11.6}
$$

When all the transmitted symbols were received and the branch metric of each legitimate transition was calculated, the maximum likelihood sequence estimator invokes the Viterbi Algorithm (VA) in order to find the maximum likelihood path associated with the best accumulated metric.

11.2.2 Other Space-time Trellis Codes

In Section 11.2.1, we have shown the encoding and decoding process of the simple 4-state, 4PSK space-time trellis code. More sophisticated 4PSK space-time trellis codes were designed by increasing the number of trellis states [70], which are reproduced in Figures 11.6 to 11.8. With an increasing number of trellis states the number of tailing symbols required for terminating the trellis at the end of a transmitted frame is also increased. Two zero symbols are needed to force

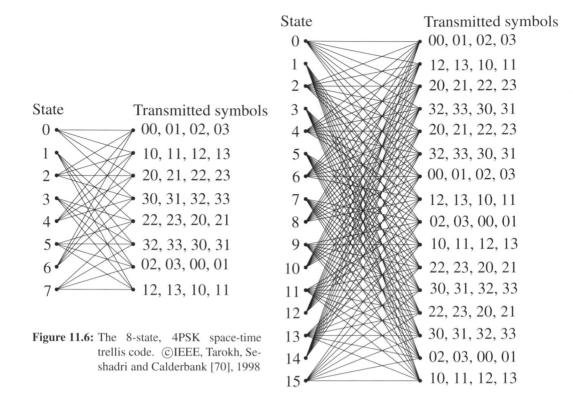

State   Transmitted symbols
0 ●————————————● 00, 01, 02, 03

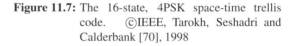

Figure 11.6: The 8-state, 4PSK space-time
trellis code. ©IEEE, Tarokh, Seshadri and Calderbank [70], 1998

Figure 11.7: The 16-state, 4PSK space-time trellis
code.   ©IEEE, Tarokh, Seshadri and
Calderbank [70], 1998

the trellis back to state zero for the space-time trellis codes shown in Figures 11.6 and 11.7. By
contrast, three zero symbols are required for the space-time trellis code shown in Figure 11.8.

Space-time trellis codes designed for the higher-order modulation scheme of 8PSK were
also proposed in [70]. In Figure 11.9, we show the constellation points employed in [70]. The
proposed 8-state, 16-state and 32-state 8PSK space-time trellis codes are reproduced from [70]
in Figures 11.10, 11.11 and 11.12, respectively. One zero symbol is required to terminate the
8-state, 8PSK space-time trellis code, whereas two zero symbols are needed for both the 16-state
and 32-state 8PSK space-time trellis codes.

11.3 Space-time-coded Transmission over
Wideband Channels

In Section 11.2, we have detailed the concept of space-time trellis codes. Let us now elaborate
further by investigating the performance of space-time codes over dispersive wideband fading
channels. As mentioned in Section 11.1, Bauch's approach [74, 75] of using turbo equalisation
for mitigating the ISI exhibits a considerable complexity. Hence we argued that using space-time-
coded OFDM constitutes a more favourable approach to transmission over dispersive wireless

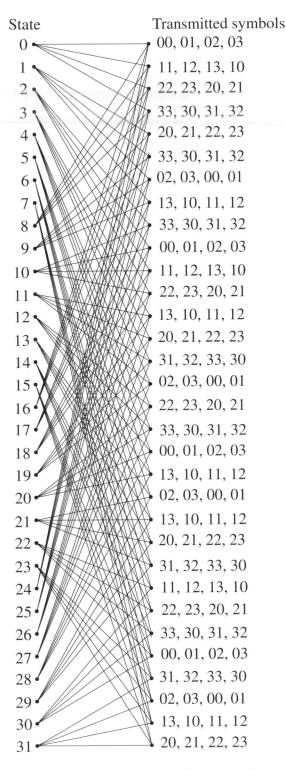

State   Transmitted symbols

State	Transmitted symbols
0	00, 01, 02, 03
1	11, 12, 13, 10
2	22, 23, 20, 21
3	33, 30, 31, 32
4	20, 21, 22, 23
5	33, 30, 31, 32
6	02, 03, 00, 01
7	13, 10, 11, 12
8	33, 30, 31, 32
9	00, 01, 02, 03
10	11, 12, 13, 10
11	22, 23, 20, 21
12	13, 10, 11, 12
13	20, 21, 22, 23
14	31, 32, 33, 30
15	02, 03, 00, 01
16	22, 23, 20, 21
17	33, 30, 31, 32
18	00, 01, 02, 03
19	13, 10, 11, 12
20	02, 03, 00, 01
21	13, 10, 11, 12
22	20, 21, 22, 23
23	31, 32, 33, 30
24	11, 12, 13, 10
25	22, 23, 20, 21
26	33, 30, 31, 32
27	00, 01, 02, 03
28	31, 32, 33, 30
29	02, 03, 00, 01
30	13, 10, 11, 12
31	20, 21, 22, 23

Figure 11.8: The 32-State, 4PSK space-time trellis code. ©IEEE, Tarokh, Seshadri and Calderbank [70], 1998

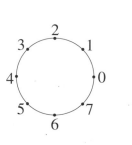

Figure 11.9: The 8PSK constellation points.

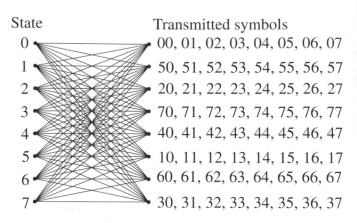

Figure 11.10: The 8-state, 8PSK space-time trellis code. ©IEEE, Tarokh, Seshadri and Calderbank [70], 1998

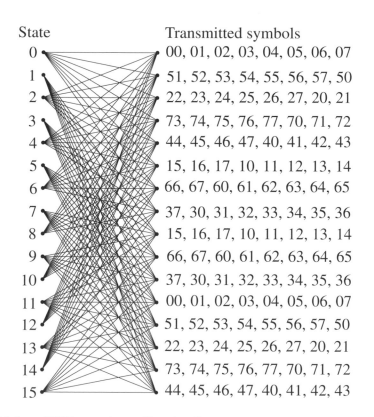

Figure 11.11: The 16-State, 8PSK space-time trellis code. ©IEEE, Tarokh, Seshadri and Calderbank [70], 1998

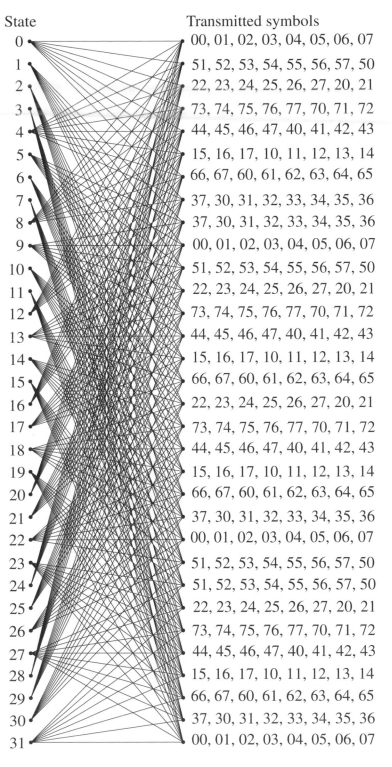

Figure 11.12: The 32-State, 8PSK space-time trellis code. ©IEEE, Tarokh, Seshadri and Calderbank [70], 1998

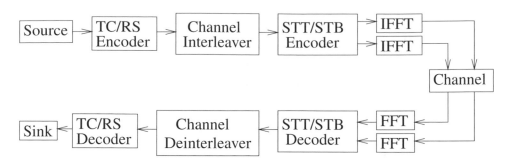

Figure 11.13: System overview.

channels, since the associated decoding complexity is significantly lower. Therefore, in this chapter OFDM is employed for mitigating the effects of dispersive channels.

It is widely recognised that space-time trellis codes [70] perform well at the cost of high complexity. However, Alamouti's $\mathbf{G}_2$ space-time block code [71] could be invoked instead of space-time trellis codes. The space-time block code $\mathbf{G}_2$ is appealing in terms of its simplicity, although there is a slight loss in performance. Therefore, we concatenate the space-time block code $\mathbf{G}_2$ with Turbo Convolutional (TC) codes in order to improve the performance of the system. The family of TC codes was favoured because it was shown in Section 10.5.3 of Chapter 10 and in [295,296] that TC codes achieve an enormous coding gain at a moderate complexity, when compared to convolutional codes, turbo BCH codes, trellis-coded modulation and turbo trellis-coded modulation. The performance of concatenated space-time block codes and TC codes will then be compared to that of space-time trellis codes. Conventionally, Reed–Solomon (RS) codes have been employed in conjunction with the space-time trellis codes [76, 78, 79] for improving the performance of the system. In our forthcoming discussion, we will concentrate on comparing the performance of space-time block and trellis codes in conjunction with various channel coders.

11.3.1 System Overview

Figure 11.13 shows the schematic of our system. At the transmitter, the information source generates random information data bits. The information bits are then encoded by TC codes, RS codes or left uncoded. The coded or uncoded bits are then channel interleaved, as shown in Figure 11.13. The output bits of the channel interleaver are then passed to the Space-Time Trellis (STT) or Space-Time Block (STB) encoder. We will investigate all the previously mentioned space-time trellis codes proposed in [70], where the associated state diagrams are shown in Figures 11.3, 11.6, 11.7, 11.10, 11.11 and 11.12. The modulation schemes employed are 4PSK as well as 8PSK and the corresponding trellis diagrams were shown in Figures 11.2 and 11.9, respectively. On the other hand, from the family of space-time block codes only Alamouti's $\mathbf{G}_2$ code is employed in our system, since we have shown in Figure 10.11 and in [296] that the best performance is achieved by concatenating the space-time block code $\mathbf{G}_2$ with TC codes. For convenience, the transmission matrix of the space-time block code $\mathbf{G}_2$ is reproduced here as follows:

$$\mathbf{G}_2 = \begin{pmatrix} x_1 & x_2 \\ -\bar{x}_2 & \bar{x}_1 \end{pmatrix}. \tag{11.7}$$

The reader is referred to Chapter 10 for an in-depth discussion on space-time block codes. Different modulation schemes could be employed [49], such as Binary Phase Shift Keying (BPSK), Quadrature Phase Shift Keying (QPSK), 16-level Quadrature Amplitude Modulation (16QAM) and 64-level Quadrature Amplitude Modulation (64QAM). Gray mapping of the bits to symbols was applied and this resulted in different protection classes in higher-order modulation schemes [207]. The mapping of the data bits and parity bits of the TC encoder was chosen such that it yielded the best achievable performance along with the application of the random separation channel interleaver [295] seen in Figure 10.14. The output of the space-time encoder was then OFDM [207] modulated and transmitted by the corresponding antenna. The number of transmit antennas was fixed to two, while the number of receive antennas constituted a design parameter. Dispersive wideband channels were used and the associated channels' profiles will be discussed later.

At the receiver the signal of each receive antenna is OFDM demodulated. The demodulated signals of the receiver antennas are then fed to the space-time trellis or space-time block decoder. The space-time decoders apply the MAP [11] or Log-MAP [52, 283] decoding algorithms for providing soft outputs for the channel decoders. If no channel codecs are employed in the system, the space-time decoders apply the VA [9, 70], which gives similar performance to the MAP decoder at a lower complexity. The decoded bits are finally passed to the sink for the calculation of the Bit Error Rate (BER) or Frame Error Rate (FER).

11.3.2 Space-time and Channel Codec Parameters

In Figure 11.13, we have given an overview of the proposed system. In this section, we present the parameters of the space-time codes and the channel codecs employed in the proposed system. We will employ the set of various space-time trellis codes shown in Figures 11.3, 11.6, 11.7, 11.8, 11.10, 11.11 and 11.12. The associated space-time trellis coding parameters are summarised in Table 11.2. On the other hand, from the family of space-time block codes only Alamouti's

Modulation scheme	BPS	Decoding algorithm	No. of states	No. of transmitters	No. of termination symbols
4PSK	2	VA	4	2	1
			8	2	2
			16	2	2
			32	2	3
8PSK	3	VA	8	2	1
			16	2	2
			32	2	2

Table 11.2: Parameters of the space-time trellis codes shown in Figures 11.3, 11.6, 11.7, 11.8, 11.10, 11.11 and 11.12.

G_2 code is employed, since we have shown in Section 10.5.3 of the previous chapter and in [296] that the best performance in the set of investigated schemes was yielded by concatenating the space-time block code G_2 with TC codes. The transmission matrix of the code is shown in Equation 11.7, while the number of transmitters used by the space-time block code G_2 is two, which is identical to the number of transmitters in the space-time trellis codes shown in Table 11.2.

Let us now briefly consider the TC channel codes used. The reader is referred to Chapters 7 and 10 for further information on the Log-MAP decoding algorithm and for a brief introduction to various TC codes, respectively. In this chapter, we will concentrate on using the simple half-rate TC(2, 1, 3) code. Its associated parameters are shown in Table 11.3. As seen in Table 11.4,

Code	Octal generator polynomial	No. of states	Decoding algorithm	Puncturing pattern	No. of iterations
TC(2, 1, 3)	7,5	4	Log-MAP	10,01	8

Table 11.3: The associated parameters of the TC(2, 1, 3) code.

different modulation schemes are employed in conjunction with the concatenated space-time block code G_2 and the TC(2,1,3) code. Since the half-rate TC(2,1,3) code is employed, higher-order modulation schemes such as 16QAM and 64QAM were chosen, so that the throughput of the system remained the same as that of the system employing the space-time trellis codes without channel coding. It is widely recognised that the performance of TC codes improves upon increasing the turbo interleaver size and near-optimum performance can be achieved using large interleaver sizes exceeding 10000 bits. However, this performance gain is achieved at the cost of high latency, which is impractical for a delay-sensitive real-time system. On the other hand, space-time trellis codes offer impressive coding gains [70] at low latency. The decoding of the space-time trellis codes is carried out on a transmission burst-by-burst basis. In order to make a fair comparison between the systems investigated, the turbo interleaver size was chosen such that all the coded bits were hosted by one transmission burst. This enables burst-by-burst turbo decoding at the receiver. Since we employ an OFDM modem, latency may also be imposed by a high number of subcarriers in an OFDM symbol. Therefore, the turbo interleaver size was increased, as the number of subcarriers increased in our investigations. In Table 11.4, we summarised the

Code	Code rate R	Modula-tion mode	BPS	Random turbo interleaver depth	Random separation interleaver depth
		128 carriers			
TC(2, 1, 3)	0.50	16QAM	2	256	512
		64QAM	3	384	768
		512 carriers			
		QPSK	1	512	1024
		16QAM	2	1024	2048

Table 11.4: The simulation parameters associated with the TC(2, 1, 3) code.

modulation schemes and interleaver sizes used for different number, of OFDM subcarriers in the proposed system. The random separation-based channel interleaver of Figure 10.14 was used. The mapping of the data bits and parity bits into different protection classes of the higher-order modulation scheme was carried out such that the best possible performance was attained. This issue was addressed in Section 10.5.2.

RS codes were employed in conjunction with the space-time trellis codes. Hard-decision decoding was utilised and the coding parameters of the RS codes employed are summarised in

Code	Galois field	Rate	Correctable symbol errors
RS(105,51)	2^{10}	0.49	27
RS(153,102)	2^{10}	0.67	25

Table 11.5: The coding parameters of the RS codes employed.

Table 11.5.

11.3.3 Complexity Issues

In this section, we will address the implementational complexity issues of the proposed system. We will, however, focus mainly on the relative complexity of the proposed systems, rather than attempting to quantify their exact complexity. In order to simplify our comparative study, several assumptions were stipulated. In our simplified approach, the estimated complexity of the system is deemed to depend only on that of the space-time trellis decoder and turbo decoder. In other words, the complexity associated with the modulator, demodulator, space-time block encoder and decoder as well as that of the space-time trellis encoder and turbo encoder are assumed to be insignificant compared to the complexity of space-time trellis decoder and turbo decoder.

In Section 10.4.3, we have detailed our complexity estimates for the TC decoder and the reader is referred to this section for further details. The estimated complexity of the TC decoder is assumed to depend purely on the number of trellis transitions per information data bit and this simple estimated complexity measure was also used in Section 10.4.3 as the basis of our comparisons. Here, we adopt the same approach and evaluate the estimated complexity of the space-time trellis decoder on the basis of the number of trellis transitions per information data bit.

In Figures 11.3, 11.6, 11.7, 11.8, 11.10, 11.11 and 11.12, we have shown the state diagrams of the 4PSK and 8PSK space-time trellis codes. From these state diagrams, we can see that the number of trellis transitions leaving each state is equivalent to 2^{BPS}, where BPS denotes the number of transmitted bits per modulation symbol. Since the number of information bits is equal to BPS, we can approximate the complexity of the space-time trellis decoder as:

$$comp\{\text{STT}\} = \frac{2^{BPS} \times \text{No. of States}}{BPS}$$
$$= 2^{BPS-1} \times \text{No. of States} . \qquad (11.8)$$

Applying Equation 11.8 and assuming that the Viterbi decoding algorithm was employed, we obtained the approximated complexities of the space-time trellis decoder given in Table 11.6.

11.4 Simulation Results

In this section, we will present our simulation results characterising the proposed OFDM-based system. As mentioned earlier, we will investigate the proposed system over dispersive wideband Rayleigh fading channels. We will commence our investigations using a simple two-ray Channel Impulse Response (CIR) having equal tap weights, followed by a more realistic Wireless Asynchronous Transfer Mode (WATM) channel [207]. The CIR of the two-ray model is shown in Figure 11.14. From the figure we can see that the reflected path has the same amplitude as the

Modulation scheme	BPS	No. of states	Complexity
4PSK	2	4	8
		8	16
		16	32
		32	64
8PSK	3	8	21.33
		16	42.67
		32	85.33

Table 11.6: Estimated complexity of the space-time trellis decoders shown in Figures 11.3, 11.6, 11.7, 11.8, 11.10, 11.11 and 11.12.

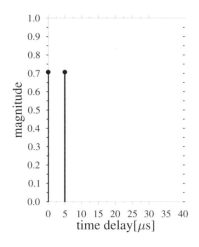

Figure 11.14: Two-ray channel impulse response having equal amplitudes.

Line Of Sight (LOS) path, although arriving 5 μs later. However, in our simulations we also present results over two-ray channels separated by various delay spreads, up to 40 μs. Jakes' model [286] was adapted for modelling the fading channels. In Figure 11.15, we portray the 128-subcarrier OFDM symbol employed, having a guard period of 40 μs. The guard period of 40 μs or cyclic extension of 32 samples was employed to overcome the inter-OFDM symbol interference due to the channel's memory.

In order to obtain our simulation results, several assumptions were stipulated:

- The average signal power received from each transmitter antenna was the same.

- All multipath components undergo independent Rayleigh fading.

- The receiver has a perfect knowledge of the CIR.

We note that the above assumptions are unrealistic, yielding the best-case performance, but

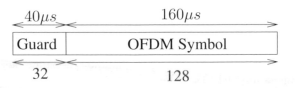

Figure 11.15: Stylised plot of 128-subcarrier OFDM time-domain signal using a cyclic extension of 32 samples.

nonetheless, facilitating the performance comparison of the various techniques under identical circumstances.

11.4.1   Space-time Coding Comparison: Throughput of 2 BPS

In Figure 11.16, we show our Frame Error Rate (FER) performance comparison between 4PSK space-time trellis codes and the space-time block code $\mathbf{G}_2$ concatenated with the TC(2,1,3) code using one receiver and the 128-subcarrier OFDM modem. The CIR had two equal-power rays separated by a delay spread of 5 μs and the maximum Doppler frequency was 200 Hz. The TC(2, 1, 3) code is a half-rate code and hence 16QAM was employed in order to support the same 2 BPS throughput as the 4PSK space-time trellis codes using no channel codes. We can

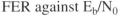

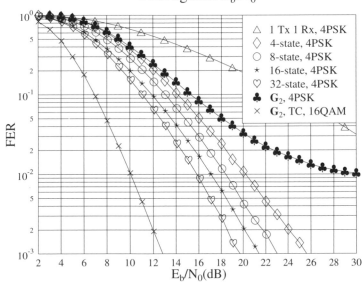

Figure 11.16: FER performance comparison between various 4PSK space-time trellis codes and the space-time block code $\mathbf{G}_2$ concatenated with the TC(2,1,3) code using **one receiver** and the 128-subcarrier OFDM modem over a channel having a CIR characterised by two equal-power rays separated by a delay spread of 5 μs. The maximum Doppler frequency was 200 Hz. The effective throughput was **2 BPS** and the coding parameters are shown in Tables 11.2, 11.3 and 11.4.

clearly see that at a FER of 10^{-3} the performance of the concatenated scheme is at least 7 dB better, than that of the space-time trellis codes.

The performance of the space-time block code $\mathbf{G}_2$ without TC(2, 1, 3) channel coding is also shown in Figure 11.16. It can be seen in the figure that the space-time block code $\mathbf{G}_2$ does not perform well, exhibiting a residual BER. Moreover, at high E_b/N_0 values, the performance of the single-transmitter, single-receiver system is better than that of the space-time block code $\mathbf{G}_2$. This is because the assumption that the fading is constant over the two consecutive transmission instants is no longer valid in this situation. Here, the two consecutive transmission instants are associated with two adjacent subcarriers in the OFDM symbol and the fading variation is relatively fast in the frequency domain. Therefore, the orthogonality of the space-time code has been destroyed by the frequency-domain variation of the fading envelope. At the receiver, the combiner can no longer separate the two different transmitted signals, namely x_1 and x_2. More explicitly, the signals interfere with each other. The increase in SNR does not improve the performance of the space-time block code $\mathbf{G}_2$, since this also increases the power of the interfering signal. We will address this issue more explicitly in Section 11.4.4. By contrast, the TC(2, 1, 3) channel codec succeeds in overcoming this problem. However, we will show later in Section 11.4.4 that the concatenated channel-coded scheme exhibits the same residual BER problem if the channel's variation becomes more rapid.

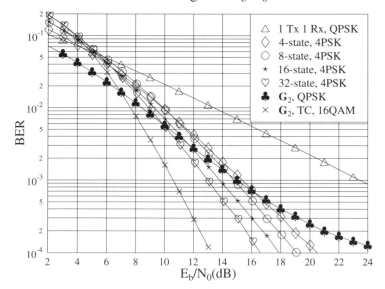

Figure 11.17: **BER** performance comparison between various 4PSK space-time trellis codes and the space-time block code $\mathbf{G}_2$ concatenated with the TC(2,1,3) code using **one receiver** and the 128-subcarrier OFDM modem over a channel having a CIR characterised by two equal-power rays separated by a delay spread of 5 μs. The maximum Doppler frequency was 200 Hz. The effective throughput was **2 BPS** and the coding parameters are shown in Tables 11.2, 11.3 and 11.4.

In Figure 11.17, we provide the corresponding BER performance comparison between the 4PSK space-time trellis codes and the space-time block code $\mathbf{G}_2$ concatenated with the TC(2,1,3) code using one receiver and the 128-subcarrier OFDM modem over a channel characterised by two equal-power rays separated by a delay spread of 5 μs and having a maximum Doppler frequency of 200 Hz. Again, we show in the figure that the 2 BPS throughput concatenated G_2/TC(2,1,3) scheme outperforms the 2 BPS space-time trellis codes using no channel coding. At a BER of 10^{-4}, the concatenated channel-coded scheme is at least 2 dB superior in SNR terms to the space-time trellis codes using no channel codes. At high E_b/N_0 values, the space-time block code G_2 again exhibits a residual BER. On the other hand, at low E_b/N_0 values the latter outperforms the concatenated G_2/TC(2,1,3) channel-coded scheme as well as the space-time trellis codes using no channel coding.

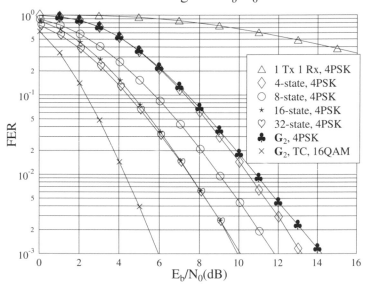

Figure 11.18: FER performance comparison between various 4PSK space-time trellis codes and the space-time block code G_2 concatenated with the TC(2,1,3) code using **two receivers** and the 128-subcarrier OFDM modem over a channel having a CIR characterised by two equal-power rays separated by a delay spread of 5 μs. The maximum Doppler frequency was 200 Hz. The effective throughput was **2 BPS** and the coding parameters are shown in Tables 11.2, 11.3 and 11.4.

Following the above investigations, the number of receivers was increased to two. In Figure 11.18, we show our FER performance comparison between the various 4PSK space-time trellis codes and the space-time block code G_2 concatenated with the TC(2,1,3) code using two receivers and the 128-subcarrier OFDM modem. As before, the CIR had two equal-power rays separated by a delay spread of 5 μs. Again, we can see that the concatenated G_2/TC(2,1,3) channel-coded scheme outperforms the space-time trellis codes using no channel coding. However, the associated difference is lower and at a FER of 10^{-3} the concatenated channel-coded scheme is about 4 dB better in E_b/N_0 terms than the space-time trellis codes using no channel codes. On the other hand, by employing two receivers the performance of the space-time block code G_2 improved and the performance flattening effect happens at a lower FER.

In Sections 10.4.3 and 11.3.3, we have derived the complexity estimates of the TC decoders and space-time trellis decoders, respectively. By employing Equations 10.44 and 11.8, we compare the performance of the proposed schemes by considering their approximate complexity.

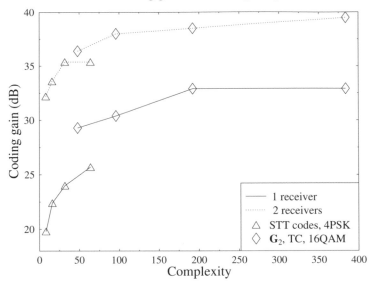

Figure 11.19: Coding gain versus estimated complexity for the various 4PSK space-time trellis codes and the space-time block code G_2 concatenated with the TC(2,1,3) code **using one as well as two receivers** and the 128-subcarrier OFDM modem over a channel having a CIR characterised by two equal-power rays separated by a delay spread of 5 μs. The maximum Doppler frequency was 200 Hz. The effective throughput was **2 BPS** and the coding parameters are shown in Tables 11.2, 11.3 and 11.4.

Our performance comparison of the various schemes was carried out on the basis of the coding gain defined as the E_b/N_0 difference, expressed in decibels (dB), at a FER of 10^{-3} between the proposed schemes and the uncoded single-transmitter, single-receiver system having the same throughput of 2 BPS. In Figure 11.19, we show our coding gain versus estimated complexity comparison for the various 4PSK space-time trellis codes and the space-time block code G_2 concatenated with the TC(2,1,3) code using one as well as two receivers. The 128-subcarrier OFDM modem transmitted over the channel having a CIR of two equal-power rays separated by a delay spread of 5 μs and a maximum Doppler frequency of 200 Hz. The estimated complexity of the space-time trellis codes was increased by increasing the number of trellis states. By constrast, the estimated complexity of the TC(2, 1, 3) code was increased by increasing the number of turbo iterations. The coding gain of the concatenated G_2/TC(2,1,3) scheme using one, two, four and eight iterations is shown in Figure 11.19. It can be seen that the concatenated scheme outperforms the space-time trellis codes using no channel coding, even though the number of turbo iterations was only one. Moreover, the improvement in coding gain was obtained at an estimated complexity comparable to that of the 32-state 4PSK space-time trellis code using no channel coding. From the figure we can also see that the performance gain of the concatenated G_2/TC(2,1,3) channel-coded scheme over the space-time trellis codes becomes lower when the number of receivers is increased to two.

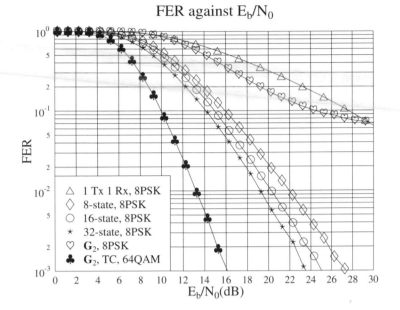

Figure 11.20: **FER** performance comparison between various 8PSK space-time trellis codes and the space-time block code G_2 concatenated with the TC(2,1,3) code using **one receiver** and the 128-subcarrier OFDM modem over a channel having a CIR characterised by two equal-power rays separated by a delay spread of 5 μs. The maximum Doppler frequency was 200 Hz. The effective throughput was **3 BPS** and the coding parameters are shown in Tables 11.2, 11.3 and 11.4.

11.4.2 Space-time Coding Comparison: Throughput of 3 BPS

In Figure 11.20, we show our FER performance comparison between the various 8PSK space-time trellis codes of Table 11.2 and space-time block code G_2 concatenated with the TC(2,1,3) code using one receiver and the 128-subcarrier OFDM modem. The CIR exhibited two equal-power rays separated by a delay spread of $5\mu s$ and a maximum Doppler frequency of 200 Hz. Since the TC(2, 1, 3) scheme is a half-rate code, 64QAM was employed in order to ensure the same 3 BPS throughput, as the 8PSK space-time trellis codes using no channel coding. We can clearly see that at a FER of 10^{-3} the performance of the concatenated channel-coded scheme is at least 7 dB better in terms of the required E_b/N_0 than that of the space-time trellis codes. The performance of the space-time block code G_2 without the concatenated TC(2, 1, 3) code is also shown in the figure. In Table 11.4, we can see that although there is an increase in the turbo interleaver size, due to employing a higher-order modulation scheme, but nonetheless, no performance gain is observed for the concatenated G_2/TC(2, 1, 3) scheme over the space-time trellis codes using no channel coding. We speculate that this is because the potential performance gain achieved with the advent of the increased interleaver size has been offset by the performance loss of the vulnerable 64QAM scheme.

We also show in Figure 11.20 that the performance of the 3 BPS 8PSK space-time block code G_2 without the concatenated TC(2, 1, 3) scheme is worse than that of the other proposed schemes. It exhibits the previously noted flattening effect, which becomes more pronounced near a FER of 10^{-1}. The same phenomenon was observed near a FER of 10^{-2} for the corresponding

G_2-coded 4PSK scheme, which has a throughput of 2 BPS.

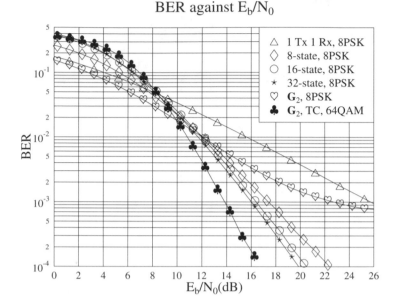

BER against E_b/N_0

Figure 11.21: **BER** performance comparison between various 8PSK space-time trellis codes and the space-time block code G_2 concatenated with the TC(2,1,3) code using **one receiver** and the 128-subcarrier OFDM modem over a channel having a CIR characterised by two equal-power rays separated by a delay spread of 5 μs. The maximum Doppler frequency was 200 Hz. The effective throughput was **3 BPS** and the coding parameters are shown in Tables 11.2, 11.3 and 11.4.

In Figure 11.21, we portray our BER performance comparison between the various 8PSK space-time trellis codes and the space-time block code G_2 concatenated with the TC(2,1,3) scheme using one receiver and the 128-subcarrier OFDM modem. The CIR exhibited two equal-power rays separated by a delay spread of 5 μs and the maximum Doppler frequency was 200 Hz. Again, we observe in the figure that the concatenated G_2/TC(2,1,3)-coded scheme outperforms the space-time trellis codes using no channel coding. At a BER of 10^{-4}, the concatenated scheme is at least 2 dB better in terms of its required E_b/N_0 value than the space-time trellis codes. The performance of the space-time block code G_2 without TC(2,1,3) channel coding is also shown in Figure 11.21. As before, at high E_b/N_0 values, the space-time block code G_2 exhibits a flattening effect. On the other hand, at low E_b/N_0 values it outperforms the concatenated G_2/TC(2,1,3) scheme as well as the space-time trellis codes.

In Figure 11.22, we compare the FER performance of the 8PSK space-time trellis codes and the space-time block code G_2 concatenated with the TC(2,1,3) channel codec using two receivers and the 128-subcarrier OFDM modem. As before, the CIR has two equal-power rays separated by a delay spread of 5 μs and exhibits maximum Doppler frequency of 200 Hz. Again, with the increase in the number of receivers the performance gap between the concatenated channel-coded scheme and the space-time trellis codes using no channel coding becomes smaller. At a FER of 10^{-3} the concatenated channel-coded scheme is only about 2 dB better in terms of its required E_b/N_0 than the space-time trellis codes using no channel coding.

With the increase in the number of receivers, the previously observed flattening effect of the

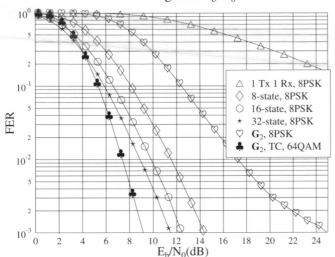

FER against E_b/N_0

Legend:
△ 1 Tx 1 Rx, 8PSK
◇ 8-state, 8PSK
○ 16-state, 8PSK
⋆ 32-state, 8PSK
♡ G_2, 8PSK
♣ G_2, TC, 64QAM

Figure 11.22: FER performance comparison between various 8PSK space-time trellis codes and the space-time block code G_2 concatenated with the TC(2,1,3) code using **two receivers** and the 128-subcarrier OFDM modem over a channel having a CIR characterised by two equal-power rays separated by a delay spread of 5 μs. The maximum Doppler frequency was 200 Hz. The effective throughput was **3 BPS** and the coding parameters are shown in Tables 11.2, 11.3 and 11.4.

space-time block code G_2 has been substantially mitigated, dipping to values below a FER of 10^{-3}. However, it can be seen in Figure 11.22 that its performance is about 10 dB worse than that of the 8-state 8PSK space-time trellis code. In our previous system characterised in Figure 11.18, which had an effective throughput of 2 BPS, the performance of the space-time block code G_2 was only about 1 dB worse in E_b/N_0 terms than that of the 4-state 4PSK space-time trellis code, when the number of receivers was increased to two. This observation clearly shows that higher-order modulation schemes have a tendency to saturate the channel's capacity and hence result in a poorer performance than the identical-throughput space-time trellis codes using no channel coding.

Similarly to the 2 BPS schemes of Figure 11.19, we compare the performance of the proposed 3 BPS throughput schemes by considering their approximate decoding complexity. The derivation of the estimated complexity has been detailed in Sections 10.4.3 and 11.3.3. As mentioned earlier, the performance comparison of the various schemes was made on the basis of the coding gain defined as the E_b/N_0 difference, expressed in decibels, at a FER of 10^{-3} between the proposed schemes and the uncoded single-transmitter, single-receiver system having a throughput of 3 BPS. In Figure 11.23, we show the associated coding gain versus estimated complexity curves for the 8PSK space-time trellis codes using no channel coding and the space-time block code G_2 concatenated with the TC(2,1,3) code using one and two receivers and the 128-subcarrier OFDM modem. For the sake of consistency, the CIR, again, exhibited two equal-power rays separated by a delay spread of 5 μs and a maximum Doppler frequency of 200 Hz. Again, the estimated

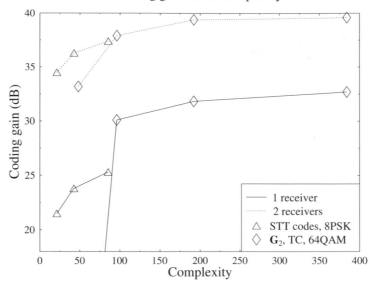

Figure 11.23: Coding gain versus estimated complexity for the various 8PSK space-time trellis codes and
the space-time block code G_2 concatenated with the TC(2,1,3) code **using one and two receivers** and the 128-subcarrier OFDM modem over a channel having a CIR characterised by
two equal-power rays separated by a delay spread of 5 μs. The maximum Doppler frequency
was 200 Hz. The effective throughput was **3 BPS** and the coding parameters are shown in
Tables 11.2, 11.3 and 11.4.

complexity of the space-time trellis codes was increased by increasing the number of states. On
the other hand, the estimated complexity of the TC(2, 1, 3) code was increased by increasing the
number of iterations. The coding gain of the concatenated channel-coded scheme invoking one,
two, four and eight iterations is shown in Figure 11.23. Previously in Figure 11.19 we have shown
that the concatenated TC(2,1,3)-coded scheme using one iteration outperformed the space-time
trellis codes using no channel coding. However, in Figure 11.23 the concatenated scheme does
not exhibit the same performance trend. For the case of one receiver, the concatenated scheme
using one iteration has a negative coding gain and exhibits a saturation effect. This is, again,
due to the employment of the high-order 64QAM scheme, which has a preponderance to exceed
the channel's capacity. Again, we can also see that the performance gain of the concatenated
G_2/TC(2,1,3)-coded scheme over the space-time trellis codes using no channel coding becomes
smaller when the number of receivers is increased to two. Having studied the performance of the
proposed schemes over the channel characterised by the two-path, 5 μs dispersion CIR at a fixed
Doppler frequency of 200 Hz, let us in the next section study the effects of varying the Doppler
frequency.

11.4.3   The Effect of Maximum Doppler Frequency

In our further investigations we have generated the FER versus E_b/N_0 curves similar to those
in Figure 11.16, when the Doppler frequency was fixed to 5, 10, 20, 50 and 100 Hz. In order
to present these results in a compact form, we then extracted the required E_b/N_0 values for

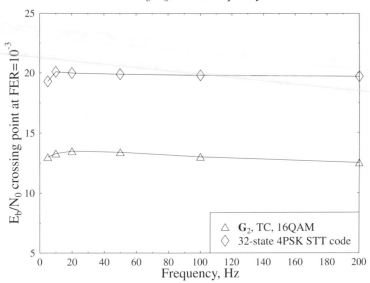

E_b/N_0 versus frequency

Figure 11.24: The E_b/N_0 value required for maintaining a FER of 10^{-3} versus the maximum Doppler frequency for the 32-state 4PSK space-time trellis code and for the space-time block code $\mathbf{G}_2$ concatenated with the TC(2,1,3) code using **one receiver** and the 128-subcarrier OFDM modem. The CIR exhibited two equal-power rays separated by a delay spread of 5 μs. The effective throughput was **3 BPS** and the coding parameters are shown in Tables 11.2, 11.3 and 11.4.

maintaining a FER of 10^{-3}. In Figure 11.24, we show the E_b/N_0 crossing point at a FER of 10^{-3} versus the maximum Doppler frequency for the 32-state 4PSK space-time trellis code using no channel coding and for the space-time block code $\mathbf{G}_2$ concatenated with the TC(2,1,3) code using one receiver and the 128-subcarrier OFDM modem. As before, the CIR exhibited two equal-power rays separated by a delay spread of 5 μs. We conclude from the near-horizontal curves shown in the figure that the maximum Doppler frequency does not significantly affect the performance of the space-time trellis codes and the concatenated scheme. Furthermore, the performance of the concatenated scheme is always better than that of the space-time trellis codes using no channel coding. Having studied the effects of various Doppler frequencies, let us now consider the impact of varying the delay spread.

11.4.4   The Effect of Delay Spreads

In this section, we will study how the variation of the delay spread between the two paths of the channel affects the system performance. By varying the delay spread, the channel's frequency-domain response varies as well. In Figure 11.25, we show the fading amplitude variation of the 128 subcarriers in an OFDM symbol for a delay spread of (a) 5 μs, (b) 10 μs, (c) 20 μs and (d) 40 μs. It can be seen from the figure that the fading amplitudes vary more rapidly when the delay spread is increased. For the space-time block code $\mathbf{G}_2$ we have shown in Section 10.3.1 that the fading envelopes of the two consecutive transmission instants of antennas Tx 1 and Tx 2 are assumed to be constant. In Figure 11.25(d), we can see that the variation of the frequency-

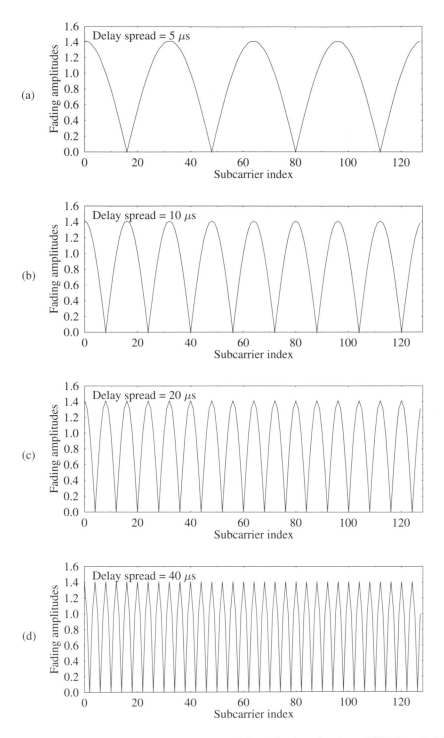

Figure 11.25: Frequency-domain fading amplitudes of the 128 subcarriers in an OFDM symbol for a delay spread of (a) 5 μs, (b) 10 μs, (d) 20 μs and (c) 40 μs.

domain fading amplitudes is so dramatic that we can no longer assume that the fading envelopes are constant for two consecutive transmission instants. The variation of the frequency-domain fading envelope will eventually destroy the orthogonality of the space-time block code $\mathbf{G}_2$.

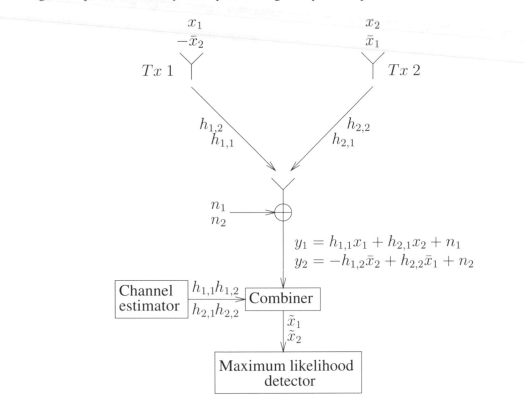

Figure 11.26: Baseband representation of the simple twin-transmitter space-time block code $\mathbf{G}_2$ of Equation 11.7 using one receiver over varying fading conditions.

We reproduce Figure 10.2 in Figure 11.26 with the modification that the two transmission instants are no longer assumed to be associated with the same complex transfer function values. The figure shows the baseband representation of the simple twin-transmitter space-time block code $\mathbf{G}_2$ of Equation 11.7 using one receiver. At the receiver, we have:

$$y_1 = h_{1,1}x_1 + h_{2,1}x_2 + n_1 \tag{11.9}$$
$$y_2 = -h_{1,2}\bar{x}_2 + h_{2,2}\bar{x}_1 + n_2 , \tag{11.10}$$

where y_1 is the first received signal and y_2 is the second. Both signals y_1 and y_2 are passed to the combiner in order to extract the signals x_1 and x_2. Aided by the channel estimator, which in this example provides perfect estimation of the diversity channels' frequency-domain transfer functions, the combiner performs simple signal processing in order to separate the signals x_1 and x_2. Specifically, in order to extract the signal x_1, it combines signals y_1 and y_2 as follows:

$$
\begin{aligned}
\tilde{x}_1 &= \bar{h}_{1,1}y_1 + h_{2,2}\bar{y}_2 \\
&= \bar{h}_{1,1}h_{1,1}x_1 + \bar{h}_{1,1}h_{2,1}x_2 + \bar{h}_{1,1}n_1 - h_{2,2}\bar{h}_{1,2}x_2 + h_{2,2}\bar{h}_{2,2}x_1 + h_{2,2}\bar{n}_2 \\
&= \left(|h_{1,1}|^2 + |h_{2,2}|^2\right)x_1 + \left(\bar{h}_{1,1}h_{2,1} - h_{2,2}\bar{h}_{1,2}\right)x_2 + \bar{h}_{1,1}n_1 + h_{2,2}\bar{n}_2 . \tag{11.11}
\end{aligned}
$$

Similarly, for signal x_2 the combiner generates:

$$
\begin{aligned}
\tilde{x}_2 &= \bar{h}_{2,1}y_1 - h_{1,2}\bar{y}_2 \\
&= \bar{h}_{2,1}h_{1,1}x_1 + \bar{h}_{2,1}h_{2,1}x_2 + \bar{h}_{2,1}n_1 + h_{1,2}\bar{h}_{1,1}x_2 - h_{1,2}\bar{h}_{2,2}x_1 - h_{1,2}\bar{n}_2 \\
&= \left(|h_{2,1}|^2 + |h_{1,2}|^2\right)x_2 + \left(\bar{h}_{2,1}h_{1,1} - h_{1,2}\bar{h}_{2,2}\right)x_1 + \bar{h}_{2,1}n_1 - h_{1,2}\bar{n}_2 \, . \quad (11.12)
\end{aligned}
$$

In contrast to the perfect cancellation scenario of Equations 10.15 and 10.16, we can see from Equations 11.11 and 11.12 that the signals x_1 and x_2 now interfere with each other. We can no longer cancel the cross-coupling of signals x_2 and x_1 in Equations 11.11 and 11.12, respectively, unless the fading envelopes satisfy the condition of $h_{1,1} = h_{1,2}$ and $h_{2,1} = h_{2,2}$.

At high SNRs the noise power is insignificant compared to the transmitted power of the signals x_1 and x_2. Therefore, we can ignore the noise terms n in Equations 11.11 and 11.12. However, the interference signals' power increases as we increase the transmission power. Assuming that both the signals x_1 and x_2 have an equivalent signal power, we can then express the Signal-to-Interference Ratio (SIR) for signal x_1 as:

$$
SIR = \frac{|h_{1,1}|^2 + |h_{2,2}|^2}{\bar{h}_{1,1}h_{2,1} - h_{2,2}\bar{h}_{1,2}} \, , \quad (11.13)
$$

and similarly for signal x_2 as:

$$
SIR = \frac{|h_{2,1}|^2 + |h_{1,2}|^2}{\bar{h}_{2,1}h_{1,1} - h_{1,2}\bar{h}_{2,2}} \, . \quad (11.14)
$$

In Figure 11.27, we show the FER performance of the space-time block code $\mathbf{G}_2$ concatenated with the TC(2,1,3) code using one receiver and the 128-subcarrier 16QAM OFDM modem. The CIR has two equal-power rays separated by various delay spreads and a maximum Doppler frequency of 200 Hz. As we can see in Equations 11.13 and 11.14, we have $SIR \rightarrow \infty$, if $h_{1,1} = h_{1,2}$ and $h_{2,1} = h_{2,2}$. On the other hand, we encounter $SIR \rightarrow 1$, if $h_{1,1} = \delta h_{1,2}$ and $h_{2,1} = \delta h_{2,2}$, where $\delta \rightarrow \infty$. Since the SIR decreases, when the delay spread increases owing to the rapidly fluctuating frequency-domain fading envelopes, as shown in Figure 11.25, we can see in Figure 11.27 that the performance of the concatenated scheme degrades when increasing the delay spread. When the delay spread is more than 15 μs, we can see from the figure that the concatenated scheme exhibits the previously observed flattening effect. Furthermore, the error floor of the concatenated scheme becomes higher, as the delay spread is increased.

Similarly to Figure 11.24, where the Doppler frequency was varied, we show in Figure 11.28 the E_b/N_0 value required for maintaining a FER of 10^{-3} versus the delay spread for the 32-state 4PSK space-time trellis code and for the space-time block code $\mathbf{G}_2$ concatenated with the TC(2,1,3) code using one receiver and the 128-subcarrier OFDM modem. The CIR exhibited two equal-power rays separated by various delay spreads. The maximum Doppler frequency was 200 Hz. We can see in the figure that the performance of the 32-state 4PSK space-time trellis code does not vary significantly with the delay spread. However, the concatenated TC(2,1,3)-coded scheme suffers severe performance degradation upon increasing the delay spread, as evidenced by the associated error floors shown in Figure 11.27. The SIR associated with the various delay spreads was obtained using computer simulations and the associated SIR values are also shown in Figure 11.28, denoted by the hearts. As we expect, the calculated SIR decreases with the delay spread. We can see in the figure that the performance of the concatenated $\mathbf{G}_2$/TC(2,1,3) scheme suffers severe degradation when the delay spread is in excess of 15 μs, as indicated by the near-vertical curve marked by triangles. If we relate this curve to the SIR curve marked by

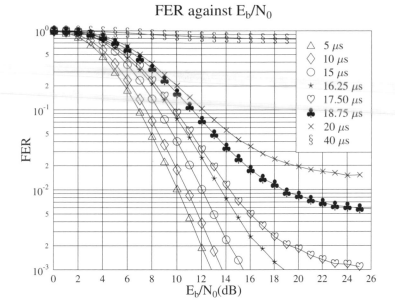

Figure 11.27: **FER** performance of the space-time block code $\mathbf{G}_2$ concatenated with the TC(2,1,3) code using one receiver, the 128-subcarrier OFDM modem and 16QAM. The CIR exhibits two equal-power rays separated by various delay spreads and a maximum Doppler frequency of 200 Hz. The coding parameters are shown in Tables 11.2, 11.3 and 11.4.

the hearts, we can see from the figure that the SIR is approximately 10 dB. Hence the SIR of the concatenated $\mathbf{G}_2$/TC(2,1,3) scheme has to be more than 10 dB, in order for it to outperform the space-time trellis codes using no channel coding.

11.4.5 Delay Non-sensitive System

Previously, we have provided simulation results for a delay-sensitive, OFDM symbol-by-symbol decoded system. More explicitly, the received OFDM symbol had to be demodulated and decoded on a symbol-by-symbol basis, in order to provide decoded bits for example for a low-delay source decoder. Therefore, the two transmission instants of the space-time block code $\mathbf{G}_2$ had to be in the same OFDM symbol. They were allocated to the adjacent subcarriers in our previous studies. Moreover, we have shown in Figure 11.25 that the variation of the frequency-domain fading amplitudes along the subcarriers becomes more severe as we increase the delay spread of the two rays. In Figure 11.29 we show both the frequency-domain and time-domain fading amplitudes of the channels' fading amplitudes for a fraction of the subcarriers in the 128-subcarrier OFDM symbols over the previously used two-path channel having two equal-power rays separated by a delay spread of 40 μs. The maximum Doppler frequency was set here to 100 Hz. It can be clearly seen from the figure that the fading amplitude variation versus time is slower than that versus the subcarrier index within the OFDM symbols. This implies that the SIR attained would be higher if we were to allocate the two transmission instants of the space-time block code $\mathbf{G}_2$ to the same subcarrier of consecutive OFDM symbols. This increase in SIR is achieved by doubling the delay of the system, since in this scenario two consecutive OFDM symbols have to be decoded, before all the received data become available.

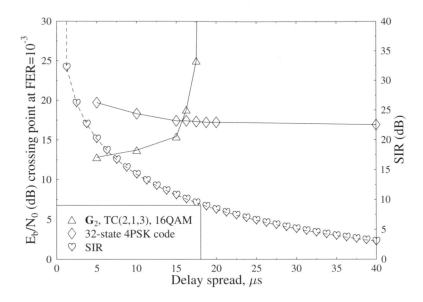

Figure 11.28: The E_b/N_0 values required for maintaining a FER of 10^{-3} versus delay spreads for the 32-state 4PSK space-time trellis code and for the space-time block code $\mathbf{G}_2$ concatenated with the TC(2,1,3) code using **one receiver** and the 128-subcarrier OFDM modem. The CIR exhibited two equal-power rays separated by various delay spreads and a maximum Doppler frequency of 200 Hz. The effective throughput was **2 BPS** and the coding parameters is shown in Tables 11.2, 11.3 and 11.4. The SIR of various delay spreads are shown as well.

In Figure 11.30, we show our FER performance comparison for the above two scenarios, namely using two adjacent subcarriers and the same subcarrier in two consecutive OFDM symbols for the space-time block code $\mathbf{G}_2$ concatenated with the TC(2,1,3) code using one 128-subcarrier 16QAM OFDM receiver. As before, the CIR exhibited two equal-power rays separated by a delay spread of $40\ \mu s$ and a maximum Doppler frequency of 100 Hz. It can be seen from the figure that there is a severe performance degradation if the two transmission instants of the space-time block $\mathbf{G}_2$ are allocated to two adjacent subcarriers. This is evidenced by the near-horizontal curve marked by diamonds across the figure. On the other hand, upon assuming that having a delay of two OFDM symbol durations does not pose any problems in terms of real-time interactive communications, we can allocate the two transmission instants of the space-time block code $\mathbf{G}_2$ to the same subcarrier of two consecutive OFDM symbols. From Figure 11.30, we can observe a dramatic improvement over the previous allocation method. Furthermore, the figure also indicates that by tolerating a two OFDM symbol delay, the concatenated $\mathbf{G}_2$/TC(2,1,3) scheme outperforms the 32-state 4PSK space-time trellis code by approximately 2 dB in terms of the required E_b/N_0 value at a FER of 10^{-3}.

Since the two transmission instants of the space-time block code $\mathbf{G}_2$ are allocated to the same subcarrier of two consecutive OFDM symbols, it is the maximum Doppler frequency that would affect the performance of the concatenated scheme more gravely, rather than the delay spread. Hence we extended our studies to consider the effects of the maximum Doppler frequency on the performance of the concatenated $\mathbf{G}_2$/TC(2,1,3) scheme. Specifically, Figure 11.31 shows the E_b/N_0 values required for maintaining a FER of 10^{-3} versus the Doppler frequency for the

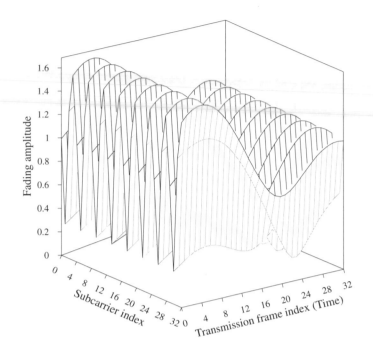

Figure 11.29: The fading amplitude versus time and frequency for various 128-subcarrier OFDM symbols over the two-path channel exhibiting two equal-power rays separated by a delay spread of 40 μs and maximum Doppler frequency of 100 Hz.

32-state 4PSK space-time trellis code, and for the space-time block code G_2 concatenated with the TC(2,1,3) code using one 128-subcarrier 16QAM OFDM receiver, when mapping the two transmission instants to the same subcarrier of two consecutive OFDM symbols. The channel exhibited two equal-power rays separated by a delay spread of 40 μs and various maximum Doppler frequencies. The SIR achievable at various maximum Doppler frequencies is also shown in Figure 11.31. Again, we can see that the performance of the concatenated G_2/TC(2,1,3) scheme suffers severely if the maximum Doppler frequency is above 160 Hz. More precisely, we can surmise that in order for the concatenated scheme to outperform the 32-state 4PSK space-time trellis code, the SIR should be at least 15 dB, which is about the same as the required SIR in Figure 11.28. From Figures 11.28 and 11.31, we can conclude that the concatenated G_2/TC(2,1,3) scheme performs better if the SIR is in excess of about 10–15 dB.

11.4.6   The Wireless Asynchronous Transfer Mode System

We have previously investigated the performance of different schemes over two-path channels having two equal-power rays. In this section, we investigate the performance of the proposed systems over indoor Wireless Asynchronous Transfer Mode (WATM) channels. The WATM system uses 512 subcarriers and each OFDM symbol is extended with a cyclic prefix of length 64. The sampling rate is 225 Msamples/s and the carrier frequency is 60 GHz. In [207] two WATM CIRs were used, namely a five-path and a three-path model, where the latter one was referred to as the shortened WATM CIR. This CIR was also used in our investigations here.

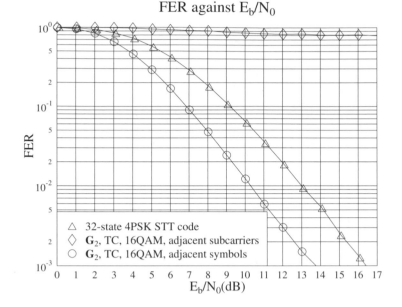

Figure 11.30: FER performance comparison between adjacent subcarriers and adjacent OFDM symbols allocated for the space-time block code G_2 concatenated with the TC(2,1,3) code using one receiver, the 128-subcarrier OFDM modem and 16QAM over a channel having a CIR characterised by two equal-power rays separated by a delay spread of 40 μs. The maximum Doppler frequency was 100 Hz. The coding parameters are shown in Tables 11.2, 11.3 and 11.4.

The shortened WATM CIR is depicted in Figure 11.32, where the longest-delay path arrived at a delay of 48.9 ns, which corresponds to 11 sample periods at the 225 Msamples/s sampling rate. The 512-subcarrier OFDM time-domain transmission frame having a cyclic extension of 64 samples is shown in Figure 11.33.

11.4.6.1   Channel-coded Space-time Codes: Throughput of 1 BPS

Previously, we have compared the performance of the space-time trellis codes to that of the TC(2,1,3)-coded space-time block code G_2. We now extend our comparisons to RS coded space-time trellis codes, which were used in [76, 78, 79]. In Figure 11.34 we show our FER performance comparison between the TC(2,1,3)-coded space-time block code G_2 and the RS(102,51) $GF(2^{10})$-coded 16-state 4PSK space-time trellis code using one 512-subcarrier OFDM receiver over the shortened WATM CIR of Figure 11.32 at an effective throughput of 1 BPS. We can see from the figure that the TC(2,1,3)-coded space-time block code G_2 outperforms the RS(102,51) $GF(2^{10})$-coded 16-state 4PSK space-time trellis code by approximately 5 dB in E_b/N_0 terms at a FER of 10^{-3}. The performance of the RS(102,51) $GF(2^{10})$-coded 16-state 4PSK space-time trellis code would be improved by about 2 dB if the additional complexity of maximum likelihood decoding were affordable. However, even assuming this improvement, the TC(2,1,3)-coded space-time block code G_2 would outperform the RS(102,51) $GF(2^{10})$-coded 16-state 4PSK space-time trellis code.

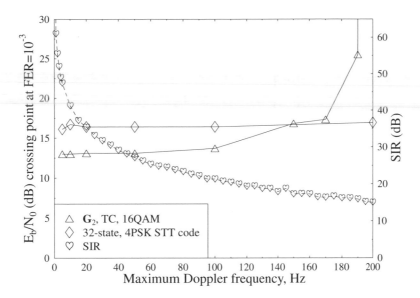

Figure 11.31: The E_b/N_0 value required for maintaining a FER of 10^{-3} versus the maximum Doppler frequency for the 32-state 4PSK space-time trellis code and for the adjacent OFDM symbols allocated for the space-time block code $\mathbf{G}_2$ concatenated with the TC(2,1,3) code using **one receiver** and the 128-subcarrier OFDM modem. The CIR exhibited two equal-power rays separated by a delay spread of 40 μs. The effective throughput was **2 BPS** and the coding parameters are shown in Tables 11.2, 11.3 and 11.4. The SIR of various maximum Doppler frequencies are shown as well.

11.4.6.2 Channel-coded Space-time Codes: Throughput of 2 BPS

In our next experiment, the throughput of the system was increased to 2 BPS by employing a higher-order modulation scheme. In Figure 11.35 we show our FER performance comparison between the TC(2,1,3)-coded space-time block code $\mathbf{G}_2$ and the RS(153,102) $GF(2^{10})$-coded 16-state 8PSK space-time trellis code using one 512-subcarrier OFDM receiver over the shortened WATM channel of Figure 11.32 at an effective throughput of 2 BPS. Again, we can see that the TC(2,1,3)-coded space-time block code $\mathbf{G}_2$ outperforms the RS(153,102) $GF(2^{10})$-coded 16-state 8PSK space-time trellis code by approximately 5 dB in terms of E_b/N_0 at a FER of 10^{-3}. The corresponding performance of the 32-state 4PSK space-time trellis code is also shown in the figure. It can be seen that its performance is about 13 dB worse in E_b/N_0 terms than that of the TC(2,1,3)-coded space-time block code $\mathbf{G}_2$. Let us now continue our investigations by considering whether channel-quality controlled adaptive space-time-coded OFDM is capable of providing further performance benefits.

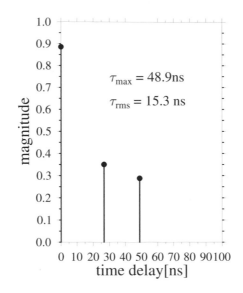

Figure 11.32: Short WATM CIR.

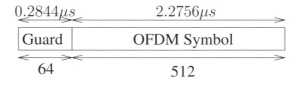

Figure 11.33: Short WATM plot of 512-subcarrier OFDM time-domain signal with a cyclic extension of 64 samples.

11.5   Space-time-coded Adaptive Modulation for OFDM

11.5.1   Introduction

Adaptive modulation was proposed by Steele and Webb [240, 297], in order to combact the time-variant fading of mobile channels. The main idea of adaptive modulation is that when the channel quality is favourable, higher-order modulation modes are employed in order to increase the throughput of the system. On the other hand, more robust but lower-throughput modulation modes are employed, if the channel quality is low. This simple but elegant idea has motivated a number of researchers to probe further [207, 242, 247, 298–303].

Recently adaptive modulation was also proposed for OFDM, which was termed adaptive OFDM (AOFDM) [207, 262, 300, 301]. AOFDM exploits the variation of the signal quality both in the time domain as well as in the frequency domain. In what is known as sub-band AOFDM transmission, all subcarriers in an AOFDM symbol are split into blocks of adjacent subcarriers, referred to as sub-bands. The same modulation scheme is employed for all subcarriers of the

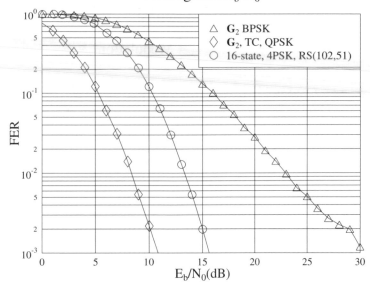

FER against E_b/N_0

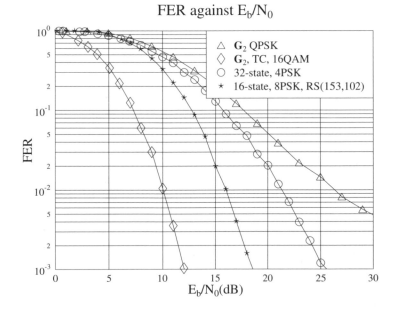

Figure 11.34: **FER** performance comparison between the TC(2,1,3)-coded space-time block code $\mathbf{G}_2$ and the RS(102,51) $GF(2^{10})$-coded 16-state 4PSK space-time trellis code using one 512-subcarrier OFDM receiver over the shortened WATM channel at an effective throughput of **1 BPS**. The coding parameters are shown in Tables 11.2, 11.3, 11.4 and 11.5.

Figure 11.35: **FER** performance comparison between the TC(2,1,3)-coded space-time block code $\mathbf{G}_2$ and the RS(153,102) $GF(2^{10})$-coded 16-state 8PSK space-time trellis code using one 512-subcarrier OFDM receiver over the shortened WATM channel at an effective throughput of 2 BPS. The coding parameters are shown in Tables 11.2, 11.3, 11.4 and 11.5.

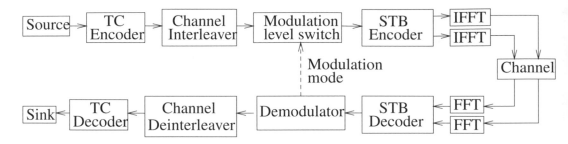

Figure 11.36: System overview of the turbo-coded and space-time-coded AOFDM.

same sub-band. This substantially simplifies the task of signalling the modulation modes, since there are typically four modes and for example 32 sub-bands, requiring a total of 64 AOFDM mode signalling bits.

11.5.2   Turbo-coded and Space-time-coded AOFDM

In this section, the adaptive OFDM philosophy proposed by Keller *et al.* [207, 300, 301] is extended, in order to exploit the advantages of multiple transmit and receive antennas. Besides, turbo coding is also employed in order to improve the performance of the system. In Figure 11.36, we show the system schematic of the turbo-coded and space-time-coded AOFDM system. Similarly to Figure 11.13, random data bits are generated and encoded by the TC(2,1,3) encoder using an octal generator polynomial of (7, 5). Various TC(2,1,3) coding rates were used for the different modulation schemes. The encoded bits were channel interleaved and passed to the modulator. The choice of the modulation scheme to be used by the transmitter for its next OFDM symbol is determined by the channel-quality estimate of the receiver based on the current OFDM symbol. In this study, we assumed perfect channel-quality estimation and perfect signalling of the required modem mode of each sub-band based on the channel-quality estimate acquired during the current OFDM symbol. Aided by the perfect channel-quality estimator, the receiver determines the highest-throughput modulation mode to be employed by the transmiter for its next transmission while maintaining the system's target BER. Five possible transmission modes were employed in our investigations, which are no transmission (NoTx), BPSK, QPSK, 16QAM and 64QAM. In order to simplify the task of signalling the required modulation modes, we employed the sub-band AOFDM transmission scheme proposed by Keller *et al.* [207, 300, 301]. The modulated signals were then passed to the encoder of the space-time block code G_2. The space-time encoded signals were OFDM modulated and transmitted by the corresponding antennas. The shortened WATM channel was used, where the CIR profile and the OFDM transmission frame as shown in Figures 11.32 and 11.33, respectively.

The number of receivers invoked constitutes a design parameter. The received signals were OFDM demodulated and passed to the space-time decoders. Log-MAP [283] decoding of the received space-time signals was performed, in order to provide soft outputs for the TC(2,1,3) decoder. Assuming that the demodulator of the receiver has perfect knowledge of the instantaneous channel quality, this information is passed to the transmitter in order to determine its next AOFDM mode allocation. The received bits were then channel de-interleaved and passed to the TC decoder, which, again, employs the Log-MAP decoding algorithm [52]. The decoded bits were finally passed to the sink for calculation of the BER.

11.5.3   Simulation Results

As mentioned earlier, all the AOFDM-based simulation results were obtained over the shortened WATM channel. The channels' profile and the OFDM transmission frame structure are shown in Figures 11.32 and 11.33, respectively. Again, Jakes' model [286] was adopted for modelling the fading channels.

In order to obtain our simulation results, several assumptions were stipulated:

- The average signal power received from each transmitter antenna was the same.

- All multipath components undergo independent Rayleigh fading.

- The receiver has a perfect knowledge of the CIR.

- Perfect signalling of the AOFDM modulation modes.

Again, we note that the above assumptions are unrealistic, yielding the best-case performance, but nonetheless, they facilitate the performance comparison of the various techniques under identical circumstances.

11.5.3.1   Space-time-coded AOFDM

In this section, we employ the fixed threshold-based modem mode selection algorithm, which was also used in [207], adapting the technique proposed by Torrance [298, 304, 305] for serial modems. Torrance assumed that the channel quality is constant for all the symbols in a transmission burst, i.e. that the channel's fading envelope varied slowly across the transmission burst. Under these conditions, all the transmitted symbols are modulated using the same modulation mode, chosen according to the predicted SNR. Torrance optimised the modem mode switching thresholds [298, 304, 305] for the target BERs of 10^{-2} and 10^{-4}, which will be appropiate for a high-BER speech system and for a low-BER data system, respectively. The resulting SNR

System	NoTx	BPSK	QPSK	16QAM	64QAM
Speech	$-\infty$	3.31	6.48	11.61	17.64
Data	$-\infty$	7.98	10.42	16.76	26.33

Table 11.7: Optimised switching levels quoted from [298] for adaptive modulation over Rayleigh fading channels for the speech and data systems, shown in instantaneous channel SNR (dB).

switching thresholds for activating a given modulation mode in a slowly Rayleigh fading narrowband channel are given in Table 11.7 for both systems. Assuming perfect channel-quality estimation, the instantaneous channel SNR is measured by the receiver and the information is passed to the modulation mode selection switch at the transmitter, as shown in Figure 11.36, using the system's control channel. This side-information signalling does not constitute a problem, since state-of-the-art wireless systems, such as for example IMT-2000 [49], have a high-rate, low-delay signalling channel. This modem mode signalling feedback information is utilised by the transmitter for selecting the next modulation mode. Specifically, a given modulation mode is selected if the instantaneous channel SNR perceived by the receiver exceeds the corresponding switching levels shown in Figure 11.7, depending on the target BER.

As mentioned earlier, the proposed adaptation algorithm [298, 304, 305] assumes constant instantaneous channel SNR over the whole transmission burst. However, in the case of an OFDM system transmitting over frequency-selective channels, the channels' quality varies across the different subcarriers. Keller *et al.* proposed employing the lowest-quality subcarrier in the sub-band for controlling the adaptation algorithm based on the switching thresholds given in Table 11.7. Again, this approach significantly simplifies the signalling and therefore it was also adopted in our investigations.

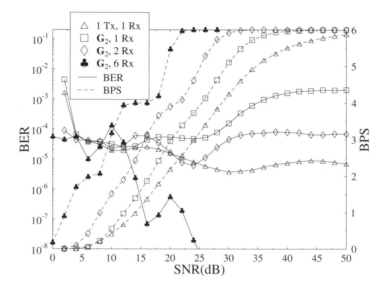

Figure 11.37: BER and BPS performance of 16 sub-band AOFDM employing the space-time block code G_2 using multiple receivers for a target BER of 10^{-4} over the shortened WATM channel shown in Figure 11.32 and the transmission format of Figure 11.33. The switching thresholds are shown in Table 11.7.

In Figure 11.37, we show the BER and BPS performance of the 16 sub-band AOFDM scheme employing the space-time block code G_2 in conjunction with multiple receivers and a target BER of 10^{-4} over the shortened WATM channel shown in Figure 11.32. The switching thresholds are shown in Table 11.7. The performance of the conventional AOFDM scheme using no diversity [207] is also shown in the figure. From Figure 11.37, we can see that the BPS performance of the space-time-coded AOFDM scheme using one receiver is better than that of the conventional AOFDM scheme. The associated performance gain improves as the throughput increases. At a throughput of 6 BPS, the space-time-coded scheme outperforms the conventional scheme by at least 10 dB in E_b/N_0 terms. However, we notice in Figure 11.37 that as a secondary effect, the BER performance of the space-time-coded AOFDM scheme using one receiver degrades as we increase the average channel SNR. Again, this problem is due to the interference of signals x_1 and x_2 caused by the rapidly varying frequency-domain fading envelope across the subcarriers. At high SNRs, predominantly 64QAM was employed. Since the constellation points in 64QAM are densely packed, this modulation mode is more sensitive to the 'cross-talk' of the signals x_1 and x_2. This limited the BER performance to 10^{-3} even at high SNRs. However, at SNRs lower than 30 dB typically more robust modulation modes were employed and hence the target BER of 10^{-4} was readily met. We will show in the next section that this problem can be overcome by employing turbo channel coding in the system.

In Figure 11.37 we also observe that the BER and BPS performance improves as we increase the number of AOFDM receivers, since the interference between the signals x_1 and x_2 is eliminated. Upon having six AOFDM receivers, the BER of the system drops below 10^{-8} when the average channel SNR exceeds 25 dB and there is no sign of the BER flattening effect. At a throughput of 6 BPS, the space-time-coded AOFDM scheme using six receivers outperforms the conventional system by more than 30 dB.

Figure 11.38 shows the probability of each AOFDM sub-band modulation mode for (a) conventional AOFDM and for space-time-coded AOFDM using (b) one, (c) two and (d) six receivers over the shortened WATM channel shown in Figure 11.32. The transmission format obeyed Figure 11.33. The switching thresholds were optimised for the data system having a target BER of 10^{-4} and they are shown in Table 11.7. By employing multiple transmitters and receivers, we increase the diversity gain and we can see in the figure that this increases the probability of the most appropriate modulation mode at a certain average channel SNR. This is clearly shown by the increased peaks of each modulation mode at different average channel SNRs. As an example, in Figure 11.38(d) we can see that there is an almost 100% probability of transmitting in the QPSK and 16QAM modes at an average channel SNR of approximately 6 dB and 15 dB, respectively. This strongly suggests that it is a better solution if fixed modulation-based transmission is employed in space-time coded OFDM, provided that we can afford the associated complexity of using six receivers. We shall investigate these issues in more depth at a later stage.

On the other hand, the increased probability of a particular modulation mode at a certain average channel SNR also means that there is less frequent switching amongst the various modulation modes. For example, we can see in Figure 11.38(b) that the probability of employing 16QAM increased to 0.8 at an average channel SNR of 25 dB compared to 0.5 in Figure 11.38(a). Furthermore, there are almost no BPSK transmissions at SNR=25 dB in Figure 11.38(b). This situation might be an advantage in the context of the AOFDM system, since most of the time the system will employ 16QAM and only occasionally switches to the QPSK and 64QAM modulation modes. This fact can be potentially exploited for reducing the AOFDM modem mode signalling traffic and hence for simplifying the system.

The characteristics of the modem mode probability density functions in Figure 11.38 in conjunction with multiple transmit antennas can be further explained with the aid of Figure 11.39. In Figure 11.39 we show the instantaneous channel SNR experienced by the 512-subcarrier OFDM symbols for a single-transmitter, single-receiver scheme and for the space-time block code $\mathbf{G}_2$ using one, two and six receivers over the shortened WATM channel. The average channel SNR is 10 dB. We can see in Figure 11.39 that the variation of the instantaneous channel SNR for a single transmitter and single receiver is fast and severe. The instantaneous channel SNR may become as low as 4 dB owing to deep fades of the channel. On the other hand, we can see that for the space-time block code $\mathbf{G}_2$ using one receiver the variation in the instantaneous channel SNR is slower and less severe. Explicitly, by employing multiple transmit antennas as shown in Figure 11.39, we have reduced the effect of the channels' deep fades significantly. This is advantageous in the context of adaptive modulation schemes, since higher-order modulation modes can be employed in order to increase the throughput of the system. However, as we increase the number of receivers, i.e. the diversity order, we observe that the variation of the channel becomes slower. Effectively, by employing higher-order diversity, the fading channels have been converted to AWGN-like channels, as evidenced by the space-time block code $\mathbf{G}_2$ using six receivers. Since adaptive modulation only offers advantages over fading channels, we argue that using adaptive modulation might become unnecessary as the diversity order is increased.

To elaborate a little further, from Figures 11.38 and 11.39 we surmise that fixed modulation

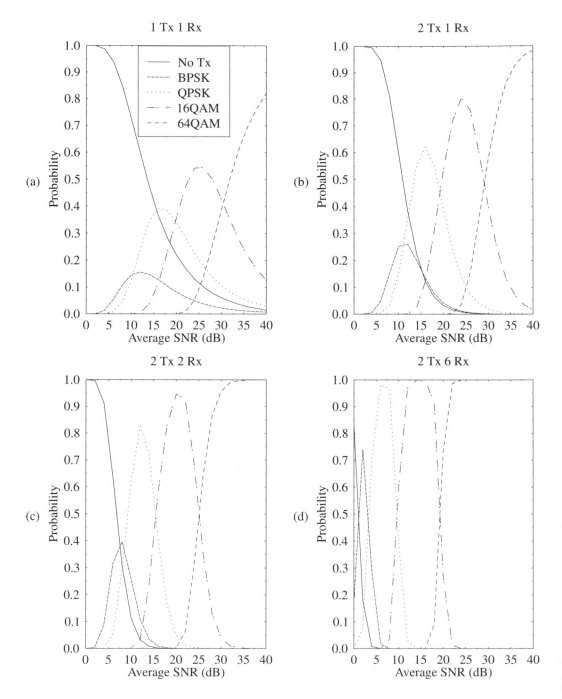

Figure 11.38: Probability of each modulation mode for (a) conventional AOFDM and for space-time-coded
AOFDM using (b) one, (c) two and (d) six receivers over the shortened WATM channel
shown in Figure 11.32 and using the transmission frame of Figure 11.33. The thresholds
were optimised for the data system and they are shown in Table 11.7. All diagrams share the
legends seen in Figure 11.38 (a).

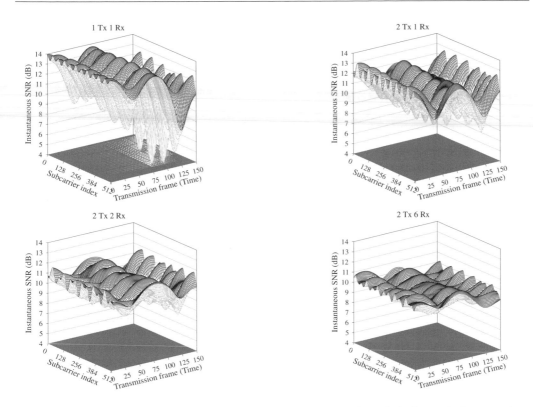

Figure 11.39: Instantaneous channel SNR of 512-subcarrier OFDM symbols for a single-transmitter, single-receiver scheme and for the space-time block code $\mathbf{G}_2$ using one, two and six receivers over the shortened WATM channel shown in Figure 11.32 and using the transmission format of Figure 11.33. The average channel SNR is 10 dB.

schemes might become more attractive when the diversity order increases, which is achieved in this case by employing more receivers. This is because for a certain average channel SNR, the probability of a particular modulation mode increases. In other words, the fading channel has become an AWGN-like channel as the diversity order is increased. In Figure 11.40 we show our throughput performance comparison between AOFDM and fixed modulation-based OFDM in conjunction with the space-time block code $\mathbf{G}_2$ employing (a) one receiver and (b) two receivers over the shortened WATM channel. The throughput of fixed OFDM was 1, 2, 4 and 6 BPS and the corresponding E_b/N_0 values were extracted from the associated BER versus E_b/N_0 curves of the individual fixed-mode OFDM schemes. It can be seen from Figure 11.40(a) that the throughput performance of the adaptive and fixed OFDM schemes is similar for a 10^{-2} target BER system. However, for a 10^{-4} target BER system, there is an improvement of 5–10 dB in E_b/N_0 terms at various throughputs for the AOFDM scheme over the fixed OFDM scheme. At high average channel SNRs the throughput performance of both schemes converged, since 64QAM became the dominant modulation mode for AOFDM.

On the other hand, if the number of receivers is increased to two, we can see in Figure 11.40(b) that the throughput performance of both adaptive and fixed OFDM is similar for both the 10^{-4} and 10^{-2} target BER systems. We would expect similar trends as the number of receivers is increased, since the fading channels become AWGN-like channels. From Fig-

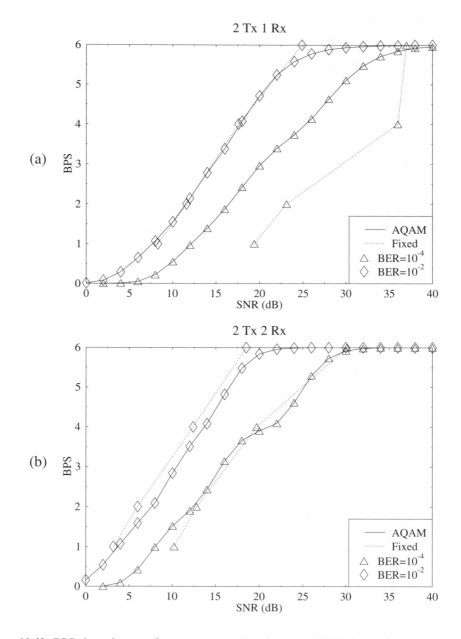

Figure 11.40: BPS throughput performance comparison between AOFDM and fixed modulation-based OFDM using the space-time block code $\mathbf{G}_2$ employing (a) one receiver and (b) two receivers over the shortened WATM channel shown in Figure 11.32 and the transmission format of Figure 11.33.

ure 11.40, we conclude that AOFDM is only beneficial for the space-time block code G_2 using one receiver in the context of the 10^{-2} target BER system.

11.5.3.2 Turbo- and Space-time-coded AOFDM

In the previous section we have discussed the performance of space-time-coded AOFDM. Here we extend our study by concatenating turbo coding with the space-time-coded AOFDM scheme in order to improve both the BER and BPS performance of the system. As earlier, the TC(2,1,3) code having a constraint length of 3 and an octal generator polynomial of $(7,5)$ was employed. Since the system was designed for high-integrity, low-BER data transmission, it was delay non-sensitive. Hence a random turbo interleaver size of approximately 10000 bits was employed. The random separation channel interleaver [295] of Figure 10.14 was utilised in order to disperse the bursty channel errors. The Log-MAP [52] decoding algorithm was employed, using eight iterations.

We propose two TC-coded schemes for the space-time-coded AOFDM system. The first scheme is a fixed half-rate turbo- and space-time-coded AOFDM system. It achieves a high BER performance, but at the cost of a maximum throughput limited to 3 BPS owing to half-rate channel coding. The second one is a variable-rate turbo- and space-time-coded AOFDM system. This scheme sacrifices the BER performance in exchange for an increased system throughput. Different puncturing patterns are employed for the various code rates R. The puncturing patterns were optimised experimentally by simulations. The design procedure for punctured turbo codes was proposed by Acikel and Ryan [64] in the context of BPSK and QPSK. The optimum AOFDM mode switching thresholds were obtained by computer simulations over the shortened WATM channel of Figure 11.32 and they are shown in Table 11.8.

	NoTx	BPSK	QPSK	16QAM	64QAM
Half-rate TC(2,1,3)					
Rate	—	0.50	0.50	0.50	0.50
Thresholds (dB)	$-\infty$	−4.0	−1.3	5.4	9.8
Variable-rate TC(2,1,3)					
Rate	—	0.50	0.67	0.75	0.90
Thresholds (dB)	$-\infty$	−4.0	2.0	9.70	21.50

Table 11.8: Coding rates and switching levels (dB) for TC(2,1,3) and space-time-coded AOFDM over the shortened WATM channel of Figure 11.32 for a target BER of 10^{-4}.

In Figure 11.41, we show the BER and BPS performance of 16 sub-band AOFDM employing the space-time block code G_2 concatenated with both half-rate and variable-rate TC(2,1,3) coding at a target BER of 10^{-4} over the shortened WATM channel of Figure 11.32. We can see in the figure that by concatenating fixed half-rate turbo coding with the space-time-coded AOFDM scheme, the BER performance of the system improves tremendously, indicated by a steep dip of the associated BER curve marked by the continuous line and diamonds. There is an improvement in the BPS performance as well, exhibiting an E_b/N_0 gain of approximately 5 dB and 10 dB at an effective throughput of 1 BPS, compared to space-time-coded AOFDM and conventional AOFDM, respectively. However, again, the maximum throughput of the system is limited to 3 BPS, since half-rate channel coding was employed. In Figure 11.41, we can see that at an E_b/N_0 value of about 30 dB the maximum throughput of the turbo-coded and space-time AOFDM system is increased from 3.0 BPS to 5.4 BPS by employing the variable-rate TC(2,1,3)

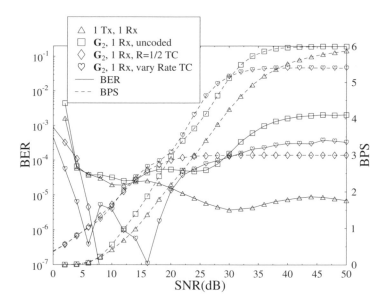

Figure 11.41: BER and BPS performance of 16 sub-band AOFDM employing the space-time block code G_2 concatenated with both half-rate and variable-rate TC(2,1,3) at a target BER of 10^{-4} over the shortened WATM channel shown in Figure 11.32 and using the transmission format of Figure 11.33. The switching thresholds and coding rates are shown in Table 11.7.

code. Furthermore, the BPS performance of the variable-rate turbo-coded scheme is similar to that of the half-rate turbo-coded scheme at an average channel SNR below 15 dB. The BER curve marked by the continuous line and clubs drops as the average channel SNR is increased from 0 dB to 15 dB. Owing to the increased probability of the 64QAM transmission mode, the variable-rate turbo-coded scheme was overloaded by the plethora of channel errors introduced by the 64QAM mode. Therefore, we can see in Figure 11.41 that the BER increases and stabilises at 10^{-4}. Again, the interference of the signals x_1 and x_2 in the context of the space-time block code G_2 prohibits further improvements in the BER performance as the average channel SNR is increased. However, employing the variable-rate turbo codec has lowered the BER floor, as demonstrated by the curve marked by the continuous line and squares.

11.6   Summary and Conclusions

Space-time trellis codes [70, 80, 278–281] and space-time block codes [71–73] constitute state-of-the-art transmission schemes based on multiple transmitters and receivers. Both codes have been introduced in Section 11.1. Since we have detailed the encoding and decoding processes of the space-time block codes in Chapter 10, the detailed description of the codes was left out of this chapter. Instead, space-time trellis codes were introduced in Section 11.2 by utilising the simplest possible 4-state, 4PSK space-time trellis code as an example. The state diagrams for other 4PSK and 8PSK space-time trellis codes were also given. The branch metric of each trellis transition was derived in order to facilitate its maximum likelihood decoding.

In Section 11.3, we proposed employing an OFDM modem for mitigating the effects of dispersive multi-path channels due to its simplicity compared to other approaches [74,75]. Turbo codes and Reed–Solomon codes were invoked in Section 11.3.1 for concatenation with the space-time block code G_2 and the various space-time trellis codes, respectively. The estimated complexity of the various space-time trellis codes was derived in Section 11.3.3.

We presented our simulation results for the proposed schemes in Section 11.4. The first scheme studied was the $TC(2, 1, 3)$-coded space-time block code G_2, whereas the second one was based on the family of space-time trellis codes. It was found that the FER and BER performance of the $TC(2, 1, 3)$ coded space-time block G_2 was better than that of the investigated space-time trellis codes at a throughput of 2 and 3 BPS over the channel exhibiting two equal-power rays separated by a delay spread of 5 μs and having a maximum Doppler frequency of 200 Hz. Our comparison between the two schemes was performed by also considering the estimated complexity of both schemes. It was found that the concatenated $G_2/TC(2, 1, 3)$ scheme still outperformed the space-time trellis codes using no channel coding, even though both schemes exhibited a similar complexity.

The effect of the maximum Doppler frequency on both schemes was also investigated in Section 11.4.3. It was found that the maximum Doppler frequency had no significant impact on the performance of both schemes. By contrast, in Section 11.4.4 we investigated the effect of the delay spread on the system. Initially, the delay-spread-dependent SIR of the space-time block code G_2 was quantified. It was found that the performance of the concatenated $G_2/TC(2,1,3)$ scheme degrades as the delay spread increases owing to the decrease in the associated SIR. However, varying the delay spread had no significant effect on the space-time trellis codes. We proposed in Section 11.4.5 an alternative mapping of the two transmission instants of the space-time block code G_2 to the same subcarrier of two consecutive OFDM symbols, a solution which was applicable to a delay non-sensitive system. By employing this approach, the performance of the concatenated scheme was no longer limited by the delay spread, but by the maximum Doppler frequency. We concluded that a certain minimum SIR has to be maintained for attaining the best possible performance of the concatenated scheme.

The shortened WATM channel was introduced in Section 11.4.6. In this section, space-time trellis codes were concatenated with Reed–Solomon codes, in order to improve the performance of the system. Once again, both channel-coded space-time block and trellis codes were compared at a throughput of 1 and 2 BPS. It was also found that the TC(2,1,3)-coded space-time block code G_2 outperforms the RS-coded space-time trellis codes.

Space-time block-coded AOFDM was proposed in Section 11.5, which is the last section of this chapter. It was shown in Section 11.5.3.1 that only the space-time block code G_2 using one AOFDM receiver outperformed the conventional single-transmitter, single-receiver AOFDM system designed for a data transmission target BER of 10^{-4} over the shortened WATM channel. We also confirmed that upon increasing the diversity order, the fading channels become AWGN-like channels. This explains why fixed-mode OFDM transmission constitutes a better trade-off than AOFDM when the diversity order is increased. In Section 11.5.3.2, we continued our investigations into AOFDM by concatenating turbo coding with the system. Two schemes were proposed: half-rate turbo and space-time-coded AOFDM as well as variable-rate turbo- and space-time-coded AOFDM. Despite the impressive BER performance of the half-rate turbo- and space-time-coded scheme, the maximum throughput of the system was limited to 3 BPS. However, by employing the variable-rate turbo- and space-time-coded scheme, the BPS performance improved, achieving a maximum throughput of 5.4 BPS. However, the improvement in BPS performance was achieved at the cost of a poorer BER performance.

In this chapter the channel-induced signal dispersion was mitigated with the aid of OFDM-based modulation techniques. By contrast, in the next chapter we will use a decision feedback channel equaliser in conjunction with adaptive modulation.

Chapter 12

Turbo-coded Adaptive Modulation versus Space-time Trellis Codes for Transmission over Dispersive Channels [1]

12.1 Introduction

In this chapter we will comparatively study a range of sophisticated, adaptive modulation and adaptive channel coding techniques in the light of space-time trellis-coded transceivers. A modest background in both turbo coding as well as modulation techiques and wireless propagation issues is asumed.

Adaptive Quadrature Amplitude Modulation (AQAM) has been documented in depth in [306], hence here only a rudimentary introduction to the basic philosophy of AQAM is provided. AQAM employs a high-throughput modulation scheme, when the channel quality is favourable, in order to increase the number of Bits Per Symbol (BPS) transmitted and, conversely, a more robust but lower-throughput modulation scheme, when the channel exhibits a reduced instantaneous quality, in order to improve the mean Bit Error Ratio (BER) performance. AQAM can be conveniently applied if the channel in the uplink and downlink can be considered reciprocal. This reciprocity can be exploited by employing Time-Division Duplexing (TDD) [289], where both the uplink and downlink carrier frequencies are the same. Consequently, the short-term channel quality can be estimated at one end of the TDD link and by utilising this estimate, a suitable modulation scheme is chosen for the reverse TDD link. However, reciprocity over wideband channels rarely holds. Another scenario where adaptive modulation can be applied is when there is a reliable, low-delay feedback path between the transmitter and receiver, an issue addressed for example in [245]. Recent developments in AQAM over narrowband channels include contributions by Webb and Steele [297], Sampei *et al.* [241], Torrance and Hanzo [305] as

---

[1]This chapter is based on T. H. Liew, B. L. Yeap, C. H. Wong, L. Hanzo: Turbo Coded Adaptive Modulation versus Space-Time Trellis Codes for Transmission over Dispersive Channels submitted to IEEE Transactions on Wireless Communications, 2002

well as by Chua and Goldsmith [303].

When applying AQAM over wideband channels, the Decision Feedback Equaliser (DFE) employed will eliminate most of the Inter-Symbol Interference (ISI). Consequently, the associated pseudo-SNR at the output of the DFE, which will be discussed in the context of Equation 12.1, is calculated and used as a metric for switching the modulation modes. This ensures that the performance is optimised by employing equalisation and AQAM switching jointly, in order to combat both the signal power and ISI fluctuations experienced in a wideband channel. Furthermore, the error correction and the error detection capability of the coding scheme invoked is exploited for improving the BER and BPS throughput performance of wideband AQAM [197, 307, 308]. Recent work on combining conventional channel coding with adaptive modulation has been conducted also by Matsuoka *et al.* [302], where punctured convolutional coding with and without an outer Reed–Solomon (RS) code was invoked in a TDD environment. Convolutional coding was also used in conjunction with adaptive modulation in [247], where results were presented in a Frequency-Division Multiple Access (FDMA) and Time-Division Multiple Access (TDMA) environment, when assuming the presence of a feedback path between the receiver and transmitter. In [309], Lau and Maric proposed to vary the coding rate of the convolutional codes used, when the channel quality varied. Chua and Goldsmith [310] demonstrated that in adaptive coded modulation the simulation and theoretical results confirmed a 3 dB coding gain at a BER of 10^{-6} for a 4-state trellis code and 4 dB by an 8-state trellis code over Rayleigh fading channels, while a 128-state code performed within 5 dB of the Shannon capacity limit. On the other hand, Liu *et al.* [311] employed Bit-Interleaved Coded Modulation (BICM) in adaptive modulation.

The third-generation (3G) mobile communication standards are expected to offer a wide range of bearer services, ranging from voice to high-rate data services supporting rates of at least 144 kbit/s in vehicular, 384 kbit/s in outdoor-to-indoor and 2 Mbit/s in indoor as well as in picocellular applications [266]. In an effort to support such high rates, the capacity of band-limited wireless channels can be increased by employing multiple antennas. Recently, different transmit diversity techniques have been introduced in order to provide diversity gain for Mobile Stations (MSs) by upgrading the Base Stations (BSs). In [70], Tarokh *et al.* proposed space-time trellis coding by jointly designing the channel coding, modulation, transmit diversity and the optional receiver diversity. Space-time block codes were subsequently proposed by Tarokh *et al.* in [70, 71, 73].

The performance of both space-time trellis and block codes was investigated by numerous researchers [70, 71, 73, 79, 80] when communicating over narrowband Rayleigh fading channels. The investigation of space-time codes was then also extended to the class of more practical wideband fading channels. The effect of multiple paths on the performance of space-time trellis codes was studied in [281] for transmission over slowly varying Rayleigh fading channels. It was shown in [281] that the presence of multiple paths does not decrease the diversity order guaranteed by the design criteria used for constructing space-time trellis codes. The evidence provided in [281] was then also extended to rapidly fading dispersive and non-dispersive channels. As a further performance improvement, turbo equalisation was employed in [74] in order to mitigate the effects of dispersive channels. As an alternative solution, the effect of ISI could also be eliminated by employing Orthogonal Frequency-Division Multiplexing (OFDM) [207]. A system using space-time trellis-coded OFDM is attractive, since the decoding complexity can be reduced, as demonstrated by the recent surge of research interests [76–79]. Similarly, the performance of space-time block codes was also investigated when communicating over frequency-selective Rayleigh fading channels. In [292], a Multiple-Input Multiple-Output (MIMO) equaliser was utilised for equalising the effects of dispersive multi-path channels. In this chapter the MIMO equaliser was employed in conjunction with space-time trellis codes for equalising the wideband channel. The Minimum Mean Squared Error (MMSE) MIMO equaliser was first studied by Tidestav *et al.* [312, 313] in the transform domain. It was then employed by Tidestav *et al.* in [314] for multi-user detection. In Space-Time Trellis (STT) codecs there are multiple transmitters and receivers at both the transmitting and receiving ends. The received symbol at each receiver is constituted by the superposition of all transmitted symbols, which undergo independent wideband fading. The MIMO equaliser was then employed for equalising and estimating the transmitted symbols of each transmitter.

In [315], we have shown that as we increase the diversity order of STT coding, the fluctuation of channel quality becomes less severe, since the fading channels have been converted to AWGN-like channels. By contrast, AQAM seeks to accommodate the temporal variation of the quality of fading channels by appropriately adjusting the turbo-coded AQAM modes. Therefore in this chapter we will investigate the achievable throughput of adaptive-coded AQAM and STT codes.

12.2   System Overview

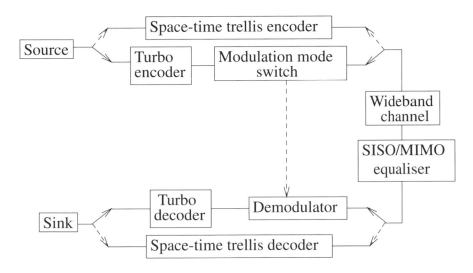

Figure 12.1: System overview of the turbo-coded AQAM and STT schemes.

Figure 12.1 shows the system overview of the proposed turbo-coded AQAM scheme as well as the STT codes employed. At the transmitter, the information source generates random information data bits. In our proposed turbo-coded AQAM scheme, the information data bits are encoded by the turbo encoder. Assuming that the next transmission burst's short-term channel quality can be estimated as in a TDD scheme, a suitable AQAM mode will be chosen by the modulation mode switch shown in the figure. The turbo-coded bits are subsequently mapped to the AQAM symbols. By contrast, the information bits will be directly encoded by the STT encoder, which produces the modulated symbols depending on the specific STT code employed. The modulated symbols are transmitted over the COST207 TU wideband channel. At the receiver, the received symbols are equalised by the appropriate Single-Input Single-Output (SISO) or Multiple-Input Multiple-Output (MIMO) DFE, respectively. Depending on the specific AQAM mode chosen at the transmitter, the demodulator provides soft outputs for the turbo decoder based on the equalised symbols output by the SISO equaliser. The turbo-decoded bits are passed to the information sink for the calculation of the BER, as shown in Figure 12.1. By contrast, the equalised symbols output by the MIMO equaliser are passed to the STT decoder.

12.2.1   SISO Equaliser and AQAM

At the receiver, the Channel Impulse Response (CIR) is estimated, which is then used for calculating the SISO DFE coefficients [207]. Subsequently, the DFE is used for equalising the ISI-corrupted received signal. Additionally, both the channel quality estimate and the SISO DFE coefficients are utilised for computing the pseudo-SNR at the output of the SISO DFE. By assuming that the residual ISI is Gaussian distributed and that the probability of decision feedback errors is negligible, the pseudo-SNR at the output of the SISO DFE, γ_{dfe}, can be calculated as [316]:

$$
\gamma_{dfe} = \frac{\text{Wanted Signal Power}}{\text{Residual ISI Power} + \text{Effective Noise Power}}
$$

$$
= \frac{E\left[|S_k \sum_{m=0}^{N_f} C_m h_m|^2\right]}{S_{k-q} \sum_{q=-(N_f-1)}^{-1} E\left[|\sum_{m=0}^{N_f-1} C_m h_{m+q}|^2\right] + N_o \sum_{m=0}^{N_f} |C_m|^2}, \qquad (12.1)
$$

where C_m and h_m denote the DFE's feedforward coefficients and the CIR, respectively. The transmitted signal and the noise spectral density are represented by S_k and N_o, respectively. Lastly, the number of DFE feedforward coefficients is denoted by N_f.

The pseudo-SNR, γ_{dfe}, is then compared against a set of AQAM switching threshold levels, f_n and subsequently the appropriate modulation mode is selected for the next transmission burst, assuming the availability of a reliable, low-delay feedback message, which can be transmitted to the remote receiver by superimposing it — after strongly protecting it — on the reverse-direction information [245]. However, in this treatise we will dispense with explicit AQAM mode signalling. The AQAM modes that are utilised in this system are Binary Phase Shift Keying (BPSK), Quadrature Phase Shift Keying (QPSK), 16-level Quadrature Amplitude Modulation (16QAM), 64-level Quadrature Amplitude Modulation (64QAM) and a 'No Transmission' (No TX) mode.

12.2.2   MIMO Equaliser

In the context of the MIMO equaliser, similarly to the AQAM receiver, the CIR is estimated, which is then used for calculating the MIMO DFE coefficients [314]. At receiver Rx, the received signal is given by the superposition of all the transmitters' signals. For each channel delay m and each combination of the specific transmitters and receivers of the MIMO scheme, we have a specific CIR. Hence we can write the corresponding channel matrix as:

$$
\mathbf{H}^m = \begin{pmatrix} h_{1,1}^m & h_{1,2}^m & \cdots & h_{1,Tx}^m \\ h_{2,1}^m & h_{2,2}^m & \cdots & h_{2,Tx}^m \\ \cdot & \cdot & \cdots & \cdot \\ \cdot & \cdot & \cdots & \cdot \\ \cdot & \cdot & \cdots & \cdot \\ h_{Rx,1}^m & h_{Rx,2}^m & \cdots & h_{Rx,Tx}^m \end{pmatrix}, \qquad (12.2)
$$

where $h_{Rx,Tx}^m$ is the CIR of the transmission link from transmitter Tx to receiver Rx at a channel delay of m. Consequently, we can write the channel matrix of the MIMO wideband channel as:

$$
\mathbf{H} = (\mathbf{H}^0, \mathbf{H}^1, \dots, \mathbf{H}^{N_c}), \qquad (12.3)
$$

where N_c is the maximum delay of the wideband channel.

The minimum mean squared error MIMO DFE feedforward filter matrix is given by solving the following system of equations for $\mathbf{S}$ [314]:

$$(\mathbf{FF}^H + \Psi) \cdot \begin{pmatrix} \mathbf{S}_0^H \\ \mathbf{S}_1^H \\ \cdot \\ \cdot \\ \cdot \\ \mathbf{S}_{N_f}^H \end{pmatrix} = \begin{pmatrix} \mathbf{H}_{N_f} \\ \mathbf{H}_{N_f-1} \\ \cdot \\ \cdot \\ \cdot \\ \mathbf{H}_0 \end{pmatrix}, \tag{12.4}$$

where $\mathbf{S}^H$ denotes the Hermitian transpose of $\mathbf{S}$ and N_f is the forward filter length, $\mathbf{F}$ is defined by:

$$\mathbf{F} = \begin{pmatrix} \mathbf{H}^0 & \mathbf{H}^1 & \mathbf{H}^2 & \cdots & \mathbf{H}^{N_f-1} \\ 0 & \mathbf{H}^0 & \mathbf{H}^1 & \cdots & \mathbf{H}^{N_f-2} \\ 0 & 0 & \mathbf{H}^0 & \cdots & \mathbf{H}^{N_f-3} \\ \cdot & \cdot & \cdot & \cdot & \cdot \\ \cdot & \cdot & \cdot & \cdot & \cdot \\ \cdot & \cdot & \cdot & \cdot & \cdot \\ 0 & 0 & 0 & \cdots & 0 \end{pmatrix}, \tag{12.5}$$

and Ψ is defined by:

$$\Psi = \begin{pmatrix} N_0 & 0 & \cdots & 0 \\ 0 & N_0 & \cdot & 0 \\ \cdot & \cdot & \cdot & \cdot \\ \cdot & \cdot & \cdot & \cdot \\ \cdot & \cdot & \cdot & \cdot \\ 0 & 0 & \cdots & N_0 \end{pmatrix}, \tag{12.6}$$

where N_0 is the noise spectral density.

After the feedforward coefficients have been calculated using Equation 12.4, we can find the N_b backward coefficients with the aid of [314]:

$$Q_n = \sum_{m=\max(0, n-N_c+N_f-1)}^{\min(N_f, N_f+n+1)} \mathbf{S}_m \cdot \mathbf{H}_{N_f+1-m+n} . \tag{12.7}$$

12.3 Simulation Parameters

The STT codes employed in this treatise were proposed by Tarokh et al. in [70]. We chose 4PSK, 8PSK and 16PSK STT codes, since they provide the throughputs of 2, 3 and 4 BPS, respectively. The associated parameters of the STT employed are summarised in Table 12.1.

As we can see in Figure 12.1, turbo codes, which were introduced by Berrou and Glavieux [13], are concatenated with the AQAM scheme. Both convolutional and block codes can be used as component codes in a turbo code. It was shown by Hagenauer et al. [61] that turbo block codes outperform punctured turbo convolutional codes when the coding rate is higher than $R = \frac{2}{3}$. It was also observed that a rate $R = 0.98$ turbo Hamming code using BPSK modulation for communicating over the non-dispersive Gaussian channel is capable of operating

Modulation scheme	BPS	Decoding algorithm	No. of states	No. of transmitters	No. of termination symbols
4PSK	2	VA	16	2	2
8PSK	3	VA	16	2	2
16PSK	4	VA	16	2	1

Table 12.1: Parameters of the STT codes proposed in [70].

Code	Octal generator polynomial	No. of states	Decoding algorithm	No. of iterations
Turbo Convolutional (TC) Code				
TC(2,1,3)	7,5	4	Log-Map	8
Turbo BCH (TBCH) Code				
TBCH(31,26)	45	32	Log-Map	8
TBCH(63,57)	103	64	Log-Map	8
TBCH(127,120)	211	128	Log-Map	8

Table 12.2: Parameters of the different turbo codes used in Figure 12.1.

within 0.27 dB of the Shannon limit [1]. In this chapter we will investigate the performance of Turbo BCH (TBCH) and Turbo Convolutional (TC) codes, when using various fixed modulation schemes and AQAM. Table 12.2 shows the parameters of each turbo codec used in the system.

As we can see in Table 12.2, different BCH component codes are employed and this results in different coding rates. On the other hand, puncturing is applied to the TC code, which also results in different coding rates. We show in Table 12.3 the puncturing patterns employed and the resulting coding rates. The puncturing patterns seen in the table consist of two parts. Specifically, the associated different puncturing patterns represent the puncturing patterns of the parity bits emanating from the first and second decoders, respectively. Note that the puncturing patterns employed in these systems have been determined experimentally, but they are not claimed to be optimal. For procedures on designing high-rate turbo codes with the aid of puncturing the interested reader is referred to [64]. Different modulation schemes were employed, which consequently results in different effective BPS throughputs as shown in Table 12.3. We have conducted prior research in terms of mapping the data and parity bits onto the different protection classes of the QAM constellations, for example in [207] by Hanzo et al. In our recent work in [295, 317] we found that the corresponding best mapping rule also depends on the error correcting power, i.e. the contraint length of the convolutional constituent codes of the turbo codec used. However, in these investigations we randomly mapped the data and parity bits to the QAM constellation points.

The random turbo interleaver and random channel interleaver depths are chosen such that each AQAM transmission burst can be individually decoded, i.e. burst-by-burst turbo decoding is used. Additionally, in an effort to quantify the best possible performance of the system, we also study a long-delay system, where the channel interleaver depth is chosen to be approximately 10^5 bits in the context of all AQAM modes. The associated channel interleaver depth and the corresponding turbo interleaver depth is indicated by the 'Long delay' phrase in Table 12.3.

Code	Code rate R	Puncturing Pattern	Modula-tion mode	BPS	Random turbo interleaver depth	Random channel interleaver depth
TC(2,1,3)	0.33	1, 1	BPSK	0.33	228	684
			QPSK	0.66	456	1368
			Long delay		34198	102600
	0.50	10, 01	BPSK	0.50	342	684
			QPSK	1.00	684	1368
			16QAM	2.00	1368	2736
			Long delay		51300	102600
	0.75	100000001000, 001000000010	BPSK	0.75	513	684
			QPSK	1.50	1026	1368
			16QAM	3.00	2052	2736
			64QAM	4.50	3078	4104
			Long delay		76950	102600
	0.83	1000000000, 0000000001	BPSK	0.83	570	684
			QPSK	1.66	1140	1368
			16QAM	3.32	2280	2736
			64QAM	4.98	3420	4104
			Long delay		85500	102600
	0.90	000001000000000000 000000000000100000, 100000000000000000 000001000000000000	BPSK	0.90	615	684
			QPSK	1.80	1230	1368
			16QAM	3.60	2462	2736
			64QAM	5.40	3694	4104
			Long delay		92340	102600

Table 12.3: Simulation parameters associated with the TC channel codec in Figure 12.1.

Similarly to Table 12.3, we show in Table 12.4 the corresponding coding rates and the achievable BPS throughput of the TBCH codes employing various BCH component codes. Again, the turbo interleaver sizes were chosen for enabling burst-by-burst turbo decoding. Additionally, both the turbo and channel interleaver depths of the delay non-sensitive system are also shown in Table 12.4 again indicated as the 'Long delay' scenario.

Our studies were conducted for transmission over the COST207 TU [265] channel. The corresponding CIR is shown in Figure 12.2. Each path was faded independently according to the Rayleigh distribution and the corresponding normalised Doppler frequency was 3.27×10^{-5}. Both the SISO and the MIMO DFE incorporated 35 forward taps and 7 feedback taps and the transmission burst structure used in our treatise is shown in Figure 12.3. The following assumptions were stipulated. First, the CIR was time-invariant for the duration of a transmission burst, but varied from burst to burst, which corresponds to assuming that the CIR is slowly varying. Second, perfect channel estimation and perfect knowledge of the AQAM mode used were assumed at the receiver. In practice a simple repetition code can be used for conveying the AQAM mode to the receiver [2]. Furthermore, the modulation mode can also be detected blindly [301].

---

[2]For example, three bits are requited for signalling one of the five possible AQAM modes, which may be repeated three times for the sake of adding protection. The receiver's Majority Logic Decision (MLD) scheme is capable of recovering the correct AQAM mode, even if three out of nine bits are erroneous, which corresponds to a channel BER of

Code	Code rate R	Puncturing pattern	Modula-tion mode	BPS	Random turbo interleaver depth	Random channel interleaver depth
TBCH(31,26)	0.72	1,1	BPSK	0.72	494	684
			QPSK	1.44	988	1368
			16QAM	2.88	1976	2736
			64QAM	4.32	2964	4104
			Long delay		74100	102600
TBCH(63,57)	0.83	1,1	BPSK	0.83	513	621
			QPSK	1.66	1083	1311
			16QAM	3.32	2223	2691
			64QAM	4.98	3363	4071
			Long delay		77976	94392
TBCH(127,120)	0.90	1,1	BPSK	0.90	600	670
			QPSK	1.80	1200	1340
			16QAM	3.60	2400	2680
			64QAM	5.40	3600	4020
			Long delay		82080	91656

Table 12.4: Simulation parameters associated with the TBCH channel codec in Figure 12.1.

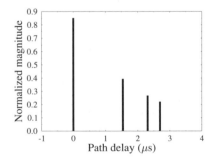

Figure 12.2: The impulse response of a COST207 TU channel [265].

Third, the pseudo-SNR at the output of the equaliser is estimated perfectly prior to transmission, which, again, tacitly assumes the existence of a reliable, low-delay feedback path between the transmitter and the receiver. In practice, there will be error propagation in the DFE's feedback loop, which will degrade the performance of the system [3].

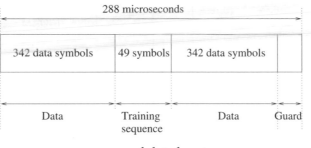

non-spread data burst

Figure 12.3: Transmission burst structure of the FRAMES Mode A (FMA1) non-spread data scheme [252].

12.4 Simulation Results

12.4.1 Turbo-coded Fixed Modulation Mode Performance

Before analysing the proposed AQAM schemes, we shall investigate the BER performance of the turbo-coded system utilising the fixed modulation modes of BPSK, QPSK, 16QAM and 64QAM. The effect of the channel interleaver length, turbo component codes, coding rate and DFE error propagation on the BER performance will be investigated.

Figure 12.4 shows the BER performance of the rate $R = 0.75$ TC-coded fixed modulation schemes for both the short channel interleaver, which enables burst-by-burst decoding, and for the long channel interleaver, which has a delay of 10^5 bits. As expected, in Figure 12.4 the BER performance associated with using the longer channel interleaver was superior compared to that using the shorter channel interleaver, although naturally at the cost of an associated higher transmission delay. Except for 64QAM, there was more than 5 dB 'interleaver gain' at a BER of 10^{-4}, when using the longer channel interleaver. The poor performance of the system when 64QAM was used was due to the error propagation of the DFE, which significantly degraded the performance of the system and hence overloaded the error correction capability of the turbo codes.

In Figure 12.5, we show the BER performance of the TC codes in conjunction with various coding rates, which were derived by puncturing using the experimentally determined patterns given in Table 12.3. Again, for procedures on designing high-rate turbo codes using puncturing the interested reader is referred to [64]. The BER performance plotted against the E_b/N_0 values in Figure 12.5 improves as the coding rate is reduced to $R = 0.33$.

In Figure 12.5, we have compared the performance of different-rate TC codes. By contrast, in Figure 12.6 we compare the performance of TC and TBCH codes having similar coding rates.

33%. Alternatively, Torrance's unequal protection scheme [318] can be used, which requires less redundancy.

[3]We note, however, that the AQAM switching thresholds can be appropriately increased, in order to compensate for the error propagation phenomenon of the DFE at the cost of a slight reduction of the system's BPS throughput.

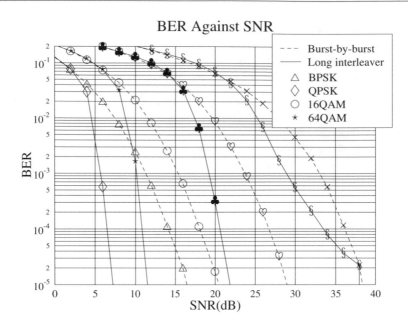

Figure 12.4: BER performance comparison between rate $R = 0.75$ TC burst-by-burst decoding and TC coding using the long channel interleaver for each individual modulation mode, when communicating over the Rayleigh fading COST207 TU channel of Figure 12.2. The associated coding parameters are shown in Tables 12.2 and 12.3.

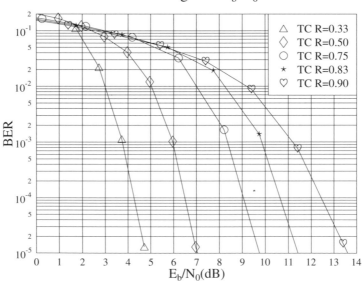

Figure 12.5: BER performance of TC codes using various coding rates and QPSK modulation, when communicating over the Rayleigh fading COST207 TU channel of Figure 12.2. The associated coding parameters are shown in Tables 12.2 and 12.3.

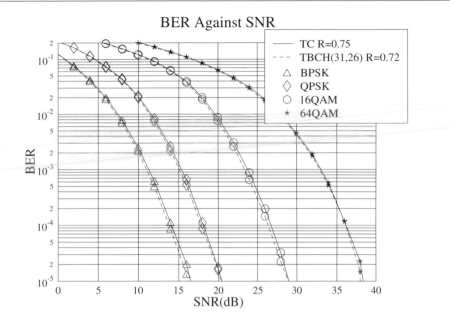

Figure 12.6: BER performance comparison between the rate $R = 0.75$ TC and the $R = 0.72$ TBCH codes using burst-by-burst decoding for each individual modulation mode for transmission over the Rayleigh fading COST207 TU channel of Figure 12.2. The associated coding parameters are shown in Tables 12.2, 12.3 and 12.4.

The channel interleaver length was chosen such that it enables burst-by-burst decoding at the receiver. As we can see in Figure 12.6, there is virtually no performance difference between the TC and TBCH codes in conjunction with the various modulation schemes studied.

Similarly to Figure 12.6, in Figure 12.7 we compare the performance of TC and TBCH codes, although in Figure 12.7 the channel interleaver's length was chosen to be approximately 10^5 bits. Unlike in Figure 12.6, we can clearly see that the performance of the rate $R = 0.72$ TBCH code becomes better than that of the TC code. At a BER of 10^{-4} we can observe from Figure 12.7 that there is a SNR gain of 5 dB for the TBCH code compared to the TC code. From Figures 12.6 and 12.7 we can conclude that the rate $R = 0.72$ TBCH code outperforms the rate $R = 0.75$ TC code when a channel interleaver length of 10^5 bits is employed in the system.

We now continue our investigations into the impact of error propagation in the SISO DFE equaliser. In Figure 12.8, we show the BER performance of the rate $R = 0.75$ TC-coded system both with and without error propagation in the SISO DFE equaliser employed. At a BER of 10^{-4}, there is a channel SNR improvement of 1–2 dB, if perfect knowledge of the transmitted symbols is assumed and hence error-free decoding is fed back to the backward filter. However, when we employ higher-order modulation schemes, we can clearly see that the performance of the TC-coded system drastically degrades when there is error propagation in the DFE equaliser. For example, if 64QAM is employed in the system, there is a SNR degradation of 10 dB at a BER of 10^{-4}, if the potentially error-prone decoded symbols are fed back to the backward filter, as a consequence of the densely-packed constellation points in 64QAM. These precipitated errors may subsequently overload the TC decoder.

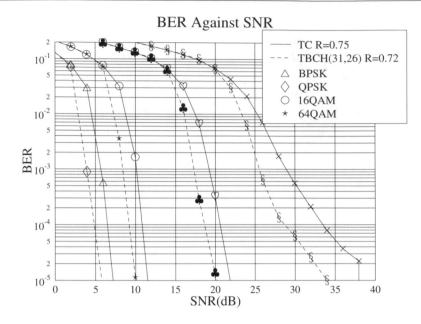

Figure 12.7: BER performance comparison between the rate $R = 0.75$ TC and the rate $R = 0.72$ TBCH codes using the 10^5 bit delay channel interleaver for each individual modulation mode for transmission over the Rayleigh fading COST207 TU channel of Figure 12.2. The associated coding parameters are shown in Tables 12.2, 12.3 and 12.4.

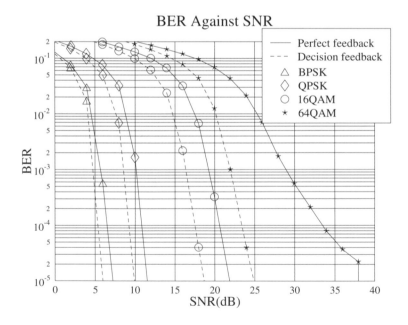

Figure 12.8: BER performance comparison between the rate $R = 0.75$ TC-coded modem both with and without error propagation in the DFE equaliser for each individual modulation mode for transmission over the Rayleigh fading COST207 TU channel of Figure 12.2. The associated coding parameters are shown in Tables 12.2 and 12.3.

12.4.2 Space-time Trellis Code Performance

In the previous section, we have presented our simulation results for various turbo-coded systems utilising fixed modulation modes. Let us now extend our investigations into the behaviour of STT codes. As argued earlier in Section 12.2.2, since multiple transmitters and receivers are employed, a MIMO equaliser will be employed for equalising the multiple-input and multiple-output channels. Since two transmitters are employed for the STT code used, there has to be a minimum of two receivers employed [314].

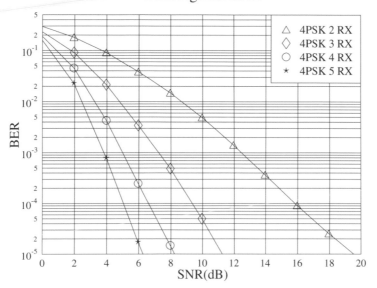

Figure 12.9: BER performance comparison between various 4PSK space-time trellis codes using two, three, four and five receivers for transmission over the Rayleigh fading COST207 TU channel of Figure 12.2. The associated coding parameters are shown in Table 12.1.

In Figure 12.9, we show the BER performance of a range of 4PSK STT codes using two to five receivers [70]. At a BER of 10^{-4}, we observe that there is a huge SNR improvement of more than 15 dB when the number of receivers is increased from two to three. As the number of receivers is further increased to four and five, the incremental performance gain becomes smaller and smaller.

Similarly, in Figure 12.10 we show the BER performance of various 8PSK STT codes using two to five receivers. Again, there is a SNR gain of 10 dB at a BER of 10^{-4} due to increasing the number of receivers from two to three. Further increasing the number of receivers reduces the achievable incremental SNR performance gain to 4 and 2 dB, respectively.

In Figures 12.9 and 12.10 we have provided BER performance results for the 4PSK and 8PSK STT codes, which have a throughput of 2 and 3 BPS, respectively. Let us now increase the throughput of the system further to 4 BPS. In Figure 12.11 we portray the BER performance of various 16PSK STT codes using two to five receivers. At a BER of 10^{-4} we observe a significant SNR gain of 12 dB due to increasing the number of receivers from two to three. Again, further increasing the number of receivers results in reduced incremental performance gains.

BER Against SNR

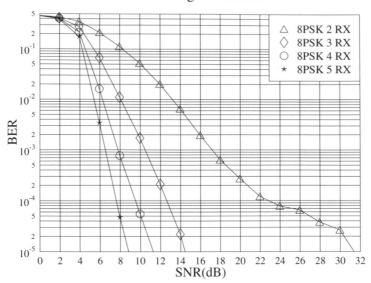

Figure 12.10: BER performance comparison between various 8PSK STT codes using two, three, four and five receivers for transmission over the Rayleigh fading COST207 TU channel of Figure 12.2. The associated coding parameters are shown in Table 12.1.

BER Against SNR

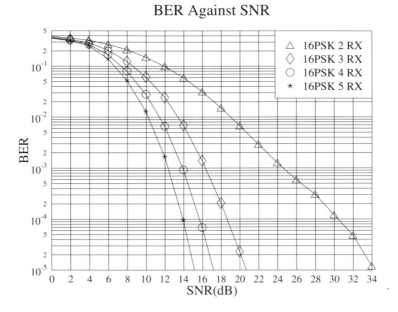

Figure 12.11: BER performance comparison between various 16PSK STT codes using two, three, four and five receivers for transmission over the Rayleigh fading COST207 TU channel of Figure 12.2. The associated coding parameters are shown in Table 12.1.

12.4.3   Adaptive Quadrature Amplitude Modulation Performance

In the previous sections, we have presented simulation results for turbo-coded systems communicating using fixed modulation modes. Furthermore, we also presented performance results for the STT codes employed in conjunction with a MIMO equaliser. Let us now study the required SNR of the turbo-coded fixed modem mode system and that of the STT codes comparatively in terms of their achievable BPS throughput, when aiming for a target BER of 10^{-4}. These results will then be compared to the BPS performance of the proposed AQAM schemes at a BER of 10^{-4} in Figure 12.12.

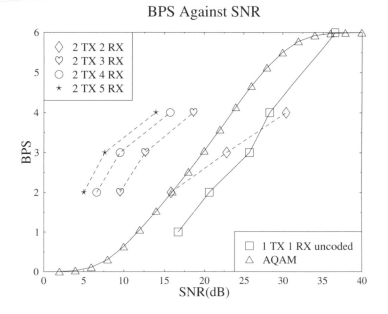

Figure 12.12: BPS performance comparison between STT codes, uncoded AQAM and the uncoded fixed modulation modes, when communicating over the Rayleigh fading COST207 TU channel of Figure 12.2. The associated coding parameters are shown in Table 12.1 and the target BER is 10^{-4}.

Explicitly, there are two set of curves in Figure 12.12. The curves drawn as continuous lines represent the system having no diversity gain, i.e. that employing one transmitter and one receiver. By contrast, the curves drawn as broken lines represent the system having diversity gain due to employing STT codes and multiple receivers. The curve marked with squares represents the performance of the uncoded system using 1, 2, 3, 4 and 6 BPS fixed modulation schemes. In [245], an uncoded wideband burst-by-burst adaptive modem was characterised and its performance is reproduced in Figure 12.12, where it is denoted by the triangles. We can see in Figure 12.12 that by employing STT codes in conjunction with two receivers, we have a SNR gain of 5 dB at a throughput of 2 BPS as compared to the uncoded system. The achievable SNR performance gain reduces to 3 dB, when the required throughput of the system is increased to 3 BPS, and there is a negative SNR gain, when the required system throughput is 4 BPS. When employing AQAM, the single-transmitter and single-receiver system outperforms the STT code using two receivers. By contrast, the AQAM BPS performance remains inferior to that of the STT code using three receivers, as shown in Figure 12.12.

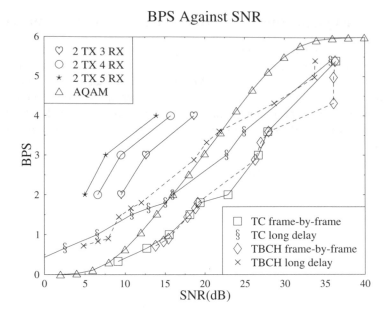

Figure 12.13: BPS performance comparison between STT codes, uncoded AQAM, TC and TBCH codes in conjunction with both burst-by-burst decoding and long-delay channel interleaving for transmission over the Rayleigh fading COST207 TU channel of Figure 12.2. The associated coding parameters are shown in Tables 12.1, 12.2, 12.3 and 12.4 and the target BER is 10^{-4}.

In Figure 12.12, we have shown the performance of the uncoded one-transmitter and one-receiver fixed modulation scheme, which was outperformed by the burst-by-burst AQAM scheme proposed in [245]. We will now investigate the achievable BPS versus SNR gain due to employing turbo codes. In Figure 12.13 we show the BPS performance of both TC and TBCH codes for the short and long channel interleaver scenarios [4]. Recall that the length of the short channel interleaver was chosen such that it enabled burst-by-burst turbo decoding, while the long channel interleaver depth was approximately 10^5 bits. We have shown in Figure 12.6 that there is no significant performance difference between the TC and TBCH codes for burst-by-burst turbo decoding, which is also reflected in Figure 12.13. More explicitly, it is shown in Figure 12.13 that both burst-by-burst TC and TBCH decoding exhibit a similar BPS performance, as evidenced by the overlapping curves marked with diamonds and squares, respectively. It can be seen that burst-by-burst turbo decoding in conjunction with fixed modulation is inferior to the uncoded adaptive modulation scheme. If, however, the delay permitted by the system allows the employment of a channel interleaver length of approximately 10^5 bits, we can clearly see in Figure 12.13 that there is a substantial SNR improvement over the burst-by-burst scenario using the same modulation schemes. For TC codes, we can see that the system outperforms uncoded AQAM for throughput values below 2 BPS. At 0.5 BPS, we can see in Figure 12.13 that the TC code outperforms the uncoded AQAM scheme by approximately 10 dB. On the other hand, if TBCH codes are employed we can see that the system outperforms the uncoded AQAM scheme up to throughput values as high as 3.5 BPS.

In Figure 12.13, we have investigated the performances of TC and TBCH codes in conjunc-

---

[4]The associated curve was generated by combining all the different throughput fixed-mode modulation schemes with all the different coding-rate schemes

tion with fixed modulation-based systems for both low and high channel interleaver lengths. Let us now fully exploit the error correction capability of turbo codes in the context of AQAM. In our earlier research we used a set of fixed AQAM thresholds, which was determined with the aid of Powell's optimisation [305]. However, the achievable throughput of the system can be further increased when we optimise the AQAM switching thresholds for each individual channel SNR value using Lagrangian optimisation [319]. We note, however, that the solutions in [319] use no channel coding during the optimisation of the AQAM thresholds. Since at the time of writing no analytical formulae exist for the BER of turbo codes, employing Lagrangian optimisation might not be feasible for AQAM systems employing turbo codes. Hence in our proposed turbo-coded system we propose a simple procedure for determining the switching thresholds for each average channel SNR value.

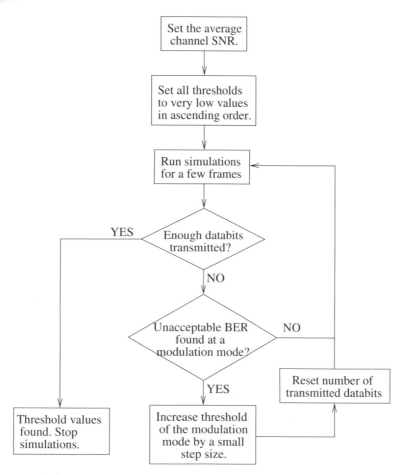

Figure 12.14: Procedure for determining the AQAM switching thresholds.

In Figure 12.14, we show the flowchart of the simple procedure proposed for determining the AQAM switching thresholds. As we can see in the figure, we commence the procedure by setting the channel SNR, where the switching thresholds will be determined. Then we assign the lowest possible SNR switching threshold to each AQAM mode in an ascending order. At this stage, the switching thresholds are set so low that the simulated BER will be significantly higher than the targeted BER. When the simulations are started and a few frames have been received, we check whether sufficient data bits were transmitted, such that a statistically relevant BER result can be obtained. If not, the BER of each AQAM mode is checked with the lowest-throughput modulation mode. Once a statistically relevant BER is found in a particular AQAM mode, the switching threshold of the modulation mode will be incremented by a small step size. Then, the number of transmitted data bits is reset to zero and the simulations are restarted. On the other hand, if there is no modulation mode having an unacceptable BER, more data bits will be transmitted. The entire process is repeated until enough data bits have been transmitted, which generate a statistically pertinent average result.

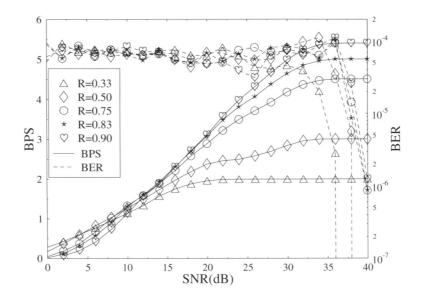

Figure 12.15: BER and BPS performance of burst-by-burst TC decoding using various coding rates and adaptive modulation when communicating over the Rayleigh fading COST207 TU channel of Figure 12.2. The associated coding parameters are shown in Tables 12.2 and 12.3 and the target BER is 10^{-4}.

In Figure 12.15 we show the BER and BPS performance of burst-by-burst TC decoding using various coding rates and adaptive modulation. The above-mentioned simple procedure was used for determining the switching thresholds. As we can see in Figure 12.15, the coded BER of the AQAM modem using various TC coding rates was maintained at an average of 10^{-4}. If the same fixed switching thresholds were employed for every channel SNR, we would have an undulating coded BER curve, which would clearly deviate from the targeted BER. In Figure 12.15 we also plotted the BPS performance associated with each TC coding rate, which is scaled on the right-hand-side axis. As we can see in Figure 12.15, at low channel SNRs, the schemes having lower TC coding rates will outperform those associated with the higher TC coding rates. For example, TC codes having a rate of $R \leq 0.5$ have a BPS throughput in excess of 0.5 for channel SNRs above 4 dB. However, as we increase the channel SNR, the TC-coded AQAM schemes having higher coding rates outperform their lower coding-rate counterparts. Furthermore, the throughput of the AQAM schemes having lower TC coding rates will saturate faster and at a lower value. For example, the TC code associated with $R = 1/3$ has a maximum BPS throughput of $6 \times 1/3 = 2$.

From Figure 12.15 we can conclude that when the channel SNR is low, lower TC coding rates have to be employed and R should be increased as the channel SNR increases.

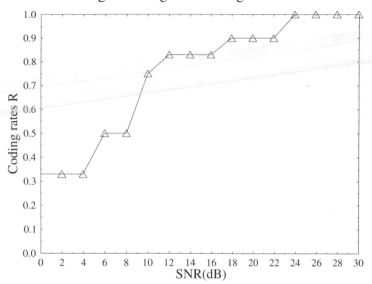

Figure 12.16: TC coding rates activated in conjunction with AQAM for different channel SNRs when communicating over the Rayleigh fading COST207 TU channel of Figure 12.2. The associated coding parameters are shown in Tables 12.2 and 12.3 and the target BER is 10^{-4}.

Based on Figure 12.15, in our adaptive regime for every channel SNR we select the specific TC coding rate which gives the maximum throughput in the context of the TC-coded AQAM system. In Figure 12.16 we portray the TC coding rates selected at the various channel SNRs encountered. As we can see in the figure, the selected TC coding rate increases as the channel SNR increases. For a channel SNR in excess of 24 dB, we can see that the TC coding rate was set to unity, which implies that the BPS throughput of the uncoded AQAM scheme becomes better than that of the TC-coded AQAM scheme. Hence, no TC codes are activated for channel SNRs in excess of 24 dB. Previously in Figure 12.14, we proposed optimising the AQAM switching thresholds individually for every channel SNR for turbo-coded AQAM. Let us now extend our proposal for optimising not only the thresholds, but also the TC coding rates for every channel SNR.

Let us hence apply the joint switching threshold and TC coding-rate optimisation techniques to the TC-coded AQAM system, which we refer to this scheme in our future discourse as ORTC AQAM. In Figure 12.17 the curve marked with diamonds represents the BPS performance of the jointly optimised ORTC AQAM with burst-by-burst decoding. As we can see in the figure, the employment of the ORTC AQAM substantially improves the performance of the system when the channel SNR is low. For example, at 0.5 BPS we observe a 5 dB SNR gain over the uncoded AQAM system. However, as the channel SNR is increased, the SNR gain due to TC coding is reduced and uncoded AQAM will be employed for channel SNRs in excess of 20 dB. For channel SNRs below 13 dB we can observe in Figure 12.17 that the performance of the ORTC AQAM scheme is similar to that of the TC-coded fixed modulation schemes using the long-delay 10^{5}-bit channel interleaver. Hence, with the aid of the ORTC AQAM we can achieve as

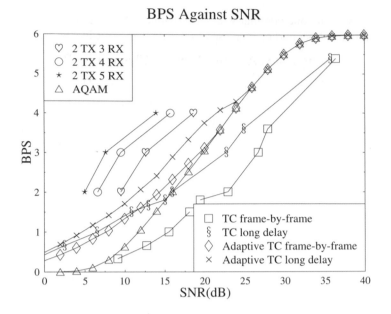

Figure 12.17: BPS performance comparison between STT codes, uncoded AQAM, adaptive-rate TC codes
using burst-by-burst decoding and long-delay channel interleaver both with fixed and adaptive modulation, when communicating over the Rayleigh fading COST207 TU channel of Figure 12.2. The associated coding parameters are shown in Tables 12.1, 12.2 and 12.3 and the target BER is 10^{-4}.

good a performance as the long-delay optimised-rate TC-coded fixed modulation scheme without incurring a long delay. More explicitly, with the aid of the AQAM system we were able to reduce the interleaving delay without compromising the BPS throughput.

On the other hand, if a long delay is tolerable in the context of the ORTC AQAM system, the 10^5-bit channel interleaver can be employed in the system. The corresponding BPS performance is shown in Figure 12.17 by the curve marked with crosses. As we can see in the figure, there is a significant BPS performance improvement over the short-delay burst-by-burst ORTC AQAM system. Observe also in Figure 12.17 that uncoded AQAM will only be employed for channel SNRs in excess of 25 dB. The corresponding performance also approaches that of the STT codes using three receivers. At a throughput of 2 BPS we can see that the long-delay AQAM SNR performance is about 2–3 dB inferior to that of the STT codes using three receivers.

In Figures 12.15 and 12.17, we have characterised the BPS performance of the ORTC AQAM system. Let us now present the corresponding results for the ORTBCH AQAM system and compare them to the ORTC AQAM system. As we can see in Figure 12.18, the performance of the ORTC and ORTBCH codes is similar for the short-delay burst-by-burst turbo-decoded AQAM scheme, which we have also observed in Figure 12.6. However, for lower channel SNRs the performance of ORTBCH AQAM is inferior to that of ORTC AQAM, since in this SNR region lower coding-rate TC codes were employed.

On the other hand, we also show in Figure 12.18 the performance of ORTC and ORTBCH AQAM using large channel interleavers. Similarly to burst-by-burst decoding, the TC codes outperform the TBCH codes for low channel SNRs. However, as the channel SNR is increased, we can see that the TBCH codes have an edge over the TC codes for channel SNRs in excess

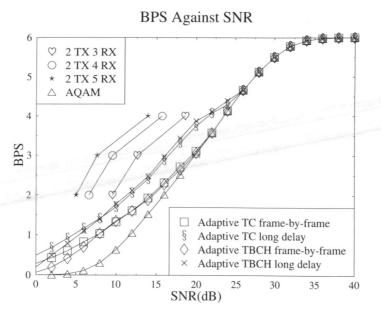

Figure 12.18: BPS performance comparison between STT codes, uncoded AQAM, adaptive-rate TC and
TBCH codes using both burst-by-burst decoding and a long-delay channel interleaver employing fixed and adaptive modulation, when communicating over the Rayleigh fading
COST207 TU channel of Figure 12.2. The associated coding parameters are shown in Tables 12.1, 12.2, 12.3 and 12.4 and the target BER is 10^{-4}.

of 7 dB, although their performance gap is insignificant as compared to those observed in Figure 12.7. Therefore, we conclude that the employment of TBCH codes does not improve the
performance of the AQAM system, although it increases the system's complexity.

12.5   Summary and Conclusions

In this chapter we comparatively studied the performance of STT codes and AQAM schemes.
When multiple transmitters and receivers are employed, a MIMO equaliser has to be used for
equalising the multiple transmitted signals. On the other hand, a SISO equaliser was employed
for the AQAM scheme, which used a single transmitter and single receiver. We then proposed to
concatenate powerful turbo codes with the AQAM system.

The simulation results were presented in Section 12.4. In Section 12.4.1 we investigated the
performance of the system using fixed modulation schemes. The performance of both TC and
TBCH codes was investigated and compared using both short and long channel interleavers. The
length of the short channel interleaver was chosen such that burst-by-burst turbo decoding was
used. The performance of TC codes in conjunction with various coding rates was presented and
we found that the performance of TC codes improved as the coding rate was reduced. Then in
Section 12.4.2 we characterised the performance of 4PSK, 8PSK and 16PSK STT codes.

Finally, in Section 12.4.3 we compiled all the results and compared them to the performance
of the proposed turbo-coded AQAM system. Initially, we showed that the uncoded AQAM system outperforms the STT codes using two receivers. We then showed that uncoded AQAM also

outperforms burst-by-burst TC-coded fixed modulation. However, in conjunction with a large channel interleaver the TC-coded fixed modulation scheme outperforms the uncoded AQAM scheme for channel SNRs below 15 dB at the cost of a high delay. Turbo-coded AQAM system was then proposed and we jointly optimised the AQAM switching thresholds and the coding rates for the turbo-coded AQAM system. The performance of the ORTC AQAM scheme using a short channel interleaver was found to be similar to that of the TC-coded long-delay fixed modulation scheme. Finally, at the expense of a high channel interleaver delay, we were able to further improve the performance of the ORTC AQAM scheme. It was found that its performance was 2 dB inferior in comparison to the STT codes using three receivers. The performance of the ORTBCH AQAM scheme was then compared to that of the ORTC AQAM scheme. It was found that there was no significant improvement over the ORTC AQAM scheme, despite the extra complexity incurred by the system. **In conclusion, ORTC AQAM can be viewed as a lower complexity measure of mitigating the channel quality fluctuations of wireless channels in comparison to multiple transmitter and multiple receiver based space-time codes. If the complexity of the latter schemes is affordable, the performance advantages of AQAM erode.**

In the next chapter our system optimisation efforts evolve further into the field of turbo-coded partial-response modulation, which constitutes the required background for the appreciation of joint, iterative channel equalisation and channel decoding techniques. These joint detection schemes have been termed turbo equalisation, which will be the subject of the next part of the book, commencing with Chapter 13.

Part V

Turbo Equalisation

Chapter 13

Turbo-coded Partial-response Modulation

13.1 Motivation

In recent years there has been an increasing demand for high-rate wireless communications services. However, there are numerous problems associated with high data-rate transmissions and with their employment in high area-spectral-efficiency systems. When transmitting at high bit rates, such as 2 Mbit/s, typically a high grade of channel-induced dispersion is experienced, which inflicts Inter-Symbol Interference (ISI). Hence, the channel equalisers employed must be capable of mitigating the effects of ISI. Furthermore, in order to achieve a high bandwidth efficiency, multilevel linear modulation techniques — such as 16-level Quadrature Amplitude Modulation (QAM) and 64-level QAM [207] — can be invoked. Partial-response modulation techniques — such as Continuous Phase Modulation (CPM) [320] schemes, which include Gaussian Minimum Shift Keying (GMSK) [321] — can also be employed. The above QAM-based systems are susceptible to noise and fading, since the signal constellation points become closer in the constellation space. CPM schemes, which spread the effects of a bit over time in order to decrease the bandwidth of the signals, additionally suffer from intentionally introduced Controlled ISI (CISI). Hence, the equaliser design must also be able to cope with the additional CISI. In conjunction with equalisation, channel decoding can also be employed in order to further improve the performance of the systems. Powerful error correction schemes — such as turbo codes [12] — have been shown to yield performances close to Shannonian performance limits.

To elaborate a little further, the scope of this chapter encompasses a range of equalisation and decoding issues with the aim of enhancing the performance of partial-response modulation schemes, such as the GMSK scheme of the GSM system. Since its standardisation in the mid-1980s, the GSM system has been evolving throughout the past decade. This process has been hallmarked by the introduction of the High-Speed Circuit Switched Data (HSCSD) service [322] and the General Packet Radio Service (GPRS) [323]. Further important developments were the introduction of the 5.6 kbit/s half-rate [324] and the Enhanced Full-Rate (EFR) speech codecs [325]. Turbo-equalised [326] and turbo-coded [327] performance improvements were proposed by Bauch and Franz and Burkert *et al.*, respectively. Motivated by these trends, in this treatise we also set out to develop a range of powerful performance enhancement techniques.

The outline of this chapter is as follows. A brief overview of the mobile radio channel is presented in Section 13.2. In Sections 13.3 and 13.4, the basic Continuous Phase Modulation (CPM) theory and the Digital Frequency Modulation (DFM) theory are presented. Subsequently, the state representation of Minimum Shift Keying (MSK) and Gaussian MSK (GMSK) is discussed in Section 13.5. Section 13.6 presents the spectral performance of MSK and GMSK, while Section 13.7 describes the principles of constructing trellis-based equaliser states. With the motivation of introducing turbo coding into GMSK modulation, in Sections 13.8.3 and 13.8.4 we briefly summarise the Log-Maximum A-Posteriori (Log-MAP) algorithm, before amalgamating it with the soft-output GMSK modem. More in-depth details of the MAP and Log-MAP algorithms can be found in Sections 5.3.3 and 5.3.5, respectively. This is followed by Section 13.8.5, which gives an overview of the complexity associated with employing turbo decoding and convolutional decoding in this context. Finally, in Section 13.8.7 the performance of the turbo-coded GSM system and the convolutional-coded GSM scheme is presented.

13.2   The Mobile Radio Channel

Mobile radio channels have been characterised in depth for example in references [49, 328–330], hence here we restrict ourselves to a rudimentary introduction. A dispersive mobile radio channel has as an impulse response, which exhibits both time-domain delay spread and Doppler frequency-domain spreading [49, 207]. The delay spread introduces time dispersion and, as a consequence of the time–frequency duality, frequency-selective fading as well. Again, owing to duality, the frequency-domain Doppler spread results in time-selective fading [49]. Although all four effects are present in mobile radio channels, their dominance is dependent on the nature of the transmitted signal, i.e. on the modulation technique and Baud rate employed. Hence, the channel can be further categorised based on the dominant effects perceived by the system. We will briefly elaborate on this below.

frequency dispersion and time-selective fading occur when the channel is time-variant. In time-selective fading the frequency-domain response of the channel varies while the signal is being transmitted, hence linearly distorting the signal. Furthermore, owing to the time–frequency duality, frequency dispersion occurs and results in the signal bandwidth being stretched. In other words, the signal bandwidth at the receiver is wider than the transmitted signal's bandwidth owing to the frequency dispersion over the channel. However, if the signal has a high bit rate and a short transmission frame length, the frame may propagate through the channel before any significant changes in channel characteristics take place. This may improve the system's resistance against time-selective fading. However, owing to the reduced bit duration the signal now becomes more susceptible to frequency-selective fading or time dispersion. More explicitly, at high bit rates, the bit period will be shorter than the channel's delay spread in a time-dispersive or frequency-selective channel. Therefore, the effect of the previous transmitted bits will be imposed on the present data bit, hence introducing ISI.

In time-dispersive channels, the transmitted signal is stretched in time, since the signal is convolved with the channel's impulse response, hence prolonging the duration of the signal. This occurs when the transmitted signal, which is reflected and refracted, reaches the receiver along a number of different paths. Each of these signal paths arrives at the receiver with different delays and possesses a different power. As a consequence, the impulse response of a wireless channel can be represented as a series of pulses, as illustrated on the left of Figure 13.1. Frequency-selective fading, portrayed on the right in Figure 13.1, arises, since the Fourier transform of this impulse response is typically a non-flat frequency-domain transfer function. When the signal bandwidth is narrow compared to the channel's so-called coherence bandwidth, B_c — which is

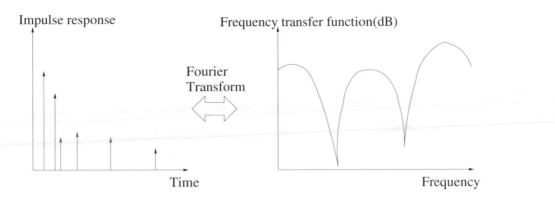

Figure 13.1: Stylised channel impulse response and frequency transfer function of a multi-path channel.

inversely proportional to the delay spread τ_d related to it by $B_c = \frac{1}{2\pi\tau_d}$ [49] — the transmitted frequency components will experience similar attenuations or frequency-flat fading. Conversely, as the delay spread increases, B_c becomes narrower. If the transmission bandwidth exceeds B_c, the transmitted signal's frequency components will begin to encounter different attenuations, unlike in the flat fading scenario described previously. Subsequently, the channel imposes a filtering effect and linearly distorts the waveform, hence resulting in frequency-selective fading. For digital systems having large transmission bandwidths, the receiver will experience distinct time-domain echoes of the previous transmitted signals, resulting in ISI.

For example, in the Global System of Mobile Communications known as GSM [49], GMSK [331] modulation is employed at a transmission rate of 270.833 KBaud, thereby giving a bit duration of 3.69 μs. For this system, the bit duration is short with respect to the typical mobile radio channel delay spread, which can be as high as 18 μs in hilly terrain environment and therefore the signal does not experience time-selective fading and frequency dispersion. Instead, it experiences frequency-selective fading and time dispersion due to its large transmission bandwidth for a static Mobile Station (MS). However, when the MS starts to move, Doppler spreading occurs, since the channel characteristics begin to change more significantly with time. Therefore, the mobile radio channel's impulse response can be modelled as a sequence of impulses, where each impulse is attenuated by values satisfying the Rician or Rayleigh distribution statistics [332]. Under these circumstances, the above-mentioned GMSK signals will be exposed to time dispersion and the receiver will observe contributions from the previous transmitted symbols. Furthermore, the GMSK signal's spectrum will be distorted, since the channel characteristics vary with time as a consequence of Doppler spreading. It is therefore necessary to perform channel equalisation in order to resolve the transmitted signals, which have been degraded by the channel-induced impairments, in addition to the deliberately introduced CISI. Another channel model employed in our simulations is the so-called narrowband fading channel. The corresponding channel impulse response is modelled as a single impulse, which is attenuated by fading values conforming to Rayleigh or Rician statistics.

Having discussed the channel models used, we proceed by describing the basic CPM principles and subsequently present two CPM schemes, namely Minimum Shift Keying (MSK) and Gaussian Minimum Shift Keying (GMSK).

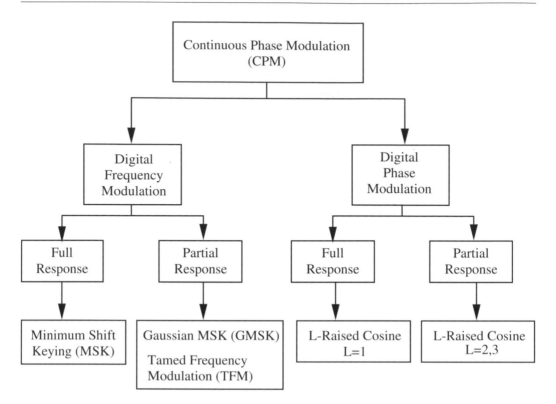

Figure 13.2: Classification tree of Continuous Phase Modulation (CPM) schemes.

13.3 Continuous Phase Modulation Theory

The classification tree of CPM schemes is shown in Figure 13.2. It is observed that there are two subclasses of CPM, namely Digital Frequency Modulation (DFM) and Digital Phase Modulation (DPM). Examples of DPM are L-Raised Cosine (L-RC) modulation, where L denotes the length of spreading in the modulator filter. Therefore, 1-RC is a full response scheme, while 2-RC and 3-RC are examples of partial-response modulation. For an in-depth treatment on the subject of DPM the interested reader is referred to references [49, 320]. In this section, the basic DFM theory, specifically MSK and GMSK, is elaborated.

13.4 Digital Frequency Modulation Systems

In DFM the transmitted radio signal can be described as:

$$s(t, a) = \sqrt{\frac{2E_s}{T}} \cos\left(2\pi f_c t + \phi(t, a) + \phi_c\right), \tag{13.1}$$

where E_s is the energy within the symbol period T. Note that in a binary system a symbol consists of only one bit. Therefore, the terms *symbol* and *bit* have the same meaning. The frequency of the carrier is f_c, while $\phi(t, a)$ is the phase contribution dependent on the information bit a. Also, the term ϕ_c is used to denote the phase offset, which can be set to 0 without loss

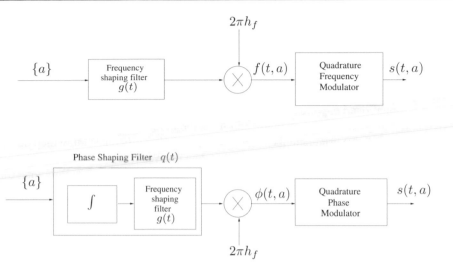

Figure 13.3: An illustration of two possible DFM implementations, where a, $f(t,a)$, $\phi(t,a)$ and $s(t,a)$ are the sequence of uncorrelated bits, the frequency of the DFM signal, the phase of the DFM signal and the modulated DFM signal, respectively. ©John Wiley UK (Ltd), Aulin and Sundberg [333], 1984

of generality. Here a represents an infinitely long and uncorrelated data sequence, which, for M-ary signalling, can be defined as:

$$a \in \big\{\, \pm 1, \pm 3, \ldots, \pm(M-1) \big\}.$$

Equation 13.1 is also often presented as:

$$s(t,a) = \sqrt{\frac{2E_s}{T}} \cos\left[\phi(t,a)\right] \cdot \cos\left[2\pi f_c t + \phi_c\right] - \sqrt{\frac{2E_s}{T}} \sin\left[\phi(t,a)\right] \cdot \sin\left[2\pi f_c t + \phi_c\right],$$

$$(13.2)$$

leading to the so-called quadrature representation, where the in-phase and quadrature-phase carriers are modulated independently by the terms $\cos\left[\phi(t,a)\right]$ and $\sin\left[\phi(t,a)\right]$. The modulated radio signal, $s(t,a)$, can be produced through two different quadrature-component-based implementations, as illustrated in Figure 13.3. The first method applies the input data, a, to a frequency shaping filter, $g(t)$, which constitutes an impulse response that spreads each data bit over a finite number of bit intervals. Subsequently, the output of the filter is multiplied by $2\pi h_f$, where h_f is the modulation index, in order to give the frequency $f(t,a)$ of the DFM signal. The resultant DFM signal $f(t,a)$ is then directed to a frequency modulator, which generates varying harmonics of frequency, depending on the transmitted data bits, thereby producing the DFM signal.

The alternative approach followed in the lower branch of Figure 13.3 is to integrate the frequency shaping filter's impulse response, $g(t)$, prior to the filtering of the input data stream. Subsequently, the response of the filter is multiplied with the scaled modulation-index term $2\pi h_f$, in order to yield the data-dependent phase, $\phi(t,a)$, of the transmitted data bits, which is applied to a phase modulator. Mathematically, this scheme can be described by representing the phase, $\phi(t,a)$, as the integral of the partial-response pulse-shaped data signal, summed over the signalling interval index $-\infty \leq i \leq \infty$ and scaled by $2\pi h_f$, yielding:

$$\phi(t, a) = 2\pi h_f \sum_{i=-\infty}^{+\infty} \int_{-\infty}^{t} a_i \cdot g(\tau - iT) d\tau, \qquad (13.3)$$

where — as mentioned previously — a is an infinitely long sequence of uncorrelated bits, $g(t)$ is the frequency shaping impulse response and h_f is the modulation index. The modulation index, h_f, is a ratio of relative prime numbers and also a proportionality constant, which determines the magnitude of the phase change upon receiving a data bit. We will elaborate further on the choice of h_f in Section 13.5. The phase, $\phi(t, a)$, of a DFM signal in the nth signalling interval can then be expressed in terms of the phase shaping function, $q(t)$ as:

$$\phi(t, a) = 2\pi h_f \sum_{i=-\infty}^{n} a_i \cdot q(t - iT) \qquad \text{for } nT \leqslant t \leqslant (n + 1)T, \qquad (13.4)$$

since:

$$q(t) = \int_{-\infty}^{t} g(\tau) d\tau. \qquad (13.5)$$

In practical implementations the integration of $g(t)$ in Equation 13.5 is performed within the limits of $t = 0$ to $t < LT$. When $t < 0$ and $t \geqslant LT$, we have:

$$q(t) = 0 \qquad \text{for } t < 0 \qquad (13.6)$$

and:

$$q(t) = 0.5 \qquad \text{for } t \geqslant LT, \qquad (13.7)$$

respectively. The frequency shaping function $g(t)$ can be of any arbitrary causal shape having a width of L bit intervals:

$$g(t) = \begin{cases} 0 & \text{for } t < 0 \text{ and } t > LT \\ \text{any arbitrary function} & \text{for } 0 \leqslant t \leqslant LT, \end{cases} \qquad (13.8)$$

where $L = 1$ for full-response systems, while for partial-response schemes we have $L > 1$. Therefore, the phase shaping impulse response, $q(t)$, is also spread over a finite number of bit intervals, L, and the effective output phase $\phi(t, a)$ has contributions from partially overlapping pulses. The intentional introduction of ISI due to partial-response pulse shaping results in a tight spectrum, which reduces the spectral spillage into adjacent channels. However, in order to mitigate the effects of ISI, an equaliser is needed at the receiver, even if the channel is non-dispersive. For the purpose of illustration, in Figure 13.4 we portray a pair of stylised impulse responses, namely $q(t)$ and $g(t)$.

13.5 State Representation

The partial-response DFM signal can be viewed as a signal exhibiting memory. This is the consequence of spreading the response of the filter, $q(t)$, over a finite number of bit intervals. For modulated signals exhibiting memory the optimal form of detection is Maximum Likelihood Sequence Estimation (MLSE) [9, 49]. MLSE-type detection can be employed by observing the

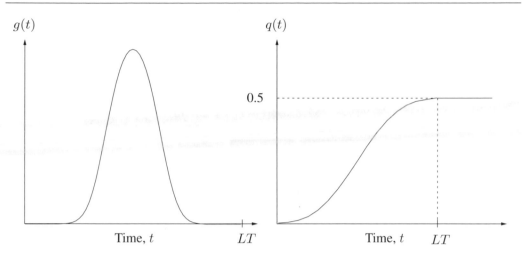

Figure 13.4: Stylised $g(t)$ and $q(t)$ impulse responses. The impulse response $q(t)$ is obtained by integrating $g(t)$, as shown in Equation 13.5.

development of the so-called phase tree over an arbitrary number of symbols, before making a decision on the oldest symbol within that time span. The number of branches in the phase tree increases exponentially with each data bit received. Hence, the number of correlations required, in order to identify the most likely received sequence, will also increase exponentially. However, upon employing a trellis structure or state representation of the DFM system, this exponential increase in complexity can be avoided. In the following discussion, the underlying principles for constructing a DFM trellis structure are established.

By using the relationship expressed in Equations 13.6 and 13.7 and by rearranging Equation 13.4, the phase of the DFM modulated radio signal at instant n can be written as [320]:

$$\phi(t,a) = 2\pi h_f \sum_{i=-\infty}^{n-L} a_i q(t-iT) + 2\pi h_f \sum_{i=n-L+1}^{n} a_i q(t-iT)$$

$$= 2\pi h_f \sum_{i=-\infty}^{n-L} a_i(0.5) + 2\pi h_f \sum_{i=n-L+1}^{n} a_i q(t-iT)$$

$$= \pi h_f \sum_{i=-\infty}^{n-L} a_i + 2\pi h_f \sum_{i=n-L+1}^{n} a_i q(t-iT)$$

$$= \theta_n + \theta(t,a),$$

where:

$$\theta(t, a) = 2\pi h_f \sum_{i=n-L+1}^{n} a_i q(t - iT)$$

$$= 2\pi h_f \underbrace{\sum_{i=n-L+1}^{n-1} a_i q(t - iT)}_{\text{Correlative state vector}} + \underbrace{2\pi h_f a_n q(t - nT)}_{\text{Current bits' effect}}, \tag{13.9}$$

and:

$$\theta_n = \pi h_f \sum_{i=\infty}^{n-L} a_i \tag{13.10}$$

is the *phase state*, which is periodic with respect to 2π and represents the accumulated phase due to the previous bits that have passed through the phase shaping filter. Equation 13.9 describes the relationship between three parameters, namely $\theta(t, a)$, the most recent information bit, a_n, and the so-called *correlative state vector*, which is the first term on the right.

Let us now establish the relationship between the number of phase states needed and the modulation index, h_f. Since the phase state, θ_n, is $\pi h_f \sum_{i=-\infty}^{n-L} a_i$ and recalling that h_f is a ratio of relative prime numbers and θ_n is periodic with respect to 2π, the four phase states for a modulation index $h_f = \frac{1}{2}$ are $0, \frac{\pi}{2}, \pi$ and $\frac{3\pi}{2}$. Similarly, for $h_f = \frac{1}{3}$, there are six phase states, namely $0, \frac{\pi}{3}, \frac{2\pi}{3}, \pi, \frac{4\pi}{3}$ and $\frac{5\pi}{3}$. When the modulation index, h_f, is $\frac{2}{3}$, the three phase states are $0, \frac{2\pi}{3}$ and $\frac{4\pi}{3}$. We observe that for modulation indices which possess even numerators, the number of phase states is equal to the denominator, as in the example of $h_f = \frac{2}{3}$. By contrast, when the numerator of h_f is odd, as in $h_f = \frac{1}{2}$ and $h_f = \frac{1}{3}$, the number of phase states is double that of the denominator. Therefore, we can express the modulation index, h_f, as [320]:

$$h_f = \frac{2 \cdot k}{p}, \tag{13.11}$$

where k and p are two integers with no common factor, such that the denominator, p, determines the number of phase states for the DFM system. For example, for $k = 1$ and $p = 4$ we arrive at $h_f = \frac{1}{2}$ and four phase states. A general expression for the p legitimate phase states , θ_n, is as follows:

$$\theta_n \in \left\{ k \cdot \frac{2\pi}{p} \right\} \qquad \text{for } k = 0 \ldots (p - 1). \tag{13.12}$$

For a full-response DFM system with $L = 1$, i.e. when employing no partial-response spreading, Equation 13.5 becomes:

$$\phi(t, a) = 2\pi h_f a_n q(t - nT) + \pi h_f \sum_{i=-\infty}^{n-1} a_i. \tag{13.13}$$

The first term on the right depends on the most recent information bit a_n and the second term is the phase state θ_n.

In a partial-response system, there are correlative state vector contributions from overlapping pulses, as we have seen in Equation 13.9. The message of Equation 13.9 will be made more explicit below by introducing the so-called trellis structure. By studying Equations 13.5 and 13.9, it is observed that the overall phase is dependent on the correlative state vector, the phase state, θ_n, and the latest data bit, a_n. Therefore, we define the trellis state, S_n, as a combination of the correlative state vector and phase state, θ_n:

$$S_n = \{\theta_n, \text{Correlative state vector}\},$$

and we will illustrate this relationship in Figure 13.5. However, since the correlative state vector relies on the information bits $a_{n-1}, a_{n-2}, \ldots, a_{n-L+1}$, as seen in Equation 13.9, the trellis state S_n in the partial-response DFM system at time nT can be defined as:

$$S_n \triangleq \{\theta_n, a_{n-1}, a_{n-2}, \ldots, a_{n-L+1}\}. \tag{13.14}$$

For a modulation index of $h_f = \frac{2k}{p}$ and M-ary signalling, where $\upsilon = \log_2 M$ number of bits are input simultaneously to the modulator, there are p number of phase states and M^{L-1} correlative state vectors for each phase state. Hence, at time nT the total number of states is $2^{\upsilon(L-1)}$. Upon receiving an information bit a_n within the interval nT and $(n+1)T$, the state S_n changes to S_{n+1}, depending on the value of a_n. By using the state representation established in Equation 13.14, S_{n+1} becomes:

$$S_n \xrightarrow{a_n} S_{n+1} \tag{13.15}$$
$$(\theta_n, \{a_{n-1}, a_{n-2}, \ldots, a_{n-L+1}\}) \xrightarrow{a_n} (\theta_{n+1}, \{a_n, a_{n-1}, \ldots, a_{n-L+2}\}),$$

where the phase state of Equation 13.12 at time $(n+1)T$ is given by:

$$\theta_{n+1} = [\theta_n + \pi h_f a_{n-L+1}] \qquad \text{modulo } 2\pi, \tag{13.16}$$

since the phase state is accumulative.

Let us consider the example of a binary partial-response DFM system spread over $L = 2$ bit intervals and with a modulation index of $h_f = \frac{2 \cdot k}{p} = \frac{1}{2}$, where k and p in Equation 13.11 will be 1 and 4, respectively. Note that the value of the modulation index h_f in this system is not necessarily an attractive setting. It was only chosen to illustrate the principles discussed previously. In this DFM system the legitimate phase states, θ_n, are as follows:

$$\theta_n \in \left\{0, \frac{\pi}{2}, \pi, \frac{3\pi}{2}\right\}.$$

For each phase state θ_n in the trellis, there are 2^{L-1} associated correlative state vectors, where L indicates the extent of the spreading. Hence, for pulses spread over two bit intervals, there will be a total of $4 \cdot 2^{2-1} = 8$ states in the trellis. The transitions from each of these states are illustrated in Figure 13.5 and governed by Equation 13.15. Specifically, for the trellis state $S_n = (\theta_n, a_{n-1}) = (0, 1)$ at the top of Figure 13.5, there are two legitimate trellis transitions, depending on the current incoming bit a_n. Note that the broken lines indicate transitions due to $a_n = +1$, while the continuous lines represent transitions due to $a_n = -1$. Hence, for each state S_n there are two legitimate trellis transitions, and similarly, there are two transitions merging in each state S_{n+1}. We note at this early stage that the MLSE receiver technique [9,49] is efficient in partial-response systems, since certain transitions are illegitimate, such as a transition from state $(0, 1)$ to $(0, 1)$ in Figure 13.5, hence allowing the receiver to eliminate the associated estimated received bits from its decisions.

In summary, DFM systems can be viewed as a form of Frequency Shift Keying (FSK) with a constant signal envelope and phase continuity at bit intervals and boundaries. They may also

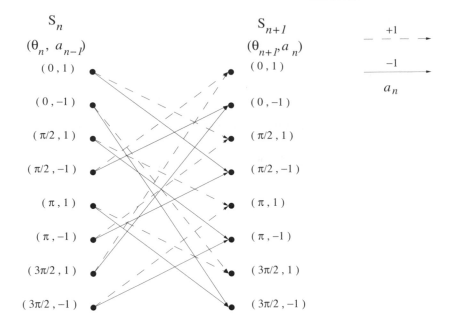

Figure 13.5: The states in a partial response DFM trellis system. The length of spreading is $L = 2$ and the modulation index used is $h_f = \frac{1}{2}$.

be further classified as full-response or partial-response systems, depending on the extent of spreading, L. For full-response systems we have $L = 1$, while partial response schemes have $L > 1$. Spreading the frequency shaping function, $g(t)$, over a higher number of bit intervals improves the spectral efficiency, but at the expense of higher complexity at the demodulator due to the increased contribution of ISI. As a consequence of the spreading in the phase and frequency shaping functions, the transmitted signals exhibit memory, since the bits in the data sequence are interdependent. In this situation, the optimum form of detection is MLSE [9, 49]. By constructing the DFM demodulator as a system of trellis states, the complexity incurred in implementing MLSE is reduced in comparison to a demodulator with a phase tree structure. We now proceed by describing two DFM systems, namely minimum shift keying [334, 335] and Gaussian minimum shift keying [321], in more detail.

13.5.1   Minimum Shift Keying

The basic theory behind Minimum Shift Keying (MSK) can be described through a brief discussion of two other modulation schemes, Quadrature Phase Shift Keying (QPSK) and Offset Quadrature Phase Shift Keying (OQPSK) [86].

QPSK is a modulation technique, which imposes a set of patterns on the phase of the carrier signal, depending on the incoming data bits. Figure 13.6(a) is an illustration of a QPSK modulator. The binary stream seen at the left of the figure is demultiplexed into odd and even index data bits, which subsequently modulate a cosine and sine carrier, hence giving:

$$s(t) = \frac{1}{\sqrt{2}} a_I \cos\left(2\pi f_c t + \frac{\pi}{4}\right) + \frac{1}{\sqrt{2}} a_Q \sin\left(2\pi f_c t + \frac{\pi}{4}\right), \qquad (13.17)$$

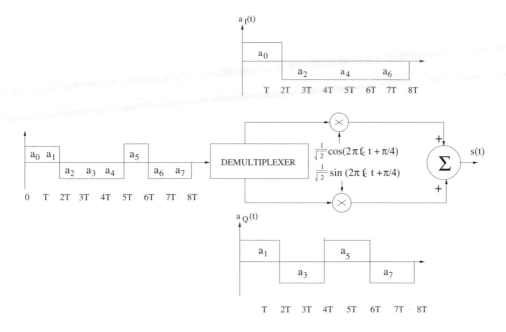

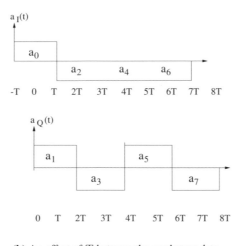

(a) A QPSK modulator. The data streams in quadrature are time-synchronous.

(b) An offset of T between the quadrature data streams is introduced.

Figure 13.6: The difference in bit-stream alignment for QPSK and OQPSK is highlighted. Owing to the staggering of data streams in OQPSK, the large phase difference of π radians is avoided. ©IEEE, Pasupathy [335], 1979

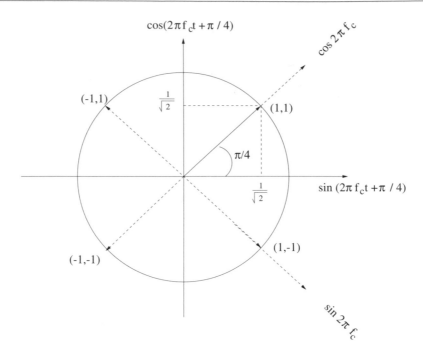

Figure 13.7: Signal space diagram for QPSK and OQPSK. The possible phase transitions for QPSK are 0, $\frac{\pi}{2}$, $-\frac{\pi}{2}$ and π radians, whereas in OQPSK the possible phase changes are 0, $\frac{\pi}{2}$ and $-\frac{\pi}{2}$ radians.

where f_c is the carrier frequency, while a_I and a_Q are the demultiplexed bits in the in-phase and quadrature arms, respectively.

In QPSK, the data bits of both quadrature arms are aligned and they are clocked synchronously. Therefore, the phase can only change once at every interval of $2T$. From Equation 13.17, the signal space diagram of Figure 13.7 can be constructed, in order to illustrate the possible phase transitions at every two-bit interval. The signal space diagram shows a phase transition of $\pm\frac{\pi}{2}$ radians when only one of the data bits in the quadrature arms changes, while a transition of π radians results when both information bits change simultaneously. In OQPSK, an offset of one bit period, T, is introduced between the two binary quadrature streams. As a consequence, only one of the quadrature bits can change in every bit interval, T, hence preventing the phase difference of π radians between consecutive phase values. Consequently, the large and abrupt signal envelope swing traversing through the origin of the coordinate system of Figure 13.7 is avoided. Hence, the bandwidth required is reduced, as compared to that of QPSK. This large amplitude swing in QPSK would result in substantial out-of-band emission, unless a perfectly linear amplifier is used [207].

The shift in time alignment between the quadrature arms does not produce different power spectra. Therefore, OQPSK and QPSK have the same power spectral density. However, when band limiting and signal envelope hard limiting are performed, both systems respond differently. Upon band limiting, the QPSK signal envelope becomes more distorted and at phase transitions of π radians the amplitude of the signal changes abruptly from unity to zero and back again, during the transition from $(1,1)$ to $(-1,-1)$ in Figure 13.7. Consequently, owing to hard limiting by finite-dynamic amplifiers, for example, large and rapid changes in phase are produced. Hence, the high-frequency components, which were removed by band limiting, are regenerated.

For OQPSK, band limiting also distorts the signal. However — in contrast to QPSK — the signal amplitude no longer falls to zero, since the phase transition of π radians has been prevented. This is because the transitions from any phasor position in Figure 13.7 are now limited to the adjacent phasors, hence limiting the envelope swing. Therefore, when the signal envelope is hard-limited, the rate of signal change is limited, thereby band limiting the high-frequency components. The improvement achieved in spectral efficiency through the minimisation of the phase difference suggests that further reduction in out-of-band emission and adjacent channel interference can be obtained if phase continuity is preserved at bit intervals. This is the motivation which led to the development of Minimum Shift Keying (MSK) [334, 335].

MSK can be viewed as OQPSK, but with the in-phase and quadrature-phase components shaped by a half-cycle sinusoid, as highlighted below. The MSK modulated signal can therefore be expressed as:

$$s(t) = \frac{1}{\sqrt{2}} a_I \cos\left(\frac{\pi t}{2T}\right) \cos\left(2\pi f_c t\right) + \frac{1}{\sqrt{2}} a_Q \sin\left(\frac{\pi t}{2T}\right) \sin\left(2\pi f_c t\right)$$

$$= \frac{1}{\sqrt{2}} \cos\left(2\pi f_c t + b_k \frac{\pi t}{2T} + \phi_k\right)$$

$$= \frac{1}{\sqrt{2}} \cos\left(2\pi t \left[f_c + \frac{b_k}{4T}\right] + \phi_k\right),$$

where:

$$\phi_k = \begin{cases} \pi \text{ radians} & \text{for } a_I = -1 \\ 0 \text{ radians} & \text{for } a_I = +1, \end{cases}$$

and:

$$a_I, a_Q = \pm 1, \, b_k = -a_I \cdot a_Q.$$

Equation 13.5.1 shows that MSK has a constant envelope and signalling frequencies of $f_c + \frac{1}{4T}$ and $f_c - \frac{1}{4T}$. The difference between the signalling frequencies is $\frac{1}{2T}$, which is half the bit rate and also equal to the minimum frequency spacing required for two signals to be so-called coherently orthogonal [86].

MSK can also be implemented as a DFM system by setting $g(t)$ in Figure 13.3 according to:

$$g(t) = \begin{cases} \frac{1}{2T} & \text{for } 0 \leqslant t \leqslant T \\ 0 & \text{for } t < 0 \text{ and } t > T. \end{cases} \tag{13.18}$$

The impulse response $g(t)$ yields $q(t)$ in Figure 13.8(b) by using the relationship in Equation 13.5. MSK is a full-response binary system with a modulation index of $h_f = \frac{1}{2}$. A value of $h_f = \frac{1}{2}$ is employed in order to ensure that the minimum frequency spacing of $\frac{1}{2T}$ — which is required for two signals to be coherently orthogonal — can be achieved so that orthogonal detection can be implemented. We note, however, that the rectangular $g(t)$ time-domain pulse shaping function of Figure 13.8 implies a sinc-shaped spectrum extending theoretically over an infinite bandwidth and hence resulting in substantial spectral side-lobes, which will be illustrated at a later stage in Figure 13.14. As described in Section 13.5, the states in the trellis structure are represented by the four phase states, $0, \frac{\pi}{2}, \pi, \frac{3\pi}{2}$. Assuming that the initial phase of the system

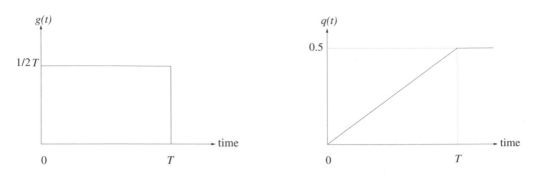

(a) The impulse response of the MSK frequency shaping filter g(t).

(b) The corresponding MSK impulse response of the phase shaping filter, q(t).

Figure 13.8: The MSK impulse response $q(t)$ is evaluated by integrating $g(t)$.

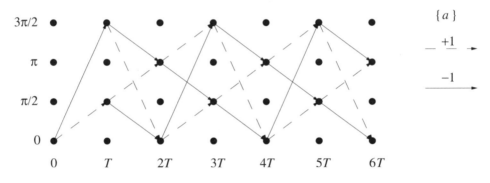

Figure 13.9: Stylised MSK trellis constructed with four phase states, 0, $\frac{\pi}{2}$, π and $\frac{3\pi}{2}$.

is 0, a data bit of $+1$ would cause the overall phase, $\phi(t, a)$, to change linearly to $\frac{\pi}{2}$. Meanwhile, a logical -1 will result in a transition from 0 to $-\frac{\pi}{2}$. Although $-\frac{\pi}{2}$ is not one of the four states in the trellis, an equivalent representation would be a transition from state 0 to $\frac{3\pi}{2}$, since the phase state is periodic with respect to 2π. By using the relationships introduced in Equations 13.15 and 13.16, the trellis diagram in Figure 13.9 can be constructed. Let us now consider how the undesirable spectral side-lobes of MSK can be reduced by invoking a partial-response pulse shaping technique in Gaussian minimum shift keying.

13.5.2 Gaussian Minimum Shift Keying

Gaussian Minimum Shift Keying (GMSK) [321] is a form of partial-response DFM, which incorporates a so-called Gaussian-shaped pulse in Figure 13.10(a) as the impulse response for $g(t)$. A so-called Non-Return To Zero (NRZ) data stream of unity amplitude can be convolved with a Gaussian low-pass filter, whose impulse response is [49]:

$$h_t(t) = \sqrt{\frac{2\pi}{\ln}} 2B_b \exp\left(-\frac{2\pi^2 B_b^2}{\ln} 2t^2\right) \tag{13.19}$$

to give [336]:

$$g(t) = \frac{1}{2T} \left[Q \left(2\pi B_b \frac{t - T/2}{\sqrt{\ln 2}} \right) - Q \left(2\pi B_b \frac{t + T/2}{\sqrt{\ln 2}} \right) \right] \qquad 0 \leqslant B_b T \leqslant \infty, \qquad (13.20)$$

where the parameter B_b is the bandwidth of the Gaussian low-pass filter, and the Gaussian Q-function, $Q(x)$ [337], is defined as:

$$Q(x) \triangleq \frac{1}{\sqrt{2\pi}} \int_x^\infty \exp \left(-\frac{v^2}{2} \right) dv. \qquad (13.21)$$

For a bit-stream having a period of T, the normalised bandwidth, B_n, can be expressed as:

$$B_n = B_b T.$$

The impulse response, $g(t)$, is illustrated in Figure 13.10(a) for different values of normalised bandwidth, B_n. It is observed that GMSK with $B_n = \infty$ is the full-response MSK system. In order to implement a GMSK modulator, the impulse response $g(t)$ must be symmetrically truncated to L bit periods. For B_n of 0.5 the impulse response, $g(t)$, has a pulse width of approximately two bit periods. Therefore, $g(t)$ may be truncated to two bit intervals, i.e. $L = 2$. With different limits of truncation depending on the normalised bandwidth B_n, the corresponding impulse response $q(t)$ in Figure 13.10(b) is obtained using the integrals of Equation 13.5. Typically, for partial-response GMSK, the filter impulse response $g(t)$ is truncated by $\lfloor \frac{1}{B_n} \rfloor$ bit periods, where $\lfloor \cdot \rfloor$ represents the nearest integer floor. For example, when $B_n = 0.3$, $g(t)$ will be truncated to $\lfloor \frac{1}{B_n} = 3.33 \rfloor = 3$ bit periods.

The objective of introducing a Gaussian filter is to prevent the instantaneous changes in frequency, which exist in MSK, when switching from $f_c + \frac{1}{4T}$ to $f_c - \frac{1}{4T}$. Hence, as depicted in Figure 13.14 in conjunction with Gaussian filtering, the power spectrum obtained is more tight and has lower side-lobes than MSK. Further improvement in spectral performance is achieved by spreading $q(t)$ over L bit periods. However, the spreading also introduces Controlled Inter-Symbol Interference (CISI), since the resulting phase in each bit interval is now dependent on the previous $L-1$ bits, as well as on the current bit. The resulting phase tree is shown in Figure 13.11 for a spreading over two bit intervals. We note that the expected $\pm\frac{\pi}{2}$ phase development in this example occurs over $2T$. However, these phase changes are not explicitly observable in the figure because of the superposition of consecutive phase trajectory changes.

Partial-response DFM signals can be equalised by constructing the GMSK modulator as a system consisting of a finite number of states and subsequently implementing trellis-based MLSE equalisers such as the Viterbi equaliser [49]. For a GMSK modulator with a modulation index of $h_f = \frac{1}{2}$ and a spreading over L bit intervals, there are four phase states, namely 0, $\frac{\pi}{2}$, π and $\frac{3\pi}{2}$, and 2^{L-1} correlative states per phase state, giving a total of $4 \cdot 2^{L-1}$ states in the trellis structure. The possible transitions are governed by the relationships established in Equations 13.15 and 13.16, and the resulting state representation for the modulator is equivalent to the trellis structure illustrated in Figure 13.5. Each transition corresponds to a particular quadrature-component baseband signal waveform. Therefore, a library of legitimate signal waveforms can be stored in a look-up table for waveform comparison with the received signal. For a binary DFM signal, the number of signal waveforms that must be stored is twice the total number of states in the trellis.

Both the full-response MSK and partial-response GMSK systems have been discussed. MSK can be viewed as a DFM system with phase continuity but frequency discontinuity at bit boundaries or as an OQPSK system, where consecutive phase tree transitions are shaped by a half-cycle sinusoid. By preventing phase discontinuities at bit boundaries, the power spectrum obtained has

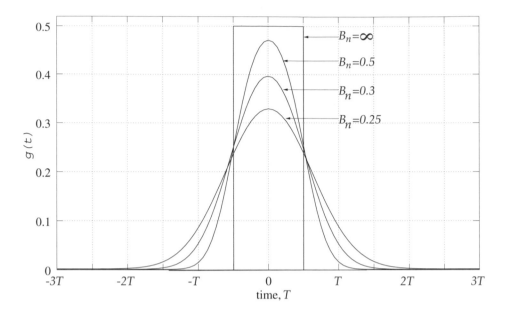

(a) Frequency-shaping impulse response, $g(t)$, for different values of B_n.

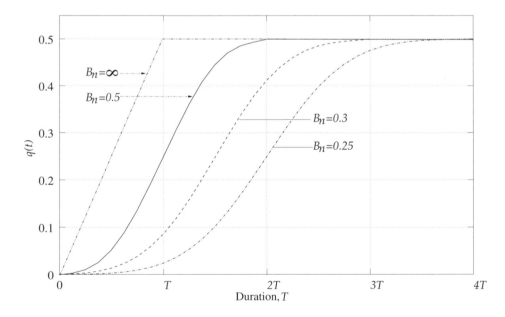

(b) The corresponding pulse-shaping impulse response, $q(t)$, is obtained by integrating $g(t)$ after symmetrically truncating $g(t)$, to L bit intervals. The origin in the time axis, is used as a reference point and represents the point of truncation.

Figure 13.10: Plots of various $g(t)$ and $q(t)$ GMSK shaping functions. The normalised bandwidth, B_n, determines the extent of the partial-response spreading. In practice, $g(t)$ is symmetrically truncated to $\lfloor \frac{1}{B_n} \rfloor$ bit intervals, where $\lfloor \cdot \rfloor$ denotes the closest integer floor.
Wiley & Sons Ltd, Steele and Hanzo [49], 1999

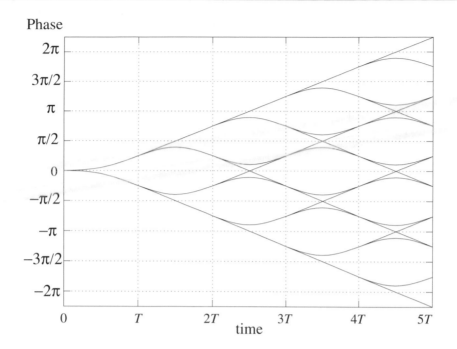

Phase

Figure 13.11: GMSK phase tree, where the impulse response $g(t)$ is spread over two bit intervals. The modulation index used is $\frac{1}{2}$.

lower side-lobes compared to OQPSK and QPSK. However, at bit transitions, there are instantaneous changes in frequency. In GMSK, a Gaussian filter is employed, in order to prevent these instantaneous changes in frequency. Furthermore, by spreading the impulse response of the filter over a finite number of bit intervals, a more tight spectrum than that of MSK is obtained. The improved spectral performance is, however, achieved at the expense of considerable CISI. Hence, equalisation must be performed at the demodulator, even in the absence of channel-induced dispersion. Let us now consider the spectral-domain properties of these signals in more detail in the next section.

13.6 Spectral Performance

With increasing demand for spectrum, the spectral occupancy of the modulated signal becomes an important consideration in the design of modulation schemes. In the following discussion, the spectral characteristics of the modulated signals considered in this treatise are evaluated by computing the power spectral density and the fractional out-of-band power, P_{ob}.

13.6.1 Power Spectral Density

The (PSD) power spectral density of the modulated signal is a measure of the distribution of power with respect to the frequency. Here, the so-called Welch method [338] is employed in order to estimate the spectra of the DFM signals, which are characterised as random processes. In these experiments a pseudo-random data sequence was directed into the MSK and GMSK

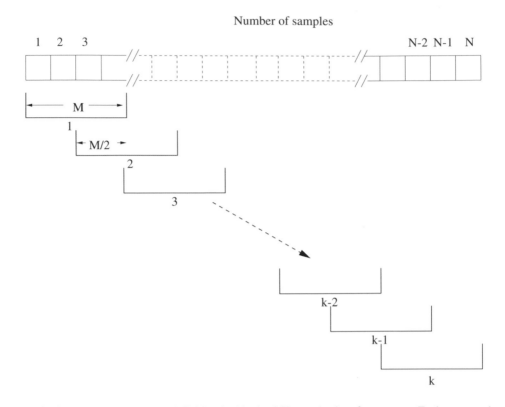

Figure 13.12: The Welch method subdivides the block of N samples into k segments. Each segment has a length M and overlaps with the successive segments by 50%.

modulator, which produced the in-phase and quadrature-phase baseband samples. Before transformation in the frequency domain, the time-domain quadrature-phase samples of the modulated signals were decomposed into k number of segments, which were overlapped by 50% between successive segments, as shown in Figure 13.12. Each segment of M samples was then multiplied with a time-domain window function. Different window functions yield subtle enhancements and degradations in terms of the sharpness of the peaks or narrowness of the computed spectral estimation. Since the purpose of this experiment is to compare the PSD of different modulation schemes, the specific choice of window function is of less significance, if the PSD of these schemes is evaluated with the same window. For our investigations, the Bartlett window [338] was chosen, which is shown in Figure 13.13 along with other commonly used windows. The Bartlett window can be expressed as:

$$w_j = 1 - \left| \frac{j - \frac{1}{2}M}{\frac{1}{2}M} \right|, \qquad \text{for } 0 \leq j \leq M, \tag{13.22}$$

where j is the sampling index. The so-called periodogram in Equation 13.23 is an estimate of the PSD, which is computed by applying the Fast Fourier Transform (FFT) to the windowed segments having M samples, in order to obtain $X(f)$. Subsequently, the magnitude of $X(f)$ is

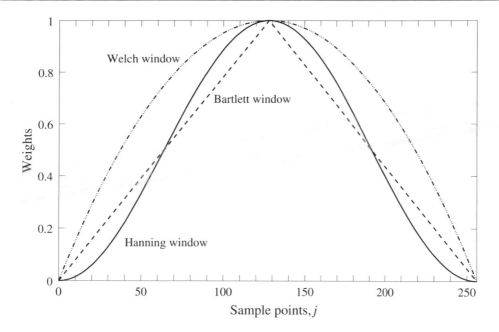

Figure 13.13: Commonly used window functions for computing the PSD of the modulated signal.

squared and normalised.

$$\text{Periodogram} = \frac{1}{M} |X(f)|^2. \tag{13.23}$$

All the k periodograms are averaged, in order to give an estimate of the PSD. A detailed mathematical treatment of PSD estimation can be found in reference [338]. The spectral density plot for MSK and GMSK having different normalised bandwidths, B_n, in Figure 13.14, was obtained by subdividing the data samples into $k = 100$ segments, each containing $M = 256$ samples.

From Figure 13.14, it is observed that the power spectrum of MSK contains significant side-lobe leakage compared with the GMSK spectra. This is due to the instantaneous changes in frequency at symbol interval boundaries. In GMSK, these rapid frequency transitions are prevented by the use of a Gaussian filter, hence giving a spectrum with lower side-lobes. For $B_n = 0.5$ the impulse response of the filter, $g(t)$, is extended approximately over two symbol periods. As a consequence, a more compact spectrum is obtained. Further reduction in the normalised bandwidth, B_n, produces power spectra, which occupy a further reduced bandwidth. However, this is at the expense of considerable CISI.

In Figure 13.14, the power spectrum for the modulated GMSK signal with $B_n = 0.3$ and $B_n = 0.5$ exhibits a bend at a normalised frequency of 1.2 and 1.6, respectively. This is due to the time-domain truncation of the impulse response $g(t)$ [320], which corresponds to multiplying $g(t)$ by a rectangular window, and hence equivalent to convolving the frequency-domain transfer function $G(f)$ with the sinc-shaped window spectrum. Note that $G(f)$ is obtained through the Fourier transform of the impulse response $g(t)$.

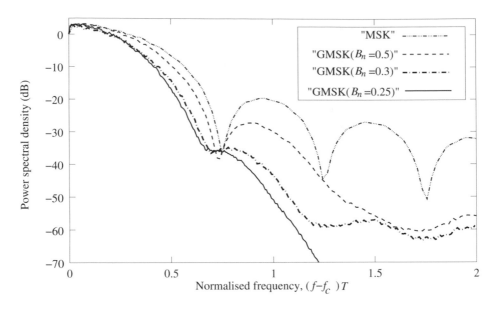

Figure 13.14: Normalised PSD for MSK and GMSK for different normalised bandwidths, B_n.

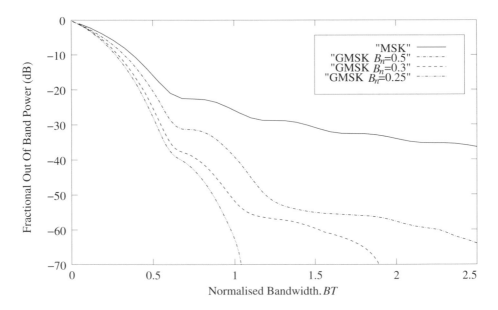

Figure 13.15: The out-of-band power for MSK and GMSK for different normalised bandwidths.

CSV$_{\text{EQ}}$ in non-dispersive channels	CSV$_{\text{EQ}}$ in truncated channel $h_w(t)$ of length L_w
$a_{n-1}, \cdots, a_{n-L+1}$	$a_{n-1}, \cdots, a_{n-L+1}, \cdots, a_{n-L-L_w+1}$

Table 13.1: Correlative State Vector (CSV$_{\text{EQ}}$) used over non-dispersive channels and dispersive radio channels with delay spread L_w at time nT.

13.6.2 Fractional Out-of-Band Power

The fractional out-of-band power, P_{ob}, represents the power in the region exceeding a particular bandwidth, B. An expression for P_{ob} is given by:

$$P_{ob}(B) = \frac{\int_{-\infty}^{\infty} S(f)df - \int_{-B}^{B} S(f)df}{\int_{-\infty}^{\infty} S(f)df}$$

$$= 1 - \frac{\int_{-B}^{B} S(f)df}{\int_{-\infty}^{\infty} S(f)df}.$$

The function $P_{ob}(B)$ for MSK and GMSK in Figure 13.15 was obtained by numerically integrating the PSDs in Figure 13.14. It shows the fraction of power — expressed in dB — that falls outside the range $-BT \leqslant fT \leqslant BT$, where B is a variable. As expected, the P_{ob} for MSK is much higher owing to the side-lobe leakage in the power spectrum. In GMSK at $B_n = 0.25$, the fraction of power leakage is -70 dB for a normalised frequency of 1.0, while MSK has a greater power leakage of -25 dB at the same normalised frequency. Since the power spectra for GMSK associated with $B_n = 0.5$ and $B_n = 0.3$ suffer from an anomalous bend due to the time-domain truncation of $g(t)$, the function P_{ob} for these systems also displays an irregularity. This occurs at a normalised frequency of approximately 1.1 and 1.3 for GMSK associated with $B_n = 0.5$ and $B_n = 0.3$, respectively.

Having described the fundamental principles and properties of the DFM schemes, specifically for MSK and GMSK, we now proceed by describing the basic principles of constructing trellis-based equaliser states.

13.7 Construction of Trellis-based Equaliser States

In order to describe the construction of the associated equaliser trellis states, we must first extend the principles of modulator state representation, which were introduced in Section 13.5. When the signal is transmitted over a Gaussian channel, a trellis-based equaliser has the same state representation as the modulator. However, in dispersive wideband radio channels the trellis-based equaliser must consider a larger set of possible signals, in order to remove the effects of the ISI introduced by the channel's multi-path components. Hence, more states will be needed in the trellis-based equaliser in order to perform the equalisation process. We will show later how each of these trellis-based equaliser states, which incorporate information needed to mitigate the CISI introduced by the modulator filter and to remove the channel-induced ISI, can be decomposed into L_w number of modulator states, in order to regenerate the received signal estimates corresponding to a trellis-based equaliser state transition.

Let us introduce the notations CSV$_{\text{mod}}$ and CSV$_{\text{EQ}}$ to represent the Correlative State Vector (CSV) of the modulator and the trellis-based equaliser, respectively. The states in the trellis-based equaliser are represented by the phase state, θ_n and the correlative state vector CSV$_{\text{EQ}}$,

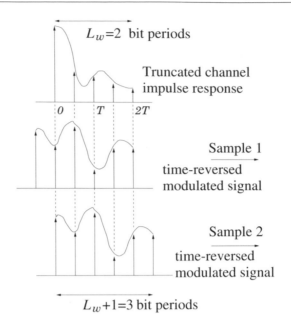

Figure 13.16: The truncated channel impulse response is over $L_w = 2$ bit periods, while the modulated signal extends over $L_w + 1 = 3$ bit periods. We obtain the regenerated signal by convolving the modulated signal with the truncated channel impulse response, in order to get the estimate of the received signal over one bit period. An oversampling by a factor of two was used in this example.

as shown in Equation 13.14. Over dispersive multi-path channels, the number of phase states remains the same. However, more data bits are required in order to represent the CSV_{EQ} and it is dependent on $v_w = L + L_w$, which is the sum of the spreading interval L in the modulator and the length of the channel window, L_w. Table 13.1 highlights the number of bits needed in the CSV_{EQ} for non-dispersive channels and dispersive wideband radio channels. At this stage, we have established that the CSV in the demodulator must be represented with the aid of L_w additional data bits, namely bits $a_{n-L}, \ldots, a_{n-L-L_w+1}$, when compared to the trellis-based equaliser designed for non-dispersive channels. The additional information is needed in order to mitigate the effects of the ISI introduced by the dispersive wideband channel.

We will now show explicitly how the information in the trellis-based equaliser state is used to regenerate the estimated received signal, which is then compared with the received signal in order to give the branch metric. At each trellis state, the phase state θ_n and the correlative state vector CSV_{EQ} are used to reproduce the GMSK-modulated signals over $L_w + 1$ bit periods. As illustrated in Figure 13.16, the modulated signal over $L_w + 1$ bit periods is convolved with the estimated channel impulse response of length L_w, hence giving the expected received signal, which extends over a single bit period. There is sufficient information stored in each trellis state in order to regenerate the GMSK-modulated signal over $L_w + 1$ bit periods, because each equaliser trellis state describes $L_w + 1$ modulator states, as depicted in Figure 13.17. Since the transition between each of these modulator states releases a modulated signal transition over one bit period, we obtain a resultant modulated signal, which extends over $L_w + 1$ bit periods. As mentioned previously, the estimated modulated signal is then convolved with the measured channel impulse response in order to give an estimate of the received signal.

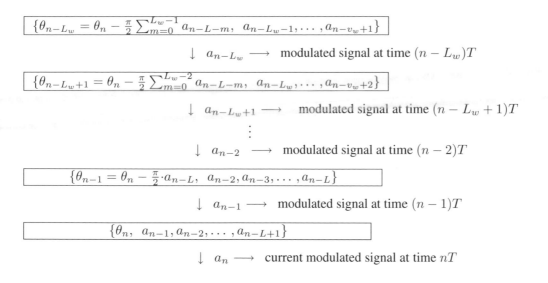

$$\{\theta_{n-L_w} = \theta_n - \tfrac{\pi}{2} \sum_{m=0}^{L_w-1} a_{n-L-m}, \ a_{n-L_w-1}, \dots, a_{n-v_w+1}\}$$

$$\downarrow \ a_{n-L_w} \longrightarrow \text{modulated signal at time } (n-L_w)T$$

$$\{\theta_{n-L_w+1} = \theta_n - \tfrac{\pi}{2} \sum_{m=0}^{L_w-2} a_{n-L-m}, \ a_{n-L_w}, \dots, a_{n-v_w+2}\}$$

$$\downarrow \ a_{n-L_w+1} \longrightarrow \text{modulated signal at time } (n-L_w+1)T$$

$$\vdots$$

$$\downarrow \ a_{n-2} \longrightarrow \text{modulated signal at time } (n-2)T$$

$$\{\theta_{n-1} = \theta_n - \tfrac{\pi}{2} \cdot a_{n-L}, \ a_{n-2}, a_{n-3}, \dots, a_{n-L}\}$$

$$\downarrow \ a_{n-1} \longrightarrow \text{modulated signal at time } (n-1)T$$

$$\{\theta_n, \ a_{n-1}, a_{n-2}, \dots, a_{n-L+1}\}$$

$$\downarrow \ a_n \longrightarrow \text{current modulated signal at time } nT$$

Figure 13.17: Each of the equaliser trellis state $S_n = \{\theta_n, a_{n-1}, \dots, a_{n-L+1}, \dots, a_{n-L-L_w+1}\}$ can be broken down into $L_w + 1$ modulator states, in order to regenerate the modulated signal $s(t, \bar{a})$, which extends over $L_w + 1$ bit periods.

Next, we describe the steps taken to decompose each trellis-based equaliser state into $L_w + 1$ modulator states. First, we determine the CSV_{mod} for each modulator state and then the corresponding phase states. The current CSV_{mod} can be obtained by choosing the first $L - 1$ bits in CSV_{EQ}. Subsequently, we skip the first bit in CSV_{EQ} and select the next $L - 1$ bits in order to determine the previous CSV_{mod}. This procedure is essentially a sliding window process, where the width of the window is $L - 1$ bits and the window is shifted by one bit position to the right each time we want to extract the next CSV_{mod} for the previous modulator states. This is illustrated in Figure 13.18.

For example, let the trellis state S_n be $\{\theta_n, a_{n-1}, \dots, a_{n-L+1}, \dots, a_{n-L-L_w+1}\}$.

The current CSV_{mod} at instant nT is $a_{n-1}, a_{n-2}, \dots, a_{n-L+1}$ and by sliding the window one bit position to the right, it will encompass the $(n-1)$th CSV_{mod}, which consists of the data bits $a_{n-2}, a_{n-3}, \dots, a_{n-L}$. By repeating this sliding window operation, we are able to identify the L_w+1 CSV_{mod}s. At the same time, we are capable of determining the previously transmitted bits from each modulator state since for each CSV_{mod}, the first bit on the left is the previous bit transmitted. Therefore, if the CSV_{mod} is $-1, +1, +1$, the previous data bit was a logical -1.

We have described how the CSV_{mod} can be obtained from CSV_{EQ}. Now we have to determine the modulator phase states for each of these CSV_{mod}s, in order to complete the description of the modulator states and hence to regenerate the modulated signal over $L_w + 1$ bit intervals. This estimated modulated signal is then convolved with the channel impulse response, in order to generate the estimated received signal. The phase state θ_n in the trellis-based equaliser state is also the current modulator phase state. Since the current modulator phase state, θ_n, is mathematically expressed as $\theta_{n+1} = [\theta_n + \pi h_f a_{n-L+1}]$ modulo 2π, in Equation 13.16, we can rearrange this equation to give:

$$\theta_n = \theta_{n+1} - \pi h_f a_{n-L+1}. \tag{13.24}$$

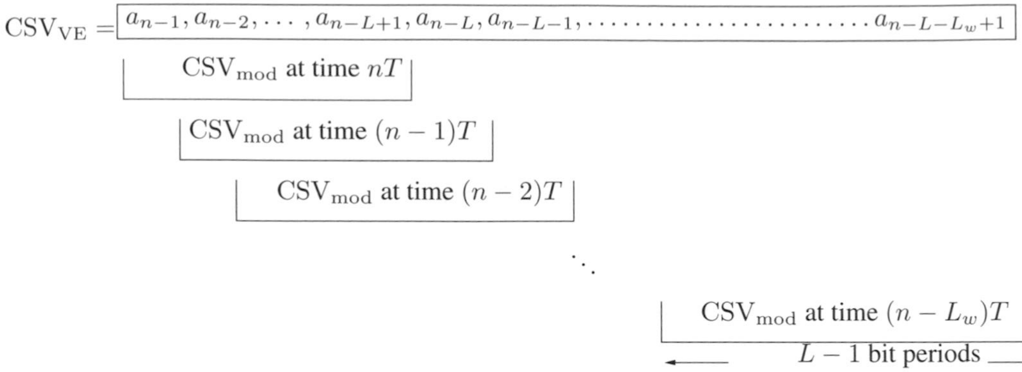

$$\mathrm{CSV_{VE}} = \boxed{a_{n-1}, a_{n-2}, \cdots, a_{n-L+1}, a_{n-L}, a_{n-L-1}, \cdots\cdots\cdots\cdots\cdots\cdots\cdots a_{n-L-L_w+1}}$$

$$\boxed{\mathrm{CSV_{mod}} \text{ at time } nT}$$

$$\boxed{\mathrm{CSV_{mod}} \text{ at time } (n-1)T}$$

$$\boxed{\mathrm{CSV_{mod}} \text{ at time } (n-2)T}$$

$$\ddots$$

$$\boxed{\mathrm{CSV_{mod}} \text{ at time } (n-L_w)T}$$

$$\longleftarrow \quad L - 1 \text{ bit periods} \quad \longrightarrow$$

Figure 13.18: Sliding window employed to extract the L_w previous $\mathrm{CSV_{mod}}$. The current $\mathrm{CSV_{mod}}$ is $a_{n-1}, \cdots, a_{n-L+1}$. By sliding the window one position to the right, the $(n-1)$th $\mathrm{CSV_{mod}}$, which is $a_{n-2}, \cdots, a_{n-L}$, is determined. This sliding operation is repeated until $L_w + 1$ $\mathrm{CSV_{mod}}$ are obtained.

In order to evaluate the previous modulator phase state, we substitute n in Equation 13.24 by $n - 1$, yielding:

$$\theta_{n-1} = \theta_n - \pi h_f a_{n-L}, \tag{13.25}$$

while for the $(n-2)$th phase state, we get:

$$\theta_{n-2} = \theta_{n-1} - \pi h_f a_{n-L-1}. \tag{13.26}$$

Substituting θ_{n-1} from Equation 13.25 into Equation 13.26, we obtain:

$$\theta_{n-2} = \theta_n - \pi h_f \left(a_{n-L} + a_{n-L-1} \right). \tag{13.27}$$

Repeating this, we can derive a closed-form equation for the L_w previous phase states, giving:

$$\theta_{n-1-k} = \theta_n - \pi h_f \sum_{m=0}^{k} a_{n-L-m} \qquad \text{for } k = 0 \ldots L_w - 1. \tag{13.28}$$

In summary, with the aid of the sliding window operation illustrated in Figure 13.18 and Equation 13.28, we can regenerate the modulated signal over $L_w + 1$ bit periods for each of the trellis-based equaliser states, by first subdividing the trellis state $S_n = \{\theta_n, a_{n-1}, \cdots, a_{n-L+1}, \cdots, a_{n-L-L_w+1}\}$ into $L_w + 1$ modulator states and the previous L_w number of bit transitions, as seen in Figure 13.17. The purpose of decomposing each of the trellis states into $L_w + 1$ modulator states is to regenerate the modulated signal over $L_w + 1$ bit periods, which is then convolved with the estimate of the channel impulse response in order to reconstruct the estimated received signal. Together, the estimate of the received signal and the actual received signal is then used to calculate the branch metric by evaluating the correlation or the squared Euclidean distance between these signals.

For example, in a GMSK system with $L = 3$ and $L_w = 2$, the total spreading, v_w, is 5. There will be four phase states as before, while the number of $\mathrm{CSV_{EQ}}$ is $2^{v_w - 1}$ or 16 for each phase

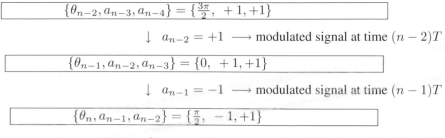

$\{\theta_{n-2}, a_{n-3}, a_{n-4}\} = \{\frac{3\pi}{2}, +1, +1\}$

$\downarrow \quad a_{n-2} = +1 \longrightarrow$ modulated signal at time $(n-2)T$

$\{\theta_{n-1}, a_{n-2}, a_{n-3}\} = \{0, +1, +1\}$

$\downarrow \quad a_{n-1} = -1 \longrightarrow$ modulated signal at time $(n-1)T$

$\{\theta_n, a_{n-1}, a_{n-2}\} = \{\frac{\pi}{2}, -1, +1\}$

$\downarrow \quad a_n = \pm 1 \longrightarrow$ current modulated signal at time nT

Figure 13.19: The trellis-based equaliser state $S_n = \{\frac{\pi}{2}, -1, +1, +1, +1\}$ contains information on the $L_w = 2$ previous bit transitions and $L_w+1 = 3$ modulator states. From the information used to represent state S_n, we can deduce that the current nth modulator state is $\{\frac{\pi}{2}, -1, +1\}$, while the $(n-1)$th and $(n-2)$th modulator state are $\{0, +1, +1\}$ and $\{\frac{3\pi}{2}, +1, +1)$, respectively.

state, giving a total of $4 \cdot 16 = 64$ states in the trellis-based equaliser. Let us assume that we are currently in the equaliser trellis state $S_n = \{\theta_n = \frac{\pi}{2}, \text{CSV}_{\text{EQ}} = -1, +1, +1, +1\}$, where $\theta_n = \frac{\pi}{2}, a_{n-1} = -1, a_{n-2} = +1, a_{n-3} = +1$ and $a_{n-4} = +1$. By substituting these values into the decomposed modulator states in Figure 13.17, we can deduce that the $(n-1)$th modulator state was $\{\theta_{n-1}, a_{n-2}, a_{n-3}\} = \{0, +1, +1\}$ and that the $(n-1)$th bit transmitted was $a_{n-1} = -1$, hence reaching the current nth modulator state $\{\theta_n, a_{n-1}, a_{n-2}\} = \{\frac{\pi}{2}, -1, +1\}$, as illustrated in Figure 13.19. Note that the phase state is periodic with respect to 2π. Therefore, the phase state at instant $n-2$ is $\frac{3\pi}{2}$ and not $-\frac{\pi}{2}$. For each trellis-based equaliser state in this system we can determine the previous modulator states and bit transitions. Therefore, being able to extract this information and by considering the latest data bit, a_n, we can regenerate the modulated sequence, which extends over $L_w + 1 = 3$ bit periods. The estimated received signal can be produced by convolving the modulated signal with the truncated channel impulse response.

In summary, over dispersive channels, the number of states in the trellis-based equaliser must be increased to mitigate the effects of the ISI introduced by the multi-path channel. Each equaliser trellis state contains the associated information of the L_w previous bit transitions and modulator states. Therefore, with the knowledge of the current data bit a_n leaving an equaliser trellis state, all combinations of the modulated signal can be generated over $L_w + 1$ bit periods. Subsequently, the received signal can be estimated by convolving the modulated signal with the windowed channel impulse response.

13.8 Soft-output GMSK Equaliser and Turbo Coding

13.8.1 Background and Motivation

Having presented an overview of the GMSK modulation scheme, we now proceed to describe how turbo coding can be employed in GMSK systems. We commence with a brief introduction to turbo codes followed by a discussion of the soft-in/soft-out decoding algorithms. The main operations of these techniques have been detailed in Chapter 5 but they are summarised here

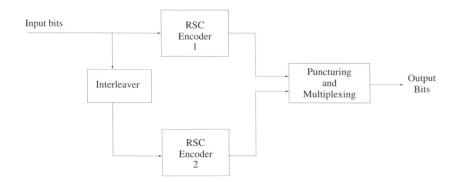

Figure 13.20: Structure of turbo encoder with RSC as component codes.

again cast in the context of the current problem of turbo-coded GMSK modulation.

The concept of turbo coding was introduced in 1993 by Berrou, *et al.* [12, 13], who demonstrated that its performance approximates the Shannonian limit. This section investigates the performance of turbo coding in terms of its BER performance when communicating over Rayleigh fading channels as well as its potential applications in enhanced versions of the GMSK-modulated GSM system [49].

Recall from Chapter 5 that turbo encoders encode the data or information sequence twice, typically using a half-rate Recursive Systematic Convolutional (RSC) encoder. As illustrated in Figure 13.20, the input data sequence is directed towards the first RSC encoder, but the same data sequence is also input into an interleaver prior to passing it to the second encoder. This is to ensure that the encoded sequences will be statistically independent of each other. The interleaver introduces time diversity and therefore, when a particular bit is erroneously decoded in the first decoder owing to channel impairments, it can still be corrected in the second decoder, since this particular bit has been mapped to a different time instant by the permutations introduced by the turbo interleaver. Hence, as the size of the interleaver increases, the performance of the turbo codes improves, since the bit sequences which are passed into the decoders become more independent. Both decoders can then exchange information in the form of the reliability of the bits, hence aiding the error correction process. In order to achieve the required coding rate, the outputs of both RSC encoders are punctured and multiplexed.

A commonly used structure, which performs turbo decoding through a series of iterations, consists of two RSC decoders. The schematic of this turbo decoder is shown in Figure 13.21. Both decoders accept soft channel values and provide soft outputs. These soft outputs express the likelihood or the probability that the data bit was decoded correctly, given the received channel values. They are usually in the form of the so-called log likelihood ratio (LLR) [126], which assists in reducing the computational complexity. As mentioned previously, the turbo decoding process is performed in a series of iterations, where information is passed from one decoder to the other in cycles, until a certain termination criterion is satisfied [61, 339]. A more detailed treatment of turbo codes can be found for example in reference [92].

Having provided a brief introduction to turbo coding, we now describe the soft-output equaliser used in the turbo-coded GSM system. Subsequently, the performance of turbo codes is characterised in the context of GSM data systems.

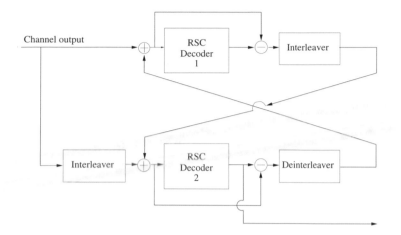

Figure 13.21: Structure of iterative turbo decoder with RSC as component decoders.

13.8.2   Soft-output GMSK Equaliser

In GSM, the most commonly used equaliser is the Viterbi Equaliser (VE) [49, 340] although Decision Feedback Equalisers (DFE) have also been used in the past [316]. The conventional VE performs MLSE by observing the development of the accumulated path metrics in the equaliser's trellis, which are evaluated recursively over several bit intervals. The length of the observation interval depends on the afforded complexity. Hard decisions are then released at the end of the equalisation process. However, since Log Likelihood Ratios (LLRs) [126] are required by the turbo decoders, we must employ specific Soft-In/Soft-Out (SISO) algorithms in place of the conventional VE. A range of SISO algorithms have been proposed in the literature such as the Maximum A-Posteriori (MAP) [11] algorithm, the Log-MAP [52], the Max-Log-MAP [50, 51], or the Soft Output Viterbi Algorithm (SOVA) [53, 54, 122]. We opted for employing the Log-MAP algorithm, since it yielded identical performance to the optimal MAP algorithm, but at a significantly lower complexity. Additionally, the Log-MAP algorithm is also numerically more stable. Other schemes — such as the Max-Log-MAP and SOVA — are computationally less intensive, but provide sub-optimal performance.

Next, we will present a brief description of the Log-MAP algorithm. For a more indepth treatment of the MAP and Log-MAP algorithms, we refer the interested reader to Sections 5.3.3 and 5.3.5, respectively.

13.8.3   The Log-MAP Algorithm

We commence by highlighting how the Log-MAP algorithm generates the LLR $L(u_m|y)$ for each input bit u_m. The LLR $L(u_m|y)$ can be defined as the ratio of the probability that the transmitted turbo-coded bit u_m was a logical '+1' to the probability that $u_m = -1$, given that y was the symbol sequence received. This is expressed as:

$$L(u_m|y) \triangleq \ln\left(\frac{P(u_m = +1|y)}{P(u_m = -1|y)}\right). \tag{13.29}$$

Depending on the specific implementation of the Log-MAP algorithm, the input bit u_m may represent the channel-coded or channel-decoded information bit. More specifically, when the

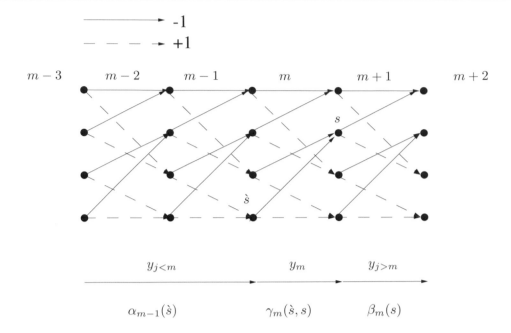

Figure 13.22: Example of binary ($M = 2$) system's trellis structure.

Log-MAP algorithm is used in an equaliser, u_m represents the channel-coded bit. However, when the MAP algorithm is used in a channel decoder, then u_m represents the source data bit. In this section, we will describe the Log-MAP algorithm in the context of the channel equaliser employed, in order to provide soft values for turbo channel decoding.

By using Bayes' theorem [341] of $P(a \wedge b) = P(a|b) \cdot P(b)$, where the symbol $\wedge$ denotes the logical 'and', Equation 13.29 can be rewritten to give the LLR in the form of:

$$L(u_m|y) = \ln \left(\frac{P(u_m = +1 \wedge y)}{P(u_m = -1 \wedge y)} \right). \tag{13.30}$$

For a binary system's trellis associated with χ number of trellis states at each trellis stage, there will be χ sets of $u_m = +1$ and $u_m = -1$ bit transitions, which are mutually exclusive. This is because only one transition could have occurred in the modulator, depending on the value of u_m. Therefore, the probability that u_m was transmitted can also be expressed as the sum of the individual probabilities of the associated χ number of trellis transitions. We can then express the LLR of Equation 13.30 as:

$$L(u_m|y) = \ln \left(\frac{\sum\limits_{(\grave{s},s)\Rightarrow u_m=+1} P(\grave{s} \wedge s \wedge y)}{\sum\limits_{(\grave{s},s)\Rightarrow u_m=-1} P(\grave{s} \wedge s \wedge y)} \right), \tag{13.31}$$

where $(\grave{s}, s) \Rightarrow u_m = +1$ denotes the set of transitions from state $\grave{s}$ to s caused by the bit $u_m = +1$ and similarly, $(\grave{s}, s) \Rightarrow u_m = -1$ for the bit $u_m = -1$. In Figure 13.22 we show an example of a binary system, which has χ states at each stage m in the trellis. Since this is an example of a binary system's trellis, at each time interval, there are $M = 2$ transitions leaving each state and $\chi = 4$ transitions due to the bit $u_m = +1$. Each of these transitions belongs to the set $(\grave{s}, s) \Rightarrow u_m = +1$. Similarly, there will be $\chi = 4$ number of $u_m = -1$ transitions, which belong to the set $(\grave{s}, s) \Rightarrow u_m = -1$. In our forthcoming discussion, we will introduce the other notations in Figure 13.22, with emphasis on $\alpha_{m-1}(\grave{s})$, $\beta_m(s)$ and $\gamma_m(\grave{s}, s)$.

Bahl *et al.* [11] showed that by using Bayes' theorem and by splitting the received sequence y into $y_{j<m}$, y_m and $y_{j>m}$, where the term y_m represents the present continuous-valued received channel output amplitude, while $y_{j<m}$ and $y_{j>m}$ are the previous and future received channel output value sequences, respectively, Equation 13.31 can be expressed as the product of three terms:

$$L(u_m|y) = \ln \left(\frac{\sum_{(\grave{s},s)\Rightarrow u_m=+1} P(y_{j<m}\wedge\grave{s}) \cdot P(y_m \wedge s|\grave{s}) \cdot P(y_{j>m}|s)}{\sum_{(\grave{s},s)\Rightarrow u_m=-1} P(y_{j<m}\wedge\grave{s}) \cdot P(y_m \wedge s|\grave{s}) \cdot P(y_{j>m}|s)} \right)$$

$$= \ln \left(\frac{\sum_{(\grave{s},s)\Rightarrow u_m=+1} \alpha_{m-1}(\grave{s}) \cdot \gamma_m(\grave{s}, s) \cdot \beta_m(s)}{\sum_{(\grave{s},s)\Rightarrow u_m=+1} \alpha_{m-1}(\grave{s}) \cdot \gamma_m(\grave{s}, s) \cdot \beta_m(s)} \right).$$

In the derivation of this equation, we assumed that the channel is memoryless. The symbol:

$$\alpha_{m-1}(\grave{s}) = P(y_{j<m}\wedge\grave{s}) \qquad (13.32)$$

is the probability that we are in state $\grave{s}$ at instant $m - 1$ and that the previous continuous-valued received channel output sequence is $y_{j<m}$. The notation:

$$\gamma_m(\grave{s}, s) = P(y_m \wedge s|\grave{s}) \qquad (13.33)$$

indicates the probability that given that we are in state $\grave{s}$ at trellis stage $m - 1$, we receive the channel output y_m at instant m, resulting in a transition to state s. Lastly:

$$\beta_m(s) = P(y_{j>m}|s) \qquad (13.34)$$

is the probability that at instant m we will receive the future channel output value sequence $y_{j>m}$ with the condition that the present state at instant m is s. By using Equations 13.8.3, 13.32, 13.33 and 13.34 we can calculate the LLR for each input bit, u_m.

In the MAP algorithm described in Section 5.3.3, the LLR $L(u_m|y)$ at instant m is calculated by evaluating $\alpha_{m-1}(\grave{s})$, $\beta_m(s)$ and $\gamma_m(\grave{s}, s)$ in Equations 5.49, 5.50 and 5.51 by performing computationally exhaustive calculations, involving multiplications and evaluating natural logarithms, $\ln(\cdot)$. However, in the Log-MAP algorithm, the complexity of these recursive calculations is reduced by transforming $\alpha_{m-1}(\grave{s})$, $\gamma_m(\grave{s}, s)$ and $\beta_m(s)$ into the logarithmic domain and then invoking the so-called Jacobian logarithmic relationship, which is defined as [52]:

$$\ln(e^{x_1} + e^{x_2}) = \max(x_1, x_2) + \ln(1 + e^{-|x_1-x_2|})$$
$$= \max(x_1, x_2) + f_c(|x_1 - x_2|)$$
$$= \max(x_1, x_2) + f_c(\delta)$$
$$= J(x_1, x_2),$$

where $\delta = |x_1 - x_2|$ and $f_c(\delta)$ can be viewed as a correction term, while $\max(x_1, x_2)$ is:

$$\max(x_1, x_2) = \begin{cases} x_1 & \text{if } x_1 > x_2 \\ x_2 & \text{if } x_2 > x_1. \end{cases} \tag{13.35}$$

The transformation to the logarithmic domain is done by defining:

$$A_m(s) \triangleq \ln(\alpha_m(s)), \tag{13.36}$$

$$B_m(s) \triangleq \ln(\beta_m(s)), \tag{13.37}$$

$$\Gamma_m(\grave{s}, s) \triangleq \ln(\gamma_m(\grave{s}, s)). \tag{13.38}$$

We will show subsequently how the transformation to the logarithmic domain and the use of the Jacobian logarithmic relationship in Equation 13.8.3 will allow the evaluation of the LLRs through additions and table look-up operations, instead of the more complex multiplication operations, natural logarithm $\ln(\cdot)$ calculations and exponential computations.

Let us consider the forward recursive calculations for $\alpha_m(s)$. By transforming $\alpha_{m-1}(\grave{s})$ to the logarithmic domain, and using Equations 13.36 and 5.49, the term $A_m(s)$ can be expressed as:

$$A_m(s) \triangleq \ln(\alpha_m(s))$$

$$= \ln \left(\sum_{\text{all } \grave{s}} \alpha_{m-1}(\grave{s}) \cdot \gamma_m(\grave{s}, s) \right)$$

$$= \ln \left(\sum_{\text{all } \grave{s}} \exp \left[A_{m-1}(\grave{s}) + \Gamma_m(\grave{s}, s) \right] \right)$$

$$= \ln \left(\sum_{\text{all } \grave{s}} \exp \left[\Upsilon_f(\grave{s}, s) \right] \right),$$

where $\Upsilon_f(\grave{s}, s)$ is:

$$\Upsilon_f(\grave{s}, s) = A_{m-1}(\grave{s}) + \Gamma_m(\grave{s}, s), \tag{13.39}$$

representing the accumulated metric for a forward transition from state $\grave{s}$ to s in Figure 13.22. Explicitly, for each transition from state $\grave{s}$ to s, we add the branch metric $\Gamma_m(\grave{s}, s)$ to the previous value $A_{m-1}(\grave{s})$, in order to obtain the current value $\Upsilon_f(\grave{s}, s)$ for that path. In a binary trellis there are two branches leaving each state, and as a consequence, two possible transitions merge to every state, as illustrated in Figure 13.22. Therefore, we will have two values of $\Upsilon_f(\grave{s}, s)$ arising from the two branches reaching state s, which we will refer to as $\Upsilon_{f1}(\grave{s}, s)$ and $\Upsilon_{f2}(\grave{s}, s)$. Exploiting this property — which is inherent in a binary system — Equation 13.8.3 can be written as:

$$A_m(s) = \ln \left(\sum_{\text{all } \grave{s}} \exp \left[\Upsilon(\grave{s}, s) \right] \right)$$

$$= \ln \left(\exp \left[\Upsilon_{f1}(\grave{s}, s) \right] + \exp \left[\Upsilon_{f2}(\grave{s}, s) \right] \right).$$

Since $A_m(s)$ can be expressed as a natural logarithm of the sum of exponentials, the Jacobian logarithmic relationship in Equation 13.8.3 can be employed, hence allowing Equation 13.8.3 to be rewritten as:

$$A_m(s) = \ln \left(\exp \left[\Upsilon_{f1}(\grave{s}, s) \right] + \exp \left[\Upsilon_{f2}(\grave{s}, s) \right] \right)$$

$$= \max(\Upsilon_{f1}(\grave{s}, s), \Upsilon_{f2}(\grave{s}, s)) + \ln \left(1 + \exp[-|\Upsilon_{f1}(\grave{s}, s) - \Upsilon_{f2}(\grave{s}, s)|] \right).$$

Although the transformation of $\alpha_m(s)$ to the logarithmic domain term and the use of the Jacobian logarithmic relationship enable us to evaluate the $A_m(s)$ term through recursive additions rather than using the more complex multiplication operations, we still need to compute the natural logarithm term, $\ln(1 + \exp[-|\Upsilon_{f1}(\grave{s}, s) - \Upsilon_{f2}(\grave{s}, s)|])$ or $f_c(\delta_f)$, as in Equation 13.8.3, where $\delta_f = \Upsilon_{f1}(\grave{s}, s) - \Upsilon_{f2}(\grave{s}, s)$. Fortunately, $f_c(\delta_f)$ can be stored as a look-up table to avoid the need for computing it repeatedly. Furthermore, Robertson and Wörz [342] have shown that it is sufficient to store only eight values of δ_f, ranging from 0 to 5. Therefore, in the Log-MAP algorithm the evaluation of $A_m(s)$, which is the logarithmic domain representation of $\alpha_m(s)$, is less complex, compared to the MAP algorithm, since it only involves additions and a look-up table search. This look-up table is specified later on in Section 14.5 in conjunction with a numerical example of iterative joint equalisation and decoding, known as turbo equalisation.

Employing the same approach for $B_{m-1}(s)$, and using Equations 5.50 and 13.37, we obtain:

$$B_{m-1}(\grave{s}) = \ln\left(\sum_{\text{all } s} \beta_m(s) \cdot \gamma_m(\grave{s}, s)\right)$$

$$= \ln\left(\sum_{\text{all } s} \exp[B_m(s) + \Gamma_m(\grave{s}, s)]\right)$$

$$= \ln\left(\sum_{\text{all } s} \exp[\Upsilon_b(\grave{s}, s)]\right),$$

where $\Upsilon_b(\grave{s}, s)$ is:

$$\Upsilon_b(\grave{s}, s) = B_m(s) + \Gamma_m(\grave{s}, s), \tag{13.40}$$

representing the accumulated metric for the backward transition from state s to $\grave{s}$. We observe in Equation 13.40 that $\Upsilon_b(\grave{s}, s)$ for each path from state s to $\grave{s}$ is evaluated by adding $\Gamma(\grave{s}, s)$ to $B_m(s)$. Since we are considering a binary system here, there are only two possible transitions or branches from state s to states $\grave{s}$. As before, by using this information and representing the accumulated metric from these two branches as $\Upsilon_{b1}(\grave{s}, s)$ and $\Upsilon_{b2}(\grave{s}, s)$, we can rewrite Equation 13.8.3 as:

$$B_{m-1}(s) = \ln\left(\sum_{\text{all } s} \exp[\Upsilon_b(\grave{s}, s)]\right)$$

$$= \ln\left(\exp[\Upsilon_{b1}(\grave{s}, s)] + \exp[\Upsilon_{b2}(\grave{s}, s)]\right).$$

Exploiting the Jacobian logarithmic relationship in Equation 13.8.3, Equation 13.8.3 becomes:

$$B_{m-1}(s) = \max(\Upsilon_{b1}(\grave{s}, s), \Upsilon_{b2}(\grave{s}, s)) + \ln(1 + \exp[-|\Upsilon_{b1}(\grave{s}, s) - \Upsilon_{b2}(\grave{s}, s)|])$$

$$= \max(\Upsilon_{b1}(\grave{s}, s), \Upsilon_{b2}(\grave{s}, s)) + f_c(\delta_b),$$

where $\delta_b = \Upsilon_{b1}(\grave{s}, s) - \Upsilon_{b2}(\grave{s}, s)$. Similarly to $A_m(s)$, we see that the recursive calculations for $B_{m-1}(\grave{s})$ are also based on additions and the evaluation of $f_c(\delta_b)$, which — as mentioned before — can be stored in a look-up table in order to reduce the computational complexity.

From Equations 13.8.3 and 13.8.3, we observe that $A_m(s)$ and $B_{m-1}(\grave{s})$ can be evaluated, once $\Gamma_m(\grave{s}, s)$ was obtained. In order to evaluate the branch metric $\Gamma_m(\grave{s}, s)$, we use Equations 5.32 and 13.38, which yields:

$$\Gamma_m(\grave{s}, s) \triangleq \ln(\gamma_m(\grave{s}, s))$$

$$= \ln P(u_m) \cdot P(y_m | \grave{s} \wedge s),$$

where $P(u_m)$ is the a-priori probability of bit u_m and $P(y_m|\grave{s} \wedge s)$ represents the probability that the signal y_m was received given that the transition from state $\grave{s}$ to state s occurred at instant m. By assuming that the signal estimate $\hat{y}_m$ was received over a Gaussian channel, we can express $P(y_m|\grave{s} \wedge s)$ as the noise-contaminated transmitted signal given by:

$$P(y_m|\grave{s} \wedge s) = \frac{1}{\sqrt{2\pi\sigma^2}} \exp\left[-\frac{1}{2\sigma^2}(y_m - \hat{y}_m)^2\right], \tag{13.41}$$

where σ^2 is the variance of the zero-mean complex white Gaussian noise. Hence, by substituting Equation 13.41 into Equation 13.8.3, we obtain:

$$\Gamma_m(\grave{s}, s) = \ln\left[\left(\frac{1}{\sqrt{2\pi\sigma^2}} \exp\left[-\frac{1}{2\sigma^2}(y_m - \hat{y}_m)^2\right]\right) \cdot P(u_m)\right]$$

$$= \ln\left(\frac{1}{\sqrt{2\pi\sigma^2}}\right) - \frac{1}{2\sigma^2}(y_m - \hat{y}_m)^2 + \ln(P(u_m)).$$

The term $\ln\left(\frac{1}{\sqrt{2\pi\sigma^2}}\right)$ on the right-hand side of Equation 13.8.3 is independent of the bit u_m and therefore can be considered as a constant and neglected. Consequently, $\Gamma_m(\grave{s}, s)$ can be evaluated as:

$$\Gamma_m(\grave{s}, s) = -\frac{1}{2\sigma^2}(y_m - \hat{y}_m)^2 + \ln(P(u_m)). \tag{13.42}$$

Furthermore, in a non-iterative equaliser we do not have a-priori information concerning the bit u_m. Therefore, we can make the assumption that all bits have equal probabilities of being transmitted. Hence, $P(u_m)$ is a constant and can be neglected, enabling the branch metric $\Gamma_m(\grave{s}, s)$ to be written as:

$$\Gamma_m(\grave{s}, s) = -\frac{1}{2\sigma^2}(y_m - \hat{y}_m)^2, \tag{13.43}$$

which is equivalent to that used in the Viterbi algorithm.

By transforming $\alpha_m(s)$, $\gamma_m(\grave{s}, s)$ and $\beta_m(s)$ into the logarithmic domain in the Log-MAP algorithm, the expression for the LLR, $L(u_m|y)$ in Equation 13.8.3, is also modified to give:

$$L(u_m|y) = \ln\left(\frac{\displaystyle\sum_{(\grave{s},s)\Rightarrow u_m=+1} \alpha_{m-1}(\grave{s}) \cdot \gamma_m(\grave{s}, s) \cdot \beta_m(s)}{\displaystyle\sum_{(\grave{s},s)\Rightarrow u_m=-1} \alpha_{m-1}(\grave{s}) \cdot \gamma_m(\grave{s}, s) \cdot \beta_m(s)}\right)$$

$$= \ln\left(\frac{\displaystyle\sum_{(\grave{s},s)\Rightarrow u_m=+1} \exp\left(A_{m-1}(\grave{s}) + \Gamma_m(\grave{s}, s) + B_m(s)\right)}{\displaystyle\sum_{(\grave{s},s)\Rightarrow u_m=-1} \exp\left(A_{m-1}(\grave{s}) + \Gamma_m(\grave{s}, s) + B_m(s)\right)}\right)$$

$$= \ln\left(\sum_{(\grave{s},s)\Rightarrow u_m=+1} \exp\left(A_{m-1}(\grave{s}) + \Gamma_m(\grave{s}, s) + B_m(s)\right)\right)$$

$$- \ln\left(\sum_{(\grave{s},s)\Rightarrow u_m=-1} \exp\left(A_{m-1}(\grave{s}) + \Gamma_m(\grave{s}, s) + B_m(s)\right)\right).$$

By considering a trellis with χ number of states at each trellis stage, there will be χ transitions which belong to the set $(\grave{s}, s) \Rightarrow u_m = +1$, and similarly, χ number of $u_m = -1$ transitions belonging to $(\grave{s}, s) \Rightarrow u_m = -1$. Therefore, we must evaluate the natural logarithm $\ln(\cdot)$ of the sum of χ exponential terms. This expression can be simplified by extending and generalising the Jacobian logarithmic relationship of Equation 13.8.3 in order to cope with a higher number of exponential summations. This can be achieved by nesting the $J(x_1, x_2)$ operations as follows:

$$\ln \left(\sum_{k=1}^{V} e^{x_k} \right) = J(x_V, J(x_{V-1}, \ldots, J(x_3, J(x_2, x_1)))), \qquad (13.44)$$

where $V = \chi$ for our χ-state trellis. Hence, by using this relationship and Equation 13.8.3, the LLR values for each input bit u_m — where x_k is equal to the sum of $A_{m-1}(\grave{s})$, $\Gamma_m(\grave{s}, s)$ and $B_m(s)$ for the kth path at instant m — can be evaluated.

13.8.4   Summary of the Log-MAP Algorithm

The Log-MAP algorithm transforms $\alpha_m(s)$, $\beta_m(s)$ and $\gamma_m(\grave{s}, s)$, in Equations 13.32, 13.34, and 13.33, to the corresponding logarithmic domain terms $A_m(s)$, $B_m(s)$ and $\Gamma_m(\grave{s}, s)$, using Equations 13.36, 13.37 and 13.38. These logarithmic domain variables are evaluated after receiving the entire burst of discretised received symbols, y_m. Figure 13.23 illustrates the sequence of key operations required for determining the LLR $L(u_m|y)$ of bit u_m in a trellis having χ number of states at each trellis stage. Once $A_{m-1}(s)$, $B_m(s)$ and $\Gamma_m(\grave{s}, s)$ have been calculated recursively, we evaluate the sum of these three terms — as seen in Equation 13.8.3 — for each path or branch caused by the bit $u_m = +1$, i.e. all the transitions belonging to the set $(\grave{s}, s) \Rightarrow u_m = +1$ at the trellis stage m between the previous state $\grave{s}$ and the current state s in Figure 13.22. We repeat the process for all paths $k = 1 \ldots \chi$ in the set $(\grave{s}, s) \Rightarrow u_m = -1$, at the trellis stage m between the previous state $\grave{s}$ and the current state s in Figure 13.22 as well. In order to evaluate the LLR for bit u_m, we use Equation 13.8.3 and the generalised expression of the Jacobian logarithmic relationship, as expressed in Equation 13.44, where k denotes the kth path ($k = 1 \ldots \chi$), which belongs to the set $(\grave{s}, s) \Rightarrow u_m = +1$ or $(\grave{s}, s) \Rightarrow u_m = -1$. Both terms on the right-hand side of the LLR expression in Equation 13.8.3 can be determined by substituting the sum $A_{m-1}(s) + B_m(s) + \Gamma_m(\grave{s}, s)$ into the term x_k in the generalised Jacobian logarithmic relationship of Equation 13.44. In comparison with the MAP algorithm, the Log-MAP algorithm is less complex since we do not need to evaluate the LLRs using multiplications, natural logarithm and exponential calculations. Instead, we only need subtractions, additions and a look-up table. This reduction in complexity — while retaining an identical BER performance to the MAP algorithm — renders the Log-MAP algorithm an attractive option for the demodulator or decoder. There are also other algorithms exhibiting a lower complexity, such as the so-called Max-Log-MAP [50, 51] and the Soft Output Viterbi Algorithm (SOVA) [53, 54, 122], which are, however, sub-optimal. In our work, we have used the optimal Log-MAP algorithm in order to obtain the best possible or upper bound performance of the system.

Having described the underlying principles of the Log-MAP algorithm, which was used in our GMSK equaliser, we now proceed by providing an overview of the complexity associated with turbo decoding and convolutional decoding.

13.8.5   Complexity of Turbo Decoding and Convolutional Decoding

In this section, the complexity of the Turbo Decoder (TC) and Convolutional Decoder (CC) is quantified in terms of the number of trellis states. The convolutional decoder employs the Viterbi algorithm, while the turbo decoder consists of two constituent decoders employing the Log-MAP

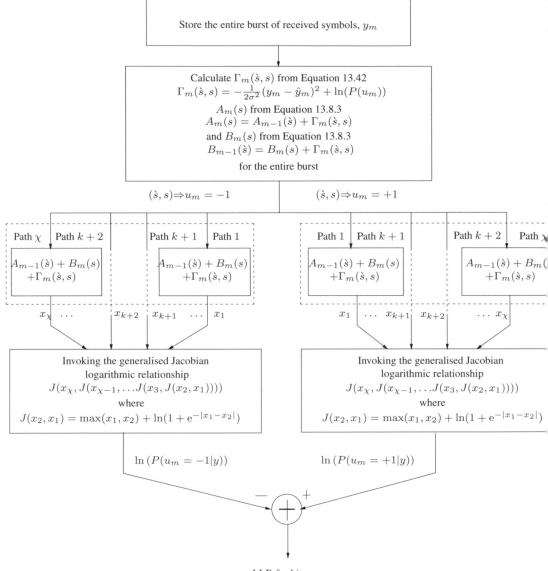

Figure 13.23: Summary of key operations in the Log-MAP algorithm, in order to evaluate the LLR for bit u_m in a trellis having χ number of states, at instant m for a non-iterative equaliser.

Typical Urban	
Position (μs)	Weights
0.00	0.622
0.92	0.248
2.30	0.062
2.76	0.039
3.22	0.025
5.06	0.004

Table 13.2: The weight and delay of each path in the COST207 typical urban channel illustrated in Figure 13.24.

algorithm. As described previously, the Log-MAP algorithms determines the LLR values which represent the reliability of the transmitted source bits. In comparison to the Viterbi algorithm, the Log-MAP algorithm is approximately three times more complex. This is because the Log-MAP algorithm has to evaluate the forward and backward recursion values and subsequently employ these values to determine the LLRs of the source bits. By contrast, the Viterbi algorithm only performs an operation similar to the forward recursion, with the exception that only the most reliable trellis path is stored and in contrast to the Jacobian logarithmic relationship, no correction term has to be computed.

First, the number of trellis states in the convolutional decoder and the turbo decoder is determined. The number of states in the convolutional decoder is dependent on the constraint length K of the convolutional code according to the relationship of 2^{K-1}. For example, a rate $R = \frac{1}{2}$ $K = 5$ convolutional decoder requires $2^{K-1} = 16$ trellis states at each trellis interval. For the turbo decoder, the number of states required is $2 \cdot 2^{K-1} = 2^K$, since the turbo decoder consists of two constituent convolutional decoders. In addition, the complexity evaluation of the turbo decoder must also account for the number of turbo decoding iterations. Therefore, the complexity of the turbo decoder can be expressed as a function of the number of iterations, and the number of states in each trellis interval, yielding:

$$\text{Complexity of TC} = 3 \cdot 2 \cdot \text{Number of iterations} \cdot \text{Number of TC states}$$
$$\text{Complexity of CC} = \text{Number of CC states},$$

$$(13.45)$$

where the factor of '3' was introduced in the complexity of the turbo decoder, since the Log-MAP algorithm employed by the constituent decoders is assumed to be three times more complex than the Viterbi algorithm utilised in the convolutional decoder.

For example, turbo codes possessing a constraint length of $K = 3$ in each constituent decoder and performing eight iterations will have a complexity of $3 \cdot 2 \cdot 8 \cdot 2^{3-1} = 192$ states. When compared with the complexity of the $R = \frac{1}{2}$ $K = 5$ convolutional decoder, it is observed that the convolutional decoder has a complexity of 16 and is 12 times less complex than the turbo decoder.

Having given an estimate of the complexity associated with turbo decoding and convolutional decoding, in our following discussion we will investigate the performance of the turbo codec employed in a GMSK-modulated GSM-like system.

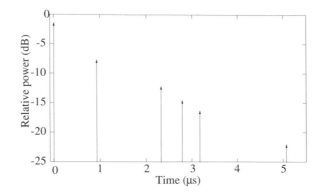

Figure 13.24: The impulse response of the COST207 typical urban channel used.

13.8.6 System Parameters

The turbo-encoded bits are passed to a channel interleaver and the interleaved bits are modulated using GMSK modulation associated with a normalised bandwidth of $B_n = 0.3$ and transmitted at 270.833 kbit/s over the COST207 [251] Typical Urban (TU) channel. Figure 13.24 portrays the TU channel model used, where each path is fading independently, obeying Rayleigh statistics, for a vehicular speed of 50 km/h or 13.89 m/s and transmission frequency of 900 MHz. The GMSK Log-MAP demodulator equalises the received signal, which has been degraded by the wideband or narrowband fading channels with the assumption that perfect channel impulse response information is available. Subsequently, the soft outputs from the demodulator are de-interleaved and passed to the turbo decoder in order to perform channel decoding. Let us now briefly describe the channel coder, the interleaver/de-interleaver scheme and the GSM-like transceiver used in our investigations.

We compare two channel coding schemes, namely the constraint length $K = 5$ convolutional codec as used in the GSM [49, 343] system, and turbo coding. Again, the turbo codec uses two $K = 3$ so-called RSC component encoders, while the decoder employs the optimal Log-MAP [52] decoding algorithm and performs eight decoding iterations. The transmission frame also contains three trellis termination bits. In GSM, the channel interleaving scheme used is known as inter-burst interleaving [49, 344].

The mapping parameters for the interburst interleaving [49, 344] have been chosen to introduce pseudo-random displacements of the coded input bits. Therefore, this scheme has the advantage of randomising the periodic fading-induced errors, since the coded input bursts $C(n, k)$ are mapped into interleaved sub-blocks $I(B, j)$ with irregular offset positions. Having briefly described the turbo codec and channel interleaving scheme used, we proceed with the description of the GMSK transceiver.

A GMSK modulator having $B_n = 0.3$, which is employed in the current GSM mobile radio standard [49], is used in our system. As detailed earlier in this chapter, GMSK belongs to the class of Continuous Phase Modulation (CPM) [320] schemes, and possesses high spectral efficiency as well as constant signal envelope, hence allowing the employment of non-linear power-efficient class-C amplifiers. However, this spectral efficiency is achieved at the expense of CISI, and therefore a channel equaliser, typically a VE is used in conventional implementations. The conventional VE [49, 340] performs MLSE by observing the development of the accumulated path metrics in the trellis, which are evaluated recursively, over several bit intervals. The length of the observation interval depends on the complexity affordable. The results of the bit-related hard decisions are then released at the end of the equalisation process. However, since LLRs [126] are required by the turbo decoders, in the proposed system we employ soft-output algorithms instead of the VE, such as the MAP [11] or the Log-MAP [52] algorithm, which were described earlier in Section 5.3.3 and Section 5.3.5, respectively. However, other techniques such as the reduced-complexity, but sub-optimum Max-Log-MAP [50, 51], and the SOVA [53, 54, 122] algorithms can also be used. We chose to invoke the Log-MAP algorithm, since it yielded identical performance to the optimal MAP algorithm at a much lower complexity.

13.8.7   Turbo Coding Performance Results

One of the key design factors in turbo codes is the choice of interleaver in the channel encoder of Figure 13.20. As described briefly in the introductory section 13.8.1 as well as in more depth in Chapter 5, the input sequence is encoded twice, where the first encoder operates upon the actual input bit sequence, while the second parallel encoder acts on the interleaved bit sequence. The performance of turbo codes improves as the interleaved and non-interleaved sequences become more statistically independent. Since they exchange information between them, these information sources supply two-near independent 'opinions' concerning the transmitted information. Therefore, as the depth of the turbo interleaver increases, the performance improves as well. It was also shown by Barbulescu and Pietrobon [68] that block interleavers with an odd number of rows and columns perform better than a block interleaver having an even number of rows and columns. This is also true when compared with block interleavers having an odd number of rows and an even number of columns. Another parameter which affects the performance of turbo coding is the number of iterations invoked at the decoder. In our work, the decoder is set to perform eight iterations, since the investigations of Chapter 5 have shown that no significant BER performance improvement is based upon using a higher number of iterations.

We commence by presenting simulation results over Gaussian channels in order to highlight the effect of different interleaver depths. The turbo encoder used consists of two $\frac{1}{2}$-rate, constraint length, $K = 3$ RSC encoders, where the generator polynomials, G_0 and G_1, are $1 + D + D^2$ and $1 + D^2$, or 7 and 5 correspondingly in the octal representation. Note that we have adopted the notation R×C to indicate the dimensions of the block interleaver, which has R rows and C columns. From Figure 13.25, we observe that the BER performance improves by approximately 0.5 dB at a BER of 10^{-3} as the size of the block interleaver is increased from 9×9 to 15×15. Another 0.25 dB improvement is gained by using a 31×31 interleaver, when compared to the performance of the 15×15 turbo encoder at a BER of 10^{-3}. This performance is expected, since a longer interleaver will render the two encoded sequences more statistically independent, hence improving the code's performance. However, this will incur long delays, since we must wait for the interleaver matrix to be filled before the turbo encoding process can begin. Therefore, there must be a compromise between the delay afforded by the system and the BER performance desired.

In the GSM data channel, which is not delay-sensitive — more specifically in the TCH/F9.6 standard [49] — we can afford a delay of 19 times 114-bit data bursts. Therefore, we used a turbo encoder with an interleaving depth of $19 \cdot 114 = 2166$ bits. A random turbo interleaver was employed as it has been shown in [68] that for long turbo coding frames, the random interleaver has superior performance over block interleavers. However, block interleavers performed better for short frames [133]. The coded BER over the narrowband Rayleigh fading channel and the COST207 TU channel are illustrated in Figures 13.26(a) and 13.26(b), respectively. Here, we observed an E_b/N_0 improvement of approximately 1.5 dB at a BER of 10^{-4} for both channels.

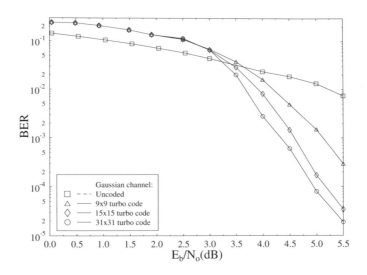

Figure 13.25: BER performance of GMSK when communicating over AWGN channels without coding and with turbo codes using 9×9 and 31×31 interleavers, half-rate, $K = 3$ RSC encoders using the octal generator polynomial of 7 and 5.

By increasing the coding frame length further, to the order of 50000 bits, we obtained an improvement of approximately 3.7 dB at a BER of 10^{-4} over the standard convolutional codes, when transmitting over the so-called perfectly interleaved narrowband Rayleigh fading channel — which is modelled as a so-called complex Gaussian channel having a Doppler frequency or vehicular speed of infinity — as illustrated in Figure 13.27. This confirmed the superiority of turbo codes over convolutional codes, when the coding frame length was sufficiently long.

13.9   Summary and Conclusions

In this chapter we have introduced the fundamental theory and properties of Digital Frequency Modulation (DFM), specifically Minimum Shift Keying (MSK) and Gaussian Minimum Shift Keying (GMSK). The basic principles of constructing the trellis states of the equaliser have also been presented.

Furthermore, in Section 13.8.3, the Log-MAP algorithm was summarised, with a view to incorporate it in a GMSK-modulated turbo-coded GSM-like system. Simulations were carried out in conjunction with the GSM data channel, where the required E_b/N_0 of turbo codes was approximately 1.5 dB lower than that of the standard constraint-length five convolutional code at a BER of 10^{-4}, when communicating over the COST 207 TU channel. When the turbo coding frame length was increased to the order of 50000 bits, it was observed that a gain of approximately 3.7 dB was achieved in E_b/N_0 terms at a BER of 10^{-4} as compared to the standard convolutional code over the so-called perfectly interleaved narrowband Rayleigh fading channel, as illustrated in Figure 13.27.

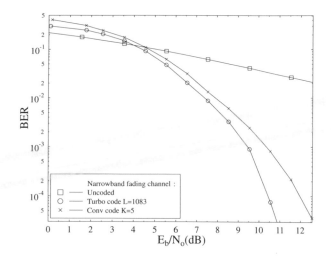

(a) Coded BER versus E_b/N_0 over the narrowband Rayleigh fading channel.

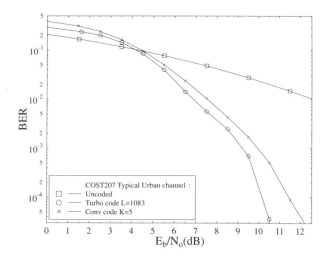

(b) Coded BER versus E_b/N_0 over the COST207 typical urban channel.

Figure 13.26: The coded BER performance of turbo code with random interleaving possessing an interleaving depth of 1083 bits, which was approximately 1.5 dB better in E_b/N_0 terms at a
BER of 10^{-4} than that of the constraint length $K = 5$ convolutional code over the narrowband Rayleigh fading channel and the COST207 TU channel for full-rate GSM data channel
TCH/F9.6 [49].

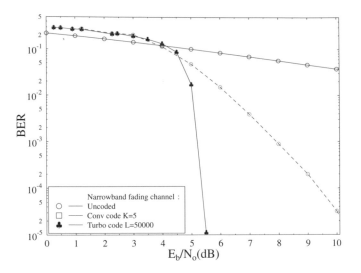

Figure 13.27: Coded BER performance of the $\frac{1}{2}$-rate, $K = 5$ turbo coding in conjunction with a random
turbo interleaver of depth 50000 bits, which was approximately 3.7 dB better in E_b/N_0
terms at a BER of 10^{-4} than convolutional codes over the so-called perfectly interleaved,
uncorrelated Rayleigh fading narrowband channel.

This chapter provided the necessary background for the appreciation of the concepts of turbo
equalisation, which will be detailed in Chapter 14.

Chapter 14

Turbo Equalisation for Partial-response Systems

The previous chapter introduced the basics of partial-response modulation techniques and channel equalisation using Soft-In/Soft-Out (SISO) algorithms, such as the Maximum A-Posteriori (MAP) algorithm and the Log-MAP algorithm. With this background, in this chapter we embark on the study of their turbo equalisation. Hence a good background in turbo decoding and SISO algorithms constitutes the foundations required for the appreciation of this chapter, which were discussed in Chapters 5 and 13.

turbo equalisation was first proposed by Douillard *et al.* in 1995 in reference [105] for a serially concatenated convolutional-coded BPSK system, as shown in Figure 14.1. In this contribution turbo equalisation employing the structure illustrated in Figure 14.2 was shown to mitigate the effects of ISI, when having perfect channel impulse response information. Instead of performing the equalisation and decoding independently, in order to overcome the channel's frequency selectivity, better performance can be obtained by the turbo equaliser, which considers the discrete channel's memory and performs both the equalisation and decoding iteratively.

Figure 14.1: Serially concatenated convolutional-coded BPSK system using the turbo equaliser, which performs the equalisation, demodulation and channel decoding iteratively.

The basic philosophy of the original turbo equalisation technique stems from the iterative turbo decoding algorithm consisting of two SISO decoders, a structure which was proposed by Berrou *et al.* [12, 13]. Before proceeding with our in-depth discussion, let us redefine below the terms a-priori, and extrinsic a-posteriori information, which we employ throughout this treatise.

A-priori The a-priori information associated with a bit v_m is the information known before equalisation or decoding commences, from a source other than the received sequence or the code constraints. It is also often referred to as intrinsic information, in order to contrast it with the extrinsic information, which is described next.

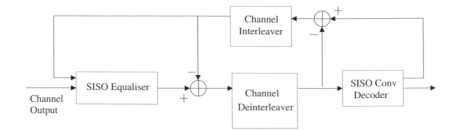

Figure 14.2: Structure of original turbo equaliser introduced by Douillard *et al.* [105] ©1995, European Transactions on Telecommunications.

Extrinsic The extrinsic information associated with a bit v_m is the information provided by the equaliser or decoder based on the received sequence and on the a-priori information of all bits with the exception of the received and a-priori information explicitly related to that particular bit v_m. To elaborate a little further, extrinsic information related to a specific bit exists because of a number of factors. First, when using an error correction encoder exhibiting memory, each input bit influences the encoder's output sequence over a long string of bits. Since a conventional convolutional code has an infinite duration impulse response [49], in theory each input bit influences the output bit sequence over an infinite number of coded bits. In practice, however, this interval is curtailed at about five times the code's constraint. In the context of RSC-coding-based turbo codes, typically a long-memory turbo interleaver is employed, which substantially extends the number of coded bits over which an input bit has an influence, since the turbo code's impulse response is typically prolonged. This justifies their high error correction power and the requirement of a high-depth turbo interleaver. Furthermore, since an input bit influences the encoded sequence over a high number of encoded bits, even when our confidence in a particular bit decision is low substantial amount of extrinsic information related to this was 'smeared' across a high number of encoded bits. With the aid of this extrinsic information the turbo decoder can iteratively enhance our confidence in the originally unreliable bit decision. We note, however, that the above arguments require that the intrinsic and extrinsic information related to a bit is treated separately by the decoder and hence remains uncorrelated, in order to be able to enhance each other's estimates in consecutive turbo decoding iterations.

As mentioned above — apart from the turbo encoder's memory — there are a number of other mechanisms generating extrinsic information due to their inherent memory, such as the channel's memory inducing dispersion over time. The deliberately introduced CISI of the GMSK modulators investigated constitutes another source of extrinsic information. These issues will be treated in more depth throughout our further discourse.

A-posteriori The a-posteriori information associated with a bit is the information that the SISO algorithm provides taking into account all available sources of information about the bit u_k.

As mentioned previously, the original turbo equaliser consists of a SISO equaliser and a SISO decoder. The SISO equaliser in Figure 14.2 generates the a-posteriori probability upon receiving the corrupted transmitted signal sequence and the a-priori probability provided by the SISO decoder. However, at the initial iteration stages — i.e. at the first turbo equalisation iteration — no a-priori information is supplied by the channel decoder. Therefore, the a-priori probability is set to $\frac{1}{2}$, since the transmitted bits are assumed to be equiprobable. Before passing the a-posteriori information generated by the SISO equaliser to the SISO decoder of Figure 14.2, the contribution of the decoder — in the form of the a-priori information — accruing from the previous iteration must be removed, in order to yield the combined channel and extrinsic information. This also minimises the correlation between the a-priori information supplied by the decoder and the a-posteriori information generated by the equaliser. The term 'combined channel and extrinsic information' indicates that they are inherently linked — in fact they are typically induced by mechanisms which exhibit memory — and hence they cannot be separated. The removal of the

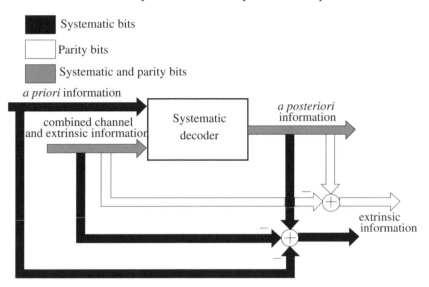

Figure 14.3: Schematic of a component decoder employed in a turbo equaliser, showing the input information received and outputs corresponding to the systematic and parity bits.

a-priori information is necessary, in order to prevent the decoder from receiving its own information, which would result in the so-called 'positive feedback' phenomenon, overwhelming the decoder's current reliability estimation of the coded bits, i.e. the extrinsic information.

The combined channel and extrinsic information is channel de-interleaved and directed to the SISO decoder, as depicted in Figure 14.2. Subsequently, the SISO decoder computes the a-posteriori probability of the coded bits. Note that the latter steps are different from those in turbo decoding. As illustrated previously in Figure 5.9, the component decoders in a turbo decoder only produce the a-posteriori probability of the source bits rather than those of all channel-coded bits. By contrast, the component decoder of the turbo equaliser computes the a-posteriori LLR values for both the parity and systematic bits, as shown in Figure 14.3. The combined channel and extrinsic information is then removed from the a-posteriori information provided by the decoder in Figure 14.2 before channel interleaving, in order to yield the extrinsic information. This is achieved by subtracting the systematic channel and extrinsic LLR values and the a-priori LLRs from the systematic bits of the a-posteriori LLRs, as illustrated in Figure 14.3. However, this only provides the extrinsic LLRs corresponding to the systematic bits. As shown in Figure 14.3, for the extrinsic LLRs of the parity bits, only the channel and extrinsic LLRs of the parity bits are subtracted from the a-posteriori LLRs corresponding to the parity bits. The aim of subtracting these input LLRs from the a-posteriori LLRs is to prevent the channel equaliser from receiving information based on its own decisions, which was generated in the previous turbo equalisation iteration. The extrinsic information computed is then employed as the a-priori input information of the equaliser in the next channel equalisation process. This constitutes the first turbo equalisation iteration. The iterative process is repeated until the required termination criteria are met [345]. At this stage, the a-posteriori information of the source bits, which has been generated by the decoder, is utilised to estimate the transmitted bits. A more in-depth treatment of the turbo equalisation principles will be presented in Section 14.2.

14.1   Motivation

Following the pioneering turbo equalisation research of Douillard *et al.* [105], further work was conducted by Gertsman and Lodge [346], who demonstrated that the iterative process of turbo equalisation can compensate for the performance degradations due to imperfect channel impulse response estimation. In the context of non-coherent detection, Marsland *et al.* demonstrated in reference [347] that turbo equalisation offered better performance, than Dai and Shwedyk's non-coherent, hard-decision-based receiver using a bank of Kalman filters [348]. Turbo codes have also been used in conjunction with turbo equalisers by Raphaeli and Zarai [106], giving increased improvement due to the performance gain of turbo coding, as well as due to the ISI mitigation achieved by employing turbo equalisation. Bauch and Franz [326] as well as Jordan and Kammeyer [349] then demonstrated how the turbo equaliser can be applied in the context of GSM — employing GMSK modulation — although having to make modifications due to the interburst interleaving scheme used in GSM. Turbo equalisation was also utilised, in order to improve the performance of a convolutional-coded GMSK system employing a differential phase detector and decoder [350]. Recent work by Narayanan and Stüber [351] demonstrated the advantage of employing turbo equalisation in the context of coded systems invoking recursive modulators, such as Differential Phase Shift Keying (DPSK). Narayanan and Stuber emphasised the importance of a recursive modulator and showed that high iteration gains could be achieved, even when there was no ISI in the channel, i.e. for transmission over the non-dispersive Gaussian channel. The above-mentioned advantages of turbo equalisation as well as the importance of a recursive modulator motivated our research on turbo equalisation of coded GMSK systems, since GMSK is also recursive in its nature. The recursive nature of GMSK modulation will be made explicit during our further discourse in Chapter 15.

The outline of this chapter is as follows. In Section 14.2, we will describe the functionality of the turbo equaliser in the context of convolutional-coded partial-response GMSK systems and portray how the original concept of turbo equalisation is extended to multiple encoder assisted systems, such as turbo-coded schemes. Subsequently, Sections 14.3 and 14.4 describe how the Log-MAP algorithm is employed in the context of channel equalisation and channel decoding, respectively. The basic principles of turbo equalisation are then highlighted with the aid of a simple convolutional-coded BPSK system — employing turbo equalisation as an example — in Section 14.5. In Section 14.6 the turbo equalisation operations are summarised. Subsequently, Section 14.7 presents the turbo equalisation performance of the convolutional-coded GMSK system, that of the convolutional-coding-based turbo-coded GMSK scheme and the Bose–Chaudhuri–Hocquengham (BCH) [15] coding-based turbo-coded GMSK system, followed by a discussion in Section 14.8.

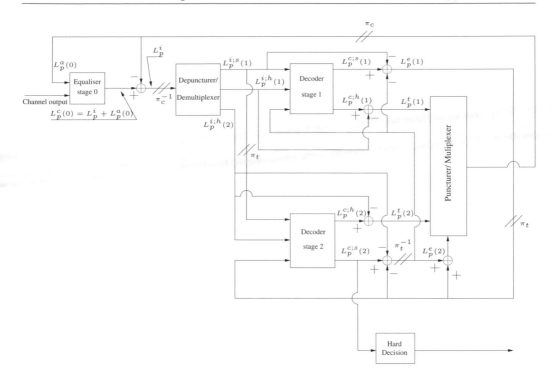

Figure 14.4: Structure of the turbo equaliser using $N_d = 2$ component decoders. For conceptual simplicity, we have omitted the interleavers and only marked the interleaver positions, where π_c and π_t represent the channel interleaver and turbo interleaver, respectively. The superscript '-1' is used to denote a de-interleaver.

14.2   Principle of Turbo Equalisation Using Single/Multiple Decoder(s)

With reference to Figure 14.4, in this section we describe the turbo equalisation principle for a baseband receiver consisting of an equaliser and N_d component decoders. This is an extension of the original turbo equalisation scheme [105], illustrated in Figure 14.2. For simplicity, we have omitted the channel interleaver π_c and turbo interleavers π_t and have only marked their positions. The superscript '-1' is used to represent a de-interleaver. Typically, for turbo codes there are $N_d = 2$ component decoders, whereas for non-iterative convolutional decoding we have $N_d = 1$. A vital rule for such turbo equalisers is that the input information to a particular block in the current iteration must not include the information contributed by that particular block from the previous iteration, since then the information used in consecutive iterations would be dependent on each other [346]. We will elaborate on this issue later in more depth.

The equaliser and decoder in Figure 14.4 employ a SISO algorithm, such as the optimal MAP algorithm [11], the Log-MAP algorithm [52] or the Soft Output Viterbi Algorithm (SOVA) [53, 54, 122], which yields the a-posteriori information. As defined previously, the a-posteriori information concerning a bit is the information that the SISO block generates taking into account all available sources of information about the bit. When using the MAP algorithm described in Chapter 5 or the Log-MAP algorithm of Section 13.8.1, we express the a-posteriori information in terms of its LLR [126]. Again, the LLR L^{v_m} of a bit v_m is defined as the natural logarithm of the ratio of the probabilities of the bit taking its two possible values of $+1$ and -1:

$$L^{v_m} \triangleq \ln \frac{P(v_m = +1)}{P(v_m = -1)}. \tag{14.1}$$

For clarity, we have employed the approach used by Gertsman and Lodge [346] and expressed the LLR of the equaliser and decoder using vector notation. The associated superscript denotes the nature of the LLR, namely

L^c: composite a-posteriori LLR consisting of source or information and parity bit a-posteriori information.

$L^{c;s}$: a-posteriori information of the source bit.

$L^{c;h}$: a-posteriori information of the parity bit.

L^i: combined channel and the so-called extrinsic information, defined previously, for the source and parity bits.

$L^{i;s}$: combined channel and extrinsic information for the source bits.

$L^{i;h}$: combined channel and extrinsic information for the parity bits.

L^e: extrinsic information corresponding to the source bit.

L^t: extrinsic information of the parity bit.

$L^{a;s}$: a-priori information of the source bit.

$L^{a;h}$: a-priori information of the parity bit.

Furthermore, the subscript notation is used to represent the iteration index, while the argument within the brackets ( ) is the index of the receiver stage.

At the equaliser — which is denoted by stage 0 in Figure 14.4 — the composite output a-posteriori LLR $L_p^c(0)$ at iteration p is generated by the sum of the LLR of the a-priori information $L_p^a(0)$ and the combined LLR of the channel and extrinsic information L_p^i, giving:

$$L_p^c(0) = L_p^i + L_p^a(0). \tag{14.2}$$

As noted before in Section 13.8.1, at the commencement of the iterative detection there is no a-priori information about the bit to be decoded and hence $L_p^a(0) = 0$ is used, indicating an equal probability of a binary one and zero. During the forthcoming iterations, however, our estimation concerning this bit can be fed back to the input of the equaliser in Figure 14.4, in order to aid its further iterations. We also have to augment the concept of a-priori and extrinsic information. Recall that the former is the information related to a specific bit, while the latter — as suggested by the terminology — indicates the information indirectly related to a bit, for example gained from the parity bits or redundant bits generated by the turbo encoder of Figure 13.20. We are unable to separate the channel's output information and extrinsic information at the output of the equaliser, which is denoted as L_p^i, since the channel impulse response can be viewed as that of a non-systematic code [326] 'smearing' the effects of the channel's input bits over time due to the associated convolution operation. Consider stage 0 of Figure 14.5, which is a detailed illustration of the equaliser, with emphasis on the input and output information. It is noted that the composite a-posteriori information $L_p^c(0)$, the a-priori information $L_p^a(0)$ of the equaliser as

Figure 14.5: SISO equaliser illustration with emphasis on the input and output information at the pth iteration.

well as the combined channel and extrinsic LLR L_p^i reflect the reliability of not only the source bits, but also the parity bits. The term L_p^i can be written as a vector, given as:

$$L_p^i \in \left\{ \underbrace{L_p^{i;s}}_{\text{source bit}} ; \underbrace{L_p^{i;h}}_{\text{parity bits}} \right\}. \tag{14.3}$$

As for the stage 0 a-priori information $L_p^a(0)$, this vector is obtained from the other N_d decoder stages — namely stages 1 and 2 in Figure 14.4 — and contains the a-priori information of the encoded bits. Therefore, like vector L_p^i, $L_p^a(0)$ is also a vector consisting of the a-priori information for the source and parity bits, hence giving:

$$L_p^a(0) \in \left\{ \underbrace{L_p^{a;s}(0); L_p^{a;h}(0)}_{\text{source bit}\quad\text{parity bits}} \right\}$$

$$\in \left\{ \underbrace{\sum_{j=1}^{N_d} L_{p-1}^e(j)}_{\text{source bit}}; \underbrace{L_{p-1}^t(1); \dots ; L_{p-1}^t(N_d)}_{\text{parity bits}} \right\}$$

where $L_{p-1}^e(j)$ and $L_{p-1}^t(j)$ are the source and parity extrinsic LLRs of the jth decoder stage from the previous iteration $p-1$ while N_d is the number of component decoders in Figure 14.4, as before. Using Equations 14.2, 14.3 and 14.2, the composite a-posteriori LLRs of the encoded bits $L_p^c(0)$ at the output of the equaliser in Figure 14.4 can be expressed as:

$$L_p^c(0) \in \left\{ \underbrace{L_p^{c;s}(0); L_p^{c;h}(0)}_{\text{source bit}\quad\text{parity bits}} \right\}$$

$$\in \left\{ \underbrace{L_p^{i;s} + \sum_{j=1}^{N_d} L_{p-1}^e(j)}_{\text{source bit}}; \underbrace{[L_p^{i;h} + L_{p-1}^t(1); \dots ; L_p^{i;h} + L_{p-1}^t(N_d)]}_{\text{parity bits}} \right\}.$$

The stage b decoder in Figure 14.4 is illustrated here again in more detail in Figure 14.6, with emphasis on its input and output information. It receives the sum of extrinsic information from

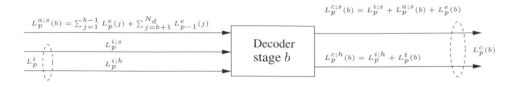

Figure 14.6: Schematic of the SISO decoder with emphasis on the input and output information at the pth iteration.

all the other decoders excluding the stage b decoder — namely $\sum_{j=1}^{b-1} L_p^e(j) + \sum_{j=b+1}^{N_d} L_{p-1}^e(j)$ — as the a-priori information and also the combined channel and extrinsic information of the source and parity LLR values, namely $L_p^{i;s}$ and $L_p^{i;h}$, respectively. The augmented source a-priori information of the stage b decoder at the pth iteration can be expressed as:

$$L_p^{a;s}(b) = \sum_{j=1}^{b-1} L_p^e(j) + \sum_{j=b+1}^{N_d} L_{p-1}^e(j). \tag{14.4}$$

From Equation 14.4 we observe that the a-priori information $L_p^{a;s}(b)$ of the source bits consists of the extrinsic information from the decoders of the **previous stages** — namely for $j < b$ in the **current iteration** p — and from the decoders of the **later stages** — namely for $j > b$ from the **previous iteration** $p-1$ — but does not include any extrinsic information from stage b. Note that unlike the equaliser, which receives the a-priori information of the source and parity bits, the N_d decoders accept a-priori LLRs of source bits only. This is because the source bit is the only common information to all decoders, whether in the interleaved order or the original sequence, whereas the parity bits are exclusive to a particular decoder, hence they are unable to contribute in the other decoders' decoding operation.

At the stage b systematic decoder — for example, at the outputs of the stages 1 and 2 in Figure 14.4 and in greater detail in Figure 14.6 — the composite a-posteriori LLR $L_p^c(b)$ consists of two LLR vectors, namely the source bit a-posteriori information $L_p^{c;s}(b)$ and the parity bit a-posteriori information $L_p^{c;h}(b)$:

$$L_p^c(b) \in \left\{ L_p^{c;s}(b); L_p^{c;h}(b) \right\}. \tag{14.5}$$

The composite source a-posteriori LLR $L_p^{c;s}(b)$ — which is the first component in Equation 14.5 — can be expressed as the sum of the augmented a-priori information $L_p^{a;s}(b)$, the extrinsic information $L_p^e(b)$ produced by the decoder stage and the combined channel and extrinsic LLR $L_p^{i;s}$ of the source bit from the equaliser:

$$L_p^{c;s}(b) = L_p^{a;s}(b) + L_p^e(b) + L_p^{i;s}, \tag{14.6}$$

which can be rewritten by using Equations 14.4 and 14.6 as:

$$\begin{aligned} L_p^{c;s}(b) &= L_p^{a;s}(b) + L_p^e(b) + L_p^{i;s} \\ &= \sum_{j=1}^{b-1} L_p^e(j) + \sum_{j=b+1}^{N_d} L_{p-1}^e(j) + L_p^e(b) + L_p^{i;s} \\ &= \sum_{j=1}^{b} L_p^e(j) + \sum_{j=b+1}^{N_d} L_{p-1}^e(j) + L_p^{i;s}, \end{aligned} \tag{14.7}$$

while the parity a-posteriori LLR $L_p^{c;h}(b)$ of Equation 14.5 can be expressed as:

$$L_p^{c;h}(b) = L_p^{i;h} + L_p^t(b). \tag{14.8}$$

The term $L_p^{i;h}(b)$ is the combined channel and extrinsic information of the parity bit and $L_p^t(b)$ is the extrinsic LLR corresponding to the parity bit.

In general, only two component encoders are used in practical turbo encoders. Therefore, by substituting $N_d = 2$ into Equations 14.2, 14.7 and 14.8, we can determine the a-posteriori LLRs of the equaliser and decoders, whereas Equations 14.2 and 14.4 can be used to determine the corresponding a-priori inputs to the equaliser and decoders. Again, Figure 14.4 illustrates the structure of the turbo equaliser, which employs the principles discussed above.

In the following sections we will concentrate on the main components of Figure 14.4, namely on the equaliser and the decoder(s).

14.3 Soft-in/Soft-out Equaliser for Turbo Equalisation

In Section 13.8.3 we discussed how the Log-MAP algorithm can be used as a soft-output GMSK equaliser in conjunction with turbo decoding. However, in this subsection, the iterations were only performed in the turbo decoder. Therefore, the equaliser did not receive any a-priori values from an independent source. However, when implementing turbo equalisation, the SISO equaliser will receive not only channel outputs, but also a-priori information from the SISO decoder. We can still employ the Log-MAP algorithm described in Section 13.8.3 for the soft-output GMSK equaliser, but we must now consider additionally the a-priori information from the SISO decoder at the turbo equaliser's input. Explicitly, the term $P(u_m)$ — which represents the a-priori probability of the code bits — required when evaluating $\Gamma_m(\check{k}, \kappa)$ in Equation 13.42 is no longer $\frac{1}{2}$, since the probability of the bit u_m being $+1$ or -1 is no longer equal. Instead, the probability $P(u_m)$ is augmented with the aid of the a-priori information from the decoder(s), as seen in Equation 14.2.

Typically, the Viterbi algorithm [8,9] — which is a maximum likelihood sequence estimation algorithm — is used to decode convolutional codes. However, in turbo equalisation we need soft outputs which represent the LLRs of the individual coded bits. In the following subsection, we will show how the LLRs of these coded bits are determined.

14.4 Soft-in/Soft-out Decoder for Turbo Equalisation

In turbo equalisation the SISO decoder block accepts soft inputs from the SISO equaliser and gives the LLR of not only the source bits, but also the coded bits, in order to implement turbo equalisation. Therefore, the decoder must employ SISO algorithms, such as the MAP algorithm, Log-Map algorithm or the SOVA. Note that since the Log-MAP algorithm yields identical performance to the optimal MAP algorithm, while incurring lower complexity, we will use this algorithm in our discussions.

Through the Log-MAP algorithm the LLRs of the source and those of the coded bits — which are output by the turbo decoder and turbo equaliser, respectively — can be computed in two key steps. Let us define the notation used before highlighting these steps. In the decoder the trellis transitions are denoted by the index d, whereas the equaliser transition intervals are represented by m. The notation v_d represents the source bit, while $c_{l,d}$ represents the lth coded bit at the dth decoder trellis interval. Relating the equaliser transition intervals to the rate $R = \frac{k}{n}$

decoder trellis intervals, we note that the lapse of $\frac{n}{K_b}$ number of equaliser intervals, where K_b is the number of bits used to represent a symbol, corresponds to a single decoder trellis interval. For example, one decoder transition interval in a rate $R = \frac{1}{2}$ decoder is equivalent to $\frac{n=2}{K_b=1} = 2$ BPSK equaliser intervals. Therefore, the bits u_m and u_{m+1} received at the BPSK equaliser correspond to the code bits $c_{1,d}$ and $c_{2,d}$ at decoder interval d. Having defined the notation used, let us now return to the discussion of the two key steps employed to determine the LLR values of the source and coded bits.

First, the forward and backward recursion values, $A_m(\kappa)$ and $B_m(\kappa)$, respectively, as well as the transition metric $\Gamma_m(\check{k}, \kappa)$ — which are the logarithmic domain counterparts of $\alpha_m(\kappa)$, $\beta_m(\kappa)$ and $\gamma_m(\check{k}, \kappa)$ in the MAP algorithm — are computed using Equations 13.8.3, 13.8.3 and 13.42. However, for the decoder the trellis transition index m is replaced by d. Second, we have to modify Equation 13.8.3 — which was discussed previously in Section 13.8.1 in the context of channel equalisation for turbo decoding — in order for the SISO decoder to give the LLR of not only the source bits v_d, but also the coded bits, namely $c_{l,d}$. Explicitly, for the rate $\frac{k}{n}$, convolutional encoder there are n number of LLR values generated at each trellis interval, whereas the binary equaliser produces only a single LLR value at each interval. Recalling Equation 13.8.3, which is repeated here for convenience:

$$L(u_m|y) = \ln \left(\frac{\displaystyle\sum_{(\check{k},\kappa)\Rightarrow u_m=+1} \exp\left(A_{m-1}(\check{k}) + \Gamma_m(\check{k}, \kappa) + B_m(\kappa)\right)}{\displaystyle\sum_{(\check{k},\kappa)\Rightarrow u_m=-1} \exp\left(A_{m-1}(\check{k}) + \Gamma_m(\check{k}, \kappa) + B_m(\kappa)\right)} \right),$$

it is noted that at the mth equaliser trellis interval, the numerator is the sum of the transition probabilities associated with the paths caused by the coded bit $u_m = +1$, whereas the denominator is the sum of the probabilities corresponding to the transitions generated by $u_m = -1$. As explained in Section 13.8.3 regarding the Log-MAP algorithm, these transition probabilities can be written as the sum of exponential terms whose arguments are the sum of $A_{m-1}(\check{k})$, $B_m(\kappa)$, and $\Gamma_m(\check{k}, \kappa)$. Therefore, at each equaliser trellis interval m, the LLR value obtained is for the coded bit u_m. However, for the $R = \frac{k}{n}$ decoder, we require the LLR values for the n number of coded bits. Therefore, in order to determine the LLR for each coded bit $c_{l,d}$ for $l = 1 \ldots n$, we must first consider the branches and the corresponding probability value — which is the summation of the term $\exp((A_{d-1}(\check{k}) + \Gamma_d(\check{k}, \kappa) + B_d(\kappa)))$ — giving the coded bit $c_{l,d} = +1$ and subsequently the branches resulting in the coded bit of $c_{l,d} = -1$. Consequently, Equation 13.8.3 is modified to:

$$L(c_{l,d}|L_p^i) = \ln \left(\frac{\displaystyle\sum_{(\check{k},\kappa)\Rightarrow c_{l,d}=+1} \exp\left(A_{d-1}(\check{k}) + \Gamma_d(\check{k}, \kappa) + B_d(\kappa)\right)}{\displaystyle\sum_{(\check{k},\kappa)\Rightarrow c_{l,d}=-1} \exp\left(A_{d-1}(\check{k}) + \Gamma_d(\check{k}, \kappa) + B_d(\kappa)\right)} \right)$$

$$= \ln \left(\displaystyle\sum_{(\check{k},\kappa)\Rightarrow c_{l,d}=+1} \exp\left(A_{d-1}(\check{k}) + \Gamma_d(\check{k}, \kappa) + B_d(\kappa)\right) \right)$$

$$- \ln \left(\displaystyle\sum_{(\check{k},\kappa)\Rightarrow c_{l,d}=-1} \exp\left(A_{d-1}(\check{k}) + \Gamma_d(\check{k}, \kappa) + B_d(\kappa)\right) \right),$$

in order to calculate the LLR of the coded bits $c_{l,d}$, where $L(c_{l,d}|L_p^i)$ is the LLR of the lth coded bit at trellis interval d, given that the combined channel and extrinsic LLR L_p^i was received by the decoder.

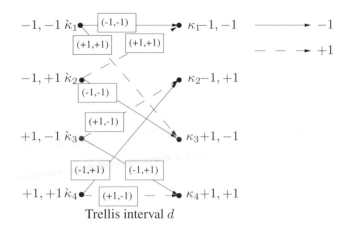

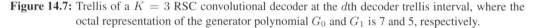

Trellis interval d

Figure 14.7: Trellis of a $K = 3$ RSC convolutional decoder at the dth decoder trellis interval, where the octal representation of the generator polynomial G_0 and G_1 is 7 and 5, respectively.

Let us now consider the example of a SISO convolutional decoder trellis for a rate $R = \frac{1}{2}$, constraint length $K = 3$ RSC encoder, as illustrated in Figure 14.7. In this example, a RSC code has been chosen instead of a non-systematic convolutional code, since we will be implementing turbo coding, which has been shown to perform better with the aid of RSC component codes [12, 13, 59]. Since the code rate is $R = \frac{1}{2}$, the output codeword consists of $k = 2$ coded bits, $c_{1,d}$ and $c_{2,d}$, at each decoder trellis interval d. The output codeword for each transition is depicted as $(c_{1,d}, c_{2,d})$ in Figure 14.7 and is governed by the generator polynomials G_0 and G_1, which can be represented in the octal form as 7 and 5, respectively. In order to determine the LLR for $c_{1,d}$, we have to consider all paths which give the coded bit $c_{1,d} = +1$. From Figure 14.7 we observe that there are four branches which fulfil this requirement, namely branches from state $\check{\kappa}_1$ to κ_3, $\check{\kappa}_2$ to κ_1, $\check{\kappa}_3$ to κ_2 and $\check{\kappa}_4$ to κ_4. Observing Equations 13.31 and 13.8.3 and by substituting Equations 13.36, 13.37 and 13.38 into Equations 13.8.3, we note that the probability that a transition from state $\check{\kappa}$ to κ occurs and that the combined channel and extrinsic LLR L_p^i was obtained is given by:

$$P(\check{\kappa} \wedge \kappa \wedge L_p^i) = \exp\left(A_{d-1}(\check{\kappa}) + \Gamma_d(\check{\kappa}, \kappa) + B_d(\kappa)\right). \tag{14.9}$$

Therefore, the individual branch probabilities associated with the coded bit of $c_{1,d} = +1$ will be:

$$
\begin{aligned}
P(\check{\kappa}_1 \wedge \kappa_3 \wedge L_p^i) &= \exp\left(A_{d-1}(\check{\kappa}_1) + \Gamma_d(\check{\kappa}_1, \kappa_3) + B_d(\kappa_3)\right) &= \exp(x(\check{\kappa}_1, \kappa_3)) \\
P(\check{\kappa}_2 \wedge \kappa_1 \wedge L_p^i) &= \exp\left(A_{d-1}(\check{\kappa}_2) + \Gamma_d(\check{\kappa}_2, \kappa_1) + B_d(\kappa_1)\right) &= \exp(x(\check{\kappa}_2, \kappa_1)) \\
P(\check{\kappa}_3 \wedge \kappa_2 \wedge L_p^i) &= \exp\left(A_{d-1}(\check{\kappa}_3) + \Gamma_d(\check{\kappa}_3, \kappa_2) + B_d(\kappa_2)\right) &= \exp(x(\check{\kappa}_3, \kappa_2)) \\
P(\check{\kappa}_4 \wedge \kappa_4 \wedge L_p^i) &= \exp\left(A_{d-1}(\check{\kappa}_4) + \Gamma_d(\check{\kappa}_4, \kappa_4) + B_d(\kappa_4)\right) &= \exp(x(\check{\kappa}_4, \kappa_4)),
\end{aligned}
\tag{14.10}
$$

where $x(\check{\kappa}_i, \kappa_j)$ represents the sum $A_{d-1}(\check{\kappa}_i) + \Gamma_d(\check{\kappa}_i, \kappa_j) + B_d(\kappa_j)$. As before, $A_{d-1}(\check{\kappa})$ and $B_d(\kappa)$ are evaluated once $\Gamma_d(\check{\kappa}, \kappa)$ has been computed. However, the expression of $\Gamma_d(\check{\kappa}, \kappa)$ for the decoder can be further simplified from Equation 13.42, which is repeated here for convenience:

$$\Gamma_m(\check{\kappa}, \kappa) = -\frac{1}{2\sigma^2}(y_m - \hat{y}_m)^2 + \ln(P(u_m)). \tag{14.11}$$

Observe that $\Gamma_m(\check{k}, \kappa)$ is related to the squared Euclidean distance between the sampled received signal y_m and the expected signal $\hat{y}_m$, which is subsequently normalised by $2\sigma^2$. The above expression can be expanded to:

$$\Gamma_m(\check{k}, \kappa) = -\frac{1}{2\sigma^2}(y_m^2 - 2 \cdot y_m \cdot \hat{y}_m + \hat{y}_m^2) + \ln(P(u_m)). \tag{14.12}$$

The expected signals $\hat{y}_m$ at the decoder are the coded bits of values $+1$ as well as -1. Therefore, the square of the expected signal $\hat{y}_m$ is always unity and hence it can be ignored. Furthermore, since the same value of y_m is considered for each trellis transition, it too can be neglected during the associated pattern matching process, leaving the decoder metric $\Gamma_d(\check{k}, \kappa)$ to be dependent on the cross-correlative term $(2 \cdot y_m \cdot \hat{y}_m)/2\sigma^2$. At the input of the decoder, the combined channel and extrinsic information L_p^i — containing the contribution of $2\sigma^2$ [326] — is received, while the locally generated expected signals are associated with the trellis transitions due to the coded bits $c_{l,d}$, for $l = 1 \ldots n$, where n is the number of codeword bits. Hence, the decoder metric $\Gamma_d(\check{k}, \kappa)$ at decoder trellis interval d can be expressed as [326]:

$$\Gamma_d(\check{k}, \kappa) = \sum_{l=1}^{n=2} \left(\frac{1}{2} L_p^i \cdot c_{l,d} \right) + \ln(P(v_d)), \tag{14.13}$$

where $P(v_d)$ is the a-priori probability of the source bit v_d. After evaluating $\Gamma_d(\check{k}, \kappa)$, the terms $A_{d-1}(\check{k})$ and $B_d(\kappa)$ can be determined using Equations 13.8.3 and 13.8.3, respectively. Therefore, the probability that $c_{1,d} = +1$ was transmitted at instant d and that L_p^i was received is determined by the sum of the $c_{1,d} = +1$ related individual branch probabilities of $P(\check{k}_1 \wedge \kappa_3 \wedge L_p^i)$, $P(\check{k}_2 \wedge \kappa_1 \wedge L_p^i)$, $P(\check{k}_3 \wedge \kappa_2 \wedge L_p^i)$ and $P(\check{k}_4 \wedge \kappa_4 \wedge L_p^i)$.

Similarly, we must consider all paths in Figure 14.7 which give $c_{1,d} = -1$, namely the branches from state $\check{k}_1$ to κ_1, $\check{k}_2$ to κ_3, $\check{k}_3$ to κ_4 and $\check{k}_4$ to κ_2. Having observed this, with the aid of Figure 14.7 we proceed to calculate the individual branch probabilities associated with the coded bit of $c_{1,d} = -1$ as follows:

$$
\begin{aligned}
P(\check{k}_1 \wedge \kappa_1 \wedge L_p^i) &= \exp\left(A_{d-1}(\check{k}_1) + \Gamma_d(\check{k}_1, \kappa_1) + B_d(\kappa_1)\right) &= \exp(x(\check{k}_1, \kappa_1)) \\
P(\check{k}_2 \wedge \kappa_3 \wedge L_p^i) &= \exp\left(A_{d-1}(\check{k}_2) + \Gamma_d(\check{k}_2, \kappa_3) + B_d(\kappa_3)\right) &= \exp(x(\check{k}_2, \kappa_3)) \\
P(\check{k}_3 \wedge \kappa_4 \wedge L_p^i) &= \exp\left(A_{d-1}(\check{k}_3) + \Gamma_d(\check{k}_3, \kappa_4) + B_d(\kappa_4)\right) &= \exp(x(\check{k}_3, \kappa_4)) \\
P(\check{k}_4 \wedge \kappa_2 \wedge L_p^i) &= \exp\left(A_{d-1}(\check{k}_4) + \Gamma_d(\check{k}_4, \kappa_2) + B_d(\kappa_2)\right) &= \exp(x(\check{k}_4, \kappa_2)).
\end{aligned}
$$
$$\tag{14.14}$$

As before, we can evaluate the probability that $c_{1,d} = -1$ was transmitted at trellis interval d given that L_p^i was received by taking the sum of the individual branch probabilities of $P(\check{k}_1 \wedge \kappa_1 \wedge L_p^i)$, $P(\check{k}_2 \wedge \kappa_3 \wedge L_p^i)$, $P(\check{k}_3 \wedge \kappa_4 \wedge L_p^i)$ and $P(\check{k}_4 \wedge \kappa_2 \wedge L_p^i)$.

With the aid of Equation 14.4 and using Equations 14.9, 14.10 and 14.14, we can express the

LLR for the first coded bit $c_{1,d}$ as:

$$
\begin{aligned}
L(c_{1,d}|L_p^i) &= \ln\left(\frac{\displaystyle\sum_{(\grave{k},\kappa)\Rightarrow c_{1,d}=+1} P(\grave{k}\wedge\kappa\wedge L_p^i)}{\displaystyle\sum_{(\grave{k},\kappa)\Rightarrow c_{1,d}=-1} P(\grave{k}\wedge\kappa\wedge L_p^i)} \right) \\
&= \ln\left(\frac{P(\grave{k}_1\wedge\kappa_3\wedge L_p^i) + P(\grave{k}_2\wedge\kappa_1\wedge L_p^i) + P(\grave{k}_3\wedge\kappa_2\wedge L_p^i) + P(\grave{k}_4\wedge\kappa_4\wedge L_p^i)}{P(\grave{k}_1\wedge\kappa_1\wedge L_p^i) + P(\grave{k}_2\wedge\kappa_3\wedge L_p^i) + P(\grave{k}_3\wedge\kappa_4\wedge L_p^i) + P(\grave{k}_4\wedge\kappa_2\wedge L_p^i)} \right) \\
&= \ln\Big(\exp(x(\grave{k}_1,\kappa_3)) + \exp(x(\grave{k}_2,\kappa_1)) + \exp(x(\grave{k}_3,\kappa_2)) + \exp(x(\grave{k}_4,\kappa_4)) \Big) \\
&\quad - \ln\Big(\exp(x(\grave{k}_1,\kappa_1)) + \exp(x(\grave{k}_2,\kappa_3)) + \exp(x(\grave{k}_3,\kappa_4)) + \exp(x(\grave{k}_4,\kappa_2)) \Big). \quad (14.15)
\end{aligned}
$$

Since the term $L(c_{1,d}|L_p^i)$ can be expressed as the natural logarithm of a sum of exponentials, the generalised Jacobian logarithmic relationship of Equation 13.44 can be employed, in order to calculate the LLR of the coded bit $c_{1,d}$ at trellis interval d.

By employing the same approach as above we can demonstrate that the LLR of the coded bit $c_{2,d}$ is given by:

$$
\begin{aligned}
L(c_{2,d}|L_p^i) &= \ln\left(\frac{\displaystyle\sum_{(\grave{k},\kappa)\Rightarrow c_{2,d}=+1} P(\grave{k}\wedge\kappa\wedge L_p^i)}{\displaystyle\sum_{(\grave{k},\kappa)\Rightarrow c_{2,d}=-1} P(\grave{k}\wedge\kappa\wedge L_p^i)} \right) \\
&= \ln\left(\frac{P(\grave{k}_1\wedge\kappa_3\wedge L_p^i) + P(\grave{k}_2\wedge\kappa_1\wedge L_p^i) + P(\grave{k}_3\wedge\kappa_4\wedge L_p^i) + P(\grave{k}_4\wedge\kappa_2\wedge L_p^i)}{P(\grave{k}_1\wedge\kappa_1\wedge L_p^i) + P(\grave{k}_2\wedge\kappa_3\wedge L_p^i) + P(\grave{k}_3\wedge\kappa_2\wedge L_p^i) + P(\grave{k}_4\wedge\kappa_4\wedge L_p^i)} \right) \\
&= \ln\Big(\exp(x(\grave{k}_1,\kappa_3)) + \exp(x(\grave{k}_2,\kappa_1)) + \exp(x(\grave{k}_3,\kappa_4)) + \exp(x(\grave{k}_4,\kappa_2)) \Big) \\
&\quad - \ln\Big(\exp(x(\grave{k}_1,\kappa_1)) + \exp(x(\grave{k}_2,\kappa_3)) + \exp(x(\grave{k}_3,\kappa_2)) + \exp(x(\grave{k}_4,\kappa_4)) \Big). \quad (14.16)
\end{aligned}
$$

The generalised Jacobian relationship in Equation 13.44 is then used to evaluate the sum of the exponential terms in Equations 14.15 and 14.16 in order to yield the LLR values $L(c_{2,d}|L_p^i)$ of the coded bit $c_{2,d}$ at trellis interval d.

14.5 Turbo Equalisation Example

In order to highlight the principle of turbo equalisation, let us now consider a simple convolutional-coded BPSK system transmitting over a three-path, symbol-spaced static channel detailed in Figure 14.8 and employing a turbo equaliser at the receiver, as shown in Figure 14.9. A rate $R = 0.5$, constraint length $K = 3$ RSC encoder using octal generator polynomials of $G_0 = 7$ and $G_1 = 5$ was utilised. The corresponding decoder trellis structure and its legitimate transitions were illustrated in Figure 14.7. Note that this turbo equaliser schematic — consisting of a trellis-based channel equaliser and a channel decoder — was first introduced in Figure 14.2, where its basic operation was highlighted. Hence here we refrain from detailing its philosophy.

Since the maximum dispersion of the channel impulse response is $\tau_d = 2$ bit periods, the trellis-based BPSK equaliser requires $2^{\tau_d=2} = 4$ states, as shown in Figure 14.10. Each of these four states can be viewed as those of a shift register possessing two memory elements and can therefore be represented by two binary bits. Consider state 1 in Figure 14.10, which

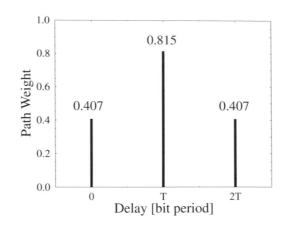

Figure 14.8: Three-path symbol-spaced static channel.

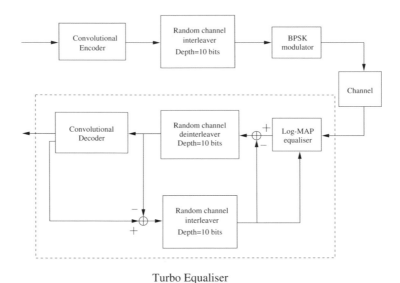

Turbo Equaliser

Figure 14.9: Schematic of the convolutional-coded BPSK system using the turbo equaliser scheme of Figure 14.2.

corresponds to the binary bits 1 and 0 in the shift register, where bit 0 is the 'older' bit. For an input bit of 1, the bit 0 is shifted out while the input bit 1 is now stored. Therefore, the new state reached is state 3, since the bits in the shift register memory elements are 1 and 1. Applying the same reasoning, all the other legitimate state transitions can be determined, producing the trellis structure of Figure 14.10.

In this example, the transmission burst consists of ten data bits and three tail bits, while the depth of the random channel interleaver is ten bits. The following discussion will describe the

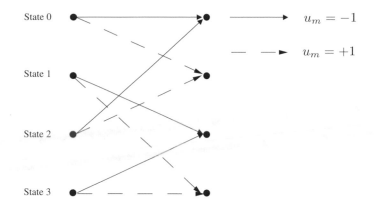

Figure 14.10: Equaliser trellis depicting the legitimate transitions.

operation of the turbo equaliser, commencing with generating the equaliser's a-posteriori LLR, which is then processed to extract the combined channel and extrinsic information to be passed to the decoder. Subsequently, it is shown how this combined channel and extrinsic information is utilised, in order to generate the decoder's a-posteriori LLR, which — as in the equaliser — is processed to obtain the extrinsic information, before passing it back to the equaliser. For simplicity, all floating point values in the following figures have been rounded to one decimal place. The trellis-based channel equaliser employs the Log-MAP algorithm described previously in Section 13.8.3. In order to evaluate the LLR value of a bit u_m, the values of $\Gamma_m(\grave{k}, \kappa)$, $A_{m-1}(\grave{k})$ and $B_m(\kappa)$ — which were described in Section 13.8.3 — must be evaluated and used jointly. This will be elaborated on numerically in the following example.

In this example, the five source bits and the corresponding ten convolutional encoded bits are specified in Table 14.1. Subsequently, these encoded bits are rearranged by a 2×5 block

Decoder trellis interval d	Source bits	Encoded bits	Interleaved encoded bits
1	+1	+1, +1	+1, −1
2	+1	+1, −1	+1, −1
3	−1	−1, −1	+1, +1
4	−1	−1, +1	−1, +1
5	+1	+1,−1	−1, −1

Table 14.1: The source bits transmitted and the corresponding encoded bits produced by a rate $R = \frac{1}{2}$ $K = 3$ recursive, systematic convolutional encoder using the octal generator polynomials of $G_0 = 7$ and $G_1 = 5$.

channel interleaver to give the new sequence detailed in Table 14.1. These interleaved bits are BPSK modulated before being transmitted through the three-path static channel of Figure 14.8. At the receiver the additive white Gaussian thermal noise corrupts the received signal. The signal is subsequently sampled and stored in a buffer, until all ten encoded bits and three tail bits arrive at the receiver. At this stage, we can begin to evaluate the transition metric $\Gamma_m(\grave{k}, \kappa)$ for all trellis intervals. Figure 14.11 illustrates the transitions in the first three trellis intervals. Specifically, starting from state 0 at trellis interval $m = 1$, the initial value of $A_0(0)$ is 0.0, while

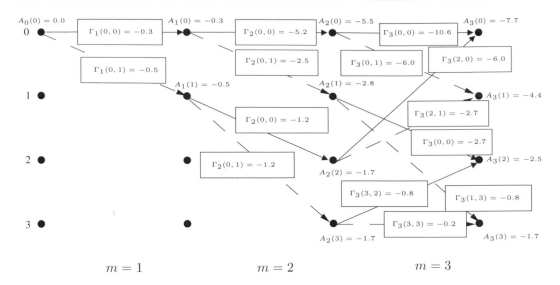

Figure 14.11: Evaluation of $A_m(\kappa)$ in the turbo equaliser of Figure 14.9 through forward recursion using a range of $\Gamma_m(\grave{\kappa}, \kappa)$ values, which were evaluated from Equations 13.38 and 5.40 by employing our simulation program over the channel of Figure 14.8.

the values of $\Gamma_1(0,0)$ and $\Gamma_1(0,1)$ are computed by using Equation 13.42, which is repeated here for convenience:

$$\Gamma_m(\grave{s}, s) = -\frac{1}{2\sigma^2}(y_m - \hat{y}_m)^2 + \ln(P(u_m)),$$

where y_m, $\hat{y}_m$, $P(u_m)$, are the sampled channel-impaired received signal, the estimated received signal and the a-priori information of the coded bits, respectively. Note that in the first iteration, no a-priori information is fed back by the channel decoder to the equaliser. Therefore, the term $P(u_m)$ is $\frac{1}{2}$, indicating that the bits $+1$ and -1 have equal probability of occurring. Recall that $A_m(\kappa)$ of Equation 13.8.3, which is repeated here for convenience, is dependent on $A_{m-1}(\grave{\kappa})$ and $\Gamma_m(\grave{\kappa}, \kappa)$:

$$A_m(\kappa) = \ln\left(\sum_{\text{all } \grave{\kappa}} \exp\left[A_{m-1}(\grave{\kappa}) + \Gamma_m(\grave{\kappa}, \kappa)\right]\right).$$

Therefore, the new values of $A_1(0)$ and $A_1(1)$ can be determined. Consider the transition from state 0 to state 0. Since we have $\Gamma_1(0,0) = -0.3$ and $A_0(0) = 0.0$, the term $A_1(0)$ becomes:

$$A_{m=1}(\kappa = 0) = \ln\left(\exp\left[A_{m=1-1}(\grave{\kappa} = 0) + \Gamma_{m=1}(\grave{\kappa} = 0, \kappa = 0)\right]\right)$$
$$A_1(0) = 0.0 + (-0.3) = -0.3, \tag{14.17}$$

while $A_1(1)$ is:

$$A_{m=1}(\kappa = 1) = \ln\left(\exp\left[A_{m=1-1}(\grave{\kappa} = 0) + \Gamma_{m=1}(\grave{\kappa} = 0, \kappa = 1)\right]\right)$$
$$A_1(1) = 0.0 + (-0.5) = -0.5. \tag{14.18}$$

At the next trellis interval of $m = 2$, the same updating process is repeated. However, in the third interval there are more than one transition reaching a particular state. Consider state 2, where there are merging transitions from state 1 and state 3. In this situation the new value of $A_3(2)$ for state 2 is:

$$A_{m=3}(\kappa = 2) = \ln(\exp\left[A_{m=2}(\check{k} = 1) + \Gamma_{m=3}(\check{k} = 1, \kappa = 2)\right]$$
$$+ \exp\left[A_{m=2}(\check{k} = 3) + \Gamma_{m=3}(\check{k} = 3, \kappa = 2)\right])$$
$$A_3(2) = \ln(\exp(-2.8 - 2.7) + \exp(-1.7 - 0.8))$$
$$= \ln(\exp(-5.5) + \exp(-2.5)),$$

which can be evaluated using the Jacobian logarithm [52] of Equation 13.8.3 and a look-up table, in order to avoid processing natural logarithm and exponential computations, yielding:

$$\ln(\exp(-5.5) + \exp(-2.5)) = \max(-5.5, -2.5) + \underbrace{\ln(1 + \exp(-|-5.5 - (-2.5)|))}_{\text{Look-up table } f_{diff} = |-5.5 - (-2.5)| = 3.0}$$

$$= -2.5 + 0.05 = -2.45,$$

where the look-up table of Reference [52] is detailed in Table 14.2. The same operation is

Range of f_{diff}	$\ln(1 + \exp(-f_{diff}))$
$f_{diff} > 3.70$	0.00
$3.70 \geq f_{diff} > 2.25$	0.05
$2.25 \geq f_{diff} > 1.50$	0.15
$1.50 \geq f_{diff} > 1.05$	0.25
$1.05 \geq f_{diff} > 0.70$	0.35
$0.70 \geq f_{diff} > 0.43$	0.45
$0.43 \geq f_{diff} > 0.20$	0.55
$f_{diff} \leqslant 0.20$	0.65

Table 14.2: Look-up table involved in order to avoid natural logarithm and exponential computations [52]. The value of f_{diff} is the absolute value of the difference between arguments of the exponential function.

performed for all the other states throughout the trellis, until all values of A_m are obtained.

The evaluation of the backward recursion term $B_m(\kappa)$ involves a similar approach. However, the starting point is at the end of the trellis, namely at trellis interval $m = 13$. Initially, the values of $B_{m=13}(\kappa)$ at trellis interval $m = 13$ are set to 0.0, as shown in Figure 14.12. Also observe that since three -1 terminating tailing bits have been inserted in the transmission burst at trellis intervals $m = 11$, $m = 12$ and $m = 13$, the only legitimate transitions at these intervals are due to bit -1. Therefore, the transitions corresponding to a transmission bit of $+1$ are illegitimate and hence are not considered. In order to evaluate the values of $B_{m=12}(\kappa)$, Equation 13.8.3 is repeated here for convenience:

$$B_{m-1}(\check{k}) = \ln\left(\sum_{\text{all } \kappa} \exp[B_m(\kappa) + \Gamma_m(\check{k}, \kappa)]\right).$$

For example, in order to determine the value of $B_{m-1=12}(\check{k} = 0)$ at $m = 13$, only the values of $B_{m=13}(\check{k} = 0)$ and $\Gamma_{m=13}(\check{k} = 0, \kappa = 0)$ are required, yielding:

$$B_{m-1=12}(\check{k} = 0) = \ln(\exp\left[B_{m=13}(\kappa = 0) + \Gamma_{m=13}(\check{k} = 0, \kappa = 0)\right])$$
$$= \ln(\exp(0.0 + 57.5))$$
$$= 57.5.$$

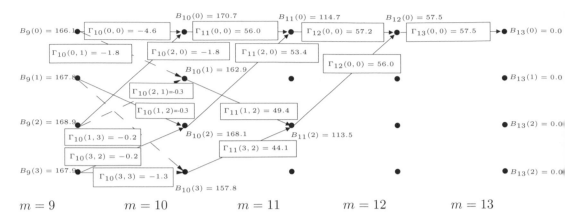

Figure 14.12: Evaluation of $B_m(\check{\kappa})$ in the turbo equaliser of Figure 14.9 through backward recursion using a range of $\Gamma_m(\check{\kappa}, \kappa)$ values, which were evaluated from Equation 13.8.3 by employing our simulation program over the channel of Figure 14.8.

This is because there is only one legitimate path, which is caused by the trellis termination bit -1. As seen in Figure 14.12, this backward recursion process is repeated in order to determine the values of $B_{m-1=11}(\check{\kappa})$ of the states $\check{\kappa} = 0 \ldots 3$ at interval $m = 12$, although owing to the tailing bits of -1 only $B_{m-1=11}(\check{\kappa} = 0)$ and $B_{m-1=11}(\check{\kappa} = 2)$ have to be computed by backtracing. At this stage, the Jacobian logarithmic relationship is not utilised, since there is only one path to be considered at state $\check{\kappa} = 0$ and state $\check{\kappa} = 2$, generated by bit -1. Consequently, $B_{m-1}(\check{\kappa})$ in Equation 13.8.3 is expressed as the natural logarithm of a single exponential term. However, at trellis interval $m = 10$ the paths associated with the bit $+1$ are also considered in Figure 14.12, since the received signal is no longer due to the -1 tailing bits. Considering state 0, it is observed that there are two transitions leaving this state during trellis interval $m = 10$, where the first is arriving at state 0, the other at state 1. Therefore, the probabilities of these transitions leaving state 0 during $m = 10$ are summed to give:

$$
\begin{aligned}
B_{m-1=9}(\check{\kappa} = 0) &= \ln(\exp\left[B_{m=10}(\kappa = 0) + \Gamma_{m=10}(\check{\kappa} = 0, \kappa = 0)\right] \\
&\quad + \exp\left[B_{m=10}(\kappa = 1) + \Gamma_{m=10}(\check{\kappa} = 0, \kappa = 1)\right]) \\
&= \ln(\exp(170.7 + (-4.6)) + \exp(162.9 + (-1.8))) \\
&= \max(166.1, 161.1) + \underbrace{\ln(1 + \exp(-|166.1 - 161.1|))}
\end{aligned}
$$

$$
\text{Look-up table:} f_{diff} = |166.1 - 161.1)| = 5.0
$$

$$
= 166.1 + 0.00 = 166.1,
$$

where the Jacobian logarithmic relationship was again employed, in order to simplify the computation of the natural logarithm of a sum of exponentials to an operation involving a maximisation process and a look-up table operation. By repeating the above backward recursion, all values of $B_m(\kappa)$ for states $\kappa = 0 \ldots 3$ and intervals $m = 1 \ldots 13$ can be determined, some of which were summarised for the sake of illustration in Figure 14.12.

Having determined the values of $A_m(\kappa)$, $B_m(\kappa)$ for all states and the $\Gamma(\check{\kappa}, \kappa)$ values associated with all transitions, the LLR of the bit u_m can be computed by using Equation 13.8.3. This involves identifying the trellis transitions caused by bit $u_m = -1$ and summing the transition probabilities of each path. Since according to Equation 13.29 the associated probabilities

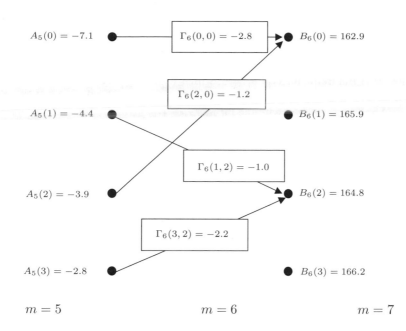

(a) Transitions corresponding to $u_m = -1$ bits

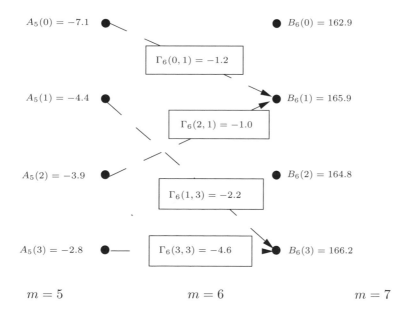

(b) Transitions corresponding to $u_m = +1$ bits

Figure 14.13: Equaliser trellis transitions, which were considered when evaluating the corresponding LLR values of bits -1 and $+1$ at trellis interval $m = 6$.

can be expressed as a sum of exponential terms, the LLR is then the natural logarithm of a sum of exponential values. Hence, the Jacobian logarithmic relationship in Equation 13.8.3 and the Look-Up Table (LUT) of Table 14.2 can again be employed in order to reduce the complexity associated with computing exponentials and logarithms. Figure 14.13(a) illustrates the set of transitions caused by $u_m = -1$, while Figure 14.13(b) depicts the set of branches corresponding to the bit $u_m = +1$ at trellis interval $m = 6$, which were extracted from the generic trellis seen in Equation 14.10. Considering this trellis interval as an example, the natural logarithm of the transition probability associated with the bit $u_m = -1$ is given by the natural logarithm of the denominator in Equation 13.8.3, yielding:

$$
\begin{aligned}
\ln[P(u_m = -1|y)] = \ln[&\exp(A_5(0) + \Gamma_6(0,0) + B_6(0)) + \exp(A_5(1) + \Gamma_6(1,2) + B_6(2)) \\
+ &\exp(A_5(2) + \Gamma_6(2,0) + B_6(0)) + \exp(A_5(3) + \Gamma_6(3,2) + B_6(2))].
\end{aligned}
$$

$$(14.19)$$

Since Equation 14.19 is the natural logarithm of the sum of four exponentials terms, the generalised Jacobian logarithmic relationship of Equation 13.44 is applied recursively, giving:

$$
\begin{aligned}
\ln[P(u_m = -1|y)] = \ln[&\exp(-7.1 + (-2.8) + 162.9) + \exp(-4.4 + (-1.0) + 164.8) \\
+ &\exp(-3.9 + (-1.2) + 162.9) + \exp(-2.8 + (-2.2) + 164.8)] \\
\approx \ln[&\exp(\overbrace{\max(153.0, 159.4) + \underbrace{0.0}_{\text{LUT}}}^{=159.4}) + \exp(-3.9 + (-1.2) + 162.9) \\
+ &\exp(-2.8 + (-2.2) + 164.8)] \\
\approx \ln[&\exp(\overbrace{\max(159.4, 157.8) + \underbrace{0.15}_{\text{LUT}}}^{=159.55}) + \exp(-2.8 + (-2.2) + 164.8)] \\
\approx &\max(159.55, 159.8) + \underbrace{0.55}_{\text{LUT}} \\
\approx &160.35.
\end{aligned}
$$

$$(14.20)$$

Similarly, the paths corresponding to bit $u_m = +1$ are grouped together with the aid of Equation 13.8.3 and Figure 14.13(b), in order to yield the natural logarithm of the probability of the bit $u_m = +1$ being received given that y was transmitted:

$$
\begin{aligned}
\ln[P(u_m = +1|y)] &= \ln[\exp(-7.1 + (-1.2) + 165.9) + \exp(-4.4 + (-2.2) + 166.2) \\
&\quad + \exp(-3.9 + (-1.0) + 165.9) + \exp(-2.8 + (-4.6) + 166.2)] \\
&\approx \ln[\exp(\overbrace{\max(157.6, 159.6)}^{=159.75} + \underbrace{0.15}_{\text{LUT}}) + \exp(-3.9 + (-1.0) + 165.9) \\
&\quad + \exp(-2.8 + (-4.6) + 166.2)] \\
&\approx \ln[\exp(\overbrace{\max(159.75, 161.0)}^{161.25} + \underbrace{0.25}_{\text{LUT}}) + \exp(-2.8 + (-4.6) + 166.2)] \\
&\approx \max(161.25, 158.8) + \underbrace{0.05}_{\text{LUT}} \\
&\approx 161.3.
\end{aligned}
$$

$$(14.21)$$

Recall from Equation 13.29, repeated here for convenience as Equation 14.5, that the LLR is the natural logarithm of the ratio of $P(u_m = +1|y)$ to $P(u_m = -1|y)$:

$$
\begin{aligned}
L(u_m|y) &\triangleq \ln\left(\frac{P(u_m = +1|y)}{P(u_m = -1|y)}\right) \\
&= \ln P(u_m = +1|y) - \ln P(u_m = -1|y).
\end{aligned}
$$

Hence, with the aid of the results from Equations 14.20 and 14.21, the a-posteriori LLR of bit u_m at trellis interval $m = 6$ is $161.3 - 160.35 = 0.95$. By applying the same approach to all trellis stages, the a-posteriori LLRs of all the 13 bits u_m can be determined. Table 14.3 shows

	Equaliser trellis interval m												
	1	2	3	4	5	6	7	8	9	10	11	12	13
APRI	0.0	0.0	0.0	0.0	0.0	0.0	0.0	0.0	0.0	0.0	0.0	0.0	0.0
APOS	2.7	2.1	−0.5	1.1	−0.1	1.0	−0.6	0.2	−0.4	−6.5	−123.0	−119.9	−116.2
EXT	2.7	2.1	−0.5	1.1	−0.1	1.0	−0.6	0.2	−0.4	−6.5	−123.0	−119.9	−116.2
TXD-C	1	−1	1	−1	1	1	−1	1	−1	−1	−1	−1	−1
DCSN	1	**1**	−1	**1**	−1	1	−1	1	−1	−1	−1	−1	−1

Table 14.3: Example of various turbo equaliser probabilities after the first iteration. Notations TXD-C, APRI, APOS, EXT and DCSN represent the transmitted encoded bits, which have been interleaved, the a-priori LLR, the a-posteriori LLR, the combined channel and extrinsic LLR, and the decision bits, respectively. In the first iteration the equaliser does not receive any a-priori information — hence APR= 0.0 — and it was observed that four bits — emphasised in bold — were corrupted.

the transmitted channel encoded bits, which have been interleaved, in order to yield the sequence (TXD-C). The associated a-priori LLRs (APRI), the a-posteriori LLRs (APOS), the combined channel and extrinsic LLRs (EXT) and the decision bits (DCSN) in the first turbo equalisation iteration are also shown in the table. At this stage, no a-priori information was received by the equaliser, yielding the values of 0.0 in the table. From the a-posteriori LLRs computed, the detected bit is determined by using a threshold of 0, i.e. a decision of $+1$ is made when the LLR value is ≥ 0 and -1 when the LLR value is < 0. The APOS value of 1.0 at $m = 6$ was due to the rounding of the previously calculated value of 0.95. It was observed in Table 14.3 that four errors — shown in bold — were made.

At this stage, we have shown how the LLR of bit u_m is determined by the Log-MAP equaliser using the example of a simple convolutional-coded BPSK system. Let us now consider the combined channel and extrinsic LLR denoted by EXT in Table 14.3, which is the only remaining quantity to be derived. At the next stage of operations, the LLR generated by the equaliser is passed to the channel decoder. However, as mentioned previously, a vital rule for turbo equalisation is that the input information to a particular block in the current iteration must not include the information contributed by that particular block from the previous iteration, since then the information used in consecutive iterations would be dependent on each other hence would fail to achieve iteration gain. Therefore, the a-posteriori LLR, which consists of the combined channel and extrinsic information plus the a-priori information provided by the channel decoder, must be processed in order to remove the previous decoder contribution in the form of the a-priori LLR, before it is passed to the decoder. This is achieved by storing the previous decoder contribution and subtracting it from the a-posteriori LLR produced by the equaliser, as was illustrated in Figure 14.9. Since in the first iteration there is no a-priori information generated by the decoder, the corresponding APRI contributions in Table 14.3 are zero and hence the a-posteriori LLR constitutes also the combined channel and extrinsic LLR, which can be passed directly to the decoder. Observe furthermore that the equaliser produces LLR values also for the three tail bits. Although these tail bit LLR values are not required by the decoder, they are employed in the equalisation process. This is because the tail bits represent contributions from the data symbols at the edge of the burst, which were spread by the dispersive channel into the adjacent symbols. Should the tail bits be ignored, low-reliability LLR values will be obtained for the bits at the edge of the received burst, since some of the symbol energy and the associated 'smeared' information is lost. Having highlighted the operation of the Log-MAP equaliser, we now proceed to describe the Log-MAP decoder in the context of turbo equalisation. As depicted in Figure 14.4, the decoder receives the combined channel and extrinsic LLR $L_p^i = \{L_p^{i;s}, L_p^{i;h}\}$, which has been de-interleaved. Here, the purpose of the de-interleaver is to minimise the correlation between the equaliser's input information and the input of the decoder. It also presents the information in the right order to the decoder, since previously the channel interleaver π_c of Figure 14.4 rearranged the bits in the order required by the equaliser. Recall that a rate $R = \frac{1}{2}$ $K = 3$ RSC encoder using the octal generator polynomials of $G_0 = 7$ and $G_1 = 5$ was employed in this example. Since the constraint length K is 3, the decoder's trellis consists of $2^{K-1} = 2^2 = 4$ states, as illustrated in Figure 14.7. As in the Log-MAP equaliser, the Log-MAP decoder also computes the values of $A_m(\kappa)$, $B_m(\kappa)$ and $\Gamma_m(\check{\kappa}, \kappa)$ using Equations 13.8.3, 13.8.3 and 13.42, respectively, for all states and all the trellis stages. However, as mentioned in Section 14.4, the calculation of $\Gamma_m(\check{\kappa}, \kappa)$ — which is evaluated using Equation 14.13 — can be simplified, since the received coded bits and the square of the expected coded bits are always constant for all trellis transitions. Note that in contrast to the equaliser trellis interval index m, here we will use d to denote the index of the decoder trellis intervals. Once these parameters are calculated, the LLRs of both the source and coded bits are determined, in contrast to stand-alone turbo decoding schemes, which only compute the LLRs of the source bits. Therefore, at each rate $R = \frac{1}{2}$ decoder trellis instant the LLR values of two coded bits are produced. In our following discussions, we will demonstrate how the LLR of each coded bit is determined. Figure 14.14(a) is a portion of the decoder trellis at trellis interval $d = 4$, which shows the set of branches corresponding to the first coded bit $c_{1,4} = -1$, while Figure 14.14(b) illustrates the group of trellis transitions caused by $c_{1,4} = +1$. Note that when a systematic encoder is employed, the LLR of the first coded bit $c_{1,d}$ constitutes the LLR of the source bit. Similarly to our approach in the Log-MAP equaliser, the

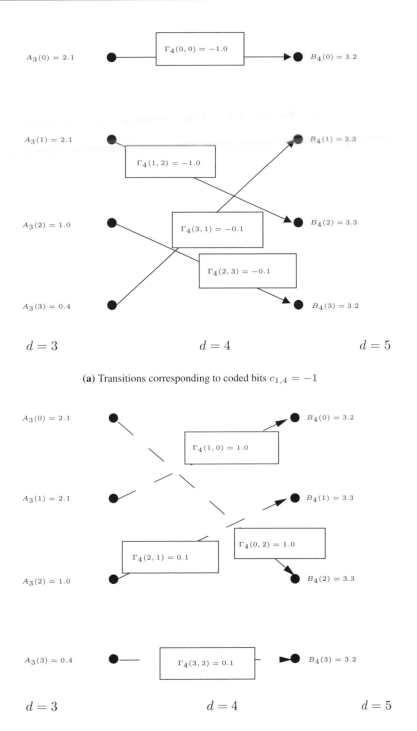

(a) Transitions corresponding to coded bits $c_{1,4} = -1$

(b) Transitions corresponding to coded bits $c_{1,4} = +1$

Figure 14.14: Half-rate, $K = 3$ RSC channel decoder trellis transitions, which were considered when evaluating the corresponding LLR values of $c_{1,4}$, which is the first coded bit at interval $d = 4$. The trellis is based on Figure 14.7.

values of $\ln[P(c_{1,4} = +1|L_p^i)]$ and $\ln[P(c_{1,4} = -1|L_p^i)]$ are determined in order to compute the LLR of $c_{1,4}$. From Figure 14.14(a) and using Equation 14.4, the term $\ln[P(c_{1,4} = -1|L_p^i)]$ is given by:

$$\ln[P(c_{1,4} = -1|L_p^i)] = \ln[\exp(2.1 + (-1.0) + 3.2) + + \exp(2.1 + (-1.0) + 3.3)$$
$$\exp(1.0 + (-0.1) + 3.2) + \exp(0.4 + (-0.1) + 3.3)]$$

$$\approx \ln[\exp(\overbrace{\max(4.3, 4.4) + \underbrace{0.65}_{\text{LUT}}}^{5.05}) + \exp(1.0 + (-0.1) + 3.2)$$
$$+ \exp(0.4 + (-0.1) + 3.3)]$$

$$\approx \ln[\exp(\overbrace{\max(5.05, 4.1) + \underbrace{0.35}_{\text{LUT}}}^{5.4}) + \exp(0.4 + (-0.1) + 3.3)]$$

$$\approx \max(5.4, 3.6) + \underbrace{0.15}_{\text{LUT}}$$

$$\approx 5.55.$$

while $\ln[P(c_{1,4} = +1|L_p^i)]$ becomes:

$$\ln[P(c_{1,4} = +1|L_p^i)] = \ln[\exp(2.1 + 1.0 + 3.3) + \exp(2.1 + 1.0 + 3.2)$$
$$+ \exp(1.0 + 0.1 + 3.3) + \exp(0.4 + 0.1 + 3.2)]$$

$$\approx \ln[\exp(\overbrace{\max(6.4, 6.3) + \underbrace{0.65}_{\text{LUT}}}^{7.05}) + \exp(1.0 + 0.1 + 3.3)$$
$$+ \exp(0.4 + 0.1 + 3.2)]$$

$$\approx \ln[\exp(\overbrace{\max(7.05, 4.4) + \underbrace{0.05}_{\text{LUT}}}^{7.1} + \exp(0.4 + 0.1 + 3.2)]$$

$$\approx \max(7.1, 3.7) + \underbrace{0.05}_{\text{LUT}}$$

$$\approx 7.15.$$

Therefore, the LLR of bit $c_{1,4}$ at trellis interval $d = 4$ is $\ln[P(c_{1,4} = +1|L_p^i)] - \ln[P(c_{1,4} = -1|L_p^i)] = 7.15 - 5.55 = 1.6$. By following the same approach, the LLR of the second coded bit $c_{2,4}$ at interval $d = 4$ can also be evaluated. Explicitly, we must calculate $\ln[P(c_{2,4} = -1|L_p^i)]$ and $\ln[P(c_{2,4} = +1|L_p^i)]$ by considering a different set of transitions obeying the trellis diagram of Figure 14.7, which were caused by the coded bits $c_{2,4} = +1$ and $c_{2,4} = -1$, respectively. Considering the set of transitions corresponding to bit $c_{2,4} = -1$ in Figure 14.15(a), the term

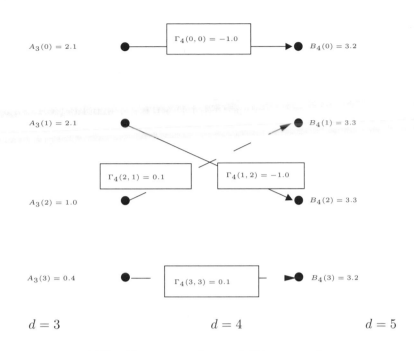

(a) Transitions corresponding to coded bits $c_{2,4} = -1$

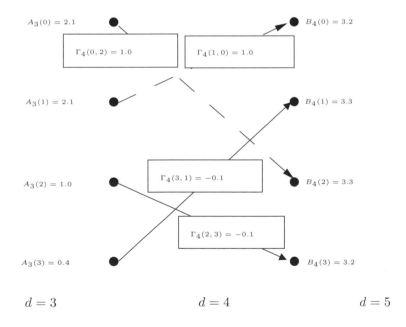

(b) Transitions corresponding to coded bits $c_{2,4} = +1$

Figure 14.15: Half-rate, $K = 3$ RSC channel decoder trellis transitions, which were considered when evaluating the corresponding LLR values of $c_{2,4}$, which is the second code bit at interval $d = 4$. The trellis is based on Figure 14.7

$\ln[P(c_{2,4} = -1|L_p^i)]$ is:

$$
\begin{aligned}
\ln[P(c_{2,4} = -1|L_p^i)] &= \ln[\exp(2.1 + (-1.0) + 3.2) + \exp(2.1 + (-1.0) + 3.3) \\
&\quad + \exp(1.0 + 0.1 + 3.3) + \exp(0.4 + 0.1 + 3.2)] \\
&\approx \ln[\exp(\overbrace{\max(4.3, 4.4) + \underbrace{0.65}_{\text{LUT}}}^{5.15}) + \exp(1.0 + 0.1 + 3.3) \\
&\quad + \exp(0.4 + 0.1 + 3.2)] \\
&\approx \ln[\exp(\overbrace{\max(5.15, 4.4) + \underbrace{0.25}_{\text{LUT}}}^{5.4}) + \exp(0.4 + 0.1 + 3.2)] \\
&\approx \max(5.4, 3.7) + \underbrace{0.15}_{\text{LUT}} \\
&\approx 5.55,
\end{aligned}
$$

while $\ln[P(c_{2,4} = +1|L_p^i)]$ can be written as:

$$
\begin{aligned}
\ln[P(c_{2,4} = +1|L_p^i)] &= \ln[\exp(2.1 + 1.0 + 3.3) + \exp(2.1 + 1.0 + 3.2) \\
&\quad + \exp(1.0 + (-0.1) + 3.2) + \exp(0.4 + (-0.1) + 3.3)] \\
&\approx \ln[\exp(\overbrace{\max(6.4, 6.3) + \underbrace{0.65}_{\text{LUT}}}^{7.05}) + \exp(1.0 + (-0.1) + 3.2) \\
&\quad + \exp(0.4 + (-0.1) + 3.3)] \\
&\approx \ln[\exp(\overbrace{\max(7.05, 4.1) + \underbrace{0.05}_{\text{LUT}}}^{7.1}) + \exp(0.4 + (-0.1) + 3.3)] \\
&\approx \max(7.1, 3.6) + \underbrace{0.05}_{\text{LUT}} \\
&\approx 7.15
\end{aligned}
$$

upon considering all legitimate trellis transitions corresponding to the bit $c_{2,4} = +1$. Therefore, the a-posteriori LLR for bit $c_{2,4}$ at interval $d = 4$ is $7.15 - 5.55 = 1.6$. This operation is applied to all trellis transition intervals in the trellis to determine the LLRs of the coded bits. In the last iteration, only the a-posteriori LLR of the source bit is computed instead of both coded bits, since only the transmitted source bits have to be determined. Note that in our example a decision concerning the source bits — based on their a-posteriori LLR — was made even after the first iteration, as shown in Table 14.4. This decision was invoked for determining the decoding performance of the convolutional code using the Log-MAP algorithm, which is equivalent to the performance of turbo equalisation after one turbo equalisation iteration. In general, however, the detected bits are only determined in the final iteration.

Various metrics characterising the convolutional decoder of Figure 14.9 were extracted from our simulations after the first turbo equalisation iteration and summarised in Table 14.4. Let us first consider the information received by the Log-MAP decoder, followed by the output information of the decoder. It is observed in Table 14.4 that the convolutional decoder never receives any source-bit-related a-priori information (APRI-S), unless this information can be extracted

	Decoder trellis interval d				
	1	2	3	4	5
APRI-S	0.0	0.0	0.0	0.0	0.0
RXD-LLR	2.7,−0.5	−0.1,−0.6	−0.4,2.1	1.1,0.8	0.2,−6.5
APOS-C	2.5,2.5	0.2,−0.3	0.1,2.3	1.6,1.6	0.2,−6.5
EXT-C	−0.2,3.0	0.3,0.3	0.5,0.2	0.6,0.8	0.0, 0.0
APOS-S	2.5	0.2	0.1	1.6	0.2
TXD-S	1	1	−1	−1	1
DCSN	1	1	**1**	**1**	1

Table 14.4: Various metrics of the convolutional decoder in Figure 14.9 extracted from our simulations after the first turbo equalisation iteration. The convolutional decoder never receives any source bit a-priori (APRI-S) information, unless this information can be extracted from another independent source, such as another convolutional decoder. The extrinsic LLR (EXT-C) is computed by subtracting the received combined channel and extrinsic LLR (RXD-LLR) from the decoder's a-posteriori LLR of the code bits (APOS-C). The decision bits (DCSN) — obtained from the the decoder source a-posteriori (APOS-S) information — is compared with the transmitted source bits (TXD-S) to determine the number of erroneous bits. It is observed that two errors were obtained in the first iteration.

from another independent source, such as another convolutional decoder in turbo-coded systems. Therefore, similarly to the a-priori LLR of the coded bit received by the equaliser, namely APRI in Table 14.3, in the first iteration the a-priori LLR of the source bit, namely APRI-S, is also initialised to 0, indicating that the bits $+1$ and -1 have equal probability of occurring. However, in contrast to the Log-MAP equaliser, in the absence of additional decoders the values of APRI-S remain at 0 in successive turbo equalisation iterations, whereas the reliability of the a-priori LLRs at the input of the equaliser APRI improves, as will be shown next in our discussion of the turbo equaliser operations during the second iteration. As mentioned previously, the decoder also received the LLR associated with the combined channel and extrinsic information, namely EXT of Table 14.3, which had been channel de-interleaved to give RXD-LLR of Table 14.4. Subsequently, with the aid of these received LLR values i.e. APRI-S as well as RXD-LLR, and by employing the principles highlighted above, the Log-MAP decoder is now ready to determine the a-posteriori LLRs of the source bits and the coded bits, namely (APOS-S) and (APOS-C), of Table 14.4, respectively. Recall that the input information to a particular block in the current iteration must not include the information contributed by that particular block from the previous iteration, in order to ensure minimum correlation between these values. Therefore, as seen at the bottom of Figure 14.9, the combined channel and extrinsic LLR RXD-LLR is subtracted from APOS-C in order to yield the extrinsic LLR (EXT-C), which is subsequently channel interleaved before being passed to the Log-MAP equaliser in the next iteration as the a-priori LLR APRI of Table 14.5. This additional information will be utilised by the equaliser in the second iteration, in order to improve the reliability of the a-posteriori LLR, i.e. to increase our confidence in the associated decision. The a-posteriori LLR of the source bits, namely APOS-S in Table 14.4, is used to detect the transmitted source bits, and is denoted by DCSN in Table 14.4. By comparing DCSN with the source bits that were actually transmitted (TXD-S), it was observed that two errors — which were emphasised in bold in Table 14.4 — were obtained in the first iteration. Continuing with our example, we will show next that the iterative exchange of information between the decoder and equaliser will improve the performance further, such that the turbo equaliser outperforms independent equalisation and decoding.

				Equaliser trellis interval m									
	1	2	3	4	5	6	7	8	9	10	11	12	13
APRI	−0.2	0.2	2.9	0.5	0.3	0.8	0.3	0.1	0.5	−0.1	−115.1	−115.1	−11
APOS	2.0	1.2	2.2	0.1	0.1	1.3	−0.6	−0.1	−0.1	−6.7	−123.1	−119.9	−11
EXT	2.2	1.0	−0.7	−0.4	−0.2	0.5	−0.9	−0.2	−0.6	−6.6	−8.0	−4.8	−1
TXD-C	1	−1	1	−1	1	1	−1	1	−1	−1	−1	−1	−
DCSN	1	**1**	1	**1**	1	1	−1	−1	−1	−1	−1	−1	−

Table 14.5: Various equaliser metrics extracted from our simulations after the second iteration. Notations TXD-C, APRI, APOS, EXT and DCSN represent the transmitted encoded bits, which have been interleaved, the a-priori LLR, the a-posteriori LLR, extrinsic LLR and the detected or decision bits. In the second iteration the equaliser now obtains a-priori information provided by the convolutional decoder, hence improving the reliability of the a-posteriori LLRs evaluated, generating only three errors, emphasised in bold.

	Decoder trellis interval d				
	1	2	3	4	5
APRI-S	0.0	0.0	0.0	0.0	0.0
RXD-LLR	2.2,−0.7	−0.2,−0.9	−0.5,1.0	−0.4,0.5	−0.1,−6.7
APOS-C	1.4,1.4	0.3,−0.6	−0.2,+0.7	−0.2,0.4	0.1,−6.7
EXT-C	−0.8,2.1	0.5,−0.3	0.3,−0.3	0.2,−0.1	0.2,0.0
APOS-S	1.4	0.3	−0.2	−0.2	0.1
TXD-S	1	1	−1	−1	1
DCSN	1	1	−1	−1	1

Table 14.6: Channel decoder metrics extracted from our simulations after the second turbo equalisation iteration. No a-priori APRI-S information is received. As in the first iteration, the extrinsic LLR EXT-C is computed by subtracting the received combined channel and extrinsic LLR RXD-LLR from the decoder's a-posteriori LLR of the code bits APOS-C. Since the decoder receives higher-reliability LLR values, it produces source a-posteriori APOS-S, which results in an error-free output.

In the second iteration, the equaliser receives a-priori information of the source bits APRI-S from the decoder output of the first iteration, as shown in Table 14.5. Recall that in the first turbo equalisation iteration the a-priori LLR shown in the top line of Table 14.3 was initialised to 0, indicating that the bits $+1$ and -1 had equal probability of being transmitted. Now this additional information assists the equaliser in providing more accurate values of the a-posteriori LLRs APOS. The improved reliability is reflected in the reduced number of errors, as compared to the results of the first iteration, seen in Table 14.3, indicating a reduction from four to three errors. As before, the a-priori information is subtracted from the a-posteriori LLR at the output of the Log-MAP equaliser of Figure 14.9, yielding the combined channel and extrinsic LLR EXT, which is subsequently de-interleaved. At the decoder the received combined channel and extrinsic LLR is processed using the Log-MAP algorithm and — as shown in Table 14.6 — the decoder corrects the two errors incurred in the previous iteration, characterised in Table 14.4. It is observed that by performing the equalisation and decoding iteratively, the dispersion induced by the channel can be overcome, as opposed to the scenario characterised in Table 14.4, which is equivalent to a situation employing independent equalisation and decoding. In order to further improve the performance of the system, more turbo equalisation iterations can be performed. However, in this example, it was sufficient to perform two turbo equalisation iterations, in order to obtain an error-free performance. Figure 14.16 illustrates the reliability of the LLRs — expressed as the product of the decoder's a-posteriori LLR and the actual source bits transmitted — versus the source bit index. A large positive reliability value indicates a high-confidence LLR value, whereas a negative value represents an erroneous decision. It is observed that in the first iteration of Figure 14.16(a) there was a large number of low-reliability LLR values. The negative reliability values in Figure 14.16(a) indicate that errors have been made. In the second turbo equalisation iteration of Figure 14.16(b) the number of low-confidence LLR values was reduced as compared to the first iteration. With increasing turbo equalisation iterations — for example, after six turbo equalisation iterations, as shown in Figure 14.16(f) — the number of low-confidence LLR values was significantly reduced, hence justifying the improved BER performance obtained with an increasing number of turbo equalisation iterations.

However, as can be seen from the results shown at a later stage, the additional gain decreases with an increasing number of turbo equalisation iterations. Iteration gains are achieved when the decoder is capable of correcting errors which were not corrected by the equaliser. When the number of turbo equalisation iterations increases, the de-interleaved output of the equaliser becomes more similar to the decoder output, yielding a high correlation. This diminishes the advantage of having information from both the equaliser and decoder, hence reducing the iteration gain.

14.6 Summary of Turbo Equalisation

In the previous sections we have shown the operation and structure of the original turbo equaliser presented by Douillard *et al.* [105] and subsequently extended the principle to turbo equalisation schemes utilising multiple decoders, as in turbo decoders. The SISO equaliser must be able to accept both the a-priori information provided by the channel decoder as well as channel outputs and to utilise both of these information sources in order to calculate the LLR of the received encoded bits. Here, the only modification to the Log-MAP algorithm of Section 13.8.3, for employment in our iterative soft-output GMSK equaliser, was to consider the a-priori probability $P(u_m)$ in the calculation of $\gamma_m(\dot{k}, \kappa)$ or $\Gamma_m(\dot{k}, \kappa)$, since $P(u_m)$ was no longer constant — in contrast to the non-iterative scenario. The decoder(s) in the turbo equaliser must also be able to provide soft outputs in the form of the source and parity LLR values. Therefore, instead of implementing a hard-decision trellis-based algorithm for the decoder — such as the Viterbi Algorithm — we have used a SISO algorithm. Specifically, the Log-MAP algorithm was employed since it guaranteed a performance identical to that of the optimal MAP algorithm, despite having lower computational complexity. We then showed in Section 14.4 that the a-posteriori LLR of both the parity and source bits can be calculated by considering all trellis branches, where the parity bit or source bit is -1 or $+1$, as expressed in Equation 14.4. The a-posteriori LLR of each turbo equalisation stage was subsequently processed, in order to yield the combined channel and extrinsic information of the equaliser or the extrinsic information of the decoder, such that the input information to a particular block in the current iteration did not include the information contributed by that particular block from the previous iteration, in order to avoid any correlation between the information used in consecutive iterations.

In the next section the performance of various coded GMSK systems employing turbo equalisation is characterised.

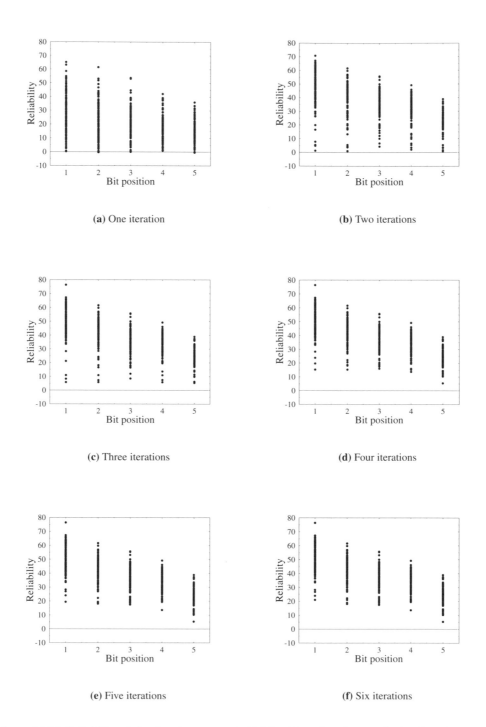

Figure 14.16: Reliability of the decoder a-posteriori LLR with each turbo equalisation iteration. The reliability is expressed as the product of the source a-posteriori LLR produced by the decoder and the transmitted source bit.

14.7 Performance of Coded GMSK Systems using Turbo Equalisation

In this section the turbo equalisation performance for three coded systems, namely that of the convolutional-coded GMSK system, the convolutional-coding-based turbo-coded GMSK scheme and the BCH-coding-based turbo-coded system, are presented. These systems were evaluated over the non-dispersive Gaussian channel and over the equally weighted and symbol-spaced five-path Rayleigh fading channel experiencing a normalised Doppler frequency of $f_d = 1.5 \times 10^{-4}$, which corresponds to a transmission frequency of 900 MHz, vehicular speed of 30 mph and a signalling rate of 270.833 KBaud. For our investigations, we have employed the transmission burst structure specified by GSM [49, 344] consisting of three tail bits at both ends of the burst and two 57-bit data segments separated by a 26-bit midamble. We have assumed that the channel impulse response was known. Furthermore, the fading magnitude and phase were kept constant for the duration of a transmission burst, a condition which we refer to as employing burst-invariant fading. In the following subsections the simulation parameters — which are specific to the class of the encoder utilised — are outlined.

Encoder type	Component code parameters
Convolutional code	Rate $R = 0.5$, Constraint length $K = 5$ RSC Octal generator polynomials $G_0 = 35$ $G_1 = 23$
Convolutional-coding-based turbo code	Rate $R = 0.5$, Constraint length $K = 5$ RSC Octal generator polynomials $G_0 = 35$ $G_1 = 23$
BCH-coding-based turbo code	Rate $R = \frac{11}{19} = 0.58$ BCH (15,11)

Table 14.7: Parameters of the encoders employed in the rate $R = 0.5$ convolutional-coded GMSK system, the $R = 0.5$ convolutional-coding-based turbo-coded GMSK scheme and the rate $R = \frac{11}{19} = 0.58$ BCH-coding-based turbo-coded GMSK system.

14.7.1 Convolutional-coded GMSK System

Non-dispersive Gaussian channel		5-path Rayleigh fading channel	
Iteration index	Iteration gain	Iteration index	Iteration gain
2	2.5 dB	2	2.9 dB
4	3.5 dB	4	3.9 dB
8	3.7 dB	8	4.2 dB

Table 14.8: The iteration gains relative to the first iteration at a BER of 10^{-4} for the $R = 0.5$ convolutional-coded GMSK system employing turbo equalisation over the non-dispersive Gaussian channel and the five-path Rayleigh fading channel using burst-invariant fading.

A code rate $R = 0.5$, constraint length $K = 5$, RSC encoder was employed. The octal generator polynomials of the encoder were $G_0 = 35$ and $G_1 = 23$ as summarised in Table 14.7.

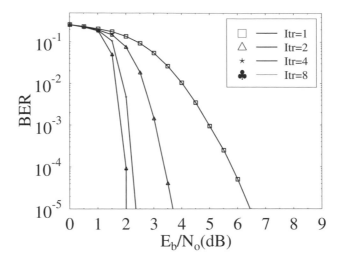

(a) Non-dispersive Gaussian channel.

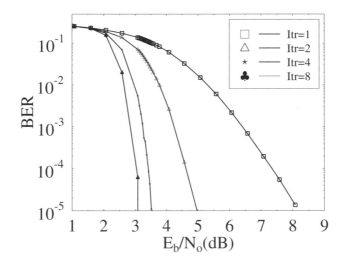

(b) Five-path Rayleigh fading channel.

Figure 14.17: Turbo equalisation performance over the non-dispersive Gaussian channel and the five-path Rayleigh fading channel using burst-invariant fading for a rate $R = 0.5$ **convolutional-coded** GMSK system using a turbo equaliser, which performs a maximum of eight turbo equalisation iterations.

No puncturing was applied to the encoded bits, which were passed to a random channel interleaver having a depth of 20000 bits. Figure 14.17(a) shows the turbo equalisation performance of the convolutional-coded GMSK system transmitting over a non-dispersive Gaussian channel after one, two, four and eight turbo equalisation iterations. Note that the convolutional decoding performance was equivalent to the first turbo equalisation iteration. It was observed that the gain obtained from iterative equalisation and decoding after eight turbo equalisation iterations was approximately 3.7 dB as compared to independent equalisation and decoding at a BER of 10^{-4}. As for the turbo equalisation performance over the five-path Rayleigh fading channel, it can be seen from Figure 14.17(b) that the gain achieved by using turbo equalisation after eight turbo equalisation iterations was 4.2 dB, compared to conventional convolutional decoding at a BER of 10^{-4}. The turbo equalisation performance was observed to improve significantly by employing more turbo equalisation iterations. At a BER of 10^{-4} and when transmitting over the non-dispersive Gaussian channel, iteration gains — i.e. gains in SNR performance with respect to the first iteration — of approximately 2.5 dB, 3.5 dB and 3.7 dB were obtained after two, four and eight turbo equalisation iterations, as displayed in Table 14.8. Over the five-path Rayleigh fading channel using burst-invariant fading the iteration gains, also summarised in this table, achieved were 2.9 dB, 3.9 dB and 4.2 dB at a BER of 10^{-4} after two, four and eight turbo equalisation iterations. Let us now consider a convolutional-coding-based turbo-coded GMSK system.

14.7.2 Convolutional-coding-based Turbo-coded GMSK System

Code rate $R = 0.5$
C1: 1 0
C2: 0 1
1 = transmitted bit
0 = non-transmitted bit

Table 14.9: Regular puncturing pattern used in order to obtain the $R = 0.5$ convolutional-coding-based turbo codes. The terms C1 and C2 represent the parity bits of the $R = 0.5$ convolutional codes of the first and second constituent codes, respectively.

Non-dispersive Gaussian channel		5-path Rayleigh fading channel	
Iteration index	Iteration gain	Iteration index	Iteration gain
2	1.6 dB	2	1.7 dB
4	2.2 dB	4	2.5 dB
8	2.4 dB	8	2.7 dB

Table 14.10: The iteration gains relative to the first iteration at a BER of 10^{-4} for the $R = 0.5$ convolutional-coding-based turbo-coded GMSK system employing turbo equalisation over the non-dispersive Gaussian channel and the five-path Rayleigh fading channel using burst-invariant fading.

The convolutional-coding-based turbo encoder consists of two rate $R = 0.5$, constraint length $K = 5$ RSC encoders, using octal generator polynomials of $G_0 = 35$ and $G_1 = 23$ as summarised in Table 14.7. A random turbo interleaver possessing a depth of 10000 bits separated the

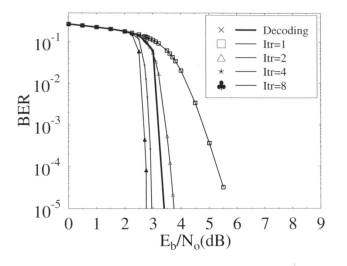

(a) Non-dispersive Gaussian channel.

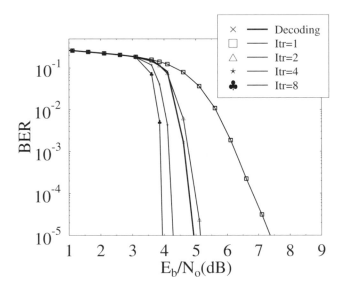

(b) Five-path Rayleigh fading channel.

Figure 14.18: Turbo equalisation performance over the non-dispersive Gaussian channel and the five-path Rayleigh fading channel using burst-invariant fading for a rate $R = 0.5$ **convolutional-coding-based turbo-coded** GMSK system employing a turbo equaliser, which performs eight turbo equalisation iterations. The turbo decoding performance after eight turbo decoding iterations is also presented.

component encoders. The encoded bits were punctured regularly using the puncturing pattern of Table 14.9 in order to obtain an overall code rate of $R = 0.5$. Subsequently, the punctured encoded bits were passed to a random channel interleaver, which had a depth of 20000 bits. The turbo equalisation performance of the convolutional-coding-based turbo-coded GMSK system transmitting over the non-dispersive Gaussian channel and the five-path Rayleigh fading channel are shown in Figures 14.18(a) and 14.18(b), respectively. In the non-dispersive Gaussian channel scenario it was observed that the iterative equalisation and decoding arrangement using eight turbo equalisation iterations outperformed the system employing eight turbo decoding iterations by approximately 0.5 dB at a BER of 10^{-4}. For the five-path Rayleigh fading channel shown in Figure 14.18(b), the gain achieved by performing turbo equalisation after eight turbo equalisation iterations over the independent equalisation and turbo decoding employing eight turbo decoding iterations was 0.8 dB at a BER of 10^{-4}. As in the convolutional-coded GMSK system, the turbo equalisation performance improved by performing more turbo equalisation iterations. When transmitting over the non-dispersive Gaussian channel, iteration gains of approximately 1.6 dB, 2.2 dB and 2.4 dB were achieved after two, four and eight turbo equalisation iterations, as displayed in Table 14.10. Over the five-path Rayleigh fading channel using burst-invariant fading, the iteration gains — also summarised in this table — achieved were 1.7 dB, 2.5 dB and 2.7 dB at a BER of 10^{-4} after two, four and eight turbo equalisation iterations. Again — as expected — including the equaliser in the iterative loop was more complex but resulted in further performance gains. Let us now consider the performance of the BCH turbo-coded systems in the context of turbo equalisation.

14.7.3   BCH-coding-based Turbo-coded GMSK System

Non-dispersive Gaussian channel		5-path Rayleigh fading channel	
Iteration index	Iteration gain	Iteration index	Iteration gain
2	2.0 dB	2	2.5 dB
4	3.8 dB	4	4.0 dB
8	4.3 dB	8	4.5 dB

Table 14.11: The iteration gains relative to the first iteration at a BER of 10^{-4} for the $R = \frac{11}{19} = 0.58$ BCH-coding based turbo-coded GMSK system employing turbo equalisation over the non-dispersive Gaussian channel and the five-path Rayleigh fading channel using burst-invariant fading.

The BCH-coding-based turbo encoders invoked consisted of two BCH (15,11) encoders — characterised in Table 14.7 — which were separated by a random turbo interleaver, having a depth of 12100 bits. No puncturing was applied, yielding an overall code rate of $R = \frac{11}{19} = 0.58$. The encoded bits were directed to a random channel interleaver possessing a depth of 20900 bits. When the BCH-coding-based turbo-coded GMSK system was transmitted over the non-dispersive Gaussian channel, it was observed from Figure 14.19(a) that by performing the equalisation and decoding iteratively — i.e. upon invoking turbo equalisation — and using eight turbo equalisation iterations, a gain of 2.8 dB was achieved over independent equalisation and BCH-coding-based turbo decoding performing eight turbo decoding iterations at a BER of 10^{-4}. The advantage of using turbo equalisation over turbo decoding was also highlighted in Figure 14.19(b). Here, it was observed that a gain of 3.0 dB was attained over the turbo decoding scheme employing eight turbo decoding iterations. The BER performance of the system

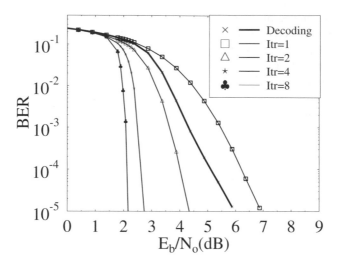

(a) Non-dispersive Gaussian channel.

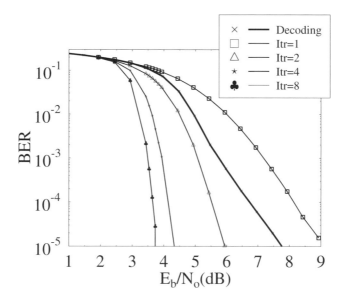

(b) Five-path Rayleigh fading channel.

Figure 14.19: Turbo equalisation performance over the non-dispersive Gaussian channel and the five-path Rayleigh fading channel using burst-invariant fading for a rate $R = 0.58$ BCH(15,11)-based turbo-coded GMSK system using a turbo equaliser, which performs eight turbo equalisation iterations. The turbo decoding performance after eight turbo decoding iterations is also plotted.

employing turbo equalisation was also improved upon subsequent iterations. Specifically, in Figure 14.19(a) and at a BER of 10^{-4}, the iteration gains obtained were 2 dB, 3.8 dB and 4.3 dB — as shown in Table 14.11 — when transmitting over the non-dispersive Gaussian channel. For the five-path fading channel scenario using burst-invariant fading, it can be seen from Figure 14.19(b) and Table 14.11 that the iteration gains achieved were 2.5 dB, 4 dB and 4.5 dB at a BER of 10^{-4}.

14.8 Discussion of Results

The key observations from the results in Sections 14.7.1, 14.7.2 and 14.7.3 were:

- The advantage of performing turbo equalisation over turbo decoding.

- Improvement with subsequent turbo equalisation iterations.

- The rate $R = 0.5$ convolutional-coded GMSK system employing turbo equalisation achieved better performance than that of the $R = 0.5$ convolutional-coding-based turbo-coded GMSK scheme.

- The $R = 0.58$ BCH-coding-based turbo-coded GMSK system employing turbo equalisation achieved better performance than that of the lower rate $R = 0.5$ convolutional-coding-based turbo-coded GMSK scheme.

It was observed that significant advantage was gained by performing the equalisation and decoding iteratively. Upon each turbo equalisation iteration the equaliser benefits from the a-priori information of the encoded bits obtained from the decoder(s). Consequently, more reliable LLR values can be generated by the equaliser, hence aiding the decoder in improving the confidence of the LLR values. In contrast, by performing independent equalisation and decoding, the decoder(s) can only operate on the initial soft values from the equaliser. Therefore, the equaliser was unable to exploit the improved decoder estimates for enhancing the reliability of the equaliser's soft values. Hence, the iterative equalisation and decoding technique outperformed the independent equalisation and decoding scheme. The disadvantage of turbo equalisation was the increased complexity due to the need of performing equalisation and decoding in each turbo equalisation iteration. For the convolutional-coded GMSK scheme, each additional iteration involved the equalisation and decoding process, hence increasing the complexity significantly, as compared to the conventional convolutional decoding technique. Upon comparing turbo decoding with turbo equalisation, it was observed that turbo decoding iterations involved exchange of information between the decoders, whereas in turbo equalisation information was exchanged not only between the decoders, but also with the equaliser in each turbo equalisation iteration. Hence, the complexity of the turbo equalisation scheme was also higher than that of turbo decoding. However, the number of iterations can be reduced by employing various termination criteria [345], similar to those used in turbo decoding [339]. This will reduce the complexity associated with turbo equalisation, making this technique a good alternative for mitigating the channel's frequency selectivity. Furthermore, the Log-MAP algorithm used for the equaliser and decoder can be substituted by lower-complexity SISO algorithms [122, 352–354], in order to practically realise the turbo equaliser. The Log-MAP algorithm can be readily parallelised [355], which speeds up the computations associated with the implementation of the Log-MAP algorithm, hence enabling the turbo equaliser — consisting of the Log-MAP equaliser and decoder — to be implemented in real-time systems.

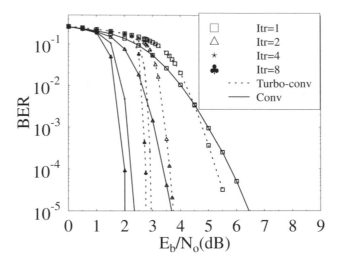

(a) Non-dispersive Gaussian channel.

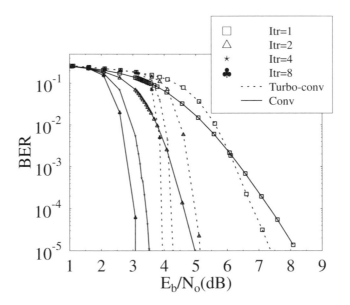

(b) Five-path Rayleigh fading channel.

Figure 14.20: Comparison of the $R = 0.5$ convolutional-coded GMSK system with $R = 0.5$ convolutional-coding-based turbo-coded GMSK scheme, transmitting over the non-dispersive Gaussian channel and the five-path Rayleigh fading channel using burst-invariant fading. Both systems employ turbo equalisation, which performs eight turbo equalisation iterations at the receiver.

As mentioned previously, by performing more iterations, the equaliser and decoder are capable of enhancing the reliability of their estimates. Therefore, the BER performance was improved as the number of turbo equalisation iterations was increased. However, as observed from the results of Sections 14.7.1, 14.7.2 and 14.7.3, the rate of increase of the iteration gain, which was again the gain in SNR performance relative to the performance achieved in the first turbo equalisation iteration, decreased with successive turbo equalisation iterations. Iteration gains were obtained when the decoder was capable of correcting errors previously inflicted by the equaliser. As the number of turbo equalisation iterations increased, the de-interleaved output of the equaliser became more similar to the decoder output. This therefore reduced the advantage of having information from the equaliser and decoder and consequently failed to achieve greater iteration gains, despite performing further turbo equalisation iterations.

Another interesting observation was the ability of the convolutional-coded GMSK system employing turbo equalisation to outperform the convolutional-coding-based turbo-coded GMSK scheme, as demonstrated by Figure 14.20. Over the non-dispersive Gaussian channel, the convolutional-coded GMSK system had a better E_b/N_0 performance than the convolutional-coding-based turbo-coded GMSK scheme by 0.8 dB at a BER of 10^{-4}, while in the five-path Rayleigh fading scenario using burst-invariant fading the advantage of the convolutional-coded system over the convolutional-coding-based turbo-coded scheme was approximately 1.0 dB, as illustrated in Figures 14.20(a) and 14.20(b), respectively. These results were surprising since the more complex turbo-coded system was expected to form a more powerful encoded system compared to the convolutional-coded scheme. Below, we offer a possible explanation, which considers the performance of both codes after the first turbo equalisation iteration. Over the non-dispersive Gaussian channel in Figure 14.20(a), the convolutional-coded scheme yielded a lower BER than that of the turbo-coded system at E_b/N_0 values less than 4.5 dB, indicating that the a-posteriori LLR of the source bits — which constitute the first coded bit of the systematic codeword — produced by the convolutional decoder had a higher reliability than that of the turbo-coded scheme. Consequently, upon receiving the higher-confidence LLR values from the decoder, the equaliser in the convolutional-coded scheme was able to produce much more reliable LLR values in the subsequent turbo equalisation iteration, as compared to the turbo-coded system. After receiving these LLR values, the decoder of the convolutional-coded system will also generate more reliable LLR values. Similarly, for the five-path Rayleigh fading channel scenario in Figure 14.20(b), it was also observed that for $E_b/N_0 < 6$ dB and after one turbo equalisation iteration, the convolutional-coded system outperformed the turbo-coded scheme. Similarly to the non-dispersive Gaussian channel scenario, the convolutional-coded system transmitting over the dispersive fading channel also produced LLR values of higher reliability as compared to those generated by the turbo-coded system, hence allowing the convolutional-coded system to outperform the turbo-coded scheme after performing eight turbo equalisation iterations. For E_b/N_0 values beyond the previously mentioned values, the performance of the convolutional-coding-based turbo-coded scheme after one turbo equalisation iteration was better than that of the convolutional-coded system. Therefore, it was predicted that although the performance of the turbo-coded system was poorer than the convolutional-coded system after eight turbo equalisation iterations for $E_b/N_0 < 4.5$ dB for the non-dispersive Gaussian channel and $E_b/N_0 < 6.0$ dB for the five-path Rayleigh fading channel, it will outperform the convolutional-coded system beyond these E_b/N_0 values, potentially yielding a lower error floor.

Finally, in Figure 14.21 the rate $R = 0.58$ BCH-coding-based turbo-coded system was also observed to outperform the $R = 0.5$ convolutional-coding-based turbo-coded GMSK scheme transmitting over the non-dispersive Gaussian channel and over the five-path Rayleigh fading channel by a margin of 0.7 dB and 0.2 dB, respectively, at a BER of 10^{-4}, when using eight

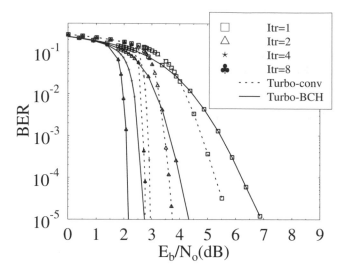

(a) Non-dispersive Gaussian channel.

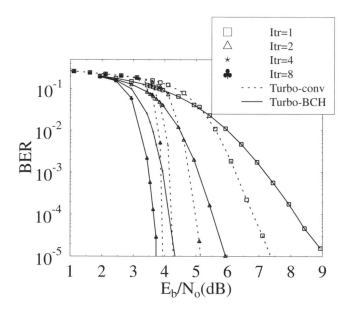

(b) Five-path Rayleigh fading channel.

Figure 14.21: Comparison of the rate $R = 0.58$ BCH-coding-based turbo-coded GMSK system with the $R = 0.5$ convolutional-coding-based turbo-coded GMSK scheme, transmitting over the non-dispersive Gaussian channel and the five-path Rayleigh fading channel using burst-invariant fading. Both systems employ turbo equalisation, which performs eight turbo equalisation iterations at the receiver.

turbo equalisation iterations. Here, the performance improvements achieved by the BCH-coding-based turbo-coded system over the convolutional-coding based turbo-coded scheme can also be attributed to the fact that the former attained a lower BER for E_b/N_0 values less than 4.0 dB and 5.5 dB over the non-dispersive channel and the five-path Rayleigh fading channel using burst-invariant fading, respectively. Therefore, the equaliser in the BCH-coding-based turbo-coded system received more reliable LLR information compared to the convolutional-coding-based turbo-coded scheme, hence allowing it to achieve better BER performance upon invoking successive turbo equalisation iterations.

In order to further justify these results, the associated Maximum Likelihood (ML) performance bound of coded systems was derived and compared in our forthcoming chapter.

14.9 Summary and Conclusions

In this chapter the principles of turbo equalisation in the context of multiple-encoder-assisted systems were described. Here, the advantage of performing the channel equalisation and channel decoding iteratively was examined and quantified in the context of GMSK-modulated systems, as compared to implementing the equalisation independently from the decoding. Gains of 3.7 dB and 4.2 dB were achieved for the rate $R = \frac{1}{2}$ convolutional-coded GMSK system transmitting over the non-dispersive Gaussian channel and the five-path Rayleigh fading channel using burst-invariant fading, respectively. Similarly, for the rate $R = \frac{1}{2}$ convolutional-coding-based turbo-coded scheme the corresponding gains achieved through turbo equalisation over independent equalisation and turbo decoding were 0.5 dB and 0.8 dB over the non-dispersive Gaussian channel and the above-mentioned dispersive Rayleigh fading channel. Finally, the BCH-coding-based turbo-coded systems employing turbo equalisation and transmitting over the non-dispersive Gaussian and the five-path fading channels exhibited gains of 2.8 dB and 3 dB, respectively. It was also observed that the rate $R = 0.5$ convolutional-coded GMSK system employing turbo equalisation obtained a better performance than that of the $R = 0.5$ convolutional-coding-based turbo-coded GMSK scheme. Over the non-dispersive Gaussian channel, the convolutional-coded GMSK system outperformed the convolutional-coding-based turbo-coded GMSK scheme by 0.8 dB at a BER of 10^{-4}, while in the five-path Rayleigh fading scenario the convolutional-coded system attained a gain of approximately 1.0 dB over the convolutional-coding-based turbo-coded scheme, as illustrated in Figures 14.20(a) and 14.20(b), respectively. These results were surprising, since the more complex turbo-coded system was expected to perform better generally, as it was deemed to be a more powerful code compared to convolutional codes. However, from Figures 14.20(a) and 14.20(b), which compared the turbo equalisation performance of the convolutional-coding-based turbo-coded system and the conventional convolutional-coded scheme over the non-dispersive Gaussian channel and five-path Rayleigh fading channel, it was observed that at E_b/N_0 values less than 4.5 dB and 6 dB, respectively, the BER of the rate $R = 0.5$ convolutional-coded GMSK system was lower than that of the $R = 0.5$ convolutional-coding-based turbo-coded GMSK scheme after the first turbo equalisation iteration. This indicated that the decoder in the convolutional-coded system mentioned above was providing more reliable LLR values to the equaliser. Consequently, the equaliser in the convolutional-coded scheme — upon receiving the higher-confidence LLR values from the decoder — was capable of producing more reliable LLR values, which was subsequently passed to the decoder in the following turbo equalisation iteration. Beyond the E_b/N_0 values of 4.5 dB and 6 dB over the non-dispersive Gaussian channel and the five-path Rayleigh fading channel, respectively, the turbo-coded scheme exhibited a better performance after one turbo equalisation iteration. Therefore, it is predicted that the convolutional-coding-based turbo-coded

GMSK system will potentially yield a lower error floor. In Figure 14.21 it was also observed that the weaker BCH-coding-based turbo-coded GMSK system yielded a better performance than that of the convolutional-coding-based turbo coded scheme after eight turbo equalisation iterations. The performance of the BCH-coding-based turbo-coded system was better than that of the convolutional-coding-based turbo-coded scheme after the first turbo equalisation iteration at E_b/N_0 values less than 4.0 dB and 5.5 dB for transmissions over the non-dispersive Gaussian channel and over the five-path Rayleigh fading channel, respectively. Therefore, the equaliser in the BCH-based turbo-coded system received more reliable LLR information compared to the convolutional-based turbo-coded scheme, hence enabling it to achieve a better BER performance after invoking successive turbo equalisation iterations.

In the next chapter we will rely on the material presented in this chapter and provide a bounding technique for characterising the performance of turbo equalisers.

15

Turbo Equalisation Performance Bound

Since the pioneering invention of turbo codes by Berrou *et al.* [12, 13], which have been shown to approach Shannonian performance limits, significant research efforts have been devoted to improving their overall performance by optimising the turbo code components. Amongst other researchers seminal work on the choice of the constituent codes has been conducted for example by the influential research team of Benedetto, Garello and Montorsi [58, 356], by Ho, Pietrobon and Giles [357], by Ushirokawa and Okamura and Kamiya [358] and others. The appropriate choice of the interleavers was the subject of research for example at NTT DoCoMo by Shibutani, Suda and Adachi [359]. The issue of combined turbo code and interleaver design was addressed by Yuan, Vucetic and Feng in [360], while He and Wang also studied various turbo interleavers [361]. A variety of decoding techniques have been studied for example in [12, 362]. Berrou *et al.* proposed the principle of turbo coding [12, 13], invoking Bahl's MAP algorithm for its iterative decoding instead of a Maximum Likelihood (ML) decoder [362], such as the one discussed in Chapter 6, in order to reduce the decoder's complexity. Since the ML decoder can only be implemented for specific codes and for rather specific interleavers [362, 363] at the cost of a high complexity, theoretical performance bounds were derived in order to observe the ability of the iterative decoder to approximate the performance of the optimal ML decoder. The theoretical bounds of the parallel concatenated convolutional code [59], namely turbo code, can be derived by employing the classic union bound technique in conjunction with the witty concept of the so-called uniform interleaver proposed by Benedetto and Montorsi [59]. More specifically, the uniform interleaver concept advocated by Benedetto and Montorsi is based on a probabilistic interleaver model, which maps the input codeword of weight w into all possible distinct permutations with equal probability. Therefore, the associated theoretical bound characterises the upper bound performance of turbo codes in conjunction with ML decoding averaged over all possible interleaver structures. The performance bound of serial concatenated convolutional codes can also be determined using this approach [104], as it was suggested by Benedetto, Divsalar, Montorsi and Pollara. Our discussions in this chapter rely on a good background in both turbo channel coding and turbo equalisation and inherently rely on the insightful seminal concept of the uniform interlaver proposed by Benedetto and Montorsi [59].

15.1 Motivation

In this chapter, the principles employed for deriving the ML performance bound for the serial concatenated convolutional coding and parallel concatenated convolutional coding schemes are adapted, in order to derive the ML bound for non-punctured convolutional-coded and for turbo-coded systems employing turbo equalisation. Previous work in this area by Narayanan and Stüber in reference [351] employed the analysis by Benedetto *et al.* [104], which emphasised the importance of possessing recursive properties for the inner encoder — constituted for example by the partial-response Gaussian Minimum Shift Keying (GMSK) modem — in serial concatenated convolutional coding schemes. This technique was utilised for determining the performance bound of the joint iterative demodulation and decoding of convolutional-coded interleaved systems in conjunction with differential modulation schemes, such as Differential Phase Shift Keying (DPSK) as proposed by Narayanan and Stüber [351] as well as by Höher and J. Lodge [364] and $\frac{\pi}{4}$-Differential Quadrature Phase Shift Keying (DQPSK) [351]. Differential modulation techniques were employed in order to show that the inherent recursive nature of these modulation schemes in conjunction with the interleaver significantly improves the distance spectrum of the transmitted signal, yielding good interleaving gains, quantified as the factor by which the bit error probability is decreased upon increasing the interleaver depth. Specifically, DPSK and DQPSK can be modelled as rate $R = 1$ convolutional codes followed by a memoryless mapper. This rate $R = 1$ recursive encoder can be viewed as an inner code in the serial concatenated convolutional coding scheme. Motivated by these research trends, we set out to employ the principles for parallel concatenated convolutional coding schemes in conjunction with serial concatenated concatenated coding, in order to evaluate the ML performance bound of non-punctured turbo-coded systems employing recursive modulators. However, the main objective of the theoretical analysis is not to obtain the exact ML performance bound, but to characterise the turbo equalisation performance trends for turbo-coded systems. Explicitly, we aim to explain the turbo equalisation results obtained for turbo-coded GMSK and convolutional-coded GMSK systems, where the convolutional-coded scheme outperformed the turbo-coded system, despite using only a single decoder in each turbo equalisation iteration. Although the union bound diverges at the so-called cut-off rate [351] giving unreliable performance bounds, this bounding technique is still capable of demonstrating the performance trends associated with the turbo-equalised systems. For our investigations we have employed DPSK modulation, since it too is inherently recursive like GMSK, but lends itself to simpler analysis.

The organisation of this chapter is as follows. Sections 15.2 and 15.3 present an analysis of the associated parallel concatenated coding and serial concatenated coding schemes, respectively. Subsequently, Section 15.4 describes the algorithm employed in determining the key parameter — namely the so-called Input Redundancy Weight Enumerating Function (IRWEF) — of an encoder. Section 15.5 models DPSK, MSK and GMSK as a recursive encoder, while Section 15.6 shows how the serial and parallel concatenated coding principles described in Sections 15.2 and 15.3 are adapted in order to determine the ML performance bound of convolutional-coded and turbo-coded systems. The results of the theoretical analysis and computer simulations are then presented in Section 15.7. This is followed by a discussion and a summary of the observations made in Section 15.8.

15.2 Parallel Concatenated Code Analysis

This section describes the fundamental principles involved in determining the performance bound for Parallel Concatenated Codes (PCCs). The PCC illustrated in Figure 15.1 consists of two en-

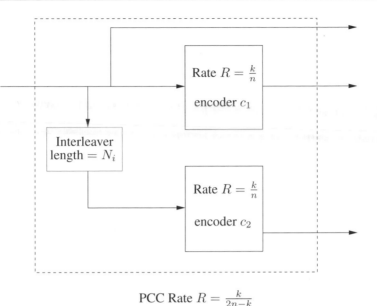

PCC Rate $R = \frac{k}{2n-k}$

Figure 15.1: Structure of the PCC considered, consisting of encoders $c1$ and $c2$ and of the turbo interleaver of depth N_i bits.

coders, namely $c1$ and $c2$, which have a common input information sequence of length N_i bits and which are linked through an interleaver, so that the information sequence entered into the second encoder is the permuted version of the original input information sequence. Therefore, the weights of the input sequences of the first and second encoder are identical, although c1's and c2's parity weight will be different. In order to determine the performance bound of the PCC, the constituent encoders are characterised by the so-called Input Redundancy Weight Enumerating Function (IRWEF) and the concept of uniform interleaving is introduced. Essential PCC notations from reference [59] are quoted and described, inherently following the insightful approach of Benedetto and Montorsi, before elaborating on the proposed turbo equalisation analysis.

The first term introduced is the IRWEF represented by $A^C(W, Z)$, where W and Z are the dummy variables whose exponents w and j represent the input weight and parity check weight, respectively. The role of these variables becomes more explicit during our further discourse. The IRWEF $A^C(W, Z)$ can be expressed with the aid of W and Z as:

$$A^C(W, Z) = \sum_{w,j} A_{w,j} W^w Z^j, \tag{15.1}$$

where $A_{w,j}$ is the number of codewords of parity weight j generated by the input sequence of weight w. The IRWEF quantifies explicitly the separate contributions of the input source bit information segment and that of the parity check segment to the total Hamming weight of the codewords. Hence, the IRWEF can be used for characterising an arbitrary encoder, since it depends on both the input information and parity bits. In order to further highlight the significance of the IRWEF, let us consider a BCH(7,4) block encoder, which accepts four input bits and produces a seven-bit codeword. Since the BCH(7,4) receives four-bit inputs, there are $2^4 = 16$ legitimate codewords.

Table 15.1 shows all the possible codewords which consist of the input word and the parity

Codeword		Input	Parity	Codeword	Term in	Term in
Input bit	Parity bit	weight w	weight	weight j	Eq. 15.2	Eq. 15.3
0000	000	0	0	0	1	1
1000	101	1	2	3	2	2
0100	111	1	3	4	3	3
1100	010	2	1	3	4	4
0010	110	1	2	3	2	2
1010	011	2	2	4	5	5
0110	001	2	1	3	4	4
1110	100	3	1	4	7	7
0001	011	1	2	3	2	2
1001	110	2	2	4	5	5
0101	100	2	1	3	4	4
1101	001	3	1	4	7	7
0011	101	2	2	4	5	5
1011	000	3	0	3	6	6
0111	010	3	1	4	7	7
1111	111	4	3	7	8	8

Table 15.1: All possible BCH(7,4) codewords which consist of the input word and parity word. The corresponding Hamming weights of the input word, parity word and codeword are also presented.

$A_{w,j}$	$A_{0,0}$	$A_{1,2}$	$A_{1,3}$	$A_{2,1}$	$A_{2,2}$	$A_{3,0}$	$A_{3,1}$	$A_{4,3}$
	1	3	1	3	3	1	3	1

Table 15.2: The number of codewords of parity weight j generated by the input sequence of weight w for the BCH(7,4) code. These $A_{w,j}$ coefficients constitute the weighting factors in Equations 15.1 and 15.2.

word. For example, it is observed that the input word 1100, whose Hamming weight — i.e. the number of 1 — is $w = 2$, produces the parity word 010 of Hamming weight $j = 1$, hence giving the codeword 1100010 of Hamming weight three. Further study of Table 15.1 reveals that there are two other codewords, namely 0110001 and 0101100, possessing input and parity Hamming weights of $w = 2$ and $j = 1$, respectively. This information is summarised in Table 15.2, which shows the total number of codewords having input and parity Hamming weights w and j and is denoted as $A_{w,j}$. Considering all input words and by using Equation 15.1 and Table 15.2, the IRWEF for the BCH(7,4) code becomes:

$$A^C(W, Z) = 1W^0 Z^0 + 3W^1 Z^2 + W^1 Z^3 + 3W^2 Z^1 + 3W^2 Z^2 + 1W^3 Z^0$$
$$+ 3W^3 Z^1 + 1W^4 Z^3$$
$$= 1 + W(3Z^2 + Z^3) + W^2(3Z + 3Z^2) + W^3(1 + 3Z) + W^4 Z^3.$$

As can be seen from Equation 15.2, each term in the IRWEF polynomial represents the input word weight w and parity segment weight j of the codewords, as well as the number of such Hamming weight codes $A_{w,j}$. For example, the second polynomial term on the right-hand side of Equation 15.2, namely $W(3Z^2 + Z^3)$, shows that there are three codewords which were generated by input bit segments having a Hamming weight $w = 1$ and resulting in parity weight

$j = 2$, while there is only a single codeword caused by a Hamming weight $w = 1$ input word and yielding a parity word of weight $j = 3$. The last but one column of Table 15.1 having the header 'Term in Eq. 15.2' refers to the BCH(7,4) codewords which contribute to the terms in the top two lines of Equation 15.2. For example, it is observed from Table 15.1 that the BCH(7,4) codewords 1000101 (having input bits 1000 and parity bits 101), 0010110 (input bits 0010 and parity bits 110) and 0001011 (input bits 0001 and parity bits 011) contribute to the second term of Equation 15.2. It is also noted that the sum of the $A_{w,j}$ coefficients is 16 in this example, which is the total number of codewords, as mentioned previously. Although we have considered block codes instead of convolutional codes in this example, we note that convolutional codes can be treated as block codes by adding all-zero tailing bits in order to terminate the encoder in the all-zero state and hence our findings can be extended to the family of convolutional codes.

Once the IRWEF of the code is determined, the bit error probability $P_b(e)$ based on ML decoding can be evaluated by using the upper bound approximation proposed by Benedetto and Montorsi in reference [59], which is given by:

$$P_b(e) \approx \frac{1}{2} \sum_{j+w=m} D_m \mathrm{erfc}\left(\sqrt{\frac{mRE_b}{N_0}}\right),$$ (15.2)

where m is the Hamming weight of the codeword generated by the input word of weight w and the weight distribution D_m is defined as:

$$D_m \triangleq \sum_{j+w=m} \frac{w}{k} A_{w,j}$$ (15.3)

while k is the number of original information bits in the n-bit codeword. Explicitly, for every codeword Hamming weight m, the weight distribution D_m is determined by adding together all weighted IRWEFs whose exponents w and j sum to m. We will show at a later stage that Equation 15.2 can also be employed for PCC and serial concatenated coding schemes.

Having defined the IRWEF and described how the BER can be subsequently evaluated, we now introduce the Conditional Weight Enumerating Function (CWEF) of the parity check bits $A_w^C(Z)$ for the purpose of evaluating the performance bound of the PCC scheme. The CWEF is the IRWEF of the encoder C corresponding to a particular input weight w. Referring to the previous BCH(7,4) example, we recall that the IRWEF in Equation 15.2 was shown to be:

$$A^C(W,Z) = 1 + W(3Z^2 + Z^3) + W^2(3Z + 3Z^2) + W^3(1 + 3Z) + W^4 Z^3.$$

Hence, the CWEF $A_w^C(Z)$ becomes:

$$\begin{aligned}
A_0^C(Z) &= 1 & w &= 0 \\
A_1^C(Z) &= 3Z^2 + Z^3 & w &= 1 \\
A_2^C(Z) &= 3Z + 3Z^2 & w &= 2 \\
A_3^C(Z) &= 1 + 3Z & w &= 3 \\
A_4^C(Z) &= Z^3 & w &= 4.
\end{aligned}$$ (15.4)

The derivation of the CWEF can be generalised according to the approach of Benedetto and Montorsi as [59]:

$$A_w^C(Z) = \sum_j A_{w,j} Z^j,$$ (15.5)

where the corresponding $A_{w,j}$ coefficients can be extracted from Table 15.1 or from Table 15.2 for the BCH(7,4) code.

Before showing the contribution of the CWEF to the evaluation of the performance bound, let us introduce a further abstract device known as the uniform interleaver. As noted before, Benedetto and Montorsi proposed the uniform interleaver [59] of depth k_i as a conceptual probabilistic device, which maps a given input word of weight w to all possible distinct $\binom{k_i}{w}$ number of permutations of it with equal probability of $1/\binom{k_i}{w}$. As a consequence of the interleaving — between the first encoder $c1$ and the second $c2$ — the CWEF of the first encoder is independent from the second. Hence, the CWEF $A_w^{C_p}(Z)$ of the entire concatenated code C_p can be expressed as the product of two CWEFs of the constituent codes, which is given by:

$$A_w^{C_p}(Z) = \frac{A_w^{c1}(Z) \times A_w^{c2}(Z)}{\binom{k_i}{w}}, \tag{15.6}$$

which is normalised by — or averaged over — the number of possible permutations $\binom{k_i}{w}$. In order to highlight the principle of computing the bit error probability of a PCC scheme, let us consider the parallel concatenated BCH(7,4) scheme employing the above abstract uniform interleaver with a depth of $k_i = 4$ bits, replacing the convolutional codec in Figure 15.1. Using Equation 15.6 and the CWEF of the BCH(7,4) code in Equation 15.4, the CWEF of the entire code becomes:

$$A_0^{C_p}(Z) = \frac{1 \times 1}{\binom{4}{0}} = 1$$

$$A_1^{C_p}(Z) = \frac{(3Z^2 + Z^3) \times (3Z^2 + Z^3)}{\binom{4}{1}} = \frac{9}{4}Z^4 + \frac{3}{2}Z^5 + \frac{1}{4}Z^6$$

$$A_2^{C_p}(Z) = \frac{(3Z + 3Z^2) \times (3Z + 3Z^2)}{\binom{4}{2}} = \frac{3}{2}Z^2 + 3Z^3 + \frac{3}{2}Z^4 \tag{15.7}$$

$$A_3^{C_p}(Z) = \frac{(1 + 3Z) \times (1 + 3Z)}{\binom{4}{3}} = \frac{1}{4} + \frac{3}{2}Z + \frac{9}{4}Z^2$$

$$A_4^{C_p}(Z) = \frac{Z^3 \times Z^3}{\binom{4}{4}} = Z^6.$$

Since $A^{C_p}(W, Z) = \sum_w A_w^{C_p}(Z)W^w$, with the aid of Equation 15.7 the IRWEF of the overall parallel concatenated BCH(7,4) code can be expressed as:

$$A^{C_p}(W, Z) = 1 + W\left(\frac{9}{4}Z^4 + \frac{3}{2}Z^5 + \frac{1}{4}Z^6\right)$$

$$+ W^2\left(\frac{3}{2}Z^2 + 3Z^3 + \frac{3}{2}Z^4\right)$$

$$+ W^3\left(\frac{1}{4} + \frac{3}{2}Z + \frac{9}{4}Z^2\right)$$

$$+ W^4\left(Z^6\right).$$

With the aid of Equation 15.2 we can now tabulate the $A_{w,j}$ values of the parallel concatenated BCH(7,4)-based code, as seen in Table 15.3. This then allows us to determine the values of the weight distribution D_m in Equation 15.3, since we can obtain $A_{w,j}$ — i.e. the number of

$A_{w,j}$	$A_{0,0}$	$A_{1,4}$	$A_{1,5}$	$A_{1,6}$	$A_{2,2}$	$A_{2,3}$	$A_{2,4}$	$A_{3,0}$	$A_{3,1}$	$A_{3,2}$	$A_{4,6}$
	1	$\frac{9}{4}$	$\frac{3}{2}$	$\frac{1}{4}$	$\frac{3}{2}$	3	$\frac{3}{2}$	$\frac{1}{4}$	$\frac{3}{2}$	$\frac{9}{4}$	1

Table 15.3: The number of codewords of parity weight j generated by the input sequence of weight w for the BCH(7,4) turbo code extracted from Equation 15.2.

codewords having a parity word of weight j generated by the input sequence of weight w — from the IRWEF $A^{C_p}(W, Z)$. Recalling from Equation 15.3 that $D_m = \sum\limits_{j+w=m} \frac{w}{k} A_{w,j}$ and using Table 15.3 for the BCH(7,4)-based turbo code, we obtain:

$$
\begin{aligned}
D_{m=0} &= D_{m=1} = D_{m=2} = 0 \\
D_{m=3} &= \frac{w=3}{k=4} \cdot A_{3,0} = \frac{3}{4} \cdot \frac{1}{4} = 0.1875 \\
D_{m=4} &= \frac{w=2}{k=4} \cdot A_{2,2} + \frac{w=3}{k=4} \cdot A_{3,1} = \frac{2}{4} \cdot \frac{3}{2} + \frac{3}{4} \cdot \frac{3}{2} = 1.875 \\
D_{m=5} &= \frac{w=1}{k=4} \cdot A_{1,4} + \frac{w=2}{k=4} \cdot A_{2,3} + \frac{w=3}{k=4} \cdot A_{3,2} = \frac{1}{4} \cdot \frac{9}{4} + \frac{2}{4} \cdot 3 + \frac{3}{4} \cdot \frac{9}{4} = 4.3125 \\
D_{m=6} &= \frac{w=1}{k=4} \cdot A_{1,5} + \frac{w=2}{k=4} \cdot A_{2,4} = \frac{1}{4} \cdot \frac{3}{2} + \frac{2}{4} \cdot \frac{3}{2} = 1.125 \\
D_{m=7} &= \frac{w=1}{k=4} \cdot A_{1,6} = \frac{1}{4} \cdot \frac{1}{4} = 0.0625 \\
D_{m=8} &= D_{m=9} = 0 \\
D_{m=10} &= \frac{w=4}{k=4} \cdot A_{4,6} = \frac{4}{4} \cdot 1 = 1.
\end{aligned}
$$

$$(15.8)$$

Consequently, with the aid of Equation 15.2 the bit error probability $P_b(e)$ is given by:

$$
P_b(e) \approx \frac{1}{2} \sum_{j+w=m} D_m \operatorname{erfc}\left(\sqrt{\frac{mRE_b}{N_0}}\right)
$$

$$
\begin{aligned}
P_b(e) \approx \frac{1}{2} \Bigg[& 0.1875 \operatorname{erfc}\left(\sqrt{\frac{3RE_b}{N_0}}\right) + 1.875 \operatorname{erfc}\left(\sqrt{\frac{4RE_b}{N_0}}\right) \\
& + 4.3125 \operatorname{erfc}\left(\sqrt{\frac{5RE_b}{N_0}}\right) + 1.125 \operatorname{erfc}\left(\sqrt{\frac{6RE_b}{N_0}}\right) \\
& + 0.0625 \operatorname{erfc}\left(\sqrt{\frac{7RE_b}{N_0}}\right) + \operatorname{erfc}\left(\sqrt{\frac{10RE_b}{N_0}}\right) \Bigg],
\end{aligned}
$$

$$(15.9)$$

which is plotted in Figure 15.2 for various E_b/N_0 values, where the overall rate of the parallel concatenated BCH(7,4) code is $R = \frac{4}{10} = 0.4$.

The PCC principles described previously can be adapted to evaluate the bit error probability of parallel concatenated convolutional codes, when the convolutional code is terminated in the all-zero state with the aid of tailing bits consisting of logical 0s.

In summary, the key parameters and principles which were required for the evaluation of the bit error probability bound were based on the seminal contributions of Benedetto and Montorsi [59], relying on the:

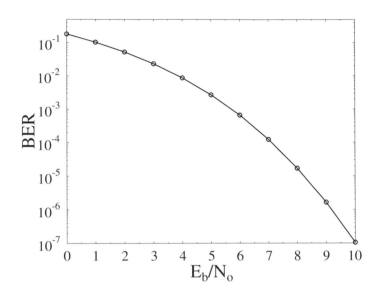

Figure 15.2: Theoretical bound of the parallel concatenated BCH(7,4) code, using a uniform interleaver and possessing an overall code rate of $R = \frac{4}{10} = 0.4$.

- Input Redundancy Weight Enumerating Function (IRWEF) of the constituent codes and that of the entire code.

- Uniform interleaver.

- Conditional Weight Enumerating Function (CWEF) of the constituent codes.

- Union bound approximation [59].

The IRWEF of each constituent code is a function of the input word and the parity word weight. As seen in Equation 15.1, the IRWEF is a polynomial of the dummy variables W and Z, whose exponents are determined by the Hamming weight w of the input word and weight j of the parity word, respectively. The polynomial coefficients $A_{w,j}$ represent the number of codewords with these weights. Consequently, the CWEF, which is constituted by the IRWEF for a particular input weight w, can be evaluated. For block codes the codeword is of finite length, while convolutional-coded codewords can be potentially infinitely long. However, when we employ terminated convolutional codes, i.e. convolutional codes using all-zero termination bits, these codes can be treated as block codes. In order to evaluate the performance bound of PCCs for a specific interleaver, the permutation of the input bit sequence by the interleaver must be considered, to allow the parity word of the second encoder to be determined. Consequently, an exhaustive enumeration of all possible cases must be performed. This high-complexity operation was circumvented by using the abstract concept of the uniform interleaver, which maps a given input into all possible distinct permutations of the input word with equal probability. Therefore, the CWEF of the overall PCC using the uniform interleaver can be evaluated as the product of the CWEF of the constituent codes and normalised by the number of possible distinct permutations.

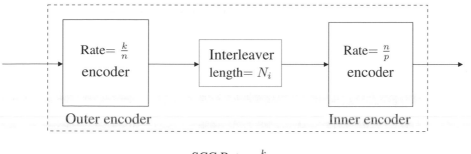

$$\text{SCC Rate} = \frac{k}{p}$$

Figure 15.3: Serial concatenated coding scheme consisting of a rate $\frac{k}{n}$ outer encoder serially cascaded with a rate $\frac{n}{p}$ inner encoder, which are separated by an interleaver.

Subsequently, the bit error probability bound can be evaluated using the union bound relationship in Equation 15.2.

Having described the key notation and principles of union bounding for the PCC scheme, we now proceed to analyse serial concatenated coding schemes.

15.3 Serial Concatenated Code Analysis

The serial concatenated code (SCC) scheme illustrated in Figure 15.3 consists of a rate $\frac{k}{n}$ outer encoder serially cascaded with a rate $\frac{n}{p}$ inner encoder, separated by an interleaver of depth N_i. Explicitly, the rate $R = \frac{k}{n}$ outer encoder generates n-bit codewords for every k input bit. These outer encoder codewords are permuted by the interleaver and subsequently directed to the inner encoder. The rate $R = \frac{n}{p}$ inner encoder then produces p-bit codewords for every n input bit. The overall rate of this SCC scheme is $\frac{k}{p}$.

Let the notations w, l, h represent the Hamming weights of the input sequence, the outer codeword and the inner codeword. Since the output of the outer encoder becomes the input of the inner encoder, the Hamming weight of the *codeword* is considered instead of the weight of the *parity* segment as in the PCC analysis. Hence, the original concept of IRWEF is modified, in order to include the weight of the codewords, resulting in another function defined as the Input Output Weight Enumerating Function (IOWEF) by Benedetto and Divsalar and Montorsi and Pollara [104]. Using the BCH(7,4) example of Section 15.2 and observing the IRWEF expressed in Equation 15.2, which is repeated here for convenience:

$$A^C(W, Z) = 1 + W(3Z^2 + Z^3) + W^2(3Z + 3Z^2) + W^3(1 + 3Z) + W^4 Z^3,$$

and with the aid of the last column in Table 15.1, the IOWEF of the BCH(7,4) code can be written as:

$$A^C(W, L) = \sum_{w,l} A_{w,l} W^w L^l$$

$$A^C(W, L) = 1W^0 L^0 + 3W^1 L^3 + W^1 L^4 + 3W^2 L^3 + 3W^2 L^4 + 1W^3 L^3$$
$$+ 3W^3 L^4 + 1W^4 L^7$$
$$= 1 + W(3L^3 + L^4) + W^2(3L^3 + 3L^4) + W^3(L^3 + 3L^4) + W^4 L^7,$$

Codeword		Input	Codeword	Parity
Input bit	Parity bit	weight w	weight	weight j
000	0	0	0	0
100	1	1	2	1
010	1	1	2	1
110	0	2	2	0
001	1	1	2	1
101	0	2	2	0
011	0	2	2	0
111	1	3	4	1

Table 15.4: All possible (4,3) parity check codewords which consist of the input word and parity word. The corresponding Hamming weights of the input word, parity word and codeword are also presented.

where W and L are dummy variables whose exponents namely w and l, are used to represent the Hamming weights of the input word and codeword, respectively. Note that the value of l in Equation 15.3 is the sum of the exponents of W and Z in Equation 15.2. Therefore — in contrast to the PCC scheme of Section 15.2 — for the purpose of the SCC analysis, all enumerations will consider the weight of the codeword, rather than that of the parity segment.

Having determined the IOWEF, the associated CWEF can be evaluated as in our previous PCC analysis. Subsequently, since the SCC scheme assumes that the interleaver permutations are of equal probability through the use of the uniform interleaver, we can obtain an expression for the IOWEF of the entire SCC using the CWEF of the constituent codes. By using the IOWEF of the overall code, the corresponding bit error probability bound can be determined. As an example, let us consider a simple Serial Concatenated Block Code (SCBC), in order to highlight the key principles in determining the performance bound of the SCC configuration concerned. Let us assume that the outer constituent encoder c_o is a (4,3) parity check code, while the inner encoder c_i employed is a BCH(7,4) code. The corresponding IOWEF of the outer encoder is derived by using Table 15.4, which details all the possible input words and the corresponding codewords for the (4,3) parity check code, yielding:

$$A^{c_o}(W, L) = 1 + W(3L^2) + W^2(3L^2) + W^3(L^4). \tag{15.10}$$

Since the output of the outer encoder forms the inner encoder inputs, in Equation 15.3 we substitute W by L, as well as L by H, hence giving the IOWEF of the BCH(7,4) inner code:

$$A^{c_i}(L, H) = 1 + L(3H^3 + H^4) + L^2(3H^3 + 3H^4) + L^3(H^3 + 3H^4) + L^4 H^7, \tag{15.11}$$

where L and H are dummy variables having exponents l and h, respectively. The exponent l represents the Hamming weights of the codeword of the outer encoder, which also constitute the input to the inner encoder, whereas h denotes the Hamming weights of the inner encoder codeword. As mentioned in Section 15.2, the CWEF is essentially the IRWEF of the constituent code for a particular weight. However, in the SCC analysis the CWEF is the IOWEF conditioned upon the Hamming weight of a word, such as the Hamming weight of the outer encoder codeword. For example, $A_l^{c_o}(W)$ is the CWEF of the outer encoder c_o obtained from the IOWEF $A^{c_o}(W, L)$, which has been conditioned upon l — namely upon the Hamming weight of the codeword — whereas $A_l^{c_i}(H)$ is the CWEF of the inner encoder c_i, which is obtained by conditioning the

IOWEF $A^{c_i}(L, H)$ upon the weight of the input word l, which constitutes also the outer encoder's codeword. Continuing with the SCBC example described previously, the corresponding values of $A_l^{c_o}(W)$ employing the (4,3) parity check code are extracted from Equation 15.10, yielding:

$$
\begin{aligned}
A_{l=0}^{c_o}(W) &= 1 \\
A_{l=1}^{c_o}(W) &= 0 \\
A_{l=2}^{c_o}(W) &= 3W + 3W^2 \\
A_{l=3}^{c_o}(W) &= 0 \\
A_{l=4}^{c_o}(W) &= W^3,
\end{aligned}
\tag{15.12}
$$

while the CWEFs $A_l^{c_i}(H)$ of the BCH(7,4) code are inferred from Equation 15.11, giving:

$$
\begin{aligned}
A_{l=0}^{c_i}(H) &= 1 \\
A_{l=1}^{c_i}(H) &= 3H^3 + H^4 \\
A_{l=2}^{c_i}(H) &= 3H^3 + 3H^4 \\
A_{l=3}^{c_i}(H) &= H^3 + 3H^4 \\
A_{l=4}^{c_i}(H) &= H^7.
\end{aligned}
\tag{15.13}
$$

Here, we have assumed uniform interleaving, i.e. each codeword of weight l is permuted at the output of the outer encoder into all of its possible distinct $\binom{k_i}{l}$ permutations, where k_i is the depth of the interleaver. Each codeword of the outer code c_o of weight l — after uniform interleaving — is then directed towards the inner encoder generating $\binom{k_i}{l}$ inner encoder codewords. In reference [104] Benedetto and Divsalar and Montorsi and Pollara have shown that the IOWEF of the entire concatenated code $A^{C_s}(W, H)$ — i.e. that of the SCC using uniform interleaver — can be expressed as the product of two CWEFs of the constituent codes, which is normalised by the number of possible permutations $\binom{k_i}{l}$:

$$
A^{C_s}(W, H) = \sum_{l=0}^{k_i} \frac{A_l^{c_o}(W) \times A_l^{c_i}(H)}{\binom{k_i}{l}}
\tag{15.14}
$$

Therefore, with the aid of Equations 15.12, 15.13 and 15.14 the IOWEF of the SCBC example using the (4,3) parity check code as the outer encoder and the BCH(7,4) as the inner encoder can be written as:

$$
\begin{aligned}
A^{C_s}(W, H) &= \sum_{l=0}^{k_i=4} \frac{A_l^{c_o}(W) \times A_l^{c_i}(H)}{\binom{k_i}{l}} \\
A^{C_s}(W, H) &= \frac{1 \times 1}{1} + \frac{0 \times (3H^3 + H^4)}{4} + \frac{(3W + 3W^2) \times (3H^3 + 3H^4)}{6} \\
&\quad + \frac{0 \times (H^3 + 3H^4)}{4} + \frac{W^3 \times H^7}{1} \\
&= 1 + W(1.5H^3 + 1.5H^4) + W^2(1.5H^3 + 1.5H^4) + W^3 H^7,
\end{aligned}
$$

where the depth of the interleaver is $N_i = k_i = 4$. Recall from Equation 15.3 that the weight distribution is $D_m = \sum_{m=j+w} \frac{w}{k} A_{w,j}$, where m — which is the sum of the input word weight w and the parity segment weight j — also represents the weight of the codeword and is denoted as

$A_{w,h}$	$A_{0,0}$	$A_{1,3}$	$A_{1,4}$	$A_{2,3}$	$A_{2,4}$	$A_{3,7}$
	1	1.5	1.5	1.5	1.5	1

Table 15.5: The number of SCBC codewords extracted from Equation 15.3.

h for the SCC configurations. Referring to Equation 15.3 and Table 15.5, it is observed that the possible codeword weights generated are $h = 3$, $h = 4$ and $h = 7$, hence giving $D_{m=h}$ values of:

$$D_{h=0} = D_{h=1} = D_{h=2} = 0$$

$$D_{h=3} = \frac{w=1}{k_i=4} \cdot A_{1,3} + \frac{w=2}{k_i=4} \cdot A_{2,3} = \frac{1}{4} \cdot 1.5 + \frac{2}{4} \cdot 1.5 = 1.125$$

$$D_{h=4} = \frac{w=1}{k_i=4} \cdot A_{1,4} + \frac{w=2}{k_i=4} \cdot A_{2,4} = \frac{1}{4} \cdot 1.5 + \frac{2}{4} \cdot 1.5 = 1.125 \qquad (15.15)$$

$$D_{h=5} = D_6 = 0$$

$$D_{h=7} = \frac{w=3}{k_i=4} \cdot A_{3,7} = \frac{3}{4} \cdot 1 = 0.75,$$

and with the aid of Equation 15.2, the bit error probability bound can be expressed as [59]:

$$P_b(e) \approx \frac{1}{2} \sum_{m=h} D_m \mathrm{erfc} \left(\sqrt{\frac{mRE_b}{N_0}} \right)$$

$$P_b(e) \approx \frac{1}{2} \left[1.125 \, \mathrm{erfc} \left(\sqrt{\frac{3RE_b}{N_0}} \right) + 1.125 \, \mathrm{erfc} \left(\sqrt{\frac{4RE_b}{N_0}} \right) + 0.75 \, \mathrm{erfc} \left(\sqrt{\frac{7RE_b}{N_0}} \right) \right]$$

$$(15.16)$$

as plotted in Figure 15.4, where R is the overall SCBC code rate, which is equal to $\frac{3}{7}$ in our example. As argued before, the same principles can also be applied for using convolutional codes as constituent codes provided that the code is terminated with all-zero tailing bits. More explicitly, this is because convolutional codes, which can potentially have infinitely long codewords, can be treated as block codes when all-zero tailing bits are inserted, such that the trellis terminates in the all-zero state.

In reference [104], Benedetto and Divsalar and Montorsi and Pollara showed that the key design criterion for serial concatenated convolutional coding schemes is that the inner encoder must possess recursive properties. This is crucial in order to obtain a large interleaving gain, which is defined as the factor by which the bit error probability is decreased with increasing interleaver length.

In summary, we have described how the ML performance of the PCC and the SCC schemes can be approximated by utilising the constituent codes' IRWEF and IOWEF, respectively. Furthermore, the application of the abstract uniform interleaver also alleviated the computational complexity associated with the evaluation of the ML bound of the PCC and SCC schemes for a specific interleaver. In order to evaluate the IRWEF of the code, all possible input words and its corresponding parity words are determined. Subsequently, the individual contribution of the input word and parity word Hamming weight to the overall codeword Hamming weight was identified, in order to yield the IRWEF. Similarly, the IOWEF of a code is determined by considering the individual contribution of the input word and codeword Hamming weight. However,

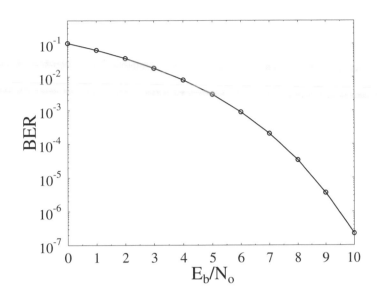

Figure 15.4: Theoretical performance bound of the serial concatenated code consisting of a (4,3) parity
check code as the outer encoder and a BCH(7,4) code as the inner encoder.

the number of input words increases exponentially as the number of bits forming the input word
increases. In parallel concatenated convolutional code or serial concatenated convolutional code
analysis, convolutional codes are assumed to be terminated and therefore the length of the input word is equivalent to the depth of the turbo interleaver for PCC schemes, or to the depth
of the interleaver separating the inner encoder and outer encoder for SCC schemes. Therefore,
the longer the interleaver depth, the greater the complexity associated with the evaluation of the
IRWEF and IOWEF. In order to overcome this complexity problem, we employed a trellis search
algorithm based on the Viterbi algorithm [8,9], which is the focus of our next discussion.

15.4   Enumerating the Weight Distribution of the
Convolutional Code

Having described the ML analysis for PCC and SCC schemes using simple BCH codes as an
example, we now employ these principles in order to enumerate the weight distribution of convolutional codes. We have considered convolutional codes here, since our aim is to characterise
the ML bound of convolutional-coded and convolutional turbo-coded DPSK systems. Furthermore, the modulation technique employed, namely DPSK, can also be modelled as a recursive
systematic convolutional code, as will be shown during our further discourse.

Recall that in Section 15.3 the IRWEF of the BCH(7,4) code was evaluated by identifying all
the possible input words and the associated parity words. This approach can also be employed for
convolutional codes. However, as the length of the input word increases, the number of possible
sequences increases exponentially, hence rendering the task in general impractical. Alternatively,

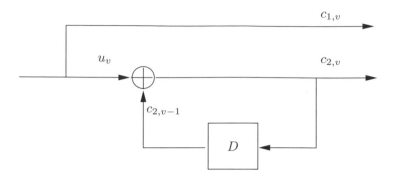

Figure 15.5: Schematic of the rate $R = \frac{1}{2}$, constraint length $K = 2$, recursive systematic convolutional code employing octal generator polynomials of $G_0 = 3$ and $G_1 = 2$. The notations D, u_v, $c_{1,v}$, $c_{2,v}$ and $c_{2,v-1}$ represent the symbol duration, the source bit at time instant v, the systematic bit of the codeword at instant v, the parity bit of the codeword at instant v and the parity bit of the codeword at the previous time instant $v - 1$, respectively.

Input word	Input weight	Parity bit	Parity weight	Codeword weight
0000	0	0000	0	0
1000	1	1111	4	5
0100	1	0111	3	4
1100	2	1000	1	3
0010	1	0011	2	3
1010	2	1100	2	4
0110	2	0100	1	3
1110	3	1011	3	6
0001	1	0001	1	2
1001	2	1110	3	5
0101	2	0110	2	4
1101	3	1001	2	5
0011	2	0010	1	3
1011	3	1101	3	6
0111	3	0101	2	5
1111	4	1010	2	6

Table 15.6: Input information, codeword and parity Hamming weight for rate $R = \frac{1}{2}$, constraint length $K = 2$ RSC encoder, using octal generator polynomials of $G_0 = 3$ and $G_1 = 2$.

the enumeration of each input word's Hamming weight and its corresponding codeword Hamming weight can be performed through a recursive trellis search algorithm, in order to reduce the associated computational complexity. The search algorithm employed is similar to the Viterbi algorithm [8, 9], where a metric is updated recursively. Below, we present a simple example of the algorithm and subsequently we generalise the key points discussed, where possible.

Consider a rate $R = \frac{1}{2}$, constraint length $K = 2$, Recursive Systematic Convolutional (RSC) encoder, using octal generator polynomials of $G_0 = 3$ and $G_1 = 2$, which is depicted in Figure 15.5. Let the length of the input sequence be four bits. As was shown in Section 15.2, the

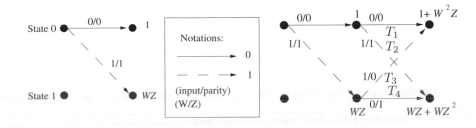

(a) Trellis transition interval $v=1$ **(b)** Trellis transition interval $v=2$

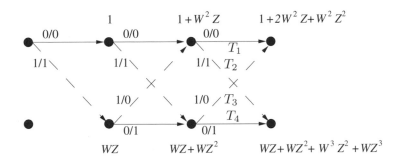

(c) Trellis transition interval $v=3$

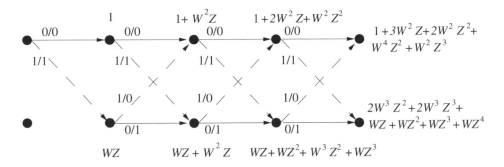

(d) Trellis transition interval $v=4$

Figure 15.6: Trellis development and the accumulated Hamming weight of the input words and the parity words for each state of the $\frac{1}{2}$-rate, constraint length $K=2$ RSC encoder, using octal generator polynomials of $G_0 = 3$ and $G_1 = 2$ illustrated in Figure 15.5.

IRWEF of the code can be determined by listing all possible input words and the corresponding parity words. Therefore, by considering all possible four-bit input sequences entered into the rate $R = \frac{1}{2}$, $K = 2$ RSC encoder of Figure 15.5 we can determine their corresponding parity segments and codeword Hamming weights, as summarised in Table 15.6. Using this approach and Table 15.6 it can be readily verified that the IRWEF and IOWEF of the code are:

$$A^C(W, Z) = 1 + W(Z + Z^2 + Z^3 + Z^4) + W^2(3Z + 2Z^2 + Z^3)$$
$$+ W^3(2Z^2 + 2Z^3) + W^4(2Z^2),$$

and:

$$A^C(W, H) = 1 + W(H^2 + H^3 + H^4 + H^5) + W^2(3H^3 + 2H^4 + H^5)$$
$$+ W^3(2H^5 + 2H^6) + W^4(2H^6),$$

respectively. Alternatively, we can employ a trellis search algorithm that updates a weight metric recursively upon traversing through the trellis. Specifically, in this case the metric updated for each state is the input and parity word Hamming weight. From this, we can easily determine the Hamming weights of the codewords. Figure 15.6 illustrates the development of the trellis until the trellis transition interval of $v = 4$, which will be detailed in the next paragraph. At each interval, there are two main trellis operations. First, the Hamming weight of the input and parity word is evaluated for each transition. This is analogous to the transition metric in the Viterbi algorithm. Second, the IRWEF associated with the previous state is updated after considering the Hamming weight of the corresponding transition, in order to yield the accumulated IRWEF of the new state. This too resembles the accumulated path metric computed by the Viterbi algorithm. One major difference between this trellis search algorithm and the Viterbi algorithm is that during the determination of the IRWEF, no paths are discarded. This will be made more explicit in our forthcoming discussion.

At trellis interval $v = 1$ in Figure 15.6(a), the accumulated IRWEF at state 0 is initialised to $W^0 Z^0 = 1$. Note that the first bit written on the transition is the input bit, which is the same as the systematic encoded bit, while the second is the associated parity bit generated by the encoder. There are two branches leaving state 0, giving the corresponding weight contributions of $W^0 Z^0 = 1$ and $W^1 Z^1$ for input bits of 0 and 1, respectively. Therefore, at the next trellis stage in state 0 we have an IRWEF of $W^0 Z^0 = 1$, while in state 1 the accumulated IRWEF is $W^1 Z^1$. In trellis interval $v = 2$ in Figure 15.6(b) there are two transitions leaving both state 0 and state 1. Consider transitions T_1 and T_3, which merge into state 0. The corresponding Hamming weights of the input word and parity bit for transition T_1 is $w = 0$ and $j = 0$, while for T_3 it is $w = 1$ and $j = 0$. Therefore, we can express the IRWEF of the transition T_1 as $W^0 Z^0 = 1$, while for transition T_3 the IRWEF is $W^1 Z^0$. Consequently, the accumulated IRWEF of state 0 at interval $v = 2$ can be expressed as:

$$
\begin{aligned}
A^C(W, Z)_{\text{of state 0 at }(v = 2)} &= A^C(W, Z)_{\text{of state 0 at } v = 1} \times A^C(W, Z)_{\text{of } T_1} \\
&\quad + A^C(W, Z)_{\text{of state 1 at }(v = 1)} \times A^C(W, Z)_{\text{of } T_3} \\
&= 1 \times 1 + WZ \times W \\
&= 1 + W^2 Z,
\end{aligned}
$$

as seen also in Figure 15.6(b). Similarly, the accumulated IRWEF of state 1 at interval $v = 2$ can be written as:

$$
\begin{aligned}
A^C(W, Z)_{\text{of state 1 at }(v = 2)} &= A^C(W, Z)_{\text{of state 0 at } v = 1} \times A^C(W, Z)_{\text{of } T_2} \\
&\quad + A^C(W, Z)_{\text{of state 1 at }(v = 1)} \times A^C(W, Z)_{\text{of } T_4} \\
&= 1 \times WZ + WZ \times Z \\
&= WZ + WZ^2,
\end{aligned}
$$

as portrayed in Figure 15.6(b). Note that — in contrast to the Viterbi algorithm — there is no selection of the winning path here and hence no paths are discarded. Instead, the Hamming weight contributions from the two merging paths contribute to the accumulated Hamming weight of the specific state reached as we have seen in Equations 15.4 and 15.4. Note that the IRWEFs of the transitions T_1, T_2, T_3 and T_4 are invariant with respect to the trellis interval v. For the sake of convenience, we will summarise the IRWEFs of these transitions here.

$$
\begin{aligned}
A^C(W, Z) \text{ of } T_1 &= 1 \\
A^C(W, Z) \text{ of } T_2 &= WZ \\
A^C(W, Z) \text{ of } T_3 &= W \\
A^C(W, Z) \text{ of } T_4 &= Z.
\end{aligned}
\tag{15.17}
$$

Following the same reasoning, at trellis interval $v = 3$ state 0 accumulates the IRWEF $A^C(W, Z)$ of:

$$
\begin{aligned}
A^C(W, Z) \text{ of state 0 at } (v = 3) &= A^C(W, Z) \text{ of state 0 at } (v = 2) \times A^C(W, Z) \text{ of } T_1 \\
&+ A^C(W, Z) \text{ of state 1 at } (v = 2) \times A^C(W, Z) \text{ of } T_3 \\
&= (1 + W^2 Z) \times 1 + (WZ + WZ^2) \times W \\
&= 1 + 2W^2 Z + W^2 Z^2,
\end{aligned}
$$

while $A^C(W, Z)$ for state 1 at $(v = 3)$ is:

$$
\begin{aligned}
A^C(W, Z) \text{ of state 0 at } (v = 3) &= A^C(W, Z) \text{ of state 0 at } (v = 2) \times A^C(W, Z) \text{ of } T_2 \\
&+ A^C(W, Z) \text{ of state 1 at } (v = 2) \times A^C(W, Z) \text{ of } T_4 \\
&= (1 + W^2 Z) \times WZ + (WZ + WZ^2) \times Z \\
&= W(Z + Z^2 + Z^3) + W^3 Z^2.
\end{aligned}
$$

These functions can also be seen in Figure 15.6(c). We can generalise the expression of $A^C(W, Z)$ for state s at trellis interval v as:

$$
A^C(W, Z) \text{ of state } s \text{ at interval } v = \sum_{\forall \grave{s} \to s} \left(A^C(W, Z) \text{ of state } \grave{s} \text{ at } v - 1 \right.
$$
$$
\left. \times\, A^C(W, Z) \text{ of transition } \grave{s} \to s \right),
$$

where $\forall \grave{s} \to s$ represents all paths from the previous state $\grave{s}$ to the current state s.

In Figure 15.6(d) the trellis ends and the sum of the accumulated IRWEF at states 0 and 1 represents the IRWEF of the entire code. This is because the IRWEF accumulated at state 0 indicates the weight of all the possible codes, which started from state 0 at trellis interval $v = 1$ and reached state 0 at trellis interval $v = 4$, while the IRWEF at state 1 encompasses all the codes that emerged from state 0 at $v = 1$ and arrived at state 1 at interval $v = 4$. Furthermore, since there are only two possible states in the trellis, the sum of the IRWEF in these states represents all the possible input words and codewords. Therefore, we infer from Figure 15.6(d) that the IRWEF of the entire code $A^C(W, Z)$ is:

$$
\begin{aligned}
A^C(W, Z) &= 1 + W(Z + Z^2 + Z^3 + Z^4) + W^2(3Z + 2Z^2 + Z^3) \\
&+ W^3(2Z^2 + 2Z^3) + W^4(Z^2),
\end{aligned}
$$

which is identical to the IRWEF in Equation 15.4 that was obtained from Table 15.6 by generating all possible input words. The advantage of employing the trellis search algorithm is that its complexity is not determined by the consideration of an exponentially increasing number of possible input words.

Although the trellis search algorithm's complexity is not determined by the exponentially increasing number of codewords, it is constrained by the memory required to store the Hamming weights of the input and parity words. Therefore, input and parity weights exceeding a certain threshold, which we term the **IRWEF weight threshold**, are not stored.

In our forthcoming discussion we aim to show that by modelling the modulator as an inner encoder, the PCC and SCC principles can be employed for the analysis of turbo equalisation performance for convolutional-coded and turbo-coded GMSK systems. However, since it is not possible to model the GMSK modulator as a binary, recursive encoder in order to exploit the SCC and PCC principles for the ML bound analysis, we employ a simpler modulator, namely DPSK, which is also inherently recursive like GMSK.

15.5   Recursive Properties of the MSK, GMSK and DPSK Modulators

As mentioned previously, one of the important design criteria for serial concatenated convolutional coding schemes is that the inner encoder must be recursive, as it was noted by Benedetto and Divsalar and Montorsi and Pollara in [104]. In this section, CPM schemes such as MSK and GMSK are shown to be recursive. However, owing to the difficulty in modelling these modulation techniques as binary recursive convolutional codes, a simpler modulation scheme is considered, namely DPSK modulation. Here, a suitable recursive convolutional code is utilised, in order to represent the DPSK modulator as an inner code, hence allowing the SCC principles to be adopted in order to derive the ML bound of coded DPSK systems.

Let us proceed by examining the first CPM scheme concerned, namely MSK. As mentioned in Section 13.5.1, the transmitted information is embedded in the phase of the signal. Recall from Equation 13.13 that the output phase of the MSK modulator at interval v is

$$\phi(t,a) = 2\pi h_f a_v q(t - vT) + \pi h_f \sum_{i=-\infty}^{v-1} a_i \qquad \text{for } vT \leqslant t \leqslant (v+1)T$$

$$= \frac{\pi}{2} a_v + \frac{\pi}{2} \sum_{i=-\infty}^{v-1} a_i \qquad \text{where } a_i = \pm 1$$

$$= \frac{\pi}{2} a_v + \theta_v,$$

since the MSK modulation index is $h_f = \frac{1}{2}$ and the phase shaping function $q(t - vT) = 0.5$ for $t > T$. The notation θ_v is the phase state and represents the accumulated phase due to the previous bits $a_i, i = -\infty \ldots v - 1$, that have passed through the filter $q(t - vT)$. Equation 15.5 can also be represented graphically by the schematic of Figure 15.7. In the MSK modulator, the interleaved encoder bits are mapped from logical 0s and 1s to -1s and $+1$s, respectively. The mapped bits are multiplied by $\frac{\pi}{2}$ and subsequently added to the phase state θ_v, in order to yield the current phase $\phi(t,a)$. We observe that there is inherent memory in MSK and the modulator is recursive its in nature. However, the MSK modulator is not readily modelled as a recursive convolutional encoder, since the range of phase values involved in the recursive operation is

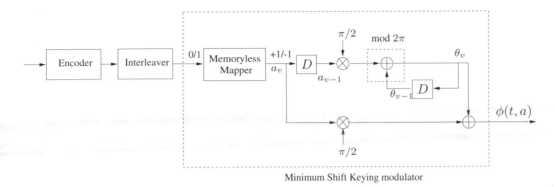

Figure 15.7: Schematic of the MSK modulator based on Figures 13.3 and 13.8, which was modified according to Equation 15.5, where D represents the symbol duration.

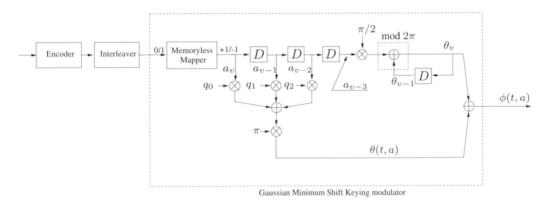

Figure 15.8: GMSK modulator model based on Figures 13.3 and 13.10 and modified according to Equation 15.5, where D represents the symbol duration.

much wider compared to the binary AND operations between bits 0 and 1, which are performed in the case of DPSK, as described at a later stage. Furthermore, the model chosen must account for the effect of the recursive summation of the phase values on the minimum distance of the modulation scheme, hence increasing the complexity of the performance analysis of turbo-coded and convolutional-coded MSK systems.

The GMSK modulator can also be represented with the aid of a model similar to that employed for MSK. Here, the additional modifications are due to the partial-response spreading introduced by the phase shaping function $q(t)$ in the time domain. For convenience, the expres-

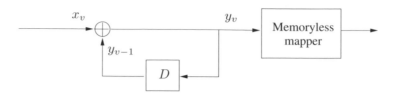

Figure 15.9: Schematic of the DPSK modulator based on Equation 15.18, where D represents the symbol duration.

sion for the phase of GMSK is repeated from Equation 13.5:

$$\phi(t,a) = 2\pi h_f \sum_{i=v-L+1}^{v} a_i q(t-iT) + \pi h_f \sum_{i=-\infty}^{v-L} a_i$$

$$= \pi \sum_{i=v-2}^{v} a_i q(t-iT) + \frac{\pi}{2} \sum_{i=-\infty}^{v-3} a_i$$

$$= \theta(t,a) + \theta_v,$$

where $\phi(t,a)$ is the overall phase of the GMSK signal, while θ_v and $\theta(t,a)$ are the phase state and the phase, which are dependent on the previously transmitted bits, also known as the correlative state vector, respectively. The phase shaping function $q(t)$ is typically spread over $L = 3$ symbol periods, as in the GSM standard [49, 344]. Taking these changes into consideration, the GMSK modulator can be schematically represented as in Figure 15.8. As before, it is observed that GMSK is also recursive in its nature and possesses memory. The recursive nature of the MSK and GMSK modulation schemes can therefore be exploited, in order to yield interleaving gains when concatenated with an outer encoder and separated by an interleaver as in the SCC schemes of Figure 15.3, for example. However, as mentioned above, the analysis of coded GMSK systems cannot readily adopt the principles invoked for SCC schemes since GMSK — like MSK — cannot be modelled as a binary recursive convolutional code because its recursive nature exhibits itself in terms of the phase, not in terms of binary bits. Let us now describe a simple differential modulation technique, namely DPSK, which is also inherently recursive.

Figure 15.9 illustrates the schematic of the recursive DPSK modulator, where D represents the symbol period delay. The DPSK modulator can be viewed as a block, which performs a binary AND operation on the current binary input bit x_v and the previous output bit y_{v-1}, in order to give y_v at trellis interval v, which is formulated as:

$$y_v = x_v \oplus y_{v-1}. \tag{15.18}$$

Subsequently, the memoryless mapper as shown in Figure 15.9 translates the resultant bit $y_j = 1$ to the phase $\theta = 0$ radians and $\theta = \pi$ radians for $y_j = 0$. Explicitly, the output of the DPSK modulator is dependent on the previous output bit, hence exhibiting recursive properties. It was observed that the schematic of the DPSK modulator of Figure 15.9 has a structure similar to the rate $R = 0.5$, constraint length $K = 2$ RSC encoder using octal generator polynomials $G_0 = 3$ and $G_1 = 2$, as illustrated in Figure 15.5. However, the DPSK modulator only generates a single discrete response for every input bit received, unlike the rate $R = 0.5$ convolutional encoder above, which produces two coded bits for every source bit received. Therefore, the DPSK modulator can be modelled as a rate $R = 1$ RSC encoder with constraint length $K = 2$

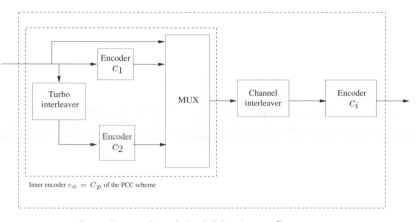

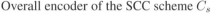

Overall encoder of the SCC scheme C_s

Figure 15.10: Modelling the turbo-coded DPSK system as a hybrid code, consisting of a parallel concatenated convolutional code, C_p, which is coupled serially via a channel interleaver to an inner encoder c_i. The DPSK modulator is modelled by a recursive, rate $R = 1$ inner convolutional encoder.

and octal generator polynomials $G_0 = 3$ and $G_1 = 2$, coupled with a memoryless mapper. The rate $R = 1$ RSC encoder is obtained by retaining the parity bit of the rate $R = 0.5$ RSC encoder and discarding the systematic bit. Furthermore, since the mapper is memoryless, the encoder state transitions are not affected. Hence, the DPSK modulator can be modelled solely by the encoder without the mapper.

At this stage, we have shown that MSK, GMSK as well as DPSK possess memory and are recursive in their nature. Hence, these modulation schemes are suitable for SCC-like schemes and capable of achieving large interleaving gains, as it was observed by Benedetto and Divsalar and Montorsi and Pollara in [104]. Owing to the difficulty of representing MSK and GMSK modulation as a binary convolutional code, in our approach DPSK modulation is employed. DPSK can be modelled as a rate $R = 1$ recursive, convolutional code using octal generator polynomials $G_0 = 3$ and $G_1 = 2$. Therefore, in our approach the theoretical analysis of turbo-coded and convolutional-coded GMSK schemes ensues employing DPSK instead of GMSK. In the next section, we discuss the analytical model of the coded DPSK system and how the associated ML bound can be determined.

15.6 Analytical Model of Coded DPSK Systems

Having identified a suitable convolutional encoder for representing the DPSK modulator, the analytical model of the coded DPSK systems can be constructed. The structure of the convolutional-coded DPSK scheme is the same as that of a SCC configuration illustrated in Figure 15.3, where the inner encoder of rate $R = \frac{n}{p}$ represents the DPSK modulator. As mentioned in the previous section, the DPSK model is a rate $R = 1$, constraint length $K = 2$ recursive convolutional code, employing octal generator polynomials of $G_0 = 3$ and $G_1 = 2$. Therefore, the ML bound of the convolutional-coded DPSK system can be determined by applying the principles of the SCC scheme of Section 15.3 directly.

For a non-punctured turbo-coded DPSK system the analytical model consists of two parallel concatenated encoders c_1 and c_2, which are separated by a turbo interleaver. This forms the outer encoder c_o — also equivalent to C_p of Section 15.2 — and is coupled serially with an inner encoder c_i, which models the DPSK modulator. The entire model C_s of the turbo-coded DPSK system is depicted in Figure 15.10. The resultant configuration is that of a hybrid between PCC and SCC and therefore the theoretical analysis of the turbo-coded DPSK system considered requires employing both the PCC and SCC principles. Initially, the IRWEF of the PCC $A^{C_p}(W, Z)$ is evaluated by employing the PCC principles described in Section 15.2. Once this IRWEF has been determined, the overall system can be viewed as a SCC scheme, consisting of the outer encoder c_o, which is the PCC now characterised by its IRWEF, and the inner encoder c_i describing the DPSK modulator. The IRWEF $A^{C_p}(W, Z)$ of Equation 15.1 is subsequently modified and treated as the IOWEF of the outer encoder $A^{c_o}(W, L)$, which is a function of the dummy variables W and L. This is achieved by substituting Z in Equation 15.1 by the dummy variable L, where the exponent of L is the associated Hamming weight of the outer codeword l obtained by taking the sum of the exponents of the dummy variables W and Z in Equation 15.1, *i.e.* $w + j$. Now, by evaluating the IOWEF of the inner encoder $A^{c_i}(L, H)$ as well, where the orders of L and H are the Hamming weights of the input words of the inner encoder (or the Hamming weight of the output encoder's codewords) and those of its output codeword, the overall IOWEF $A^{C_s}(W, H)$ of the PCC/SCC hybrid scheme can be determined by using the SCC principles of Section 15.3. Subsequently, with the aid of Equation 15.2 and using the IOWEF $A^{C_s}(W, H)$ determined, the ML bound of the turbo-coded DPSK system can be evaluated.

At this stage we have to remind the reader that there are several limitations associated with the theoretical analysis of the coded DPSK system concerned. The first limitation is due to the use of the abstract uniform interleaver in the analysis, which only yields the average ML performance instead of the exact performance of the specific interleaver employed. However, since the analysis is utilised as a tool for comparisons, all the coded systems are analysed in the same manner, and hence determining the average performance bound is adequate for these purposes. Second, the theoretical bound derived using union bound techniques produces a divergence in the ML bound at E_b/N_0 values below the so-called cut-off rate [351], leading to an inaccurate ML bound. However, as long as the systems compared are analysed using the same technique, the ML bound derived can be used — with due caution — in order to characterise the performance trends of the associated systems. Third, the recursive weight evaluation algorithm of Section 15.4 employed is constrained by the memory requirement for large Hamming weights. In order to circumvent this memory limitation, a threshold known as the IRWEF weight threshold can be employed, whereby only Hamming weights below this threshold are retained, while the others are discarded. Because of this limitation, we are unable to determine the weight distribution D_m associated with high values of the codeword Hamming weights m, which were identified as the reason for the divergence in the ML bound [59]. In such cases, where the input word is long, the ML bound can be employed in a more limited context, for example in order to determine the error floor of the system.

In the following section we will compare our simulation and theoretical results for non-punctured convolutional-coded DPSK systems and non-punctured turbo-coded DPSK schemes which employ channel interleavers of depths of 300 bits and 30000 bits between the encoder and modulator.

Encoder	Constraint length K	Octal generator polynomials
$R = \frac{1}{3}$ convolutional code	3	$G_0 = 5\ G_1 = 7\ G_2 = 7$
	5	$G_0 = 25\ G_1 = 33\ G_2 = 37$
$R = \frac{1}{3}$ convolutional-coding-based turbo code	3	$G_0 = 7\ G_1 = 5$
	5	$G_0 = 35\ G_1 = 23$

Table 15.7: Parameters of the encoders used in the $R \approx \frac{1}{3}$ rate turbo-coded and convolutional-coded DPSK systems.

15.7 Theoretical and Simulation Performance of Coded DPSK Systems

In this section, the results of our comparative study between a code rate $R = \frac{1}{3}$ turbo-coded DPSK scheme and a $R = \frac{1}{3}$ convolutional-coded DPSK system will be presented. The parameters employed are summarised in Table 15.7. Initially, a channel interleaver possessing a depth of 300 bits was employed. This corresponds to a turbo interleaver depth of 100 bits at the input of the turbo encoder, while introducing the same delay. The reason for employing such a short interleaver is to avoid the previously mentioned memory limitation associated with the theoretical analysis. In Figure 15.11(a) the simulation results of the rate $R = \frac{1}{3}$, constraint length $K = 3$ turbo-coded and convolutional-coded DPSK system employing eight turbo equalisation iterations over the non-dispersive Gaussian channel are presented. Again, a random channel interleaver with a depth of 300 bits was employed, while the random turbo interleaver had a depth of 100 bits. It was observed that iteration gains — i.e. SNR performance gains with respect to the first iteration — can be obtained by both the convolutional-coded and turbo-coded DPSK schemes. Based on our simulation results in Figure 15.11(a) the convolutional-coded DPSK system achieved an iteration gain of 2.0 dB, 2.9 dB and 3.0 dB after two, four and eight turbo equalisation iterations, respectively, at a BER of 10^{-4}. By contrast, for a BER of 10^{-4}, the turbo-coded DPSK system achieved iteration gains of 1.5 dB, 2.1 dB and 2.2 dB after two, four and eight turbo equalisation iterations respectively . After eight turbo equalisation iterations, the convolutional-coded DPSK scheme was observed to outperform the turbo-coded DPSK system by a margin of 0.5 dB at a BER of 10^{-4}.

Studying the ML bound of Figure 15.11(b) obtained through our theoretical analysis, where the same simulation parameters were employed as in our computer simulations, it was observed that at $E_b/N_0 < 3$ dB, the convolutional-coded DPSK system has a lower BER compared to the turbo-coded DPSK scheme. This result further justifies the ability of the convolutional-coded DPSK system outperforming the turbo-coded DPSK scheme. However, at $E_b/N_0 > 3$ dB the convolutional-coded system yielded a higher error floor compared to the turbo-coded scheme. Note that although the theoretical analysis only yields the average performance of the system and does not predict the BER performance accurately owing to the previously mentioned divergence, it is an adequate tool for comparing the performance of coded schemes in order to identify the more robust encoder. Note that the theoretical analysis for such short channel interleavers is not memory intensive, since the corresponding input word length, and consequently the Hamming weight, is not high. Therefore, no IRWEF weight threshold was set.

In our next experiment, random channel interleavers having depths of 30000 bits and random turbo interleavers possessing depths of 10000 bits were utilised. Here, the constraint length of the rate $R = \frac{1}{3}$ encoders was set to $K = 5$ and the octal generator polynomials of Table 15.7 were employed. The results of our computer simulations were presented in Figure 15.12(a). It was noted that the iteration gains achieved by the convolutional-coded scheme were 2.7 dB, 4.0 dB and 4.5 dB after two, four and eight turbo equalisation iterations, whereas the turbo-coded system obtained gains of 1.8 dB, 2.5 dB and 2.7 dB, respectively. Compared to the scenario where a short turbo interleaver with a depth of 100 bits was employed, the iteration gains observed in Figure 15.12(a) were higher, since the longer interleaver minimised the correlation between the input of the equaliser and the decoder(s), hence benefiting from the information provided by the other equaliser or decoder blocks. Another key observation was that after eight turbo equalisation iterations, the convolutional-coded DPSK scheme was observed to outperform the turbo-coded DPSK system by a margin of 1.2 dB at a BER of 10^{-4}.

For the theoretical analysis of Figure 15.12(b), a IRWEF weight threshold of 100 was set, since it was not feasible to accumulate Hamming weights corresponding to an input word length of 10000 bits. The analysis was based on the same $R = \frac{1}{3}$, $K = 5$ encoder parameters as tabulated in Table 15.7 and employing the abstract uniform interleaver of reference [59]. Since the Hamming weights have been truncated, we were unable to determine the weight distribution values D_m of Equation 15.3, for higher codeword Hamming weights. The distribution of the codeword Hamming weights in this high-weight region determines the divergence of the ML bound, while the D_m values of the lower-weight region are responsible for the error floor performance [59]. Therefore, the ML bound evaluated for the coded systems considered in conjunction with such long channel interleavers — with a depth of 30000 bits — only reflects the error floor performance of the coded DPSK schemes. As before, the error floor of the turbo-coded DPSK system in Figure 15.12(b) was observed to be lower than that of the convolutional-coded DPSK scheme.

In conclusion, our simulation results in Figures 15.11(a) and 15.12(a) showed that convolutional-coded DPSK systems outperformed turbo-coded DPSK schemes which employed turbo equalisation. In our simulations employing 300-bit random channel interleavers the convolutional-coded DPSK system outperformed the turbo-coded scheme by a margin of 0.5 dB at a BER of 10^{-4}, while a gain of 1.2 dB was achieved by the convolutional-coded scheme over the turbo-coded system when a channel interleaver with a depth of 30000 bits was employed. This performance trend was also observed in Chapter 14 for our coded GMSK schemes. Intuitively, the turbo-coded scheme is expected to perform better since it employs two encoders, which are concatenated in parallel, as opposed to the single encoder in convolutional codes. Therefore, the theoretical analysis of coded DPSK systems was invoked, in order to justify the simulation results obtained. Through the analysis using a 300-bit uniform channel interleaver, it was observed in Figure 15.11(b) that the convolutional-coded system performed better than the turbo-coded system at $E_b/N_0 < 3$ dB. However the former yielded a higher undesirable error floor. For longer channel interleaver depths, specifically for 30000 bits, the ML bound of the convolutional-coded system gave a worse error floor performance than that of the turbo-coded DPSK scheme. Note that in the analysis of the systems employing such interleaver depths, the input words of Hamming weights above 100 were not considered. Hence, the region where the divergence occurs could not be determined and therefore the ML bound derived reflects only the error floor performance and it is incapable of reflecting the BER performance adequately at low E_b/N_0 values. From the observations made, it can be concluded that for speech or data systems — requiring a BER of $10^{-3} - 10^{-4}$ — employing recursive modulation techniques such as DPSK or GMSK and utilising turbo equalisation, convolutional coding is the more robust choice as compared to the more complex turbo coding schemes considered.

15.8 Summary and Conclusions

This chapter described how the union bound principles of the SCC and PCC schemes can be employed to evaluate the ML bound of the convolutional-coded and turbo-coded DPSK systems investigated. Each component encoder was characterised by the Input Redundancy Weight Enumerating Function (IRWEF), which categorised the codes according to the input word and parity word Hamming weight or in terms of the input word and codeword Hamming weight quantified by the Input Output Weight Enumerating Function (IOWEF). By modelling the modulator as a convolutional encoder, the IRWEF and IOWEF were evaluated in conjunction with the abstract uniform interleaver in order to determine the ML performance of an encoder concatenated with a modulator. The aim of this study was to provide a theoretical justification of the results obtained for coded GMSK schemes employing turbo equalisation in Chapter 14. Furthermore, it was intended that the investigations showed the importance of employing inner encoders constituted by the modems in a serial concatenated scheme, which were recursive in their nature. In Chapter 14, it was observed that the convolutional-coded GMSK scheme achieved better performance than the convolutional-coding-based turbo-coded GMSK system employing turbo equalisation, despite utilising only one convolutional decoder, while the turbo-coded scheme employed two decoders. In our investigations, we have used DPSK instead of GMSK modulation. This was because DPSK — unlike GMSK — could be readily modelled as a binary recursive convolutional encoder and hence allowing the union bound SCC principles to be employed in order to evaluate the ML bound of the system. GMSK is also recursive in its nature, but in terms of phase values (radians), whereas DPSK operated recursively on binary values and hence it was modelled similarly to recursive convolutional encoders. Hence we have employed DPSK in our analysis of the associated performance bounds. From our simulation results generated for turbo-coded DPSK schemes and convolutional-coded DPSK systems, it was observed that the convolutional-coded scheme outperformed the turbo-coded system. This was further justified by the similar trends of the theoretical performance bounds. Despite the divergence of the theoretical bound below the cut-off rate, the ML bound derived was still a reliable experimental tool, since the aim of the analysis was to characterise the relative performance of the coded systems analysed with the aid of the same analytical technique. In Figure 15.11(b), for a channel interleaver depth of 300 bits, the theoretical bound of the turbo-coded DPSK system diverged at an E_b/N_0 value of approximately 3.0 dB, and showed that the performance of the convolutional-coded system was better than that of the turbo-coded system at low E_b/N_0 values. As the E_b/N_0 value increased beyond 3 dB, the theoretical bound for the convolutional-coded system began to flatten at approximately a BER of $10^{-12} - 10^{-8}$, whereas the error floor for the turbo-coded system for $E_b/N_0 > 3$ dB was approximately a BER of $10^{-14} - 10^{-10}$. For channel interleaver depths of 30000 bits the error floor of the convolutional-coded scheme was approximately a BER of $10^{-28} - 10^{-22}$, while the turbo-coded DPSK system yielded an error floor of a BER of $10^{-32} - 10^{-26}$. Therefore, it can be concluded that the turbo-coded DPSK system exhibited a better error floor than the convolutional-coded system, but at low E_b/N_0 values, its performance was poorer than that of the convolutional-coded DPSK system. These observations and conclusions were also consistent with the results obtained for convolutional-coded GMSK schemes and turbo-coded GMSK systems. Furthermore, it can be concluded that for speech and data systems — requiring a BER of 10^{-3} and a BER of 10^{-4}, respectively — employing recursive modulation techniques such as DPSK or GMSK and utilising turbo equalisation, the convolutional code is the more robust choice, compared to the more complex turbo code.

Having estimated the performance bounds of turbo equalisation schemes, in the next chapter we will embark on a comparative study of various turbo equalisers.

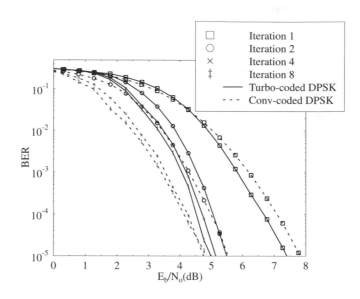

(a) Computer simulation

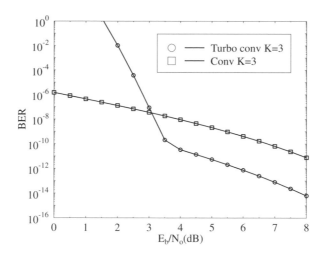

(b) Theoretical bound

Figure 15.11: Comparing the performance of the rate $R = \frac{1}{3}$ $K = 3$ convolutional-coded DPSK system with that of the $R = \frac{1}{3}$ turbo-coded DPSK scheme over the non-dispersive Gaussian channel using both computer simulations and theoretical analysis. A random channel interleaver with a depth of 300 bits was employed, while the random turbo interleaver had a depth of 100 bits.

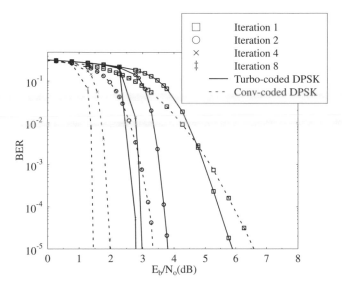

(a) Computer simulation

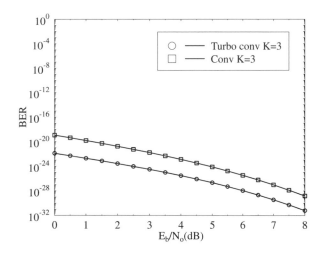

(b) Theoretical bound

Figure 15.12: Comparing the performance of the rate $R = \frac{1}{3}$ $K = 5$ convolutional-coded DPSK system with that of the $R = \frac{1}{3}$ turbo-coded DPSK scheme, over the non-dispersive Gaussian channel using both computer simulations and theoretical analysis. A random channel interleaver possessing a depth of 30000 bits and a 10000-bit random turbo interleaver were implemented.

Chapter 16

Comparative Study of Turbo Equalisers

16.1 Motivation[1]

In Chapter 14, turbo equalisation was investigated in the context of coded partial response GMSK systems. The inherent recursive nature of GMSK modulation was exploited, in order to achieve large interleaver gains. Furthermore, it was observed in Chapter 15 through computer simulations that for recursive modulation systems such as GMSK and DPSK convolutional-coded schemes outperformed convolutional-coding-based turbo-coded systems at low E_b/N_0 values and for BERs $> 10^{-5}$. The theoretical ML bounds of the convolutional-coded and turbo-coded DPSK systems in Chapter 15 also showed that the convolutional-coded scheme was more powerful than the investigated turbo-coded systems at these E_b/N_0 values. However, at higher E_b/N_0 values, it was observed that the turbo-coded scheme yielded lower error floors compared to the convolutional-coded scheme. It was therefore concluded for recursive modulation systems transmitting data and speech — i.e. upon requiring a BER of 10^{-4} and a BER of 10^{-3}, respectively, and employing turbo equalisation — that convolutional codes are more robust compared to the investigated turbo-coded schemes. In this chapter a sound appreciation of the concept presented in Chapter 14 constitutes a prerequisite, although useful system-performance-related information may be gleaned without understanding the associated iterative equalisation principles.

In this chapter, BPSK modulation is employed in order to investigate the performance of the turbo equaliser in the context of non-recursive modulation systems. In this case the modulator and dispersive channel are viewed as the inner encoder, while the channel encoder employed is perceived to be the outer encoder, as in the SCC scheme described in Section 15.3. In addition to convolutional codes and convolutional-coding-based turbo codes, block-coding-based turbo codes are also researched in conjunction with BPSK systems utilising turbo equalisation. With the ever-increasing demand for bandwidth, current systems aim to increase the spectral efficiency by invoking high-rate codes. This has been the motivation for research into block turbo codes, which have been shown by Hagenauer *et al.* [61] to outperform convolutional turbo

---

[1]This chapter is based on B. L. Yeap, T. H. Liew, J.Hámorský and L. Hanzo: Comparative study of turbo equalization schemes using convolutional, convolutional turbo and block-turbo codes, IEEE Tr. on Wireless Communications, 2002,Vol.1, No. 2, April 2002, pp 266-273

codes when the coding rate is higher than $\frac{2}{3}$. It was also observed that a rate $R = 0.981$ block turbo code using BPSK over the non-dispersive Gaussian channel can operate within 0.27 dB of the Shannon limit [63]. In reference [123] Pyndiah presented iterative decoding algorithms for BCH turbo codes. **In this chapter we construct a BPSK turbo equaliser which employs block-coding-based turbo codes, with the objective of investigating its performance in comparison to turbo equalisers employing different classes of codes for high code rates of $R = \frac{3}{4}$ and $R = \frac{5}{6}$, since known turbo equalisation results have only been presented for turbo equalisers using convolutional codes and convolutional-coding-based turbo codes for code rates of $R = \frac{1}{3}$ and $R = \frac{1}{2}$** [105, 106]. Specifically, Bose–Chaudhuri–Hocquengham (BCH) codes [14, 15] are used as the component codes of the block-coding-based turbo codec. Since BCH codes may be constructed with parameters n and k, which represent the number of coded bits and data bits, respectively, we will use the notation BCH(n,k). The BCH-coding-based turbo-coded systems are denoted as **BT**, while the convolutional-coding-based turbo-coded schemes and convolutional-coded systems are represented as **CT** and **CC**, respectively.

The organisation of this chapter is as follows. Section 16.2 provides an overview of the systems researched. Subsequently, Section 16.3 summarises the simulation parameters. Then, Section 16.4 provides results and discussions, while finally Section 16.6 summarises the systems' performance.

16.2   System overview

Again, in this comparative study three classes of encoders, namely convolutional codes, convolutional-coding-based turbo codes and BCH-coding-based turbo codes are employed, which are serially concatenated with the BPSK modulator. The encoder parameters will be specified in the following section. In addition to the channel interleaver, which separates the encoder and the modulator, turbo interleavers are also implemented for the turbo encoders. At the receiver, the equaliser and decoder(s) are configured to perform either independent equalisation and decoding or turbo equalisation, where equalisation and decoding are performed by exchanging information iteratively between the equaliser and decoder(s). Specifically, for the convolutional-coded system, the receiver implements either conventional convolutional decoding [8, 9] or turbo equalisation. The turbo equalisation operation is based on the principles described in Section 14.2 using $N_d = 1$ decoder. For convolutional-coding-based turbo codes and BCH-coding-based turbo codes, independent equalisation and decoding refer to the scenario where soft decision is passed from the equaliser to the turbo decoder, which performs its decoding by passing information between the decoders, but never with the equaliser [12, 123]. In these systems turbo equalisation is also based on the principles of Section 14.2, but in this case information is exchanged between the equaliser and the $N_d = 2$ decoders.

In the following section, we specify the parameters of the convolutional-coded BPSK system, convolutional-coding-based turbo-coded BPSK scheme and the BCH-coding-based turbo-coded BPSK system.

16.3   Simulation Parameters

In this chapter, BPSK modulation is employed in all the systems examined. The first system described is a convolutional-coded scheme, denoted by **CC**. A rate $R = \frac{1}{2}$, constraint length $K = 5$, recursive systematic convolutional code was used with octal generator polynomials of $G_0 = 35$ and $G_1 = 23$ as summarised previously in Table 14.7. In order to obtain rate $R = \frac{3}{4}$

and $R = \frac{5}{6}$ convolutional codes, we have employed the Digital Video Broadcast (DVB) puncturing pattern [365] specified in Table 16.1. For a fair comparative study, it was adequate for

Code rate $R = \frac{3}{4}$	Code rate $R = \frac{5}{6}$
G_0: 1 0 1	G_0: 1 0 1 0 1
G_1: 1 1 0	G_1: 1 1 0 1 0
1 = transmitted bit	
0 = non-transmitted bit	

Table 16.1: DVB puncturing pattern [365] applied to the coded bits of the $R = \frac{1}{2}$ convolutional code in order to obtain code rates $R = \frac{3}{4}$ and $R = \frac{5}{6}$ convolutional codes.

the turbo codes to employ a simple regular puncturing pattern, even though it was recognised
that puncturing patterns can be optimised to improve the performance of turbo codes [64]. For
the convolutional-coding-based turbo-coded system, represented by **CT**, we have used the convolutional constituent codes with the same parameters — i.e. $R = \frac{1}{2}$, $K = 5$ — as described
previously for example in Table 14.7. When no puncturing is implemented, the overall rate of the
turbo code is $R = \frac{1}{3}$. Therefore, we have applied regular puncturing — as detailed in Table 16.2
— to the turbo codes, in order to obtain rate $R = \frac{1}{2}$, $R = \frac{3}{4}$ and $R = \frac{5}{6}$ convolutional-coding-
based turbo codes. Finally, for the BCH-coding-based turbo-coded system, which we denoted
by **BT**, three different constituent BCH codes were used, namely the BCH(15,11) code, the
BCH(31,26) code and the BCH(63,57) code, in order to obtain the code rates $R = \frac{11}{19}$, $R = \frac{26}{36}$
and $R = \frac{57}{69}$, respectively. No puncturing is required for this class of turbo equalisers. A summary of all three classes of encoder parameters is shown in Table 16.3. We used random channel
interleavers for all three turbo equalisation systems and the depth was set to approximately 20000
bits. Similarly, random turbo interleavers — which have an odd-even separation [68] — were
used in the turbo equalisers employing BCH turbo codes and convolutional-coding based turbo
codes. The detailed channel and turbo interleaver depths are specified in Table 16.3.

We have assumed perfect knowledge of the channel impulse response and for the Soft-
In/Soft-Out (SISO) equaliser and SISO decoder we have used the Log-Maximum A-Posteriori
(Log-MAP) algorithm [342], since the Log-MAP algorithm achieves identical performance to
the original Maximum A-Posteriori (MAP) algorithm [11], despite having a reduced computational complexity. Furthermore, the term **decoding** refers here to the scenario where the equaliser
passes soft outputs to the decoder and there is no iterative processing between the equaliser and
decoder(s). When using **turbo decoding**, there will be decoding iterations where information
is passed iteratively between the component decoders, but not between the decoders and the
equaliser. Information is only passed iteratively between the equaliser and decoder(s) when
turbo equalisation is employed. For our work, we have used eight turbo decoding and turbo
equalisation iterations.

The complexity of the turbo equaliser for each system investigated can be characterised by
the number of states in the entire decoder trellis for each iterative step. Here, the complexity of
the equaliser is not taken into account, since the same equaliser is used in all the turbo-equalised
systems. For example, a trellis-based convolutional decoder employing the Log-MAP algorithm
has 2^{K-1} states at each time instant, where K is the code constraint length. Hence, for an
encoder input block length of 10000 bits the total number of states in the entire trellis is $10000 \cdot
2^{K-1}$. Since a turbo equaliser employing convolutional-coding-based turbo codes consists of
$N_d = 2$ convolutional decoders, its receiver complexity is twice that of the turbo equaliser using
conventional convolutional codes. For BCH(n,k) trellis decoders the total number of states in

Code rate $R = \frac{1}{2}$	Code rate $R = \frac{3}{4}$	Code rate $R = \frac{5}{6}$
C1: 1 0	C1: 1 0 0 0 0 0	C1: 1 0 0 0 0 0 0 0 0 0
C2: 0 1	C2: 0 0 1 0 0 0	C2: 0 0 0 0 1 0 0 0 0 0
	1 = transmitted bit	
	0 = non-transmitted bit	

Table 16.2: Regular puncturing pattern used in order to obtain the $R = \frac{1}{2}$, $R = \frac{3}{4}$ and $R = \frac{5}{6}$ convolutional-coding-based turbo codes. The terms C1 and C2 represent the parity bits of $R = \frac{1}{2}$ convolutional codes of the first and second constituent codes, respectively.

Encoder	Random π_t depth	Random π_c depth	Puncturing
Conv Rate $R = \frac{1}{2} = 0.5$	None	20736	None
Conv Rate $R = \frac{3}{4} = 0.75$	None	20736	See Table 16.1
Conv Rate $R = \frac{5}{6} = 0.833$	None	20736	See Table 16.1
Turbo Conv Rate $R = \frac{1}{2} = 0.5$	10368	20736	See Table 16.2
Turbo Conv Rate $R = \frac{3}{4} = 0.75$	15552	20736	See Table 16.2
Turbo Conv Rate $R = \frac{5}{6} = 0.833$	17280	20736	See Table 16.2
Turbo BCH(15,11) Rate $R = \frac{11}{19} = 0.579$	12672	21888	None
Turbo BCH(31,26) Rate $R = \frac{26}{36} = 0.722$	14976	20736	None
Turbo BCH(63,57) Rate $R = \frac{57}{69} = 0.826$	16416	19872	None

Table 16.3: Parameters of the encoders used in the $R \approx \frac{1}{2}$, $R \approx \frac{3}{4}$ and $R \approx \frac{5}{6}$ BPSK **CC**, **CT** and **BT** systems. The notations π_t and π_c represent the turbo and channel interleaver, respectively.

the trellis is approximately:

$$\text{Number of states in decoder trellis} = \frac{\text{Encoder input block length}}{k} \cdot (2k - n + 3) \cdot 2^{n-k}.$$

(16.1)

Table 16.4 shows the turbo equaliser complexity of each turbo equalisation iteration, for the three different classes of encoders.

The transmission burst structure used in all systems was the so-called FMA1 non-spread speech burst as specified in the Pan-European FRAMES proposal [252] and shown in Figure 16.1. Our comparative study was conducted over the five-path Gaussian channel and the equally weighted five-path Rayleigh fading channel using a normalised Doppler frequency of $f_d = 1.5 \times 10^{-4}$, as illustrated in Figures 16.2(a) and 16.2(b), respectively. Again, the fading

Code rate	Complexity [States]		
	Convolutional code	Convolutional-coding-based turbo code	BCH-coding-based turbo code
$R \approx \frac{1}{2}$	165888	331776	368640
$R \approx \frac{3}{4}$	248832	497664	884736
$R \approx \frac{5}{6}$	276480	552960	1990656

Table 16.4: The complexity of the BPSK turbo equalisers employing three different classes of codes after one turbo equalisation iteration, as a function of the total number of states in the trellis-based decoder(s).

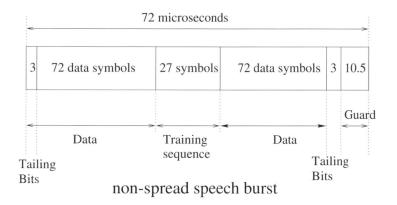

Figure 16.1: Transmission burst structure of the FMA1 non-spread speech burst of the FRAMES proposal [252].

magnitude and phase were kept constant for the duration of a transmission burst, a condition which we refer to as employing burst-invariant fading.

16.4 Results and Discussion

In this section we compare the turbo equalisation and decoding performance of the **CC**, **CT** and **BT** systems investigated. We commence by comparing the turbo equalisation performance of the **CT**, **BT** and **CC** systems, followed by a study of the turbo equalisation performance in comparison to the decoding performance of each system for code rates of $R \approx \frac{1}{2}$, $R \approx \frac{3}{4}$ and $R \approx \frac{5}{6}$ over the five-path Gaussian channel and the five-path Rayleigh fading channel using the burst-invariant fading in Figures 16.2(a) and 16.2(b), respectively.

16.4.1 Five-path Gaussian Channel

Figure 16.3 shows the BPSK turbo equalisation performance of the $R = \frac{11}{19}$ **BT** system, of the $R = \frac{1}{2}$ **CT** system and that of the $R = \frac{1}{2}$ **CC** system after one and eight turbo equalisation iterations over the five-path Gaussian channel illustrated in Figure 16.2(a). We observed that the BER performance of the $R = \frac{1}{2}$ **CT** system and the $R = \frac{11}{19}$ **BT** system was comparable and both were better than that of the $R = \frac{1}{2}$ **CC** system by approximately 0.5 dB after eight turbo

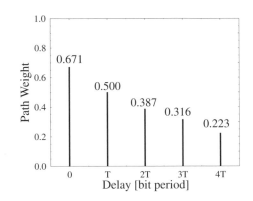

(a) Five-path Gaussian channel

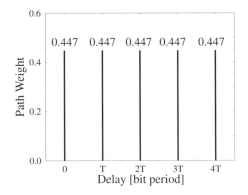

(b) Equally-weighted five-path Rayleigh fading channel

Figure 16.2: Channel impulse response of the five-path Gaussian channel and the equally weighted five-path Rayleigh fading channel.

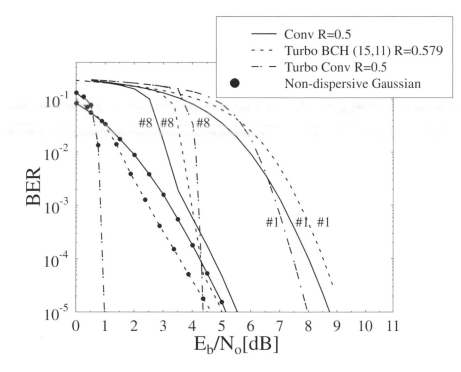

Figure 16.3: Comparing the BPSK turbo equalisation performance of the $R = \frac{1}{2}$ **CC** system, the $R = \frac{1}{2}$ **CT** scheme and the $R = \frac{11}{19}$ **BT** system for one (#1) and eight (#8) turbo equalisation iterations, over the five-path Gaussian channel of Figure 16.2(a). The decoding performance over the non-dispersive Gaussian channel — i.e the lower bound performance — is shown as well.

equalisation iterations at a BER of 10^{-4}. The same comparison was performed for $R \approx \frac{3}{4}$

Code rate	1	2	3
$R \approx \frac{1}{2}$	**BT** $\approx$ **CC** (0.0 dB)		**CC** (0.4 dB)
$R \approx \frac{3}{4}$	**BT** (0.0 dB)	**CT** (0.3 dB)	**CC** (1.1 dB)
$R \approx \frac{5}{6}$	**BT** $\approx$ **CT** (0.0 dB)		**CC** (1.0 dB)

Table 16.5: Ranking of the BPSK turbo equalisation performance for all systems over the five-path Gaussian channel from Figure 16.2(a). The notation '1' represents the system that required the lowest E_b/N_0 value and '3' for the system that needed the highest E_b/N_0 value to achieve a BER of 10^{-4}. The value within the brackets ( ) represents the E_b/N_0 loss relative to the system in column '1'.

BPSK turbo equalisers in Figure 16.4. We observed that the $R = \frac{3}{4}$ **CT** scheme achieved a gain of 0.8 dB over the $R = \frac{3}{4}$ **CC** system, whereas the $R = \frac{26}{36}$ **BT** system outperformed the $R = \frac{3}{4}$ **CT** arrangement by 0.4 dB at a BER of 10^{-4} after eight turbo equalisation iterations. Figure 16.5 shows the turbo equalisation performance of the $R = \frac{57}{69}$ **BT** system, the $R = \frac{5}{6}$ **CT** scheme and that of the $R = \frac{5}{6}$ **CC** system. For this code rate, we observed that the $R = \frac{5}{6}$ **CT** system had

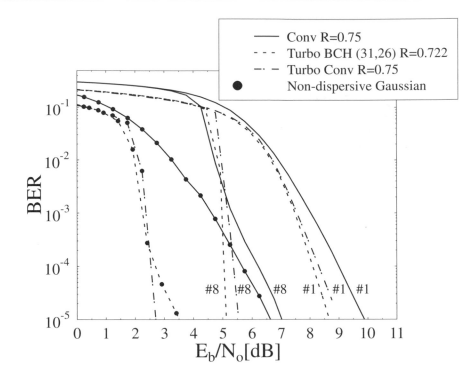

Figure 16.4: Comparing the BPSK turbo equalisation performance of the $R = \frac{3}{4}$ **CC** system, the $R = \frac{3}{4}$ **CT** scheme and the $R = \frac{26}{36}$ **BT** system for one ($\#1$) and eight ($\#8$) turbo equalisation iterations, over the five-path Gaussian channel of Figure 16.2(a). The decoding performance over the non-dispersive Gaussian channel, i.e the lower bound performance, is shown as well.

a comparable BER performance to that of the $R = \frac{57}{69}$ **BT** system after eight turbo equalisation iterations. At a BER of 10^{-4} both the $R = \frac{5}{6}$ **CT** system and the $R = \frac{57}{69}$ **BT** scheme obtained a 1 dB gain over the $R = \frac{5}{6}$ **CC** system.

The results demonstrated that at high code rates the BPSK turbo equaliser using BCH turbo codes required the lowest E_b/N_0 value of the three systems in order to achieve a BER of 10^{-4}, except at $R \approx \frac{5}{6}$, where the performance of the **BT** system and the **CT** system was similar. We summarised the above turbo equalisation performance results for the different **CC**, **CT** and **BT** systems and ranked them according to the E_b/N_0 required to achieve a BER of 10^{-4} in Table 16.5, where '1' represents the system that requires the lowest E_b/N_0 value and '3' the system that required the highest E_b/N_0 value. The value within the brackets ( ) represents the E_b/N_0 loss relative to the system in column '1'.

Next we compared the performance of the BPSK turbo equaliser with the decoding performance of each system over the five-path Gaussian channel of Figure 16.2(a). Note that for the concatenated-coded **BT** and **CT** schemes, turbo equalisation and turbo decoding have the same processing sequence when only one iteration is implemented, hence giving the same performance. From Figures 16.6(a), 16.6(b) and 16.6(c) we observed that by performing turbo equalisation using convolutional-coding-based turbo codes instead of convolutional-coding based turbo decoding, gains of 0.7 dB, 0.8 dB and 0.6 dB were achieved for code rates of $R = \frac{1}{2}$, $R = \frac{3}{4}$ and $R = \frac{5}{6}$ at a BER of 10^{-4}, respectively. In Figures 16.7(a), 16.7(b) and 16.7(c), we observed the

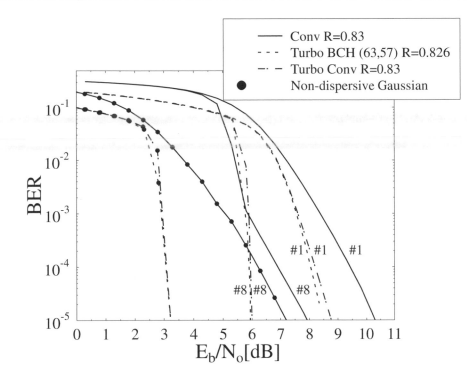

Figure 16.5: Comparing the BPSK turbo equalisation performance of the $R = \frac{5}{6}$ **CC** system, the $R = \frac{5}{6}$ **CT** scheme and the $R = \frac{57}{69}$ **BT** system for one (#1) and eight (#8) turbo equalisation iterations, over the five-path Gaussian channel of Figure 16.2(a). The decoding performance over the non-dispersive Gaussian channel, i.e. the lower bound performance, is shown as well.

same trend, where gains of 3.0 dB, 1.4 dB and 0.7 dB were achieved by the turbo equaliser using BCH turbo codes over BCH turbo decoding for code rates of $R = \frac{11}{19}$, $R = \frac{26}{36}$ and $R = \frac{57}{69}$ at a BER of 10^{-4}. Note that for the **CC** system, the performance of the turbo equaliser after one turbo equalisation iteration is the same as the convolutional decoding performance. Therefore, we have used Figures 16.3, 16.4 and 16.5 for the comparison between the turbo equalisation and the convolutional decoding performance. Here, gains of 3.2 dB, 2.8 dB and 2.5 dB were obtained by using turbo equalisation over convolutional decoding for code rates of $R = \frac{1}{2}$, $R = \frac{3}{4}$ and $R = \frac{5}{6}$ at a BER of 10^{-4}.

In summary, we can conclude from the results obtained that by performing equalisation and decoding iteratively, a better BER performance can be obtained than by performing these operations in isolation, although for the **BT** system this performance gain begins to erode as the code rate increases.

16.4.2 Equally Weighted Five-path Rayleigh Fading Channel

We now compare the turbo equalisation performance of the **CC**, **CT** and **BT** systems, for code rates of $R \approx \frac{1}{2}$, $R \approx \frac{3}{4}$ and $R \approx \frac{5}{6}$ over the five-path Rayleigh fading channel using the burst-invariant fading depicted in Figure 16.2(b).

As shown in Figure 16.8, the $R = \frac{1}{2}$ **CT** system achieved a significant gain of 2.4 dB and

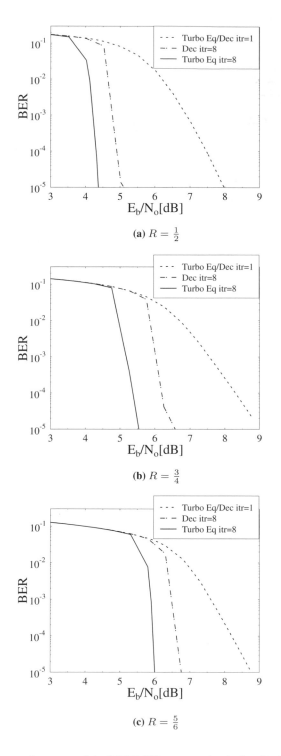

(a) $R = \frac{1}{2}$

(b) $R = \frac{3}{4}$

(c) $R = \frac{5}{6}$

Figure 16.6: Decoding performance of the BPSK **CT** system using isolated turbo decoding compared with the turbo equalisation performance after the first iteration — which is identical for both — and after the eighth iteration for code rates of $R = \frac{1}{2}$, $R = \frac{3}{4}$ and $R = \frac{5}{6}$, over the five-path Gaussian channel of Figure 16.2(a).

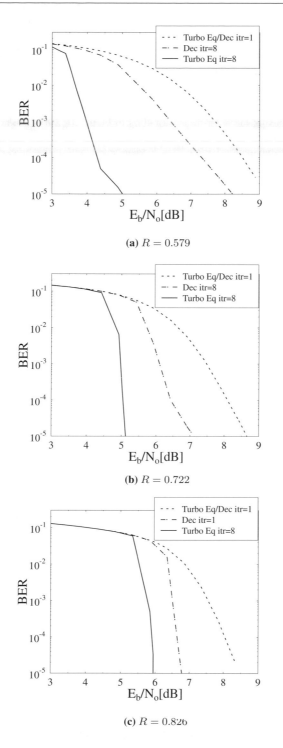

(a) $R = 0.579$

(b) $R = 0.722$

(c) $R = 0.826$

Figure 16.7: Decoding performance of the BPSK **BT** system using isolated turbo decoding compared with the turbo equalisation performance after the first iteration — which is identical for both — and after the eighth iteration for code rates of $R = \frac{11}{19}$, $R = \frac{26}{36}$ and $R = \frac{57}{69}$, over the five-path Gaussian channel of Figure 16.2(a).

Code rate	1	2	3
$R \approx \frac{1}{2}$	**CT** (0.0 dB)	**BT** (2.4 dB)	**CC** (3.0 dB)
$R \approx \frac{3}{4}$	**BT** (0.0 dB)	**CT** (0.1 dB)	**CC** (3.6 dB)
$R \approx \frac{5}{6}$	**BT** (0.0 dB)	**CT** (0.1 dB)	**CC** (3.8 dB)

Table 16.6: Ranking of the BPSK turbo equalisation performance for all systems for the five-path Rayleigh fading channel using the burst-invariant fading in Figure 16.2(b). The notation '1' represents the system that required the lowest E_b/N_0 value and '3' for the system that needed the highest E_b/N_0 value to achieve a BER of 10^{-4}. The value within the brackets ( ) represents the E_b/N_0 loss relative to the system in column '1'.

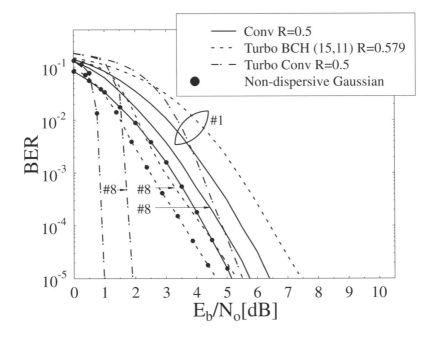

Figure 16.8: Comparing the BPSK turbo equalisation performance of the $R = \frac{1}{2}$ **CC** system, the $R = \frac{1}{2}$ **CT** scheme and the $R = \frac{11}{19}$ **BT** system for one (#1) and eight (#8) turbo equalisation iterations, over the five-path Rayleigh fading channel using the burst-invariant fading illustrated in Figure 16.2(b). The decoding performance over the non-dispersive Gaussian channel — i.e. the lower bound performance — is shown as well.

3.0 dB, when compared to the $R = \frac{11}{19}$ **BT** system and the $R = \frac{1}{2}$ **CC** system after eight turbo equalisation iterations at a BER of 10^{-4}. For a code rate of $R \approx \frac{3}{4}$ we observed in Figure 16.9 that from the set of three turbo-equalised systems, the **BT** system required the lowest E_b/N_0 value in order to achieve a BER of 10^{-4}. Relative to the **BT** system, the **CT** system exhibited an E_b/N_0 loss of 0.1 dB, while the **CC** system yielded an E_b/N_0 loss of 3.6 dB at a BER of 10^{-4} after eight turbo equalisation iterations. The same performance trend was observed in Figure 16.10 for the rate $R \approx \frac{5}{6}$ turbo equalisers, where the **BT** system obtained an E_b/N_0 gain

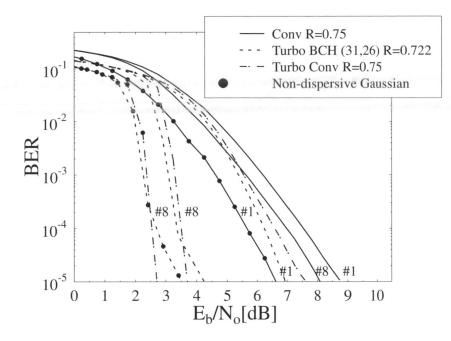

Figure 16.9: Comparing the BPSK turbo equalisation performance of the $R = \frac{3}{4}$ **CC** system, the $R = \frac{3}{4}$ **CT** scheme and the $R = \frac{26}{36}$ **BT** system for one (#1) and eight (#8) turbo equalisation iterations, over the five-path Rayleigh fading channel using the burst-invariant fading illustrated in Figure 16.2(b). The decoding performance over the non-dispersive Gaussian channel, i.e. the lower bound performance, is shown as well.

of 0.1 dB, when compared to the rate $R = \frac{5}{6}$ **CT** system, whereas a significant gain of 3.8 dB was observed, when compared to the **CC** system after eight turbo equalisation iterations at a BER of 10^{-4}.

We observed, again, in the five-path Rayleigh fading channel scenario that the turbo equaliser using high-rate — i.e. $R \approx \frac{3}{4}$ and $R \approx \frac{5}{6}$ — BCH turbo decoders outperformed the high-rate **CC** system significantly, while only a marginal improvement over the **CT** system was obtained. In Table 16.6 we ranked the **CC**, **CT** and **BT** systems according to the E_b/N_0 value required to achieve a BER of 10^{-4}, where the index '1' is used for the system that required the lowest E_b/N_0 value and '3' for the system that needed the highest E_b/N_0 value. The value within the brackets ( ) represents the E_b/N_0 loss relative to the system in column '1' for the five-path Rayleigh fading channel using the burst-invariant fading of Figure 16.2(b).

In our next endeavour a comparison of the turbo equalisation and decoding performance was conducted for the **BT**, **CT** and **CC** systems over the five-path Rayleigh fading channel. From Figures 16.11(a), 16.11(b) and 16.11(c) we observed that for all code rates investigated the turbo equaliser using convolutional-coding-based turbo codes required approximately 0.4 dB lower E_b/N_0 in order to achieve a BER of 10^{-4} when compared to isolated turbo decoding. The same performance trend was observed for the turbo-equalised systems in Figures 16.12(a), 16.12(b) and 16.12(c) using BCH turbo codes. Here, gains between 0.5 dB and 0.6 dB were achieved

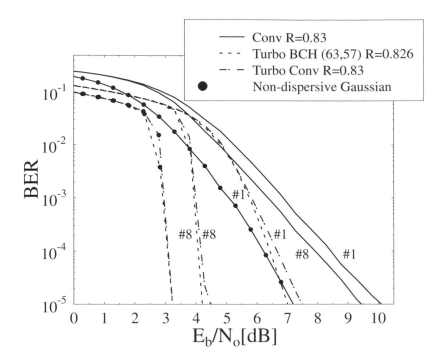

Figure 16.10: Comparing the BPSK turbo equalisation performance of the $R = \frac{5}{6}$ **CC** system, the $R = \frac{5}{6}$ **CT** scheme and the $R = \frac{57}{69}$ **BT** system for one (#1) and eight (#8) turbo equalisation iterations, over the five-path Rayleigh fading channel using the burst-invariant fading illustrated in Figure 16.2(b). The decoding performance over the non-dispersive Gaussian channel, i.e. the lower bound performance, is shown as well.

through turbo equalisation, as compared to BCH turbo decoding at a BER of 10^{-4} for all code rates investigated. Finally, Figures 16.8, 16.9 and 16.10 showed gains of 0.7 dB, 0.5 dB and 0.5 dB, which were achieved by employing turbo equalisation instead of convolutional decoding at a BER 10^{-4} for the **CC** system at code rates of $R = \frac{1}{2}$, $R = \frac{3}{4}$ and $R = \frac{5}{6}$, respectively.

In summary, we observed over the five-path Rayleigh fading channel that turbo equalisation — i.e. iterative equalisation and decoding — outperforms isolated equalisation and decoding for all code rates investigated, although for the **BT** system the gain achieved through turbo equalisation was lower than that obtained over the dispersive Gaussian channel scenario.

The turbo equalisation simulations over the five-path Rayleigh fading channel of Figure 16.2(b) also showed that the **CC** system has poor iteration gain — i.e. a modest gain in E_b/N_0 — performance with respect to the first iteration (consistent with the results presented for the $R = \frac{1}{2}$ **CC** system in references [105, 345]). For the turbo-equalised **CT** and **BT** systems, the **CT** system obtained slightly higher iteration gains. For example, the $R = \frac{5}{6}$ **CT** system obtained an iteration gain of 2.4 dB, while the $R = \frac{57}{69}$ **BT** system achieved a gain of 2.3 dB after eight turbo equalisation iterations at a BER of 10^{-4}, as shown in Figure 16.10. At this BER, the $R = \frac{5}{6}$ **CC** system only achieves an iteration gain of 0.5 dB after eight turbo equalisation iterations.

In the five-path Gaussian channel and the five-path Rayleigh fading channel scenario, the

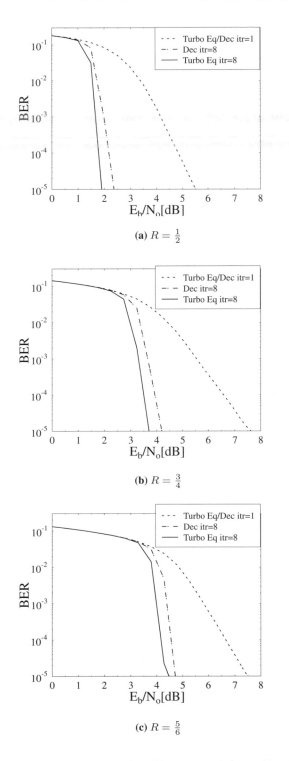

Figure 16.11: Decoding performance of the BPSK **CT** system using isolated turbo decoding compared with the turbo equalisation performance after the first iteration — which is identical for both — and after the eighth iteration for code rates of $R = \frac{1}{2}$, $R = \frac{3}{4}$ and $R = \frac{5}{6}$, over the five-path Rayleigh fading channel using the burst-invariant fading illustrated in Figure 16.2(b).

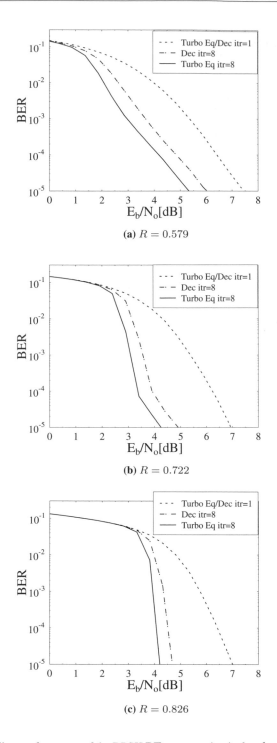

(a) $R = 0.579$

(b) $R = 0.722$

(c) $R = 0.826$

Figure 16.12: Decoding performance of the BPSK **BT** system using isolated turbo decoding compared with the turbo equalisation performance after the first iteration — which is identical for both — and after the eighth iteration for code rates of $R = \frac{11}{19}$, $R = \frac{26}{36}$ and $R = \frac{57}{69}$, over the five-path Rayleigh fading channel using the burst-invariant fading illustrated in Figure 16.2(b).

performance of the high-code-rate $R \approx \frac{3}{4}$ and $R \approx \frac{5}{6}$ **BT** system is marginally better than or comparable to the **CT** system at a BER of 10^{-4}. However, this was obtained at the cost of higher receiver complexity compared to the **CT** system as seen in Table 16.4. At high code rates the **CC** system performs poorly over the five-path Rayleigh fading channel. For example, at a BER of 10^{-4} a loss of 3.8 dB was observed, when compared to the turbo-equalised **BT** system after eight turbo equalisation iterations, and of 1 dB for the dispersive Gaussian channel, when compared to the **CC** system. This inferior performance is due to the low iteration gain, which does not exceed 0.9 dB. A reason for this marginal improvement through iterative equalisation and decoding is that the **CC** system's performance is already close to the optimum — i.e. to the decoding performance over the non-dispersive Gaussian channel — after the first iteration.

16.5 Non-Iterative Joint Channel Equalisation and Channel Decoding[2]

16.5.1 Motivation

In this section we further augment the concept of turbo equalisation by exploiting that both the convolutional encoder and the channel can be described with the aid of a trellis or a finite-state machine. These two trellises are effectively concatenated and the ML trellis path can be identified in each trellis, for example using the Viterbi algorithm or Bahl's MAP algorithm. Since both the encoder and the channel can be described by a trellis, the channel can be viewed as an additional so-called inner channel encoder, serially concatenated to an outer convolutional encoder. The iterative turbo equalisation algorithm introduced in the previous chapter can be employed for the decoding of serially concatenated codes. It can be hypothesised, however, that an amalgamated trellis decoder scheme may be able to achieve a better performance than two separate serially concatenated decoders, although at the cost of a high complexity. **Our discussions in this section will be centred around the topic of employing a serially concatenated 'super-trellis' for the joint non-iterative 'turbo equalisation' of serially concatenated codes.** In conceptual terms the associate serially concatenated 'super-trellis' can be viewed as a relative of the parallel concatenated super-trellis of Chapter 6.

To elaborate a little further, in iterative turbo equalisation and decoding there are two consecutive trellises which exchange soft-decision information concerning the encoded bits, passing the associated log-likelihood values back and forth a number of times, in order to iteratively improve the system's BER performance. The ML information sequence can be asymptotically approached by this method upon increasing the number of iterations. Let us now consider a non-iterative joint channel equalisation and channel decoding technique.

16.5.2 Non-iterative Turbo Equalisation

In Chapter 6 the optimum non-iterative turbo decoding algorithm was introduced for the decoding of parallel concatenated codes, which are also referred to as turbo codes. **By contrast, in this section we study a non-iterative joint channel decoding and equalisation scheme for the serially concatenated convolutional encoder plus the dispersive channel arrangement of Figure 16.13, which outperforms iterative turbo equalisers.** In contrast to iterative turbo equalisation algorithms, where we need a certain number of iterations in order to achieve a good

---

[2]This section is based on A. Knickenberg, B.-L. Yeap, J. Hámorský, M. Breiling, L. Hanzo: Non-iterative Joint Channel Equalisation and Channel Decoding, Proceedings of IEEE Globecom 1999, Rio de Janeiro, Brazil.

performance, according to our proposed solution we carry out equalisation and decoding in a joint, non-iterative step. We note here that conceptually a multi-path channel can be seen as a special convolutional encoder, having a coding rate of $R = 1$ and a memory length $K - 1$ which is equal to the maximum delay of the channel expressed in terms of the time units k' characteristic of the output of the decoder. We then have a system of two serial encoders, namely the convolutional encoder and the channel, separated by an interleaver, namely the channel interleaver.

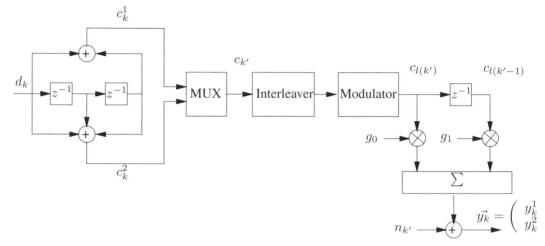

Figure 16.13: Model of the transmitter consisting of the convolutional encoder, interleaver, modulator and complex channel.

For the proposed non-iterative joint equalisation/decoding process we have to construct a finite-state machine which models the whole system. The input of this finite-state machine is constituted by the original data bits and its output is the channel output. In a finite-state machine the output $\vec{y_k}$ and the next state S_{k+1} are fully determined by the current state S_k and the current input data bit d_k. However, the interleaver renders the receiver's task complex.

Let us consider the example of Figure 16.13, where the multi-path channel has $L = 2$ symbol-spaced paths, corresponding to an inner encoder memory length of $\nu_I = L - 1 = 1$. The convolutional outer encoder of Figure 16.13 has a constraint length of $K = 3$ and a memory length of $\nu_o = K - 1 = 2$. Each encoded bit is determined by the current incoming data bit d_k and the ν_o previous ones. These encoded bits pass through the interleaver and enter the multi-path channel in the interleaved order. We denote the interleaver function by $l(k')$, which assigns the input time instant k' to the output time instant $l(k')$. Hence the output of the channel at the time instant k' depends on $c_{l(k')}, c_{l(k'-1)}, \cdots, c_{l(k'-\nu_I)}$. Each convolutionally encoded bit in turn depends on the last $\nu_o + 1$ data bits. Finally, we get the output of the channel as a function of the data bits as follows:

$$
\begin{aligned}
y_{k'} &= g_0 \left[d_{l(k)} \oplus d_{l(k)-1} \oplus \cdots \oplus d_{l(k)-\nu_o} \right] \\
&+ g_1 \left[d_{l(k-1)} \oplus d_{l(k-1)-1} \oplus \cdots \oplus d_{l(k-1)-\nu_o} \right] \\
&\vdots \\
&+ g_{\nu_I} \left[d_{l(k-\nu_I)} \oplus d_{l(k-\nu_I)-1} \oplus \cdots \oplus d_{l(k-\nu_I)-\nu_o} \right],
\end{aligned}
$$

where g_i represents the symbol-spaced dispersive channel impulse response taps. In conventional trellises, which are used to describe convolutional codes, the encoder state is determined by the bits in the encoder's shift register. In our serially concatenated convolutional encoder plus channel **we define a super-state** S_k^***, which is determined by all the bits that are used to calculate the channel output, except the actual input bit**, which is often referred to as the state transition bit.

In our forthcoming discourse we propose a non-iterative equalisation/decoding algorithm and highlight its operation using the simple example of a $2 \times N$ block interleaver and a half-rate convolutional encoder.

16.5.3 Non-iterative Joint Equalisation/Decoding Using a $2 \times N$ Interleaver

Again, Figure 16.13 shows a simple manifestation of the serially concatenated system using a non-recursive, half-rate convolutional encoder with a constraint length of $K = 3$, octal generator polynomials $G[0] = 5$ and $G[1] = 7$, a $2 \times N$ block interleaver and a two-path channel. It is readily seen that this two-column interleaver will simply separate for each encoded bit block of length $2N$ the even and the odd encoded bits. First all the odd-indexed encoded bits generated by the first generator polynomial will enter the channel, followed by the even-indexed encoded bits due to the second generator polynomial. The super-state S_k^* is constituted by all the previous bits which influence the channel output sequence, namely y_k^1 and y_k^2 in Figure 16.13. Let us denote the encoded bits emanating from the first generator polynomial at the time k by c_k^1 and those due to the second one by c_k^2. The channel outputs y_k^1 and y_k^2 in Figure 16.13 are calculated according to the following equations:

$$
\begin{aligned}
y_k^1 &= g_0 \cdot c_k^1 + g_1 \cdot c_{k-1}^1 \\
&= g_0 \cdot (d_k \oplus d_{k-2}) + g_1 \cdot (d_{k-1} \oplus d_{k-3}) \quad (16.2) \\
y_k^2 &= g_0 \cdot c_k^2 + g_1 \cdot c_{k-1}^2 \\
&= g_0 \cdot (d_k \oplus d_{k-1} \oplus d_{k-2}) + g_1 \cdot (d_{k-1} \oplus d_{k-2} \oplus d_{k-3}), \quad (16.3)
\end{aligned}
$$

where y_k^1 is transmitted during the first half of the encoded output block of length $2N$ and y_k^2 during the second half. The super-state S_k^* is determined by the data bits $(d_{k-1}, \ldots, d_{k-3})$, while d_k is the current encoder input bit. The next super-state, namely S_{k+1}^*, will be determined by $(d_k, \ldots, d_{k-2})$.

In our example the outer encoder's and the inner encoder's memory length are given by $\nu_o = 2$ and $\nu_I = 1$, respectively, hence the number of super-state bits is $\nu_o + \nu_I$ and the number of super-states is $2^{\nu_o + \nu_I}$. With each additional memory element in the channel, we have to consider one additional previous encoded bit, implying that we have to consider an additional bit in the super-state.

Let us now consider the inner working of the proposed non-iterative scheme at the commencement of operation and also near the middle of the transmitted sequence, where we change from transmitting c_k^1 to c_k^2, and lastly, at the end of each encoded block of length $2N$. Let us assume for the sake of illustration that we have a 2×10 interleaver. Then we have 20 encoded bits and hence 10 input data bits at the encoder. In order to be in a predefined state at the end of each encoded block we will use a terminated code with three zero-valued termination bits. The length of this termination sequence is equal to the number of concatenated encoder state bits, i.e. the number of bits in the super-state. This implies that in the 10-bit encoded sequence we have 7 data bits and 3 termination bits, which are, again, logical '0's, since we are using a non-recursive

Time k'	State	$c_{l(k')}$	$c_{l(k'-1)}$	Output y_k^i
1	$(0,0,0)$	$c_1^1 = d_1 \oplus 0$	0	y_1^1
2	$(d_1,0,0)$	$c_2^1 = d_2 \oplus 0$	$c_1^1 = d_1 \oplus 0$	y_2^1
$\vdots$	$\vdots$	$\vdots$	$\vdots$	$\vdots$
10	(d_9,d_8,d_7)	$\underbrace{c_{10}^1 = d_{10} \oplus d_8}_{=0}$	$c_9^1 = d_9 \oplus d_7$	y_{10}^1
11	$(0,0,0)$	$c_1^2 = d_1 \oplus 0 \oplus 0$	$\underbrace{c_{10}^1 = d_{10} \oplus d_8}_{=0}$	y_1^2
12	$(d_1,0,0)$	$c_2^2 = d_2 \oplus d_1 \oplus 0$	$c_1^2 = d_1 \oplus 0 \oplus 0$	y_2^2
$\vdots$	$\vdots$	$\vdots$	$\vdots$	$\vdots$
20	(d_9,d_8,d_7)	$\underbrace{c_{10}^2 = d_{10} \oplus d_9 \oplus d_8}_{=0}$	$c_9^2 = d_9 \oplus d_8 \oplus d_7$	y_{10}^2

Table 16.7: The contents of the concatenated encoder seen in Figure 16.13 versus time k'.

encoder. Table 16.7 shows the situation at the above-mentioned 'critical' operational points of the encoder. Observe in the example that if the input bits $d_8, \ldots, d_{10}$ are zero, then c_{10}^1 and c_{10}^2 are also zero.

The proposed decoder operates on the basis of constructing the encoder's super-trellis constituted by the previously defined super-states S_k^* described by the data bits $(d_{k-1}, \ldots, d_{k-3})$ and invoking the Log-MAP algorithm for joint channel equalisation and decoding. We note, however, that in this context the Viterbi algorithm also could have been utilised instead of the MAP algorithm, since in this particular application there is no need to produce the soft outputs provided by the MAP algorithm. More explicitly, the proposed non-iterative scheme refrains from passing soft-decision information iteratively between the separate trellises of the equaliser and channel decoder and operates on the basis of the super-trellis, rather than on the two independent trellises of the equaliser and channel decoder. Invoking the super-trellis in a non-iterative step has the potential of improving the performance of the amalgamated scheme.

16.5.4 Non-iterative Turbo Equaliser Performance

We have conducted simulations using the proposed non-iterative algorithm described above and compared its performance to that of the iterative turbo equalisation algorithm using the same parameters. In both systems we employed BPSK modulation.

In Figure 16.14 we have compared the iterative and non-iterative algorithms, when transmitting over a dispersive two-path Gaussian channel using eight iterations. We used a non-recursive rate $R = \frac{1}{2}$, constraint length $K = 5$ convolutional code with a generator polynomial of $G[0] = 37$ and $G[1] = 21$, which are expressed in octal format. The channel impulse response taps were given by $g_0 = g_1 = \frac{1}{\sqrt{2}}$ and the interleaver size was $2 \times 2052 = 4104$. As seen in the figure, the non-iterative algorithm outperforms the iterative scheme by about 0.7 dB at a BER of 10^{-3}.

In order to further quantify the proposed scheme's performance, in Figure 16.15 we compared the two algorithms over a more dispersive five-path channel, employing the following

BER Against Eb/N0

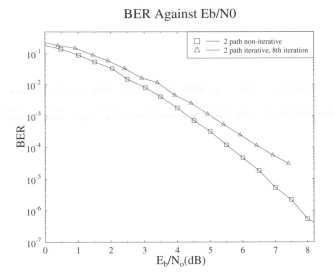

Figure 16.14: Comparison of iterative versus non-iterative joint equalisation/decoding over a two-path dispersive Gaussian channel, when using a convolutional code having the following parameters: $R = \frac{1}{2}$, $K = 5$, $G[0]$=37, $G[1]$=21, channel impulse response taps of $g_0 = g_1 = \frac{1}{\sqrt{2}}$ and an interleaver of 2×2052 bits.

parameters: $R = \frac{1}{2}$; constraint length of $K = 5$; $G[0] = 37$ and $G[1] = 21$; channel coefficients of $g_0 = g_4 = 0.227$, $g_1 = g_3 = 0.46$, $g_2 = 0.688$; interleaver size of $2 \times 2052 = 4104$; eight iterations. Over this more dispersive channel apparently the non-iterative scheme's performance becomes even more attractive, resulting in an SNR gain of ≈ 3.4 dB at a BER of 10^{-3}. We also note that the performance of the Max-Log-MAP and the Log-MAP algorithms was also compared and was found to be virtually identical. These performance improvements were achieved in the context of the $2 \times N$ interleaver at the cost of a reasonable complexity. For example, in the case of $K = 5$ and a two-path channel we have $\nu_o + \nu_I = 4 + 1$ state bits, hence there are $2^5 = 32$ states. Even for $K = 5$ and a five-path channel we have a moderate number of $2^8 = 256$ states.

16.5.4.1 Effect of Interleaver Depth

Let us now consider the effect of interleaver depth in this section. If we do not use an interleaver, then the encoded bits enter the channel in the same order as they were calculated by the encoder. We then have two channel output bits, namely y_k^1 and y_k^2, which were due to the input data bit d_k, when the super-state was S_k^*. If we consider the example of Figure 16.13, we have the following

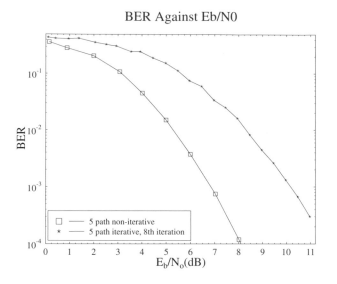

Figure 16.15: Comparison of iterative versus non-iterative turbo equalisation over a five-path Gaussian channel using a convolutional code having the following parameters: $R = \frac{1}{2}$, K=5, $G[0] = 37$, $G[1] = 21$, $g_0 = g_4 = 0.227$, $g_1 = g_3 = 0.46$, $g_2 = 0.688$, 2×2052 interleaver.

equations:

$$
\begin{aligned}
y_k^1 &= g_0 \cdot c_k^1 + g_1 \cdot c_{k-1}^2 \\
&= g_0(d_k \oplus d_{k-2}) + g_1(d_{k-1} \oplus d_{k-2} \oplus d_{k-3}) \\
y_k^2 &= g_0 \cdot c_k^2 + g_1 \cdot c_k^1 \\
&= g_0(d_k \oplus d_{k-1} \oplus d_{k-2}) + g_1(d_k \oplus d_{k-2}).
\end{aligned}
$$

(16.4)

(16.5)

Hence we have three state bits in the superstate $S_k^* = (d_{k-1}, d_{k-2}, d_{k-3})$. Owing to the lack of interleaving we now have $\nu_o + \lfloor \nu_I/2 \rfloor$ state bits, which implies that we need less state bits than the previous $\nu_o + \nu_I$ value, which was associated with the $2 \times N$ interleaver.

When using a $4 \times N$ interleaver, we have to cope with a higher detection complexity. Specifically, this interleaver produces from the sequence:

$$
\ldots, c_4^2, c_4^1, c_3^2, c_3^1, c_2^2, c_2^1, c_1^2, c_1^1 \Longrightarrow
$$

the following output sequence:

$$
\ldots, c_6^2, c_4^2, c_2^2, \ldots, c_6^1, c_4^1, c_2^1, \ldots, c_5^2, c_3^2, c_1^2, \ldots, c_5^1, c_3^1, c_1^1.
$$

In the shift-register stages of Figure 16.13 modelling the channel we now have the following encoded bits: $c_k^i, c_{k-2}^i, \ldots$. Hence, in comparison to the $2 \times N$ interleaver we now always have a displacement of two clock intervals between the indices of the coded bits which consecutively enter the channel. If we consider again the example of Figure 16.13, then the two output values are calculated as follows:

$$
\begin{aligned}
y_k^1 &= g_0 \cdot c_k^1 + g_1 \cdot c_{k-2}^1 \\
&= g_0(d_k \oplus d_{k-2}) + g_1(d_{k-2} \oplus d_{k-4}) \\
y_k^2 &= g_0 \cdot c_k^2 + g_1 \cdot c_{k-2}^2 \\
&= g_0(d_k \oplus d_{k-1} \oplus d_{k-2}) + g_1(d_{k-2} \oplus d_{k-3} \oplus d_{k-4}).
\end{aligned}
$$

(16.6)

(16.7)

This leads to the following superstate:

$$
S_k^* = (d_{k-1}, d_{k-2}, d_{k-3}, d_{k-4}).
$$

This example shows that we now need $\nu_o + 2 \cdot \nu_I$ state bits, since we have to store all the previous bits, which are either used to calculate the channel output or used to determine the next state. In this case the number of states increases rapidly with the memory or dispersion of the channel. For example, for a five-path channel and a convolutional encoder of constraint length $K = 5$ this would result in $2^{4+2 \cdot 4} = 2^{12} = 4096$ states, which is unrealistically high.

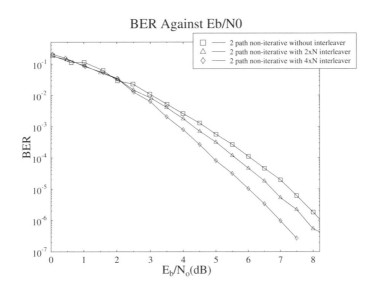

Figure 16.16: Comparison of different interleavers, using non-iterative joint equalisation/decoding over a two-path channel. $R = \frac{1}{2}$, $K{=}5$, $G[0]{=}37$, $G[1]{=}21$, $g_0 = g_1 = \frac{1}{\sqrt{2}}$.

We have conducted simulations for a two-path channel using the same parameters as in Section 16.5.3, except for the interleaver. Figure 16.16 shows the results for the two-path channel without an interleaver, and with a $2 \times N$ or a $4 \times N$ interleaver.

The study of the non-iterative turbo equaliser is extremely instructive in terms of further augmenting our exposition of the turbo equalisation principles. It clearly shows that, indeed, a dispersive channel actually acts as a serially concatenated inner channel encoder. Hence ironically, the more dispersive the channel, the better the achievable performance.

However, the performance trends emerging from our previous elaborations are somewhat misleading, since they portrayed the non-iterative turbo equaliser in a somewhat optimistic light,

BER Against E_b/N_o

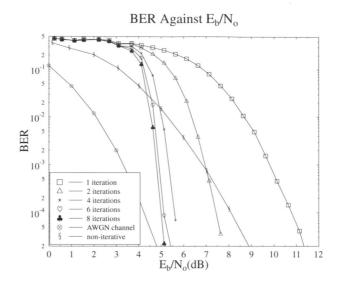

Figure 16.17: Comparison of **iterative** turbo equalisation using a **random interleaver** with the **non-iterative** system using a $2 \times$ **Ninterleaver** over a five-path channel. $R = \frac{1}{2}$, $K = 5$, $G[0] = 37, G[1] = 21$, $g_0 = g_4 = 0.227$, $g_1 = g_3 = 0.46$, $g_2 = 0.688$, output block length=4104, BPSK modulation.

which therefore outperformed the iterative turbo equaliser. This is because, as we have argued before, we have to use a random interleaver in order to achieve the best possible performance for iterative turbo equalisation systems.

Therefore, we have also compared the non-iterative turbo equalisation performance results shown above to the performance of an iterative system using a random interleaver. In this context the iterative scheme becomes capable of developing its full potential. Figure 16.17 shows this BER performance comparison, demonstrating the BER improvements as a function of the number of iterations. One iteration corresponds to a scenario employing conventional separate channel equalisation and channel decoding using soft decoding. Then from two iterations onwards the equaliser benefits from the a-priori information provided by the decoder. We observe a significant iteration gain, especially up to four iterations. In fact, at $E_b/N_0 = 5$ dB this algorithm is capable of approaching the performance of the same system communicating over a non-dispersive channel. By contrast, the non-iterative turbo equalisation algorithm using a simple $2 \times N$ interleaver has an inferior performance.

We can finally say that for non-iterative turbo equalisation the performance as well as the complexity increase with the number of rows in the interleaver. The best interleaver would probably be a random interleaver, but it appears infeasible to apply a random interleaver for the non-iterative algorithm owing to its excessive complexity.

In the next section we will employ sub-optimum decoders in the context of our joint decoding and equalisation problem, which have the advantage of being able to operate using trellises associated with a high number of states. This will potentially allow us to investigate systems associated with block interleavers having a higher number of trellis states.

16.5.4.2 The M-algorithm

In this section we examine whether we can ameliorate the performance of the non-iterative system by using interleavers having a higher number of columns than the interleavers treated in the previous sections. We have seen before that the number of states increases rapidly with the dispersion of the channel. This is the reason for employing sub-optimum decoders. The goal of the sub-optimum decoders is to reduce the complexity by refraining from considering all the possible states and transitions at the same time. Naturally, this will lead to a loss in performance.

Specifically, we have applied the *M-algorithm*, which is similar to the Viterbi algorithm, except that at each trellis stage it retains only M paths as survivors. The parameter M allows us to strike a trade-off between the achievable performance and the necessary computational complexity.

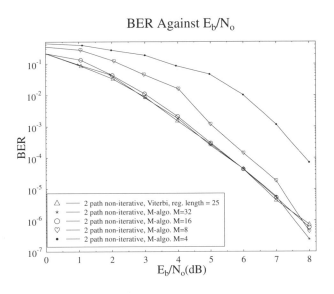

Figure 16.18: Comparison of non-iterative turbo equalisation using the **Log-MAP/Viterbi algorithm** and the **M-algorithm** over the two-path channel. $R = \frac{1}{2}$, $K = 5$, $G[0] = 37$, $G[1] = 21$, $g_0 = g_1 = \frac{1}{\sqrt{2}}$, 2×2052 interleaver, Viterbi register length = $5(\nu_o + \nu_I) = 5(4+1) = 25$, M=32,15,8,4, BPSK modulation.

If we set M to the number of states in the super-trellis, then we should achieve the same performance as in the previous simulations using the Log-MAP and Viterbi decoders. This can be verified in Figure 16.18 whre we have then decreased the parameter M. If we set M equal to 16, then the decoder retains only the 16 best states of the 32 possible states and we do not experience any significant deterioration of the performance in the simulated E_b/N_0 range. As we lower M to 8 and 4 we observe that the decoding algorithm yields a considerably degraded performance. This is due to the fact that the ML path may have been discarded, since temporarily it was not amongst the M best paths.

In order to further evaluate the potential of lowering the complexity by means of the M-algorithm, we have simulated the previously used five-path channel and compared again the performance of the M-algorithm to that of the Log-MAP and Viterbi algorithms. The results are

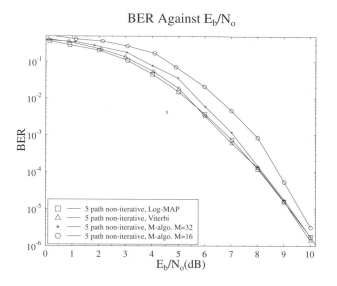

Figure 16.19: Comparison of non-iterative turbo equalisation using the **(Max-)Log-MAP/Viterbi algo-rithm** and the **M-algorithm** over a five-path channel. $R = \frac{1}{2}$, $K = 5$, $G[0] = 37$, $G[1] = 21$, $g_0 = g_4 = 0.227$, $g_1 = g_3 = 0.46$, $g_2 = 0.688$, 2×2052 interleaver, Viterbi register length $= 5(\nu_o + \nu_I)$, BPSK modulation.

shown in Figure 16.19. Now the super-trellis has 256 states. If we set M equal to 32, which corresponds to reducing the complexity by a factor of eight, we still attain nearly the same performance as in conjunction with the full-complexity algorithms which consider at each decision stage all the 256 states. By reducing M to 16 we observe a significant performance degradation over the entire simulated range of E_b/N_0 values, although the degradation was slightly reduced for high E_b/N_0 values.

This leads to the dilemma whether we may be able to achieve an overall gain if we use a more complex interleaver in conjunction with a sub-optimum decoder in comparison to the system with a $2 \times N$ interleaver and the optimum decoder. The corresponding results are shown in Figure 16.20. Specifically, we compared our simulation results when communicating over the five-path channel using on the one hand a $2 \times N$ interleaver in conjunction with the Max-Log-MAP algorithm and on the other hand a $4 \times N$ interleaver along with the M-algorithm, where M was set to 256 and 128. Previously it was unfeasible to communicate over the five-path channel in conjunction with the $4 \times N$ interleaver owing to the associated high number of states (4096), when employing the optimum decoder. Hence the achievable gain due to this interleaver could not be quantified. We can observe that the joint employment of the $4 \times N$ interleaver and the M-algorithm does not outperform the the $2 \times N$ interleaver-assisted optimum full-complexity decoder.

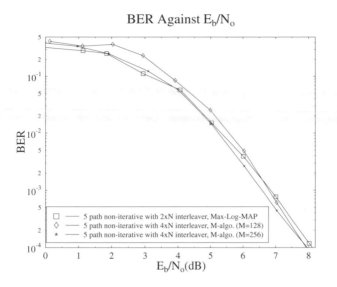

Figure 16.20: Comparison of the **M-algorithm** with a **4 × Ninterleaver** and the **Max-Log-MAP-algorithm** with a **2 × N interleaver** over a five-path channel. $R = \frac{1}{2}$, $K = 5$, $G[0] = 37$, $G[1] = 21$, $g_0 = g_4 = 0.227$, $g_1 = g_3 = 0.46$, $g_2 = 0.688$, output block length = 4104, BPSK modulation.

16.6 Summary and Conclusions

Different receiver configurations were compared for BPSK-modulated transmission systems using BCH turbo codes **BT**, convolutional-coding-based turbo codes **CT** and convolutional codes **CC**. Non-iterative and iterative equaliser/decoders operating at code rates $R \approx \frac{1}{2}$, $R \approx \frac{3}{4}$ and $R \approx \frac{5}{6}$ were studied. In the iterative cases loops containing only decoders — as in isolated turbo decoding — and loops containing joint equalisation and decoding stages — as in turbo equalisation — were implemented. The SISO equaliser and decoders employed the Log-MAP algorithm. Our comparative study of the turbo equalisers for the **BT**, **CT**, **CC** systems showed that at high code rates of $R \approx \frac{3}{4}$ and $R \approx \frac{5}{6}$ the **BT** system is marginally better or comparable to the **CT** system at a BER of 10^{-4}, at the expense of a higher complexity compared to the **CT** system. At these high code rates and over the equally weighted symbol-spaced five-path Rayleigh fading channel using burst-invariant fading of Figure 16.2(b), the **CC** system performs poorly since its iteration gain is low. This is because after the first iteration the system's performance is already close to the decoding results over the non-dispersive Gaussian channel. At $R = \frac{5}{6}$ we observed a loss of $E_b/N_0 = 1.0$ dB over the five-path Gaussian channel and a loss of $E_b/N_0 = 3.8$ dB, when compared to the **BT** system over the five-path Rayleigh fading channel at a BER of 10^{-4} after eight turbo equalisation iterations. On the whole, the turbo-equalised **CT** system is the most robust scheme, giving comparable performance within a few tenths of a dB for all code rates investigated, compared to the best system in each scenario. Furthermore, the turbo-equalised **CT** system has a lower receiver complexity when compared to the **BT** system, hence making it the best choice in most applications.

Finally, in Section 16.5 a non-iterative turbo equaliser was proposed for the sake of further augmenting the general concept of turbo equalisation. This scheme is impractical for employment in highly dispersive environments because of its high complexity, which we reduced with the aid of the M-algorithm.

Having characterised the performance of various turbo equalisers in this chapter, our discussions evolve further in the next chapter in the area of reducing the associated implementation complexity.

Chapter 17

Reduced-complexity Turbo Equaliser

17.1 Motivation[1]

In a communications system, the received signal is degraded by ISI introduced by the channel. We can mitigate the effects of ISI by employing equalisation and further reduce the bit error rate by using error correction/decoding schemes. When performing the equalisation and decoding independently, we cannot compensate effectively for the loss due to the ISI, even when soft decisions are passed from the equaliser to the decoder. Instead, by performing the equalisation and decoding jointly, as in the iterative turbo equalisation scheme proposed by Douillard *et al.* [105], the channel ISI can be mitigated effectively. Knickenberg *et al.* [366] subsequently proposed a non-iterative joint equalisation and decoding technique based on a super-trellis structure. This technique yielded an optimum performance but was restricted to simple interleavers owing to the high complexity incurred by high-depth interleavers. Therefore, Radial Basis Function (RBF) equalisers, which incur lower computational complexity as compared to conventional trellis-based equalisers, have been researched [367] in the context of turbo equalisation. Other reduced-complexity equalisers for turbo equalisation have also been proposed by Glavieux *et al.* [368]. This chapter characterises turbo equalisation implemented in the context of multi-level full-response modulation schemes, namely in M-ary Quadrature Amplitude Modulation (M-QAM) [207], where M denotes the number of constellation points in a particular modulation mode. However, owing to its complexity, turbo equalisation using trellis-based equalisers can only be realistically applied to BPSK and QPSK modulation schemes [346], since the computational complexity incurred by the trellis-based equaliser is dependent on the length of the Channel Impulse Response (CIR) and on the modulation mode utilised. Explicitly, the number of states in the M-QAM trellis-based equaliser is proportional to M^{τ_d}, where τ_d is the maximum CIR duration expressed in terms of symbol periods. Consequently, a high complexity can be incurred for long CIR durations and higher-order modulation modes. Hence, turbo equalisation research has been focused on developing reduced-complexity equalisers, such as the receiver

---

[1]This chapter is based on B. L. Yeap, C. H. Wong and L. Hanzo: Reduced Complexity In-phase/Quadrature-phase M-QAM Turbo Equalization Using Iterative Channel Estimation, submitted to IEEE Transactions on Wireless Communications

structure proposed by Glavieux *et al.* [368]. Motivated by these trends, here we propose and investigate a reduced-complexity turbo equaliser employing iterative channel estimation [369] for M-QAM [207] systems. We refer to this turbo equaliser as the In-Phase/Quadrature-Phase Turbo Equaliser (TEQ-IQ). The underlying principle of the reduced-complexity turbo equaliser is based on equalising the in-phase and quadrature-phase component of the transmitted signal independently. Therefore, the number of states required for the trellis-based channel equaliser in the TEQ-IQ scheme is reduced when compared to the afore mentioned trellis-based equaliser used in the original turbo equalisation arrangement. This is made more explicit in our forthcoming discourse. The outline of this chapter is as follows. Section 17.2 provides a brief introduction to the principles of constructing trellis-based equalisers and to the associated complexity. Subsequently, the principles of the reduced-complexity turbo equaliser are described in Sections 17.4 and 17.5. Section 17.7 then summarises the system parameters followed by the system performance in Section 17.8. Finally, Section 17.9 provides discussions and concluding remarks concerning the systems' performance.

17.2   Complexity of the Multilevel Full-response Turbo Equaliser

The original turbo equaliser proposed by Douillard *et al.* [105] for BPSK transmission can be extended to M-QAM transmissions by modifying the trellis-based equaliser. This equaliser accepts soft inputs and produces soft outputs which reflect the reliability of the equalised encoded bits. Hence, such equalisers are also known as Soft-In Soft-Out (SISO) trellis-based equalisers. In our forthcoming discussions, we will provide a brief introduction to the principles of constructing a M-QAM equaliser trellis and subsequently highlight the relationship between the equaliser complexity, the modulation mode as well as the channel's memory.

Figure 17.1(a) shows the transmitted constellation points of the 4-QAM signal using Cartesian coordinates, whereas Figure 17.1(b) depicts the corresponding signal constellation received over the two-path static channel illustrated in Figure 17.2. Similarly, when the 16-QAM signal constellation of Figure 17.1(c) is transmitted over the same dispersive Gaussian channel of Figure 17.2, the number of constellation points is determined by the number of all possible combinations of the two consecutive symbols, i.e. M^2, although some points happen to coincide as in Figure 17.1(d) owing to the regularity of the constellation. This can be explained by observing first that when the channel is non-dispersive, the output of the channel is only dependent on the transmitted symbol at that time instant. However, when there is channel dispersion, the output of the channel becomes dependent not only on the current transmitted symbol, but also on the previously transmitted symbol. Hence, the number of constellation points becomes M^2 over a two-path channel as shown in Figure 17.1(b). Again, in Figure 17.1(d) only 100 points are visible since some points coincide for the specific CIR of Figure 17.2. In order to obtain a more quantitative explanation, the design and construction of the trellis-based equaliser is briefly discussed.

Previously in Section 13.7 we described how the estimated received signal at each time instant can be regenerated when we have knowledge of the current and previous τ_d partial-response symbols, as well as the estimated CIR. Applying the same principles here, we can construct a trellis which can regenerate all possible received signals for each time instant. At instant m, information on the previous τ_d number of symbols $s_{m-\tau_d} \ldots s_{m-1}$ and the current symbol s_m is embedded in the states of the trellis and the state transitions, respectively. The symbol sequence $s_{m-\tau_d} \ldots s_{m-1}, s_m$, which is associated with each state transition, is convolved with the esti-

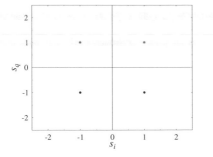

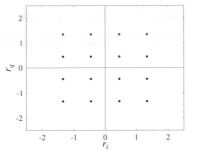

(a) 4-QAM signal constellation

(b) Signal constellation after the 4-QAM symbols are transmitted over the noiseless two-path Gaussian channel

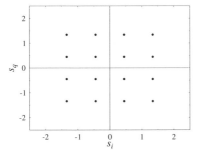

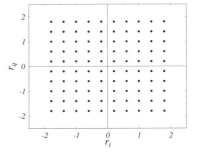

(c) 16-QAM signal constellation

(d) Signal constellation after the 16-QAM symbols are transmitted over the noiseless two-path Gaussian channel.

Figure 17.1: Plotted 4-QAM and 16-QAM signal constellations consisting of in-phase and quadrature-phase components $s_I(t)$ and $s_Q(t)$, respectively, expanded after transmission over the noiseless two-path Gaussian channel of Figure 17.2, as a consequence of the memory in the channel. The received in-phase and quadrature-phase components at the channel output are denoted by r_I and r_Q, respectively, and due to the two-path channel's memory two consecutive symbols now determine each constellation point. Therefore, the number of received points is now given by all possible combinations of two consecutive symbols, i.e. by M^2. Note, however, that in Figure 17.1(d) only 100 points — instead of $16^2 = 256$ points — are visible, since some points happen to coincide owing to the regularity of the constellation.

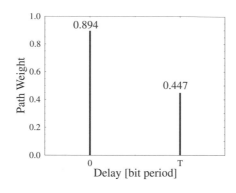

Figure 17.2: The impulse response of a two-path static channel.

mated CIR in order to yield the received signal estimate at instant m. Since there are M possible symbols in the M-QAM constellation, M^{τ_d} states are required to represent all combinations of the previous τ_d number of symbols. Furthermore, since there are M transitions leaving each state, the total number of possible received signal constellation points is $M^{\tau_d} \times M = M^{\tau_d+1}$. For 4-QAM symbols transmitted over the dispersive channel of Figure 17.2, the resulting number of possible received constellations points is $M^{\tau_d+1} = 4^{1+1} = 16$, as shown in Figure 17.1(b), since $\tau_d = 1$ symbol period. Similarly, when transmitting 16-QAM symbols over this channel, the maximum number of signal constellation points is $16^{1+1} = 256$. By studying the trellis construction, we observe that the number of states increases exponentially with the maximum CIR duration τ_d. Consequently, the practical implementation of trellis-based equalisers for higher-order full-response modulation over long dispersive channels is not feasible. In order to implement a practical turbo equaliser for M-QAM, the complexity of the SISO equaliser must be reduced. In the forthcoming sections we will discuss the operations of the reduced-complexity turbo equaliser, commencing with the description of the system model employed in our investigations.

17.3 System Model

In our investigations we have considered a coded M-QAM system employing turbo equalisation at the receiver, as illustrated in Figure 17.3. At the transmitter, the source bits u_k are convolution-

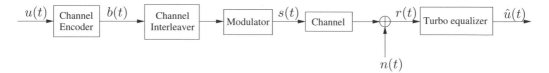

Figure 17.3: A coded M-QAM system employing a turbo equaliser at the receiver.

ally encoded, generating the coded bits b_k. Subsequently, the coded bits are channel-interleaved and passed to the modulator, which produces the modulated signal $s(t)$. The signal $s(t)$ is transmitted over the channel characterised by the CIR $h(t)$ and also corrupted by the zero-mean complex white Gaussian noise $n(t)$, having a double-sided power spectral density of $N_0/2$, yielding the received signal $r(t)$. Since the CIR $h(t)$ is complex and therefore consists of the in-phase component $h_I(t)$ and quadrature-phase component $h_Q(t)$, the resultant received signal $r(t)$ is given by:

$$\begin{aligned}
r(t) &= s(t) * h(t) + n(t) = [s_I(t) + js_Q(t)] * [h_I(t) + jh_Q(t)] + n_I(t) + jn_Q(t) \\
&= [s_I(t) * h_I(t) - s_Q(t) * h_Q(t) + n_I(t)] + j[s_I(t) * h_Q(t) + s_Q(t) * h_I(t) + n_Q(t)] \\
&= r_I(t) + jr_Q(t),
\end{aligned}$$

(17.1)

and $s_I(t)$ and $s_Q(t)$ are the in-phase and quadrature-phase components of the transmitted signal $s(t)$. Also note that $n_I(t)$ and $n_Q(t)$ are the quadrature components of the Gaussian noise $n(t)$. These equations can also be modelled as seen in Figure 17.4.

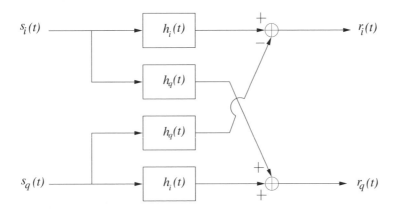

Figure 17.4: Model of the complex, dispersive channel. After transmission over the complex channel, the received signal $r(t)$ becomes dependent on the in-phase component $s_I(t)$ and quadrature-phase component $s_Q(t)$ of the transmitted signal, as expressed in Equation 17.1.

At the receiver, the CIR is estimated using an iterative technique [369,370]. Specifically, during the first turbo equalisation iteration the CIR is estimated using the Least Mean Square (LMS) algorithm [371] and the training sequence mentioned previously. The initial step size of the LMS algorithm is set to 0.10. Subsequently, the CIR estimate acquired during the first iteration is then utilised by the Soft-In/Soft-Out (SISO) [372] equaliser, which generates soft decisions in the form of the logarithmic probability ratios known as Log-Likelihood Ratios (LLRs). These soft decisions are passed to the channel decoder, which computes the reliability information referred to as a-posteriori information, corresponding to the coded bits. During the next iteration, instead of using the training sequence for re-estimating the CIR, the soft estimates of the entire transmission burst's symbols derived from the a-posteriori information of the SISO decoder are employed. Here, a smaller step size of 0.05 is utilised in the LMS algorithm. The decoder's a-posteriori information is converted from the ratio of probability values into soft estimates of the modulated symbols by computing the statistical average of the transmitted symbol probabilities [368]. Further details of this conversion are provided in Section 17.5.2. This CIR estimation process is repeated for each turbo equalisation iteration.

Although there exists a wide range of low-complexity SISO algorithms, we have opted for using the Logarithmic-Maximum A-Posteriori (Log-MAP) algorithm [52,373] for both the SISO channel equaliser and for the channel decoder, since the Log-MAP algorithm achieves optimal performance, despite having a reduced computational complexity compared to the original Maximum A-Posteriori (MAP) algorithm [11].

17.4   In-phase/Quadrature-phase Equaliser Principle

As shown in Equation 17.1, the I/Q components of the received signals, namely $r_I(t)$ and $r_Q(t)$, become dependent on $s_I(t)$ and $s_Q(t)$ after transmission over the channel described by its complex CIR. We refer to the cross-correlation between $s_I(t)$ and $s_Q(t)$ in $r_I(t)$ and $r_Q(t)$ as **cross-coupling**. This cross-coupling of the transmitted signal components requires the receiver to consider a high number of signal combinations, hence necessitating a high number of equaliser trellis states. We can reduce the number of states significantly when the cross-coupling is removed, such that the quadrature components of the decoupled channel output $r'_I(t)$ and $r'_Q(t)$ are solely dependent on $s_I(t)$ or $s_Q(t)$, respectively. The decoupling operation is performed by removing the effect of the undesired quadrature component from the received signal using the symbol and channel estimates generated by the receiver. If these estimates were perfect, perfect decoupling could be achieved. In reality, no perfect symbol and channel estimates are available. However, the initial rough estimates improve considerably during the consecutive turbo equalisation iterations, ultimately yielding a performance close to that of the system using no I/Q decoupling. This process will be elaborated on in the next section. For transmission over real-valued channels no decoupling is necessary. This is because $h_Q(t) = 0$ and from Equation 17.1 it is observed that $r_I(t)$ and $r_Q(t)$ are solely dependent on the $s_I(t)$ and $s_Q(t)$, respectively. The decoupling operation is only essential for transmissions over channels described by complex-valued CIRs.

After decoupling, we can equalise $s_I(t)$ and $s_Q(t)$ independently, hence reducing significantly the number of states in the trellis, when compared to the full-complexity Conventional Trellis-based Equaliser (CT-EQ). In the 4-QAM system $s_I(t)$ and $s_Q(t)$ can be either $+1$ or -1, whereas for 16-QAM, there are four possible values of $s_I(t)$ and $s_Q(t)$, namely $-3, -1, +1$ and $+3$. Hence, there are $\sqrt{M}$ possible values of $s_I(t)$, where, again, M is the number of constellation points for a particular modulation mode. The total number of states in the I/Q-EQ trellis is $\left(\sqrt{M}\right)^{\tau_d}$ and there are $\sqrt{M}$ number of transitions leaving each state. Once the trellis states and transitions are determined, any trellis-based SISO algorithm, such as the MAP algorithm [11] or the Log-MAP algorithm [52], can be employed. As mentioned before, each signal constellation point $s(t) = s_I(t) + js_Q(t)$ can be represented by using a given combination of bits. For example, the 16-QAM signal constellation can be represented by $K_b = 4$ bits, namely b_1, b_2, b_3 and b_4, where $b_k = \pm 1$ for $k = 1, 2, \ldots, K_b = 4$. Let us assume that bits b_1 and b_2 are used for representing $s_I(t)$, while bits b_3 and b_4 are mapped to $s_Q(t)$. Therefore, the SISO I/Q-EQ, which equalises $r'_I(t)$, will give the a-posteriori LLRs of bits b_1 and b_2. The other I/Q-EQ equalising the signal $r'_Q(t)$ is also based on the same principles. However, in this case, combinations of $s_Q(t)$ are considered instead of $s_I(t)$ and the LLRs of bits b_3 and b_4 are produced. In general, for an M-QAM scheme, the number of bits used for representing a symbol is $K_b = \log_2 M$. Therefore, the I/Q-EQ associated with $r'_I(t)$ will compute the a-posteriori LLRs of the first $\frac{K_b}{2}$ number of bits, while the other I/Q-EQ determines the LLRs of the following $\frac{K_b}{2}$ bits. Subsequently, the LLRs of both I/Q-EQs are multiplexed in the schematic of Figure 17.5 before being passed to the decoder, in order to ensure that they are rearranged in the right order, i.e. in the order of $b_1, \ldots, b_{\frac{K_b}{2}}, b_{\frac{K_b}{2}+1}, \ldots, b_{K_b}$.

Note that it is important to identify the bits which are mapped to the in-phase and quadrature-

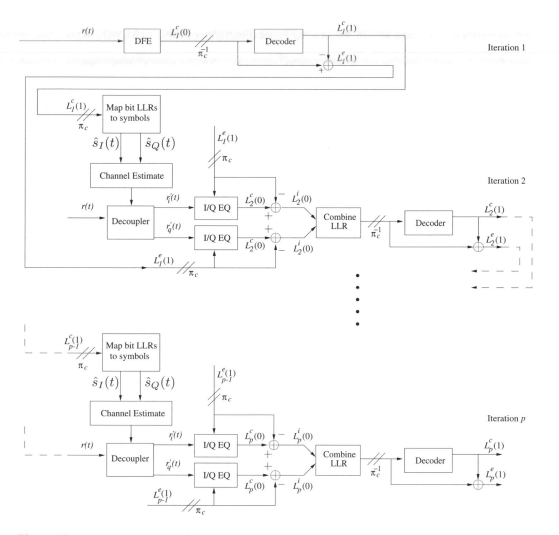

Figure 17.5: Schematic of the turbo equaliser employing a Decision Feedback Equaliser (DFE) and a SISO channel decoder in the first turbo equalisation iteration. In subsequent iterations, two reduced-complexity SISO In-phase/Quadrature-phase Equalisers (I/Q EQs) and one SISO channel decoder are employed. The notation π_c represents a channel interleaver, while π_c^{-1} is used to denote a channel de-interleaver.

phase component of the signal. For example, as in the previous 16-QAM example, bits b_1 and b_2 correspond to signal $s_I(t)$, while b_3 and b_4 map to $s_Q(t)$. This is because the channel decoder processes the received information on a coded bit basis. Therefore, the I/Q-EQ of each quadrature arm must determine the LLR values corresponding to the coded bits. If the LLRs of the quadrature symbols are determined instead, then an additional processing step is required for converting the LLR values of the quadrature symbols to the LLRs of the coded bits. Therefore, the above-mentioned principle of the I/Q-EQ is not directly applicable to M-PSK, which has M values higher than 4, such as $M = 8$, 16, 32, since the bits of the multilevel symbol do not distinctly map to $s_I(t)$ and $s_Q(t)$ as in the above square-shaped M-QAM constellations. The I/Q-EQ also applies to non-square M-QAM schemes which do not have a distinct mapping from coded bits to quadrature symbols.

17.5 Overview of the Reduced-complexity Turbo Equaliser

Figure 17.5, which is discussed in detail throughout this section, illustrates the schematic of the turbo equaliser utilising two reduced-complexity SISO equalisers. As mentioned previously, the Log-MAP algorithm [52] is employed in the I/Q-EQ and in the channel decoder blocks. Here, the soft decisions generated are in the form of LLRs. We expressed the LLR values of the equaliser and decoder using vector notations following the approach of [346], albeit using different notations. The superscript denotes the nature of the LLR, namely 'c' is used for the composite a-posteriori [11] information, 'i' [11] for the combined channel and extrinsic information and 'e' [11] for the extrinsic information. Furthermore, the subscripts in Figure 17.5 are used for representing the iteration index, while the argument within the brackets ( ) indicates the index of the receiver stage, where the equalisers are denoted as stage 0, and the channel decoder as stage 1. Furthermore, in our discussions related to multilevel QAM, the term **bit** refers to either the $+1$ or -1 bit of the M-QAM **symbols**. For 4-QAM, there are two bits in a symbol, whereas a 16-QAM symbol consists of four bits.

At the first iteration, the DFE is invoked, since it constitutes a low-complexity approach to providing initial estimates of the transmitted symbols, as compared to the more complex CT-EQ. Subsequently, the SISO channel decoder of Figure 17.5 generates the a-posteriori LLR $L_1^c(1)$, from which the extrinsic information of the encoded bits, namely $L_1^e(1)$, is extracted. In the next iteration, the a-posteriori LLR $L_1^c(1)$ is used for regenerating estimates of the in-phase and quadrature-phase components of the transmitted signal, namely $\hat{s}_I(t)$ and $\hat{s}_Q(t)$, as seen in the 'MAP bit LLRs to symbols' block of Figure 17.5. The a-posteriori information was transformed from the logarithmic domain to modulated symbols using the approach employed in [368]. The estimates of the transmitted quadrature components $\hat{s}_I(t)$ and $\hat{s}_Q(t)$ are then convolved with the estimate of the CIR $\hat{h}(t)$. At the decoupler block of Figure 17.5, the resultant signal is used for removing the cross-coupling effect — seen in Equation 17.1 — from both quadrature components of the transmitted signal, yielding $r'_I(t)$ and $r'_Q(t)$. As mentioned previously, the cross-coupling between $s_I(t)$ and $s_Q(t)$ in each quadrature arm of the received signal $r(t)$ is removed in order to reduce the number of possible signal combinations, hence reducing the number of equaliser states in the trellis. After the decoupling operation, $r'_I(t)$ and $r'_Q(t)$ are passed to the I/Q-EQ in the schematic of Figure 17.5. In addition to these received quadrature signals, the I/Q-EQ also processes the a-priori information received, which is constituted by the extrinsic LLRs $L_1^e(1)$ from the previous iteration, and generates the a-posteriori information $L_2^c(0)$. Subsequently, the combined channel and extrinsic information $L_2^i(0)$ is extracted from both I/Q-EQs in Figure 17.5 and combined, before being passed to the Log-MAP channel decoder. As in the first turbo equalisation iteration, the a-posteriori and extrinsic information of the encoded bits, namely $L_2^c(1)$

and $L_2^e(1)$, respectively, are evaluated. The following turbo equalisation iterations also obey the same sequence of operations, until the iteration termination criterion is met.

Let us now discuss in greater detail the following aspects of the schematic in Figure 17.5:

- Conversion of the DFE symbol estimates to LLR.

- Conversion of the decoder a-posteriori LLR into symbols.

- Decoupling operation.

- I/Q-EQ complexity.

17.5.1   Conversion of the DFE Symbol Estimates to LLRs

We commence by describing the DFE symbol-to-LLR conversion for a specific example, namely for 4-QAM, in order to highlight its salient steps, before generalising the concept. The associated DFE block can be found at the top left corner of Figure 17.5 at iteration 1. The output of the DFE is the estimate $\tilde{s}(t)$ of the transmitted symbol at time instant t. The squared Euclidean distance

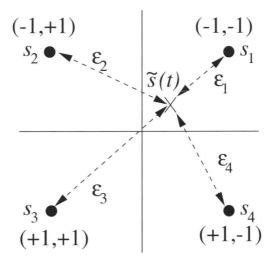

Figure 17.6: The squared Euclidean distance ε_1, ε_2, ε_3 and ε_4 between the DFE symbol estimate $\tilde{s}(t)$ and the 4-QAM signal constellation points s_1, s_2, s_3 and s_4. The values in the brackets ( ) are the bits of the 4-QAM symbol.

between the DFE symbol estimate $\tilde{s}(t)$ and the individual 4-QAM constellation points, namely s_1, s_2, s_3 and s_4, at instant t can be evaluated as:

$$\varepsilon_j = |\tilde{s}(t) - s_j|^2 \qquad j = 1, 2, \ldots, M = 4, \tag{17.2}$$

where M is the number of constellation points, which is depicted in Figure 17.6. Note that in Figure 17.6 the values in the brackets ( ) represent the bits of the 4-QAM symbol. Subsequently, the probability that symbol s_j, $j = 1, 2, \ldots, M = 4$, was transmitted given that $\tilde{s}(t)$ was received can be evaluated by using [368]:

$$P[s_j | \tilde{s}(t)] = \frac{1}{\sqrt{2\pi\sigma^2}} \exp\left(\frac{-\varepsilon_j}{2\sigma^2}\right) \qquad j = 1, 2, \ldots, M = 4, \tag{17.3}$$

where M is the number of constellation points. Before we proceed, we define new notations, in order to represent each element of the decoder's a-posteriori LLR vector $L_1^c(1)$. Information related to the time instant m and to bit b_k of a symbol, where k is the bit index, is introduced to give $L^c(m; k)$ for all bits of all symbols, which is formulated as:.

$$L_1^c(0) = [L^c(m; k = 1); L^c(m; k = 2)] \qquad m = 1, 2, \dots, B_L, \qquad (17.4)$$

where B_L is the length of the coded block, while the maximum value of k in our example is two for 4-QAM. Subsequently, having determined the probability of the transmitted symbols s_1, s_2, s_3 and s_4, we can now compute the LLR of each transmitted bit $L^c(m; k)$ at instant m, since:

$$L^c(m; k) = \ln \left(\frac{P[b_k = +1 | \tilde{s}(t)]}{P[b_k = -1 | \tilde{s}(t)]} \right) = \ln \left(\frac{\displaystyle\sum_{\forall s_j \text{ where } b_k = +1} P[s_j | \tilde{s}(t)]}{\displaystyle\sum_{\forall s_j \text{ where } b_k = -1} P[s_j | \tilde{s}(t)]} \right), \qquad (17.5)$$

where b_k is the kth bit of the M-QAM symbol. Explicitly, the numerator is the sum of all probabilities $P[s_j | \tilde{s}(t)]$, where the kth bit of the symbol s_j is $+1$. For the 4-QAM scenario we observe in Figure 17.6 that, when $k = 1$, this corresponds to signal points s_3 and s_4, and to s_2 and s_3, when $k = 2$. For the denominator, we sum all the probabilities $P[s_j | \tilde{s}(t)]$, which correspond to the kth bit of the symbol being -1. From Figure 17.6, we observed that when $k = 1$, the summation of probabilities $P(s_1 | \tilde{s}(t))$ and $P(s_2 | \tilde{s}(t))$ is performed, while for $k = 2$ $P(s_1 | \tilde{s}(t))$ is added to $P(s_4 | \tilde{s}(t))$. Therefore, at instant m the equaliser a-posteriori LLR $L^c(m; k = 1)$ for the $(k = 1)$st bit is given by:

$$L^c(m; k = 1) = \ln \left(\frac{P[s_3 | \tilde{s}(t)] + P[s_4 | \tilde{s}(t)]}{P[s_1 | \tilde{s}(t)] + P[s_2 | \tilde{s}(t)]} \right), \qquad (17.6)$$

while for the $(k = 2)$nd bit, we obtain:

$$L^c(m; k = 2) = \ln \left(\frac{P[s_2 | \tilde{s}(t)] + P[s_3 | \tilde{s}(t)]}{P[s_1 | \tilde{s}(t)] + P[s_4 | \tilde{s}(t)]} \right). \qquad (17.7)$$

The symbol estimates at the output of the DFE for the other modulation modes can be easily converted into the LLR form by substituting the value of M in Equations 17.2 and 17.3 for the corresponding modulation mode and using Equation 17.5. Having described the conversion of the DFE symbol estimates to LLRs, we now proceed with the discussion of the LLR conversion to modulated symbols.

17.5.2   Conversion of the Decoder A-Posteriori LLRs into Symbols

Let us now highlight the operation of the top left block of Figure 17.5 at iteration 2. There are two potential approaches which can be employed to convert information from LLRs into symbols. The first technique is based on the inverse of the conversion of DFE symbol estimates to LLRs — which was described in the previous subsection — while the second approach determines the statistical average of the in-phase and quadrature-phase signals, denoted as $\hat{s}_I(t)$ and $\hat{s}_Q(t)$, respectively, by using [374]:

$$E\{s_I(t)\} = \hat{s}_I(t) = \sum_{i=0}^{\sqrt{M}-1} s_{I;i} \cdot P[s_{I;i} | \tilde{s}(t)]$$

$$E\{s_Q(t)\} = \hat{s}_Q(t) = \sum_{i=0}^{\sqrt{M}-1} s_{Q;i} \cdot P[s_{Q;i} | \tilde{s}(t)], \qquad (17.8)$$

In-phase $s_I(t)$				Quadrature-phase $s_Q(t)$			
(b_1, b_2)				(b_3, b_4)			
(+1,-1)	(+1,+1)	(-1,+1)	(-1,-1)	(+1,-1)	(+1,+1)	(-1,+1)	(-1,-1)
-3	-1	+1	+3	-3	-1	+1	+3
$s_{I;3}$	$s_{I;2}$	$s_{I;1}$	$s_{I;0}$	$s_{Q;3}$	$s_{Q;2}$	$s_{Q;1}$	$s_{Q;0}$

Figure 17.7: The Gray mapping [102] of the 16-QAM mode depicting the in-phase and quadrature-phase components and the corresponding bit assignments.

where $E\{\cdot\}$ denotes the expectation or averaging operation and M is the number of constellation points in a particular modulation mode. The terms $s_{I;i}$ and $s_{Q;i}$, $i = 0 \ldots \sqrt{M} - 1$, represent the ith in-phase signal and the ith quadrature-phase signal, as illustrated in Figure 17.7 for 16-QAM. In attempting to determine the estimated symbol from the LLR values, the former technique involves solving second-order equations arising from the squared Euclidean distance term in Equation 17.2. Therefore, there are more than one possible symbol estimates at each time instant. The latter approach provides the symbol estimates based on the probabilities of symbols, which can be extracted from the LLR values. Since there is no ambiguity in the latter statistical approach — in contrast to the inverse process of the symbols-to-LLR conversion — we will adopt this technique, in order to translate LLR values to symbol estimates.

In order to highlight the associated operations, let us consider a particular example, namely 16-QAM, before generalising the concept. Each 16-QAM symbol $s(t)$ can be represented by four bits, namely b_1, b_2, b_3 and b_4. Let us assume that the first two bits (b_1, b_2) correspond to a coordinate on the in-phase axis $s_I(t)$ of the signal constellation, while the other two bits (b_3, b_4) represent a point on the quadrature-phase axis $s_Q(t)$, as depicted in Figure 17.7. Hence, the 16-QAM symbol can be expressed as a combination of the coordinates $s_I(t)$ and $s_Q(t)$:

$$s(t) = s_I(t) + js_Q(t). \tag{17.9}$$

With reference to Figure 17.7, which illustrates the Gray mapping used in each quadrature arm of the 16-QAM signal constellation, and by employing Equation 17.8, the regenerated quadrature components of $s(t)$, namely $\hat{s}_I(t)$ and $\hat{s}_Q(t)$, can be expressed as:

$$\hat{s}_I(t) = 3 \cdot P[s_{I;0}|\tilde{s}(t)] + 1 \cdot P[s_{I;1}|\tilde{s}(t)] - 1 \cdot P[s_{I;2}|\tilde{s}(t)] - 3 \cdot P[s_{I;3}|\tilde{s}(t)]$$
$$= 3 \cdot P[b_1 = -1; b_2 = -1|\tilde{s}(t)] + 1 \cdot P[b_1 = -1; b_2 = +1|\tilde{s}(t)]$$
$$- 1 \cdot P[b_1 = +1; b_2 = +1|\tilde{s}(t)] - 3 \cdot P[b_1 = +1; b_2 = -1|\tilde{s}(t)],$$

and:

$$\hat{s}_Q(t) = 3 \cdot P[s_{Q;0}|\tilde{s}(t)] + 1 \cdot P[s_{Q;1}|\tilde{s}(t)] - 1 \cdot P[s_{Q;2}|\tilde{s}(t)] - 3 \cdot P[s_{Q;3}|\tilde{s}(t)]$$
$$= 3 \cdot P[b_3 = -1; b_4 = -1|\tilde{s}(t)] + 1 \cdot P[b_3 = -1; b_4 = +1|\tilde{s}(t)]$$
$$- 1 \cdot P[b_3 = +1; b_4 = +1|\tilde{s}(t)] - 3 \cdot P[b_3 = +1; b_4 = -1|\tilde{s}(t)],$$

where the probabilities $P[s_{I;i}|\tilde{s}(t)]$ and $P[s_{Q;i}|\tilde{s}(t)]$ for $i = 0 \ldots \sqrt{M} - 1$ can be written as $P[b_1; b_2|\tilde{s}(t)]$ and $P[b_3; b_4|\tilde{s}(t)]$, respectively, since the symbol $s_{I;i}$ corresponds to bits b_1 and b_2, while the symbol $s_{Q;i}$ is represented by bits b_3 and b_4. Note also that $P[b_1; b_2|\tilde{s}(t)]$ can be expressed as $P[b_1|\tilde{s}(t)] \cdot P[b_2|\tilde{s}(t)]$, since the transmissions of bits b_1 and b_2 are statistically independent events. Similarly, the probability $P[b_3; b_4|\tilde{s}(t)]$ can be written as $P[b_3|\tilde{s}(t)] \cdot P[b_4|\tilde{s}(t)]$, yielding

$$\hat{s}_I(t) = 3 \cdot P[b_1 = -1|\tilde{s}(t)] \cdot P[b_2 = -1|\tilde{s}(t)] + 1 \cdot P[b_1 = -1|\tilde{s}(t)] \cdot P[b_2 = +1|\tilde{s}(t)]$$
$$- 1 \cdot P[b_1 = +1|\tilde{s}(t)] \cdot P[b_2 = +1|\tilde{s}(t)] - 3 \cdot P[b_1 = +1|\tilde{s}(t)] \cdot P[b_2 = -1|\tilde{s}(t)],$$

and:

$$\hat{s}_Q(t) = 3 \cdot P[b_3 = -1|\tilde{s}(t)] \cdot P[b_4 = -1|\tilde{s}(t)] + 1 \cdot P[b_3 = -1|\tilde{s}(t)] \cdot P[b_4 = +1|\tilde{s}(t)]$$
$$- 1 \cdot P[b_3 = +1|\tilde{s}(t)] \cdot P[b_4 = +1|\tilde{s}(t)] - 3 \cdot P[b_3 = +1|\tilde{s}(t)] \cdot P[b_4 = -1|\tilde{s}(t)].$$

Therefore, it is possible to infer the associated 16-QAM symbols, once the probabilities $P[b_k|\tilde{s}(t)]$ of the bits in the symbol are known. This can be extracted from the decoder's a-posteriori LLR $L_1^c(1)$ in Figure 17.5, since the LLR reflects the confidence associated with the assumption of the transmitted bit being either $+1$ or -1 in the form of a ratio, as expressed in Equation 17.5. Since $L^c(m;k)$ can be expressed as:

$$L^c(m;k) = \ln\left(\frac{P[b_k = +1|\tilde{s}(t)]}{P[b_k = -1|\tilde{s}(t)]}\right), \tag{17.10}$$

the probability that $b_k = +1$ was transmitted at instant m given that $\tilde{s}(t)$ was estimated by the DFE, namely $P[b_k = +1|\tilde{s}(t)]$, can be expressed after a few steps as:

$$P[b_k = +1|\tilde{s}(t)] = \frac{\exp(L^c(m;k))}{1 + \exp(L^c(m;k))}. \tag{17.11}$$

Subsequently, using the relationship that $P[b_k = +1|\tilde{s}(t)] + P[b_k = -1|\tilde{s}(t)] = 1$, the term $P[b_k = -1|\tilde{s}(t)]$ becomes:

$$P[b_k = -1|\tilde{s}(t)] = \frac{1}{1 + \exp(L^c(m;k))}. \tag{17.12}$$

We now have an expression for the kth bit's probability in the mth symbol, which can be extracted from the decoder's LLRs by substituting Equations 17.11 and 17.12 into Equations 17.5.2 and 17.5.2. At this stage, the probabilities in Equations 17.8 and 17.5.2–17.12 have been conditioned upon the DFE symbol estimates $\tilde{s}(t)$, since the DFE is utilised and LLR values are generated based on the DFE symbol estimates. However, when I/Q-EQs are employed instead of the DFE, the probability terms in these equations will be conditioned upon the received signal $r(t)$. Closer inspection of Equations 17.11 and 17.12 shows that the maximum value of the $\tilde{s}_I(t)$ and $\tilde{s}_Q(t)$ components is $+3$, when $P[b_1 = -1|\tilde{s}(t)] \cdot P[b_2 = -1|\tilde{s}(t)] = 1$ and $P[b_3 = -1|\tilde{s}(t)] \cdot P[b_4 = -1|\tilde{s}(t)] = 1$, respectively. Conversely, $\tilde{s}_I(t)$ and $\tilde{s}_Q(t)$ have a minimum value of -3 when $P[b_1 = +1|\tilde{s}(t)] \cdot P[b_2 = -1|\tilde{s}(t)] = 1$ and $P[b_3 = +1|\tilde{s}(t)] \cdot P[b_4 = -1|\tilde{s}(t)] = 1$, respectively. Therefore, by evaluating the average value of the in-phase and quadrature-phase components, we will obtain a constellation of estimated symbols which are within the boundaries of the maximum and minimum values of each quadrature arm.

Let us highlight this point more explicitly by considering a simple 4-QAM signal constellation as illustrated in Figure 17.8. Again, by employing Equation 17.8 and with reference to Figure 17.8, which illustrates the Gray mapping used in each quadrature arm of the 4-QAM signal constellation, the regenerated quadrature components of $s(t)$, namely $\hat{s}_I(t)$ and $\hat{s}_Q(t)$, can be expressed as:

$$\hat{s}_I(t) = 1 \cdot P[b_1 = +1|\tilde{s}(t)] - 1 \cdot P[b_1 = -1|\tilde{s}(t)] \tag{17.13}$$

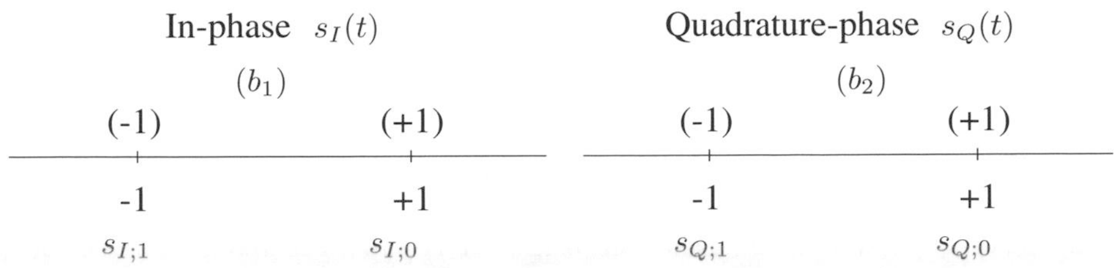

	In-phase $s_I(t)$				Quadrature-phase $s_Q(t)$	
	(b_1)				(b_2)	
(-1)		(+1)		(-1)		(+1)
-1		+1		-1		+1
$s_{I;1}$		$s_{I;0}$		$s_{Q;1}$		$s_{Q;0}$

Figure 17.8: The Gray mapping [102] of the 4-QAM mode depicting the in-phase and quadrature-phase components and the corresponding bit assignments.

and:

$$\hat{s}_Q(t) = 1 \cdot P[b_2 = +1|\tilde{s}(t)] - 1 \cdot P[b_2 = -1|\tilde{s}(t)]. \tag{17.14}$$

Since the expression $P[b_1 = +1|\tilde{s}(t)] + P[b_1 = -1|\tilde{s}(t)] = 1$ is valid, the maximum value of 1 is assumed by $\hat{s}_I(t)$ in Equation 17.13, when $P[b_1 = +1|\tilde{s}(t)] = 1$ and $P[b_1 = -1|\tilde{s}(t)] = 0$. Conversely, a minimum value of -1 is assumed by $\hat{s}_I(t)$, when $P[b_1 = -1|\tilde{s}(t)] = 1$ and $P[b_1 = +1|\tilde{s}(t)] = 0$. Employing the same reasoning as above, the maximum and minimum values of $\hat{s}_Q(t)$ in Equation 17.14 are $+1$ and -1, respectively. Hence, the average symbol values of $\hat{s}_I(t)$ and $\hat{s}_Q(t)$ will never be beyond the boundaries formed by the minimum and maximum values of these quadrature signal estimates, as will be shown in Figures 17.11, 17.13 and 17.15 for 4-QAM, 16-QAM and 64-QAM, respectively.

We summarise the associated operations described previously and generalise the concept, where possible, in order to allow the conversion of decoder's bit LLR values to M-QAM symbols in the top left block of Figure 17.5 at iteration 2, as follows:

- Initially, determine the Gray mapping for the M-QAM symbol, as shown in Figure 17.7. Knowledge of the mapping allows the in-phase and quadrature-phase components of the symbol, namely $s_I(t)$ and $s_Q(t)$, to be determined as a function of the bit probabilities, as seen in Equations 17.5.2 and 17.5.2.

- Subsequently, we evaluate the probability that bit $b_k = +1$ was transmitted at time instant m, given that $\tilde{s}(t)$ was estimated by the DFE, namely $P[b_k = +1|\tilde{s}(t)]$, by using Equation 17.11. Similarly, with the aid of Equation 17.12 we compute the probability $P[b_k = -1|\tilde{s}(t)]$ of $b_k = -1$ being transmitted at instant m, given that $\tilde{s}(t)$ was estimated by the DFE. For 4-QAM, 16-QAM and 64-QAM the maximum values of k are two, four and six, respectively.

- Finally, we substitute the above bit probability values computed into Equations 17.5.2 and 17.5.2 in order to obtain the M-QAM symbol.

So far, we have described the conversion of DFE symbol estimates $\tilde{s}(t)$ to the LLR $L_1^c(0)$. This LLR information is then passed to the decoder at iteration 1 in Figure 17.5, which subsequently computes the a-posteriori LLR $L_1^c(1)$ of the encoded bits. Next, this LLR $L_1^c(1)$ is converted into M-QAM symbols at iteration 2 in Figure 17.5 using the principles described previously. These symbols are then used in conjunction with the channel estimates $h(t)$, in order to remove the cross-coupling between $s_I(t)$ and $s_Q(t)$ in each quadrature branch, such that a lower number of states are required in the equaliser. This decoupling operation is seen at the bottom left corner of Figure 17.5 at iteration 2 and will be the focus of our following discussion.

17.5.3   Decoupling Operation

As mentioned previously, the aim of the decoupling process is to produce the resultant signals $r'_I(t)$ and $r'_Q(t)$, which are solely dependent on the I and Q signals, respectively. Let us use Equation 17.1 and consider $r_I(t)$ as our starting point in highlighting the decoupling operation. It is observed that $r_I(t)$ is dependent on signals $s_I(t)$ and $s_Q(t)$. In order to generate $r'_I(t)$, all contributions of $s_Q(t)$ must be removed. This is achieved by first generating estimates of $\hat{s}_Q(t)$, since it is unknown at the receiver, by using the previous decoder's a-posteriori information, which reflects our confidence in whether a -1 or $+1$ coded bit was transmitted. The conversion process from the decoder LLRs to M-QAM symbol estimates was described in the previous section. Subsequently, these symbols are convolved with the I and Q CIR estimates, in order to generate $\hat{s}_Q(t) * \hat{h}_I(t)$ and $j\hat{s}_Q(t) * \hat{h}_Q(t)$. These estimated signals are then removed from the received signal $r(t)$, in order to generate $r'_I(t)$:

$$
\begin{aligned}
r'_I(t) &= r(t) + \hat{s}_Q(t) * \hat{h}_Q(t) - j\hat{s}_Q(t) * \hat{h}_I(t) \\
&= s_I(t) * h_I(t) + js_I(t) * h_Q(t) + n(t) + e(\hat{s}_Q(t), \hat{h}_Q(t)) + je(\hat{s}_Q(t), \hat{h}_I(t)),
\end{aligned}
\tag{17.15}
$$

where $e(\hat{s}_Q(t), \hat{h}_Q(t))$ and $e(\hat{s}_Q(t), \hat{h}_I(t))$ are the error functions:

$$
\begin{aligned}
e(\hat{s}_Q(t), \hat{h}_Q(t)) &= r_I(t) + \hat{s}_Q(t) * \hat{h}_Q(t) \\
je(\hat{s}_Q(t), \hat{h}_I(t)) &= jr_Q(t) - j\hat{s}_Q(t) * \hat{h}_I(t),
\end{aligned}
\tag{17.16}
$$

which arise when inaccurate CIR estimates and low-confidence M-QAM symbol estimates are generated. Similarly, $r'_Q(t)$ is obtained by subtracting $\hat{s}_I(t) * \hat{h}_I(t)$ and $j\hat{s}_I(t) * \hat{h}_Q(t)$ from $r(t)$, which are generated by using the symbol estimates $\hat{s}_I(t)$ and the I and Q CIR estimates.

In the first turbo equalisation iteration the values of the error functions are high owing to the poor reliability of the regenerated signal estimates. However, through successive turbo equalisation iterations the performance of the TEQ-IQ improves, since the reliability of the symbol estimates and CIR estimates is enhanced. This can be observed in Figure 17.13 below, which shows the phasor constellation of 16-QAM signals obtained at $E_b/N_0 = 12$ dB. After the first iteration a crude approximation of the 16-QAM signal constellation was beginning to emerge. Subsequent iterations improved the reliability of the channel decoder's a-posteriori LLRs, hence yielding a more accurate mapping of the 16-QAM signal. Therefore, at the sixth turbo equalisation iteration in Figure 17.13(d) a close approximation of the transmitted 16-QAM signal constellation was achieved. The improved reliability of the regenerated symbol estimates reduces the decoupling errors, hence improving the performance of the TEQ-IQ. This will be demonstrated using our simulation results in Section 17.8.

17.6   Complexity of the In-phase/Quadrature-phase Turbo Equaliser

In order to simplify the complexity analysis of the turbo equalisers, the complexity of the channel encoder, modulator, interleaver and de-interleaver has been assumed to be negligible. Therefore, the complexity of the turbo equaliser is dependent only on the complexity of the equaliser, the decoder and the number of turbo equalisation iterations performed. Since, the complexity of the equaliser and decoder is added and subsequently multiplied by the number of turbo equalisation iterations, we must adopt the same measure of complexity for both the equaliser and decoder, which is in this case the number of associated trellis transitions per information bit. Therefore, the complexity of the equaliser, which is dependent on the number of trellis transitions per coded bit, must be normalised by the overall throughput T_r, which is the product of the number of Bits Per Symbol (BPS) and the code rate R, to become the number of transitions per information bit.

For the CT-EQ, the complexity $comp[\text{CT} - \text{EQ}]$ associated with equalising M-QAM signals transmitted over a channel having a complex CIR with a delay spread of τ_d symbols is:

$$comp[\text{CT} - \text{EQ}] = (\text{Number of states} \cdot \text{Number of transitions}) / T_r$$
$$= \left(M^{\tau_d} \cdot M \right) / T_r = \left(M^{\tau_d+1} \right) / T_r,$$

whereas for the single I/Q-EQ trellis stage:

$$comp[\text{I/Q} - \text{EQ}] = (\text{Number of states} \cdot \text{Number of transitions}) / T_r$$
$$= \left(\sqrt{M}^{\tau_d} \cdot \sqrt{M} \right) / T_r = \left(\sqrt{M}^{\tau_d+1} \right) / T_r.$$

For the rate $R = \frac{1}{2}$ and constraint length $K = 5$ convolutional decoder, the complexity $comp[CC(2, 1, K)]$ incurred is:

$$comp[CC(2, 1, K)] = \text{Number of states} \cdot \text{Number of transitions}$$
$$= 2^{K-1} \cdot 2 = 2^K,$$

since the convolutional code is a binary code, which has two branches leaving each state.

Having determined the complexity of the equaliser and decoder as a function of the number of transitions per information bit, the complexity of the Turbo EQualiser based on the Conventional Trellis, which we refer to as the TEQ-CT, can be estimated as:

$$comp[\text{TEQ} - \text{CT}] = \left(\left(M^{\tau_d+1} \right) / T_r + 2^K \right) \cdot \text{Itr[TEQ-CT]}, \tag{17.17}$$

while the complexity of TEQ-IQ is:

$$comp[\text{TEQ} - \text{IQ}] = 2^K + \left(2 \cdot \left(\sqrt{M}^{\tau_d+1} \right) / T_r + 2^K \right) \cdot \left(\text{Itr[TEQ-IQ]} - 1 \right), \tag{17.18}$$

where Itr[ ] denotes the number of iterations performed by the receiver. A factor of 2 was introduced in Equation 17.18, since two I/Q-EQs are required. In Equation 17.18, the first term on the right-hand side represents the complexity incurred in the first TEQ-IQ iteration, where a DFE and a convolutional decoder were employed. The remaining terms correspond to the complexity of the subsequent TEQ-IQ iterations. For the sake of simplicity we have assumed that the complexity of the DFE is negligible when compared to the complexity of the I/Q-EQ and CT-EQ. In terms of arithmetic operations, the DFE's complexity is approximately proportional to N_f^3 [237], where N_f is the number of feedforward filter taps. Since we have employed $N_f = 15$ feedforward filter taps in our investigations, the complexity incurred is approximately 3375 arithmetic operations per equalised M-ary symbol. By contrast, the I/Q-EQ and CT-EQ have to evaluate the trellis transition metric as well as the forward and backward recursions [11, 52] respectively, for every transition, resulting in a higher number of operations. Therefore, the complexity of the TEQ-IQ in the first iteration is only dependent on the complexity of the convolutional decoder.

Having described the salient operations of the I/Q-EQ, we now give an overview of the system parameters utilised and subsequently present performance results for our turbo equalisation scheme employing the I/Q-EQ.

17.7   System Parameters

In our forthcoming deliberations, the performance of our turbo equaliser employing the proposed reduced-complexity I/Q-EQ and a convolutional decoder is investigated in the context of square-constellation M-QAM systems having a fixed system delay. We will elaborate on our considerations related to the system delay after describing the transmission burst structure.

The rate $R = \frac{1}{2}$, constraint length $K = 5$, recursive systematic convolutional code, with octal generator polynomials of $G_0 = 35$ and $G_1 = 23$, was invoked in the turbo-equalised M-QAM systems considered. In the first iteration, the DFE of Figure 17.5 was employed, as mentioned in Section 17.5. The number of forward and backward taps in the DFE was fifteen and four, respectively. In the subsequent iterations, two SISO I/Q-EQs were employed, which utilised the Log-MAP algorithm of Section 13.8.3. The convolutional decoder used in these turbo-equalised M-QAM systems also employed the Log-MAP algorithm of Section 13.8.3 — rather than the less complex and more conventional Viterbi MLSE decoder — in order to supply the turbo equaliser with bit confidence values.

The transmission burst structure used in this system was the so-called FMA1 non-spread speech burst as specified in the Pan-European FRAMES proposal [252] and shown in Figure 16.1. In order to decide on the tolerance delay and the depth of the channel interleaver, we considered the maximum affordable delay of a speech system. This system delay is mainly determined by the latency introduced by the channel interleavers, where an entire block of bits must be received in the interleaver's buffer before their transmission can commence. Here the processing delay attributed to the channel encoding, modulation and turbo equalisation operations has been ignored although practical systems typically have a processing delay which allows them to complete their operations 'just' before they have to commence processing the incoming information block. Typically, speech systems can tolerate system delays which are less than 40 ms. Here, the acceptable delay is conservatively set to ≈ 30 ms. For example, for a Time Division Multiple Access/Time Division Duplex (TDMA/TDD) system, which employs eight uplink and eight downlink slots, one transmission slot will be available after every 16 TDMA slots. Furthermore, since each 72 μs burst of Figure 16.1 consists of 144 data symbols, the total number of symbols transmitted within ≈ 30 ms is:

$$
\begin{aligned}
\text{Number of symbols} &= \left\lfloor \frac{\text{Maximum system delay}}{\text{Burst duration} \times \text{Number of slots between transmission}} \right\rfloor \\
&\quad \times \text{Data symbols per burst} \\
&= \left\lfloor \frac{30 \text{ ms}}{72 \ \mu\text{s} \times 16} \right\rfloor \times 144 \\
&= 26 \times 144 \\
&= 3456,
\end{aligned}
$$

$$(17.19)$$

corresponding to 3456 symbols when employing BPSK transmission. The notation $\lfloor \cdot \rfloor$ represents the integer floor. Upon assuming an identical signalling rate, the corresponding number of transmitted encoded bits for 4-QAM, 16-QAM and 64-QAM is 6912, 13824 and 20736, respectively. Hence, random channel interleavers of these depths were invoked in our investigations.

A three-path, symbol-spaced fading channel of equal weights was utilised, where the Rayleigh fading statistics obeyed a normalised Doppler frequency of 3.3×10^{-5}. This corresponded to a transmission frequency of 1900 MHz, signalling rate of 2600 KBaud and a vehicular speed of 30 mph. As before, in our investigations the CIR was assumed to be known and

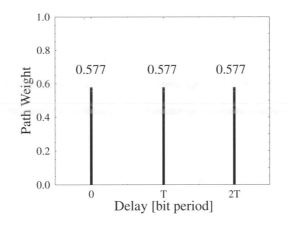

Figure 17.9: Illustration of the equally weighted, symbol spaced three-path fading channel.

the fading magnitude and phase were kept constant for the duration of a transmission burst, a condition which we refer to as employing burst-invariant fading.

17.8 System Performance

17.8.1 4-QAM System

The signals were transmitted over the equally weighted three-path Rayleigh fading channel of Figure 17.9 using burst-invariant fading. For a transmission delay of ≈ 30 ms, the depth of the channel interleaver implemented was 6912 bits. Figures 17.10(a) and 17.10(b) characterise the performance of the reduced-complexity turbo equaliser (**TEQ-IQ**) using iterative CIR estimation and the conventional trellis-based turbo equaliser (**TEQ-EQ**) having perfect CIR information for a 4-QAM system, respectively, after four turbo equalisation iterations. In Figure 17.10(a), it was observed that after two turbo equalisation iterations the performance of the TEQ-EQ did not improve significantly despite invoking further iterations. We used the term **critical number of iterations** I_t, in order to denote this iteration number, namely two in this case. When employing the TEQ-IQ receiver, the performance obtained after two and three turbo equalisation iterations was similar, as shown in Figure 17.10(b). Hence, the critical number of iterations performed by the TEQ-IQ receiver was three. The performance achieved by the TEQ-IQ receiver after three iterations was also observed to be similar to that obtained by the TEQ-EQ receiver after two iterations in Figure 17.10(b).

Let us study the complexity of the TEQ-IQ and TEQ-EQ receivers by using Equations 17.6 and 17.6. Substituting $M = 4$ and $\tau_d = 2$ symbol periods in Equation 17.18 and 17.17, the TEQ-IQ and TEQ-EQ complexity becomes 176 transitions per trellis interval and 192 transitions per trellis interval, respectively, since the critical number of iterations for the TEQ-IQ was four, whereas for TEQ-EQ it was two. Hence, a complexity reduction by a factor of 1.1 was achieved by the TEQ-IQ receiver using iterative CIR estimation, while obtaining the same performance as the TEQ-EQ receiver having perfect CIR information.

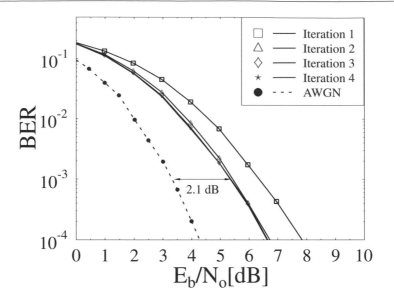

(a) Turbo equaliser using the conventional trellis-based equaliser

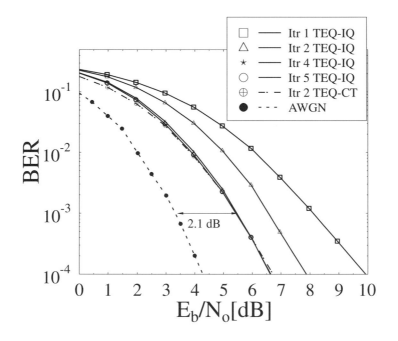

(b) Turbo equaliser using two I/Q EQs

Figure 17.10: Performance of the turbo equaliser employing two I/Q-EQs and that utilising the conventional trellis-based equaliser for a convolutional-coded 4-QAM system, possessing a channel interleaving depth of 6912 bits, transmitted over the equally weighted, three-path Rayleigh fading channel of Figure 17.9 using a normalised Doppler frequency of 3.3×10^{-5} and burst-invariant fading.

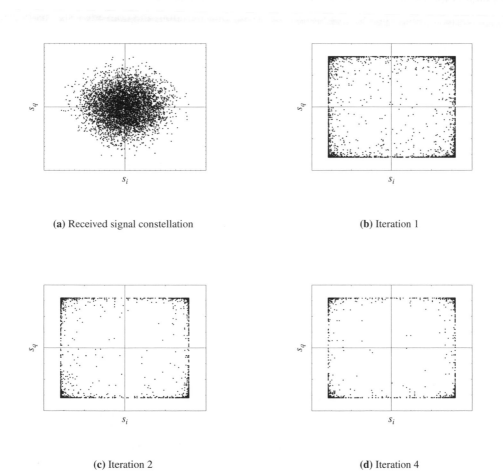

(a) Received signal constellation   (b) Iteration 1

(c) Iteration 2   (d) Iteration 4

Figure 17.11: Phasor constellation of the received signal and that of the 4-QAM signal, which was generated from the decoder's LLR after one, two and four iterations at $E_b/N_0 = 6$ dB over the three-path Rayleigh fading channel of Figure 17.9 using a normalised Doppler frequency of 3.3×10^{-5} and burst-invariant fading.

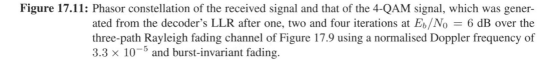

The ability of the TEQ-IQ receiver to mitigate the channel's ISI was studied as a function of the E_b/N_0 loss evaluated for the turbo equalisation scheme after the critical number of iterations with respect to the decoding performance obtained over the AWGN channel, i.e. over the ISI-free channel at a BER of 10^{-3}. In this respect, a loss of 2.1 dB was observed, as evidenced by Figure 17.10(b). In order to justify the associated E_b/N_0 loss, the constellation of the received signal and that of the signal converted from the decoder's a-posteriori LLR seen in Figure 17.5 was plotted for $E_b/N_0 = 6$ dB after one, two and four iterations in Figure 17.11 over the three-path Rayleigh fading channel of Figure 17.9. After one iteration, the converted signals began to cluster around the 4-QAM constellation points. It was observed that additional iterations did not refine the constellation of the converted signals significantly, which still exhibited substantial residual ISI. Therefore, the turbo equaliser cannot completely overcome the severe frequency selectivity of the three-path fading channel, yielding an E_b/N_0 degradation of approximately 2.1 dB when compared to the decoding performance over the non-dispersive Gaussian channel i.e. to the ISI-free channel.

In summary, our findings showed that for 4-QAM transmission over an equally weighted three-path Rayleigh fading channel, the TEQ-IQ can achieve the same performance as the TEQ-EQ after four turbo equalisation iterations. Using Equations 17.17 and 17.18, the complexity of the TEQ-IQ and TEQ-EQ receivers was found to be 176 and 192 transitions per trellis interval, respectively. The TEQ-IQ receiver achieved a complexity reduction factor of 1.1 compared to the TEQ-EQ receiver, whilst achieving an identical BER performance. It was also observed that the TEQ-IQ receiver was unable to further mitigate the channel ISI after the critical number of iterations, i.e. after four turbo equalisation iterations. This was because the signal mapped from the channel decoder's a-posteriori information to the 4-QAM symbols still retained residual ISI at this stage, as shown in Figure 17.11.

17.8.2   16-QAM System

As a further set of results, Figure 17.12(a) displays the performance of the TEQ-EQ receiver employing iterative CIR estimation for 16-QAM transmitted over the equally weighted three-path Rayleigh fading channel of Figure 17.9 using burst-invariant fading and employing a channel interleaving depth of 13824 bits in order to maintain a total system delay of approximately 30 ms. The critical number of iterations was three when employing the 16-QAM TEQ-EQ receiver, since no further significant gain was achieved by performing additional turbo equalisation iterations. In contrast, the critical number of iterations was six when employing the 16-QAM TEQ-IQ receiver over the same channel, as shown in Figure 17.12(b). Comparing the performance obtained by the 16-QAM TEQ-IQ using iterative CIR estimation and the TEQ-CT having perfect CIR information after their critical number of iterations, it was observed in Figure 17.12(b) that both receivers yielded a similar BER performance. The complexity of the 16-QAM TEQ-EQ and TEQ-IQ receivers was estimated as before, by using Equations 17.17 and 17.18, giving 6240 and 512 transitions per trellis interval, respectively. Here, the 16-QAM TEQ-IQ receiver achieved a complexity reduction by a factor of 12.2, relative to the TEQ-EQ receiver, while still obtaining the same performance.

It was also observed in Figure 17.12(b) that the performance of the 16-QAM TEQ-IQ receiver after six iterations at a BER of 10^{-3} was 2 dB from the decoding performance obtained over the ISI-free AWGN channel. Recall that the E_b/N_0 loss of the 4-QAM system was 2.1 dB, hence indicating an improved performance by the TEQ-IQ receiver for 16-QAM. The constellation of the received 16-QAM signal and the signal obtained by converting the decoder's a-posteriori LLRs to symbols was plotted in Figure 17.13 for $E_b/N_0 = 12$ dB after one, two and

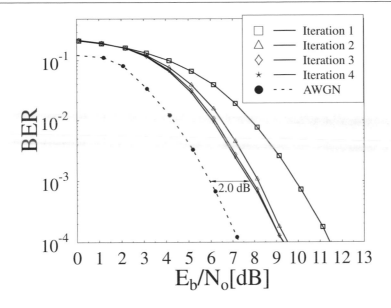

(a) Turbo equaliser using the conventional trellis-based equaliser

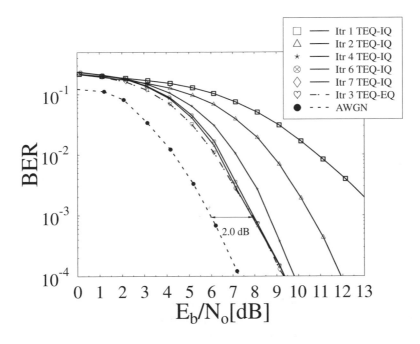

(b) Turbo equaliser using two I/Q-EQs

Figure 17.12: Performance of the turbo equaliser employing two I/Q-EQs and that utilising the conventional trellis-based equaliser for a convolutional-coded 16-QAM system possessing a channel interleaving depth of 13824 bits transmitted over the equally weighted three-path Rayleigh fading channel of Figure 17.9 using a normalised Doppler frequency of 3.3×10^{-5} and burst-invariant fading.

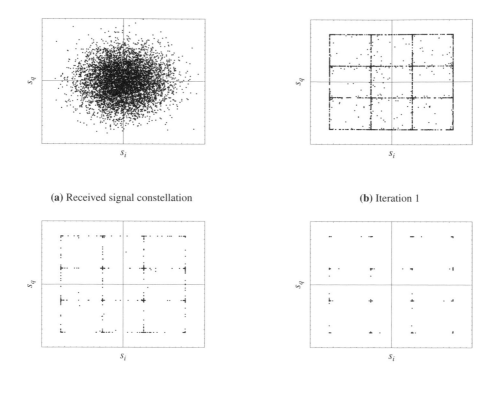

(a) Received signal constellation   (b) Iteration 1

(c) Iteration 2   (d) Iteration 6

Figure 17.13: Phasor constellation of the received signals and that of the 16-QAM signals, which were converted from the channel decoder's LLRs to 16-QAM symbols after one, two and six iterations at $E_b/N_0 = 12$ dB over the three-path Rayleigh fading channel of Figure 17.9 using a normalised Doppler frequency of 3.3×10^{-5} and burst-invariant fading.

six iterations. After the first iteration a crude approximation of the 16-QAM signal constellation was beginning to emerge. Subsequent iterations improved the reliability of the channel decoder's a-posteriori LLRs, hence yielding a more accurate mapping of the 16-QAM signal. Therefore, at the sixth turbo equalisation iteration in Figure 17.13(d) a close approximation of the transmitted 16-QAM signal constellation was achieved and no severe residual ISI persisted. Hence, an improved performance was obtained, compared to the 4-QAM system, which still exhibited residual ISI after four turbo equalisation iterations. Also worth noting is the difference in the channel interleaver depths implemented, i.e. which was 6912 bits for 4-QAM and 13824 for 16-QAM. The larger interleaving depths reduced the correlation between the bits, hence yielding a better performance.

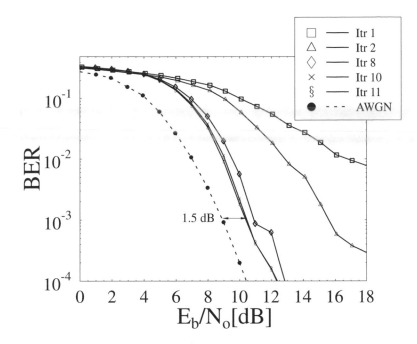

Figure 17.14: Performance of the turbo equaliser employing two I/Q-EQs for a convolutional-coded 64-QAM system possessing a channel interleaving depth of 20736 bits transmitted over the equally weighted three-path Rayleigh fading channel of Figure 17.9 using a normalised Doppler frequency of 3.3×10^{-5} and burst-invariant fading.

17.8.3 64-QAM System

Examining the performance of our 64-QAM system over the same dispersive Rayleigh fading channel in Figure 17.14, it was observed that the critical number of iterations was ten. Simulations could not be conducted for the 64-QAM TEQ-EQ system, since the trellis-based equaliser required $64^2 = 4096$ states and 64 transitions per state. However, in order to compute the complexity of the TEQ-EQ receiver, we assume — based on the 16-QAM performance — that its critical number of iteration is three, giving a complexity of 262240 number of transitions per trellis interval. By contrast, the TEQ-IQ receiver incurred a complexity of 3392 number of transitions per trellis interval, yielding a significant complexity reduction factor of 77.3. After ten turbo equalisation iterations the performance of the TEQ-IQ receiver using iterative CIR estimation at a BER of 10^{-3} was only 1.5 dB from the decoding performance curve over the non-dispersive Gaussian channel, as shown in Figure 17.14. This was an improvement, when compared to the E_b/N_0 loss of 2.1 dB and 2 dB suffered by the 4-QAM and 16-QAM systems, respectively, and can be attributed to the higher interleaving depths employed. As before, the constellation of the received 64-QAM signal and the signal obtained by converting the channel decoder's a-posteriori LLRs to 64-QAM symbols was plotted in Figure 17.15 for $E_b/N_0 = 18$ dB after one, two and ten iterations. After the first iteration the 64-QAM signal constellation was beginning to emerge. Subsequent iterations improved the reliability of the channel decoder's a-posteriori LLR values, hence yielding a more accurate mapping of the 64-QAM symbols. Therefore, by the tenth turbo equalisation iteration in Figure 17.15(d), a close approximation of the 64-QAM signal constel-

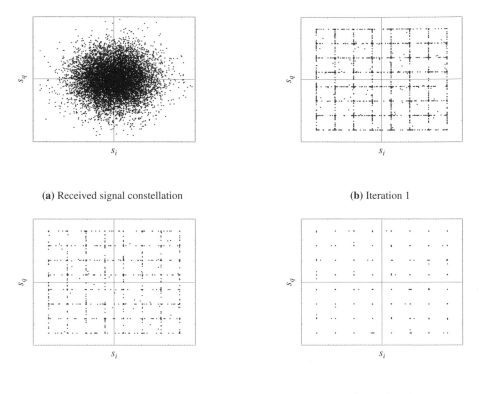

(a) Received signal constellation   (b) Iteration 1

(c) Iteration 2   (d) Iteration 10

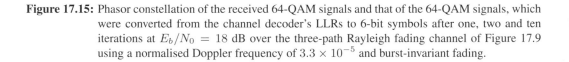

Figure 17.15: Phasor constellation of the received 64-QAM signals and that of the 64-QAM signals, which were converted from the channel decoder's LLRs to 6-bit symbols after one, two and ten iterations at $E_b/N_0 = 18$ dB over the three-path Rayleigh fading channel of Figure 17.9 using a normalised Doppler frequency of 3.3×10^{-5} and burst-invariant fading.

lation was achieved and no severe residual ISI persisted, hence justifying the lower E_b/N_0 loss obtained.

17.9   Summary and Conclusions

In this chapter, a novel reduced-complexity trellis-based turbo equaliser referred to as the TEQ-IQ using iterative CIR estimation was proposed, in order to equalise M-QAM signals and to provide soft decisions. When the M-QAM signal $s(t)$, consisting of quadrature signals $s_I(t)$ and $s_Q(t)$, was transmitted over a complex channel $h(t)$, the complex channel output contained contributions from both $s_I(t)$ and $s_Q(t)$ in each receiver quadrature arm, as expressed in Equation 17.1. Therefore, when a channel with memory τ_d was encountered, the conventional trellis-based equaliser must consider M^{τ_d+1} M-QAM signal sequence combinations at each trellis stage in order to equalise the received signal. Hence, the complexity of the conventional trellis-based equaliser increased exponentially with τ_d. However, by removing the associated cross-coupling of $s_I(t)$ and $s_Q(t)$ and rendering the channel output to be dependent on either $s_I(t)$ or $s_Q(t)$, the number of signal sequence combinations considered was reduced to $\sqrt{M}^{\tau_d+1}$. This was the motivation for the design of the TEQ-IQ. The decoupling operation was performed by using the CIR estimates and the channel decoder's a-posteriori LLRs from the previous iteration, which have been converted into M-QAM symbols. Subsequently, the cross-coupling of the complex received signals was cancelled and the signals were combined accordingly to give the resultant output signals $r'_I(t)$ and $r'_Q(t)$ of Equation 17.15. For perfect cancellation, the signals $r'_I(t)$ and $r'_Q(t)$ were no longer affected by the cross-coupling of $s_I(t)$ and $s_Q(t)$. The signals $r'_I(t)$ and $r'_Q(t)$ were then passed to the I/Q-EQs, which employed SISO algorithms — such as the Log-MAP algorithm of Section 13.8.3 — in order to provide soft decisions for turbo equalisation. The performance of the TEQ-IQ using iterative CIR estimation was compared with the conventional turbo equaliser TEQ-EQ having perfect CIR information for M-QAM transmissions over the equally weighted three-path Rayleigh fading channel of Figure 17.9 using a normalised Doppler frequency of 3.3×10^{-5} and burst-invariant fading. It was observed in Figures 17.10(b) and 17.12(b) that the TEQ-IQ receiver using iterative CIR estimation was capable of achieving the same performance as the more complex TEQ-EQ receiver having perfect CIR information for 4-QAM and 16-QAM, while achieving a corresponding complexity reduction factor of 1.1 and 12.2, respectively. For 64-QAM, the performance of the TEQ-IQ receiver at a BER of 10^{-3} is only 1.5 dB from the decoding performance curve over the non-dispersive Gaussian channel, as was shown in Figure 17.14, while reducing its complexity by a factor of 77.3 times, relative to the TEQ-EQ receiver. Note that although we were unable to simulate the performance of the TEQ-EQ receiver for 64-QAM transmissions over the same dispersive Rayleigh fading channel owing to complexity reasons, we have computed its complexity based on the assumption that its critical number of iterations was three, as in the observed performance of the TEQ-EQ receiver for the 16-QAM system. It was also concluded from the TEQ-IQ performance that despite the poor symbol mapping from low-reliability decoder LLR values in the initial turbo equalisation iterations, the employment of subsequent turbo equalisation iterations improved the symbol mapping, hence yielding a better BER performance.

Following our discourse on the various complexity reduction techiques proposed in this chapter, in the next chapter we will invoke turbo equalisation principles in the context of space-time trellis codes.

18

Turbo Equalisation for Space-time Trellis-coded Systems

18.1 Introduction

Third-generation (3G) systems aim to support full coverage and mobility for a bit rate of at least 144 kbit/s in vehicular scenarios, 384 kbit/s in scenarios providing indoor coverage from outdoors and a rate of 2 Mbit/s in indoor picocellular environments [289]. However, transmission at high data rates, such as 2 Mbit/s, leads to a high grade of channel-induced dispersion and hence the received signal suffers from Inter-Symbol Interference (ISI). Hence, the equaliser design must be capable of mitigating the effects of ISI.

In conjunction with channel equalisation, transmit diversity techniques can also be utilised for mitigating the effects of ISI. Recently, transmission diversity techniques known as Space-time Block Coding (STBC) [71] and Space-time Trellis Coding (STTC) [70] have been advocated for providing additional diversity gains for the Mobile Stations (MSs) by upgrading the Base Stations (BSs). STBC [71] is a diversity technique which utilises orthogonal code designs [72] for facilitating detection with the aid of low-complexity linear processing and simple maximum likelihood decoding. This scheme has been shown to provide considerable performance gains, while incurring a reasonable complexity. However, the employment of STBC is limited to systems which encounter slowly fading narrowband channels. For transmissions over wideband fading channels and fast fading channels, STTC is a more appropriate diversity technique. STTC [70] relies on the joint design of channel coding, modulation, transmit diversity and the optional receiver diversity schemes. The decoding operation is performed by using a maximum likelihood detector. This is an effective scheme, since it combines the benefits of Forward Error Correction (FEC) coding and transmit diversity, in order to obtain performance gains. However, the cost of this is the additional computational complexity, which increases as a function of bandwidth efficiency (bits/s/Hz) and the required diversity order. In this chapter we embark on investigating STTC systems transmitting over wideband fading channels.

Instead of implementing the channel equalisation and STTC decoding operations independently, the performance of the STTC system can be further improved by performing both operations iteratively. This philosophy is based on that of turbo equalisation [105], which was first employed in a convolutional-encoded BPSK system transmitting over dispersive channels. This technique performs channel equalisation and channel decoding iteratively and has been shown to successfully mitigate the effects of channel ISI. Bauch *et al.* [74] adapted the turbo equalisation technique and proposed an iterative equalisation and STTC decoding scheme. This approach yielded an improved performance due to the intelligent exploitation of the soft-decision-based feedback from the STTC decoder to the equaliser's input. However, owing to the high complexity of the turbo equaliser, their investigations were limited to Channel Impulse Responses (CIR) associated with a low dispersion.

Motivated by these trends, we propose a reduced-complexity turbo equaliser, in order to reduce the complexity of the STTC receiver. This is achieved by employing the I/Q signal cancellation technique proposed in [375], in conjunction with turbo equalisation.

The outline of this chapter is as follows. Section 18.2 presents an overview of the investigated system. This is followed by the description of the turbo equalisation principle in the context of our proposed STTC scheme in Section 18.3. Subsequently, the complexity of the proposed system is detailed in Section 18.4. Performance results are presented in Section 18.5 and finally, the main conclusions are summarised in Section 18.6.

18.2   System Overview

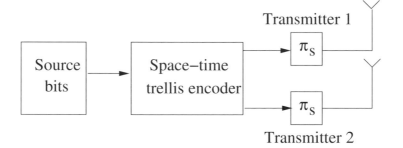

Figure 18.1: Transmitter of the STTC system, where π_s represents the STTC interleaver.

Figure 18.1 shows the transmitter of the STTC system, consisting of a STTC encoder using $T_x = 2$ transmit antennas. The source bits are passed to the STTC encoder, represented using the notation STTC(M,n), where M represents the modulation mode used and n the number of states of the STTC encoder. These codes can be characterised using trellis diagrams. For example, STTC(4,n) codes having 4-PSK constellation points seen in Figure 11.2 can be represented as trellis diagrams illustrated in Figures 11.3, 11.6, 11.7 and 11.8, which correspond to the STTC(4,4), STTC(4,8), STTC(4,16) and STTC(4,32) codes, respectively. Upon receiving an input symbol, the STTC produces an output symbol in each transmitter arm of Figure 18.1. These output symbols are displayed at the right-hand side of each trellis state in Figures 11.3, 11.6, 11.7 and 11.8, which can be treated as the STTC codewords. At the output of the STTC encoder, the encoded symbols are interleaved by a random STTC interleaver represented as π_s in Figure 18.1. A detailed description of the STTC encoder's operation was presented in Section 11.2.

As was illustrated in Figure 16.1, the transmission burst employed consists of 144 data symbols. In our investigations, two-path and five-path, symbol-spaced Rayleigh fading CIRs having equal symbol-spaced tap weights were used. The Rayleigh fading statistics obeyed a normalised

Doppler frequency of 3.3615×10^{-5}, where the fading magnitude and phase were kept constant for the duration of a transmission burst. We referred to this as burst-invariant fading. Furthermore, in order to investigate the lower bound performance of these systems, we have assumed that the CIR was perfectly estimated at the receiver. At the receiver, $R_x = 2$ receive antennas were employed. The received signal at the ith receiver can be written as:

$$
\begin{aligned}
r^i(t) &= \sum_{k=1}^{T_x=2} s^k(t) * h^{ki}(t) + n^i(t) \qquad i = 1 \ldots R_x = 2, \\
&= \sum_{k=1}^{T_x=2} \left[s_I^k(t) * h_I^{ki}(t) - s_Q^k(t) * h_Q^{ki}(t) + j \left(s_Q^k(t) * h_I^{ki}(t) + s_I^k(t) * h_Q^{ki}(t) \right) \right] + n^i(t) \\
&= r_I^i(t) + j r_Q^i(t),
\end{aligned}
\tag{18.1}
$$

where:

$$
\begin{aligned}
r_I^i(t) &= \sum_{k=1}^{T_x=2} \left(s_I^k(t) * h_I^{ki}(t) - s_Q^k(t) * h_Q^{ki}(t) \right) + n_I^i(t) \\
j r_Q^i(t) &= \sum_{k=1}^{T_x=2} j \left(s_Q^k(t) * h_I^{ki}(t) + s_I^k(t) * h_Q^{ki}(t) \right) + j n_I^i(t),
\end{aligned}
\tag{18.2}
$$

and $s^k(t)$, $r^i(t)$ and $n^i(t)$ denote the symbol transmitted from the kth transmitter, the received signal on the ith receive antenna and the AWGN imposed on the signal received by the ith receive antenna, respectively. The notation $h^{ki}(t)$ refers to the CIR corresponding to the kth transmit and ith receive link, whereas $*$ represents convolution.

18.3 Principle of In-phase/Quadrature-phase Turbo Equalisation

The basic principles of In-phase/Quadrature-phase (I/Q) turbo equalisation for single-transmitter and single-receiver systems has been outlined in Sections 17.4 and 17.5.3. Let us refresh the main points discussed in that section. After transmission over a channel exhibiting a complex CIR, the received I/Q signals, namely $r_I(t)$ and $r_Q(t)$, shown in Figure 17.4, become dependent on $s_I(t)$ and $s_Q(t)$. The inter-dependency between $s_I(t)$ and $s_Q(t)$ in the received quadrature signals $r_I(t)$ and $r_Q(t)$ is referred to here as **cross-coupling**. This cross-coupling of the transmitted signal's quadrature components requires the receiver to consider an increased number of signal combinations, hence resulting in a high number of equaliser trellis states. However, we can significantly reduce the number of trellis states to be considered when the cross-coupling is removed such that the quadrature components of the decoupled channel output $r'(t)$ are only dependent on $s_I(t)$ or $s_Q(t)$. As was shown in Equation 18.3, this is achieved by generating the estimates $\hat{s}_I(t)$ and $\hat{s}_Q(t)$ of the transmitted signal [368] using the reliability information produced by the decoder and cancelling the cross-coupling effects of the transmitted quadrature signals from both the in-phase and quadrature-phase received signal components, in order to obtain $r'_I(t)$ and $r'_Q(t)$, respectively. In the ideal scenario, where perfect signal regeneration is possible, the cross-coupling inherent in the received signal can be successfully removed. No doubt there will be errors introduced in the decoupling operation when inaccurate symbol estimates are generated from the channel-impaired low-confidence reliability values. However, as seen in the simulation results of Section 17.8, the imperfect decoupling effects are compensated through successive turbo equalisation iterations and following a number of iterations the performance achieved becomes comparable to that of the turbo equaliser utilising the conventional trellis-based equaliser.

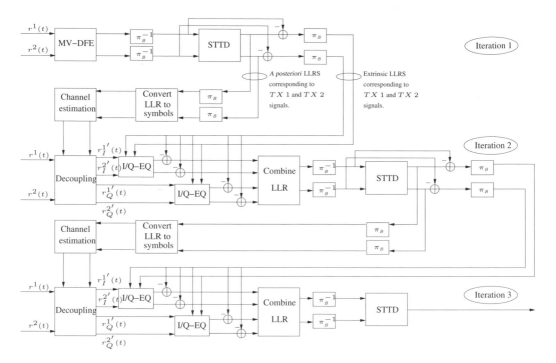

Figure 18.2: Schematic of the reduced complexity turbo equaliser in conjunction with a two-transmitter, 2 receiver based STTC scheme performing three turbo equalization iterations.

In this section, we will demonstrate how this principle has been adapted for the STTC-based system. The structure of the STTC turbo equaliser shown in Figure 18.2 is similar to that of the single-transmitter, single-receiver-based turbo equaliser, which was illustrated in Figure 17.5. However, there are a number of differences and the main differences are that the STTC turbo equaliser utilises a Multivariable Decision Feedback Equalizer (MV-DFE) [314], which simultaneously detects the signals received by all receive antennas, instead of using a conventional DFE. A further difference is that the convolutional decoder is replaced by a STTC decoder. Since the STTC schemes possess multiple transmit antennas, the signal estimates, which are passed iteratively between the component detectors, represent the transmitted symbols of all the transmitters, rather than that of a single transmit-antenna, as it would be the case in the single-transmitter, single-receiver-based turbo equalisation schemes of Figure 17.5. In the first turbo equalisation iteration the MV-DFE equalises the received signal and converts the demodulated symbols into LLR values. These LLRs are de-interleaved using STTC de-interleavers and subsequently passed to the STTC decoder. After processing the received LLR values, the STTC decoder directs the LLRs to the I/Q equaliser in the second turbo equalisation iteration. Note that in the proposed STTC turbo equaliser, we have assumed that perfect CIR knowledge was available, in order to obtain the system's upper bound performance. Apart from these differences, the STTC turbo equaliser operates in the same manner as the I/Q turbo equaliser previously described in Section 17.4.

More explicitly, as described in Section 17.4, the quadrature components of the signal estimates must be generated before the I/Q signal cancellation process can commence. In contrast to the corresponding single-transmitter and single-receiver I/Q turbo equaliser of Figure 17.5, which only generates quadrature estimates corresponding to one transmitter, the proposed reduced complexity STTC turbo equaliser requires the quadrature signal components $s_I^k(t)$ and $s_Q^k(t)$ of all transmitters to be determined. These quadrature signal estimates are obtained by converting the LLR values — generated by the STTC decoder — using Equation 17.8. Subsequently, these estimated symbols are convolved with the in-phase and quadrature-phase CIR estimates of the kth transmit and ith receive link, in order to generate $\hat{s}_I^k(t) * \hat{h}_I^{ki}(t)$, $j\hat{s}_I^k(t) * \hat{h}_Q^{ki}(t)$, $\hat{s}_Q^k(t) * \hat{h}_Q^{ki}(t)$ and $j\hat{s}_Q^k(t) * \hat{h}_I^{ki}(t)$, where $k = 1 \ldots T_x$ and $i = 1 \ldots R_x$ as seen in Figure 18.2. Let us now show how the received signals can be decoupled with the aid of linear processing, such that we obtain the received signals $r_I^{i'}(t)$ and $r_Q^{i'}(t)$, which are only dependent on a particular quadrature component, namely on $s_I^k(t)$ or $s_Q^k(t)$, rather than on both. Commencing with $r_I^{i'}(t)$, we can generate this signal by removing $\hat{s}_Q^k(t) * \hat{h}_Q^{ki}(t)$ and $j\hat{s}_Q^k(t) * \hat{h}_I^{ki}(t)$ from the in-phase component of the received signal $r^i(t)$:

$$
\begin{aligned}
r_I^{i'}(t) &= r^i(t) + \sum_{k=1}^{T_x = 2} \left(\hat{s}_Q^k(t) * \hat{h}_Q^{ki}(t) - j\hat{s}_Q^k(t) * \hat{h}_I^{ki}(t) \right) \\
&= \sum_{k=1}^{T_x = 2} \left(s_I^k(t) * h_I^{ki}(t) + js_I^k(t) * h_Q^{ki}(t) + e(\hat{s}_Q^k(t), \hat{h}_Q^{ki}(t)) + je(\hat{s}_Q^k(t), \hat{h}_I^{ki}(t)) \right) + n^i(t),
\end{aligned}
$$

$$(18.3)$$

where $e(\hat{s}_Q(t), \hat{h}_Q(t))$ and $e(\hat{s}_Q(t), \hat{h}_I(t))$ are the error terms expressed as:

$$
\begin{aligned}
e(\hat{s}_Q^k(t), \hat{h}_Q^{ki}(t)) &= r_I^i(t) + \hat{s}_Q^k(t) * \hat{h}_Q^{ki}(t) \\
je(\hat{s}_Q^k(t), \hat{h}_I^{ki}(t)) &= jr_Q^i(t) - j\hat{s}_Q^k(t) * \hat{h}_I^{ki}(t),
\end{aligned}
$$

$$(18.4)$$

which arise when inaccurate CIR estimates and low-confidence M-QAM symbol estimates are generated. Similarly, $r_Q^{i'}(t)$ is obtained by subtracting $\hat{s}_I^k(t) * \hat{h}_I^{ki}(t)$ and $j\hat{s}_I^k(t) * \hat{h}_Q^{ki}(t)$, where $k = 1 \ldots T_x$ and $i = 1 \ldots R_x$ from $r^i(t)$.

After the decoupling operation has been completed, the resultant signals corresponding to the in-phase signal components of all transmitters are directed to a SISO equaliser. We have utilised the Max-Log-MAP algorithm in our equaliser, since it constitutes a good compromise in terms of the achievable performance and the computational complexity imposed. Upon receiving signals corresponding to the in-phase signal components, the SISO equaliser determines the transition metric associated with the transition from state s to state $\grave{s}$, which can be expressed as:

$$
\Gamma(s, \grave{s}) = \sum_{i=1}^{R_x = 2} \left(r_I^{i'}(t) - \hat{r}_I^{i'} \right)^2 / 2\sigma^2,
$$

$$(18.5)$$

where $\hat{r}_I^{i'}$ is the estimate of the in-phase component of the decoupled received signal $r_I^{i'}$ in Equation 18.3. Subsequently, the forward and backward recursion values are computed, in order to generate the LLR values of the transmitted symbols. These LLR values are de-interleaved by the STTC de-interleaver of Figure 18.2 and passed to the STTC decoder. The STTC decoder computes the a-posteriori LLRs and subsequently the extrinsic LLRs, which are directed back to the I/Q equaliser of Figure 18.2 for processing in the next iteration. This iterative process is repeated until the termination criterion stipulated is satisfied. In our investigations the iterations were curtailed when no significant further performance gains could be obtained through additional iterations.

Having described the I/Q turbo equalisation operation, we now proceed further to analyse the complexity incurred by the proposed scheme. From this stage on, we will refer to the reduced-complexity turbo equaliser as TEQ-IQ and to the conventional full-complexity scheme as TEQ-CT.

18.4   Complexity Analysis

As we argued in Section 17.6, the complexity of the STTC turbo equalisers can be quantified in terms of the number of associated trellis transitions per information bit, for the sake of simplicity. Therefore, the complexity of the equaliser, which is dependent on the number of trellis transitions per coded bit, must be normalised by the overall throughput T_r, which is the product of the number of Bits Per Symbol (BPS) and the code rate R, yielding the number of transitions per information bit.

For the conventional full-complexity trellis-based equaliser CT-EQ, the complexity $comp[\text{CT-EQ}]$ associated with equalising M-QAM signals transmitted over a channel having complex weights and a delay spread of τ_d symbols is:

$$comp[\text{CT-EQ}] = (\text{Number of states} \cdot \text{Number of transitions}) / T_r$$
$$= \left(M^{T_x \tau_d} \cdot M^{T_x} \right) / T_r = \left(M^{T_x(\tau_d+1)} \right) / T_r,$$

where again M, T_x and T_r denote the number of constellation points, the number of transmitters and the overall throughput, respectively. By contrast, the complexity of a single I/Q equaliser (I/Q-EQ) trellis stage is given by:

$$comp[\text{I/Q-EQ}] = (\text{Number of states} \cdot \text{Number of transitions}) / T_r$$
$$= \left(\sqrt{M}^{\,T_x \tau_d} \cdot \sqrt{M}^{\,T_x} \right) / T_r = \left(\sqrt{M}^{\,T_x(\tau_d+1)} \right) / T_r.$$

For the STTC(M, n) decoder, where M and n denote the number of constellation points and the number of states in the encoder, the complexity $comp[\text{STTC}(M, n)]$ incurred, is:

$$comp[\text{STTC}(M, n)] = (\text{Number of states} \cdot \text{Number of transitions}) / T_r$$
$$= (n \cdot M) / T_r.$$

Having determined the complexity of the equaliser and that of the STTC decoder as a function of the number of transitions per information bit, the complexity of the TEQ-CT can be estimated as:

$$comp[\text{TEQ-CT}] = \frac{\left(\left(M^{T_x(\tau_d+1)} \right) + n \cdot M \right)}{T_r} \cdot \text{Itr}[\text{TEQ-CT}], \qquad (18.6)$$

while the complexity of TEQ-IQ is:

$$comp[\text{TEQ-IQ}] = \frac{n \cdot M + \left(2 \cdot \left(\sqrt{M}^{\,T_x(\tau_d+1)} \right) + n \cdot M \right) \cdot (\text{Itr}[\text{TEQ-IQ}] - 1)}{T_r}, \qquad (18.7)$$

where Itr[ ] denotes the number of turbo equalisation iterations. A factor of 2 was introduced in Equation 18.7, since two I/Q-EQs are required for performing the equalisation. In Equation 18.7, the first term on the right hand side represents the complexity incurred in the first TEQ-IQ iteration, where a MV-DFE and a STTC decoder were employed. The remaining terms correspond to the complexity of the subsequent TEQ-IQ iterations. For the sake of simplicity we have assumed that the complexity of the MV-DFE is negligible when compared to the complexity of the I/Q-EQ and CT-EQ. Therefore, the complexity of the TEQ-IQ in the first iteration is only dependent on the complexity of the STTC decoder.

In the next section, we will present our simulation results obtained for the turbo-equalized STTC(4,4), STTC(4,8), STTC(4,16) and STTC(4,32) schemes, which utilize STTC interleavers of varying sizes, namely 144 symbols, 576 symbols, 2304 symbols and 9216 symbols. Subsequently, we will analyse the performance versus complexity relationships of these schemes.

18.5 Results and Discussion

Figures 18.3(a), 18.3(b), 18.3(c) and 18.3(d) illustrate the performance of the STTC(4,4)-based system utilising the TEQ-IQ and the TEQ-CT schemes in conjunction with various interleaver sizes, namely the 144-symbol, 576-symbol, 2304-symbol and 9216-symbol STTC interleavers, respectively. In these figures, we observe that for all interleaver sizes, the critical number of turbo equalization iterations for the TEQ-IQ scheme is three, while the TEQ-CT technique requires two iterations. Note that the term critical number of iterations refers to the number of iterations that have to be performed, such that further iterations beyond this value do not yield significant gains in BER performance terms. After performing the critical number of iterations, the performance of the TEQ-IQ scheme approximates that of the TEQ-CT technique. Typically, in turbo detection schemes the performance of the system improves when the interleaver, which separates the SISO detectors, is long. However, it was observed from these figures that the performance of the turbo-equalised STTC(4,4) systems, which utilised longer interleaver sizes, did not improve. Although longer interleavers are capable of dispersing the bursty channel errors which have been induced by the employment of the sub-optimal MV-DFE, the STTC(4,4) decoder remains unable to correct the resultant random errors. Hence, the different-delay STTC(4,4) schemes have a comparable performance, despite employing longer STTC interleavers.

In Figures 18.4(a), 18.4(b), 18.4(c) and 18.4(d) it is also observed that the critical number of turbo equalisation iterations for the TEQ-CT and TEQ-IQ schemes is two and three, respectively. For the STTC(4,8) system employing 144-symbol or 576-symbol STTC interleavers, the TEQ-IQ is observed in Figures 18.4(a) and 18.4(b) to yield a comparable performance to that of the TEQ-CT up to a BER of 10^{-3}. Below this BER value and especially for SNR values higher than 10 dB, the BER performance curve of the TEQ-IQ tends to an error floor. For longer interleaver sizes, such as the 2304-symbol interleaver and the 9216-symbol interleaver, the TEQ-IQ is capable of attaining a performance which is comparable to that of the TEQ-CT technique for BER values above 10^{-5}. For the STTC(4,16) and the STTC(4,32) schemes characterised in Figures 18.5 and 18.6, a similar trend is observed. As in the STTC(4,8) system employing 144-symbol interleavers and 576-symbol interleavers, the TEQ-IQ of the STTC(4,16) and STTC(4,32) systems is capable of approximating the performance of the TEQ-CT scheme for BER values above 10^{-3}. For lower BER values and for SNR values higher than 9 dB, the BER curve of the TEQ-IQ begins to flatten. The performance of the STTC(4,16) and STTC(4,32) schemes improves when longer STTC interleavers, namely the 2304-symbol interleavers, and the 9216-symbol interleavers are employed

The reason for the performance degradation of the TEQ-IQ schemes using short interleavers

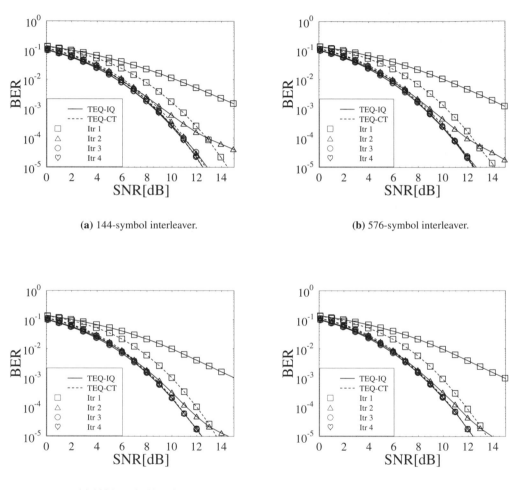

(a) 144-symbol interleaver.

(b) 576-symbol interleaver.

(c) 2304-symbol interleaver.

(d) 9216-symbol interleaver.

Figure 18.3: Performance comparison of the TEQ-CT and TEQ-IQ in conjunction with the **STTC(4,4)** scheme using various STTC interleaver sizes for transmission over a two-path, equal-weight and symbol-spaced Rayleigh fading CIR having a normalised Doppler frequency of 3.3615×10^{-5}.

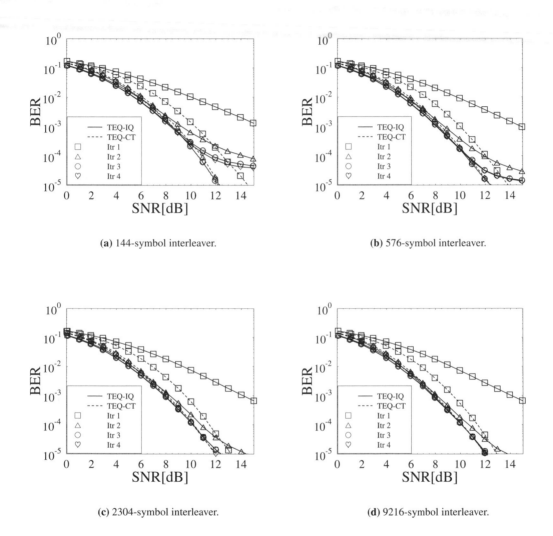

(a) 144-symbol interleaver.

(b) 576-symbol interleaver.

(c) 2304-symbol interleaver.

(d) 9216-symbol interleaver.

Figure 18.4: Performance comparison of the TEQ-CT and TEQ-IQ in conjunction with the **STTC(4,8)** scheme using various STTC interleaver sizes for transmission over a two-path, equal-weight and symbol-spaced Rayleigh fading CIR having a normalised Doppler frequency of 3.3615×10^{-5}.

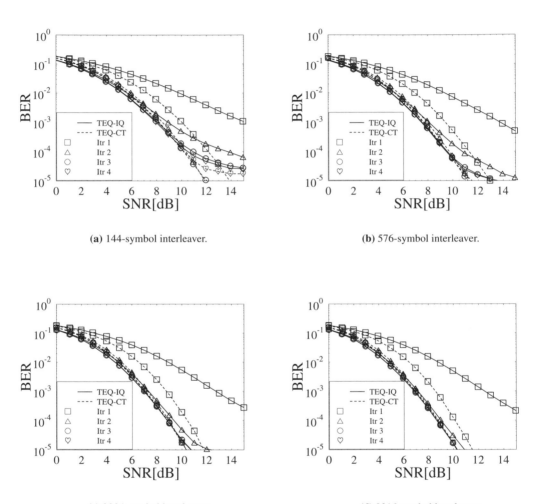

(a) 144-symbol interleaver.

(b) 576-symbol interleaver.

(c) 2304-symbol interleaver.

(d) 9216-symbol interleaver.

Figure 18.5: Performance comparison of the TEQ-CT and TEQ-IQ in conjunction with the **STTC(4,16)** scheme using various STTC interleaver sizes for transmission over a two-path, equal-weight and symbol-spaced Rayleigh fading CIR having a normalised Doppler frequency of 3.3615×10^{-5}.

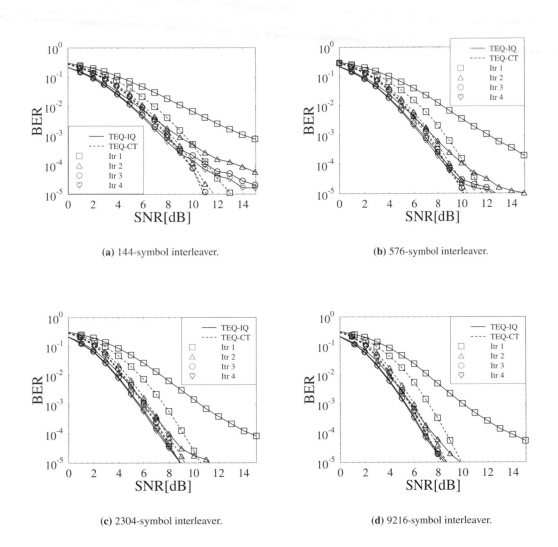

(a) 144-symbol interleaver.

(b) 576-symbol interleaver.

(c) 2304-symbol interleaver.

(d) 9216-symbol interleaver.

Figure 18.6: Performance comparison of the TEQ-CT and TEQ-IQ in conjunction with the **STTC(4,32)** scheme using various STTC interleaver sizes for transmission over a two-path, equal-weight and symbol-spaced Rayleigh fading CIR having a normalised Doppler frequency of 3.3615×10^{-5}.

is the employment of a sub-optimal MV-DFE in the first iteration. As a consequence, bursts of errors are generated when the instantaneous channel quality is poor. For short STTC interleavers, such as the 144-symbol interleaver and the 576-symbol interleaver, the bursty channel errors are insufficiently randomised. Consequently, the STTC decoders are unable to decode the MV-DFE's output information with a sufficiently high confidence, hence yielding low reliability a-priori LLR values, which are passed to the I/Q equalisers in the following turbo equalisation iteration. Since the LLR values are of low reliability, the performance of the I/Q equalisers also degrades owing to the detrimental influence of the a-priori LLRs. Another source of performance degradation is the imperfect decoupling of the quadrature components of the TEQ-IQ. Since the a-posteriori LLR values generated by the STTC decoder have a low confidence, the modulated symbols generated by using these LLR values also possess low reliabilities, hence resulting in higher decoupling errors. The combination of both degradations contributes towards the error floor observed in Figures 18.3(a), 18.3(b), 18.4(a), 18.4(b), 18.5(a), 18.5(b), 18.6(a) and 18.6(b).

In order to support these arguments, we simulated three different scenarios. The first, referred to as Scheme A, is the realistic scenario, where no modifications were made. Second, we devised Scheme B, such that the decoupling operations were performed perfectly, i.e. there were no decoupling errors. Finally, in addition to perfect decoupling Scheme C also suppressed the a-priori LLRs generated by the STTC decoder in the first turbo equalisation iteration, in order to ensure that the performance of the I/Q equalisers was not influenced by the low-reliability a-priori LLRs. Explicitly, at the end of the first turbo equalisation iteration, we replaced the values of these a-priori LLRs with LLR values, which indicated that the probability that a bit 1 was transmitted was equiprobable to that of bit 0. As observed in Figures 18.7 and 18.8, the error floor was lowered when Scheme B was utilised. This was because the TEQ-IQ was not handicapped by the poor decoupling operation. However, the performance curve of the STTC system still exhibited error floors for the STTC(4,32) system using 144-symbol interleavers and 576-symbol interleavers in Figures 18.7(d) and 18.8(d), respectively. Finally, when utilising Scheme C, there was no observable error floor in Figures 18.7(d) and 18.8(d). Therefore, we can conclude that the main cause of the associated degradation was not only the poor decoupling capability of the TEQ-IQ, but also the low-confidence a-priori LLRs generated by the STTC decoder during the first turbo equalisation iterations.

In order to overcome this error floor, longer interleavers such as the 2304-symbol interleaver and the 9216-symbol interleaver must be employed to ensure that the bursty channel errors are dispersed.

18.5.1   Performance versus Complexity Trade-off

Having observed that the performance of the TEQ-IQ becomes comparable to that of TEQ-CT, we now examine the relationship between the performance obtained and the complexity incurred by these schemes. Figure 18.9 illustrates the associated trade-off between the SNR required for achieving a BER of 10^{-3} and the complexity incurred by the different STTC systems employing various interleaver sizes. For each of these curves the complexity estimated in terms of the number of trellis transitions per information bit increases upon increasing the number of turbo equalisation iterations. The number of turbo equalisation iterations performed was four.

In Figure 18.9 it is shown that the TEQ-IQ scheme requires a comparable SNR value to that of the TEQ-CT, in order to achieve a BER of 10^{-3}, while incurring a lower computational complexity. For example, it was observed in Figure 18.9 that when employing the TEQ-IQ technique, which performed three turbo equalisation iterations in conjunction with the STTC(4,4),

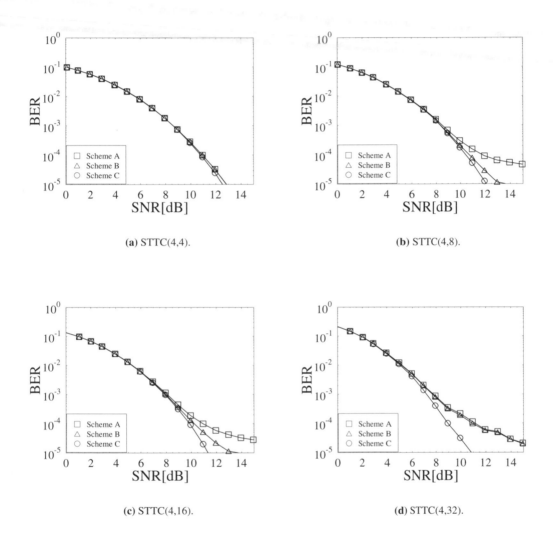

Figure 18.7: Performance of the TEQ-IQ employing a **144-symbol** random interleaver. Scheme A is the realistic scenario, Scheme B represents the system where perfect I/Q decoupling is assumed and
scheme C is the system which performs perfect decoupling and suppresses the low-reliability
a-priori LLRs produced by the STTC decoder during the first turbo equalisation iteration. The
signals were transmitted over a two-path, equal-weight and symbol-spaced Rayleigh fading
CIR having a normalised Doppler frequency of 3.3615×10^{-5}.

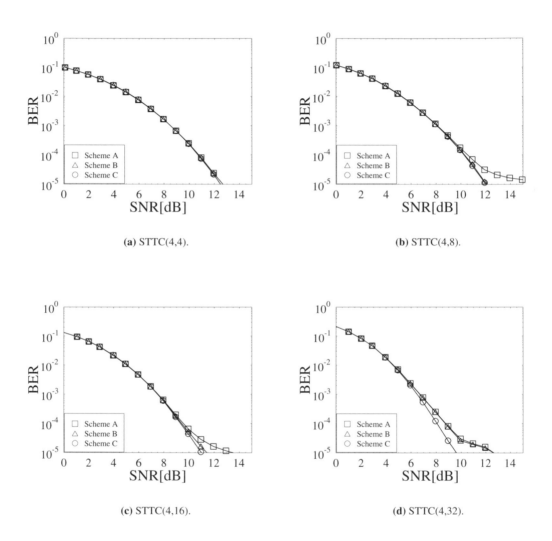

(a) STTC(4,4).

(b) STTC(4,8).

(c) STTC(4,16).

(d) STTC(4,32).

Figure 18.8: Performance of the TEQ-IQ employing a **576-symbol** random interleaver. Scheme A is the realistic scenario, Scheme B represents the system where I/Q perfect decoupling is assumed and Scheme C is the system which performs perfect decoupling and suppresses the low-reliability a-priori LLRs produced by the STTC decoder during the first turbo equalisation iteration. The signals were transmitted over a two-path, equal-weight and symbol-spaced Rayleigh fading CIR having a normalised Doppler frequency of 3.3615×10^{-5}.

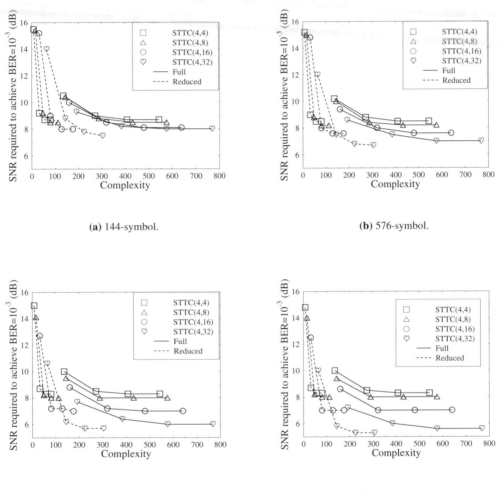

(a) 144-symbol.

(b) 576-symbol.

(c) 2304-symbol.

(d) 9216-symbol.

Figure 18.9: Comparing the complexity of the TEQ-IQ and TEQ-CT assisted STTC(4,4), STTC(4,8), STTC(4,16) and STTC(4,32) systems employing a particular interleaving size for transmission over a two-path, equal-weight and symbol-spaced Rayleigh fading CIR having a normalised Doppler frequency of 3.3615×10^{-5}. In each curve, the complexity increases upon increasing the number of turbo equalisation iterations. A maximum of four turbo equalisation iterations was performed.

STTC(4,8), STTC(4,16) and STTC(4,32) systems, the SNR value required to achieve a BER of 10^{-3} was similar to that of the corresponding TEQ-CT scheme performing two turbo equalisation iterations, while achieving a complexity reduction of a factor of 4.9, 3.6, 2.5 and 1.7, respectively.

For short interleaver sizes, such as the 144-symbol interleaver and the 576-symbol interleaver, the performance gain of the different STTC systems was modest, ranging from 0.1 dB to 0.8 dB, even when more powerful and more complex STTC decoders were employed. For example, in Figure 18.9(a), when comparing the TEQ-IQ performance of the STTC(4,4) system to that of the STTC(4,32) scheme after four turbo equalisation iterations, it was observed that the STTC(4,32) scheme required a slightly lower SNR, namely approximately 0.8 dB lower, in order to achieve the BER of 10^{-3}. This was obtained at the cost of a higher complexity, which was of a factor of 3.8. Similar trends were also observed for the TEQ-CT receiver, where after four turbo equalisation iterations the performance of the STTC(4,32) arrangement was observed to require a 0.8 dB lower SNR than the STTC(4,4), while incurring a factor of 1.4 higher complexity, in order to obtain a BER of 10^{-3}. Hence, for STTC systems employing short interleaver sizes, such as the 144-symbol interleaver and the 576-symbol interleaver, it is sufficient to use the STTC(4,4) code, since no significant performance gain is achieved by employing more complex and powerful STTC codes.

Figure 18.10 shows the trade-off between the SNR required to achieve a BER of 10^{-3} and the complexity incurred in conjunction with a particular STTC system employing various interleaver sizes. As in Figure 18.9, the complexity associated with each performance curve increases upon increasing the number of turbo equalisation iterations. The maximum number of turbo equalisation iterations performed was four. In Figures 18.10(a) and 18.10(b), it can be observed that for the weaker STTC schemes, such as for example the STTC(4,4) and STTC(4,8) arrangements, the SNR required for achieving a BER of 10^{-3} does not vary significantly, despite increasing the interleaver size. However, for the more powerful STTC schemes, such as the STTC(4,16) and STTC(4,32) arrangements, it is observed that the STTC system employing longer interleaver sizes requires lower SNR values for obtaining a BER of 10^{-3}. For example, in Figure 18.10(d), the STTC(4,32) system utilising the TEQ-IQ and using a 9216-symbol interleaver requires an SNR of 5.6 dB for achieving a BER of 10^{-3}, whereas the same system employing the 144-symbol interleaver needs 7.5 dB in order to attain the same BER. Similarly, for the STTC(4,16) and STTC(4,32) systems using TEQ-CT, it is also observed that the systems employing the 9216-symbol interleaver require 1.1 dB to 2.4 dB lower SNRs than the STTC(4,16) and STTC(4,32) systems using a 144-symbol interleaver.

As shown in Figures 18.9 and 18.10, attaining a BER performance of 10^{-3} by the various STTC systems using longer STTC interleavers is possible at a given complexity and at a lower SNR value than that of the systems utilising shorter interleaver sizes. However, this SNR improvement is achieved at the expense of increased system delay, which arises from the delay imposed by filling the interleaver buffer, before the turbo equalisation operations may commence. Before continuing our discussions on the performance versus delay relationship, we will briefly highlight our considerations when determining the system delay. The system delay is mainly determined by the latency introduced by the STTC interleavers, where an entire block of bits must be stored in the interleaver's buffer before their transmission may commence. Here, the processing delay attributed to the STTC encoding, modulation and turbo equalisation operations has been ignored, although practical systems typically have a processing delay which allows them to complete their operations 'just' before they have to commence processing the next incoming information block.

In general, speech systems can tolerate only limited system delays which are less than

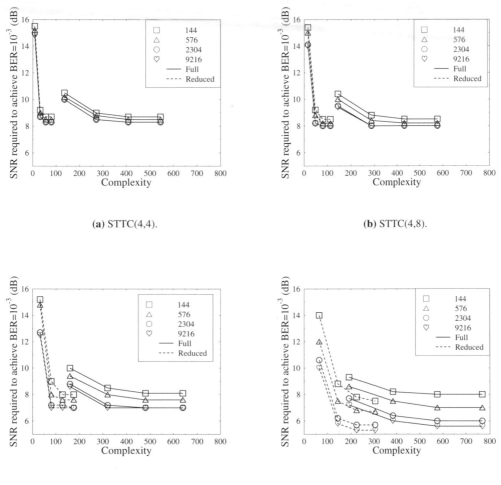

(a) STTC(4,4).   (b) STTC(4,8).

(c) STTC(4,16).   (d) STTC(4,32).

Figure 18.10: Comparing the TEQ-IQ and TEQ-CT complexity for various STTC schemes employing interleaving sizes of 144, 576, 2304 and 9216 symbols. The signals were transmitted over a two-path, equal-weight and symbol-spaced Rayleigh fading CIR having a normalised Doppler frequency of 3.3615×10^{-5}. In each curve, the complexity increases upon increasing the number of turbo equalisation iterations. A maximum of four turbo equalisation iterations was performed.

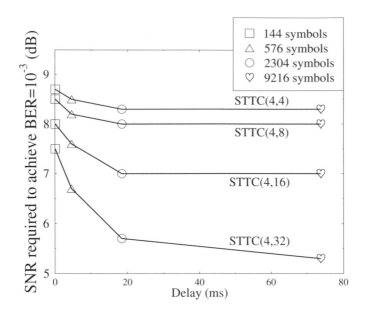

Figure 18.11: Performance of the STTC(4,4), STTC(4,8), STTC(4,16) and STTC(4,32) schemes after four turbo equalisation iterations versus the STTC interleaving delay for the TEQ-IQ arrangement.

40 ms. As an example, let us consider a Time Division Multiple Access/Time Division Duplex (TDMA/TDD) system which employs eight uplink and eight downlink slots, hence one transmission slot will be available for a specific user after every 16 TDMA slots. Since each 72 μs burst of Figure 16.1 consists of 144 data symbols, the system delay is $72\mu s$ for an interleaver size of 144 symbols. In general we have:

$$\text{System delay} = \left[\frac{\text{Interleaver size}}{\text{Data symbols per burst}}\right] \times \text{Burst duration} \times \text{Number of slots per frame}$$

$$= \left[\frac{\text{Interleaver size}}{144}\right] \times 72 \ \mu s \times 16 \qquad \text{if the interleaver size} > 144 \text{ symbols}$$

$$(18.8)$$

corresponding to 4.6 ms, 18.4 ms and 73.7 ms for STTC interleaver sizes of 576, 2304 and 9216 symbols, respectively. With the aid of Equation 18.8, the system delay of the various STTC schemes can be determined, and is illustrated in Figure 18.11 in relation to the SNR value required to achieve a BER of 10^{-3}. In Figure 18.11 it can be observed that upon increasing the interleaver size the STTC systems require lower SNR values, in order to achieve a BER of 10^{-3}. The SNR difference is marginal for the lower-complexity STTC codes, namely for the STTC(4,4) and STTC(4,8) schemes, but becomes more significant for the more powerful STTC codes, such as the STTC(4,16) and STTC(4,32) arrangements. However, the curves reach a plateau for interleaver sizes higher than 2304 symbols. Therefore, there is no substantial advantage in employing a 9216-symbol interleaver, since the achievable SNR reduction is marginal, while the system delay is significantly increased, namely by a factor of four. Hence, we will employ the 2304-symbol

STTC interleaver in our subsequent simulations.

18.5.2   Performance of STTC Systems over Channels with Long Delays

In this section, the performance of the various STTC systems communicating over channels having a CIR associated with long delay spreads, i.e. over a symbol-spaced, five-path, equal tap-weight, Rayleigh fading channel, is investigated. For this CIR, the TEQ-CT scheme cannot be implemented, since the complexity of the trellis-based equaliser alone is already associated with 10^6 transitions per trellis interval. Figure 18.12 shows the performance of the various STTC

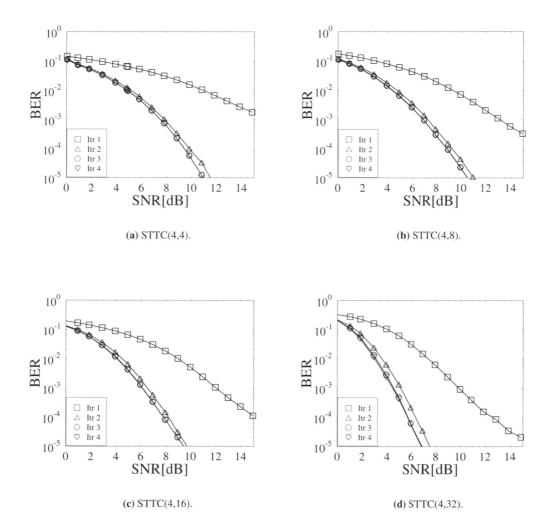

(a) STTC(4,4).   (b) STTC(4,8).

(c) STTC(4,16).   (d) STTC(4,32).

Figure 18.12: Performance of the various STTC systems employing the TEQ-IQ scheme and transmitting over a symbol-spaced, five-path, equal tap-weight, Rayleigh fading channel having a normalised Doppler frequency of 3.3615×10^{-5}.

systems utilising the TEQ-IQ scheme in conjunction with four turbo equalisation iterations. It was observed that the system employing the powerful STTC(4,32) scheme outperforms the other STTC schemes. Specifically, the STTC(4,32) system achieves a performance gain of approximately 3.5 dB, when compared to the STTC(4,4) system at a BER of 10^{-3}. This is achieved at the expense of a slight increase in complexity, corresponding to a factor of 1.08. Although the complexity of the STTC(4,32) decoder is higher than the STTC(4,4) decoder, the increase in overall turbo equaliser complexity due to the employment of the STTC(4,32) scheme is marginal. This is because the complexity of the turbo equaliser is dominated by the complexity incurred by the I/Q equaliser. Hence, it can be concluded from the results of Figure 18.12 that upon employing the TEQ-IQ, the best choice is constituted by the STTC(4,32) decoder, which outperforms the other STTC codes by 1.5 dB to 3.5 dB without significantly increasing the overall complexity of the turbo equaliser. Furthermore, it was demonstrated that the TEQ-IQ was capable of detecting signals transmitted over channels exhibiting a high delay spread, while attaining good BER performances.

18.6 Summary and Conclusions

In this chapter, we have shown that the principle of I/Q turbo equalisation, which was originally proposed for single-transmit and single-receive systems [375], can be adapted for employment in STTC systems, in order to reduce the complexity of the STTC turbo equalisation scheme. It was observed that the TEQ-IQ scheme is capable of achieving comparable performance to that of the TEQ-CT at a reduced complexity.

In Figures 18.3a-b, 18.4 (a–b), 18.5 (a–b) and 18.6 (a–b) it was shown that for STTC systems employing short interleaver sizes, namely the 144-symbol interleaver and the 576-symbol interleaver, the performance of the TEQ-IQ was comparable to that of the TEQ-CT for BERs in excess of 10^{-3}. For lower BERs and especially for SNR values higher than 9dB, an error floor was observed. When utilising longer interleavers, such as the 2304-symbol interleaver and the 9216-symbol interleaver, the TEQ-IQ approximates closely the performance of the TEQ-CT up to a BER of 10^{-5}. The performance degradation of TEQ-IQ schemes employing short STTC interleavers is due to the use of a sub-optimal MV-DFE in the first TEQ-IQ iteration, which generates low-reliability outputs, resulting in bursty channel errors. Short interleavers are unable to disperse these burst errors and therefore the STTC decoder is unable to perform the decoding operations efficiently. Consequently, the STTC decoder generates low-confidence outputs, which result in poor I/Q decoupling and perturb the equaliser's operation.

In this chapter, we have also examined the trade-offs between performance and complexity. In Figure 18.9 it was observed that when employing the TEQ-IQ scheme, which performed three turbo equalisation iterations in the context of the STTC(4,4), STTC(4,8), STTC(4,16) and STTC(4,32) systems, the SNR value required for achieving a BER of 10^{-3} was comparable to that of the TEQ-CT scheme performing two turbo equalisation iterations. Moreover, the TEQ-IQ scheme attained a complexity reduction of a factor of 4.9, 3.6, 2.5 and 1.7, respectively. It was also shown in Figure 18.9 that for short interleavers, namely for the 144-symbol interleaver and for the 576-symbol interleaver, the BER performance improvement obtained by using the more complex STTC decoders, i.e. the STTC(4,16) and STTC(4,32), was marginal. Therefore, for these interleaver lengths it is sufficient to employ the lowest-complexity STTC(4,4) scheme. However, for systems where the maximum system delay is not constrained, long STTC interleavers can be used. In conjunction with long interleavers, the STTC(4,32) scheme is the best choice, since the performance gain becomes more significant. This was further highlighted in Figure 18.10(d), where it was observed that the performance of the STTC(4,32) scheme improved upon increasing the interleaver size.

However, this performance gain was achieved at the expense of longer system delays. From Figure 18.11 it was observed that the performance achieved by the schemes utilising the 9216-symbol interleaver was only marginally better than that of the systems using the 2304-symbol interleaver, despite incurring a significantly longer system delays of 73.7 ms. Therefore, an interleaver associated with a size of approximately 2304 symbols constitutes an attractive compromise between the achievable performance gain and the system delay imposed.

In conclusion, it is observed that the TEQ-IQ constitutes a promising approach which can reduce the complexity of the transmit diversity system without sacrificing the achievable performance. For systems associated with stringent delay constraints, a 144-symbol interleaver can be employed and the best scheme to use in terms of the performance versus complexity trade-off, from the set of schemes studied, is the STTC(4,4) arrangement. For schemes which are not delay-sensitive, an interleaving size of 2304 symbols is sufficient and the recommended scheme is the STTC(4,32).

The systems studied above demonstrated the power of the proposed iterative detection technique. However, the best configuration of the system depends on the propagation environment encountered. Hence, the best performance, delay and complexity trade-off can only be met by a system that is capable of reconfiguring itself in any of the transceiver modes studied in this chapter, which can be implemented with the aid of software-controlled transceivers.

Chapter 19

Summary and Conclusions

19.1 Summary of the Book

In this chapter a brief summary of the contents of the book is provided. Specifically, in Chapter 1 of the book we commenced with a brief historical perspective on channel coding following Shannon's pioneering work [1]. Thanks to the invention of turbo codes by Berrou *et al.* [12,13], numerous researchers have been able to approach the Shannon limit within a fraction of a dB [63]. Hence one could view the past 50 years of research in channel coding as a tour motivated by Shannon's predictions concerning the capacity of a *Gaussian channel* and culminating in the invention of turbo codes, which were capable of attaining a performance asymptotically approaching Shannon's predictions. Recent research is endeavouring to extend this tour into the future by defining the achievable performance limits of narrowband and dispersive wideband *wireless channels* and creating schemes that are capable of approaching these performance limits.

We commence with a brief a summary of Part I of the book. In **Chapter 2** we provided a rudimentary introduction to the concept of convolution coding with the aid of simple worked examples. The philosophy of Viterbi decoding was highlighted and we considered various scenarios, both in conjunction with and without decoding errors.

In **Chapter 3**, after a rudimentary introduction to finite fields we briefly characterised the maximum–minimum–distance family of non-binary RS discussed various decoding algorithms designed for RS and BCH codes. The *Peterson–Gorenstein–Zierler* decoder, which is based on matrix inversion techniques, and the *Berlekamp–Massey–Forney* algorithm were considered. The latter technique constitutes a computationally more effective iterative solution to the determination of the error positions and error magnitudes. The number of multiplications is proportional to t^3 in the case of the matrix inversion method, as opposed to $6t^2$ in the case of the BM algorithm, which means that whenever $t > 6$, the BM algorithm requires less computation than the matrix-inversion-based PGZ decoder. Although for the sake of completeness we have provided a rudimentary introduction to RS and BCH codes, as well as to their algebraic decoding, for more detailed discussions on the topic the interested reader is referred to the literature [31,32,49,85,88,90,96].

In **Chapter 4**, we commenced our discussions with a brief introduction to conventional BCH codes. This was followed by the description of the Viterbi algorithm [8] using the simple BCH(7,4,3) code, following a similar rudimentary approach as in **Chapter 2**. Several examples based on both hard-decision and soft-decision Viterbi decoding were given. Then in

Section 4.3.5, we presented simulation results for both the Viterbi and the BM decoding algorithms. It was found that the performance of the soft-decision Viterbi decoding algorithm is about 2 dB better than that of the corresponding hard decision decoding algorithm. The coding gain of various BCH codes using the same codeword length n was plotted in Figures 4.19 and 4.20. We observed from the figures that the maximum coding gain of the various BCH codes was achieved typically at coding rates of 0.5–0.7. Since a high complexity is incurred by employing the Viterbi decoding of BCH codes, the reduced-complexity Chase algorithm was introduced in Section 4.4.2, which offers a significantly lower complexity. It was found that the Chase algorithm gives effectively the same performance as Viterbi decoding. We note, however, that the complexity of the Chase algorithm increases exponentially with the minimum free distance d_{min} of the code, as well as with the number of test patterns.

Following the introductory Part I of the book, in Part II we focused our attention on the family of turbo codes. In **Chapter 5** we have detailed the operation of turbo codes using convolutional constituent codes and characterised their performance using both BPSK and QPSK modulation constellations for transmissions over both AWGN and Rayleigh channels. As expected, the turbo codes have been shown to perform significantly better than convolutional codes. We have demonstrated the effects of the various decoding algorithms, the constraint length and generator polynomials of the constituent codes, as well as the influence of the transmission frame length on the achievable performance. Furthermore, based on our detailed investigations we have demonstrated the importance of the choice of the interleaver in the context of turbo codes. More explicitly, we have reached the following conclusions regarding the choice of interleavers:

- When block interleavers are used in conjunction with half-rate codes, an odd number of rows and columns should be used.

- For long frame-length systems random interleavers perform better than block interleavers, but for shorter frame-length systems, such as those that might be used for speech transmission, block interleavers perform better.

Finally in Section 5.5 we provided performance results when using turbo codes in conjunction with BPSK and QPSK modulation for transmissions over Rayleigh fading channels.

In **Chapter 6** we have shown that turbo codes can be described by means of a super-trellis, if the trellises of the component scrambler and the interleaver structure are known. For rectangular interleavers one can model this super-trellis as time-invariant, so that it resembles the trellis of a conventional convolutional code. We have also argued that a 'good' conventional convolutional code of the same trellis complexity can be more powerful than a turbo code. On the basis of simulations we have found that iterative decoding is sub-optimal, at least for the investigated simple rectangular interleavers having a low number of columns. We have presented an upper bound for the super-trellis complexity of turbo codes based on rectangular interleavers and an upper bound for the super-trellis complexity averaged over all possible interleavers. The second bound gives rise to the supposition that the complexity of a turbo code having a random interleaver is of the same magnitude as that of a random code.

Based on our introduction to conventional BCH codes in **Chapter 4** and to convolutional constituent code-based turbo codes in **Chapter 5**, in **Chapter 7** we embarked on investigating the family of novel turbo BCH codes. The structure of the turbo BCH encoder and BCH decoder was outlined and the principle of iterative decoding was highlighted, with reference to the turbo convolutional decoding MAP principles outlined in Section 5.3.3. The SOVA [53,54] turbo BCH decoding algorithm was outlined in Section 7.3.2, while a simple example of the iterative decoding cast in the context of the turbo BCH(7,4,3) code was given in Section 7.4. We highlighted how iterative decoding can improve the reliability of the decoded bits and eventually correct an increased number of errors. In Section 7.5 the MAP algorithm was modified in order to incorporate the parity check bit of extended BCH codes for enhancing the decoder's performance. This was vital in the iterative decoding of the family of extended turbo BCH codes.

Various parameters may affect the performance of turbo BCH codes and these effects were investigated in Section 7.6. We first studied the effect of iterative decoding on the performance of turbo BCH codes. It was found that the performance of turbo BCH codes does not improve significantly after four iterations. Various decoding algorithms were compared and it was found that the Log-MAP and Max-Log-MAP algorithms give similar performance. However, the Log-MAP algorithm performs badly if the estimate of the channel reliability value L_c is imperfect. Puncturing was found to degrade the performance of turbo BCH codes. Hence we concluded that no puncturing should be applied to turbo BCH codes. The performance of turbo BCH codes employing various interleaver lengths was investigated as well. It was found that the best possible performance can be achieved by employing an interleaver depth of 5000 bits over AWGN channels. In Section 7.6.6 a new interleaver design was proposed. This interleaver was referred to as the random-in-column block interleaver and it was shown to outperform other conventional block and random interleavers, when the interleaver dimension is $k \times k$. Furthermore, turbo BCH codes employing different BCH component codes were investigated and it was found that the turbo BCH(63,51,5) code performs within about 0.8 dB of the Shannon limit. The proposed modified MAP algorithm [376] was then employed, invoking extended BCH codes, and it was shown in Section 7.6.10 that the extended turbo BCH(32,26,4) code outperforms the turbo BCH(31,26,3) code by approximately 0.5 dB at a BER of 10^{-5}.

In **Chapter 8** we extended our research to the novel class of non-binary block codes referred to as Redundant Residue Number System (RRNS) codes. An RRNS code is a maximum-minimum distance block code, exhibiting identical distance properties to RS codes. In order to understand the basic principles of RRNS codes, we commenced the chapter with a brief introduction to the ancient theory of the Residue Number System (RNS). This theory provide the basis for a promising way of supporting fast arithmetic operations [40, 41] such as addition, subtraction and multiplication. After a brief introduction to the RNS, we showed in Section 8.2.6 that a RRNS can be obtained by incorporating extra moduli into the existing RNS. Then the coding theory of the RRNS codes was developed. We showed that the minimum free distance d_{min} of the RRNS code could be determined using Equation 8.39. The procedure for multiple-error correction was laid out in Section 8.4. In Section 8.5, we introduced different bit-to-residue mapping methods resulting in systematic and non-systematic RRNS codes. Furthermore, we proposed a novel systematic mapping method, which results in higher code rates for systematic RRNS codes than the conventional mapping. Then we modified the Chase algorithm so that the RRNS decoder became capable of processing soft inputs and providing soft outputs. This facilitated the iterative decoding of turbo RRNS codes. Our simulation results for the proposed RRNS codes using BPSK over AWGN channels were then given in Section 8.9. It was found that systematic RRNS codes outperform their non-systematic counterparts. The proposed mapping method which results in higher code rates for systematic RRNS codes gave a similar performance to that of the corresponding conventional systematic RRNS codes. Besides hard-decision decoding, soft-decision decoding of the proposed RRNS codes was also employed in order to improve the performance of the codes. It was found that the soft decoding performance of RS codes is similar to that of the RRNS codes. Performance results were also provided for turbo RRNS codes. We began with the performance comparison between the SISO Chase algorithm and other trellis decoding algorithms, using the turbo BCH(63,57) code as an example. It was shown in Figure 8.18 that the performance of the SISO Chase algorithm suffers only a small performance degradation compared to the Log-MAP algorithm at a significantly reduced complexity. Then, similarly to

Chapter 7, the effects of various parameters affecting the performance of turbo RRNS codes were investigated. We concluded that the best compromise in terms of the number of decoding iterations was four. In **Chapter 8**, we concluded that the performance of RRNS codes was similar to that of RS codes, while offering some advantages over RS codes. Specifically, short non-binary block codes can be readily designed using RRNS codes without having to shorten long codes, which is not an option in RS codes. Besides, in ARQ systems stronger RRNS codes can be obtained by transmitting more redundant residues without having to retransmit the whole codeword. In [198, 205, 207], we exploited this principle in designing adaptive-rate RRNS codes for near-instantaneously adaptive OFDM transmission over mobile communication channels.

In **Chapter 9**, which constitutes Part III of the book, the family of coded modulation schemes was studied in conjunction with adaptive modulation. More specifically, in Sections 9.2–9.6 we have studied the conceptual differences between four coded modulation schemes in terms of their coding structure, signal labelling philosophy, interleaver type and decoding philosophy. The symbol-based non-binary MAP algorithm was also highlighted, when operating in the logarithmic domain. Furthermore, in Section 9.7 the performance of the above-mentioned four coded modulation schemes was evaluated for transmissions over narrowband channels in Section 9.7.2 and over wideband channels in Section 9.7.3. Over the dispersive Rayleigh fading channels a DFE was utilised for supporting the operation of the coded modulation schemes, as investigated in Section 9.7.3.3. TCM was found to perform better than BICM in AWGN channels owing to its higher Euclidean distance, while BICM faired better in fading channels since it was designed for fading channels, with a higher grade of time diversity in mind. BICM-ID gave better performance both in AWGN and fading channels compared to TCM and BICM, although it exhibited a higher complexity due to employing several decoding iterations, while TTCM struck the best balance between performance and complexity. The performance of the BbB DFE-assisted adaptive coded modulation scheme was investigated in Section 9.7.3.3, which attained an improved performance in comparison to the fixed-mode-based coded modulation scheme in terms of both its BER and BPS performance. OFDM was also invoked for assisting the operation of coded modulation schemes in highly dispersive propagation environment as investigated in Section 9.7.3.5 in a multi-carrier transmission scenario. Again, TTCM/OFDM struck the most attractive trade-off in terms of its performance versus complexity balance in these investigations, closely followed by BICM-ID/OFDM.

In the next two chapters, namely in **Chapters 10** and **11**, we expanded our research into a new area. The design of the channel codes was no longer considered to be independent from the modulation and transmit diversity scheme used. Instead, channel coding, modulation and transmit diversity were jointly designed, resulting in space-time codes. We studied two distinct types of space-time codes: space-time block codes [71–73] and space-time trellis codes [70, 80, 278–281]. Despite the slight associated performance penalty of space-time block codes in comparison to space-time trellis codes, space-time block codes are appealing in terms of their simplicity and performance. Utilising the associated low complexity offered by the space-time block code G_2, we concatenated it with turbo convolutional codes, in order to improve its performance. The performance of the concatenated scheme was then compared to that of space-time trellis codes. Specifically, in **Chapter 10** we investigated space-time block codes, whereas space-time trellis codes were detailed in **Chapter 11**.

We commenced our elaborations in **Chapter 10** with a rudimentary introduction to Maximum Ratio Combining MRC [71], which constituted the basis of our further studies into space-time block codes. This was followed by an introduction to Alamouti's simple space-time block code G_2. Examples were given for the space-time block code G_2 using one and two receivers, followed by various other space-time block codes. In Section 10.4 we proposed a system which consists of the concatenation of the above-mentioned space-time block codes and a range of different channel codes. The channel coding schemes investigated were convolutional codes, turbo convolutional codes, turbo BCH codes, trellis-coded modulation and turbo trellis-coded modulation. The estimated complexity and memory requirements of the channel decoders were summarised in Section 10.4.3. Simulation results characterising the proposed concatenated system were presented in Section 10.5. We first compared the performance of the space-time codes G_2, G_3, G_4, H_3 and H_4 without using channel codecs. It was found that as we increased the effective throughput of the system, the performance of the half-rate space-time codes G_3 and G_4 degraded in comparison to that of the unity-rate space-time code G_2. This was because in order to maintain the same effective throughput, higher-order modulation schemes had to be employed in conjunction with the half-rate space-time codes G_3 and G_4, which were more prone to errors and hence degraded the performance of the system. On the other hand, for the sake of maintaining the same diversity gain and same effective throughput, we found that the performance of the space-time codes H_3 and H_4 was better than that of the space-time codes G_3 and G_4, respectively. Since the space-time code G_2 has a code rate of unity, we were able to concatenate it with half-rate TC codes, while maintaining the same effective throughput as that of the half-rate space-time code using no channel coding. We found that for the same effective throughput, the unity-rate G_2 space-time-coded and half-rate TC-coded scheme provided substantial performance improvement over other space-time block codes. We concluded that the reduction in coding rate was best invested in turbo channel codes, rather than in space-time block codes.

In the second category of our investigations of channel-coded space-time systems, we studied the effect of the binary channel codes' data and parity bits being mapped into different protection classes of multilevel modulation schemes. It was found that TC codes having different constraint lengths K require different mapping methods, as evidenced by Figure 10.12. By contrast, in the turbo BCH codes studied mapping of the parity bits to the higher-integrity protection class of a multilevel modulation scheme yielded a better performance. The so-called random separation-based interleaver as shown in Figure 10.14 was proposed, in order to improve the performance of the system. The third set of results compared the performance of all proposed channel codes in conjunction with the space-time code G_2. It was then found that the performance of the half-rate TC codes was better than that of the CC, TBCH, TCM and TTCM codes. The chapter was then concluded by comparing the G_2 space-time-coded channel codes upon taking their complexity into consideration. In Figures 10.29 and 10.30, we can clearly see that the half-rate TC codes give the best coding gain at a moderate complexity.

Our discussions in **Chapter 11** evolved further and we explored the power of various space-time trellis codes, where the performance of the space-time trellis codes was compared to that of the TC-coded space-time block code of **Chapter 10**. Space-time trellis codes were introduced in Section 11.2 by utilising the simplest possible 4-state, 4PSK space-time trellis code as an example. The state diagrams for other 4PSK and 8PSK space-time trellis codes were also provided. The branch metric of each trellis transition was derived, in order to facilitate maximum likelihood decoding. In Section 11.3, we proposed to employ an OFDM modem for mitigating the effects of dispersive multi-path channels. This technique was invoked owing to its simplicity compared to other approaches. TC codes and RS codes were invoked in Section 11.3.1 for concatenation with the space-time block code G_2 and the various space-time trellis codes, respectively. The estimated complexity of the various space-time trellis codes was derived in Section 11.3.3. We then presented our simulation results for the proposed schemes in Section 11.4. The first scheme studied was the TC(2,1,3)-coded space-time block code G_2, whereas the second one was based on the family of space-time trellis codes. It was found that the FER and BER performance of the TC(2,1,3)-coded space-time block G_2 was better than that of the investigated space-time trellis codes at a throughput of 2 and 3 BPS over the channel exhibiting two equal-power rays separated by a delay spread of 5 μs and having a maximum Doppler frequency of 200 Hz. Our comparison between the two schemes was performed by also considering the estimated complexity of both schemes. It was found that the concatenated G_2/TC(2,1,3) scheme still outperformed the space-time trellis codes using no channel coding, even though both schemes exhibited a similar complexity. The effect of the maximum Doppler frequency on both schemes was also investigated in Section 11.4.3. It was found that the maximum Doppler frequency had no significant impact on the performance of either scheme. By contrast, in Section 11.4.4, it was found that the performance of the concatenated G_2 TC(2,1,3) scheme degrades when the Doppler frequency is increased. Then, the Doppler-dependent channel-induced interference of the G_2 space-time-coded system was quantified in terms of the SIR. It was found that the SIR decreases as the delay spread increases and this phenomenon degrades the performance of the concatenated scheme. We proposed in Section 11.4.5 an alternative mapping of the two transmission instants of the space-time block code G_2 to the same subcarrier of two consecutive OFDM symbols, a solution which was applicable to a delay non-sensitive system. By employing this approach, the performance of the concatenated scheme was no longer limited by the delay spread, but rather by the maximum Doppler frequency. We concluded that a certain minimum SIR has to be maintained for attaining the best possible performance of the concatenated scheme.

Space-time block-coded adaptive OFDM was proposed in Section 11.5. It was shown in Figure 11.40 that the space-time block code G_2 using only one AOFDM receiver outperformed the conventional single-transmitter, single-receiver AOFDM system designed for a data transmission target BER of 10^{-4} over a WATM channel. We also confirmed that upon increasing the diversity order, the fading channels become AWGN-like channels. In Section 11.5.3.2, we continued our investigations into AOFDM by concatenating turbo coding with the system. Two schemes were proposed: half-rate turbo- and space-time-coded AOFDM as well as variable-rate turbo- and space-time-coded AOFDM. Despite the impressive BER performance of the half-rate turbo- and space-time-coded scheme, the maximum throughput of the system was limited to 3 BPS. However, by employing the proposed variable-rate turbo- and space-time-coded scheme, the BPS performance improved, achieving a maximum throughput of 5.4 BPS. This improvement in BPS performance terms was achieved at the cost of a poorer BER performance.

The discussions of **Chapters 10 and 11** were centred around the topic of employing multiple-transmitter, multiple-receiver (MIMO) based transmit and receive-diversity assisted space-time coding schemes. These arrangements have the potential of significantly mitigating the hostile fading wireless channel's near-instantaneous channel quality fluctuations. Hence these space-time codecs can be advantageously combined with powerful channel codecs originally designed for Gaussian channels. As a lower-complexity design alternative, **Chapter 12** introduced the concept of near-instantaneously Adaptive Quadrature Amplitude Modulation (AQAM), combined with near-instantaneously adaptive turbo channel coding. These adaptive schemes are capable of mitigating the wireless channel's quality fluctuations by near-instantaneously adapting both the modulation mode used, as well as the coding rate of the channel codec invoked. The design and performance study of these novel schemes constituted the topic of **Chapter 12**.

More specifically, in **Chapter 12** we comparatively studied the performance of space-time trellis codes and AQAM schemes. When multiple transmitters and receivers are employed, a MIMO equaliser has to be used for equalising the multiple transmitted signals. On the other hand, a SISO equaliser was employed for the AQAM scheme, which used a single transmitter and single receiver. We then proposed to concatenate powerful turbo codes with the AQAM system. The simulation results were presented in Section 12.4. In Section 12.4.1 we investigated the performance of the system using fixed modulation schemes. The performance of both TC and TBCH codes was investigated and compared using both short and long channel interleavers. The length of the short channel interleaver was chosen such that burst-by-burst turbo decoding was used. The performance of TC codes in conjunction with various coding rates was presented and we found that the performance of TC codes improved as the coding rate was reduced. Then in Section 12.4.2 we characterised the performance of 4PSK, 8PSK and 16PSK space-time trellis codes.

In order to conclude the investigations of **Chapter 12**, finally, in Section 12.4.3 we compiled all the results and compared them to the performance of the proposed turbo-coded AQAM system. Initially, we showed that the uncoded AQAM system outperforms the space-time trellis codes using two receivers. We then showed that uncoded AQAM also outperforms burst-by-burst TC-coded fixed modulation. However, in conjunction with a large channel interleaver the TC-coded fixed modulation scheme outperforms the uncoded AQAM scheme for channel SNRs below 15 dB at the cost of a high delay. The turbo-coded AQAM system was then proposed and we jointly optimised the AQAM switching thresholds and the coding rates for the turbo-coded AQAM system. The performance of the ORTC AQAM scheme using a short channel interleaver was found to be similar to that of the TC-coded long-delay fixed modulation scheme. Finally, at the expense of a high channel interleaver delay, we were able to further improve the performance of the ORTC AQAM scheme. It was found that its performance was 2 dB inferior in comparison to the space-time trellis codes using three receivers. The performance of the ORTBCH AQAM scheme was then compared to that of the ORTC AQAM scheme. It was found that there was no significant improvement over the ORTC AQAM scheme, despite the extra complexity incurred by the system. In conclusion, ORTC AQAM can be viewed as a lower-complexity measure of mitigating the channel quality fluctuations of wireless channels in comparison to multiple transmitter and multiple receiver based space-time codes. If the complexity of the latter schemes is affordable, the performance advantages of AQAM erode.

In Part IV we focused our attention on interative joint channel equalisation and channel decoding schemes. In **Chapter 13** we briefly characterised the family partial-response modulation schemes, since their employment typically requires a channel equaliser for removing the effects of the intentionally introduced Controlled Inter-Symbol Interference (CISI). The Viterbi equalisation principles were augmented with the aid of worked examples and combined with turbo coding/decoding. For the GSM data channel, the required channel SNR of turbo codes was approximately 1.5 dB lower, than that of the standard convolutional code at a BER of 10^{-4} for both of these channels. By increasing the coding frame length further to the order of 50000 bits, we obtained a gain of approximately 3.7 dB as compared to the standard convolutional code over the so-called perfectly interleaved narrowband Rayleigh fading channel, as illustrated in Figure 13.27. This confirmed the superiority of turbo codes over convolutional codes when the coding frame length was sufficiently long.

In **Chapter 14** the advantage gained by performing the equalisation and decoding jointly was examined in the context of GMSK-modulated systems, as compared to implementing the equalisation independently from the decoding, as in Section 13.8.1. This joint equalisation and decoding technique is known as turbo equalisation [105] and is performed iteratively. We then quantified the advantage gained by turbo equalisation over independent equalisation and decoding. Gains of 3.7 dB and 4.2 dB were achieved for the rate $R = \frac{1}{2}$ convolutional-coded GMSK system transmitting over the non-dispersive Gaussian channel and the five-path Rayleigh fading channel, respectively. Similarly, for the rate $R = \frac{1}{2}$ convolutional-coding-based turbo-coded scheme, the corresponding gains achieved through turbo equalisation over independent equalisation and turbo decoding were 0.5 dB and 0.8 dB, over the non-dispersive Gaussian channel and the dispersive Rayleigh fading channel considered. Finally, the BCH-coding-based turbo-coded systems employing turbo equalisation and transmitting over the non-dispersive Gaussian and the five-path fading channel also demonstrated gains of 2.8 dB and 3 dB, respectively. Second, it was observed that the rate $R = 0.5$ convolutional-coded GMSK system employing turbo equalisation obtained a better performance than that of the $R = 0.5$ convolutional-coding-based turbo-coded GMSK scheme. For transmission over the non-dispersive Gaussian channel the convolutional-coded GMSK system outperformed the convolutional-coding-based turbo-coded GMSK scheme by 0.8 dB at a BER of 10^{-4}, while in the five-path Rayleigh fading scenario the convolutional-coded system attained a gain of approximately 1.0 dB over the convolutional-coding-based turbo-coded scheme, as illustrated in Figures 14.20(a) and 14.20(b), respectively. These results were surprising, since the turbo-coded system — which consists of two convolutional encoders — was expected to perform better, as it was a more powerful coding system compared to convolutional codes. However, from Figures 14.20(a) and 14.20(b), which compared the turbo equalisation performance of the rate $R = 0.5$, $K = 5$ conventional convolutional-coded GMSK system and $R = 0.5$ convolutional-coding-based turbo-coded GMSK scheme over both the non-dispersive Gaussian channel and five-path Rayleigh fading channel, it was observed that at E_b/N_0 values less than 4.5 dB and 6 dB, respectively, the BER of the convolutional-coded system was lower than that of the turbo-coded scheme after one turbo equalisation iteration. This indicated that the decoder in the $R = 0.5$, $K = 5$ convolutional-coded GMSK system was providing more reliable LLR values for the equaliser. Consequently, the equaliser in the convolutional-coded scheme — upon receiving the higher confidence LLR values from the decoder — was capable of producing more reliable LLR values, which were subsequently passed to the decoder in the following turbo equalisation iteration. Beyond the E_b/N_0 values of 4.5 dB and 6 dB, the turbo-coded scheme exhibited a better performance after one turbo equalisation iteration. Therefore, the turbo-coded system was expected to obtain a lower BER. However, since speech and data systems have a target BER of 10^{-3} and 10^{-4}, respectively, convolutional codes constitute a better choice for coded-GMSK systems employing turbo equalisation. A similar phenomenon was also observed in Figure 14.21 in conjunction with the weaker BCH-coding-based turbo codes, which also yielded a better performance than the convolutional-coding-based turbo codes, when turbo equalisation was performed. Here, the same trend was observed as before, where the performance of the BCH-coding-based turbo-coded system was better than that of the convolutional-coding-based turbo-coded scheme after one turbo equalisation iteration. This occurred at E_b/N_0 values less than 4.0 dB and 5.5 dB for transmissions over the non-dispersive Gaussian channel and over the five-path Rayleigh fading channel, respectively. Therefore, the equaliser in the BCH-coding-based turbo-coded system received more reliable LLR information compared to the convolutional-coding-based turbo-coded scheme, hence enabling it to achieve a better BER performance after implementing successive turbo equalisation iterations.

Chapter 15 described the fundamental principles required for deriving the ML union bound performance for non-punctured convolutional-coded and convolutional-coding-based turbo-coded DPSK systems. The aim was to characterise the relative performance bounds of coded GMSK systems and to justify our observation in **Chapter 14** that the convolutional-coded GMSK scheme achieved better turbo equalisation performance than the convolutional-coding-based turbo-coded GMSK system. The ML performance analysis was based on modelling the modulator as an inner encoder and viewing the entire system as a serial concatenated coded scheme. Subsequently, the union bound principles of SCCCs [104] was employed for the convolutional-coded system, whereas the analysis for PCCCs [59] was used in conjunction with the above SCCC principles for turbo-coded schemes. However, since GMSK modulation could not be readily modelled as a binary inner encoder, it was substituted by DPSK, which is also inherently recursive. Our simulation results in Figure 15.11(a) demonstrated that the rate $R = \frac{1}{3}$

convolutional-coded scheme outperformed the rate $R = \frac{1}{3}$ turbo-coded system by a margin of 0.5 dB at a BER of 10^{-4}. The trends of our simulation results were also corroborated by the theoretical performance bound in Figure 15.11(b). As the E_b/N_0 value increased beyond 3 dB, the theoretical bound for the convolutional-coded system began to flatten at approximately a BER of 10^{-8} to a BER of 10^{-12}, whereas the error floor for the turbo-coded system was approximately at a BER of 10^{-10} to a BER of 10^{-14}. It was concluded that the turbo-coded DPSK system exhibited a lower error floor than the convolutional-coded system, but at low E_b/N_0 values, the turbo-coded system yielded poorer performance than the convolutional-coded DPSK system. These observations were also consistent with the results obtained for convolutional-coded GMSK schemes and turbo-coded GMSK systems in **Chapter 14**. Therefore, for speech and data systems — requiring a BER of 10^{-3} and a BER of 10^{-4}, respectively — employing recursive modulation techniques, such as DPSK or GMSK, and utilising turbo equalisation, convolutional coding is the more robust choice, compared to the more complex turbo coding schemes.

Having studied the turbo equalisation performance of various coded systems employing recursive modulators, **Chapter 16** investigated the different receiver configurations of BPSK modulated systems using BCH-coding-based turbo codes, convolutional-coding-based turbo codes, and convolutional codes, denoted as **BT**, **CT** and **CC**, respectively. Non-iterative and iterative equaliser/decoders operating at code rates of $R \approx \frac{1}{2}, R \approx \frac{3}{4}$ and $R \approx \frac{5}{6}$ were studied. In the iterative schemes, loops containing decoders only, as in turbo decoding, and loops containing joint equalisation and decoding stages — i.e. employing turbo equalisation — were investigated. The comparative study of the turbo equalisers for the above **BT**, **CT** and **CC** systems demonstrated that at high code rates of $R \approx \frac{3}{4}$ and $R \approx \frac{5}{6}$ the **BT** system was marginally better than or comparable to the **CT** system at a BER of 10^{-4}, at the expense of higher complexity compared to the **CT** system. At these high code rates and over the equally weighted symbol-spaced five-path Rayleigh fading channel of Figure 16.2(b), the **CC** system performed poorly since its turbo equalisation iteration gain was low. This was because after the first iteration the system's performance was already close to the decoding results over the non-dispersive Gaussian channel. At a code rate $R = \frac{5}{6}$ we observed a loss of $E_b/N_0 = 1.0$ dB over the five-path Gaussian channel and a loss of $E_b/N_0 = 3.8$ dB over the five-path Rayleigh fading channel at a BER of 10^{-4} after eight turbo equalisation iterations when compared to the **BT** system. On the whole, the turbo-equalised **CT** system was the most robust scheme, giving comparable performance — within a few tenths of a dB — for all code rates investigated, when compared to the best system in each scenario. Furthermore, the turbo-equalised **CT** system had a lower receiver complexity, when compared to the **BT** system, hence making it the best choice in most applications.

Having introduced a family of powerful, but implementationally complex, iterative joint channel equalisation and channel decoding schemes, we turned our attention towards reducing the implementational complexity imposed. Hence in **Chapter 17** a novel reduced-complexity trellis-based equaliser-referred to as the I/Q-EQ, was proposed for the turbo equalisation of M-QAM signals. When the M-QAM signals consisting of quadrature components were transmitted over a complex channel, the complex channel output contained contributions from each quadrature arm, as expressed in Equation 17.1. Therefore, when a channel with memory τ_d was encountered, the trellis-based equaliser must consider M^{τ_d+1} M-QAM signal sequence combinations, in order to equalise the received signal. Hence, the complexity of the complex-valued trellis-based equaliser increases rapidly with τ_d. However, by removing the associated cross-coupling of the in-phase and quadrature-phase signal components and hence rendering the channel output to be only dependent on either quadrature component, the number of signal–sequence combinations considered was reduced to $\sqrt{M}^{\tau_d+1}$. This was our motivation for the design of the I/Q-EQ. The performance of the turbo equaliser using two I/Q-EQs — denoted as TEQ-IQ — was compared with that of the complex-valued turbo equaliser TEQ-EQ for M-QAM transmissions over the equally weighted three-path Rayleigh fading channel using a normalised Doppler frequency of 3.3×10^{-5}, where fading was assumed to be invariant throughout the transmission burst. It was observed in Figures 17.10(b) and 17.12(b) that the TEQ-IQ receiver was capable of achieving the same performance as the TEQ-EQ receiver for 4-QAM and 16-QAM, while maintaining a complexity reduction factor of 2.67 and 16, respectively. For 64-QAM, the performance of the TEQ-IQ receiver at a BER of 10^{-3} was only 1.5 dB from the decoding performance curve over the non-dispersive Gaussian channel, as shown in Figure 17.14, while reducing its complexity by a factor of 128, relative to the TEQ-EQ receiver. Note that although we were unable to simulate the performance of the TEQ-EQ receiver for 64-QAM transmissions over the same dispersive Rayleigh fading channel, we have computed its complexity based on the assumption that its critical number of iterations was three, as for the 16-QAM TEQ-EQ system. It was also noticed from the TEQ-IQ performance that despite the poor symbol mapping from low-reliability channel decoder LLR values to M-QAM symbols in the initial stages, the employment of subsequent turbo equalisation iterations improved the symbol mapping, hence yielding a better BER performance. The iteration gain achieved after the critical number of iterations increased with M, achieving 2 dB, 5 dB and 7 dB for 4-QAM, 16-QAM and 64-QAM, respectively. This was expected, since the Euclidean distance between signal points in the constellation was reduced for higher-order modulation modes, hence increasing the likelihood of making an error when the channel ISI and noise were dominant. Therefore, the performance of the TEQ-IQ after the first iteration was poor. Furthermore, longer interleaving depths could be employed for higher modulation modes, while maintaining the target channel transmission delay of ≈ 30 ms. The higher interleaver depths reduced the correlation between the bits, hence improving the turbo equaliser's performance in mitigating the effects of ISI and approximating the ISI-free performance. Overall, it has been demonstrated that the turbo equaliser employing the reduced-complexity I/Q-EQs — namely, the TEQ-IQ receiver — is capable of achieving the same performance as the complex-valued turbo equaliser, while incurring lower computational complexity. Furthermore, the TEQ-IQ can also perform equalisation and decoding for high-order modulation modes such as 64-QAM, which is not feasible by using the complex-valued trellis-based equaliser owing to its complexity constraints.

In our concluding chapter, namely in **Chapter 18**, we have shown that the principle of I/Q turbo equalization, which was originally proposed for single-transmit and single-receive systems [375], can be adapted for employment in STTC systems, in order to reduce the complexity of the STTC turbo equalization scheme. It was observed that the TEQ-IQ scheme is capable of achieving comparable performance to that of the TEQ-CT at a reduced complexity.

In Figures 18.3a-b, 18.4(a–b), 18.5(a–b) and 18.6(a–b) it was shown that for STTC systems employing short interleaver sizes, namely the 144-symbol interleaver and the 576-symbol interleaver, the performance of the TEQ-IQ was comparable to that of the TEQ-CT for BERs in excess of 10^{-3}. For lower BERs and especially for SNR values higher than 9dB, an error floor was observed. When utilising longer interleavers, such as the 2304-symbol interleaver and the 9216-symbol interleaver, the TEQ-IQ approximates closely the performance of the TEQ-CT up to a BER of $= 10^{-5}$. The performance degradation of TEQ-IQ schemes employing short STTC interleavers is due to the use of a sub-optimal MV-DFE in the first TEQ-IQ iteration, which generates low-reliability outputs, resulting in bursty channel errors. Short interleavers are unable to disperse these burst errors and therefore the STTC decoder is unable to perform the decoding operations efficiently. Consequently, the STTC decoder generates low-confidence outputs, which result in poor I/Q decoupling and perturb the equaliser's operation.

In **Chapter 18**, we have also examined the trade-offs between performance and complexity. In Figure 18.9 it was observed that when employing the TEQ-IQ scheme, which performed three turbo equalisation iterations in the context of the STTC(4,4), STTC(4,8), STTC(4,16) and STTC(4,32) systems, the SNR value required for achieving a BER of 10^{-3} was comparable to that of the TEQ-CT scheme performing two turbo equalisation iterations. Moreover, the TEQ-IQ scheme attained a complexity reduction of a factor of 4.9, 3.6, 2.5 and 1.7, respectively. It was also shown in Figure 18.9 that for short interleavers, namely for the 144-symbol interleaver and for the 576-symbol interleaver, the BER performance improvement obtained by using the more complex STTC decoders, i.e. the STTC(4,16) and STTC(4,32), was marginal. Therefore, for these interleavers lengths it is sufficient to employ the lowest-complexity STTC(4,4) scheme. However, for systems where the maximum system delay is not constrained, long STTC interleavers can be used. In conjunction with long interleavers, the STTC(4,32) scheme is the best choice, since the performance gain becomes more significant. This was further highlighted in Figure 18.10(d), where it was observed that the performance of the STTC(4,32) scheme improved upon increasing the interleaver size.

However, this performance gain was achieved at the expense of longer system delays. From Figure 18.11 it was observed that the performance achieved by the schemes utilising the 9216-symbol interleaver was only marginally better than that of the systems using the 2304-symbol interleaver, despite incurring a significantly longer system delays of 73.7 ms. Therefore, an interleaver associated with a size of approximately 2304 symbols constitutes an attractive compromise between the achievable performance gain and the system delay imposed.

In conclusion, it is observed that the TEQ-IQ constitutes a promising approach, which can reduce the complexity of the transmit diversity system without sacrificing the achievable performance. For systems associated with stringent delay constraints, a 144-symbol interleaver can be employed and the best space-time trellis code to use in terms of the performance versus complexity trade-off, from the set of schemes studied, is the STTC(4,4) arrangement. For schemes which are not delay sensitive, an interleaving size of 2304 symbols is sufficient and the recommended space-time trellis code is the STTC(4,32).

The systems studied above demonstrated the power of the proposed iterative detection technique. However, the best configuration of the system depends on the propagation environment encountered. Hence, the best performance, delay and complexity trade-off can only be met by a system that is capable of reconfiguring itself in any of the transceiver modes studied in this chapter, which can be implemented with the aid of software-controlled transceivers.

19.2   Concluding Remarks

Since the establishment of Shannon's information theory researchers have endeavoured to design communications systems capable of attaining a performance near the Shannonian information theoretic limits [91]. About 45 years after Shannon's pioneering work Berrou and Glavieux [13] have succeeded in devising the so-called turbo channel coding scheme, which is capable of reaching the information theoretic limits over Gaussian channels within a fraction of a dB, although this is achieved at the cost of typically high turbo interleaving delays. Turbo coding revitalised channel coding research and opened a new chapter in the design of iterative detection-assisted communications systems, such as turbo trellis coding schemes, turbo channel equalisers, etc. The portrayal and study of these schemes has been the main topic of this monograph.

At the time of writing researchers are endeavouring to further enhance the performance of wireless systems communicating over hostile, dispersive fading channels. However, wireless channels require a high number of parameters for their characterisation and hence the capacity of wireless channels is typically characterised for specific scenarios only [377]. The ultimate goal of wireless communications is to devise transmission schemes which are capable of communicating near the **capacity of the wireless channel**.

Research has been also dedicated to the study of channel coding schemes specifically designed for dispersive, fading wireless channels. Some representatives of this channel coding family are BICM and BICM-ID studied in Chapter 9. Even though space-time coding schemes are often regarded as transmit diversity arrangements, they can also be viewed as specific codes designed for attaining the highest possible diversity gain. We have demonstrated in the various space-time coding chapters, namely in Chapters 10, 11 and Chapter 18, that if the diversity order is sufficiently high and the channel-induced dispersion is eliminated with the aid of a channel equaliser, the wireless channel may be rendered 'near-Gaussian' for the purposes of channel coding. Since these space-time-coded systems require multiple transmitters and receivers, they exhibit a relatively high complexity. Nonetheless, if this increased complexity is deemed acceptable in the light of the performance gains attained, space-time codecs can be advantageously combined with a range of well-known channel codecs originally designed for Gaussian channels, because they render the channel 'near-Gaussian'.

In an effort to identify attractive channel and space-time coding combinations for various wireless systems convolutional coding, turbo convolutional and turbo-BCH coding, trellis-coded modulation as well as turbo trellis-coded modulation were combined with various space-time block codes having different effective throughputs. The various schemes exhibited dramatically different implementational complexities. These investigations concluded that for the scenarios investigated the highest overall coding gain was achieved by the turbo convolutional-coded two-antenna-based unity-rate space-time block-coded scheme proposed by Alamouti. Furthermore, it was found disadvantageous to invoke space-time codecs having a less than unity coding rate. This was a consequence of having to increase the number of bits per modulation symbol for maintaining the same throughput as that of the unity-rate space-time-coded bench-markers, which rendered the space-time coding schemes having a coding rate below unity prone to transmission errors, while also increasing their associated complexity. By contrast, invoking turbo convolutional coding in conjunction with the unity-rate space-time code was found to be the most attractive scheme in complexity and throughput terms.

If, however, the implementational complexity of the space-time-coded systems presented in Chapters 10, 11 and Chapter 18 and requiring multiple transmitters and receivers is deemed excessive, *near-instantaneously adaptive modulation and channel coding schemes* may be invoked as a lower-complexity single-transmitter single-receiver design alternative. Such adaptive modulation and channel coding arrangements [306] have been studied in Chapters 9, 11 and 12.

To elaborate a little further, since the instantaneous channel quality of wireless systems fluctuates over a wide range, it is unrealistic to expect that conventional fixed-mode transceivers would maintain wireline-like transmission qualities. At the time of writing many existing wireless systems already exhibit some grade of adaptivity, as was documented for example in the excellent overview article by Nanda *et al.* [378]. The third-generation (3G) systems UTRA/IMT200 [49, 379] are also amenable to agile, Burst-by-Burst (BbB) reconfiguration [306] on the basis of 10 ms transmission bursts with the advent of their sufficiently high-rate and suitably low-delay control channels.

Let us now relate the above-mentioned issues of channel-quality-motivated transceiver adaptivity to the emerging 3G standard systems. Although the well-known Orthogonal Variable Spreading Factor (OVSF) spreading codes used in the 3G IMT2000/UTRA system [49] wsere originally not proposed for BbB-based reconfiguration, the activation of these OVSF codes can be potentially adapted on a near-instantaneous basis in an effort to counteract the near-instantaneous channel quality fluctuations of the wireless channel. Controlling the OVSF codes can also be potentially combined with near-instantaneous or BbB-adaptive modulation mode control. The advantage of counteracting the near-instantaneous channel quality fluctuations in this way instead of the extensive employment of agile power control is that power control may inflict excessive co-channel interference in an effort to maintain the channel quality of specific users. By contrast, BbB-adaptive transmission constitutes a 'non-intrusive' way of mitigating the effects of transmission bursts, since only the user of interest adjusts its transmission mode appropriately. A further advantage of this principle is that the error distribution at the output of BbB-adaptive transceivers is expecetd to be less bursty than in case of a fixed-mode transceiver. This allows channel codecs that were originally designed for Gaussian channels exhibiting random, rather than bursty channel error statistics to operate more efficiently.

Another widespread standard system is the High PERformance Local Area Network standard known as HIPERLAN2, which was also designed with a similar baseband BbB-adaptive modulation and channel coding tool-box concept in mind. However, similarly to the IMT2000/UTRA system, the BbB employment of different transmission modes was not originally foreseen. Nonetheless, the agile BbB-based application of adaptive source and channel coding, as well as modulation [380–382], is emerging as a promising new system design paradigm in conjunction with the powerful channel quality prediction techniques proposed for example by Duel-Hallen, *et al.* [254]. Again, these BbB-adaptive modulation and channel coding principles have been elaborated on in detail in Chapters 9, 11 and 12.

Fuelled by the imminent emergence of wireless multimedia communications services there is an urgent demand for flexible, bandwidth-efficient transceivers, which are also capable of multi-standard operation. The concept of BbB adaptive transceivers is capable of providing an improved Quality of Service (QoS) in comparison to their statically reconfigured counterparts, since BbB-adaptive schemes constitute a powerful means of combating the effects of hostile wireless channels [306, 379–382].

In conclusion, wireless systems of the near future are likely to witness the co-existence of space-time-coded transmit diversity and BbB-adaptive transmission schemes for years to come, where intelligent learning algorithms will configure the transceivers in the appropriate mode that ultimately provides the best trade-off in terms of satisfying the user's preference in the context of the service requested [306, 379, 380].

Another important benefit of employing BbB-adaptive transceivers is that with their advent the network's teletraffic capacity typically also improves. This is a consequence of the transceiver's ability to invoke a robust, but lower throughput transceiver mode, when the instantaneous channel quality degrades, in which case a conventional fixed-mode transceiver would have to drop the call supported, unless the QoS constraint was violated. This issue was documented in depth in conjunction with adaptive antenna-assisted dynamic channel allocation schemes in [379].[1]

∗   ∗

Throughout this monograph we have attempted to portray the range of contradictory system design trade-offs associated with the conception of channel coding arrangements for a variety

---

[1] A range of related research papers and book chapters can be found at http://www-mobile.ecs.soton.ac.uk.

of applications. We endeavoured to present the material in an unbiased fashion and sufficiently richly illustrated in terms of the associated trade-offs so that readers will be able to glean useful information from it in order to solve their own particular channel coding and communications problem. In this rapidly evolving field it is a challenge to complete a comprehensive treatise, since new advances are contrived at an ever-increasing pace, which one would like to report on. Our sincere hope is that you, the readers, have found the book a useful source of information.

Bibliography

[1] C. E. Shannon, "A mathematical theory of communication," *Bell System Technical Journal*, pp. 379–427, 1948.

[2] R. Hamming, "Error detecting and error correcting codes," *Bell System Technical Journal*, vol. 29, pp. 147–160, 1950.

[3] P. Elias, "Coding for noisy channels," *IRE Convention Record, pt.4*, pp. 37–47, 1955.

[4] J. Wozencraft, "Sequential decoding for reliable communication," *IRE Natl. Conv. Rec.*, vol. 5, pt.2, pp. 11–25, 1957.

[5] J. Wozencraft and B. Reiffen, *Sequential Decoding*. Cambridge, MA, USA: MIT Press, 1961.

[6] R. Fano, "A heuristic discussion of probabilistic coding," *IEEE Transactions on Information Theory*, vol. IT-9, pp. 64–74, April 1963.

[7] J. Massey, *Threshold Decoding*. Cambridge, MA, USA: MIT Press, 1963.

[8] A. Viterbi, "Error bounds for convolutional codes and an asymptotically optimum decoding algorithm," *IEEE Transactions on Information Theory*, vol. IT-13, pp. 260–269, April 1967.

[9] G. Forney, "The Viterbi algorithm," *Proceedings of the IEEE*, vol. 61, pp. 268–278, March 1973.

[10] J. H. Chen, "High-quality 16 kb/s speech coding with a one-way delay less than 2 ms," in *Proceedings of International Conference on Acoustics, Speech, and Signal Processing, ICASSP'90*, vol. 1, (Albuquerque, New Mexico, USA), pp. 453–456, IEEE, 3–6 April 1990.

[11] L. R. Bahl, J. Cocke, F. Jelinek, and J. Raviv, "Optimal Decoding of Linear Codes for Minimising Symbol Error Rate," *IEEE Transactions on Information Theory*, vol. 20, pp. 284–287, March 1974.

[12] C. Berrou, A. Glavieux, and P. Thitimajshima, "Near Shannon Limit Error-Correcting Coding and Decoding: Turbo Codes," in *Proceedings of the International Conference on Communications*, (Geneva, Switzerland), pp. 1064–1070, May 1993.

[13] C. Berrou and A. Glavieux, "Near optimum error correcting coding and decoding: Turbo codes," *IEEE Transactions on Communications*, vol. 44, pp. 1261–1271, October 1996.

[14] A. Hocquenghem, "Codes correcteurs d'erreurs," *Chiffres (Paris)*, vol. 2, pp. 147–156, September 1959.

[15] R. Bose and D. Ray-Chaudhuri, "On a class of error correcting binary group codes," *Information and Control*, vol. 3, pp. 68–79, March 1960.

[16] R. Bose and D. Ray-Chaudhuri, "Further results on error correcting binary group codes," *Information and Control*, vol. 3, pp. 279–290, September 1960.

[17] W. Peterson, "Encoding and error correction procedures for the Bose-Chaudhuri codes," *IEEE Transactions on Information Theory*, vol. IT-6, pp. 459–470, September 1960.

[18] J. Wolf, "Efficient maximum likelihood decoding of linear block codes using a trellis," *IEEE Transactions on Information Theory*, vol. IT-24, pp. 76–80, January 1978.

[19] B. Honary and G. Markarian, *Trellis Decoding of Block Codes*. Dordrecht, The Netherlands: Kluwer Academic, 1997.

[20] S. Lin, T. Kasami, T. Fujiwara, and M. Fossorier, *Trellises and Trellis-based Decoding Algorithms for Linear Block Codes*. Norwell, MA, USA: Kluwer Academic, 1998.

[21] G. Forney, "Coset Codes-Part II: Binary Lattices and Related Codes," *IEEE Transactions on Information Theory*, vol. 34, pp. 1152–1187, September 1988.

[22] H. Manoukian and B. Honary, "BCJR trellis construction for binary linear block codes," *IEE Proceedings, Communications*, vol. 144, pp. 367–371, December 1997.

[23] B. Honary, G. Markarian, and P. Farrell, "Generalised array codes and their trellis structure," *Electronics Letters*, vol. 29, pp. 541–542, March 1993.

[24] B. Honary and G. Markarian, "Low-complexity trellis decoding of Hamming codes," *Electronics Letters*, vol. 29, pp. 1114–1116, June 1993.

[25] G. M. B. Honary and M. Darnell, "Low-complexity trellis decoding of linear block codes," *IEE Proceedings Communications*, vol. 142, pp. 201–209, August 1995.

[26] T. Kasami, T. Takata, T. Fujiwara, and S. Lin, "On complexity of trellis structure of linear block codes," *IEEE Transactions on Information Theory*, vol. 39, pp. 1057–1937, May 1993.

[27] T. Kasami, T. Takata, T. Fujiwara, and S. Lin, "On the optimum bit orders with respect to the state complexity of trellis diagrams for binary linear codes," *IEEE Transactions on Information Theory*, vol. 39, pp. 242–245, January 1993.

[28] D. Chase, "A class of algorithms for decoding block codes with channel measurement information," *IEEE Transactions on Information Theory*, vol. IT-18, pp. 170–182, January 1972.

[29] D. Gorenstein and N. Zierler, "A class of cyclic linear error-correcting codes in p^m synbols," *Journal of the Society of Industrial and Applied Mathematics.*, vol. 9, pp. 107–214, June 1961.

[30] I. Reed and G. Solomon, "Polynomial codes over certain finite fields," *Journal of the Society of Industrial and Applied Mathematics.*, vol. 8, pp. 300–304, June 1960.

[31] E. Berlekamp, "On decoding binary Bose-Chaudhuri-Hocquenghem codes," *IEEE Transactions on Information Theory*, vol. 11, pp. 577–579, 1965.

[32] E. Berlekamp, *Algebraic Coding Theory*. New York, USA: McGraw-Hill, 1968.

[33] J. Massey, "Step-by-step decoding of the Bose-Chaudhuri-Hocquenghem codes," *IEEE Transactions on Information Theory*, vol. 11, pp. 580–585, 1965.

[34] J. Massey, "Shift-register synthesis and BCH decoding," *IEEE Transactions on Information Theory*, vol. IT-15, pp. 122–127, January 1969.

[35] M. Oh and P. Sweeney, "Bit-level soft-decision sequential decoding for Reed Solomon codes," in *Workshop on Coding and Cryptography*, (Paris, France), January 1999.

[36] M. Oh and P. Sweeney, "Low complexity soft-decision sequential decoding using hybrid permutation for RS codes," in *Seventh IMA Conference on Cryptography and Coding*, (Royal Agricultural College, Cirencester, UK), December 1999.

[37] D. Burgess, S. Wesemeyer, and P. Sweeney, "Soft-decision decoding algorithms for RS codes," in *Seventh IMA Conference on Cryptography and Coding*, (Royal Agricultural College, Cirencester, UK), December 1999.

[38] Consultative Committee for Space Data Systems, *Blue Book: Recommendations for Space Data System Standards: Telemetry Channel Coding*, May 1984.

[39] European Telecommunication Standard Institute (ETSI), *Digital Video Broadcasting (DVB); Framing structure, channel coding and modulation for MVDS at 10GHz and above*, ETS 300 748 ed., October 1996. http://www.etsi.org/.

[40] F. Taylor, "Residue arithmetic: A tutorial with examples," *IEEE Computer Magazine*, vol. 17, pp. 50–62, May 1984.

[41] N. Szabo and R. Tanaka, *Residue Arithmetic and Its Applications to Computer Technology*. New York, USA: McGraw-Hill, 1967.

[42] R. Watson and C. Hastings, "Self-checked computation using residue arithmetic," *Proceedings of the IEEE*, vol. 54, pp. 1920–1931, December 1966.

[43] H. Krishna, K. Y. Lin, and J. D. Sun, "A coding theory approach to error control in redundant residue number systems - part I: Theory and single error correction," *IEEE Transactions on Circuits and Systems*, vol. 39, pp. 8–17, January 1992.

[44] J. D. Sun and H. Krishna, "A coding theory approach to error control in redundant residue number systems — part II: Multiple error detection and correction," *IEEE Transactions on Circuits and Systems*, vol. 39, pp. 18–34, January 1992.

[45] T. H. Liew, L. L. Yang, and L. Hanzo, "Soft-decision Redundant Residue Number System Based Error Correction Coding," in *Proceedings of IEEE VTC'99*, (Amsterdam, The Netherlands), pp. 2546–2550, September 1999.

[46] G. Ungerboeck, "Treliis-coded modulation with redundant signal sets part 1: Introduction," *IEEE Communications Magazine*, vol. 25, pp. 5–11, February 1987.

[47] G. Ungerboeck, "Treliis-coded modulation with redundant signal sets part II: State of the art," *IEEE Communications Magazine*, vol. 25, pp. 12–21, February 1987.

[48] C. Schlegel, *Trellis Coding*. New York, USA: IEEE Press, 1997.

[49] R. Steele and L. Hanzo, eds., *Mobile Radio Communications: Second and Third Generation Cellular and WATM Systems*. New York, USA: IEEE Press - John Wiley & Sons, 2nd ed., 1999.

[50] W. Koch and A. Baier, "Optimum and sub-optimum detection of coded data disturbed by time-varying intersymbol interference," *IEEE Globecom*, pp. 1679–1684, December 1990.

[51] J. Erfanian, S. Pasupathy, and G. Gulak, "Reduced complexity symbol dectectors with parallel structures for ISI channels," *IEEE Transactions on Communications*, vol. 42, pp. 1661–1671, 1994.

[52] P. Robertson, E. Villebrun, and P. Höher, "A Comparison of Optimal and Sub-Optimal MAP Decoding Algorithms Operating in the Log Domain," in *Proceedings of the International Conference on Communications*, (Seattle, USA), pp. 1009–1013, June 1995.

[53] J. Hagenauer and P. Höher, "A Viterbi algorithm with soft-decision outputs and its applications," in *IEEE Globecom*, pp. 1680–1686, 1989.

[54] J. Hagenauer, "Source-controlled channel decoding," *IEEE Transactions on Communications*, vol. 43, pp. 2449–2457, September 1995.

[55] S. L. Goff, A. Glavieux, and C. Berrou, "Turbo-codes and high spectral efficiency modulation," in *Proceedings of IEEE International Conference on Communications*, pp. 645–649, 1994.

[56] U. Wachsmann and J. Huber, "Power and bandwidth efficient digital communications using turbo codes in multilevel codes," *European Transactions on Telecommunications*, vol. 6, pp. 557–567, September–October 1995.

[57] P. Robertson and T. Wörz, "Bandwidth-Efficient Turbo Trellis-Coded Modulation Using Punctured Component Codes," *IEEE Journal on Selected Areas in Communications*, vol. 16, pp. 206–218, February 1998.

[58] S. Benedetto and G. Montorsi, "Design of parallel concatenated convolutional codes," *IEEE Transactions on Communications*, vol. 44, pp. 591–600, May 1996.

[59] S. Benedetto and G. Montorsi, "Unveiling turbo codes: Some results on parallel concatenated coding schemes," *IEEE Transactions on Information Theory*, vol. 42, pp. 409–428, March 1996.

[60] L. Perez, J. Seghers, and D. Costello, "A distance spectrum interpretation of turbo codes," *IEEE Transactions on Information Theory*, vol. 42, pp. 1698–1709, November 1996.

[61] J. Hagenauer, E. Offer, and L. Papke, "Iterative decoding of binary block and convolutional codes," *IEEE Transactions on Information Theory*, vol. 42, pp. 429–445, March 1996.

[62] R. M. Pyndiah, "Near-optimum decoding of product codes: Block turbo codes," *IEEE Transactions on Communications*, vol. 46, pp. 1003–1010, August 1998.

[63] H. Nickl, J. Hagenauer, and F. Burkett, "Approaching Shannon's capacity limit by 0.27 dB using simple Hamming codes," *IEEE Communications Letters*, vol. 1, pp. 130–132, September 1997.

[64] O. F. Açikel and W. E. Ryan, "Punctured turbo-codes for BPSK/QPSK channels," *IEEE Transactions on Communications*, vol. 47, pp. 1315–1323, September 1999.

[65] P. Jung and M. Nasshan, "Performance evaluation of turbo codes for short frame transmission systems," *IEE Electronics Letters*, vol. 30, pp. 111–112, January 1994.

[66] P. Jung, "Comparison of turbo-code decoders applied to short frame transmission systems," *IEEE Journal on Selected Areas in Communications*, pp. 530–537, 1996.

[67] P. Jung, M. Naßhan, and J. Blanz, "Application of Turbo-Codes to a CDMA Mobile Radio System Using Joint Detection and Antenna Diversity," in *Proceedings of the IEEE Conference on Vehicular Technology*, pp. 770–774, 1994.

[68] A. Barbulescu and S. Pietrobon, "Interleaver design for turbo codes," *IEE Electronics Letters*, pp. 2107–2108, December 1994.

[69] B. Sklar, "A Primer on Turbo Code Concepts," *IEEE Communications Magazine*, pp. 94–102, December 1997.

[70] V. Tarokh, N. Seshadri, and A. R. Calderbank, "Space-Time Codes for High Data Rate Wireless Communication: Performance Criterion and Code Construction," *IEEE Transactions on Information Theory*, vol. 44, pp. 744–765, March 1998.

[71] S. M. Alamouti, "A Simple Transmit Diversity Technique for Wireless Communications," *IEEE Journal on Selected Areas in Communications*, vol. 16, pp. 1451–1458, October 1998.

[72] V. Tarokh, H. Jafarkhani, and A. Calderbank, "Space-time block codes from orthogonal designs," *IEEE Transactions on Information Theory*, vol. 45, pp. 1456–1467, May 1999.

[73] V. Tarokh, H. Jafarkhani, and A. R. Calderbank, "Space-time block coding for wireless communications: Performance results," *IEEE Journal on Selected Areas in Communications*, vol. 17, pp. 451–460, March 1999.

[74] G. Bauch, A. Naguib, and N. Seshadri, "MAP Equalization of Space-Time Coded Signals over Frequency Selective Channels," in *Proceedings of Wireless Communications and Networking Conference*, (New Orleans, USA), September 1999.

[75] G. Bauch and N. Al-Dhahir, "Reduced-complexity turbo equalization with multiple transmit and receive antennas over multipath fading channels," in *Proceedings of Information Sciences and Systems*, (Princeton, USA), pp. WP3 13–18, March 2000.

[76] D. Agrawal, V. Tarokh, A. Naguib, and N. Seshadri, "Space-time coded OFDM for high data-rate wireless communication over wideband channels," in *Proceedings of IEEE Vehicular Technology Conference*, (Ottawa, Canada), pp. 2232–2236, May 1998.

[77] Y. Li, N. Seshadri, and S. Ariyavisitakul, "Channel estimation for OFDM systems with transmitter diversity in mobile wireless channels," *IEEE Journal on Selected Areas in Communications*, vol. 17, pp. 461–471, March 1999.

[78] Y. Li, J. Chuang, and N. Sollenberger, "Transmitter diversity for OFDM systems and its impact on high-rate data wireless networks," *IEEE Journal on Selected Areas in Communications*, vol. 17, pp. 1233–1243, July 1999.

[79] A. Naguib, N. Seshdri, and A. Calderbank, "Increasing Data Rate Over Wireless Channels: Space-Time Coding for High Data Rate Wireless Communications," *IEEE Signal Processing Magazine*, vol. 17, pp. 76–92, May 2000.

[80] A. F. Naguib, V. Tarokh, N. Seshadri, and A. R. Calderbank, "A Space-Time Coding Modem for High-Data-Rate Wireless Communications," *IEEE Journal on Selected Areas in Communications*, vol. 16, pp. 1459–1478, October 1998.

[81] R. Pyndiah, A. Glavieux, A. Picart, and S. Jacq, "Near optimum decoding of product codes," in *GLOBECOM 94*, (San Francisco, USA), pp. 339–343, November 1994.

[82] W. Peterson and E. Weldon Jr., *Error Correcting Codes*. Cambridge, MA, USA: MIT. Press, 2nd ed., August 1972. ISBN: 0262160390.

[83] T. Kasami, *Combinational Mathematics and its Applications*. Darham, NC, USA: University of North Carolina Press, 1969.

[84] W. Peterson, *Error Correcting Codes*. Cambridge, MA, USA: MIT Press, 1st ed., 1961.

[85] F. MacWilliams and J. Sloane, *The Theory of Error-Correcting Codes*. Amsterdam: North-Holland, 1977.

[86] B. Sklar, *Digital Communications—Fundamentals and Applications*. Englewood Cliffs, NJ, USA: Prentice Hall, 1988.

[87] I. Blake, ed., *Algebraic Coding Theory: History and Development*. Dowden, Hutchinson and Ross, 1973.

[88] G. Clark Jr. and J. Cain, *Error Correction Coding for Digital Communications*. New York, USA: Plenum Press, May 1981. ISBN: 0306406152.

[89] V. Pless, *Introduction to the Theory of Error-correcting Codes*. New York, USA: John Wiley & Sons, 1982. ISBN: 0471813044.

[90] R. Blahut, *Theory and Practice of Error Control Codes*. Reading, MA, USA: Addison-Wesley, 1983. ISBN 0-201-10102-5.

[91] C. E. Shannon, *Mathematical Theory of Communication*. University of Illinois Press, 1963.

[92] C. Heegard and S. Wicker, *Turbo Coding*. Kluwer International, 1999.

[93] M. Bossert, *Channel Coding for Telecommunications*. New York, USA: John Wiley & Sons, 1999. ISBN 0-471-98277-6.

[94] B. Vucetic and J. Yuan, *Turbo Codes Principles and Applications*. Dordrecht, The Netherlands: Kluwer Academic, 2000.

[95] R. Lidl and H. Niederreiter, *Finite Fields*. Cambridge, UK: Cambridge University Press, 1996.

[96] S. Lin and D. Constello Jr., *Error Control Coding: Fundamentals and Applications*. Englewood Cliffs, NJ, USA: Prentice Hall, October 1982. ISBN: 013283796X.

[97] A. Michelson and A. Levesque, *Error Control Techniques for Digital Communication*. New York, USA: John Wiley & Sons, 1985.

[98] D. Hoffman, D. Leonard, C. Lindner, K. Phelps, C. Rodger, and J. Wall, *Coding Theory*. New York, USA: Marcel Dekker, 1991.

[99] J. Huber, *Trelliscodierung*. Berlin, Germany: Springer Verlag, 1992.

[100] J. Anderson and S. Mohan, *Source and Channel Coding — An Algorithmic Approach*. Dordrecht, The Netherlands: Kluwer Academic, 1993.

[101] S. Wicker, *Error Control Systems for Digital Communication and Storage*. Englewood Cliffs, NJ, USA: Prentice Hall, 1994.

[102] J. Proakis, *Digital Communications*. New York, USA: McGraw Hill: International Editions, 3rd ed., 1995.

[103] P. Sweeney, *Error Control Coding: An Introduction*. New York, USA: Prentice Hall, 1991.

[104] S. Benedetto, D. Divsalar, G. Montorsi, and F. Pollara, "Serial concatenation of interleaved codes: Performance analysis, design, and iterative decoding," *IEEE Transactions on Information Theory*, vol. 44, pp. 909–926, May 1998.

[105] C. Douillard, A. Picart, M. Jézéquel, P. Didier, C. Berrou, and A. Glavieux, "Iterative correction of intersymbol interference: Turbo-equalization," *European Transactions on Communications*, vol. 6, pp. 507–511, 1995.

[106] D. Raphaeli and Y. Zarai, "Combined turbo equalization and turbo decoding," *IEEE Communications Letters*, vol. 2, pp. 107–109, April 1998.

[107] J. Heller and I. Jacobs, "Viterbi decoding for satellite and space communication," *IEEE Transactions on Communication Technology*, vol. COM-19, pp. 835–848, October 1971.

[108] K. H. H. Wong, *Transmission of Channel Coded Speech and Data over Mobile Channels*. PhD thesis, University of Southampton, UK, 1989.

[109] R. Steele and L. Hanzo, eds., *Mobile Radio Communications*. Piscataway, NJ, USA: IEEE Press, 1999.

[110] J. Makhoul, "Linear prediction: A tutorial review," *Proceedings of the IEEE*, vol. 63, pp. 561–580, April 1975.

[111] R. Blahut, *Fast Algorithms for Digital Signal Processing*. Reading, MA, USA: Addison-Wesley, 1985. ISBN 0-201-10155-6.

[112] J. Schur, "Ueber Potenzreihen, die im Innern des Einheitskreises beschraenkt sind," *Journal fuer Mathematik*, pp. 205–232. Bd. 147, Heft 4.

[113] R. Chien, "Cyclic decoding procedure for the Bose-Chaudhuri-Hocquenghem codes," *IEEE Transactions on Information Theory*, vol. 10, pp. 357–363, October 1964.

[114] A. Jennings, *Matrix Computation for Engineers and Scientists*. New York, USA: John Wiley & Sons, 1977.

[115] G. Forney Jr., "On decoding BCH codes," *IEEE Transactions on Information Theory*, vol. IT-11, pp. 549–557, 1965.

[116] Y. Sugiyama, M. Kasahara, S. Hirasawa, and T. Namekawa, "A method for solving key equation for decoding Goppa codes," *Information Control*, no. 27, pp. 87–99, 1975.

[117] S. Golomb, *Shift Register Sequences*. Laugana Hills, CA, USA: Aegean Park Press, 1982.

[118] J. Stenbit, "Table of generators for Bose-Chaudhuri codes," *IEEE Transactions on Information Theory*, vol. 10, pp. 390–391, October 1964.

[119] B. Sklar, *Digital Communications Fundamentals and Applications*, ch. 5.6.5, pp. 294–296. Englewood Cliffs, NJ, USA: Prentice Hall International Editions, 1988.

[120] R. Steele and L. Hanzo, eds., *Mobile Radio Communications*, ch. 4.4.4, pp. 425–428. Piscataway, NJ, USA: IEEE Press and Pentech Press, 1999.

[121] R. Blahut, *Theory and Practice of Error Control Codes*, ch. 6, pp. 130–160. IBM Corporation, Owego, NY 13827, USA: Addison-Wesley, 1983.

[122] C. Berrou, P. Adde, E. Angui, and S. Faudeil, "A low complexity soft-output Viterbi decoder architecture," in *Proceedings of the International Conference on Communications*, pp. 737–740, May 1993.

[123] R. Pyndiah, "Iterative decoding of product codes: Block turbo codes," in *International Symposium on Turbo Codes and related topics*, (Brest, France), pp. 71–79, September 1997.

[124] M. Breiling and L. Hanzo, "Optimum non-iterative decoding of turbo codes," *IEEE Transactions on Information Theory*, vol. 46, pp. 2212–2228, September 2000.

[125] M. Breiling and L. Hanzo, "Optimum Non-iterative Turbo-Decoding," *Proceedings of PIMRC'97, Helsinki, Finland*, pp. 714–718, September 1997.

[126] P. Robertson, "Illuminating the structure of code and decoder of parallel concatenated recursive systematic (turbo) codes," *IEEE Globecom*, pp. 1298–1303, 1994.

[127] M. Breiling, "Turbo coding simulation results," tech. rep., Universität Karlsruhe, Germany and Southampton University, UK, 1997.

[128] C. Berrou, "Some clinical aspects of turbo codes," in *International Symposium on Turbo Codes and related topics*, (Brest, France), pp. 26–31, September 1997.

[129] J. Erfanian, S. Pasupathy, and G. Gulak, "Low-complexity symbol detectors for isi channels," in *IEEE Globecom*, (San Diego), December 1990.

[130] A. Viterbi, "Approaching the Shannon limit: Theorist's dream and practitioner's challenge," in *Proceedings of the International Conference on Millimeter Wave and Far Infrared Science and Technology*, pp. 1–11, 1996.

[131] A. J. Viterbi, "An Intuitive Justification and a Simplified Implementation of the MAP Decoder for Convolutional Codes," *IEEE Journal on Selected Areas in Communications*, pp. 260–264, February 1997.

[132] J. Woodard, T. Keller, and L. Hanzo, "Turbo-coded orthogonal frequency division multiplex transmission of 8 kbps encoded speech," in *Proceedings of ACTS Mobile Communication Summit '97*, (Aalborg, Denmark), pp. 894–899, ACTS, 7–10 October 1997.

[133] P. Jung and M. Nasshan, "Dependence of the error performance of turbo-codes on the interleaver structure in short frame transmission systems," *IEE Electronics Letters*, pp. 287–288, February 1994.

[134] P. Jung and M. Naßhan, "Results on Turbo-Codes for Speech Transmission in a Joint Detection CDMA Mobile Radio System with Coherent Receiver Antenna Diversity," *IEEE Transactions on Vehicular Technology*, vol. 46, pp. 862–870, November 1997.

[135] H. Herzberg, "Multilevel Turbo Coding with Short Interleavers," *IEEE Journal on Selected Areas in Communications*, vol. 16, pp. 303–309, February 1998.

[136] T.A. Summmers and S.G. Wilson, "SNR Mismatch and Online Estimation in Turbo Decoding," *IEEE Transactions on Communications*, vol. 46, pp. 421–423, April 1998.

[137] J. Cavers, "An analysis of pilot symbol assisted modulation for Rayleigh fading channels," *IEEE Transactions on Vehicular Technology*, vol. 40, pp. 686–693, November 1991.

[138] R. Gallager, "Low-density parity-check codes," *IEEE Transactions on Information Theory*, pp. 21–28, 1962.

[139] G. Battail, "On random-like codes," *Canadian Workshop on Information Theory*, 1995.

[140] A. Khandani, "Design of turbo-code interleaver using Hungarian method," *Electronics Letters*, pp. 63–65, 1998.

[141] K. Andrews, C. Heegard, and D. Kozen, "Interleaver design methods for turbo codes," *International Symposium on Information Theory*, p. 420, 1998.

[142] A. Ambroze, G. Wade, and M. Tomlinson, "Turbo code tree and code performance," *IEE Electronics Letters*, vol. 34, pp. 353–354, February 1998.

[143] J. Sadowsky, "A maximum-likelihood decoding algorithm for turbo codes," *IEEE Communications Theory Workshop, Tucson*, 1997.

[144] P. Höher, "New iterative ("turbo") decoding algorithms," in *International Symposium on Turbo Codes and related topics*, (Brest, France), pp. 63–70, September 1997.

[145] J. Proakis, *Digital Communications*. New York, USA: McGraw-Hill, 2nd ed., 1989.

[146] A. Jimenez and K. Zigangirov, "Time-varying Periodical Convolutional Codes with Low-Density Parity-Check Matrix," *IEEE Transactions onf Information Technology*, 1999.

[147] P. Adde, R. Pyndiah, O. Raoul, and J. R. Inisan, "Block turbo decoder design," in *International Symposium on Turbo Codes and related topics*, (Brest, France), pp. 166–169, September 1997.

[148] A. Goalic and R. Pyndiah, "Real-time turbo-decoding of product codes on a digital signal processor," in *International Symposium on Turbo Codes and related topics*, (Brest, France), pp. 267–270, September 1997.

[149] F. Macwilliams and N. Sloane, *The theory of error correcting codes*, vol. 16, ch. 18, pp. 567–580. Bell Laboratories, Murray Hill, NJ 07974, USA: North-Holland, 1978.

[150] S. Ng, T. Liew, L. Yang, and L. Hanzo, "Turbo coding performance using block component codes," in *Proceedings of VTC 2000 Spring*, (Tokyo, Japan), pp. 849–853, 15–18 May 2000.

[151] C. Hastings, "Automatic detection and correction of errors in digital computers using residue arithmetic," in *Proceeding of IEEE Conference*, (Region Six), pp. 465–490, 1966.

[152] R. Watson and C. Hastings, *Residue Arithmetic and Reliable Computer Design*. Washington, DC: Spartan Books, 1967.

[153] Y. Keir, P. Cheney, and M. Tannenbaum, "Division and overflow detection in residue number systems," *IRE Transactions on Electronic Computers*, vol. EC-11, pp. 501–507, August 1962.

[154] A. Baraniecka and G. Jullien, "On decoding techniques for residue number system realizations of digital signal processing hardware," *IEEE Transactions on Circuits System*, vol. CAS-25, November 1978.

[155] W. Jenkins, "A new algorithm for scaling in residue number systems with applications to recursive digital filtering," in *IEEE International Symposium on Circuits and Systems*, (Phoenix, USA), pp. 56–59, 1977.

[156] H. Huang and F. Taylor, "High speed DFT's using residue numbers," in *IEEE International Conference on Acoustics, Speech, Signal Processing*, (Denver, USA), pp. 238–242, April 1980.

[157] W. Jenkins and B. Leon, "The use of residue number system in the design of finite impulse response filters," *IEEE Transactions on Circuits and Systems*, vol. CAS-24, pp. 191–201, April 1977.

[158] T. Vu, "Efficient implementations of the Chinese remainder theorem for sign detection and residue decoding," *IEEE Transactions on Computers*, vol. 34, pp. 646–651, July 1985.

[159] C. Su and H. Lo, "An algorithm for scaling and single residue error correction in residue number system," *IEEE Transactions on Computers*, vol. 39, pp. 1053–1064, August 1990.

[160] F. Taylor and C. Huang, "An autoscale residue multiplier," *IEEE Transactions on Computers*, vol. 31, pp. 321–325, April 1982.

[161] G. Jullien, "Residue number scaling and other operations using ROM arrays," *IEEE Transactions on Computers*, vol. C-27, pp. 325–337, April 1978.

[162] W. Jenkins, "A highly efficient residue-combinatorial architecture for digital filters," *Proceedings of the IEEE*, vol. 66, pp. 700–702, June 1978.

[163] G. Alia and E. Martinelli, "A VLSI modulo m multiplier," *IEEE Transactions on Computers*, vol. 40, pp. 873–878, July 1991.

[164] D. Radhakrishnan and Y. Yuan, "Novel Approaches to the Design of VLSI RNS Multiplier," *IEEE Transactions on Circuits and Systems-II*, vol. 39, pp. 52–57, January 1992.

[165] F. Taylor and A. Ramnarayanan, "An efficient residue-to-decimal converter," *IEEE Transactions on Circuits and Systems*, vol. 28, pp. 1164–1169, December 1981.

[166] A. Shenoy and R. Kumaresan, "Residue to binary conversion for RNS arithmetic using only modular look-up table," *IEEE Transactions on Circuits and Systems*, vol. 35, pp. 1158–1162, September 1988.

[167] G. Alia and E. Martinelli, "On the lower bound to the VLSI complexity of number conversion from weighted to residue representation," *IEEE Transactions on Computers*, vol. 42, pp. 962–967, August 1993.

[168] G. Alia and E. Martinelli, "A VLSI algorithm for direct and reverse conversion from weighted binary number system to residue number system," *IEEE Transactions on Circuits and Systems*, vol. 31, pp. 1033–1039, December 1984.

[169] F. Barsi and P. Maestrini, "Error correction properties of redundant residue number systems," *IEEE Transactions on Computers*, vol. 22, pp. 307–315, March 1973.

[170] S. Yau and Y. Liu, "Error correction in redundant residue number systems," *IEEE Transactions on Computers*, vol. C-22, pp. 5–11, January 1984.

[171] H. Krishna and J. D. Sun, "On theory and fast algorithms for error correction in residue number system product codes," *IEEE Transactions on Computers*, vol. 42, pp. 840–852, July 1993.

[172] R. Cosentino, "Fault tolerance in a systolic residue arithmetic processor array," *IEEE Transactions on Computers*, vol. 37, pp. 886–890, July 1988.

[173] F. Barsi and P. Maestrini, "Error detection and correction by product codes in residue number system," *IEEE Transactions on Computers*, vol. 23, pp. 915–924, September 1974.

[174] V. Ramachandran, "Single residue error correction in residue number systems," *IEEE Transactions on Computers*, vol. 32, pp. 504–507, May 1983.

[175] A. Shenoy and R. Kumaresan, "Fast base extension using a redundant modulus in RNS," *IEEE Transactions on Computers*, vol. 38, pp. 292–297, February 1989.

[176] E. Claudio, G. Orlandi, and F. Piazza, "A systolic redundant residue arithmetic error correction circuit," *IEEE Transactions on Computers*, vol. 42, pp. 427–432, April 1993.

[177] L. Yang and L. Hanzo, "Redundant residue number system based error correction codes," in *Proceedings of IEEE VTC'01 (Fall)*, (Atlantic City, USA), pp. 1472–1476, October 2001.

[178] M. Etzel and W. Jenkins, "Redundant residue number systems for error detection and correction in digital filters," *IEEE Transactions on Acoustics, Speech and Signal Processing*, vol. ASSP-28, pp. 538–544, October 1980.

[179] C. Huang, D. Peterson, H. Rauch, J. Teague, and D. Fraser, "Implementation of a fast digital processor using the residue number system," *IEEE Transactions on Circuits and Systems*, vol. CAS-28, pp. 32–38, January 1981.

[180] W. Jenkins, "Recent advances in residue number system techniques for recursive digital filtering," *IEEE Transactions on Acoustics, Speech and Signal Processing*, vol. ASSP-27, pp. 19–30, February 1979.

[181] M. Soderstrand, "A high-speed, low-cost, recursive digital filter using residue number arithmetic," *Proceedings of the IEEE*, vol. 65, pp. 1065–1067, July 1977.

[182] M. Soderstrand, W. Jenkins, and G. Jullien, *Residue Number System Arithmetic: Modern Applications in Digital Signal Processing*. New York, USA: IEEE Press, 1986.

[183] R. Krishnan, G. Jullien, and W. Miller, "Complex digital signal processing using quadratic residue number systems," *IEEE Transactions on Acoustics, Speech and Signal Processing*, vol. 34, pp. 166–176, February 1986.

[184] L. L. Yang and L. Hanzo, "Residue number system arithmetic assisted m-ary modulation," *IEEE Communications Letters*, vol. 3, pp. 28–30, February 1999.

[185] A. Baraniecka and G. Jullien, "Residue number system implementations of number theoretic transforms," *IEEE Transactions on Acoustics, Speech and Signal Processing*, vol. ASSP-28, pp. 285–291, June 1980.

[186] M. Soderstrand and E. Fields, "Multipliers for residue number arithmetic digital filters," *Electronics Letters*, vol. 13, pp. 164–166, March 1977.

[187] B. Tseng, G. Jullien, and W. Miller, "Implementation of FFT structure using the residue number system," *IEEE Transactions on Computers*, vol. 28, pp. 831–844, November 1979.

[188] L. L. Yang and L. Hanzo, "Performance of residue number system based DS-CDMA over multipath fading channels using orthogonal sequences," *ETT*, vol. 9, pp. 525–536, November–December 1998.

[189] L. L. Yang and L. Hanzo, "Residue number system based multiple code DS-CDMA systems," in *Proceedings of VTC'99 (Spring)*, (Houston, TX, USA), IEEE, 16–20 May 1999.

[190] L. Hanzo and L. L. Yang, "Ratio statistic test assisted residue number system based parallel communication systems," in *Proceedings of VTC'99 (Spring)*, (Houston, TX, USA), pp. 894–898, IEEE, 16–20 May 1999.

[191] K. Yen, L. L. Yang, and L. Hanzo, "Residual number system assisted CDMA – a new system concept," in *Proceedings of 4th ACTS Mobile Communications Summit'99*, (Sorrento, Italy), pp. 177–182, 8–11 June 1999.

[192] L. L. Yang and L. Hanzo, "A residue number system based parallel communication scheme using orthogonal signaling: part I - system outline," *To appear in IEEE Transactions on Vehicular Technology*, 2002.

[193] L. L. Yang and L. Hanzo, "A residue number system based parallel communication scheme using orthogonal signaling: part II - multipath fading channels," *To appear in IEEE Transactions on Vehicular Technology*, 2002.

[194] D. Mandelbaum, "Error correction in residue arithmetic," *IEEE Transactions on Computers*, vol. C-21, pp. 538–543, June 1972.

[195] L. Yang and L. Hanzo, "Minimum-distance decoding of redundant residue number system codes," in *Proceedings of IEEE ICC'2001*, (Helsinki, Finland), pp. 2975–2979, June 2001.

[196] T. Liew, L. Yang, and L. Hanzo, "Turbo decoded redundant residue number system codes," in *Proceedings of VTC 2000, Spring*, (Tokyo, Japan), pp. 576–580, 15–18 May 2000.

[197] T. Keller, T. H. Liew, and L. Hanzo, "Adaptive redundant residue number system coded multicarrier modulation," *IEEE Journal on Selected Areas in Communications*, vol. 18, pp. 2292–2301, November 2000.

[198] T. Keller, T. Liew, and L. Hanzo, "Adaptive rate RRNS coded OFDM transmission for mobile communication channels," in *Proceedings of VTC 2000, Spring*, (Tokyo, Japan), pp. 230–234, 15–18 May 2000.

[199] W. Jenkins and E. Altman, "Self-checking properties of residue number error checkers based on mixed radix conversion," *IEEE Transactions on Circuits and Systems*, vol. 35, pp. 159–167, February 1988.

[200] W. Jenkins, "The design of error checkers for self-checking residue number arithmetic," *IEEE Transactions on Computers*, vol. 32, pp. 388–396, April 1983.

[201] O. Aitsab and R. Pyndiah, "Performance of Reed-Solomon block turbo code," in *GLOBECOM '96*, (London, UK), pp. 121–125, November 1996.

[202] O. Aitsab and R. Pyndiah, "Performance of concatenated Reed-Solomon/Convolutional codes with iterative decoding," in *GLOBECOM '97*, (New York, USA), pp. 644–648, 1997.

[203] R. Pyndiah, P. Combelles, and P. Adde, "A very low complexity block turbo decoder for product codes," in *GLOBECOM '96*, (London, UK), pp. 101–105, November 1996.

[204] S. Lin, D. Constello Jr., and M. Miller, "Automatic-repeat-request error-control schemes," *IEEE Communications Magazine*, vol. 22, pp. 5–17, December 1984.

[205] T. Keller, T. Liew, and L. Hanzo, "Adaptive redundant residue number system coded multicarrier modulation," *IEEE Journal on Selected Areas in Communications*, vol. 18, pp. 2292–2301, November 2000.

[206] T. James, "Study into redundant residue number system codes," tech. rep., University of Southampton, UK, May 1999.

[207] L. Hanzo, W. Webb, and T. Keller, *Single- and Multi-Carrier Quadrature Amplitude Modulation: Principles and Applications for Personal Communications, WLANs and Broadcasting*. IEEE Press, 2000.

[208] G. Ungerböck, "Channel Coding with Multilevel/Phase Signals," *IEEE Transactions on Information Theory*, vol. IT-28, pp. 55–67, January 1982.

[209] D. Divsalar and M. K. Simon, "The design of trellis coded MPSK for fading channel: Performance criteria," *IEEE Transactions on Communications*, vol. 36, pp. 1004–1012, September 1988.

[210] D. Divsalar and M. K. Simon, "The design of trellis coded MPSK for fading channel: Set partitioning for optimum code design," *IEEE Transactions on Communications*, vol. 36, pp. 1013–1021, September 1988.

[211] P. Robertson, T. Wörz, "Bandwidth-Efficient Turbo Trellis-Coded Modulation Using Punctured Component Codes," *IEEE Journal on Selected Areas in Communications*, vol. 16, pp. 206–218, February 1998.

[212] E. Zehavi, "8-PSK trellis codes for a Rayleigh fading channel," *IEEE Transactions on Communications*, vol. 40, pp. 873–883, May 1992.

[213] X. Li and J.A. Ritcey, "Bit-interleaved coded modulation with iterative decoding," *IEEE Communications Letters*, vol. 1, November 1997.

[214] X. Li and J.A. Ritcey, "Trellis-Coded Modulation with Bit Interleaving and Iterative Decoding," *IEEE Journal on Selected Areas in Communications*, vol. 17, April 1999.

[215] X. Li and J.A. Ritcey, "Bit-interleaved coded modulation with iterative decoding — Approaching turbo-TCM performance without code concatenation," in *Proceedings of CISS 1998*, (Princeton University, USA), March 1998.

[216] S. X. Ng, T. H. Liew, L. L. Yang, and L. Hanzo, "Comparative Study of TCM, TTCM, BICM and BICM-ID schemes," *IEEE Vehicular Technology Conference*, p. 265 (CDROM), May 2001.

[217] S. X. Ng, C. H. Wong and L. Hanzo, "Burst-by-Burst Adaptive Decision Feedback Equalized TCM, TTCM, BICM and BICM-ID," *International Conference on Communications (ICC)*, pp. 3031–3035, June 2001.

[218] C. S. Lee, S. X. Ng, L. Piazzo and L. Hanzo, "TCM, TTCM, BICM and Iterative BICM Assisted OFDM-Based Digital Video Broadcasting to Mobile Receivers," *IEEE Vehicular Technology Conference*, p. 113 (CDROM), May 2001.

[219] X. Li and J.A. Ritcey, "Bit-interleaved coded modulation with iterative decoding using soft feedback," *IEE Electronics Letters*, vol. 34, pp. 942–943, May 1998.

[220] G. Ungerböck, "Trellis-coded modulation with redundant signal sets. Part 1 and 2," *IEEE Communications Magazine*, vol. 25, pp. 5–21, February 1987.

[221] J. H. Chen and A. Gersho, "Gain-adaptive vector quantization with application to speech coding," *IEEE Transactions on Communications*, vol. 35, pp. 918–930, September 1987.

[222] S. S. Pietrobon, G. Ungerböck, L. C. Perez and D. J. Costello, "Rotationally invariant nonlinear trellis codes for two-dimensional modulation," *IEEE Transactions on Information Theory*, vol. IT-40, pp. 1773–1791, November 1994.

[223] C. Schlegel, "Chapter 3: Trellis Coded Modulation," in *Trellis Coding*, pp. 43–89, New York, USA: IEEE Press, September 1997.

[224] J. K. Cavers and P. Ho, "Analysis of the Error Performance of Trellis-Coded Modulations in Rayleigh-Fading Channels," *IEEE Transactions on Communications*, vol. 40, pp. 74–83, January 1992.

[225] G. D. Forney, "The Viterbi Algorithm," in *Proceedings of the IEEE*, vol. 61, pp. 268–277, March 1973.

[226] J.G. Proakis, "Optimum Receivers for the Additive White Gaussian Noise Channel," in *Digital Communications*, pp. 260–274, New York, USA: McGraw-Hill International Edition, 3rd Ed., September 1995.

[227] K. Abend and B. D. Fritchman, "Statistical detection for communication channels with intersymbol interference," *Proceedings of the IEEE*, vol. 58, pp. 779–785, May 1970.

[228] L. Piazzo, "An Algorithm for SBS Receivers/Decoders," *IEE Electronics Letters*, vol. 32, pp. 1058–1060, June 1996.

[229] S. S. Pietrobon, R. H. Deng, A. Lafanechére, G. Ungerböck, and D. J. Costello, "Trellis-Coded Multidimensional Phase Modulation," *IEEE Transactions on Information Theory*, vol. 36, pp. 63–89, January 1990.

[230] L. F. Wei, "Trellis-coded modulation with multidimensional constellations," *IEEE Transactions on Information Theory*, vol. IT-33, pp. 483–501, July 1987.

[231] P. Robertson, "An Overview of Bandwidth Efficient Turbo Coding Schemes," in *International Symposium on Turbo Codes and related topics*, (Brest, France), pp. 103–110, September 1997.

[232] G. Caire, G. Taricco, and E. Biglieri, "Bit-Interleaved Coded Modulation," *IEEE Transactions on Information Theory*, vol. 44, pp. 927–946, May 1998.

[233] J. Hagenauer, "Rate-compatible puncture convolutional codes (RCPC) and their application," *IEEE Transactions on Communications*, vol. 36, pp. 389–400, April 1988.

[234] L. Lee, "New rate-compatible puncture convolutional codes for Viterbi decoding," *IEEE Transactions on Communications*, vol. 42, pp. 3073–3079, December 1994.

[235] S. Benedetto, D. Divsalar, G. Montorsi, and F. Pollara, "A Soft-Input Soft-Output APP Module for Iterative Decoding of Concatenated Codes," *IEEE Communications Letters*, vol. 1, pp. 22–24, January 1997.

[236] J.G. Proakis, "Chapter 10: Communication Through Band-Limited Channels," in *Digital Communications*, pp. 583–635, New York, USA: McGraw-Hill International Editions, 3rd Ed., September 1995.

[237] C. H. Wong, *Wideband Adaptive Full Response Multilevel Transceivers and Equalizers*. PhD thesis, University of Southampton, UK, November 1999.

[238] D. F. Mix, *Random Signal Processing*. Englewood Cliffs, NJ, USA: Prentice Hall, 1995.

[239] J. C. Cheung, *Adaptive Equalisers for Wideband TDMA Mobile Radio*. PhD thesis, Department of Electronics and Computer Science, University of Southampton, UK, 1991.

[240] R. Steele and W. Webb, "Variable rate QAM for data transmission over Rayleigh fading channels," in *Proceeedings of Wireless '91*, (Calgary, Alberta), pp. 1–14, IEEE, 1991.

[241] S. Sampei, S. Komaki, and N. Morinaga, "Adaptive Modulation/TDMA Scheme for Large Capacity Personal Multi-Media Communication Systems," *IEICE Transactions on Communications (Japan)*, vol. E77-B, pp. 1096–1103, September 1994.

[242] J. M. Torrance and L. Hanzo, "Latency and Networking Aspects of Adaptive Modems over Slow Indoors Rayleigh Fading Channels," *IEEE Transactions on Vehicular Technology*, vol. 48, no. 4, pp. 1237–1251, 1998.

[243] J. M. Torrance and L. Hanzo, "Interference aspects of adaptive modems over slow Rayleigh fading channels," *IEEE Vehicular Technology Conference*, vol. 48, pp. 1527–1545, September 1999.

[244] A. J. Goldsmith and S. Chua, "Variable-rate variable-power MQAM for fading channels," *IEEE Transactions on Communications*, vol. 45, pp. 1218–1230, October 1997.

[245] C. Wong and L. Hanzo, "Upper-bound performance of a wideband burst-by-burst adaptive modem," *IEEE Transactions on Communications*, vol. 48, pp. 367–369, March 2000.

[246] H. Matsuoka and S. Sampei and N. Morinaga and Y. Kamio, "Adaptive Modulation System with Variable Coding Rate Concatenated Code for High Quality Multi-Media Communications Systems," in *Proceedings of IEEE VTC'96*, vol. 1, (Atlanta, USA), pp. 487–491, IEEE, 28 April–1 May 1996.

[247] V. Lau and M. Macleod, "Variable rate adaptive trellis coded QAM for high bandwidth efficiency applications in Rayleigh fading channels," in *Proceedings of IEEE Vehicular Technology Conference (VTC'98)*, (Ottawa, Canada), pp. 348–352, IEEE, 18–21 May 1998.

[248] A. J. Goldsmith and S. Chua, "Adaptive Coded Modulation for Fading Channels," *IEEE Transactions on Communications*, vol. 46, pp. 595–602, May 1998.

[249] C. H. Wong, T. H. Liew and L. Hanzo, "Burst-by-Burst Turbo Coded Wideband Adaptive Modulation with Blind Modem Mode Detection," in *Proceedings of 4th ACTS Mobile Communications Summit 1999*, (Sorrento, Italy), pp. 303–308, June 1999.

[250] D. Goeckel, "Adaptive Coding for Fading Channels using Outdated Fading Estimates," *IEEE Transactions on Communications*, vol. 47, pp. 844–855, June 1999.

[251] "COST207: Digital land mobile radio communications, final report." Office for Official Publications of the European Communities, 1989. Luxembourg.

[252] A. Klein, R. Pirhonen, J. Skoeld, and R. Suoranta, "FRAMES Multiple Access MODE 1 — Wideband TDMA with and without Spreading," in *Proceedings of the IEEE International Symposium on Personal, Indoor and Mobile Radio Communications (PIMRC)*, vol. 1, (Helsinki, Finland), pp. 37–41, 1–4 September 1997.

[253] B. J. Choi, M. Münster, L. L. Yang, and L. Hanzo, "Performance of Rake receiver assisted adaptive-modulation based CDMA over frequency selective slow Rayleigh fading channel," *Electronics Letters*, vol. 37, pp. 247–249, February 2001.

[254] A. Duel-Hallen, S. Hu, and H. Hallen, "Long Range Prediction of Fading Signals," *IEEE Signal Processing Magazine*, vol. 17, pp. 62–75, May 2000.

[255] S. M. Alamouti and S. Kallel, "Adaptive Trellis-Coded Multiple-Phased-Shift Keying Rayleigh Fading Channels," *IEEE Transactions on Communications*, vol. 42, pp. 2305–2341, June 1994.

[256] L. Piazzo and L. Hanzo, "TTCM-OFDM over Dispersive Fading Channels," *IEEE Vehicular Technology Conference*, vol. 1, pp. 66–70, May 2000.

[257] R. W. Chang, "Synthesis of Band-Limited Orthogonal Signals for Multichannel Data Transmission," *Bell Systems Technical Journal*, vol. 46, pp. 1775–1796, December 1966.

[258] M. S. Zimmermann and A. L. Kirsch, "The AN/GSC-10/KATHRYN/Variable Rate Data Modem for HF Radio," *IEEE Transactions on Communication Technology*, vol. CCM-15, pp. 197–205, April 1967.

[259] L.J. Cimini, "Analysis and Simulation of a Digital Mobile Channel Using Orthogonal Frequency Division Multiplexing," *IEEE Transactions on Communications*, vol. 33, pp. 665–675, July 1985.

[260] F. Mueller-Roemer, "Directions in audio broadcasting," *Journal of the Audio Engineering Society*, vol. 41, pp. 158–173, March 1993.

[261] M. Alard and R. Lassalle, "Principles of modulation and channel coding for digital broadcasting for mobile receivers," *EBU Review, Technical No. 224*, pp. 47–69, August 1987.

[262] I. Kalet, "The multitone channel," *IEEE Transactions on Communications*, vol. 37, pp. 119–124, February 1989.

[263] B. L. Yeap, T. H. Liew, and L. Hanzo, "Turbo Equalization of Serially Concatenated Systematic Convolutional Codes and Systematic Space Time Trellis Codes," *IEEE Vehicular Technology Conference*, p. 119 (CDROM), May 2001.

[264] ETSI, *Digital Video Broadcasting (DVB); Framing structure, channel coding and modulation for digital terrestrial television*, August 1997. ETS 300 744.

[265] "COST207: Digital land mobile radio communications, final report," tech. rep., Luxembourg, 1989.

[266] P. Chaudhury, W. Mohr, and S. Onoe, "The 3GPP proposal for IMT-2000," *IEEE Communications Magazine*, vol. 37, pp. 72–81, December 1999.

[267] G. Foschini Jr. and M. Gans, "On limits of wireless communication in a fading environment when using multiple antennas," *Wireless Personal Communications*, vol. 6, pp. 311–335, March 1998.

[268] B. Glance and L. Greestein, "Frequency-selective fading effects in digital mobile radio with diversity combining," *IEEE Transactions on Communications*, vol. COM-31, pp. 1085–1094, September 1983.

[269] F. Adachi and K. Ohno, "BER performance of QDPSK with postdetection diversity reception in mobile radio channels," *IEEE Transactions on Vehicular Technology*, vol. 40, pp. 237–249, February 1991.

[270] H. Zhou, R. Deng, and T. Tjhung, "Performance of combined diversity reception and convolutional coding for QDPSK land mobile radio," *IEEE Transactions on Vehicular Technology*, vol. 43, pp. 499–508, August 1994.

[271] J. Winters, "Switched diversity with feedback for DPSK mobile radio systems," *IEEE Transactions on Information Theory*, vol. 32, pp. 134–150, February 1983.

[272] G. Raleigh and J. Cioffi, "Spatio-temporal coding for wireless communications," in *GLOBECOM '96*, (London, UK), pp. 533–537, November 1996.

[273] A. Wittneben, "Base station modulation diversity for digital SIMULCAST," in *Proceedings of IEEE Vehicular Technology Conference*, pp. 505–511, May 1993.

[274] N. Seshadri and J. Winters, "Two signalling schemes for improving the error performance of frequency-division-duplex (FDD) transmission systems using transmitter antenna diversity," *International Journal of Wireless Information Networks*, vol. 1, pp. 49–60, January 1994.

[275] J. Winters, "The diversity gain of transmit diversity in wireless systems with Rayleigh fading," *IEEE Transactions on Vehicular Technology*, vol. 47, pp. 119–123, February 1998.

[276] T. Hattori and K. Hirade, "Multitransmitter simulcast digital signal transmission by using frequency offset strategy in land mobile radio-telephone system," *IEEE Transactions on Vehicular Technology*, vol. 27, pp. 231–238, 1978.

[277] A. Hiroike, F. Adachi, and N. Nakajima, "Combined effects of phase sweeping transmitter diversity and channel coding," *IEEE Transactions on Vehicular Technology*, vol. 41, pp. 170–176, May 1992.

[278] N. Seshadri, V. Tarokh, and A. Calderbank, "Space-Time Codes for High Data Rate Wireless Communications: Code Construction," in *Proceedings of IEEE Vehicular Technology Conference '97*, (Phoenix, Arizona), pp. 637–641, 1997.

[279] V. Tarokh and N. Seshadri and A. Calderbank, "Space-time codes for high data rate wireless communications: Performance criterion and code construction," in *Proceedings of the IEEE International Conference on Communications '97*, (Montreal, Canada), pp. 299–303, 1997.

[280] V. Tarokh, A. Naguib, N. Seshadri, and A. Calderbank, "Space-time codes for high data rate wireless communications: Mismatch analysis," in *Proceedings of the IEEE International Conference on Communications '97*, (Montreal, Canada), pp. 309–313, 1997.

[281] V. Tarokh, A. Naguib, N. Seshadri, and A. R. Calderbank, "Space-time codes for high data rate wireless communication: Performance criteria in the presence of channel estimation errors, mobility, and multile paths," *IEEE Transactions on Communications*, vol. 47, pp. 199–207, February 1999.

[282] D. Brennan, "Linear diversity combining techniques," *Proceedings of the IRE*, vol. 47, pp. 1075–1102, 1959.

[283] G. Bauch, "Concatenation of space-time block codes and Turbo-TCM," in *Proceedings of IEEE International Conference on Communications*, (Vancouver, Canada), pp. 1202–1206, June 1999.

[284] G. Forney, "Convolutional codes: I. Algebraic structure," *IEEE Transactions on Information Theory*, vol. 16, pp. 720–738, November 1970.

[285] G. Forney, "Burst-correcting codes for the classic burst channel," *IEEE Transactions on Communication Technology*, vol. COM-19, pp. 772–781, October 1971.

[286] W. Jakes Jr., ed., *Microwave Mobile Communications*. New York, USA: John Wiley & Sons, 1974.

[287] S. Red, M. Oliphant, and M. Weber, *An Introduction to GSM*. London, UK: Artech House, 1995.

[288] 3GPP, *Multiplexing and channel coding (TDD)*. 3G TS 25.222, http://www.3gpp.org.

[289] T. Ojanperä and R. Prasad, *Wideband CDMA for Third Generation Mobile Communications*. London, UK: Artech House, 1998.

[290] S. Al-Semari and T. Fuja, "I-Q TCM: Reliable communication over the Rayleigh fading channel close to the cutoff rate," *IEEE Transactions on Information Theory*, vol. 43, pp. 250–262, January 1997.

[291] R. Horn and C. Johnson, *Matrix Analysis*. New York, USA: Cambridge University Press, 1988.

[292] W. Choi and J. Cioffi, "Space-Time Block Codes over Frequency Selective Fading Channels," in *Proceedings of VTC 1999 Fall*, (Amsterdam, Holland), pp. 2541–2545, 19–22 September 1999.

[293] Z. Liu, G. Giannakis, A. Scaglione, and S. Barbarossa, "Block precoding and transmit-antenna diversity for decoding and equalization of unknown multipath channels," in *Proceedings of the 33rd Asilomar Conference Signals, Systems and Computers*, (Pacific Grove, Canada), pp. 1557–1561, 1–4 November 1999.

[294] Z. Liu and G. Giannakis, "Space-time coding with transmit antennas for multiple access regardless of frequency-selective multipath," in *Proceedings of the 1st Sensor Array and Multichannel SP Workshop*, (Boston, USA), 15–17 March 2000.

[295] T. Liew, J. Pliquett, B. Yeap, L. L. Yang, and L. Hanzo, "Comparative study of space time block codes and various concatenated turbo coding schemes," in *PIMRC 2000*, (London, UK), pp. 741–745, 18–21 September 2000.

[296] T. Liew, J. Pliquett, B. Yeap, L. L. Yang, and L. Hanzo, "Concatenated space time block codes and TCM, turbo TCM, convolutional as well as turbo codes," in *GLOBECOM 2000*, (San Francisco, USA), 27 November – 1 December 2000.

[297] W. Webb and R. Steele, "Variable rate QAM for mobile radio," *IEEE Transactions on Communications*, vol. 43, pp. 2223–2230, July 1995.

[298] J. Torrance and L. Hanzo, "Performance upper bound of adaptive QAM in slow Rayleigh-fading environments," in *Proceedings of IEEE ICCS'96/ISPACS'96*, (Singapore), pp. 1653–1657, IEEE, 25–29 November 1996.

[299] J. M. Torrance, L. Hanzo, and T. Keller, "Interference aspects of adaptive modems over slow Rayleigh fading channels," *IEEE Transactions on Vehicular Technology*, vol. 48, pp. 1527–1545, September 1999.

[300] T. Keller and L. Hanzo, "Adaptive orthogonal frequency division multiplexing schemes," in *Proceeding of ACTS Mobile Communication Summit '98*, (Rhodes, Greece), pp. 794–799, ACTS, 8–11 June 1998.

[301] T. Keller and L. Hanzo, "Blind-detection assisted sub-band adaptive turbo-coded OFDM schemes," in *Proceedings of VTC'99 (Spring)*, (Houston, USA), pp. 489–493, IEEE, 16–20 May 1999.

[302] H. Matsuoka, S. Sampei, N. Morinaga, and Y. Kamio, "Adaptive modulation systems with variable coding rate concatenated code for high quality multi-media communication systems," in *Proceedings of IEEE VTC'96*, (Atlanta, USA), pp. 487–491, 28 April–1 May 1996.

[303] S. G. Chua and A. J. Goldsmith, "Variable-rate variable-power mQAM for fading channels," in *Proceedings of IEEE VTC'96*, (Atlanta, USA), pp. 815–819, 28 April–1 May 1996.

[304] J. Torrance and L. Hanzo, "Optimisation of switching levels for adaptive modulation in a slow Rayleigh fading channel," *Electronics Letters*, vol. 32, pp. 1167–1169, 20 June 1996.

[305] J. Torrance and L. Hanzo, "On the upper bound performance of adaptive QAM in a slow Rayleigh fading," *IEE Electronics Letters*, pp. 169–171, April 1996.

[306] L. Hanzo, C. H. Wong, and M. S. Yee, *Adaptive Wireless Transceivers*. New York, USA: John Wiley, IEEE Press, 2002. (For detailed contents, please refer to http://www-mobile.ecs.soton.ac.uk.).

[307] M. Yee, T. Liew, and L. Hanzo, "Burst-by-burst adaptive turbo coded radial basis function assisted feedback equalisation," *IEEE Transactions on Communications*, vol. 49, November 2001.

[308] V. Lau and M. Macleod, "Variable-rate adaptive trellis coded QAM for flat-fading channels," *IEEE Transactions on Communications*, vol. 49, pp. 1550–1560, September 2001.

[309] V. Lau and S. Maric, "Variable rate adaptive modulation for DS-CDMA," *IEEE Transactions on Communications*, vol. 47, pp. 577–589, April 1999.

[310] S. Chua and A. Goldsmith, "Adaptive Coded Modulation for Fading Channels," *IEEE Transactions on Communications*, vol. 46, pp. 595–602, May 1998.

[311] X. Liu, P. Ormeci, R. Wesel, and D. Goeckel, "Bandwidth-efficient, low-latency adaptive coded modulation schemes for time-varying channels," in *Proceedings of IEEE International Conference on Communications*, (Helsinki, Finland), June 2001.

[312] C. Tidestav, A. Ahlén and M. Sternad, "Realizable MIMO Decision Feedback Equalizer," *in Proceedings of the 1999 International Conference on Acoustics Speech and Signal Processing*, pp. 2591–2594, 1999.

[313] C. Tidestav, A. Ahlén and M. Sternad, "Realiazable MIMO Decision Feedback Equalizer: Structure and Design," *IEEE Transactions on Signal Processing*, vol. 49, pp. 121–133, January 2001.

[314] C. Tidestav, M. Sternad, and A. Ahlén, "Reuse within a cell - interference rejection or multiuser detection," *IEEE Transactions on Communications*, vol. 47, pp. 1511–1522, October 1999.

[315] T. Liew and L. Hanzo, "Space-time block coded adaptive modulation aided OFDM," in *IEEE Globecom 2001*, (San Antonio, USA), pp. 136–140, 25–29 November 2001.

[316] J. Cheung and R. Steele, "Soft-decision feedback equalizer for continuous-phase modulated signals in wide-band mobile radio channels," *IEEE Transactions on Communications*, vol. 42, pp. 1628–1638, February/March/April 1994.

[317] T. Liew and L. Hanzo, "Space-time codes and concatenated channel codes for wireless communications," *to appear in Proceedings of the IEEE*, February 2002.

[318] J. Torrance and L. Hanzo, "Demodulation level selection in adaptive modulation," *Electronics Letters*, vol. 32, pp. 1751–1752, 12 September 1996.

[319] B. J. Choi, T. H. Liew, and L. Hanzo, "Concatenated space-time block coded and turbo coded symbol-by-symbol adaptive OFDM and multi-carrier CDMA systems," in *Proceedings of IEEE VTC 2001-Spring*, p. 528, IEEE, May 2001.

[320] J. Anderson, T. Aulin, and C. Sundberg, *Digital phase modulation*. New York, USA: Plenum Press, 1986.

[321] K. Murota and K. Hirade, "GMSK modulation for digital mobile radio telephony," *IEEE Transactions on Communications*, vol. 29, pp. 1044–1050, July 1981.

[322] ETSI, *Digital Cellular Telecommunications System (Phase 2+); High Speed Circuit Switched Data (HSCSD) — Stage 1; (GSM 02.34 Version 5.2.1)*. European Telecommunications Standards Institute, Sophia Antipolis, Cedex, France, July 1997.

[323] ETSI, *Digital Cellular Telecommunications System (Phase 2+); General Packet Radio Service (GPRS); Overall Description of the GPRS Radio Interface, Stage 2 (GSM 03.64 Version 5.2.0)*. European Telecommunications Standards Institute, Sophia Antipolis, Cedex, France, January 1998.

[324] I. Gerson, M. Jasiuk, J. M. Muller, J. Nowack, and E. Winter, "Speech and channel coding for the half-rate GSM channel," *Proceedings ITG-Fachbericht*, vol. 130, pp. 225–233, November 1994.

[325] R. Salami, C. Laflamme, B. Besette, J. P. Adoul, K. Jarvinen, J. Vainio, P. Kapanen, T. Hankanen, and P. Haavisto, "Description of the GSM enhanced full rate speech codec," in *Proceedings of ICC'97*, 1997.

[326] G. Bauch and V. Franz, "Iterative equalisation and decoding for the GSM-system," in *Proceedings of IEEE Vehicular Technology Conference (VTC'98)*, (Ottawa, Canada), pp. 2262–2266, IEEE, 18–21 May 1998.

[327] F. Burkert, G. Caire, J. Hagenauer, T. Hidelang, and G. Lechner, "Turbo decoding with unequal error protection applied to GSM speech coding," in *Proceedings of IEEE Global Telecommunications Conference, Globecom 96*, (London, UK), pp. 2044–2048, IEEE, 18–22 November 1996.

[328] D. Parsons, *The Mobile Radio Propagation Channel*. London: Pentech Press, 1992.

[329] S. Saunders, *Antennas and Propagation for Wireless Communication Systems Concept and Design*. New York, USA: John Wiley & Sons, 1999.

[330] T. Rappaport, *Wireless Communications Principles and Practice*. Englewood Cliffs, NJ, USA: Prentice Hall, 1996.

[331] ETSI, *GSM Recommendation 05.05, Annex 3*, November 1988.

[332] A. Carlson, *Communication Systems*. New York, USA: McGraw-Hill, 1975.

[333] T. Aulin and C.-E. Sundberg, "CPM-An efficient constant amplitude modulation scheme," *International Journal of Satellite Communications*, vol. 2, pp. 161–186, 1994.

[334] R. Debuda, "Coherent demodulation of frequency shift keying with low deviation ratio," *IEEE Transactions on Communications*, vol. COM-20, pp. 429–435, June 1972.

[335] S. Pasupathy, "Minimum shift keying: A spectrally efficient modulation," *IEEE Communications Magazine*, vol. 17, pp. 14–22, July 1979.

[336] M. K. Simon and C. Wang, "Differential detection of Gaussian MSK in a mobile radio environment," *IEEE Transactions on Vehicular Technology*, vol. 33, pp. 307–320, November 1984.

[337] E. Kreyszig, *Advanced engineering mathematics*. New York, USA: John Wiley & Sons, 7th ed., 1993.

[338] J. Proakis and D. Manolakis, *Digital Signal Processing — Principles, Algorithms and Applications*. Macmillan, 1992.

[339] M. Moher, "Decoding via cross-entropy minimization," in *Proceedings of the IEEE Global Telecommunications Conference 1993*, (Houston, USA), pp. 809–813, 29 November–2 December 1993.

[340] D. A. Johnson, S. W. Wales, and P. H. Waters, "Equalisers for GSM," *IEE Colloquium (Digest)*, no. 21, pp. 1/1–1/6, 1990.

[341] A. Papoulis, *Probability, Random Variables, and Stochastic Processes*. New York USA: McGraw-Hill, 3 ed., 1991.

[342] P. Robertson and T. Wörz, "Coded modulation scheme employing turbo codes," *IEE Electronics Letters*, vol. 31, pp. 1546–1547, 31 August 1995.

[343] "GSM Recommendation 05.03: Channel coding," November 1988.

[344] M. Mouly and M. Pautet, *The GSM System for Mobile Communications*. Michel Mouly and Marie-Bernadette Pautet, 1992.

[345] G. Bauch, H. Khorram, and J. Hagenauer, "Iterative equalization and decoding in mobile communications systems," in *European Personal Mobile Communications Conference*, (Bonn, Germany), pp. 301–312, 30 September–2 October 1997.

[346] M. Gertsman and J. Lodge, "Symbol-by-symbol MAP demodulation of CPM and PSK signals on Rayleigh flat-fading channels," *IEEE Transactions on Communications*, vol. 45, pp. 788–799, July 1997.

[347] I. Marsland, P. Mathiopoulos, and S. Kallel, "Non-coherent turbo equalization for frequency selective Rayleigh fast fading channels," in *International Symposium on Turbo Codes and related topics*, (Brest, France), pp. 196–199, September 1997.

[348] Q. Dai and E. Shwedyk, "Detection of bandlimited signals over frequency selective Rayleigh fading channels," *IEEE Transactions on Communications*, pp. 941–950, February/March/April 1994.

[349] F. Jordan and K. Kammeyer, "Study on iterative decoding techniques applied to GSM full-rate channels," in *Proceedings of the IEEE International Symposium on Personal, Indoor and Mobile Radio Communications, PIMRC*, (Pisa, Italy), pp. 1066–1070, 29 September–2 October 1998.

[350] G. Qin, S. Zhou, and Y. Yao, "Iterative decoding of GMSK modulated convolutional code," *IEE Electronics Letters*, vol. 35, pp. 810–811, 13 May 1999.

[351] K. Narayanan and G. Stüber, "A serial concatenation approach to iterative demodulation and decoding," *IEEE Transactions on Communications*, vol. 47, pp. 956–961, July 1999.

[352] L. Lin and R. Cheng, "Improvements in SOVA-based decoding for turbo codes," in *Proceedings of the IEEE International Conference on Communications*, vol. 3, (Montreal, Canada), pp. 1473–1478, 8–12 June 1997.

[353] Y. Liu, M. Fossorier, and S. Lin, "MAP algorithms for decoding linear block code based on sectionalized trellis diagrams," in *Proceedings of the IEEE Global Telecommunications Conference 1998*, vol. 1, (Sydney, Australia), pp. 562–566, 8–12 November 1998.

[354] Y. V. Svirid and S. Riedel, "Threshold decoding of turbo-codes," *Proceedings of the IEEE International Symposium on Information Theory*, p. 39, September 1995.

[355] J. S. Reeve, "A parallel Viterbi decoding algorithm," *Concurrency and Computation: Practice and Experience*, vol. 13, pp. 95–102, July 2001.

[356] S. Benedetto, R. Garello, and G. Montorsi, "A search for good convolutional codes to be used in the construction of turbo codes," *IEEE Transactions on Communications*, vol. 46, pp. 1101–1105, September 1998.

[357] M. Ho, S. Pietrobon, and T. Giles, "Improving the constituent codes of turbo encoders," in *Proceedings of the IEEE Global Telecommunications Conference 1998*, vol. 6, (Sydney, Australia), pp. 3525–3529, 8–12 November 1998.

[358] A. Ushirokawa, T. Okamura, and N. Kamiya, "Principles of turbo codes and their application to mobile communications," *IEICE Transactions on Fundamentals of Electronics, Communications and Computer Science*, vol. E81-A, pp. 1320–1329, July 1998.

[359] A. Shibutani, H. Suda, and F. Adachi, "Complexity reduction of turbo decoding," in *Proceedings of the IEEE Vehicular Technology Conference 1999*, (Amsterdam, The Netherlands), pp. 1570–1574, 19–22 September 1999.

[360] J. Yuan, B. Vucetic, and W. Feng, "Combined turbo codes and interleaver design," *IEEE Transactions on Communications*, vol. 47, no. 4, pp. 484–487, 1999.

[361] B. He and M. Z. Wang, "Interleaver design for turbo codes," in *Proceedings of the International Conference on Information, Communications and Signal Processing, ICICS*, vol. 1, (Singapore, Singapore), pp. 453–455, 9-12 September 1997.

[362] M. Breiling and L. Hanzo, "Optimum non-iterative turbo-decoding," in *IEEE International Symposium on Personal, Indoor and Mobile Radio Communications, PIMRC, 1997*, vol. 2, pp. 714–718, IEEE, 1997.

[363] M. Breiling and L. Hanzo, "Non-iterative optimum super-trellis decoding of turbo codes," *Electronics Letters*, vol. 33, pp. 848–849, May 1997.

[364] P. Höher and J. Lodge, "Turbo DPSK: Iterative differential PSK demodulation and channel decoding," *IEEE Transactions on Communications*, vol. 47, pp. 837–843, June 1999.

[365] ETSI, *Digital Video Broadcasting (DVB); Framing structure, channel coding and modulation for 11/12 GHz Satellite Services*, August 1997. ETS 300 421.

[366] A. Knickenberg, B. Yeap, J. Hàmorskỳ, M. Breiling, and L. Hanzo, "Non-iterative Joint Channel Equalisation and Channel Decoding," *IEE Electronics Letters*, vol. 35, pp. 1628–1630, 16 September 1999.

[367] M. S. Yee, B. L. Yeap, and L. Hanzo, "Radial basis function assisted turbo equalisation," in *Proceedings of IEEE Vehicular Technology Conference*, (Japan, Tokyo), pp. 640–644, IEEE, 15–18 May 2000.

[368] A. Glavieux, C. Laot, and J. Labat, "Turbo equalization over a frequency selective channel," in *Proceedings of the International Symposium on Turbo Codes*, (Brest, France), pp. 96–102, 3–5 September 1997.

[369] C. Wong, B. Yeap, and L. Hanzo, "Wideband Burst-by-Burst Adaptive Modulation with Turbo Equalization and Iterative Channel Estimation," in *Proceedings of the IEEE Vehicular Technology Conference 2000*, 2000.

[370] M. Sandell, C. Luschi, P. Strauch, and R. Yan, "Iterative channel estimation using soft decision feedback," in *Proceedings of the Global Telecommunications Conference 1998*, (Sydney, Australia), pp. 3728–3733, 8–12 November 1998.

[371] S. Haykin, *Digital Communications*. New York, USA: John Wiley & Sons, 1988.

[372] S. Benedetto, D. Divsalar, G. Motorsi, and F. Pollara, "A soft-input soft-output APP module for iterative decoding of concatenated codes," *IEEE Communication Letters*, pp. 22–24, 1997.

[373] P. Robertson, P. Höher, and E. Villebrun, "Optimal and sub-optimal maximum a posteriori algorithms suitable for turbo decoding," *European Transactions on Telecommunications*, vol. 8, pp. 119–125, March/April 1997.

[374] A. Whalen, *Detection of signals in noise*. New York, USA: Academic Press, 1971.

[375] B. L. Yeap, C. H. Wong, and L. Hanzo, "Reduced complexity in-phase/quadrature-phase turbo equalisation with iterative channel estimation," in *IEEE International Communications Conference 2001*, (Helsinki, Finland), 11–15 June 2001. Accepted for publication.

[376] T. H. Liew, B. L. Yeap, J. P. Woodard, and L. Hanzo, "Modified MAP Algorithm for Extended Turbo BCH Codes and Turbo Equalisers," in *Proceedings of the First International Conference on 3G Mobile Communication Technologies, January 2000*, (London, UK), pp. 185–189, 27-29 March 2000.

[377] W. Lee, "Estimate of channel capacity in Rayleigh fading environment," *IEEE Transactions on Vehicular Technology*, vol. 39, pp. 187–189, August 1990.

[378] S. Nanda, K. Balachandran, and S. Kumar, "Adaptation techniques in wireless packet data services," *IEEE Communications Magazine*, vol. 38, pp. 54–64, January 2000.

[379] J. Blogh and L. Hanzo, *3G Systems and Intelligent Networking*. New York, USA: John Wiley and IEEE Press, 2002. (For detailed contents, please refer to http://www-mobile.ecs.soton.ac.uk.).

[380] L. Hanzo, P. Cherriman, and J. Streit, *Wireless Video Communications: From Second to Third Generation Systems, WLANs and Beyond*. Piscataway, NJ, USA: IEEE Press, 2001. (For detailed contents please refer to http://www-mobile.ecs.soton.ac.uk).

[381] L. Hanzo, C. Wong, and P. Cherriman, "Channel-adaptive wideband video telephony," *IEEE Signal Processing Magazine*, vol. 17, pp. 10–30, July 2000.

[382] L. Hanzo, P. Cherriman, and E. Kuan, "Interactive cellular and cordless video telephony: State-of-the-art, system design principles and expected performance," *Proceedings of IEEE*, September 2000.

Subject Index

—

Author Index

About the Authors

Lajos Hanzo (http://www-mobile.ecs.soton.ac.uk) received his degree in electronics in 1976 and his doctorate in 1983. During his 25-year career in telecommunications he has held various research and academic posts in Hungary, Germany and the UK. Since 1986 he has been with the Department of Electronics and Computer Science, University of Southampton, UK, where he holds the chair in telecommunications. He has co-authored eight books on mobile radio communications, published over 300 research papers, organised and chaired conference sessions, presented overview lectures and been awarded a number of distinctions. Currently he is managing an academic research team, working on a range of research projects in the field of wireless multimedia communications sponsored by industry, the Engineering and Physical Sciences Research Council (EPSRC) UK, the European IST Programme and the Mobile Virtual Centre of Excellence (VCE), UK. He is an enthusiastic supporter of industrial and academic liaison and he offers a range of industrial courses. He is also an IEEE Distinguished Lecturer. For further information on research in progress and associated publications please refer to http://www-mobile.ecs.soton.ac.uk

T.H. Liew received the B.Eng degree in Electronics Engineering from the University of Southampton, U.K. In 2001 he was awarded the degree of PhD by the same university and currently he is continuing his research as a postdoctoral research fellow. His current research interests are associated with coding and modulation for wireless channels, space-time coding, adaptive transceivers, etc. He published his research results widely.

Bee Leong Yeap graduated in Electronics Engineering from the University of Southampton, UK, with a first class honours degree in 1996. In 2000, he was awarded a PhD and since then he has been continuing his research as a postdoctoral research fellow in Southampton. His research interests include turbo coding, turbo equalisation, adaptive modulation and space-time coding.

Other Related Wiley and IEEE Press Books

- L. Hanzo, W.T. Webb, T. Keller: Single- and Multi-carrier Quadrature Amplitude Modulation: Principles and Applications for Personal Communications, WATM and Broadcasting; IEEE Press-John Wiley, 2000 [2]

- R. Steele, L. Hanzo (Ed): Mobile Radio Communications: Second and Third Generation Cellular and WATM Systems, John Wiley-IEEE Press, 2nd edition, 1999, ISBN 07 273-1406-8

- L. Hanzo, F.C.A. Somerville, J.P. Woodard: Voice Compression and Communications: Principles and Applications for Fixed and Wireless Channels; IEEE Press-John Wiley, 2001

- L. Hanzo, P. Cherriman, J. Streit: Wireless Video Communications: Second to Third Generation and Beyond, IEEE Press, 2001

- L. Hanzo, C.H. Wong, M.S. Yee: Adaptive wireless transceivers: Turbo-Coded, Turbo-Equalised and Space-Time Coded TDMA, CDMA and OFDM systems, John Wiley, 2002

- J.S. Blogh, L. Hanzo: Third-Generation Systems and Intelligent Wireless Networking - Smart Antennas and Adaptive Modulation, John Wiley, 2002

---

[2] For detailed contents please refer to http://www-mobile.ecs.soton.ac.uk

Blurb

For the sake of completeness and wide reader appeal, virtually no prior knowledge is assumed in the field of channel coding. In **Chapter 1** we commence our discourse by introducing the family of convolutional codes and the hard- as well as soft-decision Viterbi algorithm in simple conceptual terms with the aid of worked examples.

Chapter 2 provides a rudimentary introduction to the most prominant classes of block codes, namely to Reed-Solomon (RS) and Bose-Chaudhuri-Hocquenghem (BCH) codes. A range of algebraic decoding techiques are reviewed and worked examples are provided.

Chapter 3 elaborates on the trellis-decoding of BCH codes using worked examples and characterises their performance. Furthermore, the classic Chase algorithm is introduced and its performance is investigated.

Chapter 4 introduces the concept of turbo convolutional codes and gives a detailed discourse on the Maximum Aposteriory (MAP) algorithm and its computationally less demanding counterparts, namely the Log-MAP and Max-Log-MAP algorithms. The Soft Output Viterbi Algorithm (SOVA) is also highlighted and its concept is augmented with the aid of a detailed worked example. Then the effects of the various turbo codec parameters are investigated.

Chapter 5 comparatively studies the trellis structure of convolutional and turbo codes, while **Chapter 6** characterises turbo BCH codes. **Chapter 7** is a unique portrayal of the novel family of Redundant Residue Number System (RNS) based codes and their turbo decoding. **Chapter 8** considers the family of joint coding and modulation based arrangements, which are often referred to as coded modulation schemes. Specifically, Trellis Coded Modulation (TCM), Turbo Trellis Coded Modulation (TTCM), Bit-Interleaved Coded Modulation (BICM) as well as iterative joint decoding and demodulation assisted BICM (BICM-ID) are studied and compared under various narrow-band and wide-band propagation conditions.

In **Chapter 9 and 10** space-time block codes and space-time trellis codes are introduced. Their performance is studied comparative in conjunction with a whole host of channel codecs, providing guide-lines for system designers. As a lower-complexity design alternative to multiple-transmitter, multiple-receiver (MIMO) based schemes the concept of near-instantaneously Adaptive Quadrature Amplitude Modulation (AQAM), combined with near-instantaneously adaptive turbo channel coding is introduced in **Chapter 11**.

Based on the introductory concepts of **Chapter 12**, **Chapter 13** is dedicated to the detailed principles of iterative joint channel equalisation and channel decoding techniques known as turbo equalisation. **Chapter 14** provides theoretical performance bounds for turbo equalisers, while **Chapter 15** offers a wide-ranging comparative study of various turbo equaliser arrangements. The problem of reduced implemenattional complexity is addressed in **Chapter 16**. Finally, turbo equalised space-time trellis codes are the subject of **Chapter 17**.